CALCULUS

A RIGOROUS FIRST COURSE

A U R O R A

DOVER MODERN MATH ORIGINALS

Dover Publications is pleased to announce the publication of its first volumes in our new Aurora Series of original books in mathematics. In this series, we plan to make available exciting new and original works in the same kind of well-produced and affordable editions for which Dover has always been known.

Aurora titles currently available are:

Numbers: Histories, Mysteries, Theories by Albrecht Beutelspacher. (978-0-486-80348-7)

The Theory and Practice of Conformal Geometry by Steven G. Krantz. (978-0-486-79344-3)

Category Theory in Context by Emily Riehl. (978-0-486-80903-8)

Optimization in Function Spaces by Amol Sasane. (978-0-486-78945-3)

An Introductory Course on Differentiable Manifolds by Siavash Shahshahani. (978-0-486-80706-5)

Elementary Point-Set Topology: A Transition to Advanced Mathematics by André L. Yandl and Adam Bowers. (978-0-486-80349-4)

Additional volumes will be announced periodically.

The Dover Aurora Advisory Board:

John B. Little
College of the Holy Cross
Worcester, Massachusetts

Daniel S. Silver
University of South Alabama
Mobile, Alabama

CALCULUS
A RIGOROUS FIRST COURSE

Daniel J. Velleman

Amherst College
University of Vermont

DOVER PUBLICATIONS, INC.
Mineola, New York

Bibliographical Note

Calculus: A Rigorous First Course is a new work, first published by Dover Publications, Inc., in 2016.
A Solutions Manual for this book is available to instructors only. If you are an instructor and would like to receive it, please send an email to: calculus@doverpublications.com, and include the name of your institution.

Library of Congress Cataloging-in-Publication Data

Names: Velleman, Daniel J.
Title: Calculus : a rigorous first course / Daniel J. Velleman.
Description: Mineola, New York : Dover Publications, [2016] | Series: Aurora: Dover modern math originals | Includes bibliographical references and index.
Identifiers: LCCN 2016034302| ISBN 9780486809366 | ISBN 0486809366
Subjects: LCSH: Calculus—Textbooks.
Classification: LCC QA303.2 .V45 2016 | DDC 515—dc23 LC record available at https://lccn.loc.gov/2016034302

Manufactured in the United States by LSC Communications
80936602   2018
www.doverpublications.com

Contents

Preface

To the Student

My advice to students reading this book is very simple: Don't believe anything you read in this book.

Perhaps I should explain further. You may be used to studying mathematics by memorizing formulas and procedures for solving different types of problems. This method can be successful in high school math, but when you get to more advanced subjects like calculus, it doesn't work well. There are just too many different types of problems in calculus to memorize a procedure for solving each one. And for some types of calculus problems, there is no step-by-step procedure you can follow to get the answer.

An approach that works better is to grasp the concepts of calculus, so that you can understand *why* problems are done the way they are. With that understanding, when you come to a problem that is a little different from ones you have seen before, you can *figure out* how to solve it, rather than trying to apply a memorized procedure for solving it. When calculus is approached in this way, solving problems is not just a matter of calculation; it involves reasoning.

For reasoning in mathematics to be effective, it must be held to a very high standard. Our standard will be *certainty*: when reasoning about a problem, our goal will be not just to determine the answer, but to become *certain* of the answer. This pursuit of certainty should be evident throughout this book. Whenever we solve a problem, we will present a solution that is intended not merely to find the answer, but to *convince* you, with complete certainty, that the answer is correct. You should read this book with a skeptical attitude, refusing to believe that an answer is correct unless the solution is completely convincing. (That's why you shouldn't believe anything you read in this book.) And you should take a similar skeptical attitude toward your own solutions to problems: your goal is not merely to get the answer, but to be certain of the answer.

This skeptical attitude is important for success in calculus. In calculus there are often many different approaches that could be taken to a problem, some of which work and some of which don't. It is sometimes impossible to know in advance which approach will work. When trying to solve a problem, you may have to try one approach, recognize that it doesn't work, and then switch to a different approach. Thus, success in calculus requires not only the ability to find correct solutions, but also the ability to reject incorrect ones.

In your previous study of mathematics, you may have focused mainly on learning how to find correct solutions. Learning to recognize, and reject, incorrect solutions may be a new skill for you. How do you know when to reject a proposed solution? The best answer we can give is that you must insist on certainty. If your reasoning on a problem does not completely convince you of the answer, then it is insufficient and must be either improved or rejected. Your best defense against incorrect solutions is the skeptical attitude that we try to encourage in this book.

For reasoning to achieve certainty, it must be expressed with precision. We introduce many technical terms in this book, and when a term is introduced we always provide a precise definition. It is important to understand that terminology and notation in mathematics are always used to mean *exactly* what the definitions say—no more and no less. You should pay close attention to definitions, referring back to them if necessary. In many cases, the best way to get started on a problem is to be guided by the definitions of the words and notation appearing in the statement of the problem.

Often the methods we use to solve problems are based on general principles that are stated in the form of theorems. We have provided proofs of almost all of the theorems stated in this book. You may choose to skip some of these proofs, especially on a first reading. But we hope that your skeptical attitude will make you want to read them, so that you can be convinced that the theorems are true, rather than merely accepting them. These proofs demonstrate one of the fundamental principles of mathematics, without which the skeptical approach would be impossible: everything in math has a reason. Reading the proof of a theorem can not only help you understand why the theorem is true, it can also deepen your understanding of the meaning of the theorem. The proofs also provide you with good models of how mathematical reasoning should be carried out and expressed in writing.

This book requires no previous knowledge of calculus, but it does require a good background in algebra and trigonometry. Chapter 1 gives a brief review of the ideas from algebra and trigonometry that will be most important to us. If this review is not sufficient for you, then you may need to refer back to other resources on algebra and trigonometry.

To the Instructor

The topics covered in this book are the usual topics of first-year calculus: limits, derivatives, integrals, and infinite series. We have tried to give a mathematically rigorous treatment of these topics, while keeping the focus on the use of calculus to solve problems, rather than on the theoretical foundations of the subject. This is a rigorous calculus book, not an analysis book.

Some instructors may believe that rigor is an advanced topic that shouldn't be introduced until a student is taking analysis. This book is an attempt to justify the opposite point of view: rigor is a useful guide to beginning students that can help them learn to distinguish between correct mathematical reasoning and reasoning that is plausible but flawed. Calculus is full of such tempting but incorrect approaches to problems, and success requires learning to recognize them. Doing so when reasoning is kept at an intuitive level can be extremely difficult. Our point of view is that rigor is not an advanced topic; it is intuitive reasoning that is an advanced topic.

One of the most difficult ideas in calculus is the precise definition of limits, but a rigorous approach to the subject is not possible without such a definition. We devote an entire section to motivating and explaining the definition, using words, pictures, and formulas. A second section gives practice using the definition to prove limit statements. Reasoning based on this definition occurs at various points throughout the book.

Another essential theoretical idea is the completeness of the real numbers. Our fundamental completeness statement is the nested interval theorem, which says that for any nested sequence of closed intervals whose lengths approach 0, there is a unique number that is in all of the intervals. We state the nested interval theorem without proof, but a diagram makes it very plausible. We then use the theorem in a number of proofs throughout the book, including a proof of another version of completeness, the existence of least upper bounds and greatest lower bounds.

Our discussion of the nested interval theorem uses sequences and their limits, so these concepts are introduced early, in the last two sections of Chapter 2. The study of sequences also provides an opportunity to introduce the method of mathematical induction, which is used in a number of places later. For example, we use induction to prove that for every positive integer n, the derivative of x^n is nx^{n-1}.

We have chosen to restrict our discussion of definite integrals to continuous functions, and to use only uniform partitions in our definition of the integral. This is sufficient for everything we do with integrals in first-year calculus. Our definition of definite integrals also makes use of sequence limits: the definite integral is defined to be the limit of a sequence of Riemann sums that are based on uniform partitions.

We have included complete proofs of a number of theorems that are often stated without proof in calculus books, such as the extreme value theorem, the integrability of continuous functions, and the term-by-term differentiability and integrability of power series. All of these proofs use only high school algebra (and the nested interval theorem). The proof of the integrability of continuous functions is given in an optional final section in Chapter 5. That section introduces two new concepts that are used in the proof, but are not needed anywhere else in the book: uniform continuity and Riemann sums based on nonuniform partitions. Readers who don't want to be sidetracked by these topics can safely skip this section. The proof of term-by-term differentiability and integrability of power series is similarly put off until the final section of Chapter 10, since it requires the introduction of uniform convergence and summation by parts. Again, some readers may want to skip this section.

There are a number of paradoxes, such as Zeno's paradox, that are usually associated with calculus. We introduce two new paradoxes in this book; we call them the *paradox of precision through approximation* and the *paradox of generalization*.

The paradox of precision through approximation is the fact that calculus is a subject in which we use approximations to find precise answers. The paradox of generalization is the observation that in calculus, generalizing a problem often makes it easier. This paradox first arises in the study of derivatives: the rules for computing derivatives make it easier to find the formula for $f'(x)$ than to find, say, $f'(2)$. We appeal to the paradox of generalization later to motivate the shift from $\int_a^b f(t)\,dt$ to $\int_a^x f(t)\,dt$ that leads to the fundamental theorems of calculus, and again to motivate the shift from series of numbers to power series.

In order to make our presentation rigorous, we have had to take novel approaches to some topics. We summarize the most noteworthy features of our presentation here.

- Many calculus books introduce the informal notation "as $x \to a$, $f(x) \to L$" to express the idea that $\lim_{x \to a} f(x) = L$. But this notation misses a crucial distinction between the way in which x approaches a and the way in which $f(x)$ approaches L, namely, that x must never be equal to a, but $f(x)$ can equal L. We have therefore changed the notation slightly by writing "as $x \to a^{\neq}$, $f(x) \to L$," with the superscript "$\neq$" indicating that $x \neq a$. This use of superscripts is in keeping with other common variants of the arrow notation, such as "as $x \to a^+$, $f(x) \to L$," which means that $\lim_{x \to a^+} f(x) = L$. We use similar superscripts on L; for example, "as $x \to a^+$, $f(x) \to L^-$" means that when x approaches a from the positive side, $f(x)$ approaches L from the negative side. These small changes in notation turn out to be extremely valuable: they make it possible for us to give an intuitive and rigorously justifiable method for computing limits of compositions of functions in Section 2.6. And such limit computations are often useful in calculus. For example, see the calculation of the derivative of $\sqrt[n]{x}$ in Theorem 3.3.7. This calculation is a preview of the method used to find derivatives of inverse functions in Theorem 7.2.2.

- We motivate the study of derivatives with several examples of rates of change of real-world quantities. While we explain how these rates of change can be visualized as slopes of tangent lines, the focus is on rates of change in the real world, not tangent lines. Similarly, integration is motivated by real-world examples of quantities that accumulate at varying rates. Again, the connection to areas is used to help the intuition, but the emphasis is on accumulating quantities in the real world.

- Our proof of the mean value theorem is different from the one given in most books. Although our proof is probably a bit longer than the usual proof, it has the virtue that it does not depend on the extreme value theorem. This makes it possible to put off the extreme value theorem until after the discussion of curve sketching techniques, in which the idea of maximum and minimum values arises naturally.

- Differentiation is a pointwise operation, but antidifferentiation must be done on intervals. When we introduce antidifferentiation, we define one-sided derivatives, to allow for antiderivatives on closed intervals, and we state a version of the chain rule on intervals that uses one-sided derivatives at the endpoints of intervals (Theorem 4.9.5). This theorem is needed to justify integration by substitution.

To see why, consider the following integration example, using the substitution $u = 1 + \sin x$:

$$\int \sqrt{1 + \sin x} \cos x \, dx = \int u^{1/2} \, du = \frac{2}{3} u^{3/2} + C = \frac{2}{3} (1 + \sin x)^{3/2} + C.$$

This answer is correct, but the usual version of the chain rule does not suffice to justify it, because $u^{3/2}$ has only a one-sided derivative at $u = 0$. However, our chain rule on intervals justifies it.

- We present trigonometric substitutions in integrals as an instance of a general method of substitution with inverse functions. A justification of this method is given in Section 8.4.

- Integrals for arc length and surface area are introduced in the context of curves given by parametric equations, which is the most natural and general context for these topics. The formulas in the case of graphs of functions follow easily from the formulas for parametric curves.

- All of the power series representations of familiar functions that are given in most calculus books are derived without the use of any formulas for the Taylor remainder. This keeps the focus of attention on calculus with power series, rather than intricate error estimates. In each case, we show that the function in question is the unique solution to some differential equation that meets some initial condition, and then we show that the Taylor series satisfies the differential equation and initial condition. We do present the Lagrange formula for the Taylor remainder, but it is put off until near the end of the chapter on series and power series.

Chapter 1

Preliminaries

1.1 Numbers and Sets

The numbers we will use in this book are the *real numbers*. These are all the numbers that can be written in decimal notation. We often think of them as corresponding to points on a number line (see Figure 1.1). The simplest real numbers are the *integers*: the numbers 0, 1, −1, 2, −2, 3, −3, and so on. A real number is said to be *rational* if it can be written as an integer divided by an integer; for example, 2/3 and −13/5 are rational numbers. Notice that every integer is also a rational number, since, for example, we can write 3 as 3/1. Real numbers that are not rational are called *irrational*. For example, it can be shown that $\sqrt{2} = 1.41421\ldots$ and $\pi = 3.14159\ldots$ are irrational. Both rational and irrational numbers are spread throughout the number line; in fact, between any two real numbers there are infinitely many rational numbers and also infinitely many irrational numbers.

We will also often work with collections of numbers. In mathematics, a collection of objects is called a *set*, and the objects in the collection are called *elements* of the set. The simplest way to specify a set is to list the elements of the set between braces. For example, $\{-1, 0, 3/2\}$ is the set whose elements are the three numbers −1, 0, and 3/2. If we let the letter A stand for this set, then we write $3/2 \in A$ to say that 3/2 is an element of A, while $4 \notin A$ means that 4 is not an element of A.

Another way to specify a set is to give a rule for determining which objects belong to the set and which do not. For example, if we write

$$B = \{x : 2x^3 - x^2 - 3x = 0\}, \tag{1.1}$$

1

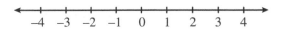

Figure 1.1: The number line.

then this means that B is the set whose elements are all values of x that satisfy the equation $2x^3 - x^2 - 3x = 0$. Equation (1.1) is read "B is equal to the set of all x such that $2x^3 - x^2 - 3x = 0$." The equation $2x^3 - x^2 - 3x = 0$ that appears in the definition of B is a statement that is true for some values of x and false for others. You should think of it as an elementhood test for the set B; those values of x that make the equation true pass the test and are elements of B, while those that make the equation false are not. To determine which numbers belong to B we simply have to solve the equation, which we can do by factoring the left-hand side. We have

$$2x^3 - x^2 - 3x = x(x+1)(2x-3),$$

so the equation can be rewritten $x(x+1)(2x-3) = 0$, and the solutions are 0, -1, and $3/2$. These are the elements of the set B. Notice that these are exactly the same as the elements of the set A defined earlier. Thus $B = A$; they are both the same collection of numbers, described in two different ways.

Although we will usually use the letter x when writing an elementhood test to define a set, as we did in the definition of B, in fact any letter can be used. For example, the set $C = \{y : 2y^3 - y^2 - 3y = 0\}$ is the set of all values of y that satisfy the equation $2y^3 - y^2 - 3y = 0$. Of course, this is the same equation that we used in the definition of B, but with x replaced by y. The values of y that satisfy the equation are therefore once again the numbers 0, -1, and $3/2$. Therefore $C = B = A$; we have the same set of numbers described in yet another way.

Here is another example of a set defined by an elementhood test: $I = \{x : 2 < x < 5\}$. This time the elementhood test is $2 < x < 5$, which is a shorthand way of saying that $2 < x$ and $x < 5$. In this case $3 \in I$, since the statement $2 < 3 < 5$ is true, but $5 \notin I$, since the statement $2 < 5 < 5$ is false. The elements of I are all the numbers strictly between 2 and 5. There are infinitely many numbers in this range, so we cannot list all the elements of I, as we did for A. But we can mark them on a number line, as in Figure 1.2.

Figure 1.2: $I = \{x : 2 < x < 5\}$. The open circles indicate that 2 and 5 are not elements of I, while the thick line shows that all numbers between 2 and 5 are elements.

The set I is an example of a kind of set called an *open interval*. For any numbers a and b with $a < b$, the set of all numbers strictly between a and b is an open interval, and it is denoted (a, b). In other words,

$$(a, b) = \{x : a < x < b\}.$$

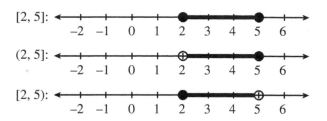

Figure 1.3: The closed interval [2, 5], and the half-open intervals (2, 5] and [2, 5). Black dots indicate points that are included in a set.

The numbers *a* and *b* are called the *endpoints* of the interval. Thus *I* = (2, 5); it is the open interval with endpoints 2 and 5.

The endpoints of an open interval are not elements of the interval. But sometimes we will want to include the endpoints, so we define

$$[a, b] = \{x : a \leq x \leq b\}.$$

This set is called a *closed interval*. For example, [2, 5] = {x : 2 ≤ x ≤ 5}. This set is exactly the same as the open interval (2, 5) considered earlier, except that it includes the endpoints 2 and 5 as elements. If we include only one endpoint, we get a *half-open interval*. As you might guess, we write half-open intervals like this:

$$(a, b] = \{x : a < x \leq b\},$$
$$[a, b) = \{x : a \leq x < b\}.$$

In general, we use a square bracket to indicate that an endpoint is included in an interval, and a parenthesis to indicate that it is not. Figure 1.3 shows examples of closed and half-open intervals.

The *interior* of an interval is the set containing all numbers in the interval except the endpoints. Thus, the interior of the closed interval [2, 5] is the open interval (2, 5). The interiors of (2, 5] and [2, 5) are also (2, 5), and we will even say that the interior of (2, 5) is (2, 5).

Finally, we sometimes want to consider intervals that extend infinitely far in some direction, so we introduce the notation:

$$(a, \infty) = \{x : x > a\},$$
$$[a, \infty) = \{x : x \geq a\},$$
$$(-\infty, b) = \{x : x < b\},$$
$$(-\infty, b] = \{x : x \leq b\}.$$

Some examples of infinite intervals are shown in Figure 1.4. The interior of [a, ∞) is (a, ∞), and the interior of (−∞, b] is (−∞, b). The set of all real numbers is often denoted ℝ, but we could also think of it as the interval (−∞, ∞). We consider any

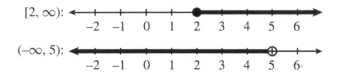

Figure 1.4: Infinite intervals.

interval that does not include its endpoints to be an open interval. Thus, the intervals (a, ∞), $(-\infty, b)$, and $(-\infty, \infty)$ are open intervals, and the interior of any interval is an open interval.

Since this is our first use of the infinity symbol ∞, it might be worthwhile to pause at this point to explain what this symbol means. The most important thing to understand about the infinity symbol is that *there is no such number as infinity.* You might wonder, then, how it can be correct to use this symbol in mathematical notation like (a, ∞). The answer is that, according to the definition we have given, this notation is simply a shorthand for $\{x : x > a\}$, and this last expression makes no mention of infinity. Every time we make a statement using the symbol ∞, it will be a similar shorthand for a statement that makes no mention of infinity. We will never use the symbol ∞ as if it stood for a number. Thus, for example, we would never set x equal to ∞ in some formula, and we would never talk about the "closed interval" $[2, \infty]$.

There are two ways of combining sets that we will sometimes make use of. If A and B are sets, then the *intersection* of A and B, denoted $A \cap B$, is the set whose elements are those objects that belong to both A and B. Thus

$$A \cap B = \{x : x \in A \text{ and } x \in B\}.$$

For example,

$$[2, \infty) \cap (-\infty, 5) = \{x : x \in [2, \infty) \text{ and } x \in (-\infty, 5)\}$$
$$= \{x : 2 \le x \text{ and } x < 5\} = [2, 5).$$

Looking at Figure 1.4, you can see that the elements of $[2, \infty) \cap (-\infty, 5)$ are those numbers that are in the overlap of the sets $[2, \infty)$ and $(-\infty, 5)$. In general, you can think of $A \cap B$ as the overlap of A and B.

The *union* of A and B, denoted $A \cup B$, is the set whose elements are all those objects that are elements of either A or B (or both). That is,

$$A \cup B = \{x : x \in A \text{ or } x \in B\}.$$

You could think of $A \cup B$ as the set you get if you take all the elements of A, and all the elements of B, and throw them together into one set. For example, if we take all the numbers in the intervals $(2, 4]$ and $[4, 5]$ and put them together into one set, we get the interval $(2, 5]$. That is,

$$(2, 4] \cup [4, 5] = \{x : x \in (2, 4] \text{ or } x \in [4, 5]\}$$
$$= \{x : 2 < x \leq 4 \text{ or } 4 \leq x \leq 5\} = \{x : 2 < x \leq 5\} = (2, 5].$$

If A and B are sets, then A is called a *subset* of B if every element of A is also an element of B. We write $A \subseteq B$ to indicate that A is a subset of B. For example, $(2, 4) \subseteq [2, 4]$, and $[2, 4] \subseteq (1, 5)$.

One reason that intervals are important in calculus is that they often come up as solution sets of inequalities. In particular, we will often be concerned with inequalities involving absolute values. Recall that the *absolute value* of a number x is defined as follows:

$$|x| = \begin{cases} x, & \text{if } x \geq 0, \\ -x, & \text{if } x < 0. \end{cases}$$

This notation means that if $x \geq 0$ then $|x| = x$, and if $x < 0$ then $|x| = -x$. For example, $|3| = 3$ and $|-4| = -(-4) = 4$.

The fact that $|x|$ is defined by cases, with one formula when $x \geq 0$ and a different formula when $x < 0$, suggests a method that can be used when solving any problem involving absolute values: reasoning by cases. As a simple example of this kind of reasoning, notice that if $x \geq 0$ then we have $|x| = x \geq 0$, and if $x < 0$ then $|x| = -x > 0$. In both cases the statement $|x| \geq 0$ is true, so we conclude that for every number x, $|x| \geq 0$. You can also use reasoning by cases to show that for every number x, $\sqrt{x^2} = |x|$ (see Exercise 15).

Here's an example of how reasoning by cases can be used to solve an inequality involving absolute values:

Example 1.1.1. Solve:

$$|x| < 3.$$

Solution. Motivated by the definition of $|x|$, we will consider $x \geq 0$ and $x < 0$ separately.

Case 1. $x \geq 0$. In this case, according to the definition of absolute value we have $|x| = x$, so the inequality $|x| < 3$ means $x < 3$.

Case 2. $x < 0$. Now the definition of absolute value says that $|x| = -x$, and substituting this into our inequality $|x| < 3$ gives us $-x < 3$. Multiplying by -1 (and remembering that when multiplying an inequality by a negative number, we must reverse the direction of the inequality) we get $x > -3$.

So what's the solution to our inequality? Is it $x < 3$, as we found in case 1, or $x > -3$, as in case 2? To answer this question, we must think about what it means to solve an inequality. To solve the inequality $|x| < 3$ means to determine which values of x make the inequality true. Our reasoning in case 1 shows that, for $x \geq 0$, the inequality means $x < 3$. Thus the inequality is true if $0 \leq x < 3$ and false if $x \geq 3$. We can't tell from this reasoning whether the inequality is true or false when $x < 0$; that's the purpose of case 2. Case 2 tells us that, for $x < 0$, the inequality will be true precisely when $x > -3$. Thus, the inequality is true if $-3 < x < 0$ and false if $x \leq -3$. Putting all this

Figure 1.5: The solution to Example 1.1.1. The parts of the number line marked with solid lines are in the solution set, and outlined areas are not. The blue lines were determined in case 1 of the solution, and the red lines in case 2.

information together, we conclude that the inequality is true if $-3 < x < 3$ and false if either $x \geq 3$ or $x \leq -3$, as shown in Figure 1.5. This means that the *solution set* of the inequality is an open interval:

$$\{x : |x| < 3\} = \{x : -3 < x < 3\} = (-3, 3).$$

Notice that in case 1 we determined that all numbers in the interval $[0, 3)$ are in the solution set, and in case 2 we determined that the numbers in $(-3, 0)$ are also in the solution set. The solution set is therefore the union of these two intervals: $[0, 3) \cup (-3, 0) = (-3, 3)$. □

Here's another way of describing the answer to Example 1.1.1. Our reasoning shows that the statements $|x| < 3$ and $-3 < x < 3$ are true for exactly the same values of x; the two statements are equivalent. In other words, for any number x, if $|x| < 3$, then $-3 < x < 3$, and if $-3 < x < 3$, then $|x| < 3$. Mathematicians usually describe this situation by saying that $|x| < 3$ is true *if and only if* $-3 < x < 3$. The phrase "if and only if" comes up often in mathematics, and we will see it many times later in this book.

Of course, there is nothing special about the number 3 in this example. Similar reasoning, using the variable y in place of the number 3, can be used to establish the following theorem. Parts 3 and 4 of the theorem follow directly from parts 1 and 2, by negating the statements involved.

Theorem 1.1.2. *For any numbers x and y, the following statements are true:*

1. $|x| < y$ *if and only if* $-y < x < y$.

2. $|x| \leq y$ *if and only if* $-y \leq x \leq y$.

3. $|x| \geq y$ *if and only if either* $x \leq -y$ *or* $x \geq y$.

4. $|x| > y$ *if and only if either* $x < -y$ *or* $x > y$.

The most important use of absolute values in calculus is to compute distances between numbers on the number line. To find the distance between two numbers, we subtract the smaller number from the larger. For example, the distance from -4 to 3 on the number line is $3 - (-4) = 7$. In general, if $a \leq b$ then the distance from a to b is $b - a$. But if $a > b$ then the distance is $a - b$. Is there a single formula that gives the distance from a to b no matter which of the numbers is larger?

It turns out that the formula $|b - a|$ does the trick. We can see this by once again using reasoning by cases.

Case 1. $a \leq b$. Then $b - a \geq 0$, so $|b - a| = b - a$, which, in this case, is the distance from a to b on the number line.

Case 2. $a > b$. Now $b - a < 0$, so $|b - a| = -(b - a) = a - b$, which is, once again, the distance from a to b in this case.

Thus, no matter which of the numbers a and b is larger, we have

$$|b - a| = \text{the distance from } a \text{ to } b \text{ on the number line.}$$

This fact provides a nice way to see why our solution to Example 1.1.1 makes sense. We have

$$|x| = |x - 0| = \text{distance from 0 to } x \text{ on the number line.}$$

With this interpretation for $|x|$, the inequality $|x| < 3$ can be thought of as saying

$$(\text{distance from 0 to } x \text{ on the number line}) < 3.$$

It is clear geometrically that the values of x for which this statement is true are the numbers in the open interval $(-3, 3)$, exactly as we found in our solution to Example 1.1.1.

Example 1.1.3. Solve:
$$|3t + 2| \leq 4.$$

Solution. By part 2 of Theorem 1.1.2, the inequality to be solved is equivalent to

$$-4 \leq 3t + 2 \leq 4.$$

Subtracting 2 all the way through the inequality gives us

$$-6 \leq 3t \leq 2,$$

and then dividing by 3 we get
$$-2 \leq t \leq 2/3.$$

Thus the solution set for this inequality is a closed interval:

$$\{t : |3t + 2| \leq 4\} = \{t : -2 \leq t \leq 2/3\} = [-2, 2/3]. \qquad \square$$

Sometimes it is useful to be able to simplify absolute values of complicated expressions. In such situations, the following theorem can be helpful.

Theorem 1.1.4. *For any numbers x and y, the following statements are true:*

1. $|xy| = |x||y|$.

2. *If $y \neq 0$ then* $\left|\dfrac{x}{y}\right| = \dfrac{|x|}{|y|}$.

Proof. We will just prove part 1; the proof of part 2 is similar. First note that if either x or y is 0 then both sides of the equation $|xy| = |x||y|$ are 0, so the equation is true.

If neither of them is 0, then each is either positive or negative. This leaves us with four cases to consider:

Case 1. $x > 0$, $y > 0$. Then by the definition of absolute value, $|x| = x$ and $|y| = y$. Also, since the product of two positive numbers is positive, $xy > 0$. Therefore $|xy| = xy = |x||y|$.

Case 2. $x > 0$, $y < 0$. Then $|x| = x$ and $|y| = -y$. Since a positive number times a negative number is negative, $xy < 0$. Thus $|xy| = -xy = x(-y) = |x||y|$.

Case 3. $x < 0$, $y > 0$. Then $|x| = -x$, $|y| = y$, and $xy < 0$, so $|xy| = -xy = |x||y|$.

Case 4. $x < 0$, $y < 0$. Then $|x| = -x$ and $|y| = -y$. Since a negative times a negative is positive, we have $xy > 0$, so $|xy| = xy = (-x)(-y) = |x||y|$. ☐

To illustrate the use of Theorem 1.1.4, we briefly revisit Example 1.1.3. Notice that $|3t + 2| = |3(t + 2/3)| = |3||t + 2/3| = 3|t - (-2/3)|$, where we have used part 1 of Theorem 1.1.4 in the second step. Thus, the inequality in Example 1.1.3 can be rewritten

$$3|t - (-2/3)| \leq 4.$$

Dividing through by 3, this is equivalent to

$$|t - (-2/3)| \leq 4/3,$$

which can be interpreted as meaning

$$(\text{distance from } t \text{ to } -2/3 \text{ on the number line}) \leq 4/3.$$

Thus the solution set of the inequality is

$$[-2/3 - 4/3, -2/3 + 4/3] = [-2, 2/3],$$

exactly as we found in Example 1.1.3.

Theorem 1.1.4 tells us how to simplify an absolute value of a product or a quotient. What about the absolute value of a sum? Is it always true that $|x + y| = |x| + |y|$? A little experimentation shows that, unfortunately, this equation is not always true. For example, $|5 + (-3)| = 2$, but $|5| + |-3| = 8$. However, we do have the following important fact, which is known as the *triangle inequality*.

Theorem 1.1.5 (Triangle Inequality). *For all numbers x and y, $|x + y| \leq |x| + |y|$.*

Proof. According to part 2 of Theorem 1.1.2, the triangle inequality is equivalent to the statement

$$-(|x| + |y|) \leq x + y \leq |x| + |y|, \tag{1.2}$$

so it will suffice to prove this inequality. We will leave it to you to verify (using reasoning by cases) that

$$-|x| \leq x \leq |x|$$

and

$$-|y| \leq y \leq |y|$$

(see Exercise 18). Adding these two inequalities gives inequality (1.2). ☐

The theorems we have proven about absolute values could be regarded as shortcuts that allow us to solve problems more easily. For example, using Theorem 1.1.2 we were able to solve Example 1.1.3 without having to resort to reasoning by cases. This is a pattern that we will see repeated many times in this book. When a new concept is introduced, we will initially rely on the definition to tell us how to solve problems involving that concept. But often solutions based on the definition will be long and complicated, so we will develop theorems that provide shortcuts that allow us to solve problems more easily.

You may wonder, why bother learning the definitions? Why not just learn the shortcuts, if they provide easier ways of solving problems? One answer is that shortcuts, while helpful, are also often somewhat limited. They usually allow us to solve only a restricted range of problems. Sometimes you come across a problem for which the shortcuts are not helpful, and then you have no choice but to return to the definitions of the relevant concepts to solve the problem.

As an illustration of this, we close this section by solving an inequality involving absolute values for which our various shortcuts don't seem to be very helpful.

Example 1.1.6. Solve:
$$2|x| - 3 \geq |x - 1|.$$

Solution. Although it is possible to use Theorem 1.1.2 to solve this inequality (see Exercise 16), the solution is not easy. Here we use the more straightforward approach of working from the definition of absolute value, which suggests reasoning by cases. Since we need to work with both $|x|$ and $|x - 1|$, we are led to consider three cases: $x < 0$, $0 \leq x < 1$, and $x \geq 1$.

Case 1. $x < 0$. In this case we also have $x - 1 < 0$, so according to the definition of absolute value, $|x| = -x$ and $|x - 1| = -(x - 1) = 1 - x$. Substituting into our inequality, we find that we must solve
$$-2x - 3 \geq 1 - x.$$

Adding $2x$ to both sides and subtracting 1, we get $-4 \geq x$. Thus, for negative values of x, the inequality is true if and only if $x \leq -4$. In other words, it is true if $x \leq -4$ and false if $-4 < x < 0$.

Case 2. $0 \leq x < 1$. We have $x \geq 0$, so $|x| = x$, and $x < 1$, so $x - 1 < 0$ and therefore $|x - 1| = 1 - x$. Thus in this case the inequality means
$$2x - 3 \geq 1 - x,$$

which can be simplified to $x \geq 4/3$. But this inequality is false for all values of x in the range $0 \leq x < 1$. Thus, none of the numbers in the interval $[0, 1)$ will be included in the solution set of our inequality.

Case 3. $x \geq 1$. Then $x \geq 0$ and $x - 1 \geq 0$, and therefore $|x| = x$ and $|x - 1| = x - 1$. Filling in these formulas in our inequality we get
$$2x - 3 \geq x - 1,$$

and simplifying leads to $x \geq 2$. Thus, the inequality is true for $x \geq 2$ and false for $1 \leq x < 2$.

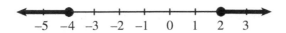

Figure 1.6: The solution set for Example 1.1.6 is $(-\infty, -4] \cup [2, \infty)$.

Combining the information from all three cases, we find that the solution set is the set containing all numbers in the interval $(-\infty, -4]$ and also all numbers in the interval $[2, \infty)$. In other words,

$$\{x : 2|x| - 3 \ge |x - 1|\} = \{x : \text{either } x \le -4 \text{ or } x \ge 2\} = (-\infty, -4] \cup [2, \infty). \quad \square$$

Exercises 1.1

1–13: Solve the inequality. If possible, write the solution set as an interval or a union of intervals.

1. $|x - 5| < 7$.

2. $|4x + 2| \le 6$.

3. $|5 - 2t| \ge 4$.

4. $|x - 4| < x$.

5. $2|x - 4| < x$.

6. $|x + 4| < x$.

7. $|x + 4| < 2x$.

8. $|6 - 2u| - 3 < u$.

9. $|3y| \le |y| + 10$.

10. $|x - 3| \ge |x| - 1$.

11. $|2x - 3| \ge |x| - 1$.

12. $|2x - 2| > |x| - 1$.

13. $|3x - 6| \le |3 - 6x|$.

14. Write the following sets as intervals or unions of intervals. Use interval notation.

   (a) $\{x : x^2 < 64\}$.

   (b) $\{x : x^3 < 64\}$.

   (c) $\{x : 4 \le x^2 < 9\}$.

   (d) $(-\infty, 5] \cap (3, \infty)$.

   (e) $(-3, 5] \cup (3, 7]$.

15. Use reasoning by cases to prove that for every number x, $\sqrt{x^2} = |x|$.

16. Solve Example 1.1.6 by using Theorem 1.1.2.

17. Prove part 1 of Theorem 1.1.2. (Hint: Treat the cases $y > 0$ and $y \leq 0$ separately.)

18. Prove that for every number x, $-|x| \leq x \leq |x|$. (This fact is used in the proof of Theorem 1.1.5.)

19. Prove that for all numbers x and y, $|x - y| \geq |x| - |y|$. (Hint: Start by applying the triangle inequality to $|(x - y) + y|$.)

20. Use the triangle inequality to show that for every number x, if $|x| \leq 2$ then $|x^3 - 7x + 3| \leq 25$.

21. Show that for every number x, if $|x| \leq 2$ then $\left| \dfrac{x^3 - 7x + 3}{9 - x^2} \right| \leq 5$. (Hint: Use Exercise 19 to show that $|9 - x^2| \geq 5$.)

1.2 Graphs in the Plane

In Section 1.1 we considered statements involving a single variable, usually x, and we illustrated these statements by marking on a number line the values of the variable for which the statement is true. Often in calculus we will work with statements involving two variables, usually x and y. To illustrate such statements we use the coordinate plane. Each point in the plane represents an ordered pair of numbers (x, y), as illustrated in Figure 1.7.[1]

For example, consider the statement

$$1 < x \leq 3 \text{ and } -2 \leq y < 1. \tag{1.3}$$

The points (x, y) whose coordinates make this statement true are shown in Figure 1.8. This set of points is called the *graph* of the statement. For example, the statement is true if $x = 3$ and $y = -2$, and therefore the point $(3, -2)$ is included in the graph. On the other hand, it is false if $x = 3$ and $y = 1$, so the point $(3, 1)$ is not included.

The statements we will be most concerned with in this book are equations involving two variables, most often x and y. The graph of such an equation is usually a curve in the plane. Among the most important examples are equations whose graphs are straight lines. We assume you are familiar with equations of straight lines, but since they will be so important in calculus we briefly review the most important facts.

For any numbers m and b, the graph of the equation $y = mx + b$ is a straight line with slope m and y-intercept b. This equation is called the *slope-intercept* equation of the line. For example, the line $y = 2x - 3$, which has slope 2 and y-intercept -3, is shown in Figure 1.9a. The fact that the y-intercept is -3 means that the line crosses the y-axis

---

[1]It is an unfortunate fact that the notation for a point in the plane is the same as the notation for an open interval, namely a pair of numbers in parentheses. You will have to tell from context whether, for example, the notation $(3, 5)$ stands for the open interval from 3 to 5 or the point in the plane whose coordinates are 3 and 5.

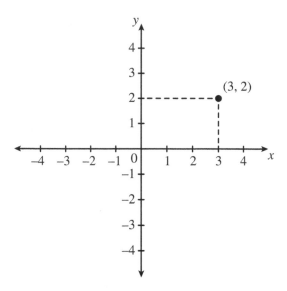

Figure 1.7: The coordinate plane. The black dot is located at the point with coordinates $(3, 2)$.

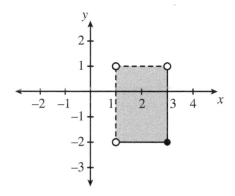

Figure 1.8: The graph of statement (1.3). The shaded region, solid lines, and black dot are included in the graph; the dotted lines and open circles are not.

at -3. In other words, the line passes through the point $(0, -3)$, as you can verify by checking that the equation $y = 2x - 3$ is true when $x = 0$ and $y = -3$. To understand the meaning of the slope, it might be helpful to imagine a point (x, y) moving along the line from left to right. The fact that the slope is 2 means that every time the x-coordinate of our moving point increases by 1, the y-coordinate increases by 2. More generally, if we add some number h to the x-coordinate of a point on the line, then we must add $2h$ to the y-coordinate to reach another point on the line. To see why this is true, suppose that some point (x_1, y_1) is on the line $y = 2x - 3$. This means that the equation

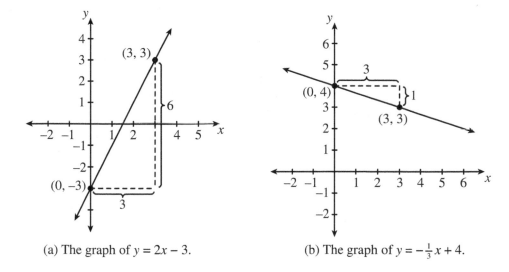

(a) The graph of $y = 2x - 3$. (b) The graph of $y = -\frac{1}{3}x + 4$.

Figure 1.9

$y = 2x - 3$ is true when $x = x_1$ and $y = y_1$, or in other words, $y_1 = 2x_1 - 3$. It follows that $y_1 + 2h = 2x_1 - 3 + 2h = 2(x_1 + h) - 3$, so the point $(x_1 + h, y_1 + 2h)$ is also on the line. For example, Figure 1.9a shows that if we move along the line from $(0, -3)$ to $(3, 3)$, x increases by 3 and y increases by $2 \cdot 3 = 6$. The change in y is always twice as big as the change in x, and we therefore say that the slope gives the *rate of change of y with respect to x*.

Lines with positive slope slant upwards as we move from left to right, and lines with negative slope slant downwards. For example, Figure 1.9b shows the line $y = -\frac{1}{3}x + 4$. This time, when we add h to x, we must add $-\frac{1}{3}h$ to y—in other words, if x increases by h, then y *decreases* by $\frac{1}{3}h$. Figure 1.9b shows that when we move from $(0, 4)$ to $(3, 3)$, x increases by 3 and y decreases by $\frac{1}{3} \cdot 3 = 1$.

A line with slope 0 has an equation of the form $y = 0 \cdot x + b$, or in other words $y = b$. The graph of this equation is a horizontal line passing through the point b on the y-axis. Similarly, the equation of a vertical line passing through the point a on the x-axis is $x = a$. Vertical lines are the only lines whose equations cannot be written in the slope-intercept form $y = mx + b$. The slope of a vertical line is undefined.

Example 1.2.1. Find an equation of the line with slope 3 that passes through the point $(-1, 2)$.

Solution. The slope-intercept equation of the line must have the form $y = 3x + b$, where b is the y-intercept. We must find b.

Since the line passes through the point $(-1, 2)$, the equation for the line must be true when $x = -1$ and $y = 2$. In other words, $2 = 3(-1) + b$, and therefore $b = 5$. So the equation of the line is $y = 3x + 5$. □

More generally, suppose we are looking for the line with slope m passing through the point (x_1, y_1). As in Example 1.2.1, we can substitute x_1 and y_1 for x and y in the equation $y = mx + b$ to conclude that $y_1 = mx_1 + b$, and therefore $b = y_1 - mx_1$. Thus, the slope-intercept equation for the line is $y = mx + (y_1 - mx_1)$. It is sometimes convenient to rearrange this equation to put it in the form

$$y - y_1 = m(x - x_1). \tag{1.4}$$

Equation (1.4) is called the *point-slope* form of the equation for the line.

If a line with equation $y = mx + b$ passes through two different points (x_1, y_1) and (x_2, y_2), then we can substitute the coordinates of both points into the equation of the line to conclude that $y_1 = mx_1 + b$ and $y_2 = mx_2 + b$. Subtracting the first equation from the second we conclude that $y_2 - y_1 = m(x_2 - x_1)$, and therefore

$$m = \frac{y_2 - y_1}{x_2 - x_1}. \tag{1.5}$$

This gives a convenient formula for the slope of a nonvertical line, given two points on the line. The formula confirms our earlier interpretation of the slope as the rate of change of y with respect to x, since it says that if you move along the line from (x_1, y_1) to (x_2, y_2), then the slope is equal to the number of units of change in y per unit of change in x.

Example 1.2.2. Find an equation of the line through the points $(-2, 3)$ and $(4, -1)$.

Solution. Applying equation (1.5), we find that the slope of the line is

$$m = \frac{-1 - 3}{4 - (-2)} = -\frac{2}{3}.$$

Plugging this slope and the given point $(-2, 3)$ into the point-slope formula (1.4), we see that one equation for the line is

$$y - 3 = -\frac{2}{3}(x + 2).$$

Of course, we could have used the other given point $(4, -1)$ in the point-slope formula, so another answer is

$$y + 1 = -\frac{2}{3}(x - 4).$$

These two answers are equivalent, as you can see by checking that both can be rearranged to give the same slope-intercept equation

$$y = -\frac{2}{3}x + \frac{5}{3}. \qquad \square$$

Example 1.2.3. Find the intersection point of the two lines

$$y = 4x + 1,$$
$$y = -2x + 3.$$

Solution. We are looking for the unique point (x, y) whose coordinates satisfy both equations. If both equations are to be true, then we must have $4x + 1 = -2x + 3$, and therefore $6x = 2$ and $x = 1/3$. Substituting this value into the first equation, we find that $y = 4(1/3) + 1 = 7/3$. (Of course, substituting into the second equation would give the same answer for y.) Thus, the intersection point is $(1/3, 7/3)$. □

Since the slope of a line measures how steeply it is inclined, parallel lines have the same slope. Figure 1.10 illustrates that if a line has slope $m \neq 0$, then a perpendicular line will have slope $-1/m$. (Of course, if a line has slope 0 then it is horizontal, so a perpendicular line will be vertical.)

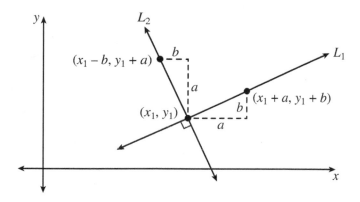

Figure 1.10: Let m be the slope of line L_1. Then $m = b/a$, and the slope of the perpendicular line L_2 is $-a/b = -1/m$.

In Section 1.1, we determined that the distance between two points x_1 and x_2 on the number line is $|x_2 - x_1|$. How do we compute the distance between two points (x_1, y_1) and (x_2, y_2) in the plane? Applying the Pythagorean theorem to the triangle in Figure 1.11, we see that the distance d satisfies the equation $d^2 = (x_2 - x_1)^2 + (y_2 - y_1)^2$, and therefore

$$d = \sqrt{(x_2 - x_1)^2 + (y_2 - y_1)^2}. \tag{1.6}$$

Equation (1.6) is known as the *distance formula*.

Example 1.2.4. Find the area of the triangle whose vertices are the points $(-1, 0), (5, 3)$, and $(1, 6)$.

Solution. See Figure 1.12. We take the side from $(-1, 0)$ to $(5, 3)$ as the base of the triangle. By the distance formula, it has length

$$b = \sqrt{(5 - (-1))^2 + (3 - 0)^2} = \sqrt{45} = 3\sqrt{5}.$$

Let L_1 be the line containing the base of the triangle. Then L_1 has slope $m = (3 - 0)/(5 - (-1)) = 1/2$, and by the point-slope formula its equation is $y - 0 = (1/2)(x - (-1))$,

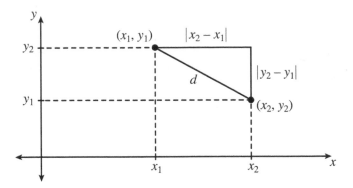

Figure 1.11: By the Pythagorean theorem, $d^2 = |x_2 - x_1|^2 + |y_2 - y_1|^2 = (x_2 - x_1)^2 + (y_2 - y_1)^2$.

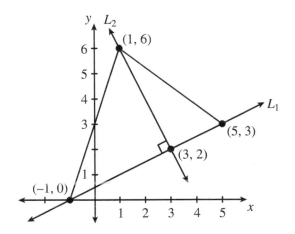

Figure 1.12: The solution to Example 1.2.4.

or equivalently $y = (1/2)x + 1/2$. To find the altitude of the triangle, we need to find the line L_2 that is perpendicular to L_1 and passes through the point $(1, 6)$. This line will have slope $-1/m = -2$, so the point-slope formula now tells us that L_2 has equation $y - 6 = -2(x - 1)$, or equivalently $y = -2x + 8$. As in Example 1.2.3, by solving the equation $(1/2)x + 1/2 = -2x + 8$ we find that L_1 and L_2 intersect at the point $(3, 2)$. Thus, the altitude of the triangle is the distance from $(3, 2)$ to $(1, 6)$, which is

$$h = \sqrt{(1 - 3)^2 + (6 - 2)^2} = \sqrt{20} = 2\sqrt{5}.$$

Finally, we can conclude that the area of the triangle is

$$\frac{1}{2}bh = \frac{1}{2} \cdot 3\sqrt{5} \cdot 2\sqrt{5} = 15. \qquad \square$$

Note that in our solution to Example 1.2.4 we left the base and altitude of the triangle in the form $b = 3\sqrt{5}$ and $h = 2\sqrt{5}$. We could have used a calculator to determine that $b \approx 6.708$ and $h \approx 4.472$, leading to the area calculation $(1/2)bh \approx 14.999$. But notice that this approach involves more work and leads to a less accurate answer. We usually find it easier and more useful to leave all of our numbers in exact form, as we did in Example 1.2.4, even when this means using an expression like $\sqrt{5}$.

So far we have focused on equations whose graphs are straight lines. Perhaps the most fundamental geometric shape other than a straight line is a circle. The circle of radius r centered at the point (a, b) consists of all points (x, y) whose distance from (a, b) is equal to r. Thus, to find an equation for this circle, we simply have to write an equation that expresses the statement

$$(\text{distance from } (x, y) \text{ to } (a, b)) = r.$$

Using the distance formula for the left-hand side of this equation, we get

$$\sqrt{(x - a)^2 + (y - b)^2} = r.$$

Squaring both sides gives the simpler, equivalent formula

$$(x - a)^2 + (y - b)^2 = r^2.$$

For example, the graph of the equation

$$x^2 + y^2 = 1$$

is the circle of radius 1 centered at the origin, $(0, 0)$.

Exercises 1.2

1–14: Graph the given statement in the plane.

1. $x = 3$ and $1 \le y \le 4$.

2. $y = x - 2$.

3. $y = -3x + 5$.

4. $y = (5 - x)/2$.

5. $2x + 3y = 5$.

6. $y = 3$.

7. $x = -2$.

8. $(y - 3)(x + 2) = 0$.

9. $x^2 - y^2 = 0$.

10. $y < x - 2$.

11. $0 \leq x \leq y \leq 1$.

12. $x^2 + (y-2)^2 = 4$.

13. $x^2 + (y-2)^2 \leq 4$.

14. $x^2 + 2x + y^2 - 6y = 6$. (Hint: Add something to both sides of the equation to get it into the form $(x-a)^2 + (y-b)^2 = r^2$.)

15. Show that the points that are equidistant from the points $(0, 3)$ and $(6, 0)$ form a straight line. What are the slope and y-intercept of this line?

16. Show that the points that are twice as far from $(6, 0)$ as they are from $(0, 3)$ form a circle. What are the center and radius of this circle?

17–23: Find an equation for the line.

17. The line passing through the points $(-1, 1)$ and $(2, 7)$.

18. The line passing through the points $(-1, 7)$ and $(2, 1)$.

19. The line passing through the points $(-1, 7)$ and $(2, 7)$.

20. The line passing through the points $(2, 1)$ and $(2, 7)$

21. The line through the point $(2, 3)$ that is parallel to the line $x - 3y = 2$.

22. The line through the point $(2, 3)$ that is perpendicular to the line $x - 3y = 2$.

23. The line tangent to the circle $(x-3)^2 + y^2 = 13$ at the point $(1, 3)$. (Hint: The line tangent to a circle at a point is perpendicular to the radius from the center of the circle to that point.)

24. Find the point of intersection of the lines $y = 4x - 5$ and $y = -2x + 4$.

25. Find all points of intersection of the line $y = -2x + 1$ and the circle $(x+1)^2 + (y-1)^2 = 4$.

26. Show that the points $(0, 0)$, $(3, 1)$, $(1, 7)$, and $(-2, 6)$ are the vertices of a rectangle. What is the area of this rectangle?

27. Give an alternative solution to Example 1.2.4 by computing the area of the gray region in Figure 1.13 and then subtracting the area of the part of that region that is striped.

1.3   Functions

High school mathematics is mostly concerned with numbers and operations on numbers, such as addition, subtraction, multiplication, and division. One of the most distinctive features of calculus is that it is primarily concerned with operations on functions rather than operations on numbers.

A *function* is a rule that associates, with every number, exactly one corresponding number. For example, the rule might associate with any number x the square of that number, x^2. Or it might associate with each number x the number $3x^5 - 7x$.

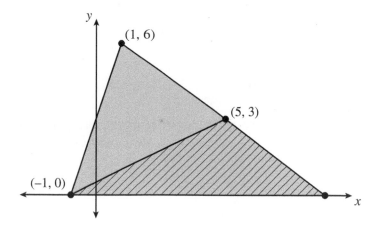

Figure 1.13: An alternative solution to Example 1.2.4.

We will often use letters to stand for functions, most often the letters f and g. If f is a function, then we write $f(x)$ to denote the number associated with x by the function f. For example, if f and g are the two functions defined in the last paragraph, then we could say that for every number x, $f(x) = x^2$ and $g(x) = 3x^5 - 7x$. The fact that these equations hold for *every* number x means that we can substitute any number we please for x. For example, we have $f(5) = 5^2 = 25$ and $g(2) = 3(2^5) - 7(2) = 82$. In other words, the number associated with 5 by the function f is 25, and the number associated with 2 by the function g is 82. We can also replace x in these formulas by a more complicated formula. For example, for any numbers a and b we have $g(a - 3b) = 3(a - 3b)^5 - 7(a - 3b)$.

In a formula like $f(x) = x^2$, it may be helpful to think of the number x as the *input* to the function f. The function can be thought of as an operation or calculation that is performed on this input to find the number associated with x by the function—in this case, x^2. We therefore sometimes speak of the function f as being *applied* to the number x, and we say that $f(x)$ is the *result of applying the function f to x*. It is also sometimes called the *value of f at x*, or simply *f of x*. Another image that may be helpful is to think of a function f as a machine, as shown in Figure 1.14. We feed a number x into the machine, and the machine spits out the number $f(x)$, which in this example is x^2. For example, if we feed the number 5 into this machine, the number 25 comes out.

Figure 1.14: The function f viewed as a machine.

We usually define a function f by giving a formula that tells us, for any number x, the value of $f(x)$. For example, here are two more definitions of functions:

$$\text{For every number } x, \quad f(x) = \frac{x^3 - 3}{x^2 + 2}.$$

$$\text{For every number } x, \quad g(x) = \sqrt[3]{x}.$$

However, a function need not be defined by a formula. Any rule that specifies unambiguously, for each number x, the corresponding value $f(x)$ counts as a function. For example, here are two more examples of functions:

For every number x, $f(x) = $ the greatest integer that is less than or equal to x.

$$\text{For every number } x, \quad g(x) = \begin{cases} x, & \text{if } x \geq 0, \\ -x, & \text{if } x < 0. \end{cases}$$

Of course, you recognize $g(x)$ as $|x|$, the absolute value of x. Thus, we may speak of the *absolute value function*. The function f is called the *greatest integer function*, and $f(x)$ is often denoted $\lfloor x \rfloor$. For example, $\lfloor 3.7 \rfloor = 3$, $\lfloor -3.7 \rfloor = -4$, and $\lfloor 6 \rfloor = 6$.

Although we usually define a function f by specifying the value of $f(x)$ for every x, it is important to understand the distinction between the expressions f and $f(x)$. The letter f is the name of a function, and the expression $f(x)$ denotes the number associated with x by f. In particular, f is a function, while $f(x)$ is a number.

Can we define a function f by saying that for every number x, $f(x) = \sqrt{x}$? There is a problem with this definition: the formula for $f(x)$ can't be used for *every* number x, since $\sqrt{x}$ is undefined if $x < 0$. But we will certainly want to be able to work with formulas like $\sqrt{x}$ in calculus. We are thus led to extend our definition by allowing a function to be a rule that applies to only *some* numbers. The set of numbers to which the function applies is called the *domain* of the function, and it could be any subset of $\mathbb{R}$, the set of all real numbers. Thus, our new definition is that if $D \subseteq \mathbb{R}$, then a function f with domain D is a rule that associates, with each number $x \in D$, exactly one corresponding number $f(x)$. We write $f : D \to \mathbb{R}$ to indicate that f is a function with domain D; in other words, it is a rule that associates, with each element of D, a corresponding element of $\mathbb{R}$. In all of our previous examples we had $D = \mathbb{R}$, but sometimes our functions will have smaller domains. For example, the equation $f(x) = \sqrt{x}$ can be used to define a function with domain $\{x : x \geq 0\} = [0, \infty)$. For this example we would say $f : [0, \infty) \to \mathbb{R}$.

Here are two more examples of functions whose domains are not all of $\mathbb{R}$. In each case, we define the function by first specifying the domain and then giving the result of applying the function to any element of the domain.

The domain of f is $[-1, 1]$, and for all $x \in [-1, 1]$, $f(x) = \sqrt{1 - x^2}$.

The domain of g is $(0, \infty)$, and for all $x \in (0, \infty)$, $g(x) = 1/x$.

For example, $f(1/2) = \sqrt{3}/2$ and $g(7) = 1/7$. However, $g(-7)$ is undefined, because -7 is not in the domain of g; the equation $g(x) = 1/x$ was declared to hold only for $x \in (0, \infty)$, so it does not apply to $x = -7$.

There are certain shortcuts that mathematicians often take when defining a function. First, mathematicians often don't specify the domain of the function they are defining. If no domain is specified, it is understood that the domain is the set of all values for which the definition of the function makes sense. Second, when defining a function f by giving a formula for $f(x)$, mathematicians usually don't say explicitly that this formula applies to all x in the domain of the function. Again, this is to be understood. Finally, mathematicians sometimes write things like "the function $f(x) = x^2$," when what they really mean is "the function f defined by the equation $f(x) = x^2$."

Example 1.3.1. Find the domains of the functions f, g, and h defined by the following formulas:

$$f(x) = \frac{x^2 - 4}{x - 2},$$
$$g(x) = x + 2,$$
$$h(x) = \sqrt{x} + \frac{1}{\sqrt{3 - x}}.$$

Solution. The formula for $f(x)$ makes sense for all values of x except $x = 2$, for which we have a 0 in the denominator. Thus, the domain of f is $\{x : x \neq 2\}$; we could also write this set as the union of two intervals, $(-\infty, 2) \cup (2, \infty)$. Of course, the formula for $g(x)$ makes sense for all values of x, so the domain of g is $\mathbb{R}$.

The formula for $h(x)$ involves the expressions $\sqrt{x}$ and $\sqrt{3 - x}$, and these are undefined for some values of x. In order for $\sqrt{x}$ to make sense we must have $x \geq 0$, and in order for $\sqrt{3 - x}$ to make sense we must have $3 - x \geq 0$, and therefore $x \leq 3$. But there is one more restriction; if $x = 3$ then the formula for $h(x)$ will involve division by 0, so we must exclude 3 from the domain. Thus, the domain of h is the half-open interval $[0, 3)$. □

The formula for $f(x)$ in Example 1.3.1 can be simplified by factoring the numerator and then canceling:

$$f(x) = \frac{x^2 - 4}{x - 2} = \frac{(x - 2)(x + 2)}{x - 2} = x + 2.$$

You might think, therefore, that the functions f and g in Example 1.3.1 are really the same. However, the algebraic simplification we have done is correct only for $x \neq 2$; when $x = 2$, the fraction $(x - 2)(x + 2)/(x - 2)$ is undefined, whereas $x + 2$ is equal to 4. In other words, for all $x \neq 2$ we have $f(x) = g(x)$, but $f(2)$ is undefined, whereas $g(2) = 4$. Thus, we will insist that f and g are different functions. It may seem that we are being unnecessarily fussy about this, but as we will see in the next chapter, maintaining this distinction will be crucial to understanding some of the central concepts of calculus.

The *graph* of a function f is the graph of the equation $y = f(x)$. For example, if we define a function f by the equation $f(x) = 2x - 3$, then the graph of f is the graph of the equation $y = 2x - 3$, which is the line in Figure 1.9a. In the equation $y = f(x)$, the variable x is sometimes called the *independent variable*, and y is the *dependent variable*. The reason for this terminology is that we are free to choose any value for x (as long as the value is in the domain of f), but once a value for x has been chosen, the equation $y = f(x)$ determines a corresponding value for y. Thus, the value of y *depends* on the value of x, but the value of x doesn't depend on anything. We may sometimes use other letters for the independent and dependent variables, but the letters x and y are used most often.

You can learn a lot about a function f very easily by looking at its graph, and we will therefore use graphs of functions extensively throughout this book. Perhaps most important is the fact that the value of $f(a)$ for any particular number a can be read off from the graph. To see how, simply note that a point (a, b) will be on the graph of f if and only if the assignment of values $x = a$, $y = b$ makes the equation $y = f(x)$ true, that is, if and only if $b = f(a)$. Thinking of the function f as a machine, this means that a point (a, b) will be on the graph of f if and only if when you feed the number a into the machine, the number b comes out. Thus, if a is in the domain of f then there will be exactly one point on the graph of f whose x-coordinate is a; the y-coordinate of that point will be $f(a)$. If a is not in the domain, then there will be no point on the graph with x-coordinate a.

Another way to say this is that if a is in the domain of f then the vertical line $x = a$ intersects the graph of f exactly once, at the point $(a, f(a))$, and if a is not in the domain of f then the vertical line $x = a$ does not intersect the graph of f at all. Thus, no vertical line intersects the graph of f more than once. Conversely, if some curve does not intersect any vertical line more than once, then the curve is the graph of some function f. We can define f by saying that the domain of f is the set of numbers a such that the line $x = a$ intersects the curve, and for each such number a, $f(a)$ is the y-coordinate of the unique point where the line $x = a$ intersects the curve. Thus, we can say that a curve is the graph of a function if and only if no vertical line intersects the curve more than once. This is sometimes called the *vertical line test*.

We have seen that the domain of a function f is equal to the set of all numbers that appear as the x-coordinate of a point on the graph of f. There is also a name for the set of numbers appearing as the y-coordinate of a point on the graph of f; it is called the *range* of f. Thinking of f as a machine, we could say that the domain of f is the set of all numbers that can be fed into the machine, and the range is the set of all numbers that come out.

Let's try out these ideas in some examples. The curve in Figure 1.15a passes the vertical line test, so it is the graph of a function f. Although we don't have a formula for $f(x)$, we can see that $f(-1) = 3$, $f(1) = 2$, $f(3) = 4$, and $f(4) = 5$, because the points $(-1, 3)$, $(1, 2)$, $(3, 4)$, and $(4, 5)$ are on the graph. We can also see that the domain of f is the interval $[-1, 5)$, and the range is the interval $[2, 5]$. On the other hand, the black curve in Figure 1.15b is not the graph of a function, because it fails the vertical line test. The line $x = 2$, shown in red, crosses the curve three times.

Figure 1.16 shows the graphs of the functions $g(x) = |x|$ and $h(x) = \lfloor x \rfloor$. The graph of g is the graph of the equation $y = |x|$, and filling in the definition of $|x|$ this means

$$y = \begin{cases} x, & \text{if } x \geq 0, \\ -x, & \text{if } x < 0. \end{cases}$$

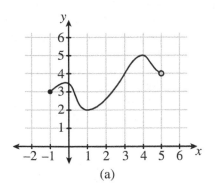
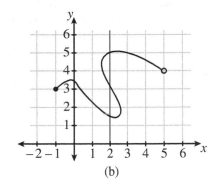

(a)   (b)

Figure 1.15: The curve in (a) is the graph of a function, but the curve in (b) is not.

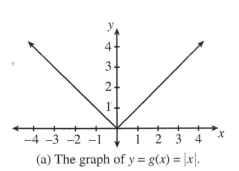
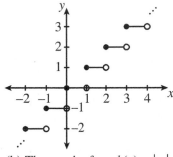

(a) The graph of $y = g(x) = |x|$.   (b) The graph of $y = h(x) = \lfloor x \rfloor$.

Figure 1.16: Graphs of the functions $g(x) = |x|$ and $h(x) = \lfloor x \rfloor$.

For $x \geq 0$ this is the same as the graph of $y = x$, which is a line with slope 1 and passing through the origin. For $x < 0$, it is the same as the graph of $y = -x$, a line through the origin with slope -1. Putting these two pieces together gives us the graph shown in Figure 1.16a. Similarly, the equation $y = h(x) = \lfloor x \rfloor$ means that $y = 0$ when $0 \leq x < 1$, $y = 1$ when $1 \leq x < 2$, and so on. This leads to the infinitely many horizontal line segments that make up the graph of h in Figure 1.16b. The domains of g and h are both $\mathbb{R}$. The range of g is the interval $[0, \infty)$, and the range of h is the set of all integers, which is usually denoted $\mathbb{Z}$.

To illustrate the usefulness of graphs of functions, let us return to Example 1.1.6, in which we solved the inequality $2|x| - 3 \geq |x - 1|$. Define functions f and g by the formulas

$$f(x) = 2|x| - 3, \qquad g(x) = |x - 1|.$$

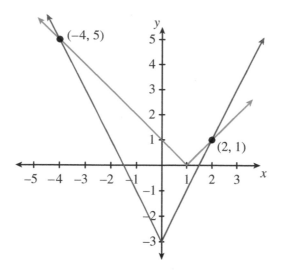

Figure 1.17: The red curve is the graph of $y = f(x) = 2|x| - 3$, and the blue curve is the graph of $y = g(x) = |x - 1|$.

To graph f, we graph the equation $y = 2|x| - 3$. As in Figure 1.16a, this consists of the line $y = 2x - 3$ for $x \geq 0$ together with the line $y = -2x - 3$ for $x < 0$. Thus, the graph of f is the red curve in Figure 1.17, and similar reasoning shows that the graph of g is the blue curve. We will leave it to you to compute the intersection points that are shown in Figure 1.17. The inequality from Example 1.1.6 is $f(x) \geq g(x)$, and since $f(x)$ and $g(x)$ give the y-coordinates of points on the red and blue curves, the solution set of the inequality is just the set of x values for which the red curve is at least as high as the blue curve. Looking at Figure 1.17, it is clear that this solution set is $(-\infty, -4] \cup [2, \infty)$, exactly as we found in Example 1.1.6.

Since graphs of functions are so useful, given a function it will be important to be able to draw its graph. We will see later how calculus can be used to draw accurate graphs of functions, but for now we will just mention a method that can be used to get started on drawing the graph of a function f. Pick a collection of values of x in the domain of the function and compute, for each value of x, the corresponding value of $y = f(x)$. This will give you the coordinates of a collection of points on the graph of f.

For example, in Figure 1.18 we have computed a table of values for the function f defined by the equation $f(x) = x^3 - 4x^2 + 2x + 2$ and plotted the corresponding points. The points seem to lie along a smooth curve, so we have filled in such a curve passing through the plotted points as the graph of f. Filling in this curve is, at this point, guesswork. The only points we can be sure are on the graph are the ones whose coordinates we computed. The graph of f might have extra wiggles between the plotted points, and it might have further bends beyond the part of the graph we have drawn. Of course, we could plot more points to try to detect further features of the graph, but no matter how many points we plot we will always be guessing if we fill in a curve connecting these points. One important application of calculus is that it will allow us to get many important features of the graph of a function right *without guesswork*.

x	$y = f(x)$
-1	-5
$-1/2$	$-1/8$
0	2
$1/2$	$17/8$
1	1
$3/2$	$-5/8$
2	-2
$5/2$	$-19/8$
3	-1
$7/2$	$23/8$

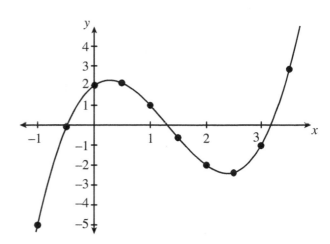

Figure 1.18: Graphing the function $f(x) = x^3 - 4x^2 + 2x + 2$.

You might think that another way to get an accurate graph of a function without guesswork is to use a computer or graphing calculator. But what does a computer do when you ask it to graph a function? The answer is that it simply plots some points and then draws a smooth curve through those points! Thus, the computer-produced graph is also based on guesswork. In practice, computer-drawn graphs are usually fairly accurate, because the computer is able to plot many more points than you could plot by hand. For example, in Figure 1.18 we plotted 10 points, but a computer might plot hundreds of points when graphing a function. But computers do sometimes miss important features of graphs of functions. In contrast, the calculus-based methods we will learn later *never* miss important features of graphs of functions, because they involve no guesswork.

Exercises 1.3

1–9: Let f, g, and h be the functions defined by the following equations:

$$f(x) = x^2 - 4, \qquad g(x) = \sqrt{x^2 + 5}, \qquad h(x) = \lfloor 5x - 2 \rfloor.$$

Evaluate each expression, simplifying as much as possible.

1. $f(3)$.

2. $g(2)$.

3. $h(2.7)$.

4. $f(2 - a)$.

5. $g(2x + 1)$.

6. $h(\lfloor x \rfloor + 1)$.

7. $f(g(x))$.

8. $g(f(x))$.

9. $h(h(x))$.

10–18: Find the domain of each function. Write your answer as a union of intervals, if possible.

10. $f(x) = \dfrac{2x - 5}{5x - 2}$.

11. $g(x) = \dfrac{x^2 - 9}{x^2 + x - 6}$.

12. $f(x) = \sqrt{2 - x}$.

13. $g(x) = \sqrt{16 - x^2}$.

14. $h(x) = \dfrac{\sqrt{16 - x^2}}{\sqrt{2 - x}}$.

15. $h(x) = \sqrt{\dfrac{16 - x^2}{2 - x}}$.

16. $f(x) = \sqrt{2 - |x|}$.

17. $g(x) = \sqrt{2 - \lceil x \rceil}$.

18. $f(x) = \sqrt{x^4 - x^2}$.

19. Which of the graphs in Figure 1.19 are graphs of functions?

20–30: Graph each function.

20. $f(x) = x - 2$.

21. $g(x) = 2 - x$.

22. $h(x) = \dfrac{x + 4}{3}$.

23. $f(x) = \begin{cases} 3 - x, & \text{if } x > 2, \\ 1, & \text{if } -2 \le x \le 2, \\ x + 4, & \text{if } x < -2. \end{cases}$

24. $f(x) = 2|x| - 1$.

25. $g(x) = |2x - 1|$.

26. $h(x) = |2|x| - 1|$.

27. $f(x) = 2\lfloor x \rfloor - 1$.

28. $g(x) = \lfloor 2x - 1 \rfloor$.

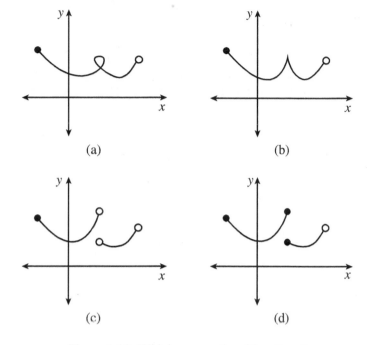

Figure 1.19: Which are graphs of functions?

29. $f(x) = |\lfloor x \rfloor|$.

30. $g(x) = \lfloor |x| \rfloor$.

31. The graph of a function f is shown in Figure 1.20.

    (a) What is the domain of f?

    (b) What is the range of f?

    (c) What is $f(2)$?

    (d) What is $f(5)$?

    (e) Solve the inequality: $f(x) > 1$. Write the solution set as a union of intervals.

    (f) For what value of x, if any, is $f(x)$ smallest?

    (g) For what value of x, if any, is $f(x)$ largest?

32. Let $f(x) = x^5 - 30x^{5/3} + 36x - 8\sqrt[3]{x}$.

    (a) Use a calculator to evaluate $f(x)$ at $x = -2, -1, 0, 1, 2$. Plot the corresponding points and draw a smooth curve through these points to get a guess at the shape of the graph of f.

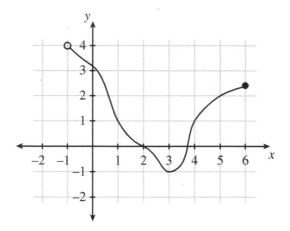

Figure 1.20: The graph of a function f.

(b) Evaluate $f(x)$ at $x = -3/2, -1/2, 1/2, 3/2$ and plot the corresponding points. Do these points lie on the curve you drew in part (a)? Adjust your graph to accommodate these new points.

(c) Evaluate $f(-1/40)$ and $f(1/40)$. Do you need to change your graph again?

1.4   Combining Functions

Sometimes we will want to combine functions to create new functions. For example, suppose f and g are functions. We can define a new function h by the equation

$$h(x) = f(x) + g(x). \tag{1.7}$$

The function h is called the *sum* of f and g, and it is denoted $f + g$. Thus, we can write $h = f + g$, and substituting $f + g$ for h in equation (1.7) we have

$$(f + g)(x) = f(x) + g(x). \tag{1.8}$$

It is important to understand that the two plus signs in equation (1.8) have different meanings. On the right-hand side of the equation, the plus sign appears between $f(x)$ and $g(x)$, which are numbers, and the plus sign stands for ordinary addition of numbers. But on the left-hand side, it appears between f and g, which are functions, and it stands for a new kind of addition, addition of functions. The expression $f + g$ is the name of a function—the function that we originally called h in equation (1.7). Equation (1.8) tells us how to apply the function $f + g$ to a number x. Figure 1.21 shows the function $f + g$ as a machine, constructed from machines for the functions f and g.

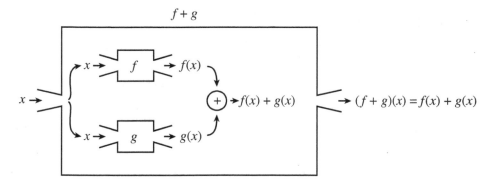

Figure 1.21: The function $f + g$, viewed as a machine.

For example, suppose f and g are the functions defined by the equations

$$f(x) = \sqrt{x}, \qquad g(x) = \frac{1}{\sqrt{3-x}}.$$

Then $f + g$ is the function defined by

$$(f+g)(x) = f(x) + g(x) = \sqrt{x} + \frac{1}{\sqrt{3-x}}.$$

This is just the function that we called h in Example 1.3.1. Notice that in order for the formula for $(f + g)(x)$ to make sense for some particular value of x, both $f(x)$ and $g(x)$ must make sense; in other words, x must be in the domains of both f and g. It follows that the domain of $f + g$ is the intersection of the domains of f and g. We will leave it for you to verify that in this example, the domain of f is $[0, \infty)$ and the domain of g is $(-\infty, 3)$, so the domain of $f + g$ is $[0, \infty) \cap (-\infty, 3) = [0, 3)$. Of course, this is in agreement with what we got for the domain of h in Example 1.3.1.

The ideas we have just discussed for addition of functions can also be applied to subtraction, multiplication, and division. We summarize these ideas with the following definition.

Definition 1.4.1. Suppose f and g are functions. Then the functions $f + g$, $f - g$, $f \cdot g$, and f/g are defined as follows:

$$(f + g)(x) = f(x) + g(x),$$
$$(f - g)(x) = f(x) - g(x),$$
$$(f \cdot g)(x) = f(x) \cdot g(x),$$
$$(f/g)(x) = f(x)/g(x).$$

Notice that if the domains of f and g are D_f and D_g, then the domains of $f + g$, $f - g$, and $f \cdot g$ are all $D_f \cap D_g$. For f/g, we must add the further restriction that $g(x)$

must not be 0. Thus, the domain of f/g is

$$D_{f/g} = \{x : x \in D_f \cap D_g \text{ and } g(x) \neq 0\}.$$

There is one more important way of combining functions that we will make extensive use of:

Definition 1.4.2. Suppose f and g are functions. Then the *composition* of f and g is the function $f \circ g$ defined by the formula

$$(f \circ g)(x) = f(g(x)).$$

Thus, to apply the function $f \circ g$ to a number x, we first apply the function g to x to get $g(x)$, and then we apply f to $g(x)$ to get $f(g(x))$. Figure 1.22 shows how machines for the functions f and g can be combined to construct a machine for $f \circ g$.

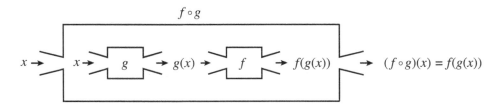

Figure 1.22: The function $f \circ g$, viewed as a machine.

For example, let $f(x) = 3x + 1$ and $g(x) = x^2 - 2$. To compute $(f \circ g)(3) = f(g(3))$, we first compute $g(3) = 3^2 - 2 = 7$, and then we apply f to get $f(g(3)) = f(7) = 3(7) + 1 = 22$. More generally, for every number x we have

$$(f \circ g)(x) = f(g(x)) = f(x^2 - 2) = 3(x^2 - 2) + 1 = 3x^2 - 5.$$

Notice that

$$(g \circ f)(x) = g(f(x)) = g(3x + 1) = (3x + 1)^2 - 2 = 9x^2 + 6x - 1,$$

which is an entirely different function. For example, we have seen that $(f \circ g)(3) = 22$, but $(g \circ f)(3) = 9(3^2) + 6(3) - 1 = 98$. Thus, $f \circ g$ is not the same as $g \circ f$.

For any functions f and g, in order for the formula $(f \circ g)(x) = f(g(x))$ to make sense, x must be in the domain of g, so that $g(x)$ is defined, and then $g(x)$ must be in the domain of f, so that $f(g(x))$ is defined. Thus, if the domains of f and g are D_f and D_g, as before, then the domain of $f \circ g$ is

$$D_{f \circ g} = \{x : x \in D_g \text{ and } g(x) \in D_f\}.$$

Example 1.4.3. Let f, g, and h be the following functions:

$$f(x) = \frac{1}{x-3}, \qquad g(x) = \frac{x}{x^2-9}, \qquad h(x) = \sqrt{x+4}.$$

Find $(f-g)(x)$, $(f/g)(x)$, $((f-g)\circ h)(x)$, and $(f-(g\circ h))(x)$. What are the domains of these functions?

Solution.

$$(f-g)(x) = f(x) - g(x) = \frac{1}{x-3} - \frac{x}{x^2-9} = \frac{x+3}{x^2-9} - \frac{x}{x^2-9} = \frac{3}{x^2-9},$$

where we got a common denominator in order to be able to combine the fractions and simplify. We also do some algebraic simplifications in our next calculation:

$$(f/g)(x) = \frac{f(x)}{g(x)} = \frac{1/(x-3)}{x/(x^2-9)} = \frac{1}{x-3} \cdot \frac{(x-3)(x+3)}{x} = \frac{x+3}{x} = 1 + \frac{3}{x}.$$

Using our answer for $(f-g)(x)$, we can compute that

$$((f-g)\circ h)(x) = (f-g)(h(x)) = (f-g)(\sqrt{x+4}) = \frac{3}{(\sqrt{x+4})^2 - 9} = \frac{3}{x-5}.$$

Finally, the difference in grouping in the third function leads to a different answer:

$$(f-(g\circ h))(x) = f(x) - (g\circ h)(x) = f(x) - g(h(x)) = f(x) - g(\sqrt{x+4})$$

$$= \frac{1}{x-3} - \frac{\sqrt{x+4}}{(\sqrt{x+4})^2 - 9} = \frac{1}{x-3} - \frac{\sqrt{x+4}}{x-5}.$$

To find the domains of these functions, we first note that the domains of f, g, and h are

$$D_f = \{x : x \neq 3\} = (-\infty, 3) \cup (3, \infty),$$
$$D_g = \{x : x \neq \pm 3\} = (-\infty, -3) \cup (-3, 3) \cup (3, \infty),$$
$$D_h = \{x : x + 4 \geq 0\} = \{x : x \geq -4\} = [-4, \infty).$$

The domain of $f - g$ is

$$D_{f-g} = D_f \cap D_g = \{x : x \neq \pm 3\} = (-\infty, -3) \cup (-3, 3) \cup (3, \infty).$$

The domain of f/g is the same, except that we must also rule out any values of x for which $g(x) = 0$. Since the only such value is $x = 0$, we find that

$$D_{f/g} = \{x : x \neq \pm 3, 0\} = (-\infty, -3) \cup (-3, 0) \cup (0, 3) \cup (3, \infty).$$

You might think that, since our formula for $(f/g)(x)$ simplified to $1 + 3/x$, the only number missing from the domain of f/g should be 0, but this is incorrect. As in Example 1.3.1, our algebraic simplifications have hidden the fact that the formula $f(x)/g(x)$ makes no sense for $x = \pm 3$.

Similarly, since our formula for $((f - g) \circ h)(x)$ simplified to $3/(x - 5)$, you might think at first that the domain of $(f - g) \circ h$ is $\{x : x \neq 5\}$, but this would be incorrect. The correct calculation is

$$D_{(f-g) \circ h} = \{x : x \in D_h \text{ and } h(x) \in D_{f-g}\} = \{x : x \geq -4 \text{ and } \sqrt{x+4} \neq \pm 3\}.$$

Now, $\sqrt{x+4}$ can never be equal to -3, since the notation $\sqrt{x+4}$ means the *nonnegative* square root of $x + 4$, so we always have $\sqrt{x+4} \geq 0$. But solving $\sqrt{x+4} = 3$ we get $x + 4 = 9$, or $x = 5$. So the domain of $(f - g) \circ h$ is

$$D_{(f-g) \circ h} = \{x : x \geq -4 \text{ and } x \neq 5\} = [-4, 5) \cup (5, \infty).$$

Finally, the domain of $f - (g \circ h)$ is

$$\begin{aligned} D_{f-(g \circ h)} &= D_f \cap D_{g \circ h} = D_f \cap \{x : x \in D_h \text{ and } h(x) \in D_g\} \\ &= \{x : x \neq 3\} \cap \{x : x \geq -4 \text{ and } \sqrt{x+4} \neq \pm 3\} \\ &= \{x : x \neq 3, x \geq -4, \text{ and } x \neq 5\} = [-4, 3) \cup (3, 5) \cup (5, \infty). \qquad \square \end{aligned}$$

In Example 1.4.3, we started with formulas for the functions f, g, and h and then worked out combinations of these functions. Sometimes it will be useful to be able to go in the opposite direction: given a complicated function, can we figure out how it could have been constructed by combining simpler functions?

Example 1.4.4. Let f, g, and h be defined as follows:

$$f(x) = \frac{x-3}{x^2+1}, \qquad g(x) = \sqrt[3]{x^3+1}, \qquad h(x) = \lfloor (x+7)^5 \rfloor.$$

Write each of these functions as a combination of simpler functions.

Solution. A good way to approach this problem is to imagine that you are computing the value of one of these functions at some particular value of x, and think about what the *last* step in this computation would be. For example, if you were computing $f(x)$ for some particular value of x, you would first compute $x - 3$ and $x^2 + 1$, and then you would divide the first by the second. The last step in this computation is division, which suggests that f could be written as a quotient of two functions. Indeed, if we define $f_1(x) = x - 3$ and $f_2(x) = x^2 + 1$, then

$$(f_1/f_2)(x) = \frac{f_1(x)}{f_2(x)} = \frac{x-3}{x^2+1} = f(x),$$

so $f = f_1/f_2$.

For the function g, to compute $g(x)$ you would first compute $x^3 + 1$ and then take the cube root of that number. The last step is taking the cube root, which suggests that we could write $g = g_1 \circ g_2$, where $g_1(x) = \sqrt[3]{x}$. The quantity that we must take the cube root of is $x^3 + 1$, so we define $g_2(x) = x^3 + 1$ and then check that

$$(g_1 \circ g_2)(x) = g_1(g_2(x)) = g_1(x^3 + 1) = \sqrt[3]{x^3 + 1} = g(x).$$

Finally, the last step in computing $h(x)$ is to apply the greatest integer function, which suggests that $h = h_1 \circ h_2$, where $h_1(x) = \lfloor x \rfloor$. You can check that if $h_2(x) = (x + 7)^5$ then $h = h_1 \circ h_2$. But notice that h_2 can itself be written as a composition of two simpler functions: to compute $h_2(x)$ we first compute $x + 7$ and then raise that number to the fifth power, so $h_2 = h_3 \circ h_4$, where $h_3(x) = x^5$ and $h_4(x) = x + 7$. So we have $h = h_1 \circ h_2 = h_1 \circ (h_3 \circ h_4)$. □

Example 1.4.4 shows that a complicated function can sometimes be written as a combination of simpler functions, and these simpler functions might themselves be expressible as combinations of still simpler functions. This suggests that we ask the question: What are the simplest possible functions?

Surely the simplest functions are those functions f defined by equations of the form $f(x) = c$, for some real number c. For example, consider the function f defined by the equation

$$f(x) = 13.$$

You might think at first that this can't be a definition of a function, because the formula for $f(x)$ doesn't have an x in it. But remember that what this function definition really means is:

For every real number x, $f(x) = 13$.

Thus, f is the rule that associates, with any number x, the number 13. If we think of this function as a machine, then when a number x is fed into the machine, the machine simply ignores the input x and outputs the number 13. The graph of f is the graph of the equation $y = 13$, which is a horizontal line passing through the number 13 on the y-axis. The graph of any function defined by an equation of the form $f(x) = c$, where c is a number, will be a horizontal line. Such functions are called *constant functions*.

The simplest function that is not a constant function is the function g defined by the formula

$$g(x) = x.$$

A machine for computing $g(x)$ would take its input x and simply spit it back out unchanged. The output of this machine is identical to its input, so the function g is called the *identity function*. Its graph is the line $y = x$, a line with slope 1 passing through the origin $(0, 0)$.

Starting with just constant functions and the identity function, we can construct many more complex functions. For example, if g is the identity function then

$$(g \cdot g)(x) = g(x) \cdot g(x) = x \cdot x = x^2,$$
$$((g \cdot g) \cdot g)(x) = (g \cdot g)(x) \cdot g(x) = x^2 \cdot x = x^3,$$
$$(((g \cdot g) \cdot g) \cdot g)(x) = ((g \cdot g) \cdot g)(x) \cdot g(x) = x^3 \cdot x = x^4,$$

etc.

Thus, any function of the form $f(x) = x^n$, where n is a positive integer, can be obtained by multiplying the identity function by itself repeatedly. Multiplying such a function by a constant function, we can obtain any function of the form $h(x) = cx^n$, where c is a real number. For example, the functions $f_1(x) = 3x^5$, $f_2(x) = -5x^2$, and $f_3(x) = 7x$ can all be obtained by multiplying constant functions and the identity function. Adding these functions, we obtain the function $((f_1 + f_2) + f_3)(x) = 3x^5 - 5x^2 + 7x$. In fact, it should be clear now that if we start with constant functions and the identity function and use multiplication and addition of functions, we can obtain any function of the form

$$p(x) = c_n x^n + c_{n-1} x^{n-1} + \cdots + c_1 x + c_0,$$

where $c_0, c_1, \ldots, c_n$ are all real numbers. Functions of this form are called *polynomials*, and the number c_k is called the *coefficient* of x^k in the polynomial. The highest power of x appearing in the formula with a nonzero coefficient is called the *degree* of the polynomial. For example, $p(x) = -2x^4 + 8x^2 + 3$ is a polynomial of degree 4. Functions of the form $h(x) = mx + b$ are called *linear functions*, because their graphs are straight lines. If $m \neq 0$ then h is a polynomial of degree 1. If $m = 0$, then it is a constant function, which can also be thought of as a polynomial of degree 0. Polynomials of degree 2 have the form $h(x) = ax^2 + bx + c$, where $a \neq 0$. Such functions are called *quadratic functions*, and their graphs are parabolas.

Why does it matter that polynomials can be built up from constant functions and the identity function by addition and multiplication? Later in this book we will introduce several new operations on functions, and we will want to learn how to apply these operations to many functions. A strategy we will use repeatedly is to first learn how to apply a new operation to constant functions and the identity function, and then learn how to apply the operation when functions are combined, for example by addition or multiplication. What we have just discovered is that this strategy will allow us to apply our new operation to any polynomial. In fact, we will push this idea much further in order to deal with a very wide range of functions.

If we divide two polynomials, we get a function of the form

$$f(x) = \frac{p(x)}{q(x)},$$

where p and q are polynomials. Functions of this form are called *rational functions*. For instance, the function f in Example 1.4.4 is a rational function. We leave it as an exercise for you to check that if you add, subtract, multiply, or divide two rational functions, the result can be simplified so that it is once again in the form of a rational function. In fact, even a composition of two rational functions can be expressed as a rational function. Thus, we have reached the end of the line: starting with constant functions and the identity function, the functions you can construct by addition, subtraction, multiplication, division, and composition are the rational functions.

Of course, there are many functions that are not rational functions. For example, the function $f(x) = \sqrt{x}$ is not a rational function. Combining this function with previously constructed functions, we can construct functions like these:

$$g(x) = \sqrt{\frac{x+1}{x-1}}, \qquad h(x) = \frac{x^2 - \sqrt{x^2 + 1}}{x - 7}, \qquad j(x) = \sqrt{x + \sqrt{x+1}}.$$

As we continue to study more functions, we will always be able to combine any new function we introduce with functions we have already studied to create and investigate more and more complicated functions.

Before closing this chapter we briefly mention a family of functions that we will want to include in all of our investigations: the trigonometric functions. We assume that you have studied trigonometric functions before, but here we review their definitions and some of their most important properties.

When defining the trigonometric functions, it is traditional to use the Greek letter θ (theta) as the independent variable. Some people define the trigonometric functions in terms of right triangles, but we prefer to base the definitions on the unit circle—the circle of radius 1 centered at the origin, whose equation is $x^2 + y^2 = 1$. Let P be the point $(1, 0)$, which is on the unit circle. For any number θ, to compute the cosine and sine of θ, we begin at the point P and travel a distance of θ units counterclockwise along the circle, reaching a point Q. We define $\cos(\theta)$ to be the x-coordinate of Q, and $\sin(\theta)$ the y-coordinate, as shown in Figure 1.23. (A negative value of θ means that we should go in the opposite direction. In other words, if θ is negative, we travel clockwise $|\theta|$ units.) When working with trigonometric functions we often leave out the parentheses and write $\cos\theta$ and $\sin\theta$.

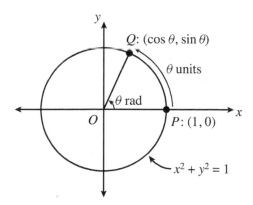

Figure 1.23: The definitions of $\cos\theta$ and $\sin\theta$.

We have specified the point Q in Figure 1.23 by giving the length of the arc PQ, but we could also think of Q as being determined by the measure of the angle POQ, where O is the origin. In calculus, we always measure angles in radians, not degrees. In Figure 1.23, the measure of angle POQ is θ radians; in other words, the radian measure of an angle is, by definition, the length of the arc of the unit circle cut off by the angle.

Thus, in Figure 1.23 the number θ has two meanings: it is the length of the arc PQ, and it is also the measure, in radians, of the angle POQ. Since the circumference of the unit circle is 2π, 2π radians is the same as 360 degrees. Thus, an angle of π radians is 180 degrees, and an angle of $\pi/2$ radians is 90 degrees, a right angle.

Some facts about the cosine and sine functions are clear from the definition. Since the point $(\cos\theta, \sin\theta)$ is always on the unit circle, we have the identity

$$\cos^2\theta + \sin^2\theta = 1,$$

where $\cos^2\theta$ and $\sin^2\theta$ mean $(\cos\theta)^2$ and $(\sin\theta)^2$. It should also be clear that $-1 \leq \cos\theta \leq 1$ and $-1 \leq \sin\theta \leq 1$. If we start at the point Q in Figure 1.23 and travel an additional distance 2π counterclockwise around the unit circle, then we just go all the way around the circle and end up back at Q. It follows that $\cos(\theta + 2\pi) = \cos\theta$ and $\sin(\theta + 2\pi) = \sin\theta$. We say that cos and sin are *periodic*, with *period* 2π. And if we reflect Q across the x-axis we reach the point $(\cos(-\theta), \sin(-\theta))$. Therefore $\cos(-\theta) = \cos\theta$ and $\sin(-\theta) = -\sin\theta$; we say that cos is an *even* function, and sin is an *odd* function. The graphs of the functions cos and sin are shown in Figure 1.24, and some important values of these functions are given in Table 1.1.

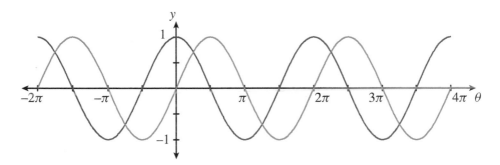

Figure 1.24: The graphs of $y = \cos\theta$ (red) and $y = \sin\theta$ (blue).

θ	$\cos\theta$	$\sin\theta$
0	1	0
$\pi/6$	$\sqrt{3}/2$	$1/2$
$\pi/4$	$\sqrt{2}/2$	$\sqrt{2}/2$
$\pi/3$	$1/2$	$\sqrt{3}/2$
$\pi/2$	0	1

Table 1.1: Some important values of the sine and cosine functions.

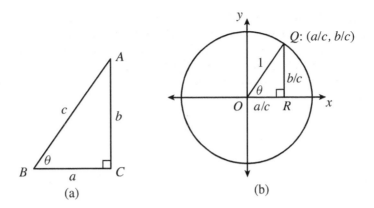

Figure 1.25: Triangle ABC in (a) is similar to triangle QOR in (b).

The other trigonometric functions are tangent, cotangent, secant, and cosecant, and they are defined as follows:

$$\tan \theta = \frac{\sin \theta}{\cos \theta}, \qquad \cot \theta = \frac{\cos \theta}{\sin \theta},$$

$$\sec \theta = \frac{1}{\cos \theta}, \qquad \csc \theta = \frac{1}{\sin \theta}.$$

Although we have not used right triangles to define the trigonometric functions, it is important to know how the functions can be used to relate the sides and angles in a right triangle. In the right triangle in Figure 1.25a, we have $\cos \theta = a/c$, $\sin \theta = b/c$, and $\tan \theta = b/a$. You can see this by observing that triangle ABC in Figure 1.25a is similar to triangle QOR in Figure 1.25b. You can now read off $\cos \theta$ and $\sin \theta$ from the coordinates of the point Q, using the definitions of the cosine and sine functions.

Exercises 1.4

1–8: Let f, g, and h be the functions defined by the following equations:

$$f(x) = 3 - \frac{2}{x}, \qquad g(x) = \frac{x^2 - 4}{x + 2}, \qquad h(x) = \sqrt{3 - x}.$$

Find formulas for the following functions, simplifying as much as possible. Also, specify the domain of each function.

1. $(f + g)(x)$.

2. $(g \cdot h)(x)$.

3. $(g/f)(x)$.

4. $(f \circ g)(x)$.

5. $(g \circ f)(x)$.

6. $(h \circ f)(x)$.

7. $(g \circ h)(x)$.

8. $((g/f) \circ h)(x)$.

9–14: Let f, g, and h be the functions defined by the following equations:

$$f(x) = \cot x, \qquad g(x) = \frac{\pi}{x}, \qquad h(x) = \lfloor x \rfloor.$$

Find formulas for the following functions, simplifying as much as possible. Also, specify the domain of each function.

9. $(f - g)(x)$.

10. $(f \circ g)(x)$.

11. $(g \circ h)(x)$.

12. $(f \circ h)(x)$.

13. $(g \circ g)(x)$.

14. $(f \circ (g \circ h))(x)$.

15–19: Write each functions as a combination of simpler functions.

15. $f(x) = \sin x + \cos x$.

16. $g(x) = \dfrac{\tan x}{\sqrt{x}}$.

17. $h(x) = \sqrt{9 - x^2}$.

18. $f(x) = \sec(\sqrt{x^3 - 7})$.

19. $g(x) = \sqrt{\dfrac{x + 7}{\cos x}}$.

20. The graph of a function f is shown in Figure 1.26. Let g be the function defined as follows:

$$g(x) = \begin{cases} x^2 - 4, & \text{if } x < 2, \\ x + 1, & \text{if } x \geq 2. \end{cases}$$

(a) Find $(f \circ g)(2)$.

(b) Find $(g \circ f)(2)$.

(c) Find $(f \circ (f + g))(3)$.

(d) Find all x such that $f(x) = 2$.

(e) Find all x such that $(f \circ g)(x) = 2$.

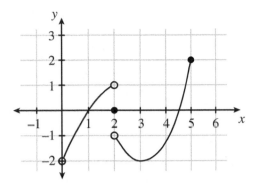

Figure 1.26: The graph of f.

21. Suppose that f, g, and h are functions from $\mathbb{R}$ to $\mathbb{R}$.

    (a) Show that $(f + g) \circ h = (f \circ h) + (g \circ h)$. (Hint: Show that for any number x, $((f + g) \circ h)(x) = ((f \circ h) + (g \circ h))(x)$.)

    (b) Is it necessarily the case that $h \circ (f + g) = (h \circ f) + (h \circ g)$? Give either a proof or a counterexample.

    (c) Is it necessarily the case that $f \circ (g \circ h) = (f \circ g) \circ h$? Give either a proof or a counterexample.

22. Show that for every angle α,

$$\tan^2 \alpha + 1 = \sec^2 \alpha.$$

23. In Figure 1.27, some angles and lengths are labeled.

    (a) Find the angles and lengths labeled with question marks in the figure.

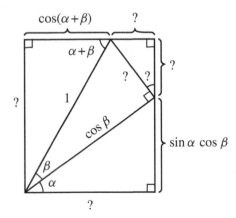

Figure 1.27: Deriving formulas for $\sin(\alpha + \beta)$ and $\cos(\alpha + \beta)$.

(b) Use the figure to derive the identities:

$$\sin(\alpha + \beta) = \sin \alpha \cos \beta + \cos \alpha \sin \beta,$$
$$\cos(\alpha + \beta) = \cos \alpha \cos \beta - \sin \alpha \sin \beta.$$

24. Derive the identities

$$\sin(2\alpha) = 2 \sin \alpha \cos \alpha, \qquad \cos(2\alpha) = \cos^2 \alpha - \sin^2 \alpha.$$

(Hint: Use Exercise 23.)

25. Derive the identity

$$\tan(\alpha + \beta) = \frac{\tan \alpha + \tan \beta}{1 - \tan \alpha \tan \beta}.$$

(Hint: Use Exercise 23.)

Chapter 2

Limits

2.1 What is Calculus About?

What is calculus about? One way to answer this question is to list topics that are studied in calculus: slopes of tangent lines, velocities and other rates of change, areas, volumes, and so on. But this answer doesn't explain what these topics have in common, and what sets them off from the topics studied in other branches of mathematics such as algebra.

In fact, there is something that all of these topics have in common, and that makes calculus different from algebra. All of the topics studied in calculus involve computing something that, it seems at first, can only be computed approximately and not exactly. For every topic we study, we will begin with a method that can be used to find an approximate answer to the problem we are trying to solve. And there is a second ingredient that will be a part of every calculus topic: given an approximate answer to the problem, there will be a way to make the answer more accurate, although still not exactly correct.

Thus, in calculus we study problems for which we can find better and better approximate answers, but we have no direct way of finding an exact answer. It might seem at first that this is all we can accomplish: that we will have to settle for approximate answers that are good enough for all practical purposes, and we will never know the exact answers to our problems. But in fact we can do better. The central idea in calculus is a very clever method of finding exact answers in these situations in which it seems impossible to do so. The idea is to compute the exact answer by finding *the unique number that the approximations get closer and closer to as they get more accurate.*

There are many paradoxical features of calculus, and we will mention some others as we come to them, but this is the first and most central paradox: calculus is a subject in which we find exact answers by means of approximation. We will call this the *paradox of precision through approximation.*

To explain how it is possible to achieve precision through approximation, let us consider an example. Suppose you throw a ball straight up with an initial velocity of 160 feet per second. Physicists have determined that the height of the ball after t seconds will be $160t - 16t^2$ feet, for $0 \leq t \leq 10$. Of course, this is an example of a function. If we let $h(t)$ stand for the height of the ball, in feet, after t seconds, then h is a function with domain [0, 10], and for $0 \leq t \leq 10$ we have

$$h(t) = 160t - 16t^2.$$

If we think of the ball as moving along a vertical number line, with the origin at ground level and units marked off vertically in feet, then we could also think of $h(t)$ as giving the position of the ball on this number line at time t. The graph of h is shown in Figure 2.1. (Note that this is the graph of h, not a picture of the trajectory of the ball; the ball is thrown straight up, so it simply goes up and then down along our vertical number line, without traveling horizontally at all.)

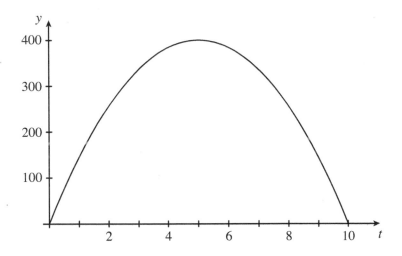

Figure 2.1: The graph of $y = h(t)$.

A number of properties of this function can be seen immediately in the graph. For example, we can see that $h(0) = 0$, $h(5) = 400$, and $h(10) = 0$. As t increases from 0 to 5 to 10, the value of $h(t)$ increases from 0 to 400 and decreases back to 0. In terms of the behavior of the ball, this means that the ball begins at ground level, rises to a height of 400 feet after 5 seconds, and then comes back down, hitting the ground after 10 seconds.

A closer examination of the graph reveals more details of the ball's motion. The first part of the graph—say, for $0 \leq t \leq 2$—slopes up steeply. This tells us that as t increases through the interval [0, 2], $h(t)$ increases quickly, which means that the ball is rising quickly. In the time interval [3, 5], the graph is still sloping upwards, but more gently.

Thus, during this interval, $h(t)$ is increasing more slowly, and therefore the ball is rising more slowly. Similarly, for $t \in [5, 7]$ the ball is falling slowly, and for $t \in [8, 10]$ it is falling more quickly.

Can we quantify these statements? Let us define the *displacement* of the ball during a time interval $[a, b]$ to be $h(b) - h(a)$. The displacement tells us how much the position of the ball has changed over this time interval, with a positive value indicating that it is higher at the end of the time interval than at the beginning, and a negative value indicating that it is lower. The *average velocity* of the ball during this time interval is the displacement divided by $b - a$, the length of the time interval, and it tells us how far the ball has moved per unit of time that has elapsed. Since we are measuring distances in feet and time in seconds in this example, the average velocity will be measured in feet per second, abbreviated ft/sec.

For example, the average velocity of the ball during the time interval $[0, 2]$ is

$$\frac{h(2) - h(0)}{2 - 0} = \frac{256 - 0}{2} = 128 \text{ ft/sec.}$$

But during the interval $[3, 5]$ the average velocity is only

$$\frac{h(5) - h(3)}{5 - 3} = \frac{400 - 336}{2} = 32 \text{ ft/sec.}$$

This confirms our informal observation that the ball was rising more quickly during the time interval $[0, 2]$ than during the interval $[3, 5]$. Notice that these average velocities could also be interpreted as the slopes of two lines. The first is the slope of the line passing through the points $(0, h(0)) = (0, 0)$ and $(2, h(2)) = (2, 256)$, and the second is the slope of the line through $(3, h(3)) = (3, 336)$ and $(5, h(5)) = (5, 400)$. All of these points are on the graph of h. Thus, each line is determined by two points on the graph of h, as shown in Figure 2.2; such lines are called *secant lines* for the curve $y = h(t)$. Of course, the secant line for the interval $[0, 2]$ slants up more steeply than the one for the interval $[3, 5]$, so its slope is larger.

The average velocities during the intervals $[5, 7]$ and $[8, 10]$ are

$$\frac{h(7) - h(5)}{7 - 5} = \frac{336 - 400}{2} = -32 \text{ ft/sec,} \qquad \frac{h(10) - h(8)}{10 - 8} = \frac{0 - 256}{2} = -128 \text{ ft/sec.}$$

The negative values here mean that the height of the ball is decreasing, and it is falling more quickly during the time interval $[8, 10]$ than during the interval $[5, 7]$. Once again these average velocities can be interpreted as slopes of secant lines, as illustrated in Figure 2.2.

So far we have found exact answers for all of the questions we have asked, but now let us ask a more difficult question: What is the velocity of the ball at the time $t = 3$? Since we are asking for the velocity at a single instant, rather than over an interval of time, this is called the *instantaneous velocity* of the ball at $t = 3$.

As a first step, we note that 3 is the beginning of the interval $[3, 5]$, and we have already computed that the average velocity of the ball during this time interval is 32 ft/sec. So 32 ft/sec might be a good approximation to the velocity at the time $t = 3$.

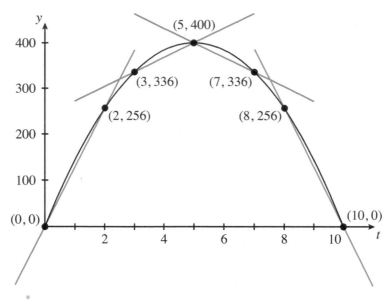

Figure 2.2: Average velocities interpreted as slopes of secant lines. The slopes of the four blue secant lines shown in the graph are 128, 32, −32, and −128. These are the average velocities of the ball during the time intervals [0, 2], [3, 5], [5, 7], and [8, 10].

But in Figure 2.1 it appears that the ball is slowing down during this time interval, so the velocity is probably more than 32 ft/sec at the beginning of the time interval and less than 32 ft/sec at the end. Indeed, if we split the interval in half and compute the average velocity in the first and second halves of the interval, we find that the average velocity during the time interval [3, 4] is

$$\frac{h(4) - h(3)}{4 - 3} = \frac{384 - 336}{1} = 48 \text{ ft/sec},$$

while the average velocity during the interval [4, 5] is

$$\frac{h(5) - h(4)}{5 - 4} = \frac{400 - 384}{1} = 16 \text{ ft/sec}.$$

Thus, 48 ft/sec is probably a better estimate of the instantaneous velocity of the ball at $t = 3$. However, it is still only an approximation, since the velocity of the ball changes during the time interval [3, 4] as well.

Figure 2.3 shows the secant lines whose slopes give the average velocity of the ball during the intervals [3, 5] and [3, 4]. We see in the figure that as the right endpoint of the time interval decreases from 5 to 4, the secant line rotates counterclockwise and its slope increases. It appears that if we continue to decrease the right endpoint, the secant line will continue to rotate, coming closer and closer to the line that just touches the graph of h at the point $(3, h(3))$. This line is said to be *tangent* to the graph at $t = 3$.

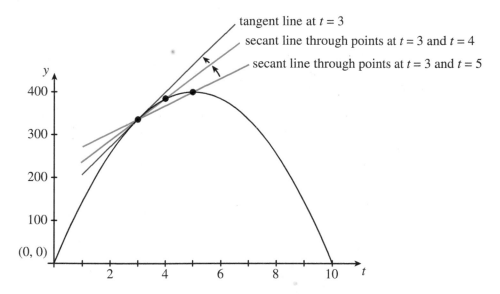

Figure 2.3: Secant lines whose slopes give the average velocity of the ball during the intervals [3, 5] and [3, 4]. As the right endpoint of the interval decreases, the secant line rotates counterclockwise and comes closer to the tangent line at $t = 3$, shown in red.

Interval	Average Velocity
[3, 3.1]	62.4
[3, 3.01]	63.84
[3, 3.001]	63.984
[2.9, 3]	65.6
[2.99, 3]	64.16
[2.999, 3]	64.016

Table 2.1: Average velocity of the ball during some short time intervals.

Perhaps to get a really good approximation to the velocity of the ball at $t = 3$ we should compute the average velocity over a time interval so short that the velocity doesn't have time to change much. Table 2.1 shows the average velocity during the intervals [3, 3.1], [3, 3.01], and [3, 3.001]. We could also consider time intervals that come before $t = 3$, so the table shows the average velocity during the intervals [2.9, 3], [2.99, 3], and [2.999, 3]. If we were to add the secant lines corresponding to these average velocities to Figure 2.3, they would be almost indistinguishable from the tangent line.

It appears from the table that the instantaneous velocity at $t = 3$ is close to 64 ft/sec, and it seems reasonable to say that this is also the slope of the line tangent to the graph of h at the point $(3, h(3))$. However, the data in the table are consistent with other answers, such as 63.99 ft/sec or 64.01 ft/sec. For many purposes this approximate answer might be good enough. But remember that the central idea of calculus is to achieve *precision through approximation*. The question we want to ask is, can we determine the *exact* velocity at $t = 3$, even though all of our calculations give only approximate answers?

We could compute even better approximations by using even shorter time intervals, but that won't solve our problem. If we are to discover a precise answer, we need to see if there is a pattern to how the approximations change as they improve. As a step in this direction, let's find the general formula for the average velocity of the ball during the time interval between $t = 3$ and $t = 3 + x$, for any number $x \neq 0$. For $x > 0$ this would be the interval $[3, 3 + x]$, and the average velocity would be

$$\frac{h(3+x) - h(3)}{3 + x - 3} = \frac{h(3+x) - h(3)}{x}.$$

For $x < 0$ we have $3 + x < 3$, so the time interval would be $[3 + x, 3]$, with average velocity

$$\frac{h(3) - h(3+x)}{3 - (3+x)} = \frac{h(3) - h(3+x)}{-x} = \frac{h(3+x) - h(3)}{x}.$$

Thus, for both $x > 0$ and $x < 0$ the average velocity is given by the function

$$f(x) = \frac{h(3+x) - h(3)}{x}.$$

What is the domain of the function f? In order for the term $h(3 + x)$ that appears in the formula for $f(x)$ to make sense, $3 + x$ must be in the domain of h, which is the interval $[0, 10]$. Thus, we must have $0 \leq 3 + x \leq 10$, or equivalently $-3 \leq x \leq 7$. But the formula for $f(x)$ is also undefined when $x = 0$, since we would have a 0 in the denominator. Therefore the domain of f is $[-3, 0) \cup (0, 7]$.

We have already computed several values of $f(x)$. For example, the first three rows of Table 2.1 show that $f(0.1) = 62.4$, $f(0.01) = 63.84$, and $f(0.001) = 63.984$. By computing $f(x)$ for values of x closer and closer to 0, we should be able to get better and better approximations to the velocity of the ball at time $t = 3$. It seems that if only we could find the value of $f(x)$ at $x = 0$ we would have the exact answer. But 0 is the one value of x in the interval $[-3, 7]$ that we cannot use, because it is not in the domain of f.

Maybe we can learn something by examining the graph of f. To find the graph, we first simplify the formula for $f(x)$. We have already computed that $h(3) = 336$, and we have

$$h(3 + x) = 160(3 + x) - 16(3 + x)^2$$
$$= 480 + 160x - 16(9 + 6x + x^2)$$
$$= 336 + 64x - 16x^2.$$

Thus, for all $x \in [-3, 0) \cup (0, 7]$,

$$f(x) = \frac{h(3+x) - h(3)}{x} = \frac{(336 + 64x - 16x^2) - 336}{x} = \frac{64x - 16x^2}{x} = 64 - 16x.$$

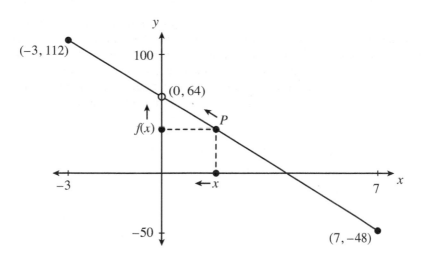

Figure 2.4: The graph of the function f. As x approaches 0, $f(x)$ approaches 64.

The graph of $y = f(x)$ is therefore the same as the graph of $y = 64 - 16x$, except that we must include only points whose x-coordinates are in the set $[-3, 0) \cup (0, 7]$, since this is the domain of f. The graph of $y = 64 - 16x$ is a straight line with slope -16 and y-intercept 64, so the graph of f is a line segment with one point removed, as shown in Figure 2.4. The line $y = 64 - 16x$ passes through the point $(0, 64)$, but this point is missing from the graph of f, as indicated by the open circle in Figure 2.4, because 0 is not in the domain of f.

We can think of a number x that is getting closer and closer to 0 as a point on the x-axis that is moving toward the origin. Directly above this point is a point P on the graph of f. The coordinates of P are $(x, f(x))$, so a horizontal line through P crosses the y-axis at the number $f(x)$. As x gets closer and closer to 0, the point P moves along the graph of f toward the point $(0, 64)$, and therefore $f(x)$ gets closer and closer to 64, as shown in Figure 2.4. The figure shows x approaching 0 from the right, but the situation would be similar if x were approaching 0 from the left. Thus, as x gets closer and closer to 0 from either side, $f(x)$ gets closer and closer to 64. We can also see this from the formula $f(x) = 64 - 16x$: as x gets closer and closer to 0, the term $16x$ gets closer and closer to 0, and the value of $64 - 16x$ approaches 64. We say that 64 is the *limit of $f(x)$ as x approaches 0*, and we write $\lim_{x \to 0} f(x) = 64$. We will give a more careful explanation of this statement later in this chapter, but for the moment we simply rely on intuition.

Recall that our strategy for finding the exact velocity of the ball at time $t = 3$ was to determine the value that the approximations get closer and closer to as they are made more and more accurate. We have now carried out this strategy: we can make the approximations more and more accurate by taking x closer and closer to 0, and when we do this the approximations approach 64 ft/sec. Thus, the exact velocity of the ball at $t = 3$ is 64 ft/sec. We can also say that the slope of the line tangent to the curve $y = h(t)$ at the point $(3, h(3))$ is 64.

In the end, the problem of finding the instantaneous velocity of the ball turned into the problem of evaluating $\lim_{x \to 0} f(x)$. For the rest of this chapter we set aside problems like the ball velocity problem and address the issue of evaluating limits of functions. Once we have learned how to evaluate limits, we will be ready to develop a systematic approach to solving problems involving velocities, tangent lines, and other related concepts. In all of these problems, the use of limits is what will allow us to achieve precision through approximation.

In the example above, once we found a simple formula for $f(x)$ and drew its graph, the value of the limit became intuitively clear. But this is not always the case. For example, consider the limit

$$\lim_{\theta \to 0} \frac{\sin \theta}{\theta}.$$

To investigate this limit, we must see if $(\sin \theta)/\theta$ gets closer and closer to some particular number as θ gets closer and closer to 0. We cannot actually set $\theta = 0$ in the formula $(\sin \theta)/\theta$, since we would be dividing by 0. But plugging in values of θ close to 0 gives the values shown in Table 2.2. In this table, it appears that $(\sin \theta)/\theta$ is getting very close to 1 as θ approaches 0. But can we be sure that this trend will continue? Can we be sure that 1 is the *exact* answer, rather than, say, 1.001? In this case, in contrast to the case of the function f, we have no explanation for *why* the values of $(\sin \theta)/\theta$ are getting close to 1. We don't know anything about the pattern of how $(\sin \theta)/\theta$ will continue to change as θ gets even closer to 0, so we cannot be confident that we have found the value of the limit.

Or consider the limit

$$\lim_{x \to 2} \frac{\sqrt{x} - \sqrt{2}}{x - 2}.$$

Notice that the expression $(\sqrt{x} - \sqrt{2})/(x - 2)$ is undefined not only when $x = 2$, but also when x is negative. However, it is defined for all values of x that are sufficiently close to 2 but not equal to 2. More specifically, if the distance between x and 2 is less than 2 but not 0, then x is positive and $x \neq 2$, so $f(x)$ is defined. Since the distance between x and 2 is $|x - 2|$, we can express this by saying that if $0 < |x - 2| < 2$ then $f(x)$ is defined. In general, we will only study a limit of the form $\lim_{x \to a} f(x)$ if there is a number $d > 0$ such that for all x, if $0 < |x - a| < d$ then $f(x)$ is defined. It is only

θ	$\frac{\sin \theta}{\theta}$	θ	$\frac{\sin \theta}{\theta}$
0.5	0.9588511	−0.5	0.9588511
0.1	0.9983342	−0.1	0.9983342
0.01	0.9999833	−0.01	0.9999833
0.001	0.9999998	−0.001	0.9999998

Table 2.2: Values of $(\sin \theta)/\theta$ for θ close to 0. (Note that θ is measured in radians here.)

x	$\frac{\sqrt{x}-\sqrt{2}}{x-2}$	x	$\frac{\sqrt{x}-\sqrt{2}}{x-2}$
2.5	0.33385	1.5	0.37894
2.1	0.34924	1.9	0.35809
2.01	0.35311	1.99	0.35400
2.001	0.35351	1.999	0.35360

Table 2.3: Values of $(\sqrt{x} - \sqrt{2})/(x - 2)$ for x close to 2.

in this situation that it makes sense to ask what happens to the value of $f(x)$ as x gets closer and closer to a.

Some values of $(\sqrt{x} - \sqrt{2})/(x - 2)$ for x close to 2 are given in Table 2.3. The values seem to be settling on a number a bit larger than 0.35, but are they really settling on some particular number, or are they just wandering around a certain small region on the number line? Is there a way to discover an *exact* value for this limit?

In fact, the value of $\lim_{\theta \to 0}(\sin\theta)/\theta$ is *exactly* 1, and the *precise* value of $\lim_{x \to 2}(\sqrt{x} - \sqrt{2})/(x - 2)$ turns out to be $\sqrt{2}/4$; your calculator will tell you that the decimal value of this number is about 0.3535534. The graphs of the equations $y = (\sin\theta)/\theta$ and $y = (\sqrt{x} - \sqrt{2})/(x - 2)$ are shown in Figure 2.5. By the end of this chapter, we will have developed the tools necessary to determine the values of both of these limits, and many more.

Exercises 2.1

1. An object moves along a number line, marked in meters, in such a way that its position at any time t, measured in seconds, is $s(t) = t^2 - 3$.

   (a) Find the average velocity of the object over the time interval $[2, 5]$. Be sure to specify the units.

   (b) Let $f(x)$ be the average velocity of the object during the time interval between $t = 2$ and $t = 2 + x$. Find a formula for $f(x)$, for all $x \neq 0$.

   (c) Draw the graph of f.

   (d) What is the velocity of the object at time $t = 2$?

2. An object moves along a number line, marked in feet, in such a way that its position at any time t, measured in minutes, is $s(t) = \sin(\pi t)$.

   (a) Find the average velocity of the object over the time interval $[1, 3]$.

   (b) Let $f(x)$ be the average velocity of the object during the time interval between $t = 1$ and $t = 1 + x$. Find a formula for $f(x)$, for all $x \neq 0$.

   (c) Guess the velocity of the object at $t = 1$ by evaluating $f(x)$ for x close to 0.

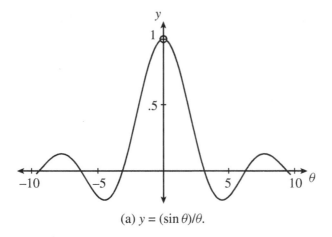

(a) $y = (\sin \theta)/\theta$.

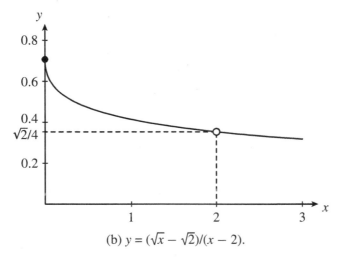

(b) $y = (\sqrt{x} - \sqrt{2})/(x - 2)$.

Figure 2.5: The graphs of $y = (\sin \theta)/\theta$ and $y = (\sqrt{x} - \sqrt{2})/(x - 2)$.

3–12: Guess the value of $\lim_{x \to a} f(x)$ by evaluating $f(x)$ at values of x close to a.

3. $\lim\limits_{x \to 3} \dfrac{3 - x}{x^2 - x - 6}$.

4. $\lim\limits_{x \to 1} \dfrac{(x + 2)^2 - 9}{(x + 1)^2 - 4}$.

5. $\lim\limits_{x \to 0} \dfrac{\sqrt{x + 1} - 1}{x}$.

6. $\lim\limits_{x \to 0} \dfrac{\sqrt[3]{x + 1} - 1}{x}$.

7. $\lim\limits_{x \to 64} \dfrac{\sqrt[3]{x} - 4}{\sqrt{x} - 8}$.

8. $\lim\limits_{x \to 0} \left(\dfrac{1}{x} + \dfrac{1}{x^2 - x} \right)$.

9. $\lim\limits_{x \to 2} \left[\dfrac{1}{x - 2} \cdot \left(\dfrac{1}{x} - \dfrac{1}{2} \right) \right]$.

10. $\lim\limits_{x \to \pi/2} \dfrac{\sec x - \tan x}{\cos x}$.

11. $\lim\limits_{x \to \pi/2} \left(\dfrac{2}{2x - \pi} + \dfrac{1}{\cos x} \right)$.

12. $\lim\limits_{x \to 0} \dfrac{x^3}{\sin x - x}$.

2.2 What Does "Limit" Mean?

It is surprisingly difficult to put one's finger on exactly what the word "limit" means. In the last section we suggested that "$\lim_{x \to a} f(x) = L$" means that as x gets closer and closer to a, $f(x)$ gets closer and closer to L. While this description may give you an intuitive feeling for what we mean by the word "limit," it turns out to be unhelpful when it comes to developing the kind of precise understanding of the limit concept that is necessary for solving difficult limit problems.

To begin to develop a better understanding of limits, let's consider another example. Let

$$g(x) = \frac{2x^2 - 5x + 2}{x - 2}. \tag{2.1}$$

Of course, the domain of g is $\{x : x \neq 2\} = (-\infty, 2) \cup (2, \infty)$. We can't compute $g(2)$, since 2 is not in the domain of g, but we can investigate $\lim_{x \to 2} g(x)$.

Factoring the numerator of $g(x)$, we find that for all $x \in (-\infty, 2) \cup (2, \infty)$,

$$g(x) = \frac{(2x - 1)(x - 2)}{x - 2} = 2x - 1. \tag{2.2}$$

It follows that the graph of g is the same as the graph of the equation $y = 2x - 1$, except that we must leave out the point with x-coordinate 2. In other words, the graph of g is the line with slope 2 and y-intercept -1, but with the point $(2, 3)$ removed. The graph of g is shown in Figure 2.6.

It seems clear from Figure 2.6 that $\lim_{x \to 2} g(x) = 3$. But now we would like to develop a more precise understanding of exactly what this means. We begin with the idea that when x is close to 2, $g(x)$ should be close to 3. Of course, we can't actually let $x = 2$, since $g(2)$ is undefined, so a more careful statement is that when x is close

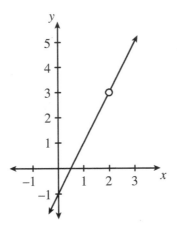

Figure 2.6: The graph of g.

to 2, but not equal to 2, $g(x)$ should be close to 3. To say that two numbers are close is the same as saying that the distance between them is small, and we know that the distance between two numbers a and b is $|a - b|$. So another way to express the idea we are interested in is that whenever $|x - 2|$ is small, but not 0, $|g(x) - 3|$ should be small. In other words, statements of the following form should be true:

$$\text{For every number } x, \text{ if } 0 < |x - 2| < \underline{\ ?\ } \text{ then } |g(x) - 3| < \underline{\ ?\ }. \qquad (2.3)$$

It remains to determine what numbers to put in the blanks. Of course, these numbers will measure how close x is to 2, and how close $g(x)$ is to 3.

As a first step in understanding statement (2.3), let's try putting particular numbers in the blanks. A little experimentation with different numbers leads to the following fact:

Proposition 2.2.1. *For the function g defined in equation* (2.1), *the following statement is true:*

For every number x, if $0 < |x - 2| < 1/2$ *then* $|g(x) - 3| < 1$.

Proof. Suppose $0 < |x - 2| < 1/2$. Since $|x - 2| > 0$, $x \neq 2$, so $g(x)$ is defined, and by equation (2.2), $g(x) = 2x - 1$. Therefore

$$|g(x) - 3| = |(2x - 1) - 3| = |2x - 4| = |2(x - 2)| = |2||x - 2| = 2|x - 2|,$$

where we have used part 1 of Theorem 1.1.4 in the fourth step. We know that $|x - 2| < 1/2$, and multiplying both sides of this inequality by 2, we find that $2|x - 2| < 1$. Since $|g(x) - 3| = 2|x - 2|$, we conclude that $|g(x) - 3| < 1$. $\qquad \square$

Figure 2.7 illustrates Proposition 2.2.1. The values of x for which $0 < |x - 2| < 1/2$ are those that are within 1/2 of 2, except for 2 itself. In other words, they are the values that lie between the vertical dashed lines in Figure 2.7. Looking at the part of the graph

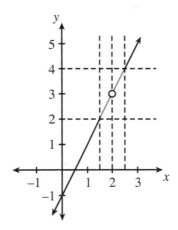

Figure 2.7: For every number x, if $0 < |x - 2| < 1/2$ then $|g(x) - 3| < 1$.

of g between these vertical dashed lines, which is shown in blue in the figure, we see that for all these values of x, $g(x)$ is between 2 and 4, which means that $|g(x) - 3| < 1$.

Proposition 2.2.1 shows that when x gets close enough to 2, $g(x)$ gets fairly close to 3—specifically, $g(x)$ gets to be within 1 of 3. But our hope is to achieve precision through approximation, so we need to do better than this. As at least a step in the right direction, can we show that $g(x)$ will get closer to 3—say, within 1/10 of 3? The answer is yes, but to make it happen we will need to require that x be closer to 2. Perhaps you can guess how close x will have to be to 2. Imitating the proof of Proposition 2.2.1, you should be able to establish the following fact:

$$\text{For every number } x, \text{ if } 0 < |x - 2| < 1/20, \text{ then } |g(x) - 3| < 1/10. \tag{2.4}$$

There is a pattern here, and rather than working out more particular examples let's state the general pattern. It is customary to use the Greek letter ϵ (epsilon) to measure how close $g(x)$ is to 3. Using this notation, we can state the general pattern like this:

Proposition 2.2.2. *Let g be the function defined in equation (2.1). Then for every positive number ϵ, the following statement is true:*

For every number x, if $0 < |x - 2| < \epsilon/2$ then $|g(x) - 3| < \epsilon$.

Proof. Let ϵ be any positive number, and suppose that $0 < |x - 2| < \epsilon/2$. Since $|x - 2| > 0$, $x \neq 2$, so $g(x)$ is defined and $g(x) = 2x - 1$. Therefore

$$|g(x) - 3| = |2x - 4| = |2(x - 2)| = |2||x - 2| = 2|x - 2| < 2(\epsilon/2) = \epsilon.$$

Notice that the second-to-last step above is justified by multiplying both sides of the known inequality $|x - 2| < \epsilon/2$ by 2. $\square$

Setting $\epsilon = 1$ in Proposition 2.2.2 we get Proposition 2.2.1, and setting $\epsilon = 1/10$ we get statement (2.4). Proposition 2.2.1 achieves a fairly low level of accuracy about the value of $g(x)$ when x is close to 2; statement (2.4) achieves a somewhat higher level of accuracy. But Proposition 2.2.2 achieves complete precision by allowing for *all possible* positive numbers ϵ. Proposition 2.2.2 tells us that $g(x)$ will get within 0.001 of 3, or within 0.0001, or within whatever positive distance you might choose, if only we make sure that x is sufficiently close to 2. (Notice that only positive numbers make sense for ϵ. It would make no sense to talk about $g(x)$ being within, say, -0.01 of 3!) We have therefore achieved our goal of precision through approximation.

We will express the fact established in Proposition 2.2.2 by saying that as x approaches 2, without being equal to 2, $g(x)$ approaches 3. And it will be convenient to have a shorthand notation for this, so we write: as $x \to 2^{\neq}$, $g(x) \to 3$. We should warn you that, although many calculus books use similar arrow notation, the use of the superscript "$\neq$" is not common. However, we prefer to include this in our notation to remind you that although x gets close to 2, it is never equal to 2.

In the future, we will leave out the phrase "for every number x" in statement (2.3), leaving it as understood that we are making a general statement about all values of x. And it will be convenient to have names for the numbers that go in *both* of the blanks in this statement. We have already introduced the use of the letter ϵ for the number that goes in the second blank. There is also a customary letter that is used in the first blank: the Greek letter δ (delta). Using these conventions, we could restate the conclusion of Proposition 2.2.2 as follows. For every $\epsilon > 0$, if we let $\delta = \epsilon/2$; then the following statement is true:

$$\text{If } 0 < |x - 2| < \delta \text{ then } |g(x) - 3| < \epsilon.$$

Notice that the equation $\delta = \epsilon/2$ implies that for different values of ϵ, we will have different values of δ. For each value of ϵ there is a corresponding value of δ.

Much of what we have said about this example will also apply to other cases. In general, when we are investigating a limit of the form $\lim_{x \to a} f(x)$, we will want to find a number L such that if x is close to a, but not equal to a, then $f(x)$ is close to L. Thus, we will be interested in the statement "if $0 < |x - a| < \delta$ then $|f(x) - L| < \epsilon$." To achieve complete precision we will want this statement to be true for every positive number ϵ, as long as we make a suitable choice for δ. But the choice $\delta = \epsilon/2$ that worked for our example g might not be the right choice in other cases. We are therefore led to make the following definition.

Definition 2.2.3. For any function f and numbers a and L, we write "as $x \to a^{\neq}$, $f(x) \to L$" to mean that for every number $\epsilon > 0$, there is some corresponding number $\delta > 0$ such that if $0 < |x - a| < \delta$ then $|f(x) - L| < \epsilon$.

To put it more informally, the statement "as $x \to a^{\neq}$, $f(x) \to L$" means that we can get $f(x)$ to be as close as we please to L by making x sufficiently close to a (but not equal to a).

Perhaps you have guessed by now that we are going to define $\lim_{x \to a} f(x)$ to be the number L such that as $x \to a^{\neq}$, $f(x) \to L$. But there is a potential problem with this definition. Might there be more than one number L such that as $x \to a^{\neq}$, $f(x) \to L$?

If so, the notation $\lim_{x \to a} f(x)$ would be ambiguous. Fortunately, this never happens:

Theorem 2.2.4. *There cannot be more than one number L such that as $x \to a^{\neq}$, $f(x) \to L$.*

You might regard Theorem 2.2.4 as confirming that the statement "as $x \to a^{\neq}$, $f(x) \to L$" pins down the value of L with complete precision. We will see in the next section why Theorem 2.2.4 is true. For now, we ask you to accept it so that we can make the following definition.

Definition 2.2.5. If there is a number L such that as $x \to a^{\neq}$, $f(x) \to L$, then we define $\lim_{x \to a} f(x)$ to be the unique such number L. If there is no such number L, then $\lim_{x \to a} f(x)$ is undefined.

We have seen that Definitions 2.2.3 and 2.2.5 accurately describe our example $\lim_{x \to 2} g(x) = 3$. But do they agree with our intuitions about *all* limits? To try to answer this question, let's try out these definitions on a few more examples. It will be most illuminating to examine these examples graphically, so we begin by getting a better idea of the graphical meaning of Definitions 2.2.3 and 2.2.5. According to Definition 2.2.5, to say that $\lim_{x \to a} f(x) = L$ means that as $x \to a^{\neq}$, $f(x) \to L$. The meaning of this statement is given by Definition 2.2.3, which we might regard as describing a challenge and response. The challenge is given by the positive number ϵ: can we ensure that $f(x)$ is within ϵ of L? And the response is the positive number δ. The response succeeds in meeting the challenge if whenever x is within δ of a, but not equal to a, $f(x)$ is within ϵ of L. In other words, the response succeeds if the following statement, which we will call the ϵ-δ *criterion*, is true:

$$\text{If } 0 < |x - a| < \delta \text{ then } |f(x) - L| < \epsilon. \qquad (\epsilon\text{-}\delta \text{ criterion})$$

To say that as $x \to a^{\neq}$, $f(x) \to L$ means that every challenge can be met successfully.

What is the graphical meaning of the ϵ-δ criterion? By part 1 of Theorem 1.1.2, the inequality $|x - a| < \delta$ is equivalent to $-\delta < x - a < \delta$, or in other words $a - \delta < x < a + \delta$. So to say that $0 < |x - a| < \delta$ means that x is between $a - \delta$ and $a + \delta$, but $x \neq a$. Similarly, to say that $|f(x) - L| < \epsilon$ means that $L - \epsilon < f(x) < L + \epsilon$. Thus, the ϵ-δ criterion tells us to look at the part of the graph of f that is between $x = a - \delta$ and $x = a + \delta$, except for $x = a$. For the criterion to be met, the y-coordinates of the points on this part of the graph must be between $L - \epsilon$ and $L + \epsilon$.

Perhaps it would be helpful to see what it would mean for the ϵ-δ criterion to be *false*. For the ϵ-δ criterion to fail, there would have to be a number x such that $a - \delta < x < a + \delta$ and $x \neq a$, but $|f(x) - L| \geq \epsilon$. This number would correspond to a point on the graph of f that is either between the vertical lines $x = a - \delta$ and $x = a$ or between the vertical lines $x = a$ and $x = a + \delta$, but *not* between the horizontal lines $y = L - \epsilon$ and $y = L + \epsilon$. This point would be in the region that is shaded red in Figure 2.8. We call this region the *forbidden region*. Thus, for the ϵ-δ criterion to fail, the graph must include a point in the forbidden region. The ϵ-δ criterion could be thought of as saying that the graph of f does not go into the forbidden region.

Figure 2.8 shows the graph of a function f and some examples of values of ϵ and δ that might be used to test whether or not $\lim_{x \to a} f(x) = L$. In Figure 2.8a, the graph

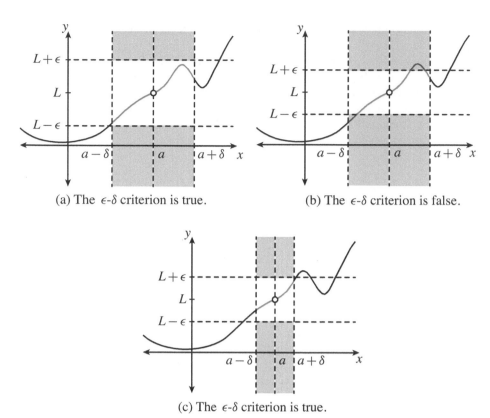

(a) The ϵ-δ criterion is true.   (b) The ϵ-δ criterion is false.

(c) The ϵ-δ criterion is true.

Figure 2.8: The graph of a function f for which $\lim_{x \to a} f(x) = L$. The red shaded region is the forbidden region. In (a), the graph does not enter the forbidden region, and the ϵ-δ criterion is true. In (b), ϵ has been decreased, and the ϵ-δ criterion is false. In (c), δ has also been decreased, and the ϵ-δ criterion is true again.

does not go into the forbidden region, and the ϵ-δ criterion is true. In Figure 2.8b, ϵ has been decreased, creating a more demanding challenge. The graph now goes into the forbidden region (where the graph is colored red), so the ϵ-δ criterion is false and the challenge has not been met. But the challenge *could* have been met, by using a smaller value of δ. Figure 2.8c shows the same challenge with a choice for δ that is small enough to meet the challenge. In fact, if you imagine ϵ being decreased further in Figure 2.8, you should be able to see that δ could always be decreased enough to make the ϵ-δ criterion true. Thus, in Figure 2.8 every challenge can be met, and therefore $\lim_{x \to a} f(x) = L$.

On the other hand, Figure 2.9 shows a function f for which $\lim_{x \to a} f(x)$ is undefined. In Figure 2.9a the ϵ-δ criterion is satisfied, but in Figure 2.9b the value of ϵ is smaller and the ϵ-δ criterion is false. Notice that the point at $(a, f(a))$ is *not* in the forbidden region; the forbidden region includes only points whose x-coordinates are between $a - \delta$ and $a + \delta$ *but not equal to a*. However, there are points on the graph of f just to the left of $(a, f(a))$ that are in the forbidden region. Furthermore, this part of the graph goes right up to $(a, f(a))$, so no matter how small we make δ, we will not be able to avoid having points on the graph of f just to the left of $(a, f(a))$ that are in the forbidden region. Thus, for the value of ϵ shown in Figure 2.9b there is no corresponding value of δ for which the ϵ-δ criterion is true, and therefore according to Definition 2.2.3 it is not the case that as $x \to a^{\neq}$, $f(x) \to L$. Similar reasoning shows that in fact there is no value of L for which as $x \to a^{\neq}$, $f(x) \to L$; you should try picking other values for L in the figure and checking that, in each case, you can draw an ϵ challenge that cannot be met. Therefore $\lim_{x \to a} f(x)$ is undefined. This should make intuitive sense: When x is close to a but less than a, $f(x)$ has one value, but when x is close to a but greater than a it has a different value. Thus, there is no *single* number that $f(x)$ gets close to when x is close to a.

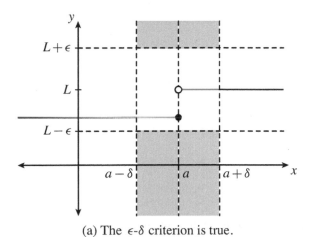

(a) The ϵ-δ criterion is true.

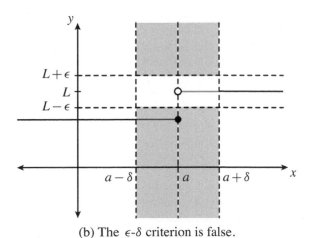

(b) The ϵ-δ criterion is false.

Figure 2.9: The graph of a function f for which $\lim_{x \to a} f(x)$ is undefined.

x	$\frac{\pi}{2x}$	$\sin\left(\frac{\pi}{2x}\right)$	x	$\frac{\pi}{2x}$	$\sin\left(\frac{\pi}{2x}\right)$
1	$\pi/2$	1	-1	$-\pi/2$	-1
1/2	π	0	$-1/2$	$-\pi$	0
1/3	$3\pi/2$	-1	$-1/3$	$-3\pi/2$	1
1/4	2π	0	$-1/4$	-2π	0
1/5	$5\pi/2$	1	$-1/5$	$-5\pi/2$	-1
1/6	3π	0	$-1/6$	-3π	0
1/7	$7\pi/2$	-1	$-1/7$	$-7\pi/2$	1
1/8	4π	0	$-1/8$	-4π	0
$\vdots$	$\vdots$	$\vdots$	$\vdots$	$\vdots$	$\vdots$

Table 2.4: Values of $\sin(\pi/(2x))$ for $x = \pm 1, \pm 1/2, \pm 1/3, \ldots$.

An interesting function that is helpful in elucidating the limit concept is the function h_1 defined by the formula

$$h_1(x) = \sin\left(\frac{\pi}{2x}\right).$$

Clearly $h_1(x)$ is defined for all $x \neq 0$, but $h_1(0)$ is undefined.

Table 2.4 gives values of $h_1(x)$ for $x = 1, 1/2, 1/3, \ldots$ and also for $x = -1, -1/2, -1/3, \ldots$. It is not hard to see that the pattern in this table will continue, and as x approaches 0 from either the positive or the negative side, the value of $h_1(x)$ swings back and forth infinitely many times between -1 and 1. For values of x between the values shown in the table, $h_1(x)$ takes on all values between -1 and 1. It never takes values outside of this range, because the value of the sine function is always in the interval $[-1, 1]$. The graph of h_1 is shown in Figure 2.10. The black area in the middle of the figure is actually infinitely many up-and-down wiggles that are crowded so close together that they are indistinguishable from each other.

Another interesting function related to h_1 is the function h_2 defined by the formula

$$h_2(x) = |x| \sin\left(\frac{\pi}{2x}\right).$$

The graph of h_2, like that of h_1, wiggles up and down infinitely many times. However, multiplying the inequality $-1 \leq \sin(\pi/(2x)) \leq 1$ by $|x|$, we see that

$$-|x| \leq |x| \sin\left(\frac{\pi}{2x}\right) \leq |x|.$$

Thus, although the wiggles in the graph of h_1 go down to $y = -1$ and up to $y = 1$, the wiggles in the graph of h_2 stay between the curves $y = -|x|$ and $y = |x|$. The graph of h_2 is shown in Figure 2.11.

What can we say about $\lim_{x \to 0} h_1(x)$ and $\lim_{x \to 0} h_2(x)$? We see in Figure 2.11 that $h_2(x)$ appears to home in on 0 as x approaches 0, so intuitively it seems that

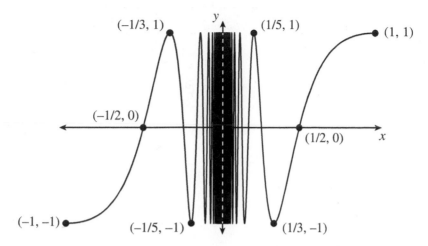

Figure 2.10: The graph of h_1. A few of the points from Table 2.4 are labeled in the graph.

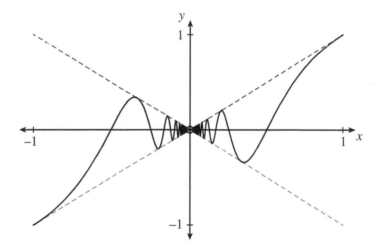

Figure 2.11: The graph of h_2. The dashed lines are the graphs of $y = |x|$ (red) and $y = -|x|$ (blue).

$\lim_{x \to 0} h_2(x)$ should be 0. Do Definitions 2.2.3 and 2.2.5 lead us to this conclusion? Figure 2.12 shows an ϵ challenge for this limit that has been successfully met by a suitable δ response. The graph does not go into the forbidden region, and therefore the ϵ-δ criterion is true. Furthermore, since the wiggles in the graph of h_2 get smaller and smaller as the curve approaches the origin, it appears that we could meet any similar ϵ challenge by using a sufficiently small δ. We will be able to confirm this more carefully later, but for the moment we will simply say that in the graph it appears to be the case that as $x \to 0^{\ne}$, $h_2(x) \to 0$, and therefore $\lim_{x \to 0} h_2(x) = 0$.

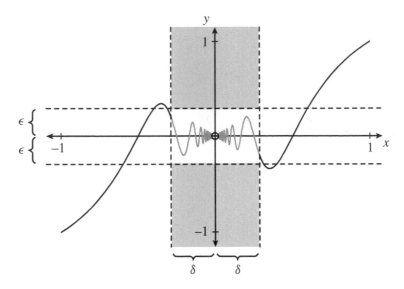

Figure 2.12: Testing whether or not $\lim_{x \to 0} h_2(x) = 0$. For this choice of ϵ and δ, the ϵ-δ criterion is true.

Can we also say that as $x \to 0^{\neq}$, $h_1(x) \to 0$? Figure 2.13 shows an ϵ challenge for this statement that has not been successfully met. And it should be clear from the figure that in fact this challenge *cannot* be met: since the wiggles continue right up to the y-axis, no positive number δ, no matter how small, can make the graph avoid the forbidden region and make the ϵ-δ criterion true. Thus, according to Definition 2.2.3, it is not the case that as $x \to 0^{\neq}$, $h_1(x) \to 0$. And similar reasoning leads to the conclusion that there is no number L such that as $x \to 0^{\neq}$, $h_1(x) \to L$: no matter what value we choose for L, if ϵ is small enough then the range $L - \epsilon < y < L + \epsilon$ will be too small to contain the wiggles in the graph of h_1, which extend from $y = -1$ to $y = 1$ (see Exercise 9). Thus, $\lim_{x \to 0} h_1(x)$ is undefined. Looking at Figure 2.10, this should make intuitive sense: unlike $h_2(x)$, $h_1(x)$ does not seem to home in on any number as x approaches 0.

We will consider one more function that is related to h_1 and h_2. Let h_3 be the function defined by the formula

$$h_3(x) = (|x| + 0.01) \sin\left(\frac{\pi}{2x}\right).$$

The graph of h_3 is shown in Figure 2.14. This time the wiggles stay between the curves $y = -|x| - 0.01$ and $y = |x| + 0.01$. At first glance, the graph of h_3 looks very similar to the graph of h_2, so you might be tempted to say that $\lim_{x \to 0} h_3(x) = 0$. But a closer look shows that this is incorrect.

Figure 2.15 shows a closeup of the graph of h_3 near $x = 0$. In this figure we can see that, although the wiggles in the graph get small as x approaches 0, they don't go away altogether. In fact, the tops of the wiggles are at $y = |x| + 0.01 > 0.01$ and the

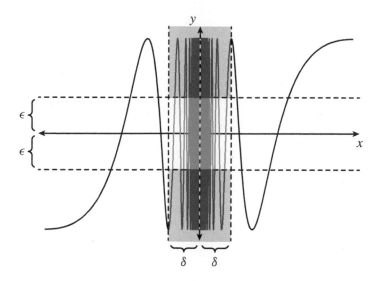

Figure 2.13: Testing whether or not $\lim_{x \to 0} h_1(x) = 0$. For this choice of ϵ and δ, the ϵ-δ criterion is false.

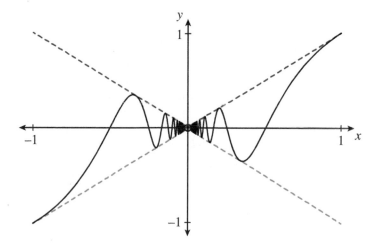

Figure 2.14: The graph of h_3. The dashed lines are the graphs of $y = |x| + 0.01$ (red) and $y = -|x| - 0.01$ (blue).

bottoms are at $y = -|x| - 0.01 < -0.01$, so the wiggles always extend more than 0.01 above and below 0. As a result, the challenge $\epsilon = 0.05$ can be met, as you can see in Figure 2.15a, but the challenge $\epsilon = 0.01$ cannot, as shown in Figure 2.15b. We conclude that $\lim_{x \to 0} h_3(x)$ is not 0, and in fact, as in the case of h_1, the limit is undefined. This should agree with your intuition. As x gets close to 0, $h_3(x)$ narrows down to a very small range of values, but it doesn't home in on a single number. Since we expect limits to be determined with complete precision, it makes sense that $\lim_{x \to 0} h_3(x)$ is undefined.

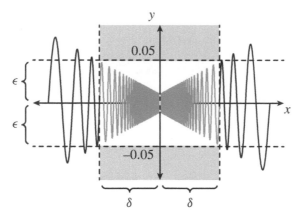

(a) When $\epsilon = 0.05$, there is a corresponding value of δ for which the ϵ-δ criterion is true.

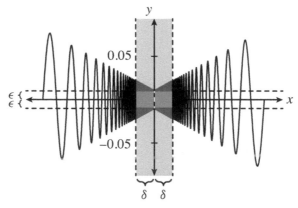

(b) When $\epsilon = 0.01$, there is no corresponding value of δ for which the ϵ-δ criterion is true.

Figure 2.15: A closer look at the graph of h_3 near $x = 0$.

Exercises 2.2

1. Let $f(x) = 3x - 2$.

   (a) Prove that the following statement is true:

   $$\text{If } 0 < |x - 2| < 1/6 \text{ then } |f(x) - 4| < 1/2.$$

   (b) Fill in the blank with a positive number that makes the statement true, and prove the resulting statement:

   $$\text{If } 0 < |x - 2| < \underline{\ ?\ } \text{ then } |f(x) - 4| < 1/10.$$

2. Let $g(x) = 5 - 2x$.

    (a) Prove that the following statement is true:

    $$\text{If } 0 < |x - 1| < 1/2 \text{ then } |g(x) - 3| < 1.$$

    (b) Prove that for every number $\epsilon > 0$, the following statement is true:

    $$\text{If } 0 < |x - 1| < \epsilon/2 \text{ then } |g(x) - 3| < \epsilon.$$

3. Let $h(x) = \dfrac{3x^2 + 8x + 5}{x + 1}$.

    (a) Prove that the following statement is true:

    $$\text{If } 0 < |x - (-1)| < 1/30 \text{ then } |h(x) - 2| < 1/10.$$

    (b) Fill in the blank with a positive number that makes the statement true for every
    number $\epsilon > 0$, and prove the resulting statement:

    $$\text{If } 0 < |x - (-1)| < \epsilon/\underline{\ ?\ } \text{ then } |h(x) - 2| < \epsilon.$$

4. Let $f(x) = x \cos x$. Prove that for every number $\epsilon > 0$, the following statement
    is true:

    $$\text{If } 0 < |x - 0| < \epsilon \text{ then } |f(x) - 0| < \epsilon.$$

5. Figure 2.16 shows the graphs of four curves $y = f(x)$. In which graphs is the ϵ-δ
    criterion true?

6. Figure 2.17 shows the graphs of four curves $y = f(x)$. In which graphs is there a
    number $\delta > 0$ such that the ϵ-δ criterion is true?

7. Figure 2.18 shows the graphs of four curves $y = f(x)$. In which graphs is it the
    case that $\lim_{x \to a} f(x) = L$?

8. Figure 2.19 shows the graphs of four curves $y = f(x)$. In which graphs is it the
    case that $\lim_{x \to a} f(x) = L$?

9. Draw a figure similar to Figure 2.13 to show that it is not the case that as $x \to 0^{\neq}$,
    $h_1(x) \to 1$.

10. Let f be the function defined by the formula

    $$f(x) = \begin{cases} 1, & \text{if } x \text{ is rational,} \\ -1, & \text{if } x \text{ is irrational.} \end{cases}$$

    (a) Sketch the graph of f. (Note: You will not be able to make your sketch very
    accurate, but you should be able to give some idea of what the graph looks
    like.) Is $\lim_{x \to 0} f(x)$ defined?

    (b) Let $g(x) = |x| f(x)$. Sketch the graph of g. Is $\lim_{x \to 0} g(x)$ defined?

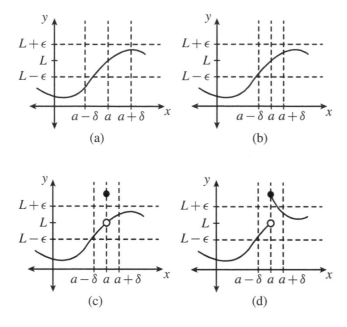

Figure 2.16: In which graphs is the ϵ-δ criterion true?

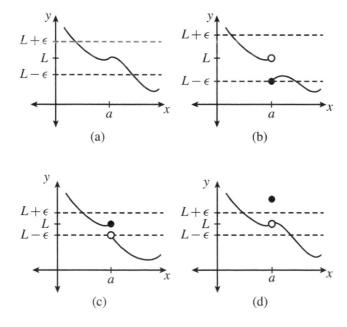

Figure 2.17: Is there a number $\delta > 0$ such that the ϵ-δ criterion is true?

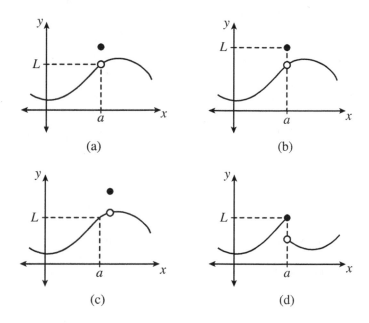

Figure 2.18: Is it the case that $\lim_{x \to a} f(x) = L$?

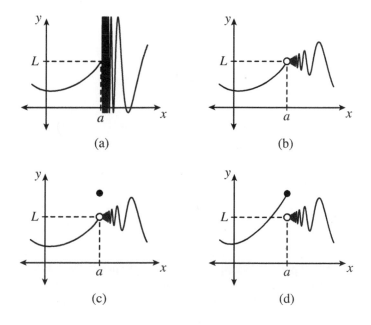

Figure 2.19: Is it the case that $\lim_{x \to a} f(x) = L$?

11. Suppose that for some function f and some numbers $a, L, \epsilon > 0$, and $\delta > 0$, the ϵ-δ criterion is true.

    (a) If we increase the value of ϵ, will the ϵ-δ criterion necessarily still be true?

    (b) If we decrease the value of ϵ, will the ϵ-δ criterion necessarily still be true?

    (c) If we increase the value of δ, will the ϵ-δ criterion necessarily still be true?

    (d) If we decrease the value of δ, will the ϵ-δ criterion necessarily still be true?

2.3   Limits by the Definition

Now that we have a careful definition of what the word "limit" means, we are ready to begin applying this definition to answer questions about limits. We begin by using the definition to verify the values of some particular limits. In all of the examples in this section the value of the limit in question will be given, and we will simply practice using the definition to prove that this value is correct. In the next section we will begin to learn how to find the values of limits.

According to Definitions 2.2.3 and 2.2.5, to verify a statement of the form $\lim_{x \to a} f(x) = L$, we must show that for every $\epsilon > 0$ there is some corresponding $\delta > 0$ such that if $0 < |x - a| < \delta$ then $|f(x) - L| < \epsilon$. This definition guides the form our solutions will take. When we want to prove that such a limit statement is true, we will usually specify, for any given ϵ, how to choose the corresponding δ, and then verify that with this choice for δ, if $0 < |x - a| < \delta$ then $|f(x) - L| < \epsilon$. Thus, we will generally write our proofs in the following form:

> Suppose $\epsilon > 0$.
> Let $\delta = \underline{\ ?\ } > 0$.
> Suppose $0 < |x - a| < \delta$.
> Then $|f(x) - L| = \cdots < \epsilon$.

There are just two parts of these proofs that will vary from one example to another. First, for different limits we will use different values for δ. For example, for the limit statement $\lim_{x \to 2} g(x) = 3$ in the last section, the choice $\delta = \epsilon/2$ worked, as we showed in Proposition 2.2.2, but other limit statements will require different choices for δ. Usually we will give a formula expressing δ in terms of ϵ, and according to the definition of "limit" we must make sure that this formula gives a positive value for δ. Second, there will usually be some algebraic steps involved in working out a formula for $|f(x) - L|$ and verifying that it is less than ϵ, and this algebra will vary from one example to the next.

Often the hardest part of working out a proof of a limit statement is finding the right formula to use for δ. One strategy we sometimes use for this is simply to leave the formula for δ blank, continue with the proof, and then, when we reach a point in the proof where we can see what value would work for δ, go back and fill in the blank. We illustrate this strategy in our next example.

Example 2.3.1. Show that

$$\lim_{x \to -1} \frac{2 - x - 3x^2}{x + 1} = 5.$$

Solution. We begin by leaving the formula for δ blank, filling in the particular function f and numbers a and L in this example, and trying to work out the algebra at the end of the proof:

Suppose $\epsilon > 0$.
Let $\delta = \underline{\ ?\ } > 0$.
Suppose $0 < |x - (-1)| < \delta$.
Then $x \neq -1$, so $(2 - x - 3x^2)/(x + 1)$ is defined and

$$\left| \frac{2 - x - 3x^2}{x + 1} - 5 \right| = \left| \frac{(x + 1)(2 - 3x)}{x + 1} - 5 \right| = |(2 - 3x) - 5| = |-3x - 3|$$
$$= |(-3)(x + 1)| = |-3||x + 1| = 3|x - (-1)|.$$

Now, we know that $|x - (-1)| < \delta$, so $3|x - (-1)| < 3\delta$, and we wanted to show that $|(2 - x - 3x^2)/(x + 1) - 5| < \epsilon$. If only 3δ were equal to ϵ, the proof would work out perfectly. This suggests that we go back and fill in the blank for the value of δ with $\epsilon/3$, which is positive because ϵ is positive. Here is the resulting proof:

Proof. Suppose $\epsilon > 0$. Let $\delta = \epsilon/3 > 0$. Suppose $0 < |x - (-1)| < \delta$. Then $x \neq -1$, so $(2 - x - 3x^2)/(x + 1)$ is defined and

$$\left| \frac{2 - x - 3x^2}{x + 1} - 5 \right| = \left| \frac{(x + 1)(2 - 3x)}{x + 1} - 5 \right| = |(2 - 3x) - 5| = |-3x - 3|$$
$$= |(-3)(x + 1)| = |-3||x + 1| = 3|x - (-1)| < 3\delta = 3(\epsilon/3) = \epsilon.$$
$\square$

In most of our examples so far we have studied limits $\lim_{x \to a} f(x)$ in which $f(a)$ was undefined. For example, this was the case when we computed the velocity of the ball in Section 2.1, the example that motivated our study of limits. But there is nothing in the definition of limits that requires that $f(a)$ be undefined, and it will turn out to be useful to study limits in which $f(a)$ is defined. Our next example is such a limit.

Example 2.3.2. Show that

$$\lim_{x \to 2} \frac{2x^2 + x}{x^2 + 1} = 2.$$

Solution. Once again, we begin by leaving the choice for δ blank and trying to work out the algebra at the end of the proof:

> Suppose $\epsilon > 0$.
> Let $\delta = \underline{\ ?\ } > 0$.
> Suppose $0 < |x - 2| < \delta$. Then
>
> $$\left| \frac{2x^2 + x}{x^2 + 1} - 2 \right| = \left| \frac{2x^2 + x}{x^2 + 1} - \frac{2x^2 + 2}{x^2 + 1} \right| = \left| \frac{x - 2}{x^2 + 1} \right| = \frac{1}{x^2 + 1} |x - 2|.$$

Note that in the last step we used part 2 of Theorem 1.1.4, and we used the fact that $x^2 \geq 0$, so $x^2 + 1 \geq 1$ and therefore $|x^2 + 1| = x^2 + 1$. Dividing both sides of the inequality $1 \leq x^2 + 1$ by the positive number $x^2 + 1$ we conclude that $1/(x^2 + 1) \leq 1$, and multiplying this inequality by the positive number $|x - 2|$ we conclude that

$$\frac{1}{x^2 + 1} |x - 2| \leq |x - 2| < \delta.$$

Thus $|(2x^2 + x)/(x^2 + 1) - 2| < \delta$. Since we wanted to end up with $|(2x^2 + x)/(x^2 + 1) - 2| < \epsilon$, it appears that the choice $\delta = \epsilon$ would work. Filling this in, we get the following proof:

Proof. Suppose $\epsilon > 0$. Let $\delta = \epsilon > 0$. Suppose $0 < |x - 2| < \delta$. Then

$$\left| \frac{2x^2 + x}{x^2 + 1} - 2 \right| = \left| \frac{x - 2}{x^2 + 1} \right| = \frac{1}{x^2 + 1} |x - 2| \leq |x - 2| < \delta = \epsilon. \qquad \square$$

    Finding a suitable choice for δ in our next example is much harder than it was in the previous ones.

Example 2.3.3. Show that

$$\lim_{x \to 1} \frac{1}{x} = 1.$$

Solution. As usual, we begin by leaving the choice for δ blank:

> Suppose $\epsilon > 0$.
> Let $\delta = \underline{\ ?\ } > 0$.
> Suppose $0 < |x - 1| < \delta$.

Next we would like to compute $|1/x - 1|$ and show that it is less than ϵ, but our first worry is that x could be 0, giving us a 0 in the denominator. One way to deal with this worry would be to choose $\delta = 1$. Then from $|x - 1| < \delta = 1$ we would get $0 < x < 2$, so x can't be 0. Now we can go ahead and compute

$$\left| \frac{1}{x} - 1 \right| = \left| \frac{1 - x}{x} \right| = \frac{|1 - x|}{|x|} = \frac{1}{x} |x - 1|,$$

where we have used the fact that $x > 0$ to allow us to rewrite $|x|$ as just x. Unfortunately, the factor $1/x$ will be much harder to deal with than the factor $1/(x^2 + 1)$ in the previous example. The reason is that if x is close to 0 then $1/x$ will be very large, leading to very large values for $|1/x - 1|$. For example,

$$\text{if } x = 0.1, \quad \left|\frac{1}{x} - 1\right| = \frac{1}{x}|x - 1| = \frac{1}{0.1}|0.1 - 1| = 10(0.9) = 9;$$

$$\text{if } x = 0.01, \quad \left|\frac{1}{x} - 1\right| = \frac{1}{x}|x - 1| = \frac{1}{0.01}|0.01 - 1| = 100(0.99) = 99;$$

$$\text{if } x = 0.001, \quad \left|\frac{1}{x} - 1\right| = \frac{1}{x}|x - 1| = \frac{1}{0.001}|0.001 - 1| = 1000(0.999) = 999.$$

It appears that if we want to be able to prove that $|1/x - 1|$ is small—smaller than ϵ, which might itself be very small—then we have to make sure not only that x isn't equal to 0, but also that it isn't too close to 0. So let's try a smaller value of δ—say, $\delta = 1/2$. Now from $|x - 1| < \delta = 1/2$ we can conclude that $1/2 < x < 3/2$. From this we can show that $1/x$ won't be very large: multiplying the inequality $1/2 < x$ by 2 and then dividing by x, we conclude that $1/x < 2$, and therefore

$$\left|\frac{1}{x} - 1\right| = \frac{1}{x}|x - 1| < 2|x - 1| < 2\delta.$$

Since we want $|1/x - 1| < \epsilon$, we now see that if $\delta = \epsilon/2$, then we will be able to finish the proof.

But wait! We seem to be using two different choices for δ. In the first part of the proof we assumed $\delta = 1/2$, and at the end of the proof we decided we wanted to have $\delta = \epsilon/2$. Which one should we choose? The number δ can have only one value.

In situations like this, when we have two choices for δ and we don't know which to use, there is a trick that is often helpful: use whichever candidate for δ is smaller. We introduce the notation $\min(a, b)$ to denote the smaller of the two numbers a and b. In other words,

$$\min(a, b) = \begin{cases} a, & \text{if } a \leq b, \\ b, & \text{if } a > b. \end{cases}$$

Notice that in general, $\min(a, b) \leq a$ and $\min(a, b) \leq b$. Also, if $a > 0$ and $b > 0$ then $\min(a, b) > 0$. Our choice for δ will be $\delta = \min(1/2, \epsilon/2)$.

Proof. Suppose $\epsilon > 0$. Let $\delta = \min(1/2, \epsilon/2)$. Notice that we have $\delta > 0$, $\delta \leq 1/2$, and $\delta \leq \epsilon/2$. Now suppose that $0 < |x - 1| < \delta$. Then $|x - 1| < \delta \leq 1/2$, so $1/2 < x < 3/2$, and therefore $1/x$ is defined and $1/x < 2$. Thus,

$$\left|\frac{1}{x} - 1\right| = \left|\frac{1 - x}{x}\right| = \frac{|1 - x|}{|x|} = \frac{1}{x}|x - 1| < 2|x - 1| < 2\delta \leq 2(\epsilon/2) = \epsilon. \qquad \square$$

In later examples we will usually just give the final proof, without discussing how we came up with the choice for δ. We suggest that on a first reading of such a proof you simply try to follow the steps without worrying about how δ was chosen. After you have been through the proof once, you can then go back and try to figure out what motivated the choice of δ.

So far we have used x as the variable in all of our limits, but sometimes it is convenient to use other variables. In our next example, we use the variable t. This time we present the proof first, and then explain how we chose δ afterwards.

Example 2.3.4. Show that
$$\lim_{t \to 3} t^2 = 9.$$

Solution. Suppose $\epsilon > 0$. Let $\delta = \min(1, \epsilon/7)$. Then $\delta > 0$, $\delta \le 1$, and $\delta \le \epsilon/7$. Suppose $0 < |t - 3| < \delta$. Note that

$$|t^2 - 9| = |(t + 3)(t - 3)| = |t + 3||t - 3|.$$

Since $|t - 3| < \delta \le 1$, we have $2 < t < 4$. Thus, $5 < t + 3 < 7$, so $-7 < t + 3 < 7$ and therefore $|t + 3| < 7$. Hence

$$|t^2 - 9| = |t + 3||t - 3| < 7|t - 3| < 7\delta \le 7(\epsilon/7) = \epsilon. \qquad \square$$

What is the motivation for the formula we used for δ in Example 2.3.4? We start by considering the quantity $|t^2 - 9|$, which will have to come out less than ϵ at the end of the proof. Since $|t^2 - 9| = |t + 3||t - 3|$, we should try to make sure that both $|t + 3|$ and $|t - 3|$ are small. We know we will have $|t - 3| < \delta$, but what about $|t + 3|$? Since t will be close to 3, $|t + 3|$ will be close to 6, but how close? How much larger than 6 could it be? It depends on how close t is to 3, which in turn depends on the value of δ. So we seem to be going in circles: we need to know something about δ in order to know how big $|t + 3|$ could be, in order to decide how to choose δ! The solution to this dilemma is to consider a formula for δ of the form $\delta = \min(1, \underline{\ ?\ })$. (The choice of the number 1 here was completely arbitrary; see Exercise 14 to see how the proof might have come out if we had used the number 2 instead of 1.) Although this doesn't pin down δ completely, it does tell us that we will have $\delta \le 1$. This allows us to show that if $0 < |t - 3| < \delta$ then $|t + 3| < 7$, and therefore $|t^2 - 9| = |t + 3||t - 3| < 7\delta$. Finally, in order to end up with ϵ at the end of the proof, we fill in the blank in our formula for δ with $\epsilon/7$.

We end this section with one more example of a proof of a limit statement. This time you're on your own to figure out the motivation for our choice of δ!

Example 2.3.5. Show that
$$\lim_{w \to 4} \sqrt{w} = 2.$$

Solution. Suppose $\epsilon > 0$. Let $\delta = \min(4, 2\epsilon)$. Then $\delta > 0$, $\delta \le 4$, and $\delta \le 2\epsilon$. Suppose $0 < |w - 4| < \delta$. Then since $\delta \le 4$ we have $|w - 4| < 4$ and therefore $0 < w < 8$, so $\sqrt{w}$ is defined. Note that since $\sqrt{w}$ denotes the *nonnegative* square root of w, we always have $\sqrt{w} \ge 0$. To complete the proof we begin with a tricky algebraic step, multiplying by $(\sqrt{w} + 2)/(\sqrt{w} + 2)$:

$$|\sqrt{w} - 2| = \left| \frac{(\sqrt{w} - 2)(\sqrt{w} + 2)}{\sqrt{w} + 2} \right| = \left| \frac{w - 4}{\sqrt{w} + 2} \right|$$

$$= \frac{1}{\sqrt{w} + 2}|w - 4| \le \frac{1}{2}|w - 4| < \frac{1}{2}\delta \le \frac{1}{2}(2\epsilon) = \epsilon. \qquad \square$$

Exercises 2.3

1–7: Use the definition of limits to prove the limit statement.

1. $\lim\limits_{x \to 3} (2x - 4) = 2.$

2. $\lim\limits_{t \to 1} (5 - 3t) = 2.$

3. $\lim\limits_{x \to -2} \dfrac{x^2 - 4}{x + 2} = -4.$

4. $\lim\limits_{x \to 2} \dfrac{x^2 - x - 2}{x - 2} = 3.$

5. $\lim\limits_{r \to 3} \dfrac{3r^2 - 8r - 3}{2r - 6} = 5.$

6. $\lim\limits_{x \to 2} \dfrac{x^2 + 2x - 8}{3x - 6} = 2.$

7. $\lim\limits_{u \to -1} \left(\dfrac{u^2 + 4u + 3}{u + 1} + u \right) = 1.$

8–13: Use the given formula for δ to prove the limit statement.

8. $\lim\limits_{x \to 2} (x^2 - 3x + 5) = 3, \delta = \min(1, \epsilon/2).$

9. $\lim\limits_{v \to 1} \dfrac{4}{3v - 1} = 2, \delta = \min(1/2, \epsilon/12).$

10. $\lim\limits_{t \to 1} \dfrac{2 - 2t}{t^2 - 1} = -1, \delta = \min(1, \epsilon).$

11. $\lim\limits_{x \to 3} \lfloor x/2 \rfloor = 1, \delta = 1.$

12. $\lim\limits_{x \to 2} x^3 = 8, \delta = \min(2, \epsilon/28).$

13. $\lim\limits_{w \to 2} \sqrt{2w + 5} = 3, \delta = \min(1, 3\epsilon/2).$

14. Redo Example 2.3.4 using a formula for δ of the form $\delta = \min(2, \underline{\ ?\ })$. You need to find something you can put in the blank so that the proof can be completed.

15–24: Use the definition of limits to prove the limit statement.

15. $\lim\limits_{x \to 1} (x^2 + x) = 2.$

16. $\lim\limits_{x \to 2} (3x^2 - 6x) = 0.$

17. $\lim\limits_{x \to -1} \dfrac{x^3 + x^2 - 3x - 3}{x + 1} = -2.$

18. $\lim\limits_{u \to 3} \dfrac{4}{u-1} = 2.$

19. $\lim\limits_{t \to 5} \dfrac{3t - 10}{t} = 1.$

20. $\lim\limits_{z \to 1} \dfrac{z-1}{2z^2 + z - 3} = \dfrac{1}{5}.$

21. $\lim\limits_{y \to 1} (y^3 - y^2 + 7) = 7.$

22. $\lim\limits_{y \to 0} (y^3 - y^2 + 7) = 7.$

23. $\lim\limits_{x \to 4} \dfrac{1}{\sqrt{x}} = \dfrac{1}{2}.$

24. $\lim\limits_{x \to 0} \sqrt{x^2 + 1} = 1.$

2.4   Limit Theorems

The examples in the last section show that using the ϵ-δ definition to verify limits of even fairly simple functions can be quite difficult. How will we ever be able to deal with limits of complicated functions? Fortunately, there is a better way. We already hinted at this better way in Section 1.4, when we discussed the fact that many complicated functions can be built up from simple functions. Our strategy will be to figure out limits of the simplest functions, and then work out what happens to limits of functions when the functions are combined. As we saw in Section 1.4, the simplest functions are constant functions and the identity function, so we begin with their limits. Fortunately, they are quite easy:

Theorem 2.4.1. *Let a be any number.*

1. *For any number c, $\lim_{x \to a} c = c$.*

2. *$\lim_{x \to a} x = a$.*

Proof. The first statement says that if x is close to a, but not equal to a, then c is close to c. The second says that if x is close to a, but not equal to a, then x is close to a. Both statements should be intuitively obvious, but they can also be proven easily by the ϵ-δ method:

1. Suppose $\epsilon > 0$. Let $\delta = 1 > 0$. Suppose $0 < |x - a| < \delta$. Then $|c - c| = 0 < \epsilon$.

2. Suppose $\epsilon > 0$. Let $\delta = \epsilon > 0$. Suppose $0 < |x - a| < \delta$. Then since $\delta = \epsilon$, $|x - a| < \epsilon$. $\square$

For limits of combinations of functions, we have the following theorem:

Theorem 2.4.2. *Suppose* $\lim_{x \to a} f(x) = L$ *and* $\lim_{x \to a} g(x) = M$. *Then:*

1. $\lim_{x \to a} (f(x) + g(x)) = L + M$.

2. *For any number c,* $\lim_{x \to a} cf(x) = cL$.

3. $\lim_{x \to a} (f(x) - g(x)) = L - M$.

4. $\lim_{x \to a} (f(x) \cdot g(x)) = LM$.

5. *If* $M \neq 0$ *then* $\lim_{x \to a} (f(x)/g(x)) = L/M$.

Once again, these statements should make intuitive sense. For example, if $f(x)$ is close to L and $g(x)$ is close to M, it should make sense that $f(x) + g(x)$ will be close to $L + M$. But does this intuitive reasoning provide the complete precision that we expect in our limit statements? The only way to be sure that these statements are correct is to give a careful proof based on the definition of limits.

Before giving the proof, we make a few remarks about the strategy for the proof. We are told in the statement of the theorem that $\lim_{x \to a} f(x) = L$ and $\lim_{x \to a} g(x) = M$, so we know that the following statements are true:

(a) For every $\epsilon_1 > 0$ there is some $\delta_1 > 0$ such that if $0 < |x - a| < \delta_1$ then $|f(x) - L| < \epsilon_1$.

(b) For every $\epsilon_2 > 0$ there is some $\delta_2 > 0$ such that if $0 < |x - a| < \delta_2$ then $|g(x) - M| < \epsilon_2$.

(We have added subscripts to the letters ϵ and δ here so that we won't confuse the ϵ's and δ's for f with those for g, or with the ϵ's and δ's we will use to prove the limit statements in parts 1–5 of the theorem.) Notice that we are not trying to prove statements (a) and (b); these are statements that we *know* are true, and that we can use in the course of the proof. Since statement (a) says that something is true about *every* $\epsilon_1 > 0$, when using statement (a) we are free to assign any positive value we please to ϵ_1. Similarly, we can choose any positive value for ϵ_2 in statement (b). One of the key ideas in proving all parts of this theorem will be making appropriate choices for ϵ_1 and ϵ_2.

In the proof of part 1, as usual we will start with "Suppose $\epsilon > 0$," and then we will specify a choice for δ and assume that $0 < |x - a| < \delta$. At the end of the proof we will need to show that $|(f(x) + g(x)) - (L + M)|$ is smaller than ϵ. To relate this quantity to the ones mentioned in statements (a) and (b), we begin by using the triangle inequality (Theorem 1.1.5):

$$|(f(x) + g(x)) - (L + M)| = |(f(x) - L) + (g(x) - M)| \leq |f(x) - L| + |g(x) - M|.$$

Our plan is to get both $|f(x) - L|$ and $|g(x) - M|$ to be smaller than $\epsilon/2$, since it will follow then that their sum is smaller than ϵ. Looking back at statements (a) and (b), this suggests letting $\epsilon_1 = \epsilon_2 = \epsilon/2$. Statements (a) and (b) will then give us two numbers δ_1 and δ_2, and these seem like good candidates to use for the δ in the proof of part 1. But which one should we use? The smaller one, of course!

Proof. 1. Suppose $\epsilon > 0$. Applying statement (a) above with $\epsilon_1 = \epsilon/2$, we conclude that there is some number $\delta_1 > 0$ such that

(a′) if $0 < |x - a| < \delta_1$ then $|f(x) - L| < \epsilon/2$.

Similarly, statement (b) with $\epsilon_2 = \epsilon/2$ tells us that there is a number $\delta_2 > 0$ such that

(b′) if $0 < |x - a| < \delta_2$ then $|g(x) - M| < \epsilon/2$.

Let $\delta = \min(\delta_1, \delta_2)$. Then $\delta > 0$, $\delta \leq \delta_1$, and $\delta \leq \delta_2$.

Suppose $0 < |x - a| < \delta$. Then since $\delta \leq \delta_1$, $0 < |x - a| < \delta_1$, so by statement (a′), $|f(x) - L| < \epsilon/2$. Similarly, $0 < |x - a| < \delta_2$, and statement (b′) tells us that $|g(x) - M| < \epsilon/2$. Therefore

$$
\begin{aligned}
|(f(x) + g(x)) - (L + M)| &= |(f(x) - L) + (g(x) - M)| \\
&\leq |f(x) - L| + |g(x) - M| \quad \text{(by the triangle inequality)} \\
&< \frac{\epsilon}{2} + \frac{\epsilon}{2} = \epsilon.
\end{aligned}
$$

2. If $c = 0$, then by part 1 of Theorem 2.4.1, $\lim_{x \to a} cf(x) = \lim_{x \to a} 0 = 0 = cL$. Now suppose $c \neq 0$, and suppose $\epsilon > 0$. By statement (a) with $\epsilon_1 = \epsilon/|c| > 0$, there must be some number $\delta_1 > 0$ such that if $0 < |x - a| < \delta_1$ then $|f(x) - L| < \epsilon/|c|$. Let $\delta = \delta_1$.

Suppose $0 < |x - a| < \delta$. Then $0 < |x - a| < \delta_1$, so $|f(x) - L| < \epsilon/|c|$ and therefore

$$
|cf(x) - cL| = |c(f(x) - L)| = |c||f(x) - L| < |c| \cdot \frac{\epsilon}{|c|} = \epsilon.
$$

3. This follows from parts 1 and 2. Applying part 2 to the function g with $c = -1$, we get $\lim_{x \to a}(-1)g(x) = (-1)M$. Therefore by part 1,

$$
\lim_{x \to a}(f(x) - g(x)) = \lim_{x \to a}(f(x) + (-1)g(x)) = L + (-1)M = L - M.
$$

4. As in the proof of part 1, we need to bring the quantities $f(x) - L$ and $g(x) - M$ into the proof somehow. We begin with following equation:

$$
(f(x) - L)(g(x) - M) = f(x)g(x) - Lg(x) - Mf(x) + LM.
$$

Rearranging, we get

$$
f(x)g(x) = (f(x) - L)(g(x) - M) + Lg(x) + Mf(x) - LM. \tag{2.5}
$$

Our plan now is to find the limit of $f(x)g(x)$ by finding the limit of the right-hand side of (2.5), using parts 1–3, which we have already proven. By part 2, we have $\lim_{x \to a} Lg(x) = LM$ and $\lim_{x \to a} Mf(x) = ML$, so by part 1,

$$
\lim_{x \to a}(Lg(x) + Mf(x)) = LM + ML = 2LM.
$$

Thus, by part 3, $\lim_{x \to a}(Lg(x) + Mf(x) - LM) = 2LM - LM = LM$. If we can show that $\lim_{x \to a}(f(x) - L)(g(x) - M) = 0$, then by one more application of part 1 we can conclude from (2.5) that

$$\lim_{x \to a} (f(x) \cdot g(x)) = \lim_{x \to a} [(f(x) - L)(g(x) - M) + (Lg(x) + Mf(x) - LM)]$$
$$= 0 + LM = LM,$$

as required.

Thus, we just need to prove that $\lim_{x \to a} ((f(x) - L)(g(x) - M)) = 0$. For this, we begin as usual by assuming $\epsilon > 0$. Applying statements (a) and (b) with $\epsilon_1 = \epsilon_2 = \sqrt{\epsilon}$, we conclude that there are numbers $\delta_1 > 0$ and $\delta_2 > 0$ such that

(a″) if $0 < |x - a| < \delta_1$ then $|f(x) - L| < \sqrt{\epsilon}$;

(b″) if $0 < |x - a| < \delta_2$ then $|g(x) - M| < \sqrt{\epsilon}$.

Let $\delta = \min(\delta_1, \delta_2)$. Then $\delta > 0$, $\delta \le \delta_1$, and $\delta \le \delta_2$.

Suppose $0 < |x - a| < \delta$. Then $0 < |x - a| < \delta_1$ and $0 < |x - a| < \delta_2$, so by statements (a″) and (b″) we have $|f(x) - L| < \sqrt{\epsilon}$ and $|g(x) - M| < \sqrt{\epsilon}$. Therefore

$$|(f(x) - L)(g(x) - M)| = |f(x) - L||g(x) - M| < \sqrt{\epsilon} \cdot \sqrt{\epsilon} = \epsilon.$$

5. We leave it as an excrcise for you to verify that $\lim_{x \to a} (1/g(x)) = 1/M$ (see Exercise 28). Applying part 4, we conclude that

$$\lim_{x \to a} \frac{f(x)}{g(x)} = \lim_{x \to a} \left(f(x) \cdot \frac{1}{g(x)} \right) = L \cdot \frac{1}{M} = \frac{L}{M}. \qquad \square$$

Using Theorems 2.4.1 and 2.4.2, we can now find many limits very easily. For example, according to part 2 of Theorem 2.4.1, for any number a, $\lim_{x \to a} x = a$. Since $x^2 = x \cdot x$, we can use part 4 of Theorem 2.4.2 to conclude that

$$\lim_{x \to a} x^2 = \lim_{x \to a} (x \cdot x) = a \cdot a = a^2.$$

Since $x^3 = x^2 \cdot x$, applying part 4 of Theorem 2.4.2 again we get

$$\lim_{x \to a} x^3 = \lim_{x \to a} (x^2 \cdot x) = a^2 \cdot a = a^3.$$

Repeating this reasoning, we find that for every positive integer k,

$$\lim_{x \to a} x^k = a^k.$$

Next we apply part 2 of Theorem 2.4.2 to see that for any number c,

$$\lim_{x \to a} cx^k = ca^k.$$

Adding functions of this form, and applying part 1 of Theorem 2.4.2, we conclude that for any numbers $c_0, c_1, \ldots, c_n$,

$$\lim_{x \to a} (c_n x^n + c_{n-1} x^{n-1} + \cdots + c_1 x + c_0) = c_n a^n + c_{n-1} a^{n-1} + \cdots + c_1 a + c_0.$$

Recall that functions like the one in this limit are called polynomials. Thus, we have proven the following fact:

Theorem 2.4.3. *If f is a polynomial, then for any number a,*

$$\lim_{x \to a} f(x) = f(a).$$

Example 2.4.4. Find the following limits:

(a) $\lim_{x \to 3} (x^3 - 2x^2 + 4x - 1)$, (b) $\lim_{x \to 3} \dfrac{x^3 - 2x^2 + 4x - 1}{x^2 - 5x + 1}$.

Solution. (a) According to Theorem 2.4.3,

$$\lim_{x \to 3} (x^3 - 2x^2 + 4x - 1) = 3^3 - 2(3^2) + 4(3) - 1 = 20.$$

(b) Part (a) gives the limit of the numerator. Similarly, the limit of the denominator is $\lim_{x \to 3}(x^2 - 5x + 1) = 3^2 - 5(3) + 1 = -5$. Since the limit of the denominator isn't 0, by part 5 of Theorem 2.4.2,

$$\lim_{x \to 3} \frac{x^3 - 2x^2 + 4x - 1}{x^2 - 5x + 1} = \frac{20}{-5} = -4. \qquad \square$$

To appreciate the power of the theorems we have proven, we suggest that you try to verify the limits in the last example using the ϵ-δ method! This example also illustrates a principle that we will see repeatedly in this book: theorems are not just for advancing the theoretical side of mathematics; they also often provide useful methods for solving problems.

Part 5 of Theorem 2.4.2 gives us a method that we can use to compute limits of the form $\lim_{x \to a}(f(x)/g(x))$ if $\lim_{x \to a} g(x) \neq 0$. But what if $\lim_{x \to a} g(x) = 0$? Theorem 2.4.2 does not tell us what happens in this case. You might be tempted to say that $\lim_{x \to a}(f(x)/g(x))$ must be undefined in this case because we would have a 0 in the denominator, but this is not true. In fact, we saw in Example 2.3.1 that $\lim_{x \to -1}(2 - x - 3x^2)/(x + 1) = 5$, even though the limit of the denominator is $\lim_{x \to -1}(x + 1) = 0$. How is this possible? Isn't division by 0 always undefined?

To resolve this apparent paradox, we must think more carefully about what "limit" means. Remember that to compute $\lim_{x \to a}(f(x)/g(x))$ we must look at $f(x)/g(x)$ for values of x that are close to a but not equal to a. If $\lim_{x \to a} g(x) = 0$, then for such values of x, $g(x)$ will be close to 0, but it might not be equal to 0. And although division by 0 is not allowed, there is nothing wrong with dividing by a number close to 0. Thus, there is no reason why $\lim_{x \to a}(f(x)/g(x))$ couldn't be defined.

How, then, do we evaluate $\lim_{x \to a}(f(x)/g(x))$ if $\lim_{x \to a} g(x) = 0$? For such limits, our next theorem is often helpful. The idea behind this theorem is that in computing a limit as x approaches a, only values of x close to a, but not equal to a, should matter. So if f_1 and f_2 are functions such that $f_1(x)$ and $f_2(x)$ are the same for all values of x that are close to a but not equal to a, then their limits as x approaches a should also be the same. The theorem is illustrated in Figure 2.20.

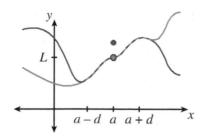

Figure 2.20: Theorem 2.4.5. The red and blue curves are the graphs of two functions f_1 and f_2. For all x, if $0 < |x - a| < d$ then $f_1(x) = f_2(x)$. Also, $\lim_{x \to a} f_1(x) = \lim_{x \to a} f_2(x) = L$.

Theorem 2.4.5. *Suppose that $d > 0$, and f_1 and f_2 are functions such that for all x, if $0 < |x - a| < d$ then $f_1(x) = f_2(x)$. Then $\lim_{x \to a} f_1(x) = \lim_{x \to a} f_2(x)$, where we interpret this equation to mean that either both limits are defined and they are equal, or both limits are undefined.*

Proof. We will show that if as $x \to a^{\neq}$, $f_1(x) \to L$, then as $x \to a^{\neq}$, $f_2(x) \to L$. Similar reasoning would show that if as $x \to a^{\neq}$, $f_2(x) \to L$, then as $x \to a^{\neq}$, $f_1(x) \to L$, and the theorem follows.

So suppose that as $x \to a^{\neq}$, $f_1(x) \to L$. To prove that as $x \to a^{\neq}$, $f_2(x) \to L$, we begin as usual by assuming that $\epsilon > 0$. Since we have assumed that as $x \to a^{\neq}$, $f_1(x) \to L$, we know that there is some number $\delta_1 > 0$ such that if $0 < |x - a| < \delta_1$ then $|f_1(x) - L| < \epsilon$.

Now let $\delta = \min(\delta_1, d)$, and suppose that $0 < |x - a| < \delta$. Then $0 < |x - a| < \delta_1$, so $|f_1(x) - L| < \epsilon$, and also $0 < |x - a| < d$, so $f_1(x) = f_2(x)$. Combining these two facts we conclude that $|f_2(x) - L| < \epsilon$, as required. □

Theorem 2.4.5 says that if you are trying to compute $\lim_{x \to a} f_1(x)$, you can get the answer by computing $\lim_{x \to a} f_2(x)$ instead, as long as f_1 and f_2 agree on values of x close to a but not equal to a. Whatever answer you get for $\lim_{x \to a} f_2(x)$—whether you find a numerical value for the limit or you discover that the limit is undefined—the answer has to be the same for $\lim_{x \to a} f_1(x)$.

Example 2.4.6. Find

$$\lim_{x \to -1} \frac{x^3 + x^2 + 4x + 4}{x^2 - 1}.$$

Solution. Since the limit of the denominator is $\lim_{x \to -1}(x^2 - 1) = 0$, we cannot apply part 5 of Theorem 2.4.2. However, for all $x \neq \pm 1$ we have

$$\frac{x^3 + x^2 + 4x + 4}{x^2 - 1} = \frac{(x^2 + 4)(x + 1)}{(x - 1)(x + 1)} = \frac{x^2 + 4}{x - 1}.$$

In other words, if we let

$$f_1(x) = \frac{x^3 + x^2 + 4x + 4}{x^2 - 1}, \qquad f_2(x) = \frac{x^2 + 4}{x - 1},$$

then for all x except ± 1, $f_1(x) = f_2(x)$. Notice that f_1 and f_2 are not the same function, since $f_1(-1)$ is undefined, but $f_2(-1) = -5/2$. Also, neither $f_1(1)$ nor $f_2(1)$ is defined. However, neither of these facts is relevant when we are computing a limit as x approaches -1. For such limits, only values of x close to -1 but not equal to -1 matter, and for these values of x, both functions are defined and they are equal. More specifically, if $0 < |x - (-1)| < 2$ then $-3 < x < 1$ and $x \neq -1$, so $x \neq \pm 1$; therefore $f_1(x)$ and $f_2(x)$ are defined and equal. We want to find $\lim_{x \to -1} f_1(x)$, and according to Theorem 2.4.5 we can get the answer by evaluating $\lim_{x \to -1} f_2(x)$ instead.

Fortunately, $\lim_{x \to -1} f_2(x)$ is easy to compute: By Theorem 2.4.3 we have $\lim_{x \to -1}(x^2 + 4) = 5$ and $\lim_{x \to -1}(x - 1) = -2$. Since the limit of the denominator is not 0, part 5 of Theorem 2.4.2 tells us that

$$\lim_{x \to -1} f_2(x) = \lim_{x \to -1} \frac{x^2 + 4}{x - 1} = \frac{5}{-2} = -\frac{5}{2}.$$

Thus, this is the answer to the original limit problem.   □

Recall that a quotient of two polynomials is called a rational function. For example, the function whose limit we computed in the last example is a rational function. When computing the limit of a rational function, if the limit of the denominator is 0 then it is often helpful to factor the numerator and denominator and see if a factor can be canceled, as in the last example.

Usually when we use Theorem 2.4.5, we won't spell out the steps in quite so much detail. For example, here is another application of Theorem 2.4.5.

Example 2.4.7. Find
$$\lim_{x \to 2} \frac{3x^2 - 7x + |x - 4|}{x - 2}.$$

Solution. The hardest part of this limit is the term $|x - 4|$. According to the definition of absolute values, $|x - 4|$ is either $x - 4$ or $-(x - 4)$, but which should we use in this problem? Since we are computing a limit as x approaches 2, we are only interested in values of x that are close to 2 but not equal to 2. For such values of x, $x - 4$ is negative, and therefore $|x - 4| = -(x - 4)$. Thus, this is the formula we should use in our calculations:

$$\lim_{x \to 2} \frac{3x^2 - 7x + |x - 4|}{x - 2} = \lim_{x \to 2} \frac{3x^2 - 7x - (x - 4)}{x - 2} \qquad \text{(by Theorem 2.4.5)}$$

$$= \lim_{x \to 2} \frac{3x^2 - 8x + 4}{x - 2}$$

$$= \lim_{x \to 2} \frac{(x - 2)(3x - 2)}{x - 2}$$

$$= \lim_{x \to 2}(3x - 2) \qquad \text{(by Theorem 2.4.5)}$$

$$= 4 \qquad \text{(by Theorem 2.4.3).} \quad □$$

Notice that in steps one and four of this solution, we changed the function whose limit is being computed, but by Theorem 2.4.5 in each case the change has no effect on the limit. In the first step this is because for all $x < 4$, $|x - 4| = -(x - 4)$, and in the fourth step it is because canceling the factor of $x - 2$ leaves the function unchanged for $x \neq 2$. Thus, in both cases the function has not been changed for values of x that are close to 2 but not equal to 2. Since the final limit turns out to equal 4, we conclude that the original limit is also 4.

You might have noticed that in Example 2.4.7, not only was the limit of the denominator equal to 0, but the limit of the numerator was as well: $\lim_{x \to 2}(3x^2 - 7x + |x - 4|) = \lim_{x \to 2}(3x^2 - 7x - (x - 4)) = 0$. And you can check that the same is true in Examples 2.3.1 and 2.4.6. Is this just a coincidence? No, it is part of a general pattern:

Theorem 2.4.8. *If* $\lim_{x \to a}(f(x)/g(x))$ *is defined and* $\lim_{x \to a} g(x) = 0$, *then* $\lim_{x \to a} f(x) = 0$ *as well. Thus, if* $\lim_{x \to a} g(x) = 0$ *but* $\lim_{x \to a} f(x) \neq 0$, *then* $\lim_{x \to a}(f(x)/g(x))$ *must be undefined.*

Proof. Suppose that $\lim_{x \to a}(f(x)/g(x))$ is defined—say it is equal to L—and $\lim_{x \to a} g(x) = 0$. Notice that for $\lim_{x \to a}(f(x)/g(x))$ to make sense, $f(x)/g(x)$ must be defined for x close to a but not equal to a. Thus, in particular, although $g(x)$ must be close to 0 for such values of x, it cannot be equal to 0. To compute $\lim_{x \to a} f(x)$, we begin with the equation

$$f(x) = \frac{f(x)}{g(x)} \cdot g(x).$$

This equation may not be true for all values of x. For example, it would not be true if $g(x) = 0$ or $g(x)$ is undefined. But as we have just observed, for x close to a but not equal to a, $f(x)/g(x)$ is defined, and therefore the equation is correct. We can therefore compute

$$\lim_{x \to a} f(x) = \lim_{x \to a}\left(\frac{f(x)}{g(x)} \cdot g(x)\right) = L \cdot 0 = 0. \qquad \square$$

Example 2.4.9. Find the following limits:

(a) $\lim\limits_{x \to 1} \dfrac{x^2 - 1}{x^2 - 2x + 1}$, (b) $\lim\limits_{x \to 1}\left(\dfrac{x}{x - 1} - \dfrac{1}{x^2 - x}\right)$.

Solution. For limit (a), we have

$$\lim_{x \to 1} \frac{x^2 - 1}{x^2 - 2x + 1} = \lim_{x \to 1} \frac{(x - 1)(x + 1)}{(x - 1)^2} = \lim_{x \to 1} \frac{x + 1}{x - 1}.$$

In the last limit, the denominator approaches 0 and the numerator approaches 2, so by Theorem 2.4.8, the limit is undefined.

For limit (b), we have $\lim_{x \to 1} x = 1$ and $\lim_{x \to 1}(x - 1) = 0$, so by Theorem 2.4.8, $\lim_{x \to 1}(x/(x - 1))$ is undefined, and similar reasoning shows that $\lim_{x \to 1}(1/(x^2 - x))$ is also undefined. You might be tempted to conclude that limit (b) is also undefined, but this would be a misuse of Theorem 2.4.2. Part 3 of Theorem 2.4.2 says that if the limits of $f(x)$ and $g(x)$ are L and M, respectively, then the limit of $f(x) - g(x)$ is $L - M$. It does *not* say that if the limits of $f(x)$ and $g(x)$ are undefined then the limit of $f(x) - g(x)$

is undefined. In fact, limit (b) is defined, as we can see by getting a common denominator and combining the fractions:

$$\lim_{x \to 1} \left(\frac{x}{x-1} - \frac{1}{x^2 - x} \right) = \lim_{x \to 1} \left(\frac{x^2}{x^2 - x} - \frac{1}{x^2 - x} \right) = \lim_{x \to 1} \frac{x^2 - 1}{x^2 - x}$$

$$= \lim_{x \to 1} \frac{(x-1)(x+1)}{x(x-1)} = \lim_{x \to 1} \frac{x+1}{x} = \frac{2}{1} = 2. \qquad \square$$

We prove one more theorem in this section to deal with functions like our example $h_2(x) = |x| \sin(\pi/(2x))$ from Section 2.2. Recall that the graph of $y = h_2(x)$ was a complicated curve squeezed between the much simpler curves $y = |x|$ and $y = -|x|$. For functions like this the following theorem is often useful:

Theorem 2.4.10 (Squeeze Theorem). *Suppose that $d > 0$, and for all x, if $0 < |x - a| < d$ then $g(x) \leq f(x) \leq h(x)$. Suppose also that $\lim_{x \to a} g(x) = \lim_{x \to a} h(x) = L$. Then $\lim_{x \to a} f(x) = L$.*

Proof. Suppose $\epsilon > 0$. Since $\lim_{x \to a} g(x) = L$, there is a number $\delta_1 > 0$ such that

(a) if $0 < |x - a| < \delta_1$ then $|g(x) - L| < \epsilon$.

Similarly, since $\lim_{x \to a} h(x) = L$, there is a number $\delta_2 > 0$ such that

(b) if $0 < |x - a| < \delta_2$ then $|h(x) - L| < \epsilon$.

Let $\delta = \min(\delta_1, \delta_2, d)$. In other words, δ is the smallest of the three numbers δ_1, δ_2, and d, and therefore $\delta > 0$, $\delta \leq \delta_1$, $\delta \leq \delta_2$, and $\delta \leq d$. Suppose $0 < |x - a| < \delta$. Then $0 < |x - a| < \delta_1$, so by statement (a), $|g(x) - L| < \epsilon$, which means that

$$L - \epsilon < g(x) < L + \epsilon.$$

Similarly, $0 < |x - a| < \delta_2$, so by statement (b),

$$L - \epsilon < h(x) < L + \epsilon.$$

And finally, $0 < |x - a| < d$, so

$$g(x) \leq f(x) \leq h(x).$$

Putting all of this information together, we get

$$L - \epsilon < g(x) \leq f(x) \leq h(x) < L + \epsilon,$$

so $|f(x) - L| < \epsilon$. $\qquad \square$

The squeeze theorem says that if the value of $f(x)$ is "squeezed" between two quantities $g(x)$ and $h(x)$ that both approach the same limit L, then $f(x)$ is forced to approach L as well. This is illustrated in Figure 2.21. As an example, recall the inequality $-|x| \leq |x| \sin(\pi/(2x)) \leq |x|$, which holds for all $x \neq 0$. We will leave it to you to use the ϵ-δ method to verify that $\lim_{x \to 0} |x| = 0$ (see Exercise 1), and therefore

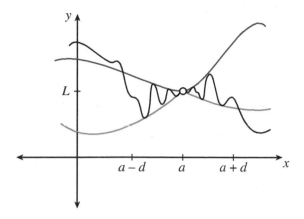

Figure 2.21: The squeeze theorem. For $0 < |x - a| < d$, the graph of f (black) is squeezed between the graphs of g (blue) and h (red).

$\lim_{x \to 0} -|x| = -1 \cdot 0 = 0$. (A second demonstration that $\lim_{x \to 0} |x| = 0$ will be given in the next section.) By the squeeze theorem, it follows that $\lim_{x \to 0} |x| \sin(\pi/(2x)) = 0$, as claimed in Section 2.2.

As another example of the use of the squeeze theorem, we prove the following useful fact. (For an alternative approach to the proof, see Exercise 2.)

Theorem 2.4.11. *If* $\lim_{x \to a} |f(x)| = 0$, *then* $\lim_{x \to a} f(x) = 0$.

Proof. Since $f(x) = \pm|f(x)|$, we have $-|f(x)| \le f(x) \le |f(x)|$. (Another way to see this is to apply Exercise 18 from Section 1.1.) If $\lim_{x \to a} |f(x)| = 0$, then $\lim_{x \to a} -|f(x)| = 0$ as well, and we can apply the squeeze theorem to conclude that $\lim_{x \to a} f(x) = 0$. $\qquad\square$

We end this section by returning to an issue we left unresolved in Section 2.2. Now that we have had some practice with the ϵ-δ definition of limits, we are ready to prove Theorem 2.2.4:

Theorem 2.2.4 (restated). There cannot be more than one number L such that as $x \to a^{\ne}$, $f(x) \to L$.

Proof. Suppose there are two different numbers L_1 and L_2 such that as $x \to a^{\ne}$, $f(x) \to L_1$ and as $x \to a^{\ne}$, $f(x) \to L_2$. Let $\epsilon = |L_1 - L_2|/2 > 0$. The idea behind the proof is that no number can be within ϵ of both L_1 and L_2, as you can see in Figure 2.22. Therefore it can never be the case that both $|f(x) - L_1| < \epsilon$ and $|f(x) - L_2| < \epsilon$ are true.

Since as $x \to a^{\ne}$, $f(x) \to L_1$, there is some $\delta_1 > 0$ such that if $0 < |x - a| < \delta_1$ then $|f(x) - L_1| < \epsilon$. Similarly, since as $x \to a^{\ne}$, $f(x) \to L_2$, there is some $\delta_2 > 0$ such that if $0 < |x - a| < \delta_2$ then $|f(x) - L_2| < \epsilon$. But now consider a value of x close enough to a that $0 < |x - a| < \delta_1$ and also $0 < |x - a| < \delta_2$. For example, we could let $x = a + \min(\delta_1, \delta_2)/2$. Then $|f(x) - L_1| < \epsilon$ and $|f(x) - L_2| < \epsilon$, which as we saw in Figure 2.22 is impossible. $\qquad\square$

Figure 2.22: If $\epsilon = |L_1 - L_2|/2$, then the solution sets of the inequalities $|y - L_1| < \epsilon$ (blue) and $|y - L_2| < \epsilon$ (red) do not overlap at all.

Exercises 2.4

1. Use the ϵ-δ definition of limits to show that $\lim_{x \to 0} |x| = 0$.

2. Use the ϵ-δ definition of limits to give an alternative proof of Theorem 2.4.11.

3–27: Evaluate the limit.

3. $\lim_{x \to 3} (x^3 - 5x^2 + 7x + 2)$.

4. $\lim_{v \to -1} (v^4 + 2v^3 + 3v^2 - v + 4)$.

5. $\lim_{v \to -1} (v^{100} + v^{99} + v^{98} + v^{97} + \cdots + v^2 + v + 1)$.

6. $\lim_{x \to 4} \dfrac{x^2 - 3x - 4}{x - 2}$.

7. $\lim_{x \to 5} \dfrac{x^2 - 3x - 10}{x^2 - 5x}$.

8. $\lim_{w \to -3} \dfrac{w^3 + 4w^2 + 3w}{w + 3}$.

9. $\lim_{x \to 2} \dfrac{x^2 + x - 6}{x^2 - x - 2}$.

10. $\lim_{t \to 3} \dfrac{t^3 - 27}{t^2 - 9}$.

11. $\lim_{x \to 2} \dfrac{x^2 - 2x}{x^2 - 4x + 4}$.

12. $\lim_{x \to -1} \dfrac{x + 4 - \frac{3}{x+2}}{x + 1}$.

13. $\lim_{u \to 1} \dfrac{u - \frac{1}{u}}{1 - \frac{1}{u}}$.

14. $\lim_{x \to -2} \left(\dfrac{x}{x + 2} - \dfrac{8}{x^2 - 4} \right)$.

15. $\lim_{x \to 7} \left(\dfrac{x-8}{x-7} + \dfrac{7}{x^2 - 7x} \right)$.

16. $\lim_{z \to 0} \left(\dfrac{1}{z} - \dfrac{1}{z^2} \right)$.

17. $\lim_{x \to 0} \left(\dfrac{1}{x} + \dfrac{1}{x^2 - x} \right)$.

18. $\lim_{u \to -3} \left[\dfrac{1}{u+3} \cdot \left(\dfrac{3}{u} + 1 \right) \right]$.

19. $\lim_{x \to 2} \left[\dfrac{1}{x-2} \cdot \left(\dfrac{1}{x} - \dfrac{1}{2} \right) \right]$.

20. $\lim_{x \to 0} \left[\dfrac{1}{x} \cdot \left(\dfrac{1}{x+1} + \dfrac{1}{x-1} \right) \right]$.

21. $\lim_{x \to 2} \dfrac{x^2 + |x| - 6}{x - 2}$.

22. $\lim_{x \to -2} \dfrac{x^2 + |x| - 6}{x + 2}$.

23. $\lim_{x \to 2} \dfrac{x^2 + |x| - 6}{x^2 + |x - 4| - 6}$.

24. $\lim_{x \to 0} \lfloor 2 - x^2 \rfloor$.

25. $\lim_{x \to 1/2} \cos(\lfloor x \rfloor)$.

26. $\lim_{x \to 0} (x^2 \sin(1/x))$.

27. $\lim_{x \to 0} \dfrac{|x|}{\cos^2(\pi/x) + 1}$.

28. (a) Show that if $\lim_{x \to a} h(x) = 1$ then $\lim_{x \to a}(1/h(x)) = 1$. (Hint: Suppose $\epsilon > 0$. Show that there is some number $\delta_1 > 0$ such that if $0 < |x - a| < \delta_1$ then $1/2 < h(x) < 3/2$, and there is some $\delta_2 > 0$ such that if $0 < |x - a| < \delta_2$ then $|h(x) - 1| < \epsilon/2$. Let $\delta = \min(\delta_1, \delta_2)$.)

  (b) Show that if $\lim_{x \to a} g(x) = M$ and $M \neq 0$ then $\lim_{x \to a}(1/g(x)) = 1/M$. (Hint: Let $h(x) = g(x)/M$ and apply part (a).)

29. Suppose that f and g are functions that agree on all values except one. In other words, there is some number c such that for all $x \neq c$, $f(x) = g(x)$, but $f(c) \neq g(c)$. Show that for every number a, $\lim_{x \to a} f(x) = \lim_{x \to a} g(x)$, where we interpret this equation to mean that either both limits are defined and they are equal, or both limits are undefined. (Note: We could state this result more informally by

saying that the value of the limit $\lim_{x \to a} f(x)$ will not be affected if we change the value of $f(c)$, for any number c. This may seem paradoxical: it suggests that none of the values of the function f are relevant to the value of $\lim_{x \to a} f(x)$!)

30. Prove the following version of the squeeze theorem: Suppose that $b < a < c$, and for all x, if $b < x < c$ and $x \neq a$ then $g(x) \leq f(x) \leq h(x)$. Suppose also that $\lim_{x \to a} g(x) = \lim_{x \to a} h(x) = L$. Then $\lim_{x \to a} f(x) = L$.

31. Prove that if $\lim_{x \to a} f(x)$ is undefined and $\lim_{x \to a} g(x)$ is defined, then $\lim_{x \to a}(f(x) + g(x))$ is undefined. (Hint: Let $L = \lim_{x \to a} g(x)$. Suppose that $\lim_{x \to a}(f(x) + g(x))$ is defined, and let $M = \lim_{x \to a}(f(x) + g(x))$. Now derive a contradiction by showing that $\lim_{x \to a} f(x)$ is defined.)

2.5   Variations on Limits

Consider the function f whose graph is shown in Figure 2.23. Clearly $\lim_{x \to a} f(x)$ is undefined, since there is no single number that $f(x)$ is close to when x is close to a but not equal to a. However, it appears that if x is close to a but less than a then $f(x)$ is close to L_1, and if x is close to a but greater than a then $f(x)$ is close to L_2. This suggests that we could define one-sided versions of limits.

We write "as $x \to a^<$, $f(x) \to L$" to indicate that if x is close to a but less than a, then $f(x)$ is close to L. Similarly, "as $x \to a^>$, $f(x) \to L$" means that if x is close to a but greater than a then $f(x)$ is close to L. For example, in Figure 2.23 we would say that as $x \to a^<$, $f(x) \to L_1$, and as $x \to a^>$, $f(x) \to L_2$. Here are the precise definitions:

Definition 2.5.1. To say that as $x \to a^<$, $f(x) \to L$ means that for every $\epsilon > 0$ there is some $\delta > 0$ such that if $a - \delta < x < a$ then $|f(x) - L| < \epsilon$. To say that as $x \to a^>$, $f(x) \to L$ means that for every $\epsilon > 0$ there is some $\delta > 0$ such that if $a < x < a + \delta$ then $|f(x) - L| < \epsilon$.

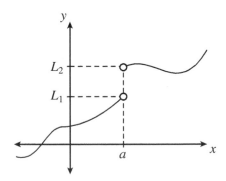

Figure 2.23: The graph of a function f for which $\lim_{x \to a} f(x)$ is undefined.

The only difference between these new definitions and our original definition of the statement "as $x \to a^{\neq}$, $f(x) \to L$" is that where the original definition contained the inequality $0 < |x - a| < \delta$, the new definitions contain the inequalities $a - \delta < x < a$ and $a < x < a + \delta$. The inequality $0 < |x - a| < \delta$ says that x is within δ of a, but not equal to a. Similarly, the inequality $a - \delta < x < a$ says that x is within δ of a but less than a, and $a < x < a + \delta$ says that x is within δ of a but greater than a. Thus, the new definitions are simply one-sided versions of the same limit concept that we captured with our original definition.

You could think of the statement "as $x \to a^{<}$, $f(x) \to L$" as meaning that when x approaches a from the negative side (i.e., from the left on the x-axis), $f(x)$ approaches L. For this reason, another notation that is often used for this statement is "as $x \to a^{-}$, $f(x) \to L$." The two notations mean exactly the same thing, and are interchangeable. Similarly, the notation "as $x \to a^{+}$, $f(x) \to L$" is an alternative way of saying that as $x \to a^{>}$, $f(x) \to L$, and it means that as x approaches a from the positive side (the right), $f(x)$ approaches L. The notations with the superscript $+$ or $-$ are more common in other books, so from now on we will usually use those notations.

As before, we can show that there cannot be more than one number L such that as $x \to a^{-}$, $f(x) \to L$, and there cannot be more than one number L such that as $x \to a^{+}$, $f(x) \to L$. We can therefore make the following definitions:

Definition 2.5.2. If there is a number L such that as $x \to a^{-}$, $f(x) \to L$, then we define $\lim_{x \to a^{-}} f(x)$ to be the unique such number L. If there is no such number L, then $\lim_{x \to a^{-}} f(x)$ is undefined. Similarly, $\lim_{x \to a^{+}} f(x)$ is the unique number L such that as $x \to a^{+}$, $f(x) \to L$, if there is such a number L.

All of the theorems about limits from the last section apply to one-sided limits as well. For example, if $\lim_{x \to a^{-}} f(x) = L$ and $\lim_{x \to a^{-}} g(x) = M$, then $\lim_{x \to a^{-}}(f(x) + g(x)) = L + M$, and similar statements hold for limits of differences, products, and quotients, and for limits as x approaches a from the right (see Exercise 1). In some cases, simple adjustments are needed in the theorems. For example, when computing $\lim_{x \to a^{-}} f(x)$, only values of x close to a but less than a should matter, so we have the following one-sided version of Theorem 2.4.5:

Theorem 2.5.3. *Suppose that $d > 0$, and f_1 and f_2 are functions such that for all x, if $a - d < x < a$ then $f_1(x) = f_2(x)$. Then $\lim_{x \to a^{-}} f_1(x) = \lim_{x \to a^{-}} f_2(x)$.*

We leave it to you to prove this theorem, and to formulate the similar theorem that holds for limits as x approaches a from the right (see Exercise 2).

Example 2.5.4. Find the following limits:

$$\text{(a)} \quad \lim_{x \to 0^{+}} |x|, \qquad \text{(b)} \quad \lim_{x \to 3^{-}} \lfloor x \rfloor.$$

Solution. For limit (a), we start with the fact that if $x > 0$, then $|x| = x$. Thus,

$$\lim_{x \to 0^{+}} |x| = \lim_{x \to 0^{+}} x = 0.$$

For limit (b), recall that $\lfloor x \rfloor$ denotes the greatest integer less than or equal to x. Therefore for $2 < x < 3$ we have $\lfloor x \rfloor = 2$. Thus, by Theorem 2.5.3 the value of limit (b) is

$$\lim_{x \to 3^-} \lfloor x \rfloor = \lim_{x \to 3^-} 2 = 2.$$

If you look back at Figure 1.16, these answers should make sense. $\square$

One reason that one-sided limits are important is that they can sometimes be used to help us compute ordinary two-sided limits. The reason is that we can study what happens to $f(x)$ when x approaches a from both sides by studying what happens when x approaches a from each side separately. Spelling this idea out more carefully proves our next theorem.

Theorem 2.5.5. *For any function f and numbers a and L, $\lim_{x \to a} f(x) = L$ if and only if both $\lim_{x \to a^-} f(x) = L$ and $\lim_{x \to a^+} f(x) = L$.*

Proof. Recall that according to the meaning of the phrase "if and only if," we must prove that if $\lim_{x \to a} f(x) = L$ then $\lim_{x \to a^-} f(x) = L$ and $\lim_{x \to a^+} f(x) = L$, and we must also prove that if $\lim_{x \to a^-} f(x) = L$ and $\lim_{x \to a^+} f(x) = L$ then $\lim_{x \to a} f(x) = L$. We will give only the second proof, leaving the first as an exercise for you (see Exercise 3).

Suppose that $\lim_{x \to a^-} f(x) = L$ and $\lim_{x \to a^+} f(x) = L$. We use the ϵ-δ method to prove that $\lim_{x \to a} f(x) = L$. Suppose that $\epsilon > 0$. Since $\lim_{x \to a^-} f(x) = L$, there must be some number $\delta_1 > 0$ such that

(a) if $a - \delta_1 < x < a$ then $|f(x) - L| < \epsilon$.

Similarly, since $\lim_{x \to a^+} f(x) = L$, there is some $\delta_2 > 0$ such that

(b) if $a < x < a + \delta_2$ then $|f(x) - L| < \epsilon$.

Of course, we let $\delta = \min(\delta_1, \delta_2)$. Now suppose $0 < |x - a| < \delta$. Then $a - \delta < x < a + \delta$ and $x \neq a$, so either $x < a$ or $x > a$. Suppose first that $x < a$. Then since $\delta \leq \delta_1$, we have $a - \delta_1 \leq a - \delta < x < a$. It follows, by statement (a), that $|f(x) - L| < \epsilon$. Similarly, if $x > a$ then $a < x < a + \delta \leq a + \delta_2$, and by (b) we again have $|f(x) - L| < \epsilon$. Thus $|f(x) - L| < \epsilon$ holds in both cases, which completes the proof. $\square$

Example 2.5.6. Find the following limits:

$$\text{(a) } \lim_{x \to 0} |x|, \quad \text{(b) } \lim_{x \to -1} \frac{x^2 - 1 + |2x + 2|}{x^2 + x}.$$

Solution. For limit (a), since the formula for $|x|$ will be different depending on whether $x > 0$ or $x < 0$, it is easier to evaluate the two one-sided limits than to evaluate the ordinary (two-sided) limit directly. Indeed, we have already computed that $\lim_{x \to 0^+} |x| = 0$

in part (a) of Example 2.5.4. For the other one-sided limit, we note that for $x < 0$, $|x| = -x$, so

$$\lim_{x \to 0^-} |x| = \lim_{x \to 0^-} (-x) = 0.$$

Since both one-sided limits are equal to 0, we conclude by Theorem 2.5.5 that $\lim_{x \to 0} |x| = 0$.

For limit (b), we again compute the two one-sided limits, because if $x < -1$ then $2x + 2 < 0$, and therefore $|2x + 2| = -(2x + 2)$, and if $x > -1$ then $2x + 2 > 0$, so $|2x + 2| = 2x + 2$. Thus, in each one-sided limit we can eliminate the absolute value signs:

$$\lim_{x \to -1^-} \frac{x^2 - 1 + |2x + 2|}{x^2 + x} = \lim_{x \to -1^-} \frac{x^2 - 1 - (2x + 2)}{x^2 + x} = \lim_{x \to -1^-} \frac{x^2 - 2x - 3}{x^2 + x}$$

$$= \lim_{x \to -1^-} \frac{(x - 3)(x + 1)}{x(x + 1)} = \lim_{x \to -1^-} \frac{x - 3}{x} = \frac{-4}{-1} = 4,$$

$$\lim_{x \to -1^+} \frac{x^2 - 1 + |2x + 2|}{x^2 + x} = \lim_{x \to -1^+} \frac{x^2 - 1 + (2x + 2)}{x^2 + x} = \lim_{x \to -1^+} \frac{x^2 + 2x + 1}{x^2 + x}$$

$$= \lim_{x \to -1^+} \frac{(x + 1)^2}{x(x + 1)} = \lim_{x \to -1^+} \frac{x + 1}{x} = \frac{0}{-1} = 0.$$

If limit (b) were defined, then by Theorem 2.5.5, the two one-sided limits would have to be equal; since they are not, we can conclude that limit (b) must be undefined. The graph of the equation $y = (x^2 - 1 + |2x + 2|)/(x^2 + x)$ near $x = -1$ is shown in Figure 2.24, where the two different one-sided limits lead to a break in the graph. $\qquad \square$

There is one more limiting process that we will be interested in. We will sometimes want to know what happens to $f(x)$, not when x gets close to some number a, but rather

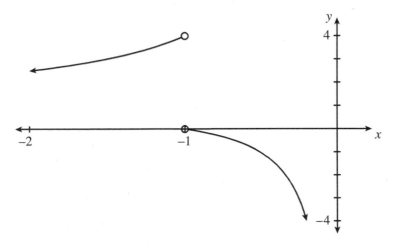

Figure 2.24: The graph of $y = (x^2 - 1 + |2x + 2|)/(x^2 + x)$ near $x = -1$. ✳

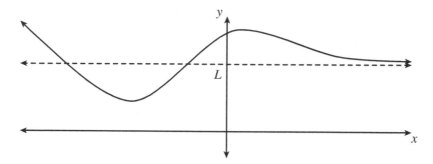

Figure 2.25: The graph of a function f for which as $x \to \infty$, $f(x) \to L$. The dashed line is the graph of $y = L$.

when x gets larger and larger. We will write "as $x \to \infty$, $f(x) \to L$" to mean that as x gets larger and larger, $f(x)$ approaches L. Figure 2.25 shows the graph of a function f for which this statement is true. The dashed line in the figure is the line $y = L$. The graph of f gets close to the line $y = L$ at the right side of the figure, and therefore when x is very large, $f(x)$ is close to L. The horizontal line $y = L$ is called an *asymptote* of the graph of f.

To state more carefully exactly what it means to say that as $x \to \infty$, $f(x) \to L$, we begin with the statement that if x is large then $f(x)$ is close to L, which we can express as follows:

$$\text{If } x > N \text{ then } |f(x) - L| < \epsilon.$$

In this statement, ϵ measures how close $f(x)$ is to L, and N measures how large x is. To ensure that $f(x)$ is very close to L we would let ϵ be a very small positive number, and to make x very large we would take N to be very large. As usual, we think of ϵ as representing a challenge, and this time the response is given by the number N. And once again our precise definition says that every challenge can be met:

Definition 2.5.7. To say that as $x \to \infty$, $f(x) \to L$ means that for every $\epsilon > 0$ there is some N such that if $x > N$ then $|f(x) - L| < \epsilon$. We define $\lim_{x \to \infty} f(x)$ to be the unique number L such that as $x \to \infty$, $f(x) \to L$, if there is such a number L.

Similarly, we can consider the limit of $f(x)$ as x gets smaller and smaller—i.e., as x gets to be a large magnitude negative number:

Definition 2.5.8. We say that as $x \to -\infty$, $f(x) \to L$ if for every $\epsilon > 0$ there is some N such that if $x < N$ then $|f(x) - L| < \epsilon$. We define $\lim_{x \to -\infty} f(x)$ to be the unique L such that as $x \to -\infty$, $f(x) \to L$, if there is such a number L.

If $\lim_{x \to -\infty} f(x) = L$, then the horizontal line $y = L$ will again be an asymptote of the graph of f. In this case the *left* side of the graph of f will approach the asymptote.

Example 2.5.9. Show that

$$\lim_{x \to \infty} \frac{1}{x} = 0, \qquad \lim_{x \to -\infty} \frac{1}{x} = 0.$$

Solution. These limit statements say that if x is a very large magnitude number, either positive or negative, then $1/x$ is close to 0. This should make intuitive sense. For example, both $1/1000000$ and $1/(-1000000)$ are close to 0. But, as usual, to be certain that our answers are exactly correct we need to have a proof. We will prove only the first limit statement, since the second is similar. We use our usual methods for proving limit statements from the definition, with adjustments for the new features of Definition 2.5.7. In particular, note that we respond to an ϵ challenge by producing a value for N.

Suppose $\epsilon > 0$. Let $N = 1/\epsilon$. Suppose $x > N$. Notice that since ϵ is positive, $N = 1/\epsilon$ is positive as well, so we have $x > N > 0$. Dividing this inequality by the positive number xN, we get $1/N > 1/x > 0$. Therefore

$$\left|\frac{1}{x} - 0\right| = \left|\frac{1}{x}\right| = \frac{1}{x} < \frac{1}{N} = \frac{1}{1/\epsilon} = \epsilon. \qquad \square$$

All of our previous limit theorems also apply to limits as x approaches ∞ or $-\infty$, with obvious minor adjustments where necessary. For example, we can combine limits as x approaches ∞ or $-\infty$ by adding, subtracting, multiplying, and dividing. Thus, we have $\lim_{x\to\infty}(1/x^2) = \lim_{x\to\infty}[(1/x)\cdot(1/x)] = 0\cdot 0 = 0$, $\lim_{x\to\infty}(1/x^3) = \lim_{x\to\infty}[(1/x^2)\cdot(1/x)] = 0\cdot 0 = 0$, and so on. In general, for any positive integer n, $\lim_{x\to\infty}(1/x^n) = \lim_{x\to-\infty}(1/x^n) = 0$.

When computing a limit as x approaches ∞ or $-\infty$ of an expression that contains a polynomial, an algebraic trick that is often helpful is to factor out of the polynomial the highest power of x appearing in the polynomial. In particular, when computing a limit as $x \to \pm\infty$ of a rational function we will usually apply this trick to both the numerator and denominator, as in the next example.

Example 2.5.10. Find the following limits:

$$\text{(a)} \lim_{x\to\infty} \frac{x^2-1}{x^2+1}, \qquad \text{(b)} \lim_{x\to-\infty} \frac{x^2-1}{x^2+1}, \qquad \text{(c)} \lim_{x\to\infty} \frac{5x^2-3}{x^3+2x-7}.$$

Solution. For (a), we factor out an x^2 from both numerator and denominator, cancel, and then apply Example 2.5.9:

$$\lim_{x\to\infty} \frac{x^2-1}{x^2+1} = \lim_{x\to\infty} \frac{x^2\cdot(1-1/x^2)}{x^2\cdot(1+1/x^2)} = \lim_{x\to\infty} \frac{1-1/x^2}{1+1/x^2} = \frac{1-0}{1+0} = 1.$$

The computation of limit (b) is exactly the same:

$$\lim_{x\to-\infty} \frac{x^2-1}{x^2+1} = \lim_{x\to-\infty} \frac{x^2\cdot(1-1/x^2)}{x^2\cdot(1+1/x^2)} = \lim_{x\to-\infty} \frac{1-1/x^2}{1+1/x^2} = \frac{1-0}{1+0} = 1.$$

The graph of the function $f(x) = (x^2-1)/(x^2+1)$ is shown in Figure 2.26. Since $\lim_{x\to-\infty} f(x) = \lim_{x\to\infty} f(x) = 1$, $f(x)$ is close to 1 when x is very large or very small. Therefore the graph of f approaches the horizontal asymptote $y = 1$ at both the left and right sides of the figure.

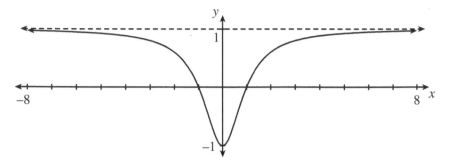

Figure 2.26: The graph of $y = (x^2 - 1)/(x^2 + 1)$. The dashed line is the graph of $y = 1$.

For limit (c), we factor out x^2 from the numerator and x^3 from the denominator:

$$\lim_{x \to \infty} \frac{5x^2 - 3}{x^3 + 2x - 7} = \lim_{x \to \infty} \frac{x^2 \cdot (5 - 3/x^2)}{x^3 \cdot (1 + 2/x^2 - 7/x^3)}$$

$$= \lim_{x \to \infty} \frac{1}{x} \cdot \frac{5 - 3/x^2}{1 + 2/x^2 - 7/x^3} = 0 \cdot \frac{5}{1} = 0. \qquad \square$$

Example 2.5.11. Find

$$\lim_{x \to \infty} \frac{\sin x}{x}.$$

Solution. For all x, we have $-1 \le \sin x \le 1$. If x is positive then we can divide this inequality through by x to conclude that

$$-\frac{1}{x} \le \frac{\sin x}{x} \le \frac{1}{x}.$$

(For negative x we would have to reverse the direction of the inequalities, but for the limit we are computing this doesn't matter; since we are taking the limit as $x \to \infty$, we are only interested in large values of x, so we can restrict attention to positive x.) We have already seen that $\lim_{x \to \infty}(1/x) = 0$, and therefore $\lim_{x \to \infty}(-1/x) = (-1) \cdot 0 = 0$. Thus, by the squeeze theorem, $\lim_{x \to \infty} \sin x/x = 0$. $\qquad \square$

So far, all of the variations on the limit idea that we have considered have involved making changes to the "$x \to a^{\neq}$" part of the definition. But we can also make similar changes to the "$f(x) \to L$" part of the definition. By now you should understand how these changes work, so rather than listing all of the possibilities we simply give a few examples in Table 2.5.

For example, consider the limit

$$\lim_{x \to 0} (3 - x^2) = 3.$$

This means that if x is close to 0 but not equal to 0 then $3 - x^2$ is close to 3, which we can express by saying that as $x \to 0^{\neq}$, $(3 - x^2) \to 3$. However, it is clear that if $x \ne 0$ then

Statement	Intuitive Meaning	Precise Definition
As $x \to a^{\neq}$, $f(x) \to L^{+}$ As $x \to a^{\neq}$, $f(x) \to L^{>}$	If x is close to a, but not equal to a, then $f(x)$ is close to L and greater than L.	For every $\epsilon > 0$ there is some $\delta > 0$ such that if $0 < |x - a| < \delta$ then $L < f(x) < L + \epsilon$.
As $x \to a^{-}$, $f(x) \to \infty$ As $x \to a^{<}$, $f(x) \to \infty$	If x is close to a but less than a, then $f(x)$ is very large.	For every number M there is some $\delta > 0$ such that if $a - \delta < x < a$ then $f(x) > M$.
As $x \to \infty$, $f(x) \to -\infty$	If x is very large then $f(x)$ is a very large magnitude negative number.	For every number M there is some number N such that if $x > N$ then $f(x) < M$.

Table 2.5: Some examples of limit statements and their meanings.

$x^2 > 0$, and therefore $3 - x^2 < 3$. So we could make our limit statement slightly more informative by saying that as $x \to 0^{\neq}$, $(3 - x^2) \to 3^{-}$. This statement says that if x is close to 0 but not equal to 0, then $3 - x^2$ is close to 3 *and less than 3*. Alternatively, we could express this idea by saying that as $x \to 0^{\neq}$, $(3 - x^2) \to 3^{<}$. We will occasionally find it useful to add such extra information to our limit statements.

Limit statements involving $f(x) \to \infty$ or $f(x) \to -\infty$ will be particularly useful.

Example 2.5.12. Show that as $x \to 0^{+}$, $1/x \to \infty$, and as $x \to 0^{-}$, $1/x \to -\infty$.

Solution. The two statements are similar, so we only prove that as $x \to 0^{+}$, $1/x \to \infty$. Our proof is guided by the meaning of the statement to be proven, which is: for every number M, there is some $\delta > 0$ such that if $0 < x < \delta$ then $1/x > M$.

Let M be any number. Let $\delta = 1/\max(M, 1)$, where of course $\max(M, 1)$ denotes the larger of the two numbers M and 1. Note that $\max(M, 1) \geq 1$, so δ is positive.

Now suppose $0 < x < \delta$. Dividing this inequality by $x\delta$, we conclude that $0 < 1/\delta < 1/x$. Thus,

$$\frac{1}{x} > \frac{1}{\delta} = \frac{1}{1/\max(M, 1)} = \max(M, 1) \geq M. \qquad \square$$

The result established in Example 2.5.12 is sometimes expressed by writing $\lim_{x \to 0^{+}}(1/x) = \infty$ and $\lim_{x \to 0^{-}}(1/x) = -\infty$. This notation can be misleading, because it appears to be giving values for these limits. However, you must remember our rule that ∞ is not a number, and any mathematical notation involving the symbol ∞ is just a shorthand for something else. In general, if $\lim_{x \to a} f(x) = \pm\infty$, then there is no *number* that $f(x)$ approaches as x approaches a, so the limit is considered to be undefined. The notation $\lim_{x \to a} f(x) = \pm\infty$ does not give a value for the limit, so it

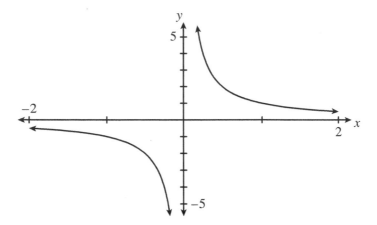

Figure 2.27: The graph of $y = 1/x$. The lines $x = 0$ and $y = 0$ are asymptotes.

does not say that the limit is defined; rather, it expresses the way in which the limit is undefined.

The graph of $y = 1/x$ is shown in Figure 2.27. Since $\lim_{x \to 0^+}(1/x) = \infty$, the graph goes up very high just to the right of the y-axis. Similarly, the fact that $\lim_{x \to 0^-}(1/x) = -\infty$ tells us that just to the left of the y-axis, the graph goes down very low. Thus, at the top and bottom of the figure the graph gets close to the y-axis, which is the line $x = 0$, so this vertical line is considered to be an asymptote of the graph. In general, for any function f, if $f(x)$ approaches either ∞ or $-\infty$ as x approaches some number a, then the vertical line $x = a$ will be an asymptote of the graph. Notice that in Figure 2.27, the horizontal line $y = 0$ (that is, the x-axis) is also an asymptote, since as we showed in Example 2.5.9, $\lim_{x \to \infty}(1/x) = \lim_{x \to -\infty}(1/x) = 0$.

Many of our previous results about limits can be adapted to the case of infinite limits. For example, we have the following version of the squeeze theorem (see Exercise 28 for the proof):

Theorem 2.5.13. *Suppose that $d > 0$, and for all x, if $0 < |x - a| < d$ then $f(x) \geq g(x)$. Suppose also that $\lim_{x \to a} g(x) = \infty$. Then $\lim_{x \to a} f(x) = \infty$.*

We can also extend our previous results about limits of sums and products to include the case of infinite limits:

Theorem 2.5.14.

1. *If $\lim_{x \to a} f(x) = \infty$ and either $\lim_{x \to a} g(x) = L$ or $\lim_{x \to a} g(x) = \infty$, then $\lim_{x \to a}(f(x) + g(x)) = \infty$.*

2. *If $\lim_{x \to a} f(x) = -\infty$ and either $\lim_{x \to a} g(x) = L$ or $\lim_{x \to a} g(x) = -\infty$, then $\lim_{x \to a}(f(x) + g(x)) = -\infty$.*

3. *If $\lim_{x \to a} f(x) = \infty$ and either $\lim_{x \to a} g(x) = L > 0$ or $\lim_{x \to a} g(x) = \infty$, then $\lim_{x \to a}(f(x) \cdot g(x)) = \infty$.*

4. *If* $\lim_{x \to a} f(x) = \infty$ *and either* $\lim_{x \to a} g(x) = L < 0$ *or* $\lim_{x \to a} g(x) = -\infty$,
   then $\lim_{x \to a}(f(x) \cdot g(x)) = -\infty$.

5. *If* $\lim_{x \to a} f(x) = -\infty$ *and either* $\lim_{x \to a} g(x) = L > 0$ *or* $\lim_{x \to a} g(x) = \infty$,
   then $\lim_{x \to a}(f(x) \cdot g(x)) = -\infty$.

6. *If* $\lim_{x \to a} f(x) = -\infty$ *and either* $\lim_{x \to a} g(x) = L < 0$ *or* $\lim_{x \to a} g(x) = -\infty$,
   then $\lim_{x \to a}(f(x) \cdot g(x)) = \infty$.

Proof. We will just prove part of statement 1, since the proofs are all quite similar. Suppose that $\lim_{x \to a} f(x) = \infty$ and $\lim_{x \to a} g(x) = L$. To prove that $\lim_{x \to a}(f(x) + g(x)) = \infty$, suppose that M is any number. Since $\lim_{x \to a} f(x) = \infty$, there is some $\delta_1 > 0$ such that

$$\text{if } 0 < |x - a| < \delta_1 \text{ then } f(x) > M - L + 1.$$

Since $\lim_{x \to a} g(x) = L$, there is some $\delta_2 > 0$ such that

$$\text{if } 0 < |x - a| < \delta_2 \text{ then } |g(x) - L| < 1, \text{ and therefore } L - 1 < g(x) < L + 1.$$

Now let $\delta = \min(\delta_1, \delta_2)$, and suppose that $0 < |x - a| < \delta$. Then $0 < |x - a| < \delta_1$, so $f(x) > M - L + 1$, and also $0 < |x - a| < \delta_2$, so $L - 1 < g(x) < L + 1$. Putting these together, we have

$$f(x) + g(x) > (M - L + 1) + (L - 1) = M. \qquad \square$$

We have stated Theorem 2.5.14 for ordinary two-sided limits, but it also applies to one-sided limits and to limits as x approaches $\pm\infty$. There are thus so many cases to consider that it would be tedious to check all of them. But we hope that by now you have had enough experience with limits that you have begun to develop some intuition about how to think about them. All of the statements in Theorem 2.5.14 should make intuitive sense. For example, part 1 of the theorem says that if $f(x)$ gets very large, and either $g(x)$ gets close to some number L or $g(x)$ also gets very large, then $f(x) + g(x)$ will get very large. You should be able to give similar informal interpretations of all parts of the theorem, and see that they make intuitive sense. Theorem 2.5.14 doesn't discuss limits of differences or quotients, but they can be handled by rewriting $f(x) - g(x)$ as $f(x) + (-g(x))$ and $f(x)/g(x)$ as $f(x) \cdot (1/g(x))$. Limits of quotients will be discussed further in the next section.

For example, let's use Theorem 2.5.14 to compute $\lim_{x \to -\infty}(5x^3 - 7)$. You should be able to verify that $\lim_{x \to -\infty} x = -\infty$ (see Exercise 29). By part 6 of Theorem 2.5.14, $\lim_{x \to -\infty} x^2 = \lim_{x \to -\infty}(x \cdot x) = \infty$, and by part 4, $\lim_{x \to -\infty} x^3 = \lim_{x \to -\infty}(x^2 \cdot x) = -\infty$. By part 5, $\lim_{x \to -\infty} 5x^3 = -\infty$, and by part 2, $\lim_{x \to -\infty}(5x^3 - 7) = \lim_{x \to -\infty}(5x^3 + (-7)) = -\infty$.

Perhaps just as important as understanding what the theorem says is understanding what it *doesn't* say. For example, the theorem doesn't say anything about the limit of $f(x) + g(x)$ in the case when the limit of $f(x)$ is ∞ and the limit of $g(x)$ is $-\infty$. The reason is that if $f(x)$ is very large, and $g(x)$ is a very large magnitude negative number, then $f(x) + g(x)$ could be anything—positive or negative, large or small—depending on the relative sizes of $f(x)$ and $g(x)$. Since $f(x)$ is getting very large, you might think of $f(x)$ as "pulling" $f(x) + g(x)$ in the positive direction. At the same time, since $g(x)$ is a large magnitude negative number, $g(x)$ is "pulling" $f(x) + g(x)$ in the negative direction. The two terms $f(x)$ and $g(x)$ are pulling in opposite directions, and the outcome will depend on which one pulls harder. For example, suppose that $f(x) = x + 17$ and $g(x) = -x$. Then $\lim_{x \to \infty} f(x) = \lim_{x \to \infty} (x + 17)$ $= \infty$ and $\lim_{x \to \infty} g(x) = \lim_{x \to \infty} (-x) = -\infty$, but $\lim_{x \to \infty} (f(x) + g(x)) =$ $\lim_{x \to \infty} (x + 17 - x) = 17$. Clearly we could replace 17 here with any other number, leading to similar examples in which the limit of $f(x) + g(x)$ is anything we please. Another case that is not covered by the theorem is the limit of $f(x) \cdot g(x)$ when the limit of $f(x)$ is ∞ and the limit of $g(x)$ is 0. You should think about how, in this case, we could also think of $f(x)$ and $g(x)$ as pulling in opposite directions. In Exercise 30 you are asked to give examples showing that the limit of $f(x) \cdot g(x)$ could be anything in this case as well.

Example 2.5.15. Find the following limits:

$$\text{(a) } \lim_{x \to 0^-} \frac{2x - 3}{x}, \qquad \text{(b) } \lim_{x \to \infty} \frac{x^3 + 2x - 7}{5x^2 - 3}, \qquad \text{(c) } \lim_{x \to \infty} (x^2 - x - 7).$$

Solution. To evaluate limit (a), we first rewrite the quotient as a product:

$$\lim_{x \to 0^-} \frac{2x - 3}{x} = \lim_{x \to 0^-} \left[(2x - 3) \cdot \frac{1}{x} \right].$$

Since $\lim_{x \to 0^-} (2x - 3) = -3$ and $\lim_{x \to 0^-} (1/x) = -\infty$ (by Example 2.5.12), we can apply part 6 of Theorem 2.5.14 to conclude that

$$\lim_{x \to 0^-} \frac{2x - 3}{x} = \infty.$$

For part (b), we begin as in Example 2.5.10 by factoring out the highest powers of x in the numerator and denominator:

$$\lim_{x \to \infty} \frac{x^3 + 2x - 7}{5x^2 - 3} = \lim_{x \to \infty} \frac{x^3 \cdot (1 + 2/x^2 - 7/x^3)}{x^2 \cdot (5 - 3/x^2)} = \lim_{x \to \infty} \left(x \cdot \frac{1 + 2/x^2 - 7/x^3}{5 - 3/x^2} \right).$$

Since

$$\lim_{x \to \infty} x = \infty \quad \text{and} \quad \lim_{x \to \infty} \frac{1 + 2/x^2 - 7/x^3}{5 - 3/x^2} = \frac{1}{5},$$

we can conclude by part 3 of Theorem 2.5.14 that

$$\lim_{x \to \infty} \frac{x^3 + 2x - 7}{5x^2 - 3} = \infty.$$

We could think of the limit in part (c) as $\lim_{x \to \infty} (f(x) + g(x))$, where $f(x) = x^2$ and $g(x) = -x - 7$, but this doesn't help: we have $\lim_{x \to \infty} f(x) = \lim_{x \to \infty} x^2 = \infty$

and $\lim_{x\to\infty} g(x) = \lim_{x\to\infty}(-x-7) = -\infty$, and this case is not covered by Theorem 2.5.14. However, once again factoring out the highest power of x leads to a solution:

$$\lim_{x\to\infty}(x^2 - x - 7) = \lim_{x\to\infty} x^2(1 - 1/x - 7/x^2).$$

Since $\lim_{x\to\infty} x^2 = \infty$ and $\lim_{x\to\infty}(1 - 1/x - 7/x^2) = 1 - 0 - 0 = 1$, part 3 of Theorem 2.5.14 tells us that $\lim_{x\to\infty}(x^2 - x - 7) = \infty$. $\qquad\square$

Exercises 2.5

1. (a) Prove that if $\lim_{x\to a^-} f(x) = L$ and $\lim_{x\to a^-} g(x) = M$ then it follows that $\lim_{x\to a^-}(f(x) + g(x)) = L + M$.

   (b) Prove that if $\lim_{x\to a^+} f(x) = L$ and $\lim_{x\to a^+} g(x) = M$ then it follows that $\lim_{x\to a^+}(f(x) - g(x)) = L - M$.

   (c) Prove that if $\lim_{x\to a^+} f(x) = L$ and $\lim_{x\to a^+} g(x) = M$ then it follows that $\lim_{x\to a^+}(f(x) \cdot g(x)) = L \cdot M$.

2. (a) Prove Theorem 2.5.3.

   (b) State and prove a similar theorem for limits as x approaches a from the right.

3. Prove the left-to-right direction of Theorem 2.5.5.

4–27: Evaluate the limit.

4. $\displaystyle\lim_{x\to\infty} \frac{3x^2 + 1}{x^2 - 3}$.

5. $\displaystyle\lim_{t\to-\infty} \frac{100t^2 + 1}{2t^{100} - 1}$.

6. $\displaystyle\lim_{x\to\infty} \frac{3 - x^2}{2x^2 + 7x}$.

7. $\displaystyle\lim_{x\to\infty} \left(x^2 - \frac{x^4}{x^2 + 3}\right)$.

8. $\displaystyle\lim_{x\to\infty} \left(\frac{x^2 - 2}{x} - \frac{x^2}{x + 3}\right)$.

9. $\displaystyle\lim_{x\to\infty} \left(x^2 \left[\frac{2}{2x - 1} - \frac{1}{x - 1}\right]\right)$.

10. $\displaystyle\lim_{r\to-\infty} \frac{r + \frac{1}{r}}{r - \frac{1}{r^2}}$.

11. $\displaystyle\lim_{x\to-\infty} (x^5 + 100x^4)$.

12. $\displaystyle\lim_{x\to\infty} \frac{\cos(x-7)}{x^2-7x}$.

13. $\displaystyle\lim_{z\to-\infty} (z^2+\sin z)$.

14. $\displaystyle\lim_{z\to-\infty} (z(\sin z+3))$.

15. $\displaystyle\lim_{x\to\infty} \frac{(2+x)\sin^2 x}{x^2(2+\sin x)}$.

16. $\displaystyle\lim_{x\to 0^+} \frac{x^2-5}{x}$.

17. $\displaystyle\lim_{x\to 0^-} \frac{1}{x^2-5x}$.

18. $\displaystyle\lim_{u\to 3^-} \frac{u^2-3u}{|u-3|}$.

19. $\displaystyle\lim_{x\to 2^+} \frac{x+|2-x|-2}{x^2+3x-10}$.

20. $\displaystyle\lim_{x\to\infty} \frac{2x}{1+|x|}$.

21. $\displaystyle\lim_{x\to-\infty} \frac{2x}{1+|x|}$.

22. $\displaystyle\lim_{t\to -1^-} \lfloor 2t+1\rfloor$.

23. $\displaystyle\lim_{w\to 2^-} \frac{w^2-4}{w-2\lfloor w\rfloor}$.

24. $\displaystyle\lim_{x\to 2^+} \frac{f(3-x)-2}{x-2}$, where $f(x)=\begin{cases} 2x, & \text{if } x\le 1, \\ x+1, & \text{if } x>1. \end{cases}$

25. $\displaystyle\lim_{x\to -2^+} \frac{|\lfloor x\rfloor|}{\lfloor |x|\rfloor}$.

26. $\displaystyle\lim_{x\to 1^+} \frac{x^2+|x-1|-1}{x^2+|x-2|-2}$.

27. $\displaystyle\lim_{x\to -2^+} \frac{|x^2+2x|}{3x^2+5x-2}$.

28. Prove Theorem 2.5.13.

29. Use the definition of limits to verify that $\lim_{x\to-\infty} x = -\infty$.

30. Show that for any number c, it is possible to find functions f and g such that $\lim_{x\to 0} f(x)=\infty$, $\lim_{x\to 0} g(x)=0$, and $\lim_{x\to 0}(f(x)\cdot g(x))=c$.

2.6 Limits of Compositions

In previous sections we have studied limits of sums, differences, products, and quotients of functions. But what about compositions of functions? How should we figure out $\lim_{x \to a}(f \circ g)(x)$?

For example, suppose that $f(x) = x^2 + 1$ and $g(x) = 5x - 2$, and we wish to compute $\lim_{x \to 1}(f \circ g)(x) = \lim_{x \to 1} f(g(x))$. It will be convenient to introduce names for the important quantities in this problem, so we let $u = g(x)$ and $y = f(g(x)) = f(u)$. Thus, the value of x determines the value of u via the equation $u = g(x)$, and the value of u determines the value of y via the equation $y = f(u)$. We want to know what happens to $y = f(g(x))$ as x approaches 1, and we will investigate this by studying what happens to u as x approaches 1, and then what this behavior of u tells us about y.

We have $\lim_{x \to 1} g(x) = \lim_{x \to 1}(5x - 2) = 3$, so we can say that as $x \to 1^{\neq}$, $g(x) \to 3$. Since we have introduced the letter u to stand for $g(x)$, another way to say this is that

$$\text{as } x \to 1^{\neq}, u \to 3. \tag{2.6}$$

Similarly, we can compute that $\lim_{u \to 3} f(u) = \lim_{u \to 3}(u^2 + 1) = 10$, and since we have let $y = f(u)$, this means that

$$\text{as } u \to 3^{\neq}, y \to 10. \tag{2.7}$$

Statement (2.6) says that if x is close to 1 but not equal to 1, then u is close to 3, and statement (2.7) says that if u is close to 3 but not equal to 3 then y is close to 10. Unfortunately, there is a slight mismatch between the right side of (2.6) and the left side of (2.7); statement (2.6) doesn't tell us that $u \neq 3$, and it seems that we need to know this to be able to apply (2.7). But let us look more closely at (2.6). We have $u = g(x) = 5x - 2$, and it follows from this equation that if $x \neq 1$ then $u \neq 3$ (indeed, if $x > 1$ then $u > 3$, and if $x < 1$ then $u < 3$). Thus, we can give a slightly more informative version of statement (2.6):

$$\text{as } x \to 1^{\neq}, u \to 3^{\neq}. \tag{2.6$'$}$$

We can now combine statements (2.6$'$) and (2.7) to conclude that

$$\text{as } x \to 1^{\neq}, y \to 10.$$

In other words, $\lim_{x \to 1} f(g(x)) = 10$.

Since this is our first time combining statements like (2.6$'$) and (2.7), perhaps we should check more carefully that our reasoning here can really be justified. We begin by writing out what these statements mean. Statement (2.6$'$) tells us that

(a) for every $\epsilon_1 > 0$ there is some $\delta_1 > 0$ such that if $0 < |x - 1| < \delta_1$ then $0 < |u - 3| < \epsilon_1$,

while by (2.7),

(b) for every $\epsilon_2 > 0$ there is some $\delta_2 > 0$ such that if $0 < |u - 3| < \delta_2$ then $|y - 10| < \epsilon_2$.

Recall that when using statements (a) and (b), we can assign whatever values we please to ϵ_1 and ϵ_2. Now let's see if we can use these two statements to prove that as $x \to 1^{\neq}$, $y \to 10$.

Suppose $\epsilon > 0$. Since we want our proof to end with $|y - 10| < \epsilon$, we first apply statement (b), plugging in ϵ for ϵ_2, to conclude that there is some number $\delta_2 > 0$ such that

(b') if $0 < |u - 3| < \delta_2$ then $|y - 10| < \epsilon$.

Next, we apply statement (a) with $\epsilon_1 = \delta_2$ to conclude that there is some $\delta_1 > 0$ such that

(a') if $0 < |x - 1| < \delta_1$ then $0 < |u - 3| < \delta_2$.

Now let $\delta = \delta_1$, and suppose that $0 < |x - 1| < \delta$. Then $0 < |x - 1| < \delta_1$, so by statement (a'), $0 < |u - 3| < \delta_2$, and therefore by (b'), $|y - 10| < \epsilon$. This shows that as $x \to 1^{\neq}$, $y \to 10$.

In the future, we'll just combine statements like (2.6') and (2.7) without providing any justification, but in every case our conclusion could be justified by similar ϵ-δ reasoning.

For the problem we just solved, there is actually another way we could have gotten the answer. We could simply have computed that $f(g(x)) = f(5x - 2) = (5x - 2)^2 + 1 = 25x^2 - 20x + 5$, and therefore

$$\lim_{x \to 1} f(g(x)) = \lim_{x \to 1} (25x^2 - 20x + 5) = 25 - 20 + 5 = 10.$$

Of course, it is reassuring to see that both methods lead to the same answer! However, in other examples there is sometimes no easy alternative to the composition method we have developed.

Example 2.6.1. Find

$$\lim_{x \to 0} \lfloor 3 - x^2 \rfloor.$$

Solution. We can write the requested limit as $\lim_{x \to 0} f(g(x))$, where $f(x) = \lfloor x \rfloor$ and $g(x) = 3 - x^2$. Imitating our last example, we let $u = g(x) = 3 - x^2$ and $y = f(u) = \lfloor u \rfloor = \lfloor 3 - x^2 \rfloor$. We begin with the fact that $\lim_{x \to 0} g(x) = \lim_{x \to 0}(3 - x^2) = 3$, which tells us that as $x \to 0^{\neq}$, $u \to 3$. But as we observed in the last section, for $x \neq 0$ we have $u = 3 - x^2 < 3$, so we can actually say that

$$\text{as } x \to 0^{\neq}, u \to 3^-. \tag{2.8}$$

This tells us that for our next step we should compute $\lim_{u \to 3^-} f(u) = \lim_{u \to 3^-} \lfloor u \rfloor$. Fortunately, we solved this limit in Example 2.5.4, where we found that the answer is 2. Thus,

$$\text{as } u \to 3^-, y \to 2. \tag{2.9}$$

Since the right side of (2.8) matches the left side of (2.9), we can combine the two limit statements to conclude that as $x \to 0^{\neq}$, $y \to 2$, and therefore $\lim_{x \to 0} \lfloor 3 - x^2 \rfloor = 2$.   □

When combining two limit statements like (2.8) and (2.9), is it really necessary to make sure the right side of the first matches the left side of the second exactly? The answer is yes; you can get wrong answers if you're not careful about this. For example, suppose f is the function defined by the equation

$$f(x) = \begin{cases} 1, & \text{if } x > 0, \\ 0, & \text{if } x = 0, \\ -1, & \text{if } x < 0. \end{cases}$$

(The function f is called the sign or signum function, and is sometimes denoted sgn.) Clearly $f(0) = 0$, $\lim_{x \to 0^+} f(x) = 1$, $\lim_{x \to 0^-} f(x) = -1$, and $\lim_{x \to 0} f(x)$ is undefined. We will consider $\lim_{x \to 0} f(g(x))$ for several different choices of the function $g(x)$.

Let $g_1(x) = x^2$, $g_2(x) = -x^2$, $g_3(x) = -x$, and $g_4(x) = 0$. Clearly all four of these functions approach 0 as x approaches 0, and this may make you think that when we compose these functions with f, the limit as x approaches 0 will be the same in every case. But in fact these four functions approach 0 in different ways, and this will lead to different limits when they are composed with f.

First we consider $\lim_{x \to 0} f(g_1(x))$. If we let $u = g_1(x) = x^2$, then for every $x \neq 0$ we have $u = x^2 > 0$. It follows that as $x \to 0^{\neq}$, $u \to 0^+$. Since $\lim_{u \to 0^+} f(u) = 1$, we can say that as $u \to 0^+$, $f(u) \to 1$. Combining our two limit statements, we conclude that as $x \to 0^{\neq}$, $f(u) = f(g_1(x)) \to 1$. In other words, $\lim_{x \to 0} f(g_1(x)) = 1$.

Similarly, if $u = g_2(x) = -x^2$, then for every $x \neq 0$ we have $u < 0$, and therefore as $x \to 0^{\neq}$, $u \to 0^-$. Since $\lim_{u \to 0^-} f(x) = -1$, we find that $\lim_{x \to 0} f(g_2(x)) = -1$.

The function g_3 is more complicated, because $g_3(x)$ approaches 0 from both sides. If $u = g_3(x) = -x$, then u is negative when x is positive, and u is positive when x is negative. This suggests that we should consider the one-sided limits as x approaches 0 from the left and the right. We see that as $x \to 0^+$, $u \to 0^-$, and as $x \to 0^-$, $u \to 0^+$, and as before this leads to the conclusions $\lim_{x \to 0^+} f(g_3(x)) = -1$ and $\lim_{x \to 0^-} f(g_3(x)) = 1$. Thus, $\lim_{x \to 0} f(g_3(x))$ is undefined.

Finally, in the case of g_4, for every x we have $g_4(x) = 0$, and therefore $f(g_4(x)) = f(0) = 0$. Thus, $\lim_{x \to 0} f(g_4(x)) = 0$. To summarize, even though all four of $g_1(x)$, $g_2(x)$, $g_3(x)$, and $g_4(x)$ approach 0 as x approaches 0, we have $\lim_{x \to 0} f(g_1(x)) = 1$, $\lim_{x \to 0} f(g_2(x)) = -1$, $\lim_{x \to 0} f(g_3(x))$ is undefined, and $\lim_{x \to 0} f(g_4(x)) = 0$. To get these answers, we had to pay careful attention to *how* u approached 0 in each case.

We can discover some useful limit facts by studying limits of compositions $f \circ g$ in which $f(x) = 1/x$. In this case, $f(g(x)) = 1/g(x)$. Suppose, for example, that g is a function such as $x \to a^{\neq}$, $g(x) \to 0^+$. Letting $u = g(x)$, we can say that as $x \to a^{\neq}$, $u \to 0^+$. We saw in the last section that as $u \to 0^+$, $1/u \to \infty$. Combining these limit statements, we conclude that as $x \to a^{\neq}$, $1/g(x) = 1/u \to \infty$. This establishes the first statement in our next theorem. The proofs of the other three are similar.

Theorem 2.6.2.

1. *If as $x \to a^{\neq}$, $g(x) \to 0^+$, then as $x \to a^{\neq}$, $1/g(x) \to \infty$.*

2. *If as $x \to a^{\neq}$, $g(x) \to 0^-$, then as $x \to a^{\neq}$, $1/g(x) \to -\infty$.*

3. If as $x \to a^{\neq}$, $g(x) \to \infty$, then as $x \to a^{\neq}$, $1/g(x) \to 0^{+}$.

4. If as $x \to a^{\neq}$, $g(x) \to -\infty$, then as $x \to a^{\neq}$, $1/g(x) \to 0^{-}$.

As usual, the theorem also applies to one-sided limits and limits as x approaches $\pm\infty$. We can use this theorem to extend our understanding of limits of quotients, as our next example shows.

Example 2.6.3. Find the following limits:

$$\text{(a) } \lim_{x \to 0} \frac{x-7}{x^3-x^2}, \qquad \text{(b) } \lim_{x \to \infty} \frac{1/x-7}{x^3-x^2}.$$

Solution. In limit (a), the numerator approaches -7 and the denominator approaches 0, so the limit is undefined. But we can give a more informative answer by using Theorem 2.6.2. We begin by asking whether the denominator approaches 0 from the right or the left. It is not easy to tell with the denominator written in the form $x^3 - x^2$, but if we rewrite it as $x^2(x-1)$, then we see that when x is close to 0 but not equal to 0, x^2 is positive and $x-1$ is negative, and therefore $x^3 - x^2 = x^2(x-1)$ is negative. Thus, as $x \to 0^{\neq}$, $x^3 - x^2 \to 0^{-}$, and therefore by part 2 of Theorem 2.6.2, as $x \to 0^{\neq}$, $1/(x^3-x^2) \to -\infty$. Combining this with the fact that as $x \to 0^{\neq}$, $x - 7 \to -7$, we can use Theorem 2.5.14 to conclude that

$$\lim_{x \to 0} \frac{x-7}{x^3-x^2} = \lim_{x \to 0}\left[(x-7) \cdot \frac{1}{x^3-x^2}\right] = \infty.$$

Notice that this answer does not contradict our earlier conclusion that the limit is undefined. We have simply said more specifically why it is undefined. Intuitively, what is happening in this limit is that when x is close to 0 but not equal to 0, $x - 7$ is close to -7 and $x^3 - x^2$ is a negative number close to 0, and therefore their quotient is a large positive number.

In limit (b), the numerator also approaches -7. It is not easy to see what the limit of the denominator is unless we factor out the highest power of x: $x^3 - x^2 = x^3(1 - 1/x)$. Now we can see that as $x \to \infty$, $x^3 \to \infty$ and $1 - 1/x \to 1$, and therefore $x^3 - x^2 = x^3(1 - 1/x) \to \infty$. By part 3 of Theorem 2.6.2, as $x \to \infty$, $1/(x^3-x^2) \to 0^{+}$, so

$$\lim_{x \to \infty} \frac{1/x-7}{x^3-x^2} = \lim_{x \to \infty}\left[(1/x-7) \cdot \frac{1}{x^3-x^2}\right] = (-7) \cdot 0 = 0.$$

See Exercise 1 for another way of computing this limit.   □

Example 2.6.4. Let f be the function whose graph is shown in Figure 2.28, and let

$$g(x) = \frac{1}{3-x}.$$

Find the following limits:

(a) $\lim_{x \to 2^{+}} f(g(x))$,   (b) $\lim_{x \to 2^{-}} f(g(x))$,

(c) $\lim_{x \to 0^{+}} g(f(x))$,   (d) $\lim_{x \to 4} g(f(x))$.

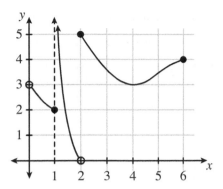

Figure 2.28: The graph of $y = f(x)$.

Solution. For limit (a), it is useful to let $u = g(x) = 1/(3 - x)$. We have $\lim_{x \to 2+}(1/(3 - x)) = 1$, so as $x \to 2^+$, $u \to 1$. But looking at Figure 2.28 we see that $\lim_{u \to 1+} f(u)$ and $\lim_{u \to 1-} f(u)$ are different, so we need to know whether u approaches 1 from the positive side or the negative side before we can tell what happens to $f(g(x)) = f(u)$. Since x is approaching 2 from the positive side, we are interested in values of x that are close to 2 but larger than 2. For $2 < x < 3$ we have $0 < 3 - x < 1$, and therefore $u = 1/(3 - x) > 1$. Thus, as $x \to 2^+$, $u \to 1^+$. The vertical asymptote in Figure 2.28 indicates that as $u \to 1^+$, $f(u) \to \infty$, so as $x \to 2^+$, $f(g(x)) = f(u) \to \infty$. In other words, $\lim_{x \to 2+} f(g(x)) = \infty$.

Limit (b) is similar, except this time we are interested in values of x that are smaller than 2. If $x < 2$ then $3 - x > 1$, and therefore $u = 1/(3 - x) < 1$. Thus, as $x \to 2^-$, $u \to 1^-$, and we see in the figure that as $u \to 1^-$, $f(u) \to 2$. Therefore $\lim_{x \to 2-} f(g(x)) = 2$.

To evaluate limit (c) we let $u = f(x)$, and we read off from Figure 2.28 that as $x \to 0^+$, $u = f(x) \to 3^-$. Thus, we must compute $\lim_{u \to 3-} g(u)$. As $u \to 3^-$, $3 - u \to 0^+$, and therefore $g(u) = 1/(3 - u) \to \infty$. Putting everything together, we conclude that $\lim_{x \to 0+} g(f(x)) = \infty$.

We also let $u = f(x)$ for limit (d), and observe that if x is close to 4 but not equal to 4, then $f(x)$ is close to 3 but larger than 3. Thus, as $x \to 4^{\neq}$, $u = f(x) \to 3^+$. As $u \to 3^+$, $3 - u \to 0^-$, and therefore $g(u) = 1/(3 - u) \to -\infty$. We conclude that $\lim_{x \to 4} g(f(x)) = -\infty$.   □

Exercises 2.6

1. Give an alternative solution to Example 2.6.3(b) by rewriting the function in the limit as a rational function and then using methods from Section 2.5.

2–18: Evaluate the limit.

2. $\displaystyle\lim_{x \to 3^-} \frac{x - 2}{x - 3}$.

3. $\lim\limits_{x\to 3^-} \dfrac{x-4}{x-3}$.

4. $\lim\limits_{t\to 0^+} \dfrac{t^2+3t}{t^3-t^2}$.

5. $\lim\limits_{x\to 1^-} \dfrac{x^2-x+2}{x^2+x-2}$.

6. $\lim\limits_{v\to 1} \dfrac{v^2-3v+1}{v^2-2v+1}$.

7. $\lim\limits_{v\to 1} \dfrac{v^2-3v+2}{v^2-2v+1}$.

8. $\lim\limits_{x\to -1^+} \dfrac{x^2+2x-3}{x^2+2|x|-3}$.

9. $\lim\limits_{x\to 2^-} \lfloor 4+3x \rfloor$.

10. $\lim\limits_{x\to 2^-} (4+\lfloor 3x \rfloor)$.

11. $\lim\limits_{x\to 2^-} (4+3\lfloor x \rfloor)$.

12. $\lim\limits_{x\to 0} \lfloor 1-x^2 \rfloor$.

13. $\lim\limits_{x\to 0} \lfloor 1-x^2+x^3 \rfloor$.

14. $\lim\limits_{x\to 0} \lfloor 1-x^2+x \rfloor$.

15. $\lim\limits_{x\to 0} (\lfloor 1-x^2 \rfloor +x)$.

16. $\lim\limits_{x\to 0} \left(\dfrac{1}{x} - \dfrac{1}{x^2} \right)$.

17. $\lim\limits_{u\to \infty} \dfrac{u^3+5}{u^2+u}$.

18. $\lim\limits_{u\to -\infty} \dfrac{u^3+5}{u^2+u}$.

19. Let f be the function whose graph is shown in Figure 2.29a, and let g be defined by the formula

$$g(x) = \begin{cases} 2-x^2, & \text{if } x \leq 2, \\ 9-2x, & \text{if } x > 2. \end{cases}$$

Find the following limits:

(a) $\lim\limits_{x\to 0} f(g(x))$. (Hint: Let $u = g(x)$. First show that as $x \to 0^{\neq}$, $u \to 2^-$.)

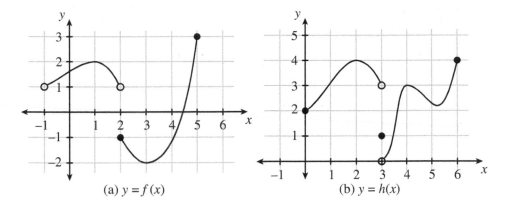

Figure 2.29: Graphs of the functions f and h.

(b) $\displaystyle\lim_{x\to 1} g(f(x))$.

(c) $\displaystyle\lim_{x\to 2^+} f(g(x))$.

(d) $\displaystyle\lim_{x\to 2^+} g(f(x))$.

20. Let h be the function whose graph is shown in Figure 2.29b. Find the following limits:

(a) $\displaystyle\lim_{x\to 3^+} h(x)$.

(b) $\displaystyle\lim_{x\to 3^+} h(\lfloor x\rfloor)$.

(c) $\displaystyle\lim_{x\to 3^-} h(x)$.

(d) $\displaystyle\lim_{x\to 3^-} h(\lfloor x\rfloor)$.

(e) $\displaystyle\lim_{x\to 3^-} h(h(x))$.

(f) $\displaystyle\lim_{x\to 4} h(h(x))$.

(g) $\displaystyle\lim_{x\to 4} h(h(h(x)))$.

21. Show that if $\lim_{x\to a} f(x) = L$ and either $\lim_{x\to a} g(x) = \infty$ or $\lim_{x\to a} g(x) = -\infty$ then $\lim_{x\to a}(f(x)/g(x)) = 0$.

22. (a) Suppose that as $x \to a^{\neq}$, $f(x) \to L > 0$ and as $x \to a^{\neq}$, $g(x) \to 0^+$. Prove that $\lim_{x\to a}(f(x)/g(x)) = \infty$.

(b) Formulate and prove similar statements for cases in which either $f(x) \to L < 0$ or $g(x) \to 0^-$, or both.

23. Show that for any number $c \geq 0$, there are functions f and g such that $\lim_{x \to 0} f(x) = \infty$, $\lim_{x \to 0} g(x) = \infty$, and $\lim_{x \to 0}(f(x)/g(x)) = c$.

2.7   Continuity

In Section 2.4, we saw that if f is a polynomial, then for every number a, $\lim_{x \to a} f(x) = f(a)$. There is a name for this important property.

Definition 2.7.1. Suppose that f is a function and a is a number. We say that f is *continuous at a* if $f(a)$ is defined, $\lim_{x \to a} f(x)$ is defined, and $\lim_{x \to a} f(x) = f(a)$.

According to this definition, f will fail to be continuous at a if either $f(a)$ is undefined, or $\lim_{x \to a} f(x)$ is undefined, or both are defined but they are not equal. These three possibilities are illustrated in Figure 2.30. We could describe the function

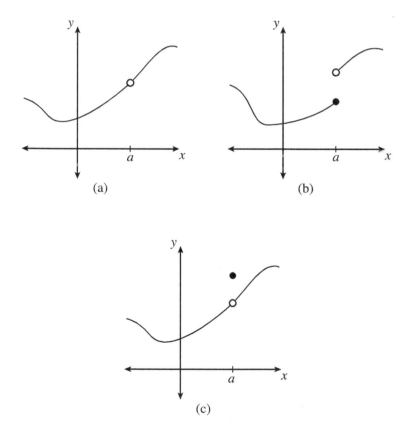

Figure 2.30: Three kinds of discontinuous functions. In (a), $f(a)$ is undefined. In (b), $\lim_{x \to a} f(x)$ is undefined. In (c), $f(a)$ and $\lim_{x \to a} f(x)$ are both defined, but they are not equal.

in Figure 2.30a as having a *missing point* at $x = a$. The function in Figure 2.30b has a *jump* at $x = a$, and the function in Figure 2.30c has a *misplaced point* at $x = a$.

You might think of $\lim_{x \to a} f(x)$ as what $f(a)$ should be, if it is to be in line with the values of $f(x)$ for x close to a. With this interpretation, we could say that f is continuous at a if the value $f(a)$ is what it should be. In Figure 2.30a, $f(a)$ is not what it should be because it is undefined. In Figure 2.30b, there is no single number that $f(a)$ should be in order to be in line with nearby points. And in Figure 2.30c, $f(a)$ is defined but not equal to what it should be.

Definition 2.7.2. We say that a function f is *continuous on an open interval I* if for every $a \in I$, f is continuous at a.

Thus, to say that f is continuous on an open interval I means that if you look at the points on the graph of f whose x-coordinates are in I, each of these points is in line with the points on either side of it. Intuitively, this means that you could draw this part of the graph of f without picking up your pencil. We restrict attention to open intervals for the moment because continuity at endpoints of an interval will require special treatment. We discuss continuity on closed and half-open intervals later in this section.

Since polynomials are continuous at all numbers, we can say that any polynomial is continuous on the interval $(-\infty, \infty)$. Thus, the graph of a polynomial is always a curve with no jumps and no missing or misplaced points; you could draw it without lifting up your pencil. For example, in Chapter 1 we graphed the function $f(x) = x^3 - 4x^2 + 2x + 2$ by plotting some points and then drawing a curve through those points (see Figure 1.18). We pointed out in Chapter 1 that by drawing this curve we were guessing at the values of f between the plotted points. While we still have not justified all aspects of the shape of the graph we drew, we have at least seen why the graph of f must be a curve with no breaks in it passing through the plotted points.

Theorem 2.4.2 allows us to see that many other functions are continuous:

Theorem 2.7.3. *Suppose f and g are continuous at a. Then so are $f + g$, $f - g$, and $f \cdot g$. If $g(a) \neq 0$, then f/g is also continuous at a.*

Proof. Since f and g are continuous at a, $\lim_{x \to a} f(x) = f(a)$ and $\lim_{x \to a} g(x) = g(a)$. Therefore by part 1 of Theorem 2.4.2,

$$\lim_{x \to a} (f + g)(x) = \lim_{x \to a} (f(x) + g(x)) = f(a) + g(a) = (f + g)(a),$$

so $f + g$ is continuous at a. The proofs of the other parts of the theorem are similar. □

According to Theorem 2.7.3, a rational function is continuous at all points except where the denominator is 0. For example, consider the function $f(x) = (x^2 - 2x + 7)/(x^3 - 3x^2)$. The denominator of $f(x)$ is 0 only when $x = 0$ or $x = 3$, so the domain of f is $(-\infty, 0) \cup (0, 3) \cup (3, \infty)$, and f is continuous on the intervals $(-\infty, 0)$, $(0, 3)$, and $(3, \infty)$. The graph of f is shown in Figure 2.31, and it consists of three pieces, each of which you can draw without picking up your pencil. We leave it as an exercise for you to check that $\lim_{x \to 0} f(x) = -\infty$, $\lim_{x \to 3^-} f(x) = -\infty$, $\lim_{x \to 3^+} f(x) = \infty$, $\lim_{x \to \infty} f(x) = 0$, and $\lim_{x \to -\infty} f(x) = 0$ (see Exercise 26). As a result, the lines $x = 0$, $x = 3$, and $y = 0$ are all asymptotes of the graph of f, as you can see in Figure 2.31.

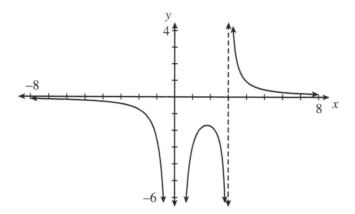

Figure 2.31: The graph of the rational function $f(x) = (x^2 - 2x + 7)/(x^3 - 3x^2)$. The dashed line is the asymptote $x = 3$.

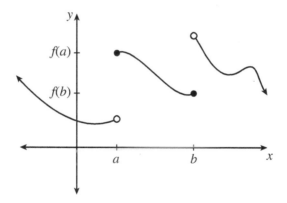

Figure 2.32: The graph of a function f that is continuous from the right at a and continuous from the left at b.

Since we have one-sided versions of limits, we can also define one-sided versions of continuity:

Definition 2.7.4. We say that f is *continuous from the right at a* if $f(a)$ is defined, $\lim_{x \to a^+} f(x)$ is defined, and $\lim_{x \to a^+} f(x) = f(a)$. We say that f is *continuous from the left at a* if $f(a)$ is defined, $\lim_{x \to a^-} f(x)$ is defined, and $\lim_{x \to a^-} f(x) = f(a)$.

It is not hard to see, using Theorem 2.5.5, that f is continuous at a if and only if it is continuous from both the left and the right at a. Of course, Theorem 2.7.3 applies to one-sided continuity as well.

For example, consider the function whose graph is shown in Figure 2.32. This function is discontinuous at a and b. However, $\lim_{x \to a^+} f(x) = f(a)$ and $\lim_{x \to b^-} f(x) = f(b)$, so f is continuous from the right at a and continuous from the left at b.

Intuitively, this means that the point $(a, f(a))$ does not seem out of place if we approach it from the right; it is only if we approach from the left that we see a sudden jump in the graph when $x = a$. Similarly, the point $(b, f(b))$ is in line with the points to its left, but not those to its right.

In Figure 2.32, consider the points on the graph of f whose x-coordinates are in the closed interval $[a, b]$. These points lie on a part of the graph that has no breaks or jumps in it; you could draw this part of the graph without lifting your pencil. It seems natural, therefore, to say that f is continuous on the interval $[a, b]$. This suggests the following definition of continuity on a closed interval.

Definition 2.7.5. We say that a function f is *continuous on a closed interval* $[a, b]$ if it is continuous on (a, b), continuous from the right at a, and continuous from the left at b.

Similarly, we would say that a function f is continuous on a half-open interval $[a, b)$ if it is continuous on (a, b) and continuous from the right at a, and it is continuous on the interval $(a, b]$ if it is continuous on (a, b) and continuous from the left at b. The general rule is that for f to be continuous on an interval I, we require that it be continuous on the interior of I, and that it have the appropriate one-sided continuity at any endpoints that are included in I. For example, the function whose graph is shown in Figure 2.33 is continuous on the intervals $(-\infty, a)$, $[a, b)$, and $[b, \infty)$. Looking back at Figure 1.16b, it should be clear that for every integer n, the function $h(x) = \lfloor x \rfloor$ is continuous on the interval $[n, n + 1)$. Notice that if f is continuous on an interval I, and J is another interval such that $J \subseteq I$, then f is also continuous on J.

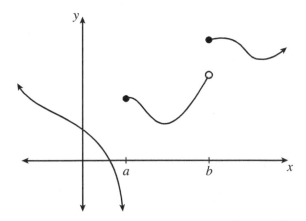

Figure 2.33: The graph of a function f that is continuous on the intervals $(-\infty, a)$, $[a, b)$, and $[b, \infty)$.

One of the reasons why continuous functions are so important in calculus is the following theorem.

Theorem 2.7.6 (Intermediate Value Theorem). *Suppose that f is continuous on the interval $[a, b]$, and either $f(a) < r < f(b)$ or $f(a) > r > f(b)$. Then there is some number c such that $a < c < b$ and $f(c) = r$.*

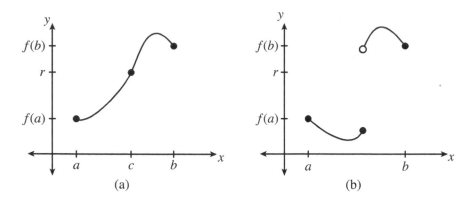

Figure 2.34: The intermediate value theorem. In (b), f is not continuous on $[a, b]$, and there is no c such that $f(c) = r$.

The intermediate value theorem is illustrated in Figure 2.34a. It may seem that the theorem is so obvious that it hardly requires proof. How could the value of the function possibly get from $f(a)$ to $f(b)$ without hitting the intermediate value r somewhere along the way? However, Figure 2.34b illustrates that it is possible for there to be no c such that $f(c) = r$, even though $f(a) < r < f(b)$. Of course, the function f in Figure 2.34b is not continuous, so it does not contradict the theorem. But it does illustrate that we have to include the assumption that f is continuous in the statement of the theorem; if that assumption were dropped, then the theorem would be incorrect. Thus, the proof will have to make use of the continuity of f, which makes the proof more subtle than one might expect. We will discuss the proof of the intermediate value theorem in the next section.

Example 2.7.7. Show that there is at least one solution to the equation

$$x^3 - 6x^2 + 5x + 2 = 0. \tag{2.10}$$

Solution. We could try to solve the equation, but there is an easier way. Let $f(x) = x^3 - 6x^2 + 5x + 2$. Table 2.6 shows some values of $f(x)$. Notice that $f(-1) = -10 < 0 < 2 = f(0)$. Since f is a polynomial, it is continuous everywhere. Therefore it is continuous on the interval $[-1, 0]$, and we can apply the intermediate value theorem to conclude that there must be some number c such that $-1 < c < 0$ and $f(c) = 0$. Such a number c is a solution to equation (2.10). Similarly, since $f(1) = 2 > 0 > -4 = f(2)$ and $f(4) = -10 < 0 < 2 = f(5)$, there must be two more solutions to the equation, one between 1 and 2 and another between 4 and 5.   □

x	-1	0	1	2	3	4	5
$f(x)$	-10	2	2	-4	-10	-10	2

Table 2.6: Some values of the function $f(x) = x^3 - 6x^2 + 5x + 2$.

As another example of the intermediate value theorem, consider the function $g(x) = x^2$. Since $g(1) = 1$, $g(2) = 4$, and g is continuous on the interval $[1, 2]$, there must be some number c such that $1 < c < 2$ and $g(c) = c^2 = 2$. In this case you may be thinking that you don't need the intermediate value theorem to show that there is such a number c; you can just let $c = \sqrt{2} \approx 1.41421$. After all, $\sqrt{2}$ is, by definition, the unique positive number whose square is 2. But how do we know that there *is* a positive number whose square is 2? When you first learned about square roots, this question was probably not discussed; it was probably simply assumed that, since $1^2 = 1 < 2$ and $2^2 = 4 > 2$, there is obviously a number between 1 and 2 whose square is exactly 2. We have now seen that this assumption is, indeed, correct, but it is not quite as obvious as you might have thought. What justifies it is the continuity of the function g and the intermediate value theorem.

Similar reasoning can be used to show that for every $r \geq 0$, there is some $c \geq 0$ such that $c^2 = r$. Furthermore, one can show that if $0 \leq a < b$ then $a^2 < b^2$, and therefore the value of c for which $c^2 = r$ is unique (see Exercise 27). This is what justifies us in defining $\sqrt{r}$ to be the unique $c \geq 0$ such that $c^2 = r$, for every r in the interval $[0, \infty)$.

In fact, the function $f(x) = \sqrt{x}$ is continuous on the interval $[0, \infty)$. Too see why, suppose $a \geq 0$. We first show that f is continuous from the right at a; in other words, we must show that $\lim_{x \to a^+} \sqrt{x} = \sqrt{a}$. We use the ϵ-δ method.

Suppose $\epsilon > 0$. Let $b = (\sqrt{a} + \epsilon)^2$, so that $\sqrt{b} = \sqrt{a} + \epsilon$, and notice that since $0 \leq \sqrt{a} < \sqrt{a} + \epsilon$,

$$b = (\sqrt{a} + \epsilon)^2 > (\sqrt{a})^2 = a.$$

Thus, we can let $\delta = b - a > 0$ (see Figure 2.35). Now suppose that $a < x < a + \delta$; to complete the proof, we must show that $|\sqrt{x} - \sqrt{a}| < \epsilon$. Since $a + \delta = b$, we have $a < x < b$. It follows (see part (c) of Exercise 27) that

$$\sqrt{a} < \sqrt{x} < \sqrt{b} = \sqrt{a} + \epsilon,$$

and therefore $|\sqrt{x} - \sqrt{a}| < \epsilon$, as required.

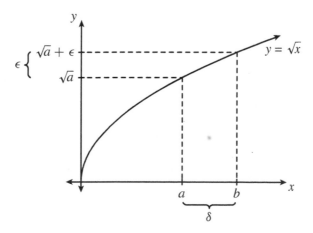

Figure 2.35: The proof that $\lim_{x \to a^+} \sqrt{x} = \sqrt{a}$.

Similar reasoning can be used to show that if $a > 0$ then f is continuous from the left at a (see Exercise 28). Thus, f is continuous on $(0, \infty)$ and continuous from the right at 0, so it is continuous on $[0, \infty)$.

Generalizing this reasoning leads to our next theorem. Recall that if n is even then $\sqrt[n]{x}$ is undefined when x is negative, but if n is odd then $\sqrt[n]{x}$ is defined for all x.

Theorem 2.7.8. *Let n be any positive integer and let* $f(x) = \sqrt[n]{x}$.

1. *If n is even, then f is continuous on* $[0, \infty)$.

2. *If n is odd, then f is continuous on* $(-\infty, \infty)$.

Many other familiar functions are continuous on their entire domains. For example, it is easy to show that the function $f(x) = |x|$ is continuous on $(-\infty, \infty)$. The trigonometric functions are also continuous everywhere they are defined. To show this we will need the following lemma. (A *lemma* is a fact that is proven just for the purpose of using it to prove something else.)

Lemma 2.7.9. *For any numbers* α *and* β,

$$|\cos \beta - \cos \alpha| \le |\beta - \alpha|, \qquad |\sin \beta - \sin \alpha| \le |\beta - \alpha|.$$

Proof. Let Q_α be the point on the unit circle with coordinates $(\cos \alpha, \sin \alpha)$, and similarly let Q_β be the point $(\cos \beta, \sin \beta)$. Recall that according to the definitions of the cosine and sine functions, we can reach Q_α by beginning at the point $(1, 0)$ and traveling α units along the unit circle, with a positive value for α indicating counterclockwise motion and a negative value indicating clockwise motion, and a similar statement holds for Q_β. Thus, if $\beta \ge \alpha$, then starting at Q_α and traveling $\beta - \alpha$ units counterclockwise along the unit circle we would reach Q_β, while if $\alpha > \beta$, then from Q_β we would travel $\alpha - \beta$ units counterclockwise to reach Q_α. In either case, we can travel $|\beta - \alpha|$ units along the unit circle to move between Q_α and Q_β, as indicated in Figure 2.36.

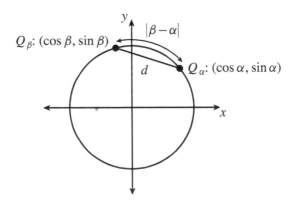

Figure 2.36: The proof of Lemma 2.7.9.

Now let d be the straight-line distance from Q_α to Q_β. Since a straight line is the shortest distance between two points, $d \le |\beta - \alpha|$. Thus, by the distance formula,

$$\sqrt{(\cos\beta - \cos\alpha)^2 + (\sin\beta - \sin\alpha)^2} = d \le |\beta - \alpha|. \qquad (2.11)$$

But we also have

$$\sqrt{(\cos\beta - \cos\alpha)^2 + (\sin\beta - \sin\alpha)^2} \ge \sqrt{(\cos\beta - \cos\alpha)^2} = |\cos\beta - \cos\alpha|, \qquad (2.12)$$

$$\sqrt{(\cos\beta - \cos\alpha)^2 + (\sin\beta - \sin\alpha)^2} \ge \sqrt{(\sin\beta - \sin\alpha)^2} = |\sin\beta - \sin\alpha|. \qquad (2.13)$$

Combining inequalities (2.11)–(2.13) proves the lemma. $\qquad\square$

Theorem 2.7.10. *All of the trigonometric functions are continuous everywhere they are defined.*

Proof. We first verify that the cosine function is continuous on $(-\infty, \infty)$. To do this, we use the ϵ-δ definition to show that for any number a, $\lim_{x\to a} \cos x = \cos a$.

Suppose $\epsilon > 0$. Let $\delta = \epsilon$. Suppose $0 < |x - a| < \delta$. Then by the lemma,

$$|\cos x - \cos a| \le |x - a| < \delta = \epsilon.$$

Similarly, the sine function is continuous on $(-\infty, \infty)$. By definition, $\tan x = \sin x / \cos x$, so it is undefined when $\cos x = 0$, which happens when $x = \pi/2, 3\pi/2, 5\pi/2, \ldots, -\pi/2, -3\pi/2, \ldots$. But by Theorem 2.7.3 it is continuous on the intervals that make up its domain: $(-\pi/2, \pi/2), (\pi/2, 3\pi/2), \ldots, (-3\pi/2, -\pi/2), \ldots$. Similarly, the other trigonometric functions are continuous everywhere they are defined. $\qquad\square$

Theorem 2.7.3 tells us that when we add, subtract, multiply, and divide continuous functions, the result is continuous. We can therefore now recognize many more functions as continuous. For example, the function $f(x) = \sqrt[3]{x} + \sin x$ is continuous on $(-\infty, \infty)$, while $g(x) = \sqrt{x} \tan x$ is continuous on $[0, \pi/2), (\pi/2, 3\pi/2), \ldots$.

What about compositions of continuous functions? Are they continuous as well? To answer this question, we begin by taking a closer look at the definition of continuity.

If a function f is continuous at a number a, then according to the definition of continuity, $\lim_{x\to a} f(x) = f(a)$, which means that as $x \to a^{\ne}$, $f(x) \to f(a)$. In other words, if x is close to a, but not equal to a, then $f(x)$ is close to $f(a)$. But notice that the restriction in this statement that x must not be equal to a is unnecessary; if x is equal to a then it is still true that $f(x)$ is close to $f(a)$, because in fact $f(x)$ is equal to $f(a)$! Thus, we can simply say that if x is close to a, then $f(x)$ is close to $f(a)$. In other words, we have established the following theorem:

Theorem 2.7.11. *Suppose that f is continuous at a. Then as $x \to a$, $f(x) \to f(a)$. In other words, for every $\epsilon > 0$ there is some $\delta > 0$ such that if $|x - a| < \delta$ then $|f(x) - f(a)| < \epsilon$.*

Notice that in this theorem we have eliminated the superscript "$\ne$" in our limit notation, which indicates that we have dropped the restriction that x cannot equal a.

A similar idea applies to one-sided continuity. If f is continuous from the left at a, then we can say that as $x \to a^<$, $f(x) \to f(a)$. But once again there is no need to exclude the possibility that $x = a$. We therefore introduce the notation "as $x \to a^\le$, $f(x) \to L$" to mean that if x is close to a and $x \le a$, then $f(x)$ is close to L; more precisely, we define this notation to mean that for every $\epsilon > 0$ there is some $\delta > 0$ such that if $a - \delta < x \le a$ then $|f(x) - L| < \epsilon$. You should be able to formulate a similar precise definition for the notation "as $x \to a^\ge$, $f(x) \to L$," to express the idea that if x is close to a and $x \ge a$ then $f(x)$ is close to L. We can now use this new notation to state a one-sided version of Theorem 2.7.11:

Theorem 2.7.12.

1. *If f is continuous from the left at a, then as $x \to a^\le$, $f(x) \to f(a)$.*

2. *If f is continuous from the right at a, then as $x \to a^\ge$, $f(x) \to f(a)$.*

Theorems 2.7.11 and 2.7.12 makes it easy to work with limits of compositions $f \circ g$ when f is continuous:

Theorem 2.7.13.

1. *Suppose that $\lim_{x \to a} g(x) = L$ and f is continuous at L. Then $\lim_{x \to a} f(g(x)) = f(L)$.*

2. *Suppose that g is continuous at a and f is continuous at $g(a)$. Then $f \circ g$ is continuous at a.*

3. *Suppose that g is continuous on an interval J, f is continuous on an interval I, and for every $x \in J$, $g(x) \in I$. Then $f \circ g$ is continuous on J.*

Proof. 1. As usual, we begin by letting $u = g(x)$. Since $\lim_{x \to a} g(x) = L$, we can say that as $x \to a^{\ne}$, $u \to L$. And according to Theorem 2.7.11, as $u \to L$, $f(u) \to f(L)$. Combining these two statements we conclude that as $x \to a^{\ne}$, $f(u) \to L$, so $\lim_{x \to a} f(g(x)) = L$.

2. Since g is continuous at a, $\lim_{x \to a} g(x) = g(a)$, and by part 1 it follows that $\lim_{x \to a} f(g(x)) = f(g(a))$. Thus, $f \circ g$ is continuous at a.

3. Suppose $a \in J$. Then by the assumptions in this part of the theorem, $g(a) \in I$. If a is in the interior of J and $g(a)$ is in the interior of I, then g is continuous at a and f is continuous at $g(a)$, and therefore $f \circ g$ is continuous at a by part 2. If either a is an endpoint of J or $g(a)$ is an endpoint of I, then similar reasoning can be used, with small adjustments to deal with one-sided limits. For example, suppose that a is in the interior of J but $g(a)$ is the left endpoint of I, and let $u = g(x)$. Then as $x \to a$, $u = g(x) \to g(a)^\ge$, and as $u \to g(a)^\ge$, $f(u) \to f(g(a))$. Therefore as $x \to a$, $f(g(x)) = f(u) \to f(g(a))$. For an alternative approach to this proof, see Exercise 31. $\square$

Example 2.7.14. On what intervals are the following functions continuous?

$$\text{(a) } h_1(x) = \sin(3x^5 - 2x^3 + 7), \quad \text{(b) } h_2(x) = \sqrt{x^2 - 1}.$$

Solution. We have $h_1 = f \circ g$, where $f(x) = \sin x$ and $g(x) = 3x^5 - 2x^3 + 7$. Since f and g are continuous at all numbers, by part 2 of Theorem 2.7.13, h_1 is continuous on $(-\infty, \infty)$.

Since $h_2(x)$ is undefined if $x^2 - 1 < 0$, the domain of h_2 is $(-\infty, -1] \cup [1, \infty)$. We have $h_2 = f \circ g$, where $f(x) = \sqrt{x}$ and $g(x) = x^2 - 1$. The function g is continuous on the interval $[1, \infty)$, f is continuous on $[0, \infty)$, and for every $x \in [1, \infty)$, $g(x) = x^2 - 1 \subset [0, \infty)$. Therefore, by part 3 of Theorem 2.7.13, $h_2 = f \circ g$ is continuous on $[1, \infty)$. Similar reasoning shows that it is also continuous on $(-\infty, -1]$. $\qquad\square$

Example 2.7.15. Find the following limits:

(a) $\displaystyle\lim_{x \to 2} \sqrt[3]{\frac{x^2 + 4x - 12}{x^2 - 5x + 6}}$,

(b) $\displaystyle\lim_{\theta \to \pi/2} (\sec\theta - \tan\theta)$,

(c) $\displaystyle\lim_{t \to \infty} \sqrt[5]{t}$,

(d) $\displaystyle\lim_{x \to \infty} (\sqrt{x^2 + x} - x)$.

Solution. For limit (a), we begin by finding the limit of the rational function inside the cube root. Since numerator and denominator both approach 0 as x approaches 2, we factor and cancel:

$$\lim_{x \to 2} \frac{x^2 + 4x - 12}{x^2 - 5x + 6} = \lim_{x \to 2} \frac{(x-2)(x+6)}{(x-2)(x-3)} = \lim_{x \to 2} \frac{x+6}{x-3} = \frac{8}{-1} = -8.$$

Since the cube root function is continuous at -8, by part 1 of Theorem 2.7.13,

$$\lim_{x \to 2} \sqrt[3]{\frac{x^2 + 4x - 12}{x^2 - 5x + 6}} = \sqrt[3]{-8} = -2.$$

We could try to find limit (b) by evaluating $\lim_{\theta \to \pi/2} \sec\theta$ and $\lim_{\theta \to \pi/2} \tan\theta$ separately. But $\lim_{\theta \to \pi/2} \sec\theta = \lim_{\theta \to \pi/2}(1/\cos\theta)$, and by the continuity of the cosine function, $\lim_{\theta \to \pi/2} \cos\theta = \cos\pi/2 = 0$. Thus, $\lim_{\theta \to \pi/2} \sec\theta$ is undefined, and similar reasoning shows that $\lim_{\theta \to \pi/2} \tan\theta$ is also undefined. You might be tempted to conclude that limit (b) is undefined, but as we saw in Example 2.4.9(b), this reasoning is not correct. In fact, the limit is defined. To find its value, we begin by writing out the definitions of the secant and tangent functions:

$$\lim_{\theta \to \pi/2} (\sec\theta - \tan\theta) = \lim_{\theta \to \pi/2} \left[\frac{1}{\cos\theta} - \frac{\sin\theta}{\cos\theta} \right] = \lim_{\theta \to \pi/2} \frac{1 - \sin\theta}{\cos\theta}.$$

By the continuity of the sine and cosine functions, the numerator and denominator of the last fraction both approach 0, so we try to find a factor we can cancel. A tricky algebraic step accomplishes this; we multiply numerator and denominator by $1 + \sin\theta$ and then apply the identity $\cos^2 \theta + \sin^2 \theta = 1$:

$$\lim_{\theta \to \pi/2} (\sec \theta - \tan \theta) = \lim_{\theta \to \pi/2} \frac{1 - \sin \theta}{\cos \theta} \cdot \frac{1 + \sin \theta}{1 + \sin \theta} = \lim_{\theta \to \pi/2} \frac{1 - \sin^2 \theta}{\cos \theta (1 + \sin \theta)}$$

$$= \lim_{\theta \to \pi/2} \frac{\cos^2 \theta}{\cos \theta (1 + \sin \theta)} = \lim_{\theta \to \pi/2} \frac{\cos \theta}{1 + \sin \theta}$$

$$= \frac{\cos \pi/2}{1 + \sin \pi/2} = \frac{0}{1 + 1} = 0.$$

For limit (c), it appears intuitively that $\lim_{t \to \infty} \sqrt[5]{t} = \infty$, but none of our theorems seem to apply. With no theorems to help us, we fall back on the definition of this limit statement, which is: for every number M, there is some number N such that if $t > N$ then $\sqrt[5]{t} > M$. To prove this statement, let M be any number. Let $N = M^5$. Now suppose $t > N$. Then $\sqrt[5]{t} > \sqrt[5]{N} = \sqrt[5]{M^5} = M$, as required. Similar reasoning can be used to show that for every positive integer n, $\lim_{x \to \infty} \sqrt[n]{x} = \infty$, and for odd n, $\lim_{n \to -\infty} \sqrt[n]{x} = -\infty$.

Finally, for limit (d) we begin with an algebraic step we've seen before that allows some terms to cancel each other out. Since this step eliminates the radical sign in the numerator of a fraction, it is sometimes called "rationalizing the numerator":

$$\lim_{x \to \infty} (\sqrt{x^2 + x} - x) = \lim_{x \to \infty} \frac{\sqrt{x^2 + x} - x}{1} \cdot \frac{\sqrt{x^2 + x} + x}{\sqrt{x^2 + x} + x}$$

$$= \lim_{x \to \infty} \frac{x^2 + x - x^2}{\sqrt{x^2 + x} + x} = \lim_{x \to \infty} \frac{x}{\sqrt{x^2 + x} + x}.$$

Next, we factor out the highest power of x in the polynomial inside the square root:

$$\lim_{x \to \infty} (\sqrt{x^2 + x} - x) = \lim_{x \to \infty} \frac{x}{\sqrt{x^2 \cdot \left(1 + \frac{1}{x}\right)} + x} = \lim_{x \to \infty} \frac{x}{\sqrt{x^2} \cdot \sqrt{1 + \frac{1}{x}} + x}$$

$$= \lim_{x \to \infty} \frac{x}{x \cdot \sqrt{1 + \frac{1}{x}} + x} = \lim_{x \to \infty} \frac{x}{x \left(\sqrt{1 + \frac{1}{x}} + 1\right)}$$

$$= \lim_{x \to \infty} \frac{1}{\sqrt{1 + \frac{1}{x}} + 1} = \frac{1}{\sqrt{1} + 1} = \frac{1}{2}.$$

Notice that we have used the continuity of the square root function to conclude that since $\lim_{x \to \infty}(1 + 1/x) = 1$, $\lim_{x \to \infty} \sqrt{1 + 1/x} = \sqrt{1} = 1$.   □

We end this section by returning to the two limits we introduced at the end of Section 2.1: $\lim_{\theta \to 0}(\sin \theta)/\theta$ and $\lim_{x \to 2}(\sqrt{x} - \sqrt{2})/(x - 2)$. We promised that by the end of this chapter we would be able to compute these limits, and now we are ready to do so.

We begin with the second, which is a little easier. Once again, rationalizing the numerator turns out to be the key:

$$\lim_{x \to 2} \frac{\sqrt{x} - \sqrt{2}}{x - 2} = \lim_{x \to 2} \frac{\sqrt{x} - \sqrt{2}}{x - 2} \cdot \frac{\sqrt{x} + \sqrt{2}}{\sqrt{x} + \sqrt{2}} = \lim_{x \to 2} \frac{x - 2}{(x - 2)(\sqrt{x} + \sqrt{2})}$$

$$= \lim_{x \to 2} \frac{1}{\sqrt{x} + \sqrt{2}} = \frac{1}{\sqrt{2} + \sqrt{2}} = \frac{1}{2\sqrt{2}} \cdot \frac{\sqrt{2}}{\sqrt{2}} = \frac{\sqrt{2}}{4}.$$

In the last step, we have rationalized the denominator, putting the answer into the form in which we gave it in Section 2.1.

Finally, for $\lim_{\theta \to 0}(\sin \theta)/\theta$, we begin with some inequalities. Figure 2.37 shows an angle of θ radians in the unit circle, with $0 < \theta < \pi/2$. In the figure, triangle OPQ is contained in the sector of the circle cut off by OP and OQ, so

$$\text{(area of triangle } OPQ) \le \text{(area of sector } OPQ). \tag{2.14}$$

Since the coordinates of Q are $(\cos \theta, \sin \theta)$, the height of triangle OPQ is $AQ = \sin \theta$, so

$$\text{area of triangle } OPQ = \frac{1}{2}(1)(\sin \theta) = \frac{\sin \theta}{2}.$$

The area of the entire unit circle is $\pi(1)^2 = \pi$, and the fraction of the circle contained in sector OPQ is $\theta/(2\pi)$ (since θ is measured in radians, and a full circle is 2π radians), so

$$\text{area of sector } OPQ = \frac{\theta}{2\pi}\pi = \frac{\theta}{2}.$$

Thus, inequality (2.14) means

$$\frac{\sin \theta}{2} \le \frac{\theta}{2},$$

and it follows that

$$\frac{\sin \theta}{\theta} \le 1. \tag{2.15}$$

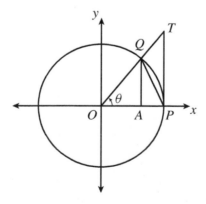

Figure 2.37: Finding $\lim_{\theta \to 0} \sin \theta / \theta$.

Next, notice that sector OPQ is contained in triangle OPT, and therefore

$$\text{(area of sector } OPQ) \leq \text{(area of triangle } OPT). \tag{2.16}$$

Triangle OPT is similar to triangle OAQ, so

$$PT = \frac{PT}{OP} = \frac{AQ}{OA} = \frac{\sin\theta}{\cos\theta} = \tan\theta.$$

Thus, the area of triangle OPT is $\frac{1}{2}(1)\tan\theta = (\tan\theta)/2$, and inequality (2.16) says

$$\frac{\theta}{2} \leq \frac{\tan\theta}{2} = \frac{\sin\theta}{2\cos\theta}.$$

Rearranging, we have

$$\cos\theta \leq \frac{\sin\theta}{\theta}, \tag{2.17}$$

and putting this together with (2.15) gives us

$$\cos\theta \leq \frac{\sin\theta}{\theta} \leq 1. \tag{2.18}$$

We have derived inequality (2.18) under the assumption that $0 < \theta < \pi/2$, but similar reasoning can be used to show that it holds for $-\pi/2 < \theta < 0$ as well (see Exercise 32).

You may not have known where we were headed with these calculations, but now you should see that inequality (2.18) is exactly what we need to apply the squeeze theorem. Since $\lim_{\theta\to 0}\cos\theta = \cos 0 = 1$ and $\lim_{\theta\to 0} 1 = 1$, the squeeze theorem implies that $\lim_{\theta\to 0}(\sin\theta)/\theta = 1$. Note that in our reasoning we have assumed that θ is measured in radians. If we were to measure angles in degrees, the answer for this limit would be different (see Exercise 29). The fact that the use of radians leads to such a simple answer for this limit is one of the reasons why we always use radians to measure angles in calculus.

Exercises 2.7

1–25: Evaluate the limit.

1. $\displaystyle\lim_{x\to 1/2} \tan(\pi x^2)$.

2. $\displaystyle\lim_{x\to 6} \frac{\sqrt{x+3}-2}{x-1}$.

3. $\displaystyle\lim_{x\to 1} \frac{\sqrt{x+3}-2}{x-1}$.

4. $\displaystyle\lim_{x\to 0} \frac{\sqrt{4+x}-\sqrt{4-x}}{x}$.

5. $\lim\limits_{t \to 1} \dfrac{\sqrt{t^2 + 3t} - 2}{t - 1}$.

6. $\lim\limits_{x \to 4} \dfrac{x - 2 - \sqrt{x}}{x - 4}$.

7. $\lim\limits_{z \to 1^+} \dfrac{z - 1}{\sqrt{z^2 - 1}}$.

8. $\lim\limits_{x \to 4} \left(\dfrac{1}{\sqrt{x} - 2} - \dfrac{4}{x - 4} \right)$.

9. $\lim\limits_{x \to 8} \dfrac{\sqrt[3]{x} - 2}{x - 8}$. (Hint: Use the fact that $(a - b)(a^2 + ab + b^2) = a^3 - b^3$.)

10. $\lim\limits_{x \to 0^+} \left(\dfrac{1}{\sqrt[3]{x}} - \dfrac{1}{\sqrt{x}} \right)$.

11. $\lim\limits_{u \to \infty} \dfrac{\sqrt{u^2 + 9}}{3u}$.

12. $\lim\limits_{x \to \infty} (\sqrt{x^2 + 100} - x)$.

13. $\lim\limits_{x \to -\infty} (\sqrt{x^2 + x} + x)$.

14. $\lim\limits_{x \to \infty} (\sqrt{x^4 + x} - x^2)$.

15. $\lim\limits_{x \to \infty} (\sqrt{x^4 + x^2} - x^2)$.

16. $\lim\limits_{x \to \infty} (\sqrt{x^4 + x^3} - x^2)$.

17. $\lim\limits_{\theta \to \pi/2} \dfrac{\sin(2\theta)}{\cos \theta}$. (Hint: Use a trigonometric identity to reexpress $\sin(2\theta)$.)

18. $\lim\limits_{\theta \to \pi/4} \dfrac{\sqrt{2} - 2\cos \theta}{1 - 2\cos^2 \theta}$.

19. $\lim\limits_{\theta \to \pi/4} \dfrac{\cos \theta - \sin \theta}{1 - 2\sin^2 \theta}$.

20. $\lim\limits_{\theta \to 0} \dfrac{1 - \cos \theta}{\theta}$.

21. $\lim\limits_{\theta \to 0} \dfrac{1 - \cos \theta}{\theta^2}$.

22. $\lim\limits_{\theta \to 0} \dfrac{2 - \sqrt{\cos \theta + 3}}{\theta^2}$.

23. $\lim\limits_{x \to 0} \dfrac{\sin(5x)}{5x}$. (Hint: Let $\theta = 5x$, and use the fact that $\lim\limits_{\theta \to 0} \frac{\sin \theta}{\theta} = 1$.)

24. $\lim\limits_{x \to 0} \dfrac{\tan(5x)}{3x}$.

25. $\lim\limits_{x \to 0} \dfrac{\tan(5x)}{x^2 - 3x}$.

26. Let

$$f(x) = \frac{x^2 - 2x + 7}{x^3 - 3x^2}.$$

Verify the following limits, thus confirming the asymptotes shown in Figure 2.31.

(a) $\lim\limits_{x \to 0} f(x) = -\infty$.

(b) $\lim\limits_{x \to 3^-} f(x) = -\infty$.

(c) $\lim\limits_{x \to 3^+} f(x) = \infty$.

(d) $\lim\limits_{x \to \infty} f(x) = 0$.

(e) $\lim\limits_{x \to -\infty} f(x) = 0$.

27. (a) Show that if $0 \le a < b$ then $a^2 < b^2$. (Hint: Show that $b^2 - a^2 > 0$ by factoring $b^2 - a^2$.)

(b) Show that if $r \ge 0$ then there is exactly one number $c \ge 0$ such that $c^2 = r$. (It follows that we can define $\sqrt{r}$ to be this unique number c.)

(c) Show that if $0 \le a < b$ then $\sqrt{a} < \sqrt{b}$.

28. Let $f(x) = \sqrt{x}$. Show that if $a > 0$ then f is continuous from the left at a. (Hint: If $\epsilon > \sqrt{a}$, then let $\delta = a$. If $\epsilon \le \sqrt{a}$, then let $b = (\sqrt{a} - \epsilon)^2$ and $\delta = a - b$.)

29. Let $f(x)$ be the sine of an angle of x degrees (not radians). Find $\lim\limits_{x \to 0}(f(x)/x)$. (Hint: Let $g(x)$ be the number of radians in an angle of x degrees. Then $f(x) = \sin(g(x))$. Now find a formula for $g(x)$.)

30. (a) Suppose that as $x \to a^{\ne}$, $g(x) \to L^{\le}$, and f is continuous from the left at L. Show that $\lim\limits_{x \to a} f(g(x)) = f(L)$. Of course, this is a one-sided version of part 1 of Theorem 2.7.13

(b) State and prove a similar statement for continuity from the right.

31. Suppose f is a function, I is an interval contained in the domain of f, and $a \in I$. We define "as $x \to a^{\in I}$, $f(x) \to L$" to mean that if x is close to a and x is in I, then $f(x)$ is close to L. More precisely, it means that for every $\epsilon > 0$ there is some $\delta > 0$ such that if $|x - a| < \delta$ and $x \in I$ then $|f(x) - L| < \epsilon$.

(a) Show that a function f is continuous on an interval I if and only if for every $a \in I$, as $x \to a^{\in I}$, $f(x) \to f(a)$.

Now consider the functions f and g and intervals I and J of part 3 of Theorem 2.7.13.

(b) Show that for any $a \in J$, as $x \to a^{\in J}$, $g(x) \to g(a)^{\in I}$. (First write out what this notation means.)

(c) Show that $f \circ g$ is continuous on J.

32. In this exercise you will use two different methods to show that inequality (2.18) holds for $-\pi/2 < \theta < 0$.

(a) Draw a diagram similar to Figure 2.37, but for $-\pi/2 < \theta < 0$. Then imitate the reasoning used in the text to show that inequality (2.18) holds in this case.

(b) Suppose that $-\pi/2 < \theta < 0$. Then $0 < -\theta < \pi/2$, so by the reasoning in the text,

$$\cos(-\theta) \leq \frac{\sin(-\theta)}{-\theta} \leq 1.$$

Use this fact to show that inequality (2.18) holds.

2.8 Sequences and the Nested Interval Theorem

As you no doubt know, $\pi = 3.14159\ldots$, where the digits go on forever. But what does it mean to specify a number with a sequence of digits that we can never completely write down? According to the meaning of decimal notation, the 1 after the decimal point stands for 1/10, the following 4 stands for 4/100, and so on, so the decimal representation of π says that

$$\pi = 3 + \frac{1}{10} + \frac{4}{100} + \frac{1}{1000} + \frac{5}{10000} + \frac{9}{100000} + \cdots .$$

But, again, it is not clear what this means: there are infinitely many terms in this sum, so we could never finish adding them up. (We will study such infinite sums in Chapter 10.)

One thing that is clear from the decimal expansion of π is that π is approximately 3.1. The complete infinite decimal expansion of π is hard to understand, but there is nothing mysterious about the finite decimal expansion 3.1, which represents the finite sum $3 + 1/10$. Unfortunately, it isn't a very good approximation of π, but if we want a better approximation we could say that $\pi \approx 3.14 = 3 + 1/10 + 4/100$, or $\pi \approx 3.141 = 3 + 1/10 + 4/100 + 1/1000$.

You should recognize that we now have the crucial ingredients for applying the methods of calculus: a way of approximating a number, and a way of making the approximations better and better. We can therefore use a limit to achieve precision through approximation: the exact value of π is the limit that the approximations approach as we take more and more digits from the decimal expansion. In fact, this is how the value of an infinite decimal expansion is defined.

Unfortunately, this limit doesn't quite fit the pattern of the limits we have studied so far in this chapter. Up until now we have studied limits of functions, but what is the function whose limit we are computing in this case? We could define $f(1) = 3.1$, $f(2) = 3.14$, and, in general,

$$f(n) = 3.14 \cdots d_n = 3 + \frac{1}{10} + \frac{4}{100} + \cdots + \frac{d_n}{10^n},$$

where d_n is the nth digit after the decimal point in the decimal expansion of π. We would then want to say that the exact value of π is $\pi = \lim_{n\to\infty} f(n)$, but this limit is slightly different from the limits we have studied previously. The reason is that $f(n)$ only makes sense if n is a positive integer, so the domain of f is the set of all positive integers, usually denoted $\mathbb{Z}^+$. (It would make no sense to take, say, the first three and a half digits of π!) A function whose domain is $\mathbb{Z}^+$ is called a *sequence*. So we need to say exactly what it means to take the limit of a sequence. Of course, we follow the pattern of our previous definitions, and just add in the restriction that only positive integer values are considered for n.

Definition 2.8.1. If f is a sequence, then to say that as $n \to \infty$, $f(n) \to L$ means that for every $\epsilon > 0$ there is some number N such that if n is a positive integer and $n > N$, then $|f(n) - L| < \epsilon$. We define $\lim_{n\to\infty} f(n)$ to be the unique L such that as $n \to \infty$, $f(n) \to L$, if there is such an L.

Figure 2.38 shows the graph of a sequence f for which $\lim_{n\to\infty} f(n) = L$. Since $f(n)$ is defined only for positive integers n, the graph consists of a sequence of dots rather than a curve. Since $f(n)$ gets close to L when n is large, the dots gets close to the line $y = L$ toward the right side of the graph.

Sometimes you will have to tell from context whether a limit is a limit of a sequence or an ordinary limit of a function. We will usually use the letter n as the independent variable for sequences to help make the distinction clear. For the most part, working with limits of sequences is very similar to working with other limits. All of our

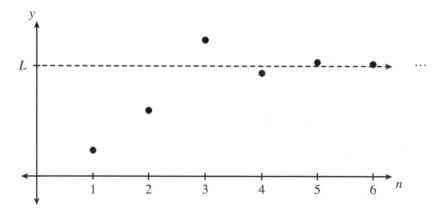

Figure 2.38: The graph of $y = f(n)$, for a sequence f such that $\lim_{n\to\infty} f(n) = L$.

previous theorems still apply, with some obvious small adjustments. We also use the notation $\lim_{n\to\infty} f(n) = \infty$ (or $\lim_{n\to\infty} f(n) = -\infty$) to mean that when n is a very large positive integer, $f(n)$ is very large (or $f(n)$ is a very large magnitude negative number).

For example, clearly $\lim_{n\to\infty} n = \infty$, and therefore $\lim_{n\to\infty} 1/n = 0$. Also,

$$\lim_{n\to\infty} \frac{3n^2 - 5n + 2}{5n^2 + 3n - 1} = \lim_{n\to\infty} \frac{n^2 \cdot (3 - 5/n + 2/n^2)}{n^2 \cdot (5 + 3/n - 1/n^2)} = \lim_{n\to\infty} \frac{3 - 5/n + 2/n^2}{5 + 3/n - 1/n^2} = \frac{3}{5}.$$

However, there are occasionally subtle differences between ordinary limits of functions and limits of sequences. For example, Figure 2.39 shows the graph of the equation $y = \cos(2\pi x)$, and it is clear from this figure that $\lim_{x\to\infty} \cos(2\pi x)$ is undefined; as x gets larger and larger, $\cos(2\pi x)$ oscillates back and forth between -1 and 1 and does not approach any single number. However, if n is any positive integer then $\cos(2\pi n) = 1$, and therefore the graph of the equation $y = \cos(2\pi n)$ (where the horizontal axis now represents n, and it is understood that n stands for a positive integer) consists of just the dots in Figure 2.39 at the points $(1, 1)$, $(2, 1)$, $(3, 1)$, ... Thus, $\lim_{n\to\infty} \cos(2\pi n) = 1$.

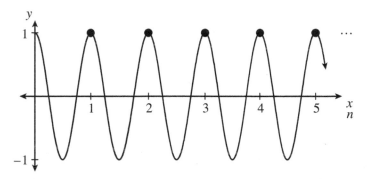

Figure 2.39: $\lim_{x\to\infty} \cos(2\pi x)$ is undefined, but $\lim_{n\to\infty} \cos(2\pi n) = 1$.

This example illustrates that it is possible for $\lim_{x\to\infty} f(x)$ to be undefined, but for $\lim_{n\to\infty} f(n)$ to be defined. However, the reverse situation is not possible. If $\lim_{x\to\infty} f(x) = L$, then for large real numbers x, $f(x)$ is close to L. This is true for all large real numbers, so in particular it is true for large integers. In other words, if n is a large integer, then $f(n)$ must be close to L, which means that $\lim_{n\to\infty} f(n) = L$. Thus, the following theorem is true.

Theorem 2.8.2. *If $\lim_{x\to\infty} f(x) = L$, then $\lim_{n\to\infty} f(n) = L$, where in the second limit n stands for an integer.*

The theorem is illustrated in Figure 2.40. Since the curve $y = f(x)$ approaches the horizontal asymptote $y = L$, the dots on the curve at integer x-coordinates, which make up the graph of $y = f(n)$, must also approach the asymptote. For a more careful proof of the theorem, see Exercise 22.

We usually think of a sequence f as representing a list of numbers $f(1)$, $f(2)$, $f(3)$, ... The number $f(n)$ is called the *nth term* of the sequence. We often write

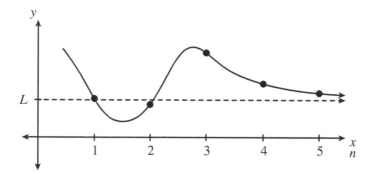

Figure 2.40: The curve is the graph of $y = f(x)$, and the dots are the graph of $y = f(n)$.

$(f(n))_{n=1}^{\infty}$ to denote the sequence $f(1)$, $f(2)$, $f(3)$, ... The subscript "$n = 1$" and superscript "∞" indicate that the terms of the sequence are generated from the expression $f(n)$ by plugging in values of n starting with 1 and running through all the positive integers. Sometimes we may want to start at a value of n other than 1. For example, $(f(n))_{n=7}^{\infty}$ would denote the sequence $f(7)$, $f(8)$, $f(9)$, ...

If $\lim_{n\to\infty} f(n) = L$, then as we go through the list $f(1)$, $f(2)$, $f(3)$, ..., the numbers in the list eventually get close to L. In this situation, we say that the sequence $(f(n))_{n=1}^{\infty}$ *converges* to L. For example, $(1/n)_{n=1}^{\infty}$ denotes the sequence 1, 1/2, 1/3, 1/4, ... Since $\lim_{n\to\infty} 1/n = 0$, this sequence converges to 0. If $\lim_{n\to\infty} f(n)$ is undefined, then we say that the sequence $(f(n))_{n=1}^{\infty}$ *diverges*. For example, since $\lim_{n\to\infty} \sqrt{n} = \infty$, the sequence $(\sqrt{n})_{n=1}^{\infty}$ diverges.

Example 2.8.3. Do the following sequences converge or diverge?

$$\text{(a) } ((-1)^n)_{n=1}^{\infty}, \qquad \text{(b) } \left(\frac{(-1)^n}{n}\right)_{n=4}^{\infty}, \qquad \text{(c) } \left(\cos\left(\frac{2\pi(n^2+1)}{n}\right)\right)_{n=1}^{\infty}.$$

Solution. For every integer n, $(-1)^n$ is equal to 1 if n is even and -1 if n is odd. Thus, $((-1)^n)_{n=1}^{\infty}$ is the sequence -1, 1, -1, 1, ... This sequence will continue to alternate between the values -1 and 1 forever, so $\lim_{n\to\infty}(-1)^n$ is undefined and the sequence diverges.

Sequence (b) is the sequence 1/4, $-1/5$, 1/6, $-1/7$, ... The numerators of these fractions alternate between -1 and 1, as in part (a). Notice that

$$\lim_{n\to\infty} \left|\frac{(-1)^n}{n}\right| = \lim_{n\to\infty} \frac{1}{n} = 0.$$

By Theorem 2.4.11 (or, more precisely, the sequence version of the theorem), it follows that $\lim_{n\to\infty}(-1)^n/n = 0$. In other words, sequence (b) converges to 0.

Finally, in part (c) we have $\lim_{n\to\infty} 2\pi(n^2+1)/n = \lim_{n\to\infty} 2\pi(n+1/n) = \infty$. Since $\cos x$ oscillates up and down between -1 and 1 as $x \to \infty$, you might be tempted to think that sequence (c) diverges. But if n is an integer then

$$\cos\left(\frac{2\pi(n^2+1)}{n}\right) = \cos\left(2\pi n + \frac{2\pi}{n}\right) = \cos\left(\frac{2\pi}{n}\right),$$

since the cosine function is periodic with period 2π. Therefore

$$\lim_{n\to\infty} \cos\left(\frac{2\pi(n^2+1)}{n}\right) = \lim_{n\to\infty} \cos\left(\frac{2\pi}{n}\right) = \cos(0) = 1,$$

so sequence (c) converges to 1. $\qquad\qquad\square$

When thinking of a sequence $(f(n))_{n=1}^{\infty}$ as a list of numbers, we often use subscript notation rather than function notation. For example, we might let $a_n = f(n)$, so that the sequence would be denoted $(a_n)_{n=1}^{\infty}$. This sequence is the list of numbers $a_1, a_2, a_3, \ldots$, and the nth term of the sequence is a_n.

Sometimes we define a sequence by specifying how each term in the sequence can be used to compute the next term. Such a definition is called a *recursive* definition. For example, we might define a sequence $(a_n)_{n=1}^{\infty}$ as follows:

$$a_1 = 0, \qquad\qquad\qquad (2.19a)$$

$$\text{for all positive integers } n, a_{n+1} = \frac{1}{2 - a_n}. \qquad\qquad (2.19b)$$

It might seem at first that this definition is circular, because a_{n+1} is defined in terms of a_n. But in fact the definition can be used to compute all the terms of the sequence in order. For example, since $a_1 = 0$, we can use equation (2.19b), with $n = 1$, to conclude that

$$a_2 = \frac{1}{2 - a_1} = \frac{1}{2 - 0} = \frac{1}{2}.$$

But then applying equation (2.19b) with $n = 2$ we see that

$$a_3 = \frac{1}{2 - a_2} = \frac{1}{2 - 1/2} = \frac{2}{3}.$$

Continuing in this way, we can compute

$$a_4 = \frac{1}{2 - 2/3} = \frac{3}{4}, \quad a_5 = \frac{1}{2 - 3/4} = \frac{4}{5}, \quad a_6 = \frac{1}{2 - 4/5} = \frac{5}{6}, \quad \ldots$$

These calculations reveal a surprising pattern: for values of n up to 6 we have $a_n = (n-1)/n$. Will this pattern continue? Is there a way to be sure?

In the last step of our calculations above, we saw that the value $a_5 = 4/5$ could be used to compute $a_6 = 5/6$. In general, will a value $a_n = (n-1)/n$ always lead to $a_{n+1} = n/(n+1)$? According to equation (2.19b), if $a_n = (n-1)/n$ then

$$a_{n+1} = \frac{1}{2 - a_n} = \frac{1}{2 - \frac{n-1}{n}} \cdot \frac{n}{n} = \frac{n}{2n - (n-1)} = \frac{n}{n+1}.$$

Thus, the pattern we have observed for n up to 6 will always be carried over from one number to the next. Since we have already checked that $a_6 = 5/6$, we can conclude that $a_7 = 6/7$. But then this statement in turn implies that $a_8 = 7/8$, which implies that $a_9 = 8/9$, and so on. We can now see that the pattern will hold for every positive integer n.

The reasoning we have just used is an example of an important kind of reasoning called *mathematical induction*. Mathematical induction is a method that can be used to prove that all positive integers have some property. There are two parts to a proof by mathematical induction. In the first part, called the *base case*, we verify that the number 1 has the property in question. In the second part, called the *induction step*, we prove that if any positive integer n has the property, then $n + 1$ does as well. For example, here is a formal proof by mathematical induction of the pattern we just verified.

Proposition 2.8.4. *Suppose that a sequence $(a_n)_{n=1}^{\infty}$ is defined recursively by equations (2.19). Then for every positive integer n, $a_n = (n - 1)/n$.*

Proof. We use mathematical induction.
Base case. If $n = 1$, then $a_n = a_1 = 0 = 0/1 = (n - 1)/n$.
Induction step. Suppose that n is a positive integer for which $a_n = (n - 1)/n$. Then

$$a_{n+1} = \frac{1}{2 - a_n} = \frac{1}{2 - \frac{n-1}{n}} = \frac{n}{n + 1}. \qquad \square$$

Why does this proof establish that every positive integer n has the property that $a_n = (n - 1)/n$? The induction step shows that if any positive integer n has the property, then so does $n + 1$. In particular, if 1 has the property, then so does 2. But the base case confirms that 1 does have the property, so we can conclude that 2 has the property. Since 2 has the property, it follows by the inductions step that 3 does as well. This implies that 4 has the property, and so on. Thus, all positive integers have the property.

By the way, now that we have a simple formula for a_n, we can see that

$$\lim_{n \to \infty} a_n = \lim_{n \to \infty} \frac{n - 1}{n} = \lim_{n \to \infty} \left(1 - \frac{1}{n}\right) = 1.$$

In other words, the sequence defined by equations (2.19) converges to 1.

In general, when writing the induction step of a proof by mathematical induction, we begin by assuming that n is a positive integer that has the property in question, and then we must prove that $n + 1$ has the property. The assumption that n has the property is known as the *inductive hypothesis*, and it is often the key step in proving that $n + 1$ has the property.

Example 2.8.5. Prove that for every positive integer n, $2^n > n$.

Solution. We use mathematical induction.
Base case. $2^1 = 2 > 1$.
Induction step. Suppose n is a positive integer and $2^n > n$. Then

$$2^{n+1} = 2 \cdot 2^n > 2 \cdot n \qquad \text{(by the inductive hypothesis)}$$
$$= n + n \geq n + 1. \qquad \square$$

Note that Example 2.8.5 implies that $\lim_{n \to \infty} 2^n = \infty$, and therefore $\lim_{n \to \infty}(1/2^n) = 0$. In other words, the sequence $(1/2^n)_{n=1}^{\infty}$ converges to 0. This fact will be useful to us later.

We give one more example of a proof by mathematical induction, this time involving the factorial sequence. For any positive integer n, we define $n!$, which is called n *factorial*, to be the product of all integers from n down to 1:

$$n! = n \cdot (n-1) \cdot (n-2) \cdots 1.$$

For example, $4! = 4 \cdot 3 \cdot 2 \cdot 1 = 24$. A useful fact about $n!$ is that for every positive integer n,

$$(n+1)! = (n+1) \cdot n \cdot (n-1) \cdots 1 = (n+1) \cdot n!.$$

By convention, $0!$ is defined to be 1. Notice that with this convention, the equation $(n+1)! = (n+1) \cdot n!$ holds when $n = 0$ as well. Thus, we could have defined the sequence $(n!)_{n=0}^{\infty}$ by using the recursive definition

$$0! = 1, \text{ and for every integer } n \geq 0, (n+1)! = (n+1) \cdot n!.$$

In many cases, if a sequence has been defined recursively, then facts about that sequence are most easily proven by mathematical induction. We illustrate this in our next example by proving a property of $n!$. We will use a slight variation on mathematical induction, but if you understand why induction works, then you should have no trouble seeing why this variation works as well. (For a generalization of this example, see Exercise 21.)

Example 2.8.6. Prove that for all integers $n \geq 9$, $n! > 4^n$.

Solution. We use mathematical induction, but we use $n = 9$ as the base case of the induction.
Base case. If $n = 9$, then $n! = 9! = 362880$ and $4^n = 4^9 = 262144$, so $n! > 4^n$.
Induction step. Suppose that n is an integer, $n \geq 9$, and $n! > 4^n$. Then

$$(n+1)! = (n+1) \cdot n! > (n+1) \cdot 4^n \quad \text{(by the inductive hypothesis)}$$
$$\geq 10 \cdot 4^n > 4 \cdot 4^n = 4^{n+1}. \qquad \square$$

Now that we know a little bit about sequences and their limits, we are ready to present the proof of the intermediate value theorem. To motivate the proof, we begin by discussing a method you might use to compute $\sqrt{2}$ if your calculator doesn't have a "$\sqrt{\ }$" key. As we have already observed, $1^2 = 1 < 2$ and $2^2 = 4 > 2$, so $\sqrt{2}$ must be between 1 and 2. Thus, in all of our calculations we can restrict attention to the interval $[u_1, v_1] = [1, 2]$.

A natural next step is to look at the midpoint of the interval $[1, 2]$, which is 1.5, and compute $1.5^2 = 2.25 > 2$. We conclude that $\sqrt{2} < 1.5$, so we now focus our attention on the interval $[u_2, v_2] = [1, 1.5]$. Testing the midpoint of this interval, we find that $1.25^2 = 1.5625 < 2$, so $\sqrt{2}$ must belong to the interval $[u_3, v_3] = [1.25, 1.5]$. The midpoint of this interval is 1.375 and $1.375^2 = 1.890625 < 2$, so we can narrow down further to the interval $[u_4, v_4] = [1.375, 1.5]$. Continuing in this way, we generate an infinite sequence of closed intervals, each one containing the next. In other words, for every n, $[u_{n+1}, v_{n+1}] \subseteq [u_n, v_n]$. We call this a *nested sequence of intervals*. As the intervals shrink, we are able to pin down $\sqrt{2}$ with greater and greater accuracy.

The endpoints of our intervals are two sequences of numbers $(u_n)_{n=1}^{\infty}$ and $(v_n)_{n=1}^{\infty}$, and since the intervals are nested we have $u_1 \le u_2 \le u_3 \le \cdots$ and $v_1 \ge v_2 \ge v_3 \ge \cdots$. The lengths of the intervals are $v_1 - u_1 = 2 - 1 = 1$, $v_2 - u_2 = 1.5 - 1 = 1/2$, $v_3 - u_3 = 1.5 - 1.25 = 1/4$, $v_4 - u_4 = 1.5 - 1.375 = 1/8$, and so on. In general, the length of the nth interval is $v_n - u_n = 1/2^{n-1}$, and therefore, by Example 2.8.5, $\lim_{n \to \infty}(v_n - u_n) = \lim_{n \to \infty}(1/2^{n-1}) = 0$. It turns out that these facts are sufficient to guarantee that the intervals shrink down to a single point on the number line. We state this precisely in the following theorem.

Theorem 2.8.7 (Nested Interval Theorem). *Suppose that $(u_n)_{n=1}^{\infty}$ and $(v_n)_{n=1}^{\infty}$ are two sequences of numbers such that $u_1 \le u_2 \le u_3 \le \cdots$ and $v_1 \ge v_2 \ge v_3 \ge \cdots$. Suppose also that for every n, $u_n \le v_n$, and $\lim_{n \to \infty}(v_n - u_n) = 0$. Then there is a unique number c such that for every n, $u_n \le c \le v_n$. Furthermore, $\lim_{n \to \infty} u_n = \lim_{n \to \infty} v_n = c$; in fact, since for every n we have $u_n \le c \le v_n$, we can say that as $n \to \infty$, $u_n \to c^{\le}$ and $v_n \to c^{\ge}$.*

The nested interval theorem is illustrated in Figure 2.41. The theorem expresses a fundamental property of the real numbers, and we will use it repeatedly in the rest of this book. We will not give a proof of the nested interval theorem. The proof of the existence of a number c belonging to all of the intervals $[u_n, v_n]$ is surprisingly difficult, because it depends on subtle facts about the real numbers that are beyond the scope of this book. However, it should be intuitively clear from the figure that such a number c must exist, and we ask you to accept it on these intuitive grounds.

Once the existence of the number c is established, it is not hard to see why the limit statements in the last sentence of the theorem must be true. For every n we have $u_n \le c \le v_n$, and therefore $0 \le v_n - c \le v_n - u_n$. But $\lim_{n \to \infty}(v_n - u_n) = 0$, so by the squeeze theorem $\lim_{n \to \infty}(v_n - c) = 0$, and therefore

$$\lim_{n \to \infty} v_n = \lim_{n \to \infty} ((v_n - c) + c) = 0 + c = c.$$

Similarly, $\lim_{n \to \infty} u_n = c$. In other words, the sequences $(u_n)_{n=1}^{\infty}$ and $(v_n)_{n=1}^{\infty}$ both converge to c. This is also intuitively clear in Figure 2.41.

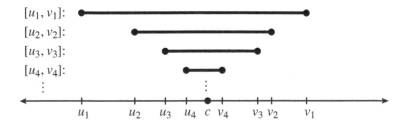

Figure 2.41: The nested interval theorem.

The nested interval theorem is the key tool we will use to prove the intermediate value theorem. We will also need the following fact:

Theorem 2.8.8. *Suppose that* $\lim_{x \to a} f(x) = L$ *and* r *is any real number.*

1. *If* $L > r$, *then there is some number* $\delta > 0$ *such that if* $0 < |x - a| < \delta$ *then* $f(x) > r$.

2. *If* $L < r$, *then there is some number* $\delta > 0$ *such that if* $0 < |x - a| < \delta$ *then* $f(x) < r$.

The theorem should make intuitive sense. For example, part 1 says that if, as $x \to a$, $f(x)$ approaches a limit L that is larger than r, then when x gets close enough to a, $f(x)$ will be larger than r. A careful proof uses the ϵ-δ definition of limits:

Proof. To prove part 1, assume that $L > r$. According to the definition of limits, for every number $\epsilon > 0$ there is some $\delta > 0$ such that if $0 < |x - a| < \delta$ then $|f(x) - L| < \epsilon$. Since this is true for *every* $\epsilon > 0$, we can assign any positive value we please to ϵ. We will use the value $\epsilon = L - r > 0$. Applying the definition of limits, we conclude that there is some number $\delta > 0$ such that if $0 < |x - a| < \delta$ then $|f(x) - L| < \epsilon$, and therefore

$$L - \epsilon < f(x) < L + \epsilon.$$

(See Figure 2.42.) The first half of this inequality tells us that $f(x) > L - \epsilon = L - (L - r) = r$, as required in part 1 of the theorem.

The proof of part 2 is similar. □

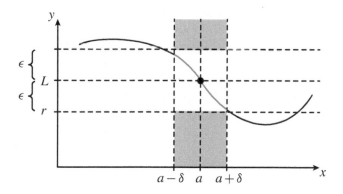

Figure 2.42: The proof of Theorem 2.8.8.

As an immediate consequence of Theorem 2.8.8, we get the following fact. A fact that follows from a theorem is often called a *corollary* of the theorem.

Corollary 2.8.9. *Suppose that* $\lim_{x \to a} f(x) = L$, r *is any real number, and* $d > 0$.

1. *Suppose that for all* x, *if* $0 < |x - a| < d$ *then* $f(x) \leq r$. *Then* $L \leq r$.

2. *Suppose that for all* x, *if* $0 < |x - a| < d$ *then* $f(x) \geq r$. *Then* $L \geq r$.

To put it more informally, Corollary 2.8.9 says that if $f(x) \le r$ for x close to a but not equal to a, then $\lim_{x \to a} f(x) \le r$, and similarly if $f(x) \ge r$ for x close to a but not equal to a, then $\lim_{x \to a} f(x) \ge r$. We say that the weak inequalities $f(x) \le r$ and $f(x) \ge r$ are *preserved by limits*. Notice that the strict inequalities $f(x) < r$ and $f(x) > r$ are *not* preserved by limits. For example, if $f(x) > r$ for x close to a but not equal to a, it does not follow that $\lim_{x \to a} f(x) > r$ (see Exercise 23).

Proof of Corollary 2.8.9. Once again, the proofs of the two parts are very similar, so we will only prove part 1. We will use the method of proof by contradiction, so we begin by assuming that the conclusion $L \le r$ is false; in other words, we assume that $L > r$. Then by part 1 of Theorem 2.8.8, there is some number $\delta > 0$ such that if $0 < |x - a| < \delta$ then $f(x) > r$. But we have also assumed in part 1 that if $0 < |x - a| < d$ then $f(x) \le r$. Thus, if we let x be a number close enough to a that $0 < |x - a| < d$ and $0 < |x - a| < \delta$, then $f(x) \le r$ and $f(x) > r$, which is impossible. Therefore it cannot be the case that $L > r$, so $L \le r$.   □

We have stated Theorem 2.8.8 and Corollary 2.8.9 for limits as $x \to a$, but similar facts hold for all of our other versions of limits: one-sided limits, limits as $x \to \pm\infty$, and limits of sequences. For example, the sequence version of part 1 of Corollary 2.8.9 says that if a sequence $(a_n)_{n=1}^{\infty}$ converges to L and there is some number N such that for all $n > N$, $a_n \le r$, then $L \le r$. In other words, weak inequalities are preserved by sequence limits. This is the fact that we will use in our proof of the intermediate value theorem.

Theorem 2.7.6 (Intermediate Value Theorem, restated). Suppose that f is continuous on the interval $[a, b]$, and either $f(a) < r < f(b)$ or $f(a) > r > f(b)$. Then there is some number c such that $a < c < b$ and $f(c) = r$.

Proof. We will assume that $f(a) < r < f(b)$; the case in which $f(a) > r > f(b)$ is similar. Our plan is to define a nested sequence of intervals $[u_1, v_1], [u_2, v_2], [u_3, v_3], \ldots$ that will shrink down to the desired number c.

We start by letting $[u_1, v_1] = [a, b]$. To get from one interval to the next, we use the following fact, which tells us how to shrink an interval $[u, v]$ to an interval $[u', v']$ that is half as big:

Lemma 2.8.10. *Suppose $a \le u < v \le b$ and $f(u) \le r \le f(v)$. Then there are numbers u' and v' such that $u \le u' < v' \le v$, $v' - u' = (v - u)/2$, and $f(u') \le r \le f(v')$.*

Proof of Lemma. Let $w = (u + v)/2$; in other words, w is the midpoint of the interval $[u, v]$. If $f(w) \le r$, then we have $f(w) \le r \le f(v)$ and $v - w = (v - u)/2$, so we can let $u' = w$ and $v' = v$. Similarly, if $f(w) > r$ then we let $u' = u$ and $v' = w$. This proves the lemma.   □

By assumption, we have $f(u_1) = f(a) < r < f(b) = f(v_1)$. Applying the lemma, we conclude that there are numbers u_2 and v_2 such that $u_1 \le u_2 < v_2 \le v_1$, $v_2 - u_2 = (v_1 - u_1)/2 = (b - a)/2$, and $f(u_2) \le r \le f(v_2)$. Applying the lemma again, we see that there are numbers u_3 and v_3 such that $u_2 \le u_3 < v_3 \le v_2$, $v_3 - u_3 = (v_2 - u_2)/2 = (b - a)/4$, and $f(u_3) \le r \le f(v_3)$. Continuing in this way, we construct a nested sequence of intervals $[u_n, v_n]$ such that for every n, $v_n - u_n = (b-a)/2^{n-1}$ and $f(u_n) \le r \le f(v_n)$. Since $\lim_{n \to \infty}(v_n - u_n) = \lim_{n \to \infty}(b-a)/2^{n-1} = 0$, we can apply the nested interval theorem to conclude that there is a unique number c such that for every n, $u_n \le c \le v_n$. In particular, $a = u_1 \le c \le v_1 = b$. We must prove now that $f(c) = r$.

Suppose that $f(c) > r$. Since $f(a) < r$, this implies that $c \ne a$. Therefore $a < c \le b$, and since f is continuous on $[a, b]$, it follows that f is continuous from the left at c. According to the nested interval theorem, as $n \to \infty$, $u_n \to c^{\le}$, and thus, by Theorem 2.7.12, $f(u_n) \to f(c)$. In other words, $\lim_{n \to \infty} f(u_n) = f(c)$. But recall that we chose the numbers u_n so that for every n, $f(u_n) \le r$. Applying the fact that weak inequalities are preserved by sequence limits (the sequence version of Corollary 2.8.9), we can conclude that $f(c) = \lim_{n \to \infty} f(u_n) \le r$. Thus, our assumption that $f(c) > r$ could not have been correct.

Similar reasoning shows that it cannot be the case that $f(c) < r$, so $f(c) = r$. Since $f(a) < r < f(b)$, this implies that $c \ne a$ and $c \ne b$. Thus, $a < c < b$, so c has all the properties required in the theorem and the proof is complete.   □

Exercises 2.8

1–3: Find the first four terms of the sequence. Then determine whether the sequence converges, and if so find the number it converges to.

1. $\left(\dfrac{50n}{n^2 - 8n + 17} \right)_{n=1}^{\infty}$.

2. $\left(\dfrac{1 + (-1)^n}{n} \right)_{n=1}^{\infty}$.

3. $(\sin(\pi n))_{n=3}^{\infty}$.

4–16: Evaluate the limit.

4. $\displaystyle \lim_{n \to \infty} \frac{3n^2 - 1}{n^2 + 3}$.

5. $\displaystyle \lim_{n \to \infty} \left(\frac{n^2}{n+1} + \frac{4 - n^3}{n^2 - 1} \right)$.

6. $\displaystyle \lim_{n \to \infty} \left[n^3 \left(\frac{1}{n^2} - \frac{1}{(n+1)^2} \right) \right]$.

7. $\displaystyle \lim_{n \to \infty} \frac{n \sin n}{n^2 + 1}$.

8. $\displaystyle \lim_{n \to \infty} \frac{n}{n^2 + \sin n}$.

9. $\displaystyle \lim_{n \to \infty} \frac{1}{(3 + (-1)^n)^n}$.

10. $\displaystyle\lim_{n\to\infty} \frac{1}{(2+(-1)^n)^n}$.

11. $\displaystyle\lim_{n\to\infty} \frac{1}{2^n+(-1)^n}$.

12. $\displaystyle\lim_{n\to\infty} (\sqrt{n^3+n} - n\sqrt{n})$.

13. $\displaystyle\lim_{n\to\infty} (\sqrt{n^3+n^2} - n\sqrt{n})$.

14. $\displaystyle\lim_{n\to\infty} \sin\left(\frac{(2n-1)\pi}{2}\right)$.

15. $\displaystyle\lim_{n\to\infty} \sin\left(\frac{(4n-1)\pi}{2}\right)$.

16. $\displaystyle\lim_{n\to\infty} \sin\left(\frac{6\pi n^2}{3n+1}\right)$.

17. Use mathematical induction to prove that for every positive integer n, $3^n > n^2+1$.

18. (a) Prove that for every positive integer n, $(3/2)^n > n$. (Hint: Use mathematical induction to prove that the inequality is true for all $n \geq 2$, and then check that it is also true when $n = 1$.)

   (b) Find $\displaystyle\lim_{n\to\infty} \frac{2^n}{3^n}$.

19. A sequence $(a_n)_{n=1}^{\infty}$ is defined recursively as follows:

$$a_1 = \frac{1}{3}, \text{ and for every positive integer } n, \ a_{n+1} = \frac{1-a_n}{3-4a_n}.$$

   Prove that for all n, $a_n = n/(2n+1)$. Does the sequence converge?

20. A sequence $(a_n)_{n=1}^{\infty}$ is defined recursively as follows:

$$a_1 = 1, \text{ and for every positive integer } n, \ a_{n+1} = \left(\frac{a_n}{n}+1\right)^2.$$

   Find a formula for a_n, and prove that your formula is correct. Does the sequence converge?

21. (a) Prove that for all positive integers m and n, $(m+n)! > m^n$. (Hint: Let m be a fixed positive integer, and then use mathematical induction to prove that for every positive integer n, $(m+n)! > m^n$.)

   (b) Prove that for every positive integer k, $(2k^2)! > k^{2k^2}$. (Hint: Use part (a).)

   (c) Suppose that k is a positive integer. Prove that there is some positive integer N such that for every integer $n \geq N$, $n! > k^n$. (Hint: Let $N = 2k^2$, and then imitate the solution to Example 2.8.6.)

22. Use Definitions 2.5.7 and 2.8.1 to prove Theorem 2.8.2.

23. Give an example of a function f and a number a such that for all $x \neq a$, $f(x) > 0$, but $\lim_{x \to a} f(x) = 0$. This shows that in Corollary 2.8.9, we cannot replace $\leq$ and $\geq$ with $<$ and $>$.

24. Show that if a sequence $(a_n)_{n=1}^{\infty}$ converges, then it is *bounded*; that is, there is some number M such that for every positive integer n, $|a_n| \leq M$. (Hint: Let $L = \lim_{n \to \infty} a_n$. Begin by applying the definition of this limit statement, with $\epsilon = 1$.)

25. (a) Show that in the nested interval theorem, we cannot strengthen the conclusion to say that for all n, $u_n < c < v_n$. (Hint: Consider the sequences $(u_n)_{n=1}^{\infty}$ and $(v_n)_{n=1}^{\infty}$ defined by the equations $u_n = 0$ and $v_n = 1/n$.)

    (b) Suppose that $(u_n)_{n=1}^{\infty}$ and $(v_n)_{n=1}^{\infty}$ are two sequences of numbers such that $u_1 < u_2 < u_3 < \cdots$ and $v_1 > v_2 > v_3 > \cdots$. Suppose also that for every n, $u_n < v_n$, and $\lim_{n \to \infty}(v_n - u_n) = 0$. Show that there is a unique number c such that for every n, $u_n < c < v_n$. Furthermore, as $n \to \infty$, $u_n \to c^-$ and $v_n \to c^+$. (Hint: First use the nested interval theorem to find a number c such that for every n, $u_n \leq c \leq v_n$. Then show that, in fact, for every n, $u_n < c < v_n$.)

2.9 Monotone Sequences and the Completeness of the Real Numbers

When studying a sequence $(a_n)_{n=1}^{\infty}$, it is sometimes useful to know how the terms compare to each other. We therefore make the following definition:

Definition 2.9.1. Let $(a_n)_{n=1}^{\infty}$ be a sequence.

1. We say that $(a_n)_{n=1}^{\infty}$ is *strictly increasing* if $a_1 < a_2 < a_3 < \cdots$.

2. We say that $(a_n)_{n=1}^{\infty}$ is *weakly increasing* if $a_1 \leq a_2 \leq a_3 \leq \cdots$.

3. We say that $(a_n)_{n=1}^{\infty}$ is *strictly decreasing* if $a_1 > a_2 > a_3 > \cdots$.

4. We say that $(a_n)_{n=1}^{\infty}$ is *weakly decreasing* if $a_1 \geq a_2 \geq a_3 \geq \cdots$.

Notice that if a sequence is strictly increasing then it is also weakly increasing, and if it is strictly decreasing then it is also weakly decreasing. If a sequence has any of these four properties, then we say that it is *monotone*.

Some books use the term "increasing" without the modifier "strictly" or "weakly," but there is some inconsistency in the literature about whether the word "increasing," when used alone, means strictly increasing or weakly increasing. To avoid confusion, we will always use one of the modifiers "strictly" or "weakly."

Example 2.9.2. Show that the following sequences are monotone:

$$\text{(a)} \left(\frac{12n}{n+1} \right)_{n=1}^{\infty}, \quad \text{(b)} \left(\frac{n!}{2^n} \right)_{n=1}^{\infty}.$$

Solution. The nth term of sequence (a) is $a_n = 12n/(n + 1)$. Setting n equal to 1, 2, 3, and 4 in this formula, we see that the sequence starts

$$6, 8, 9, 9.6, \ldots.$$

Thus, $a_1 < a_2 < a_3 < a_4$, but will this pattern continue? We need to know whether, in general, $a_n < a_{n+1}$.

One convenient way to compare a_n to a_{n+1} is to compute $a_{n+1} - a_n$ and see whether it is positive, negative, or 0. In this case we have

$$a_{n+1} - a_n = \frac{12(n + 1)}{n + 2} - \frac{12n}{n + 1} = \frac{12(n + 1)^2 - 12n(n + 2)}{(n + 2)(n + 1)} = \frac{12}{(n + 2)(n + 1)} > 0.$$

Thus, for every positive integer n we have $a_{n+1} - a_n > 0$, and therefore $a_{n+1} > a_n$. We conclude that sequence (a) is strictly increasing.

If the terms of a sequence $(a_n)_{n=1}^{\infty}$ are positive, then another way to compare a_{n+1} to a_n is to compute the ratio a_{n+1}/a_n and compare it to 1. This method often works well if the formula for a_n involves exponentiation or factorials. For sequence (b) we have $a_n = n!/2^n > 0$, so we compute

$$\frac{a_{n+1}}{a_n} = \frac{(n + 1)!/2^{n+1}}{n!/2^n} = \frac{(n + 1)!}{2^{n+1}} \cdot \frac{2^n}{n!} = \frac{(n + 1) \cdot n!}{2 \cdot n!} = \frac{n + 1}{2}.$$

Since n stands for a positive integer, we have $n \geq 1$, and therefore $(n + 1)/2 \geq 2/2 = 1$. Thus, $a_{n+1}/a_n \geq 1$, and multiplying this inequality by a_n (which is positive), we conclude that $a_{n+1} \geq a_n$. Therefore sequence (b) is weakly increasing. (It is not strictly increasing, since $a_1 = a_2$. However, $a_2 < a_3 < a_4 < \cdots$, so we could say that the sequence $(n!/2^n)_{n=2}^{\infty}$ is strictly increasing.)   □

Let's compute the limits of the two sequences in the last example. For sequence (a), we have

$$\lim_{n \to \infty} \frac{12n}{n + 1} = \lim_{n \to \infty} \frac{12}{1 + 1/n} = 12.$$

To find the limit of sequence (b), we apply Example 2.8.6, which tells us that for $n \geq 9$,

$$\frac{n!}{2^n} > \frac{4^n}{2^n} = 2^n.$$

Since $\lim_{n \to \infty} 2^n = \infty$, we conclude that $\lim_{n \to \infty} n!/2^n = \infty$.

Thus, one of the sequences in Example 2.9.2 converges, and the terms of the other approach ∞. It turns out that these are the only possibilities for a weakly increasing sequence. To see why, we will need the following concept:

Definition 2.9.3. Suppose A is a set of numbers. We say that a number b is an *upper bound* for A if for every $x \in A$, $x \leq b$. Similarly, b is a *lower bound* for A if for every $x \in A$, $x \geq b$.

For example, let A be the set whose elements are the terms of sequence (a) in Example 2.9.2. In other words,

$$A = \{x : \text{for some positive integer } n, x = 12n/(n+1)\} = \{6, 8, 9, 9.6, \ldots\}.$$

Another notation for A is $A = \{12n/(n+1) : n \text{ is a positive integer}\}$. For every positive integer n we have $12n/(n+1) \le 12n/n = 12$, so 12 is an upper bound for A. Of course, it is also true that for every integer n, $12n/(n+1) \le 13$, so 13 is also an upper bound. In fact, every number greater than 12 is an upper bound for A. But now consider a number $b < 12$. Since sequence (a) converges to 12, the sequence version of Theorem 2.8.8 implies that for sufficiently large n, $12n/(n+1) > b$, and therefore b is not an upper bound for A. Thus, 12 is the smallest number that is an upper bound for A; we say that 12 is the *least upper bound* of A. To summarize, we have shown that sequence (a) converges to the least upper bound of its set of terms.

On the other hand, suppose that $B = \{n!/2^n : n \text{ is a positive integer}\}$, the set of terms of sequence (b). Since $\lim_{n\to\infty} n!/2^n = \infty$, for any number b we might choose, there will be positive integers n for which $n!/2^n > b$. Thus, no number b will be an upper bound for B; the set of terms of sequence (b) has no upper bound.

We plan to show that every weakly increasing sequence follows the pattern of either sequence (a) or sequence (b). In other words, we will show that if the set of terms of a weakly increasing sequence has an upper bound, then the sequence converges to the least upper bound, and if there is no upper bound, then the terms approach ∞. And a similar fact is true of weakly decreasing sequences: if the set of terms of a weakly decreasing sequence has a lower bound, then the sequence converges to the greatest lower bound, and if not then the terms approach $-\infty$.

There is one difficulty we must deal with before we can carry out this plan: what if the set of terms of a weakly increasing sequence has an upper bound, but among the upper bounds there isn't a smallest one? After all, not every collection of numbers has a smallest element. For example, there is no smallest positive number: for any positive number x, $x/2$ is a smaller positive number. Could it happen that a set of numbers has an upper bound, but no least upper bound? Fortunately, we won't have to worry about this problem:

Theorem 2.9.4 (Completeness of the Real Numbers). *Suppose that A is a nonempty set of real numbers. That is, A is a set of numbers that has at least one element.*

1. *If A has an upper bound, then it has a least upper bound.*

2. *If A has a lower bound, then it has a greatest lower bound.*

Completeness is an important property of the real numbers that is closely related to the nested interval theorem. In fact, we will use the nested interval theorem to prove Theorem 2.9.4 at the end of this section. But first, let's see how it helps us understand weakly increasing and decreasing sequences.

Theorem 2.9.5. *Suppose $(a_n)_{n=1}^{\infty}$ is a sequence, and let A be the set whose elements are the terms of the sequence.*

1. If $(a_n)_{n=1}^{\infty}$ is weakly increasing and A has an upper bound, then $(a_n)_{n=1}^{\infty}$ converges to the least upper bound of A.

2. If $(a_n)_{n=1}^{\infty}$ is weakly increasing and A does not have an upper bound, then $\lim_{n \to \infty} a_n = \infty$.

3. If $(a_n)_{n=1}^{\infty}$ is weakly decreasing and A has a lower bound, then $(a_n)_{n=1}^{\infty}$ converges to the greatest lower bound of A.

4. If $(a_n)_{n=1}^{\infty}$ is weakly decreasing and A does not have a lower bound, then $\lim_{n \to \infty} a_n = -\infty$.

Proof. For part 1, suppose that $(a_n)_{n=1}^{\infty}$ is weakly increasing and A has an upper bound. Then by the completeness of the real numbers, A has a least upper bound. Let b be this least upper bound. We now use the definition of limits to show that $\lim_{n \to \infty} a_n = b$.

Suppose $\epsilon > 0$. Since b is the *least* upper bound of A, $b - \epsilon$ is not an upper bound, so there must be some positive integer N such that $a_N > b - \epsilon$. Now consider any integer $n > N$. Since the sequence is weakly increasing, $a_n \geq a_N > b - \epsilon$. And since b is an upper bound for A, $a_n \leq b$. Combining these facts, we conclude that $b - \epsilon < a_n \leq b < b + \epsilon$. Thus, for every $n > N$ we have $|a_n - b| < \epsilon$, as required.

We use the definition of limits in the proof of part 2 as well. Suppose now that A does not have an upper bound, and consider any number M. Since M is not an upper bound for A, there must be some positive integer N such that $a_N > M$. And since the sequence is weakly increasing, for any integer $n > N$, $a_n \geq a_N > M$. This shows that $\lim_{n \to \infty} a_n = \infty$.

The proofs of parts 3 and 4 are similar to the proofs of parts 1 and 2 (see Exercise 20). $\qquad\qquad\square$

Example 2.9.6. A sequence $(a_n)_{n=1}^{\infty}$ is defined recursively as follows:

$$a_1 = 1, \text{ and for every positive integer } n, \ a_{n+1} = 3 - \frac{1}{a_n}.$$

Find $\lim_{n \to \infty} a_n$.

Solution. The first few terms of the sequence are

$$a_1 = 1, \quad a_2 = 3 - \frac{1}{1} = 2, \quad a_3 = 3 - \frac{1}{2} = \frac{5}{2} = 2.5, \quad a_4 = 3 - \frac{1}{5/2} = \frac{13}{5} = 2.6.$$

So far, the terms are strictly increasing, but will they continue to increase, and do they have an upper bound? Since the sequence is defined recursively, it may be easiest to use mathematical induction to answer these questions. We will use induction to prove that for every positive integer n, $0 < a_n < a_{n+1} < 3$.
Base case. Since $a_1 = 1$ and $a_2 = 2$, we have $0 < a_1 < a_2 < 3$.
Induction step. Suppose that n is a positive integer and $0 < a_n < a_{n+1} < 3$. Then $1/a_n > 1/a_{n+1} > 0$, so

$$3 - \frac{1}{a_n} < 3 - \frac{1}{a_{n+1}} < 3.$$

According to the definition of the sequence, this means that $a_{n+1} < a_{n+2} < 3$, and we already know that $a_{n+1} > a_n > 0$. Therefore $0 < a_{n+1} < a_{n+2} < 3$. This completes the mathematical induction proof.

We now know that the sequence $(a_n)_{n=1}^{\infty}$ is strictly increasing, and that 3 is an upper bound for the terms of the sequence. By Theorem 2.9.5, the sequence converges; in other words, there is some number L such that as $n \to \infty$, $a_n \to L$. The number L is the least upper bound for the set of terms of the sequence. Since $a_4 = 2.6$ and 3 is an upper bound, we must have $2.6 \le L \le 3$, but we don't yet know the exact value of L.

We are going to use a somewhat unusual procedure to find L: we are going to compute $\lim_{n\to\infty} a_{n+1}$ in two different ways, and then set the two answers equal to each other. For the first computation, let $m = n + 1$. Then as $n \to \infty$, $m = n + 1 \to \infty$, and as $m \to \infty$, $a_m \to L$. Combining these, we conclude that as $n \to \infty$, $a_{n+1} = a_m \to L$, so $\lim_{n\to\infty} a_{n+1} = L$. For the second computation, we use the recursive definition of the sequence to see that

$$\lim_{n\to\infty} a_{n+1} = \lim_{n\to\infty} \left(3 - \frac{1}{a_n}\right) = 3 - \frac{1}{L}.$$

Both computations are correct, so the two answers must be equal to each other. Therefore $L = 3 - 1/L$. Multiplying by L, we get $L^2 = 3L - 1$, or $L^2 - 3L + 1 = 0$, and the quadratic formula gives

$$L = \frac{3 \pm \sqrt{5}}{2}.$$

Thus, we have narrowed down the possibilities for L to $(3 - \sqrt{5})/2 \approx 0.634$ and $(3 + \sqrt{5})/2 \approx 2.618$. But we already know $2.6 \le L \le 3$, so we can rule out the first possibility. Therefore $\lim_{n\to\infty} a_n = L = (3 + \sqrt{5})/2$. $\square$

Our solution to the last example had a somewhat unusual form. First we showed that the limit exists, and that entitled us to introduce the name L for the value of the limit. Then we did some further analysis to find the value of L. The lesson here is that sometimes showing that a limit exists is the first step in finding its value. Once you know that a limit is defined, you can give a name to its value. And giving a name to a number is surprisingly useful, because it makes it much easier to reason about the number. We will see more examples of this kind of reasoning later in this book.

Example 2.9.7. Find

$$\lim_{n\to\infty} \frac{(2n)!}{(n!)^2}.$$

Solution. If we let $a_n = (2n)!/(n!)^2$, then we have

$$a_1 = 2, \qquad a_2 = 6, \qquad a_3 = 20, \qquad a_4 = 70.$$

Will the numbers continue to increase? Since the formula for a_n involves factorials, a good way to answer this question is to compute

$$\frac{a_{n+1}}{a_n} = \frac{(2(n+1))!/((n+1)!)^2}{(2n)!/(n!)^2} = \frac{(2n+2)!}{((n+1)!)^2} \cdot \frac{(n!)^2}{(2n)!}.$$

We have already observed that $(n+1)! = (n+1) \cdot n!$, and it follows that $(2n+2)! = (2n+2) \cdot (2n+1)! = (2n+2) \cdot (2n+1) \cdot (2n)!$. Therefore

$$\frac{a_{n+1}}{a_n} = \frac{(2n+2) \cdot (2n+1) \cdot (2n)!}{((n+1) \cdot n!)^2} \cdot \frac{(n!)^2}{(2n)!} = \frac{(2n+2)(2n+1)}{(n+1)^2} = \frac{4n+2}{n+1} > 1.$$
(2.20)

Since $a_{n+1}/a_n > 1$, we conclude that $a_{n+1} > a_n$, so the sequence $(a_n)_{n=1}^{\infty}$ is strictly increasing. Thus, according to Theorem 2.9.5, either the sequence converges or $\lim_{n \to \infty} a_n = \infty$.

Suppose that the sequence converges—say, $\lim_{n \to \infty} a_n = L$. Then L is the least upper bound for the set of terms of the sequence, and since we have already computed $a_4 = 70$, we must have $L \geq 70$. Imitating our last example, we now compute $\lim_{n \to \infty} a_{n+1}$ in two ways. On the one hand, as $n \to \infty$ we have $n+1 \to \infty$, and therefore $\lim_{n \to \infty} a_{n+1} = L$. But on the other hand, according to equation (2.20), $a_{n+1} = (4n+2)/(n+1) \cdot a_n$, and therefore

$$\lim_{n \to \infty} a_{n+1} = \lim_{n \to \infty} \left(\frac{4n+2}{n+1} \cdot a_n \right) = \lim_{n \to \infty} \left(\frac{4+2/n}{1+1/n} \cdot a_n \right) = 4L.$$

The two answers must be equal, so $L = 4L$. Dividing by L (note that $L \geq 70$, so $L \neq 0$), we get $1 = 4$, which is obviously false. We conclude that our assumption that the sequence converges must have been wrong.

Thus, the sequence does not converge. By Theorem 2.9.5, the only remaining possibility is that $\lim_{n \to \infty} a_n = \infty$. For an alternative proof of this fact, see Exercise 18.
$\square$

We end this section with the promised proof of Theorem 2.9.4. Recall that this theorem says that if a nonempty set of real numbers A has an upper bound, then it has a least upper bound, and if it has a lower bound, then it has a greatest lower bound.

Proof of Theorem 2.9.4. We will prove only the statement about upper bounds; the proof for lower bounds is similar (see Exercise 21). Suppose that A has an upper bound; let v_1 be an upper bound for A. The assumption that A is nonempty means that there is at least one number in A, so we can let a be some element of A. Then if we let u_1 be any number smaller than a, then u_1 is not an upper bound for A. Note that $u_1 < a \leq v_1$. We will look for the least upper bound of A between u_1 and v_1. The key fact we will use to narrow down our search for the least upper bound is the following fact:

Lemma 2.9.8. *Suppose $u < v$, u is not an upper bound for A, and v is an upper bound for A. Then there are numbers u' and v' such that $u \leq u' < v' \leq v$, $v' - u' = (v-u)/2$, u' is not an upper bound for A, and v' is an upper bound for A.*

Proof. Let $m = (u+v)/2$, the midpoint of the interval $[u, v]$. If m is an upper bound for A, then we can let $u' = u$ and $v' = m$. If not, then we can let $u' = m$ and $v' = v$. $\square$

By the lemma, there must be numbers u_2 and v_2 such that $u_1 \leq u_2 < v_2 \leq v_1$, $v_2 - u_2 = (v_1 - u_1)/2$, u_2 is not an upper bound for A, and v_2 is an upper bound. Applying the lemma again, we see that there are numbers u_3 and v_3 such that

$u_2 \leq u_3 < v_3 \leq v_2$, $v_3 - u_3 = (v_2 - u_2)/2 = (v_1 - u_1)/4$, u_3 is not an upper bound for A, and v_3 is an upper bound. Continuing in this way, we get sequences $(u_n)_{n=1}^{\infty}$ and $(v_n)_{n=1}^{\infty}$ such that $u_1 \leq u_2 \leq u_3 \leq \cdots$, $v_1 \geq v_2 \geq v_3 \geq \cdots$, and for every n, $u_n \leq v_n$, $v_n - u_n = (v_1 - u_1)/2^{n-1}$, u_n is not an upper bound for A, and v_n is an upper bound for A. Since $\lim_{n\to\infty}(v_n - u_n) = \lim_{n\to\infty}(v_1 - u_1)/2^{n-1} = 0$, the nested interval theorem implies that there is a number b such that for every n, $u_n \leq b \leq v_n$, and $\lim_{n\to\infty} u_n = \lim_{n\to\infty} v_n = b$. We will show that b is the least upper bound of A.

To see that b is an upper bound for A, let x be any element of A. For every n, v_n is an upper bound for A, so $v_n \geq x$. Since weak inequalities are preserved by limits, it follows that $b = \lim_{n\to\infty} v_n \geq x$, so b is an upper bound. Finally, to see that b is the least upper bound, consider any number $c < b$. Then since $\lim_{n\to\infty} u_n = b$, by the sequence version of Theorem 2.8.8, for sufficiently large n, $u_n > c$. Since u_n is not an upper bound for A, there is some $x \in A$ such that $x > u_n$. Combining this with $u_n > c$, we have $x > c$, so c is not an upper bound for A. Thus, we see that no number smaller than b can be an upper bound for A, so b is the least upper bound of A. $\square$

Exercises 2.9

1–10: Determine whether the sequence is monotone.

1. $\left(\dfrac{2n-1}{n}\right)_{n=1}^{\infty}$.

2. $\left(\dfrac{n}{2n+1}\right)_{n=1}^{\infty}$.

3. $\left(n^2 - 2n\right)_{n=1}^{\infty}$.

4. $\left(n^2 - 3n\right)_{n=1}^{\infty}$.

5. $\left(n^2 - 4n\right)_{n=1}^{\infty}$.

6. $\left(\dfrac{2^{3n+1}}{3^{2n+1}}\right)_{n=1}^{\infty}$.

7. $\left(\dfrac{(2n)!}{4^n (n!)^2}\right)_{n=1}^{\infty}$.

8. $\left(\dfrac{(3n)! \, n!}{((2n)!)^2}\right)_{n=1}^{\infty}$.

9. $(\sin(n\pi/4))_{n=1}^{\infty}$.

10. $(\sin(\pi/(4n)))_{n=1}^{\infty}$.

11. Let $(a_n)_{n=1}^{\infty}$ be the sequence defined by the formula $a_n = (9/10)^n$.

    (a) Show that for every positive integer n, $a_{n+1} = (9/10) \cdot a_n$.

    (b) Show that $(a_n)_{n=1}^{\infty}$ is strictly decreasing.

(c) Show that $(a_n)_{n=1}^{\infty}$ has a lower bound.

(d) By parts (b) and (c), the sequence $(a_n)_{n=1}^{\infty}$ converges, so we can let $L = \lim_{n \to \infty} a_n$. Find L by computing $\lim_{n \to \infty} a_{n+1}$ in two different ways (see Example 2.9.6).

12. Let $(a_n)_{n=1}^{\infty}$ be the sequence defined by the formula $a_n = (11/10)^n$.

(a) Show that for every positive integer n, $a_{n+1} = (11/10) \cdot a_n$.

(b) Show that $(a_n)_{n=1}^{\infty}$ is strictly increasing.

(c) Show that $\lim_{n \to \infty} a_n = \infty$. (Hint: Imitate Example 2.9.7.)

13. Let $(a_n)_{n=1}^{\infty}$ be the sequence defined by the formula

$$a_n = \frac{3^n}{(n+2)!}.$$

(a) Show that for every positive integer n, $a_{n+1} = \frac{3}{n+3} \cdot a_n$.

(b) Show that $(a_n)_{n=1}^{\infty}$ is strictly decreasing.

(c) Show that $(a_n)_{n=1}^{\infty}$ has a lower bound.

(d) By parts (b) and (c), we can let $L = \lim_{n \to \infty} a_n$. Find L.

14. (a) Show that if $(a_n)_{n=1}^{\infty}$ and $(b_n)_{n=1}^{\infty}$ are strictly decreasing, then so is $(a_n + b_n)_{n=1}^{\infty}$.

(b) Show that $((3^n + 4^n)/7^n)_{n=1}^{\infty}$ is strictly decreasing.

15. A sequence $(a_n)_{n=1}^{\infty}$ is defined recursively as follows:

$$a_1 = 2, \text{ and for every positive integer } n, \ a_{n+1} = 3 + \frac{a_n}{2}.$$

(a) Show that for every positive integer n, $a_{n+1} > a_n$.

(b) Show that for every positive integer n, $a_n < 10$.

(c) Find $\lim_{n \to \infty} a_n$.

16. A sequence $(a_n)_{n=1}^{\infty}$ is defined recursively as follows:

$$a_1 = 2, \text{ and for every positive integer } n, \ a_{n+1} = \sqrt{8 + \frac{a_n^2}{2}}.$$

(a) Show that for every positive integer n, $a_{n+1} > a_n$.

(b) Show that $(a_n)_{n=1}^{\infty}$ has an upper bound.

(c) Find $\lim_{n \to \infty} a_n$.

17. A sequence $(a_n)_{n=1}^{\infty}$ is defined recursively as follows:

$$a_1 = 1, \text{ and for every positive integer } n, \ a_{n+1} = 6 - \frac{9}{a_n}.$$

   (a) Compute a_1, a_2, a_3, and a_4.

   (b) Show that for every $n \geq 3$, $a_n > a_{n+1} > 3$.

   (c) Find $\lim_{n \to \infty} a_n$.

18. Give an alternative solution to Example 2.9.7 by showing that for every positive integer n, $(2n)!/(n!)^2 \geq 2^n$. It follows that $\lim_{n \to \infty} (2n)!/(n!)^2 = \infty$.

19. Show that for every positive integer k, $\lim_{n \to \infty} n!/k^n = \infty$. (Hint: One approach is to use Exercise 21 in Section 2.8 and then imitate our proof that the limit of sequence (b) in Example 2.9.2 is ∞. Another possibility is to show that $(n!/k^n)_{n=k}^{\infty}$ is strictly increasing, and then imitate our solution to Example 2.9.7.)

20. Prove parts 3 and 4 of Theorem 2.9.5. (Note: One approach is to imitate the proofs of parts 1 and 2. Another is to apply parts 1 and 2 to the sequence $(-a_n)_{n=1}^{\infty}$.)

21. Prove part 2 of Theorem 2.9.4. (Note: One approach is to imitate the proof of part 1. Another is to apply part 1 to the set $\{-x : x \in A\}$.)

Chapter 3

Derivatives

3.1 Rates of Change and Slopes

In many applications of mathematics, an equation of the form $y = f(x)$ is used to express a relationship between real-world quantities x and y. This equation says that the value of y depends on the value of x, and therefore if the value of x changes, then the value of y will also change. Sometimes we are interested in knowing how fast y will change when x changes.

For example, consider an electric heater in which electrical current flows through a coil of wire, causing the wire to heat up. Suppose the resistance of the wire coil is 25 ohms. Then according to Joule's law, if a current of I amps flows through the coil, then the power of the heater will be P watts, where

$$P = 25I^2.$$

If we let $f(I) = 25I^2$, then we could say that $P = f(I)$. The graph of the equation $P = f(I)$ is shown in Figure 3.1.

With a current of $I = 4$ amps, the power of the heater will be $P = f(4) = 400$ watts. If we increase the current to $I = 6$, the power increases to $P = f(6) = 900$ watts. The change in the current in this case is $6 - 4 = 2$ amps, and the change in power is $f(6) - f(4) = 900 - 400 = 500$ watts. We say that the *average rate of change of P with respect to I over the interval* $[4, 6]$ is

$$\frac{f(6) - f(4)}{6 - 4} = \frac{500}{2} = 250 \text{ watts/amp.}$$

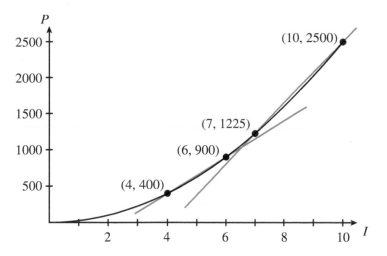

Figure 3.1: The graph of $P = f(I)$. The slopes of the two blue secant lines are the average rates of change of P with respect to I over the intervals $[4, 6]$ and $[7, 10]$.

This tells us *how fast* the power of the heater changes as we increase the current from 4 to 6 amps. On average, over this interval the power increases by 250 watts for each additional amp of current. Notice that the points $(4, 400)$ and $(6, 900)$ are on the graph of f. The line through these two points is a secant line for the graph of f, and the average rate of change of P with respect to I over the interval $[4, 6]$ can be interpreted as the slope of this secant line.

The average rate of change of P with respect to I over the interval $[7, 10]$ would be

$$\frac{f(10) - f(7)}{10 - 7} = \frac{2500 - 1225}{10 - 7} = \frac{1275}{3} = 425 \text{ watts/amp.}$$

Thus, over this interval the power increases more quickly as the current increases. You can see this in Figure 3.1. The average rate of change of P with respect to I over the interval $[7, 10]$ is the slope of the secant line through the points $(7, 1225)$ and $(10, 2500)$, and you can see in the figure that this line slopes up more steeply than the secant line through $(4, 400)$ and $(6, 900)$.

Here's another example: Consider a cylinder that is filled with a gas. Suppose that a piston can be pushed into the cylinder to decrease the volume of the gas, or pulled out to increase the volume. When the piston is pushed all the way in, the volume is 10 cubic inches (abbreviated in^3) and the pressure of the gas is 300 kilopascals (abbreviated kPa). According to Boyle's law, if the piston is pulled out, increasing the volume V, then the pressure P will decrease according to the formula

$$P = \frac{3000}{V}.$$

Once again, we have a functional relationship; we can write $P = g(V)$, where $g(V) = 3000/V$.

Suppose that the piston is pulled out enough to increase the volume from 10 in³ to 15 in³. Then the pressure will decrease from its initial value of $g(10) = 300$ kPa to $g(15) = 200$ kPa. The change in volume is $15 - 10 = 5$ in³, and the change in pressure is $g(15) - g(10) = 200 - 300 = -100$ kPa; the fact that the change in pressure is negative indicates that the pressure has decreased. The average rate of change of P with respect to V over the interval [10, 15] is therefore

$$\frac{g(15) - g(10)}{15 - 10} = \frac{-100}{5} = -20 \text{ kPa/in}^3.$$

The rate of change is negative, indicating that as the volume increases, the pressure decreases. Over this interval, the pressure decreases an average of 20 kilopascals for each cubic inch the volume increases. The average rates of change over other intervals may be different. For example, the average rate of change over the interval [20, 25] is

$$\frac{g(25) - g(20)}{25 - 20} = \frac{120 - 150}{25 - 20} = \frac{-30}{5} = -6 \text{ kPa/in}^3.$$

Once again, these rates of change can be interpreted as slopes of secant lines for the graph of g, as shown in Figure 3.2.

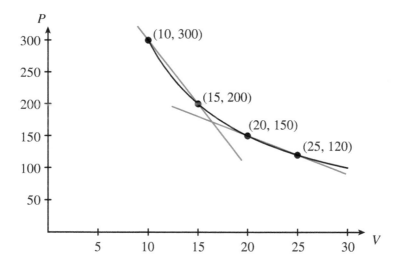

Figure 3.2: The graph of $P = g(V)$. The slopes of the two blue secant lines are the average rates of change of P with respect to V over the intervals [10, 15] and [20, 25].

Rates of change are especially useful for talking about quantities that change over time. For example, recall that in Section 2.1 we discussed a ball that was thrown up in the air. The ball moved along a vertical number line with units marked off in feet, and the position of the ball on this number line after t seconds was given by the formula $y = h(t) = 160t - 16t^2$. We saw that the average velocity of the ball during the time interval [3, 5] was

$$\frac{h(5) - h(3)}{5 - 3} = \frac{400 - 336}{5 - 3} = \frac{64}{2} = 32 \text{ ft/sec.}$$

In the terminology we are using now, this is just the average rate of change of y with respect to t over the interval [3, 5]. In other words, the average velocity of an object moving along a line is the average rate of change of its position with respect to time. This means that the average velocity tells you how fast, on average, the position of the object changes as time passes.

Generalizing from these examples, we make the following definition.

Definition 3.1.1. Suppose two quantities x and y are related by the equation $y = f(x)$. We say that the *average rate of change of y with respect to x over the interval* $[a, b]$ *is*

$$\frac{f(b) - f(a)}{b - a}.$$

If the quantity x is measured in, say, smoots, and the quantity y is measured in millihelens, then the units for the average rate of change will be millihelens per smoot; the average rate of change tells us the number of millihelens by which y changes, on average, for each smoot by which x is increased, as x increases from a to b.[1] A positive rate of change indicates that y increases, while a negative rate of change indicates that it decreases.

In our discussion in Section 2.1 of the ball thrown up in the air, we also studied the *instantaneous velocity* of the ball at time $t = 3$. This is more difficult to compute than an average velocity, because it cannot be computed by dividing distance traveled by elapsed time; in a single instant, the ball doesn't move at all, and no time passes, so we would be computing 0/0, which is undefined. Recall that our strategy in Section 2.1 was to compute the average velocity of the ball during very short time intervals either beginning or ending with $t = 3$. Our reasoning was that during a very short time interval the velocity wouldn't have time to change much, and therefore the average velocity over such a time interval would be a good approximation of the velocity at time $t = 3$. Furthermore, the approximation could be made more and more accurate by making the time interval shorter and shorter. We then found the precise velocity at time $t = 3$ by taking the limit of these approximations as they were made more and more accurate. In other words, to find the instantaneous velocity at $t = 3$, we found the average velocity over a time interval beginning or ending at $t = 3$, and then took the limit of this average velocity as the length of the time interval approached 0.

Let's see if we can generalize this idea to other situations. Consider quantities x and y related by an equation $y = f(x)$. We would like to determine how fast y changes, per unit of increase of x, when x is equal to some number a; we will call this the *instantaneous rate of change of y with respect to x at x = a*. Imitating our earlier

---

[1] The smoot and the millihelen are real, if somewhat fanciful, units. The *smoot* is a measure of length; it is 5 feet, 7 inches, which was the height of Oliver R. Smoot in 1958, when he was used by his MIT Lamba Chi Alpha brothers to measure the length of the Harvard Bridge. The *millihelen*, named for Helen of Troy, is a measure of beauty; it is the amount of beauty needed to launch one ship.

reasoning, we approximate this with the average rate of change of y with respect to x over the interval between a and $a + h$, for h close to 0 but not equal to 0. If h is positive, then this would be the interval $[a, a + h]$, and the average rate of change would be

$$\frac{f(a+h) - f(a)}{a + h - a} = \frac{f(a+h) - f(a)}{h}.$$

If h is negative, then $a + h < a$. The interval would then be $[a + h, a]$, over which the average rate of change is

$$\frac{f(a) - f(a+h)}{a - (a+h)} = \frac{f(a) - f(a+h)}{-h} = \frac{f(a+h) - f(a)}{h}.$$

Thus, we have the same formula for the average rate of change of y with respect to x over the interval between a and $a + h$, whether h is positive or negative. For h close to 0, this will be a good approximation of the instantaneous rate of change of y with respect to x at $x = a$, and the approximation will improve as h gets closer to 0. We can't actually let $h = 0$, since then our formula for the average rate of change would be $0/0$. But we can still get a precise answer for the instantaneous rate of change by taking the limit as h approaches 0.

Definition 3.1.2. Suppose two quantities x and y are related by the equation $y = f(x)$. If the limit

$$\lim_{h \to 0} \frac{f(a+h) - f(a)}{h}$$

is defined, then we say that it is the *instantaneous rate of change of y with respect to x at $x = a$*. Sometimes we will leave out the word *instantaneous* and simply call this the *rate of change of y with respect to x at $x = a$*. In other words, if a rate of change is not explicitly specified to be an average rate of change, then it is understood to be an instantaneous rate of change. Since $y = f(x)$, we may substitute $f(x)$ for y and use the phrase *rate of change of $f(x)$ with respect to x at $x = a$*.

As we will see, the limit in Definition 3.1.2 is sometimes undefined, but for the moment we will focus on cases when it is defined. In the graph of f, the fraction $(f(a+h) - f(a))/h$ represents the slope of a secant line through the point $P = (a, f(a))$ and a nearby point $Q_h = (a+h, f(a+h))$. As h approaches 0, the point Q_h approaches P, and the secant line approaches the line tangent to the graph of f at P, as shown in Figure 3.3. Thus, the instantaneous rate of change of y with respect to x at $x = a$ could also be thought of as the slope of the line tangent to the graph of f at the point P. Intuitively, the tangent line at P is the line passing through P that goes in the same direction as the curve at P, so the instantaneous rate of change tells us how steeply the curve $y = f(x)$ is inclined at P.

For example, consider again the electric heater whose power P is related to the current I by the equation $P = f(I) = 25I^2$. Let's find the instantaneous rate of change

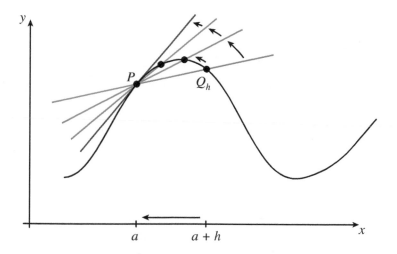

Figure 3.3: As h approaches 0, the point $a + h$ moves along the x-axis toward a, the point Q_h moves along the graph of f toward P, and the secant line through P and Q_h (blue) approaches the tangent line at P (red).

of P with respect to I at $I = 4$. According to Definition 3.1.2, this is

$$\lim_{h \to 0} \frac{f(4+h) - f(4)}{h} = \lim_{h \to 0} \frac{25(4+h)^2 - 400}{h} = \lim_{h \to 0} \frac{400 + 200h + 25h^2 - 400}{h}$$

$$= \lim_{h \to 0} \frac{200h + 25h^2}{h} = \lim_{h \to 0} (200 + 25h) = 200 \text{ watts/amp.}$$

This tells us that if the current is increasing, then just at the instant when the current is $I = 4$ amps, the power is increasing at a rate of 200 watts/amp. It also tells us that the slope of the line tangent to the graph of f at the point $(4, 400)$ is 200. Looking at Figure 3.1, this makes sense; it appears that the slope of the tangent line at $(4, 400)$ is slightly less than the slope of the secant line through $(4, 400)$ and $(6, 900)$, which, as we computed earlier, is 250.

In the case of the gas in the cylinder, the instantaneous rate of change of pressure with respect to volume when $V = 20$ would be

$$\lim_{h \to 0} \frac{g(20+h) - g(20)}{h} = \lim_{h \to 0} \frac{\frac{3000}{20+h} - 150}{h} = \lim_{h \to 0} \left(\frac{1}{h} \cdot \frac{3000 - 150(20+h)}{20+h} \right)$$

$$= \lim_{h \to 0} \left(\frac{1}{h} \cdot \frac{-150h}{20+h} \right) = \lim_{h \to 0} \frac{-150}{20+h} = -\frac{150}{20}$$

$$= -7.5 \text{ kPa/in}^3.$$

Thus, if the volume is increasing, then when it is 20 in^3, the pressure is decreasing at a rate of 7.5 kPa/in^3. We could also say that the line tangent to the graph of g at the point (20, 150) has slope -7.5. This line slopes down a little more steeply than the secant line through (20, 150) and (25, 120), whose slope we computed earlier to be -6.

Finally, for the ball thrown straight up, let's compute the instantaneous rate of change of position with respect to time when $t = 3$. Since the name of the function in this case is h, to avoid confusion we have switched to a different letter for the variable in the limit below.

$$\lim_{k \to 0} \frac{h(3+k) - h(3)}{k} = \lim_{k \to 0} \frac{(336 + 64k - 16k^2) - 336}{k}$$

$$= \lim_{k \to 0} \frac{64k - 16k^2}{k} = \lim_{k \to 0} (64 - 16k) = 64 \text{ ft/sec.}$$

Of course, this agrees with the instantaneous velocity we computed in Section 2.1.

Example 3.1.3. Suppose that the temperature in a room is 50 degrees Fahrenheit. A heater is turned on, and the temperature, in degrees Fahrenheit, t hours later is given by the formula

$$T = f(t) = 70 - \frac{20}{t+1}.$$

How fast is the temperature increasing one hour after the heater was turned on?

Solution. We compute the rate of change of T with respect to t at $t = 1$:

$$\lim_{h \to 0} \frac{f(1+h) - f(1)}{h} = \lim_{h \to 0} \frac{\left(70 - \frac{20}{(1+h)+1}\right) - 60}{h} = \lim_{h \to 0} \left[\frac{1}{h} \cdot \left(10 - \frac{20}{2+h}\right)\right]$$

$$= \lim_{h \to h} \left[\frac{1}{h} \cdot \frac{20 + 10h - 20}{2+h}\right] = \lim_{h \to 0} \left[\frac{1}{h} \cdot \frac{10h}{2+h}\right]$$

$$= \lim_{h \to 0} \frac{10}{2+h} = 5 \text{ °F/hr.}$$

Thus, the temperature is increasing at a rate of 5 °F/hr after 1 hour. $\square$

In our examples so far we have considered equations relating quantities that have physical meanings. But rates of change are useful any time one quantity depends on another, whether the quantities have a physical meaning or not.

Example 3.1.4. If $y = f(x) = \sqrt{3x + 7}$, find the rate of change of y with respect to x at $x = 6$. Also, find the equation of the line tangent to the graph of f at the point where $x = 6$.

Solution. The rate of change is

$$\lim_{h \to 0} \frac{f(6+h) - f(6)}{h} = \lim_{h \to 0} \frac{\sqrt{3(6+h) + 7} - 5}{h}$$

$$= \lim_{h \to 0} \left(\frac{\sqrt{25 + 3h} - 5}{h} \cdot \frac{\sqrt{25 + 3h} + 5}{\sqrt{25 + 3h} + 5}\right) = \lim_{h \to 0} \frac{25 + 3h - 25}{h(\sqrt{25 + 3h} + 5)}$$

$$= \lim_{h \to 0} \frac{3h}{h(\sqrt{25 + 3h} + 5)} = \lim_{h \to 0} \frac{3}{\sqrt{25 + 3h} + 5} = \frac{3}{10}.$$

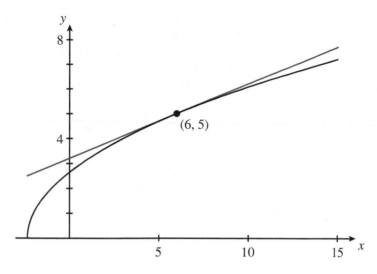

Figure 3.4: The graph of $y = \sqrt{3x + 7}$ (black), and the tangent line at the point $(6, 5)$ (red).

The graph of the equation $y = f(x)$ passes through the point $(6, f(6)) = (6, 5)$, and the calculation above shows that the slope of the line tangent to the graph at this point is $3/10$. Thus, using the point-slope equation, the equation of this tangent line is $y - 5 = (3/10)(x - 6)$, or equivalently

$$y = \frac{3}{10}x + \frac{16}{5}.$$

The graph of f and the tangent line at $(6, 5)$ are shown in Figure 3.4. $\qquad\square$

Example 3.1.5. If $g(s) = 2s/(s + 3)$, find the rate of change of $g(s)$ with respect to s at $s = 1$.

Solution. The rate of change is

$$\lim_{h \to 0} \frac{g(1 + h) - g(1)}{h} = \lim_{h \to 0} \frac{\frac{2(1+h)}{(1+h)+3} - \frac{1}{2}}{h} = \lim_{h \to 0} \left(\frac{1}{h} \cdot \frac{(4 + 4h) - (4 + h)}{2(4 + h)} \right)$$

$$= \lim_{h \to 0} \left(\frac{1}{h} \cdot \frac{3h}{8 + 2h} \right) = \lim_{h \to 0} \frac{3}{8 + 2h} = \frac{3}{8}. \qquad\square$$

Exercises 3.1

1–9: Compute the rate of change of y with respect to x at the specified value of x.

   1. $y = x^2 - 3x + 4$; $x = 7$.

   2. $y = 1/(2x + 3)$; $x = -1$.

3. $y = x^3$; $x = 2$.

4. $y = \sqrt{5 - 2x}$; $x = -2$.

5. $y = 1/x^2$; $x = 4$.

6. $y = \sin x$; $x = 0$.

7. $y = \sin x$; $x = \pi/3$. (Hint: Use the formulas derived in Exercise 2.3 of Section 1.4, and then imitate the method used in part (b) of Example 2.7.15.)

8. $y = \tan(3x)$; $x = 0$.

9. $y = 1/\sqrt{x}$; $x = 2$.

10. Find the equation of the line tangent to the curve $y = 3x^2 - 5x + 1$ at the point $(2, 3)$.

11. Find the equation of the line tangent to the curve $y = 1/x^2$ at the point $(-1, 1)$.

12. A particle moves along a number line, marked off in inches, in such a way that its position at time t, measured in seconds, is $y = t + \sqrt{t}$.

    (a) Find the average velocity of the particle from $t = 1$ to $t = 4$.

    (b) Find the average velocity of the particle from $t = 1$ to $t = 1.1$.

    (c) Find the instantaneous velocity of the particle at $t = 1$.

13. A *potentiometer* is an electrical component that can be used to introduce a variable resistance into a circuit. If a voltage of 10 volts is applied to a circuit containing a potentiometer, then according to Ohm's law, when the resistance of the circuit is R ohms, the current through the circuit, measured in amps, will be

$$I = \frac{10}{R}.$$

It follows that if the resistance R is changed by adjusting the potentiometer, then the current I will change as well. Find the rate of change of I with respect to R when $R = 5$ ohms. Be sure to specify the units. Does the current increase or decrease as the resistance increases?

14. According to Coulomb's law, if two electrons are separated by a distance of r picometers (abbreviated pm), then they will repel each other, and the repulsive force F, measured in dynes, is given by the formula

$$F = \frac{k}{r^2},$$

where $k \approx 23.07$. Find the rate of change of F with respect to r when $r = 30$ pm. Be sure to specify the units. Does the force increase or decrease as r increases? (Note: A picometer is a trillionth of a meter.)

15. According to Kepler's third law, if a planet orbits the sun in a circular orbit of radius r astronomical units (abbreviated au), then the time it takes to complete one orbit, measured in years, will be

$$P = r^{3/2}.$$

Find the rate of change of P with respect to r when $r = 4$ au. Be sure to specify the units. Does P increase or decrease as r increases? (Note: An astronomical unit is the mean distance from the earth to the sun, approximately 150 million kilometers.)

16. The *luminosity* of a star is the rate at which it radiates energy. If a star the size of the sun has a surface temperature of T kelvin (abbreviated K), then the Stefan-Boltzmann law implies that its luminosity, measured in watts, will be

$$L = kT^4,$$

where $k \approx 3.45 \times 10^{11}$. Find the rate of change of L with respect to T when $T = 2000$ K. Be sure to specify the units.

17. Suppose $y = f(x)$. You may wonder why, in the definition of the instantaneous rate of change of y with respect to x at $x = a$, we only considered intervals that have a as one endpoint. Why not consider, for example, intervals centered at a? In this exercise we explore this possibility.

   (a) Show that for any $h \neq 0$, the average rate of change of y with respect to x over the interval with endpoints $a - h$ and $a + h$ is

   $$\frac{f(a+h) - f(a-h)}{2h}.$$

   (Note that h can be either positive or negative.)

   (b) Suppose that the instantaneous rate of change of y with respect to x at $x = a$ is L. Show that

   $$\lim_{h \to 0} \frac{f(a+h) - f(a-h)}{2h} = L.$$

   (Hint: Start with the fact that $f(a+h) - f(a-h) = f(a+h) - f(a) + f(a) - f(a-h)$.)

   (c) Suppose $y = f(x) = |x|$. Show that the instantaneous rate of change of y with respect to x at $x = 0$ is undefined, but

   $$\lim_{h \to 0} \frac{f(0+h) - f(0-h)}{2h} = 0.$$

3.2   Derivatives

It is time for our next paradox of calculus. Usually, if you generalize a math problem then it gets harder. For example, finding the sum of the first 5 positive integers is not hard: $1+2+3+4+5 = 15$. But finding a general formula for the sum of the first n positive integers, for any positive integer n, is more difficult (see equation (5.4) in Section 5.1 for the answer). However, in calculus sometimes generalizing makes a problem easier, not harder. We call this the *paradox of generalization*.

    To see how this applies to rates of change, consider the equation $y = f(x)$. In the last section we saw how to find the rate of change of y with respect to x at a particular value of x. But it turns out that it will make things easier if we generalize this problem and instead find the general formula for the rate of change of y with respect to x at any value of x. Indeed, by the end of this chapter we will see that the easiest way to find the rate of change of y with respect to x at a particular value of x is to first find the general formula for the rate of change at any value of x, and then plug in the particular value of x that we are interested in.

    For example, consider the equation $y = f(x) = x^3$. Let's find the general formula for the rate of change of y with respect to x, for any value of x. To do this, we calculate the rate of change exactly as we did in the last section, but without plugging in a value for x. In our calculation we treat x as standing for a particular number, but leave the number unspecified:

$$\lim_{h\to 0} \frac{f(x+h) - f(x)}{h} = \lim_{h\to 0} \frac{(x+h)^3 - x^3}{h} = \lim_{h\to 0} \frac{(x^3 + 3x^2h + 3xh^2 + h^3) - x^3}{h}$$

$$= \lim_{h\to 0} \frac{3x^2h + 3xh^2 + h^3}{h} = \lim_{h\to 0} (3x^2 + 3xh + h^2) = 3x^2.$$

It follows that, for example, the rate of change of y with respect to x when $x = 2$ is $3(2^2) = 12$, and the rate of change when $x = 5$ is $3(5^2) = 75$.

    Notice that in this example we started with a formula involving x, namely x^3, and when we computed the rate of change we got another formula involving x, namely $3x^2$. This suggests that we could think of this calculation as an operation that can be performed on a function to produce another function.

Definition 3.2.1. For any function f, the *derivative* of f is the function f' defined by the formula

$$f'(x) = \lim_{h\to 0} \frac{f(x+h) - f(x)}{h}.$$

The domain of f' is the set of all values of x for which this limit is defined. If a is in the domain of f', then we say that f is *differentiable* at a. In that case, $f'(a)$ is the rate of change of $f(x)$ with respect to x at $x = a$, which is also the slope of the line tangent to the graph of f at the point $(a, f(a))$. The process of finding the derivative of a function is called *differentiation*.

    If I is an open interval, we say that f is *differentiable on I* if it is differentiable at a for every number $a \in I$. Recall that when we discussed continuity on closed or half-open intervals, the endpoints required special treatment. Similarly, there are complications associated with discussing differentiability at endpoints of intervals. For the moment we will avoid these complications by sticking to open intervals when discussing differentiability on intervals. Later (see Section 4.9) we will discuss differentiability on closed or half-open intervals.

For example, our earlier calculation shows that if $f(x) = x^3$ then $f'(x) = 3x^2$. Since $f'(x)$ is defined for all real numbers, f is differentiable on the interval $(-\infty, \infty)$. Let's compute a few more examples of derivates.

Example 3.2.2. Find the derivatives of the following functions:

$$f(x) = \frac{1}{x}, \qquad g(x) = \sqrt{x}.$$

What are the domains of the derivatives?

Solution. The derivative of f is the function f' defined by the formula

$$f'(x) = \lim_{h \to 0} \frac{f(x+h) - f(x)}{h}.$$

Since $f(0)$ is undefined, this formula will be undefined when $x = 0$. For $x \neq 0$, we have

$$f'(x) = \lim_{h \to 0} \frac{f(x+h) - f(x)}{h} = \lim_{h \to 0} \frac{\frac{1}{x+h} - \frac{1}{x}}{h} = \lim_{h \to 0} \left(\frac{1}{h} \cdot \frac{x - (x+h)}{x(x+h)} \right)$$

$$= \lim_{h \to 0} \left(\frac{1}{h} \cdot \frac{-h}{x(x+h)} \right) = \lim_{h \to 0} \frac{-1}{x(x+h)} = -\frac{1}{x^2}.$$

Thus, the domain of f' is $(-\infty, 0) \cup (0, \infty)$, and for all x in this domain, $f'(x) = -1/x^2$. We could also say that f is differentiable on each of the intervals $(-\infty, 0)$ and $(0, \infty)$.

Similarly, since $g(x)$ is undefined for $x < 0$, $g'(x)$ will also be undefined for $x < 0$. However, in the case of the function g it turns out that, although $g(0)$ is defined, $g'(0)$ is undefined. To see why, consider the limit that defines $g'(0)$:

$$g'(0) = \lim_{h \to 0} \frac{g(0+h) - g(0)}{h} = \lim_{h \to 0} \frac{\sqrt{0+h} - \sqrt{0}}{h} = \lim_{h \to 0} \frac{\sqrt{h}}{h} = \lim_{h \to 0} \frac{1}{\sqrt{h}}.$$

This limit is undefined, because $\sqrt{h}$ is undefined for $h < 0$. The definition of derivative calls for a two-sided limit in which h approaches 0 from both sides, and in this case the two-sided limit makes no sense. Thus, g is not differentiable at 0. (You may be thinking that we could compute a "one-sided version" of the derivative by taking the one-sided limit as $h \to 0^+$. But in this case even this one-sided limit is undefined, since the denominator approaches 0.)

For $x > 0$ we rationalize the numerator to evaluate $g'(x)$:

$$g'(x) = \lim_{h \to 0} \frac{g(x+h) - g(x)}{h} = \lim_{h \to 0} \left(\frac{\sqrt{x+h} - \sqrt{x}}{h} \cdot \frac{\sqrt{x+h} + \sqrt{x}}{\sqrt{x+h} + \sqrt{x}} \right)$$

$$= \lim_{h \to 0} \frac{x+h-x}{h(\sqrt{x+h} + \sqrt{x})} = \lim_{h \to 0} \frac{h}{h(\sqrt{x+h} + \sqrt{x})}$$

$$= \lim_{h \to 0} \frac{1}{\sqrt{x+h} + \sqrt{x}} = \frac{1}{2\sqrt{x}}.$$

Although the domain of g is $[0, \infty)$, the domain of g' is $(0, \infty)$, and for all x in this domain, $g'(x) = 1/(2\sqrt{x})$.   □

To make sure that the meaning of the calculations in Example 3.2.2 is clear, we illustrate them with a couple of examples. In the case of the function $f(x) = 1/x$, we have $f(3) = 1/3$, so the graph of the function passes through the point $(3, 1/3)$. Using the formula for $f'(x)$ that we derived in the example, we see that the tangent line to the graph at that point has slope $f'(3) = -1/3^2 = -1/9$. Thus, the equation of the tangent line is $y - 1/3 = (-1/9)(x - 3)$, or equivalently

$$y = -\frac{1}{9}x + \frac{2}{3}.$$

The graphs of the curve $y = 1/x$ and the line $y = (-1/9)x + 2/3$ are shown in Figure 3.5, where you can see that the line is tangent to the curve at the point $(3, 1/3)$. Similarly, the graph of $g(x) = \sqrt{x}$ passes through the point $(4, 2)$, the slope of the tangent line there is $g'(4) = 1/(2\sqrt{4}) = 1/4$, and therefore the equation of the tangent line is $y = (1/4)x + 1$.

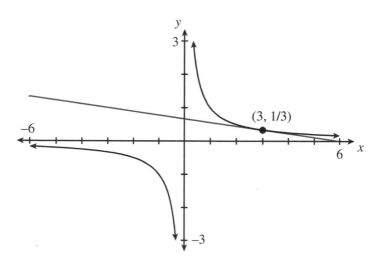

Figure 3.5: The graphs of the curve $y = 1/x$ (black) and the line $y = (-1/9)x + 2/3$ (red).

One of the lessons of Example 3.2.2 is that, since the formula that defines $f'(x)$ contains both $f(x)$ and $f(x+h)$, in order for $f'(x)$ to be defined, not only must $f(x)$ be defined, but also $f(x+h)$ must be defined for h close to 0. In other words, there must be some number $d > 0$ such that the interval $(x - d, x + d)$ is contained in the domain of f. For example, if the domain of f is an interval, then f cannot be differentiable at an endpoint of that interval. This is why the function g in Example 3.2.2 failed to be differentiable at 0: the domain of g was the interval $[0, \infty)$, so g couldn't be differentiable at the endpoint 0. (Later we will consider a one-sided version of derivatives that can be used at endpoints of intervals. But for the moment we will stick to the version of derivatives given in Definition 3.2.1, which is undefined at endpoints.)

But even if there is a number $d > 0$ such that $(x - d, x + d)$ is contained in the domain of f, the limit that defines $f'(x)$ may be undefined. For example, consider the function $f(x) = |x|$. If $x > 0$, then $x + h > 0$ for h close to 0, so $|x| = x$ and $|x + h| = x + h$. Therefore

$$f'(x) = \lim_{h \to 0} \frac{|x + h| - |x|}{h} = \lim_{h \to 0} \frac{x + h - x}{h} = \lim_{h \to 0} \frac{h}{h} = \lim_{h \to 0} 1 = 1.$$

Similarly, if $x < 0$ then

$$f'(x) = \lim_{h \to 0} \frac{|x + h| - |x|}{h} = \lim_{h \to 0} \frac{-(x + h) - (-x)}{h} = \lim_{h \to 0} \frac{-h}{h} = \lim_{h \to 0} (-1) = -1.$$

However, the computation of $f'(0)$ is more delicate. We have

$$f'(0) = \lim_{h \to 0} \frac{|0 + h| - |0|}{h} = \lim_{h \to 0} \frac{|h|}{h}.$$

In this limit, h can be either positive or negative, so $|h|$ can be either h or $-h$. The best way to deal with this is to compute the two one-sided limits:

$$\lim_{h \to 0^+} \frac{|h|}{h} = \lim_{h \to 0^+} \frac{h}{h} = \lim_{h \to 0^+} 1 = 1,$$

$$\lim_{h \to 0^-} \frac{|h|}{h} = \lim_{h \to 0^-} \frac{-h}{h} = \lim_{h \to 0^-} (-1) = -1.$$

Since the two one-sided limits are different, the two-sided limit is undefined, which means that $f'(0)$ is undefined. Thus, although $f(x)$ is defined for all x, f is not differentiable at 0. Putting all of our calculations together, we have

$$f'(x) = \begin{cases} -1, & \text{if } x < 0, \\ \text{undefined}, & \text{if } x = 0, \\ 1, & \text{if } x > 0. \end{cases}$$

If we look at the graph of f (see Figure 3.6), then we can see that this answer makes sense. If $a > 0$, then the graph of $y = |x|$ passes through the point $(a, |a|) = (a, a)$, and the line that is tangent to the graph at that point—that is, the line through (a, a) that

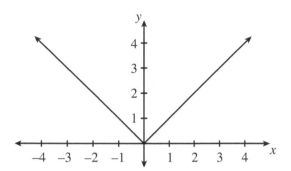

Figure 3.6: The graph of $y = f(x) = |x|$. There is no tangent line at $x = 0$, so f is not differentiable at 0.

goes in the same direction as the graph of f at that point—is the line $y = x$, which has slope 1. Similarly, for $a < 0$, the line tangent to the graph of $y = |x|$ at $(a, |a|) = (a, -a)$ is $y = -x$, which has slope -1. But at the point $(0, 0)$, the graph of f has an angle where the direction of the curve changes abruptly. Thus, the graph is not going in a particular direction at the point $(0, 0)$, and therefore there is no tangent line at that point. This, then, is another reason why a function may fail to be differentiable at a particular value of x: if there is no tangent line at that value of x, then the derivative, which gives the slope of the tangent line, will be undefined.

Another instructive example is the function $f(x) = \sqrt[3]{x}$. The derivative is

$$f'(x) = \lim_{h \to 0} \frac{f(x+h) - f(x)}{h} = \lim_{h \to 0} \frac{\sqrt[3]{x+h} - \sqrt[3]{x}}{h}.$$

A natural idea for evaluating this limit is to rationalize the numerator, but how do we do that? When we computed the derivative of the function $g(x) = \sqrt{x}$ in Example 3.2.2, we rationalized the numerator, and what made our algebraic steps work was the formula for factoring a difference of two squares,

$$(a - b)(a + b) = a^2 - b^2.$$

In the case of the function $f(x) = \sqrt[3]{x}$, it seems that we need a similar formula for factoring a difference of two cubes. In other words, we need a formula of the form

$$(a - b) \cdot \underline{\,?\,} = a^3 - b^3.$$

A little experimentation leads to a formula that works in the blank:

$$(a - b)(a^2 + ab + b^2) = a^3 - b^3.$$

Substituting $\sqrt[3]{x+h}$ for a and $\sqrt[3]{x}$ for b, we get the following formula:

$$(\sqrt[3]{x+h} - \sqrt[3]{x})((\sqrt[3]{x+h})^2 + \sqrt[3]{x+h}\sqrt[3]{x} + (\sqrt[3]{x})^2) = (\sqrt[3]{x+h})^3 - (\sqrt[3]{x})^3$$
$$= x + h - x = h.$$

This motivates our evaluation of $f'(x)$:

$$f'(x) = \lim_{h \to 0} \left(\frac{\sqrt[3]{x+h} - \sqrt[3]{x}}{h} \cdot \frac{(\sqrt[3]{x+h})^2 + \sqrt[3]{x+h}\sqrt[3]{x} + (\sqrt[3]{x})^2}{(\sqrt[3]{x+h})^2 + \sqrt[3]{x+h}\sqrt[3]{x} + (\sqrt[3]{x})^2} \right)$$

$$= \lim_{h \to 0} \frac{h}{h((\sqrt[3]{x+h})^2 + \sqrt[3]{x+h}\sqrt[3]{x} + (\sqrt[3]{x})^2)}$$

$$= \lim_{h \to 0} \frac{1}{(\sqrt[3]{x+h})^2 + \sqrt[3]{x+h}\sqrt[3]{x} + (\sqrt[3]{x})^2}.$$

Taking the limit of the denominator in the last fraction, we get

$$\lim_{h \to 0} ((\sqrt[3]{x+h})^2 + \sqrt[3]{x+h}\sqrt[3]{x} + (\sqrt[3]{x})^2) = 3(\sqrt[3]{x})^2.$$

Thus, as long as $x \neq 0$ we have

$$f'(x) = \lim_{h \to 0} \frac{1}{(\sqrt[3]{x+h})^2 + \sqrt[3]{x+h}\sqrt[3]{x} + (\sqrt[3]{x})^2} = \frac{1}{3(\sqrt[3]{x})^2}.$$

However, if $x = 0$ then the limit of the denominator is 0 and the limit of the fraction is undefined. Thus, f is not differentiable at 0.

It is worthwhile to examine more closely why $f'(0)$ is undefined. Substituting 0 for x in the definition of the derivative we see that

$$f'(0) = \lim_{h \to 0} \frac{f(0+h) - f(0)}{h} = \lim_{h \to 0} \frac{\sqrt[3]{0+h} - \sqrt[3]{0}}{h} = \lim_{h \to 0} \frac{\sqrt[3]{h}}{(\sqrt[3]{h})^3} = \lim_{h \to 0} \frac{1}{(\sqrt[3]{h})^2}.$$

Now, as $h \to 0^{\neq}$, $(\sqrt[3]{h})^2 \to 0^+$, and therefore

$$f'(0) = \lim_{h \to 0} \frac{f(0+h) - f(0)}{h} = \lim_{h \to 0} \frac{1}{(\sqrt[3]{h})^2} = \infty.$$

Of course, this doesn't change our conclusion that $f'(0)$ is undefined; if a limit is equal to ∞ then the limit is undefined, so f is not differentiable at 0. However, we see more clearly now *why* f is not differentiable at 0. As h approaches 0, the fraction $(f(0+h) - f(0))/h$ fails to approach any number because it gets very large. Recall that this fraction represents the slope of the secant line through the points $P = (0, f(0)) = (0, 0)$ and $Q_h = (0+h, f(0+h)) = (h, \sqrt[3]{h})$. Thus, this secant line slopes up very steeply, and in fact it slopes up more and more steeply as h approaches 0 and Q_h approaches P, as you can see in Figure 3.7. Our conclusion is that the secant lines approach the vertical line through P as h approaches 0. There is a line tangent to the graph of f at the point P, but it is the vertical line $x = 0$, whose slope is undefined. This is therefore another way in which a function f can fail to be differentiable: if the tangent line to the graph at the point $(a, f(a))$ is vertical, then the slope of the tangent is undefined, and the function is not differentiable at a.

There are a number of variations on the notation that mathematicians use to talk about derivatives. To explain the first of these variations, we begin with the fact that if a

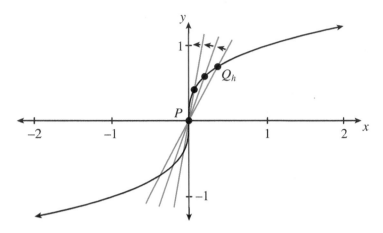

Figure 3.7: The graph of the equation $y = f(x) = \sqrt[3]{x}$. The secant lines through P and Q_h (blue) approach the vertical line $x = 0$ (red) as h approaches 0.

function f is differentiable at a number a, then we have

$$f'(a) = \lim_{h \to 0} \frac{f(a+h) - f(a)}{h}.$$

Now, suppose we let $x = a + h$. Then $h = x - a$, so as $x \to a^{\ne}$, $h \to 0^{\ne}$. And by the definition of derivative, as $h \to 0^{\ne}$,

$$\frac{f(x) - f(a)}{x - a} = \frac{f(a+h) - f(a)}{h} \to f'(a).$$

Putting these together, we see that as $x \to a^{\ne}$, $(f(x) - f(a))/(x - a) \to f'(a)$. In other words, we have proven the following theorem:

Theorem 3.2.3. *Suppose f is differentiable at a. Then*

$$f'(a) = \lim_{x \to a} \frac{f(x) - f(a)}{x - a}.$$

This gives us an alternative limit we can use to evaluate $f'(a)$. As an application of this alternative limit, we use it to prove our next theorem.

Theorem 3.2.4. *If f is differentiable at a, then f is continuous at a.*

Proof. Suppose f is differentiable at a. In order to prove that f is continuous at a, we must show that $\lim_{x \to a} f(x) = f(a)$. To evaluate this limit, we reexpress it in a way that relates it to our new formula for $f'(a)$:

$$\lim_{x \to a} f(x) = \lim_{x \to a} [f(x) - f(a) + f(a)] = \lim_{x \to a} \left[\frac{f(x) - f(a)}{x - a} \cdot (x - a) + f(a) \right]$$

$$= f'(a) \cdot 0 + f(a) = f(a). \qquad \qquad \square$$

This theorem gives us one more way in which a function can fail to be differentiable: if the function is not continuous at some number, then it cannot be differentiable at that number. For example, the function $f(x) = \lfloor x \rfloor$ is not continuous at any integer, so it is also not differentiable at any integer.

Another notational variation involves replacing the h in the definition of derivative with a different symbol. Consider two quantities x and y that are related by an equation of the form $y = f(x)$. In our definition of $f'(x)$, the fraction $(f(x+h) - f(x))/h$ represents the slope of the secant line through the points $P = (x, f(x))$ and $Q_h = (x+h, f(x+h))$. The number h is the amount by which the x-coordinate changes when we go from P to Q_h, and it is sometimes denoted Δx (the symbol "Δ" is the Greek uppercase letter delta). This is not a product of two numbers, Δ and x; rather, it should be thought of as a single symbol standing for the change in x. Similarly, the numerator $f(x+h) - f(x)$ is the amount by which y changes when we go from P to Q_h, so it can be denoted Δy. We can therefore write

$$f'(x) = \lim_{\Delta x \to 0} \frac{\Delta y}{\Delta x}.$$

This way of writing the derivative motivates another notation that is sometimes used for $f'(x)$: $\frac{dy}{dx}$, or dy/dx. Thus, we could write

$$\frac{dy}{dx} = \lim_{\Delta x \to 0} \frac{\Delta y}{\Delta x} = \lim_{h \to 0} \frac{f(x+h) - f(x)}{h} = f'(x).$$

For example, if $y = x^3$, then according to a derivative calculation we did earlier in this section, $dy/dx = 3x^2$.

It is important to remember that although dy/dx looks like a fraction, it is not a fraction. If x and y are related by the equation $y = f(x)$, then the notation dy/dx is just another name for $f'(x)$; it does not denote a fraction ih which a number dy is divided by a number dx. However, dy/dx is closely related to a fraction: according to the definition of the derivative, it is the limit of the fraction $\Delta y / \Delta x$ as Δx approaches 0. The notation dy/dx serves to remind us of the fact that, if Δx is small, then the derivative dy/dx, which is the instantaneous rate of change of y with respect to x, is close to the fraction $\Delta y / \Delta x$, which is the average rate of change of y with respect to x over a small interval.

Another way to write $\frac{dy}{dx}$ is $\frac{d}{dx}(y)$. In this version of the notation, we could think of "$\frac{d}{dx}(\ldots)$" as meaning "the rate of change of ... with respect to x." Thus,

$$\frac{dy}{dx} = \frac{d}{dx}(y) = \text{the rate of change of } y \text{ with respect to } x.$$

We also call dy/dx the *derivative of y with respect to x*. When using this notation we will sometimes replace y with the formula expressing y in terms of x. For example, if $y = x^3$ then we could write

$$\frac{dy}{dx} = \frac{d}{dx}(y) = \frac{d}{dx}(x^3) = 3x^2.$$

More generally, we can say that

$$\frac{d}{dx}(f(x)) = f'(x).$$

Sometimes we use other letters for the independent and dependent variables, in which case we make appropriate adjustments in our derivative notation. For example, if quantities w and t were related by the equation $w = g(t)$, then we would write $dw/dt = g'(t)$.

Example 3.2.5. Find

$$\frac{d}{dt}\left(\frac{1}{t^2+1}\right).$$

Solution.

$$\frac{d}{dt}\left(\frac{1}{t^2+1}\right) = \lim_{h\to 0} \frac{\frac{1}{(t+h)^2+1} - \frac{1}{t^2+1}}{h} = \lim_{h\to 0}\left(\frac{1}{h}\cdot\frac{(t^2+1)-((t+h)^2+1)}{((t+h)^2+1)(t^2+1)}\right)$$

$$= \lim_{h\to 0}\left(\frac{1}{h}\cdot\frac{-2th-h^2}{((t+h)^2+1)(t^2+1)}\right) = \lim_{h\to 0}\frac{-2t-h}{((t+h)^2+1)(t^2+1)}$$

$$= -\frac{2t}{(t^2+1)^2}. \qquad\qquad \square$$

Each of the notations for the derivative that we have introduced has its advantages and disadvantages. The f' notation emphasizes the fact that differentiation is an operation on functions: we begin with a function f, and its derivative is another function, f'. The use of functional notation also gives us a convenient way to talk about evaluating the derivative at a particular number: the rate of change of $f(x)$ with respect to x at a particular number $x = a$ can be written $f'(a)$.

The dy/dx notation does not share these advantages. But if we read "d/dx" as "rate of change with respect to x," then the dy/dx notation expresses the interpretation of the derivative as a rate of change more directly, and it reminds us that the derivative is close to the fraction $\Delta y/\Delta x$. It also gives us a convenient way of talking about a derivative without having to give a name to the function that we are differentiating. For example, in Example 3.2.5, we found the derivative of the function $f(t) = 1/(t^2+1)$, but we didn't need to introduce a name for this function.

The dy/dx notation is in a sense less explicit than the f' notation. It is only appropriate to write dy/dx in a context in which y is related to x by some equation of the form $y = f(x)$. But the notation dy/dx does not explicitly mention the function f, so it is up to the reader to identify from the context the function that relates y to x. The dy/dx notation also does not give us an easy way of talking about evaluating the derivative at a particular value of x. However, we can remedy this last deficiency by introducing a new notational convention: we write $dy/dx|_{x=a}$ to denote the derivative of y with respect to x, with the number a substituted for x. For example, we have already observed that if $y = x^3$ then $dy/dx = 3x^2$. Therefore

$$\left.\frac{dy}{dx}\right|_{x=2} = 3(2^2) = 12.$$

One of the reasons there are several different notations for derivatives is that when calculus was being developed, different mathematicians wrote derivatives in different ways. Isaac Newton (1643–1727) and Gottfried Wilhelm Leibniz (1646–1716) are generally given credit for developing calculus independently at around the same time, although they were both building on work of earlier mathematicians. The notation dy/dx was introduced by Leibniz, so it is sometimes called "Leibniz notation" for the derivative.

For Leibniz, dy/dx was actually a fraction; he thought of dx as an "infinitesimal" change in x, with dy being the corresponding infinitesimal change in y. Other mathematicians at the time also talked about infinitesimal numbers, although they did not have a good explanation of what these infinitesimal numbers were. Eventually mathematicians recognized that among the real numbers there are no infinitesimal numbers, and therefore if they wanted to put calculus on a firm logical foundation they would need to change their methods. Through the work of Augustin-Louis Cauchy (1789–1857) and Karl Theodor Wilhelm Weierstrass (1815–1897), the idea of an infinitesimal change in x was replaced with the modern idea of taking the limit as Δx approaches 0.[2]

Newton had a different notation for derivatives that involved putting dots over variables. Although his notation is still sometimes used today, it is not used as widely as Leibniz notation, and we will not use it in this book. The f' notation was introduced later, by Joseph-Louis Lagrange (1736–1813).

Exercises 3.2

1–9: Find the derivative of the function.

1. $f(x) = 3 - x^2$.

2. $g(x) = x^2 - 3x + 2$.

3. $f(x) = \dfrac{1}{\sqrt{x}}$.

4. $g(x) = \sqrt{x^2 + 1}$.

5. $f(x) = 2x - \sqrt{x}$.

6. $g(x) = x^3 - 5x$.

7. $f(x) = \sqrt[4]{x}$.

8. $g(x) = x^{3/2}$.

9. $f(x) = \dfrac{3x}{2x + 1}$.

---

[2]Much more recently, Abraham Robinson (1918–1974) showed that the set of real numbers could be enlarged to form a number system called the *hyperreal numbers* that includes infinitesimals. Robinson showed that the hyperreal numbers could be used to justify methods similar to those used at the time of Newton and Leibniz. His approach to calculus is known as *nonstandard analysis*.

10–13: Compute the derivative.

10. $\dfrac{d}{dx}((5x+2)^2)$.

11. $\dfrac{d}{dt}\left(\dfrac{1}{\sqrt{t}+1}\right)$.

12. $\dfrac{d}{dz}\left(\dfrac{2z+3}{3z-2}\right)$.

13. $\dfrac{d}{dx}((\sqrt{x}+3)^2)$.

14. Let
$$f(x) = \begin{cases} x^2 - 1, & \text{if } x \geq 1, \\ x - 1, & \text{if } x < 1. \end{cases}$$

    Is f differentiable at 1?

15. Let
$$g(x) = \begin{cases} x^2 - x, & \text{if } x \geq 1, \\ x - 1, & \text{if } x < 1. \end{cases}$$

    Is g differentiable at 1?

16. Let $f(x) = \sqrt{|x|}$. Is f differentiable at 0?

17. Let $g(x) = |x|^{3/2} = |x|\sqrt{|x|}$. Is g differentiable at 0?

3.3   Derivative Rules

We have not yet fully explained why it is easier to compute the general formula for the rate of change of $f(x)$ with respect to x—that is, to compute the derivative $f'(x)$—than it is to compute the rate of change at a particular value of x. The reason is that there are patterns in the relationship between f and f' that make it very easy to find the derivatives of many functions without having to work out any limits at all! We investigate these patterns in this section.

To discover these patterns, we will use the same method we used when we studied limits: we will begin by finding the derivatives of the simplest functions, namely constant functions and the identity function, and then we will see what happens to the derivatives of functions when the functions are combined. We will state most of our results using both Leibniz's d/dx notation and Lagrange's f' notation.

Theorem 3.3.1. *Constant functions and the identity function are differentiable at all numbers, and their derivatives are given by the following formulas:*

1. *Suppose f is a constant function; that is, there is some number c such that $f(x) = c$. Then $f'(x) = 0$. In other words, $\frac{d}{dx}(c) = 0$.*

2. *Suppose f is the identity function; that is, $f(x) = x$. Then $f'(x) = 1$. In other words, $\frac{d}{dx}(x) = 1$.*

Proof. For part 1, suppose $f(x) = c$. Then

$$f'(x) = \lim_{h \to 0} \frac{f(x+h) - f(x)}{h} = \lim_{h \to 0} \frac{c - c}{h} = \lim_{h \to 0} \frac{0}{h} = \lim_{h \to 0} 0 = 0.$$

For part 2, we have

$$f'(x) = \lim_{h \to 0} \frac{f(x+h) - f(x)}{h} = \lim_{h \to 0} \frac{x+h-x}{h} = \lim_{h \to 0} \frac{h}{h} = \lim_{h \to 0} 1 = 1. \qquad \square$$

For combinations of functions, we begin with the simplest cases: adding functions, subtracting functions, and multiplying a function by a constant. For a function f and a number c, let cf be the function defined by the equation $(cf)(x) = c(f(x))$. In other words, cf is just the product of f with the constant function whose constant value is c.

Theorem 3.3.2. *Suppose that f and g are functions and c is a number.*

1. *If f and g are differentiable at a number x, then so is $f + g$, and*

$$(f + g)'(x) = f'(x) + g'(x).$$

In other words,

$$\frac{d}{dx}(f(x) + g(x)) = \frac{d}{dx}(f(x)) + \frac{d}{dx}(g(x)).$$

2. *If f is differentiable at x, then so is cf, and*

$$(cf)'(x) = c(f'(x)).$$

In other words,

$$\frac{d}{dx}(cf(x)) = c \cdot \frac{d}{dx}(f(x)).$$

3. *If f and g are differentiable at x, then so is $f - g$, and*

$$(f - g)'(x) = f'(x) - g'(x).$$

In other words,

$$\frac{d}{dx}(f(x) - g(x)) = \frac{d}{dx}(f(x)) - \frac{d}{dx}(g(x)).$$

Proof. To prove part 1, suppose f and g are differentiable at x. According to the definition of derivative,

$$(f + g)'(x) = \lim_{h \to 0} \frac{(f + g)(x+h) - (f + g)(x)}{h}.$$

We need to be careful to interpret the first term in the numerator correctly: to evaluate $(f+g)(x+h)$, we must apply the *function* $f+g$ to the *number* $x+h$. The definition of $f+g$ tells us that $(f+g)(x+h) = f(x+h) + g(x+h)$. Thus,

$$
\begin{aligned}
(f+g)'(x) &= \lim_{h \to 0} \frac{(f+g)(x+h) - (f+g)(x)}{h} \\
&= \lim_{h \to 0} \frac{(f(x+h) + g(x+h)) - (f(x) + g(x))}{h} \\
&= \lim_{h \to 0} \frac{(f(x+h) - f(x)) + (g(x+h) - g(x))}{h} \\
&= \lim_{h \to 0} \left(\frac{f(x+h) - f(x)}{h} + \frac{g(x+h) - g(x)}{h} \right).
\end{aligned}
$$

Now, we have assumed that f and g are both differentiable at x, which means that the limits of $(f(x+h) - f(x))/h$ and $(g(x+h) - g(x))/h$ are both defined, and their values are:

$$
\lim_{h \to 0} \frac{f(x+h) - f(x)}{h} = f'(x), \qquad \lim_{h \to 0} \frac{g(x+h) - g(x)}{h} = g'(x).
$$

Therefore by part 1 of Theorem 2.4.2,

$$
(f+g)'(x) = \lim_{h \to 0} \left(\frac{f(x+h) - f(x)}{h} + \frac{g(x+h) - g(x)}{h} \right) = f'(x) + g'(x).
$$

Similarly, to prove part 2 we apply the definitions of derivative and cf. Suppose f is differentiable at x. Then

$$
\begin{aligned}
(cf)'(x) &= \lim_{h \to 0} \frac{(cf)(x+h) - (cf)(x)}{h} \\
&= \lim_{h \to 0} \frac{c(f(x+h)) - c(f(x))}{h} \\
&= \lim_{h \to 0} \left(c \cdot \frac{f(x+h) - f(x)}{h} \right) = c(f'(x)).
\end{aligned}
$$

Finally, to prove part 3 we combine parts 1 and 2. We find it more convenient to write the proof of this part in Leibniz notation:

$$
\begin{aligned}
\frac{d}{dx}(f(x) - g(x)) &= \frac{d}{dx}(f(x) + (-1)g(x)) \\
&= \frac{d}{dx}(f(x)) + \frac{d}{dx}((-1)g(x)) && \text{(by part 1)} \\
&= \frac{d}{dx}(f(x)) + (-1)\frac{d}{dx}(g(x)) && \text{(by part 2)} \\
&= \frac{d}{dx}(f(x)) - \frac{d}{dx}(g(x)). && \qquad\qquad \square
\end{aligned}
$$

To state Theorem 3.3.2 more informally, it says that to differentiate a sum or difference of two functions, just differentiate the two functions separately. To differentiate a function that has been multiplied by a constant, just leave the constant factor alone and differentiate the function. For example, combining Theorems 3.3.1 and 3.3.2 we can now see that

$$\frac{d}{dx}(5x - 7) = \frac{d}{dx}(5x) - \frac{d}{dx}(7) = 5 \cdot \frac{d}{dx}(x) - \frac{d}{dx}(7) = 5 \cdot 1 - 0 = 5.$$

Notice that we were able to compute this derivative without having to work out a limit. But this is still a fairly simple function. To differentiate more complex functions we need more rules.

Based on Theorem 3.3.2, you might expect the rule for products to be $(fg)'(x) = f'(x)g'(x)$, but it turns out that this is not correct. To discover the correct rule, we begin with the definition of derivative:

$$(fg)'(x) = \lim_{h \to 0} \frac{(fg)(x + h) - (fg)(x)}{h} = \lim_{h \to 0} \frac{f(x + h)g(x + h) - f(x)g(x)}{h}.$$

To relate this to $f'(x)$ and $g'(x)$, we will need to find an algebraic relationship between the numerator of the last fraction and the differences $f(x + h) - f(x)$ and $g(x + h) - g(x)$. There is an algebraic trick that allows us to do that: we subtract and add the same thing in the numerator.

$$\begin{aligned}
(fg)'(x) &= \lim_{h \to 0} \frac{f(x + h)g(x + h) - f(x)g(x)}{h} \\
&= \lim_{h \to 0} \frac{f(x + h)g(x + h) - f(x + h)g(x) + f(x + h)g(x) - f(x)g(x)}{h} \\
&= \lim_{h \to 0} \frac{f(x + h)(g(x + h) - g(x)) + g(x)(f(x + h) - f(x))}{h} \\
&= \lim_{h \to 0} \left(f(x + h) \cdot \frac{g(x + h) - g(x)}{h} + g(x) \cdot \frac{f(x + h) - f(x)}{h} \right).
\end{aligned}$$

If we assume that f and g are both differentiable at x, then we have

$$\lim_{h \to 0} \frac{g(x + h) - g(x)}{h} = g'(x), \qquad \lim_{h \to 0} \frac{f(x + h) - f(x)}{h} = f'(x).$$

Also, if f is differentiable at x then it is continuous at x, so since $\lim_{h \to 0}(x + h) = x$, we have $\lim_{h \to 0} f(x + h) = f(x)$. Putting it all together, we conclude that

$$\begin{aligned}
(fg)'(x) &= \lim_{h \to 0} \left(f(x + h) \cdot \frac{g(x + h) - g(x)}{h} + g(x) \cdot \frac{f(x + h) - f(x)}{h} \right) \\
&= f(x)g'(x) + g(x)f'(x).
\end{aligned}$$

Thus, we have proven the following theorem.

Theorem 3.3.3 (Product Rule). *If f and g are differentiable at x, then so is fg, and*

$$(fg)'(x) = f(x)g'(x) + g(x)f'(x).$$

In other words,

$$\frac{d}{dx}(f(x)g(x)) = f(x) \cdot \frac{d}{dx}(g(x)) + g(x) \cdot \frac{d}{dx}(f(x)).$$

For the purpose of applying the product rule in examples, it may be useful to restate the rule in words. We can state it as follows:

The derivative of a product of two terms is equal to the first term times the derivative of the second plus the second term times the derivative of the first.

For example, we have seen previously that $\frac{d}{dx}(x^3) = 3x^2$ and $\frac{d}{dx}(\sqrt{x}) = 1/(2\sqrt{x})$. The first of these derivatives is defined for all values of x, but the second is defined only for $x > 0$. Therefore $x^3\sqrt{x}$ is differentiable on $(0, \infty)$, and

$$\frac{d}{dx}(x^3\sqrt{x}) = x^3 \cdot \frac{d}{dx}(\sqrt{x}) + \sqrt{x} \cdot \frac{d}{dx}(x^3) = x^3 \cdot \frac{1}{2\sqrt{x}} + \sqrt{x} \cdot 3x^2$$

$$= \frac{x^2\sqrt{x}}{2} + 3x^2\sqrt{x} = \frac{7x^2\sqrt{x}}{2}.$$

We will use the product rule to derive our next derivative rule. We saw in Theorem 3.3.1 that $\frac{d}{dx}(x) = 1$. It follows by the product rule that

$$\frac{d}{dx}(x^2) = \frac{d}{dx}(x \cdot x) = x \cdot \frac{d}{dx}(x) + x \cdot \frac{d}{dx}(x) = x \cdot 1 + x \cdot 1 = 2x.$$

Applying the product rule again we find that

$$\frac{d}{dx}(x^3) = \frac{d}{dx}(x^2 \cdot x) = x^2 \cdot \frac{d}{dx}(x) + x \cdot \frac{d}{dx}(x^2) = x^2 \cdot 1 + x \cdot 2x = 3x^2.$$

Of course, this agrees with the answer we got for this derivative in the last section. Another application of the product rule gives us

$$\frac{d}{dx}(x^4) = \frac{d}{dx}(x^3 \cdot x) = x^3 \cdot \frac{d}{dx}(x) + x \cdot \frac{d}{dx}(x^3) = x^3 \cdot 1 + x \cdot 3x^2 = 4x^3.$$

Looking at these examples, you should see a pattern emerging. For n up to 4, we have shown that $\frac{d}{dx}(x^n) = nx^{n-1}$. Will this pattern continue for larger values of n? Let's see if we can prove it. In the calculations we have done so far, each case was derived from the previous one. This suggests the method of proof that we should try to use to show that the pattern will continue: mathematical induction.

Theorem 3.3.4 (Power Rule). *Suppose n is a positive integer, and $f(x) = x^n$. Then f is differentiable at all numbers, and $f'(x) = nx^{n-1}$. In other words, $\frac{d}{dx}(x^n) = nx^{n-1}$.*

Proof. We use mathematical induction.
Base case.[3]

$$\frac{d}{dx}(x^1) = \frac{d}{dx}(x) = 1 = 1 \cdot x^0.$$

Induction step. Suppose that n is a positive integer and $\frac{d}{dx}(x^n) = nx^{n-1}$. Then by the product rule,

$$\frac{d}{dx}(x^{n+1}) = \frac{d}{dx}(x^n \cdot x) = x^n \cdot \frac{d}{dx}(x) + x \cdot \frac{d}{dx}(x^n) = x^n \cdot 1 + x \cdot nx^{n-1} = (n+1)x^n.$$

$\square$

Using the theorems we have proven, we can now easily find the derivative of any polynomial. For example,

$$\frac{d}{dx}(3x^5 - 7x^2 + 5) = \frac{d}{dx}(3x^5) + \frac{d}{dx}(-7x^2 + 5)$$

$$= \frac{d}{dx}(3x^5) + \frac{d}{dx}(-7x^2) + \frac{d}{dx}(5)$$

$$= 3 \cdot \frac{d}{dx}(x^5) + (-7) \cdot \frac{d}{dx}(x^2) + \frac{d}{dx}(5)$$

$$= 3(5x^4) - 7(2x) + 0 = 15x^4 - 14x.$$

In the future, we won't write out so many steps when computing the derivative of a polynomial. Imitating this example, you should be able to see that polynomials are differentiable at all numbers, and that you can find the derivative of any polynomial by simply replacing each power of x by its derivative, which is given by the power rule, and replacing the constant term with 0. For example,

$$\frac{d}{dx}(6x^4 + x^3 - 5x^2 + 2x - 11) = 6(4x^3) + 3x^2 - 5(2x) + 2(1) - 0$$

$$= 24x^3 + 3x^2 - 10x + 2.$$

To find derivatives of rational functions, we need a rule for quotients:

Theorem 3.3.5 (Quotient Rule). *If f and g are differentiable at x and $g(x) \neq 0$, then f/g is differentiable at x, and*

$$(f/g)'(x) = \frac{g(x)f'(x) - f(x)g'(x)}{(g(x))^2}.$$

In other words,

$$\frac{d}{dx}\left(\frac{f(x)}{g(x)}\right) = \frac{g(x) \cdot \frac{d}{dx}(f(x)) - f(x) \cdot \frac{d}{dx}(g(x))}{(g(x))^2}.$$

---

[3]There is a small technical point that should be mentioned here. Some books do not assign any meaning to the expression 0^0. In calculus, it is best to follow the convention that $0^0 = 1$, so that the equation $x^0 = 1$ is true for all values of x.

Proof. Suppose that f and g are differentiable at x, and $g(x) \neq 0$. First we use the definition to find the derivative of $1/g(x)$:

$$\frac{d}{dx}\left(\frac{1}{g(x)}\right) = \lim_{h \to 0} \frac{\frac{1}{g(x+h)} - \frac{1}{g(x)}}{h} = \lim_{h \to 0}\left(\frac{1}{h} \cdot \frac{g(x) - g(x+h)}{g(x+h)g(x)}\right)$$

$$= \lim_{h \to 0}\left(-\frac{g(x+h) - g(x)}{h} \cdot \frac{1}{g(x+h)g(x)}\right)$$

$$= -g'(x) \cdot \frac{1}{(g(x))^2} = -\frac{g'(x)}{(g(x))^2}.$$

Now we can use this result, together with the product rule, to find the derivative of $f(x)/g(x)$:

$$\frac{d}{dx}\left(\frac{f(x)}{g(x)}\right) = \frac{d}{dx}\left(f(x) \cdot \frac{1}{g(x)}\right) = f(x) \cdot \frac{d}{dx}\left(\frac{1}{g(x)}\right) + \frac{1}{g(x)} \cdot \frac{d}{dx}(f(x))$$

$$= f(x) \cdot \left(-\frac{g'(x)}{(g(x))^2}\right) + \frac{1}{g(x)} \cdot f'(x) = -\frac{f(x)g'(x)}{(g(x))^2} + \frac{g(x)f'(x)}{(g(x))^2}$$

$$= \frac{g(x)f'(x) - f(x)g'(x)}{(g(x))^2}. \qquad\qquad \square$$

Once again, it is probably best to learn this in words:

The derivative of a quotient is equal to the denominator times the derivative of the numerator minus the numerator times the derivative of the denominator, all divided by the denominator squared.

According to the quotient rule, rational functions are differentiable at all numbers except where the denominator is 0. For example,

$$\frac{d}{dx}\left(\frac{3x^2 - 4x + 1}{x^2 - 3}\right) = \frac{(x^2 - 3) \cdot \frac{d}{dx}(3x^2 - 4x + 1) - (3x^2 - 4x + 1) \cdot \frac{d}{dx}(x^2 - 3)}{(x^2 - 3)^2}$$

$$= \frac{(x^2 - 3) \cdot (6x - 4) - (3x^2 - 4x + 1) \cdot (2x)}{(x^2 - 3)^2}$$

$$= \frac{(6x^3 - 4x^2 - 18x + 12) - (6x^3 - 8x^2 + 2x)}{(x^2 - 3)^2}$$

$$= \frac{4x^2 - 20x + 12}{(x^2 - 3)^2}.$$

This derivative is defined for all values of x except $x = \pm\sqrt{3}$.

This may be a good time to make a comment about simplifying answers. We multiplied out the products in the numerator of our last answer because when we did so we were able to combine some terms and simplify the numerator. In fact, the x^3 terms canceled out completely when we did this! We didn't bother to multiply out the square in the denominator because it would not have led to any similar simplification. It is a good idea to get in the habit of simplifying your answer as much as you can. We will eventually be using derivatives to solve problems, and it will be much easier to use a derivative if it has been simplified. But notice that simplifying does not merely mean multiplying everything out; it means looking for algebraic steps that will make the answer simpler and easier to use in further calculations.

An interesting consequence of the quotient rule is that the power rule works for negative integer exponents as well:

Theorem 3.3.6. *Suppose n is a negative integer, and for all $x \neq 0$, $f(x) = x^n$. Then f is differentiable at all $x \neq 0$, and $f'(x) = nx^{n-1}$. In other words, $\frac{d}{dx}(x^n) = nx^{n-1}$.*

Proof. Since n is a negative integer, $n = -m$ for some positive integer m. Recall that by the definition of negative exponents,

$$x^n = x^{-m} = \frac{1}{x^m}.$$

This is undefined when $x = 0$, but defined for all other values of x. We can now use the power rule for positive exponents and the quotient rule to compute the derivative, which is also defined for all $x \neq 0$:

$$\frac{d}{dx}(x^n) = \frac{d}{dx}\left(\frac{1}{x^m}\right) = \frac{x^m \cdot \frac{d}{dx}(1) - 1 \cdot \frac{d}{dx}(x^m)}{(x^m)^2} = \frac{0 - mx^{m-1}}{x^{2m}}$$

$$= -mx^{m-1-2m} = -mx^{-m-1} = nx^{n-1}. \qquad \square$$

In the last section we showed that

$$\frac{d}{dx}(\sqrt{x}) = \frac{1}{2\sqrt{x}}, \qquad \frac{d}{dx}(\sqrt[3]{x}) = \frac{1}{3(\sqrt[3]{x})^2}.$$

Recall that for a rational number m/n written in lowest terms, $x^{m/n} = (\sqrt[n]{x})^m$. Rewriting the derivatives above using fractional exponents, we have

$$\frac{d}{dx}(x^{1/2}) = \frac{1}{2}x^{-1/2} = \frac{1}{2}x^{1/2-1}, \qquad \frac{d}{dx}(x^{1/3}) = \frac{1}{3}x^{-2/3} = \frac{1}{3}x^{1/3-1}.$$

So the power rule works for the exponents $1/2$ and $1/3$ too! In fact, it works for $1/n$, for every integer $n \geq 2$. One way to see this is to imitate the method we used in the last section for the exponents $1/2$ and $1/3$ (see Exercise 24). But the algebra needed for that approach gets a little messy. A clever change of variables makes it possible to compute these derivatives with very little algebraic work.

Theorem 3.3.7. *Suppose n is an integer, $n \geq 2$, and $f(x) = x^{1/n}$. Then $f'(x) = (1/n)x^{1/n-1}$. In other words, $\frac{d}{dx}(x^{1/n}) = (1/n)x^{1/n-1}$. If n is even then the domain of f' is $(0, \infty)$, and if n is odd then it is $(-\infty, 0) \cup (0, \infty)$.*

Proof. We have $f(x) = \sqrt[n]{x}$, so

$$f'(x) = \lim_{h \to 0} \frac{\sqrt[n]{x+h} - \sqrt[n]{x}}{h}.$$

If n is even, then this limit makes sense only for $x > 0$, and we will find its value for all of these values of x. For n odd, we will see that the limit is undefined when $x = 0$, but we will find its value for all other values of x.

To compute the limit, we make a change of variables. Let $y = \sqrt[n]{x}$ and $k = \sqrt[n]{x+h} - \sqrt[n]{x}$. Then $\sqrt[n]{x+h} = \sqrt[n]{x} + k = y + k$. Raising both sides of this equation to the power n, we get $x + h = (y+k)^n$, and therefore

$$h = (y+k)^n - x = (y+k)^n - y^n.$$

For $h \neq 0$ we also have $k \neq 0$, so we can write

$$\frac{\sqrt[n]{x+h} - \sqrt[n]{x}}{h} = \frac{k}{(y+k)^n - y^n} = \frac{1}{\frac{(y+k)^n - y^n}{k}}.$$

The denominator of the last fraction is beginning to look like the derivative of y^n with respect to y. Indeed, as $k \to 0^{\neq}$,

$$\frac{(y+k)^n - y^n}{k} \to \frac{d}{dy}(y^n) = ny^{n-1}.$$

And since $k = \sqrt[n]{x+h} - \sqrt[n]{x}$, as $h \to 0^{\neq}$, $k \to 0^{\neq}$. Putting these two limit facts together, we see that if $x \neq 0$, then as $h \to 0^{\neq}$,

$$\frac{\sqrt[n]{x+h} - \sqrt[n]{x}}{h} = \frac{1}{\frac{(y+k)^n - y^n}{k}} \to \frac{1}{ny^{n-1}} = \frac{1}{n(\sqrt[n]{x})^{n-1}}. \tag{3.1}$$

Thus,

$$f'(x) = \lim_{h \to 0} \frac{\sqrt[n]{x+h} - \sqrt[n]{x}}{h} = \frac{1}{n(\sqrt[n]{x})^{n-1}} = \frac{1}{n}x^{-(n-1)/n} = \frac{1}{n}x^{1/n-1}.$$

However, if n is odd and $x = 0$, then the denominator of the second fraction in (3.1) approaches 0 as $h \to 0^{\neq}$, and therefore $f'(0)$ is undefined. $\square$

Example 3.3.8. Find

$$\frac{d}{dx}\left(\frac{x^2}{\sqrt[5]{x}+3}\right).$$

Solution.

$$\begin{aligned}
\frac{d}{dx}\left(\frac{x^2}{\sqrt[5]{x}+3}\right) &= \frac{(x^{1/5}+3)\cdot\frac{d}{dx}(x^2) - x^2\cdot\frac{d}{dx}(x^{1/5}+3)}{(\sqrt[5]{x}+3)^2} \\
&= \frac{(x^{1/5}+3)\cdot(2x) - x^2\cdot(\frac{1}{5}x^{-4/5})}{(\sqrt[5]{x}+3)^2}\cdot\frac{5}{5} \\
&= \frac{(10x^{6/5}+30x) - x^{6/5}}{5(\sqrt[5]{x}+3)^2} \\
&= \frac{9x\sqrt[5]{x}+30x}{5(\sqrt[5]{x}+3)^2}. \quad\square
\end{aligned}$$

We end this section by finding the derivatives of the trigonometric functions. To do this we will need to use the following trigonometric identities:

$$\sin(\alpha + \beta) = \sin \alpha \cos \beta + \cos \alpha \sin \beta,$$
$$\cos(\alpha + \beta) = \cos \alpha \cos \beta - \sin \alpha \sin \beta.$$

We assume you have seen these identities before, but if not you can find them in any trigonometry book (or see Exercise 2.3 in Section 1.4). We will also need the following two limits:

$$\text{(a) } \lim_{h \to 0} \frac{\sin h}{h} = 1, \quad \text{(b) } \lim_{h \to 0} \frac{\cos h - 1}{h} = 0.$$

Limit (a) is one of the examples we gave in Section 2.1, and we confirmed its value at the end of Section 2.7. To compute limit (b), we use an algebraic trick we have seen before and then apply limit (a):

$$
\begin{aligned}
\lim_{h \to 0} \frac{\cos h - 1}{h} &= \lim_{h \to 0} \left(\frac{\cos h - 1}{h} \cdot \frac{\cos h + 1}{\cos h + 1} \right) = \lim_{h \to 0} \frac{\cos^2 h - 1}{h(\cos h + 1)} \\
&= \lim_{h \to 0} \frac{\cos^2 h - (\cos^2 h + \sin^2 h)}{h(\cos h + 1)} = \lim_{h \to 0} \frac{-\sin^2 h}{h(\cos h + 1)} \\
&= \lim_{h \to 0} \left(-\frac{\sin h}{h} \cdot \frac{\sin h}{\cos h + 1} \right) = -1 \cdot \frac{0}{2} = 0.
\end{aligned}
$$

We are now ready to differentiate the trigonometric functions. We begin with the sine function. Near the end of the calculation, we apply limits (a) and (b).

$$
\begin{aligned}
\frac{d}{dx}(\sin x) &= \lim_{h \to 0} \frac{\sin(x + h) - \sin x}{h} \\
&= \lim_{h \to 0} \frac{\sin x \cos h + \cos x \sin h - \sin x}{h} \\
&= \lim_{h \to 0} \frac{\sin x (\cos h - 1) + \cos x \sin h}{h} \\
&= \lim_{h \to 0} \left(\frac{\cos h - 1}{h} \cdot \sin x + \frac{\sin h}{h} \cdot \cos x \right) \\
&= 0 \cdot \sin x + 1 \cdot \cos x = \cos x.
\end{aligned}
$$

The derivative of the cosine function is similar:

$$
\begin{aligned}
\frac{d}{dx}(\cos x) &= \lim_{h \to 0} \frac{\cos(x + h) - \cos x}{h} \\
&= \lim_{h \to 0} \frac{\cos x \cos h - \sin x \sin h - \cos x}{h} \\
&= \lim_{h \to 0} \left(\frac{\cos h - 1}{h} \cdot \cos x - \frac{\sin h}{h} \cdot \sin x \right) \\
&= 0 \cdot \cos x - 1 \cdot \sin x = -\sin x.
\end{aligned}
$$

$$\frac{d}{dx}(\sin x) = \cos x \qquad \frac{d}{dx}(\tan x) = \sec^2 x \qquad \frac{d}{dx}(\sec x) = \sec x \tan x$$

$$\frac{d}{dx}(\cos x) = -\sin x \qquad \frac{d}{dx}(\cot x) = -\csc^2 x \qquad \frac{d}{dx}(\csc x) = -\csc x \cot x.$$

Table 3.1: The derivatives of the trigonometric functions.

The remaining trigonometric functions are defined in terms of sine and cosine, and their derivatives can now be computed by the quotient rule. For example,

$$\frac{d}{dx}(\tan x) = \frac{d}{dx}\left(\frac{\sin x}{\cos x}\right) = \frac{\cos x \cdot \frac{d}{dx}(\sin x) - \sin x \cdot \frac{d}{dx}(\cos x)}{\cos^2 x}$$

$$= \frac{\cos x \cdot \cos x - \sin x \cdot (-\sin x)}{\cos^2 x} = \frac{\cos^2 x + \sin^2 x}{\cos^2 x}$$

$$= \frac{1}{\cos^2 x} = \sec^2 x.$$

We leave the other three calculations to you (see Exercise 25), and just summarize the results in Table 3.1. And we remind you that the value for limit (a) is only correct when angles are measured in radians, so the derivatives in Table 3.1 are also only correct when angles are measured in radians.

Example 3.3.9. Find

$$\frac{d}{dx}(\sqrt[4]{x}\cot x).$$

Solution.

$$\frac{d}{dx}(\sqrt[4]{x}\cot x) = \sqrt[4]{x} \cdot \frac{d}{dx}(\cot x) + \cot x \cdot \frac{d}{dx}(x^{1/4})$$

$$= \sqrt[4]{x} \cdot (-\csc^2 x) + \cot x \cdot \left(\frac{1}{4}x^{-3/4}\right)$$

$$= -\sqrt[4]{x}\csc^2 x + \frac{\cot x}{4(\sqrt[4]{x})^3} \qquad \square$$

Exercises 3.3

1–16: Find the derivative of the function.

1. $f(x) = 2x^6 - 3x^4 + x^3 - 7x^2 + 4x - 8.$

2. $g(x) = \dfrac{x^4}{2} - \dfrac{2x^3}{5} + x - 5.$

3. $f(x) = \dfrac{x^2 + 3x - 5}{3x^2 - 5x + 1}.$

4. $h(x) = (x^3 - 4x)^2.$

5. $g(x) = \dfrac{2x+3}{x^3+2}$.

6. $f(x) = x^5 + 3\sqrt[5]{x}$.

7. $g(x) = \dfrac{4x-1}{x^2+\sqrt{x}}$.

8. $h(x) = \dfrac{3+\frac{2}{x-1}}{x+1}$.

9. $f(x) = \dfrac{\sqrt{x}-1}{\sqrt[3]{x}+1}$.

10. $h(x) = x^{7/3}$. (Hint: Write this as $h(x) = x^{2+1/3} = x^2 \cdot \sqrt[3]{x}$.)

11. $h(x) = x^{2/3}$. (Hint: Write this as $h(x) = (\sqrt[3]{x})^2 = \sqrt[3]{x} \cdot \sqrt[3]{x}$.)

12. $f(x) = \sin x \cos x$.

13. $f(x) = \sqrt{x}\sin x \cos x$.

14. $g(x) = x^2 \tan x$.

15. $h(x) = \dfrac{x}{\sec x + \tan x}$.

16. $f(x) = \dfrac{\sin x}{\cos x + \sec x}$.

17. Find the equation of the line tangent to the curve $y = 3x - \dfrac{2}{x}$ at the point $(2, 5)$.

18. The lines tangent to the curve $y = x^2$ at two points P and Q pass through the point $(3, 5)$ (see Figure 3.8). Find the points P and Q.

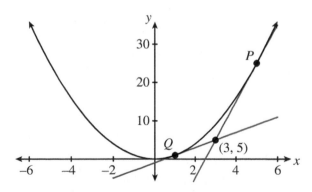

Figure 3.8: Two lines tangent to the curve $y = x^2$ pass through $(3, 5)$.

19. The line tangent to the curve $y = 1/x$ at some point P passes through the point $(4, 0)$. Find P.

20. A rocket is to be launched straight up from the point $(10, 0)$. An observer located at the origin is watching the launch, but unfortunately his view is blocked by a hill in the shape of the curve $y = -x^2 + 10x - 16$, $2 \leq x \leq 8$ (see Figure 3.9). Where will the rocket be when the observer first sees it?

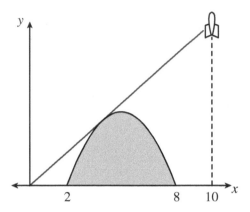

Figure 3.9: A rocket becomes visible to an observer at the origin.

21. A board is resting on top of two hills in the shape of the curves $y = 16 - x^2$, $-4 \leq x \leq 4$ and $y = -x^2 + 12x - 32$, $4 \leq x \leq 8$, as shown in Figure 3.10. Find the equation of the line along which the board lies.

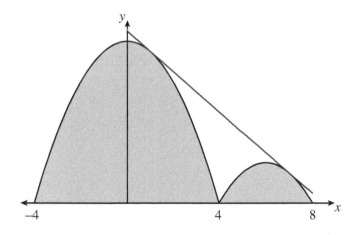

Figure 3.10: A board resting on two hills.

22. A board is resting on top of two hills in the shape of the curves $y = 17 - x^2$, $-\sqrt{17} \leq x \leq \sqrt{17}$, and $y = -2x^2 + 26x - 80$, $5 \leq x \leq 8$. (You may find it helpful to draw a figure similar to Figure 3.10.) Find the equation of the line along which the board lies.

23. Compute the derivative of the function $f(x) = (x^2+3)(2x-5)$ in two different ways: by applying the product rule to the given formula for $f(x)$, and by multiplying out the formula for $f(x)$ before differentiating. Verify that the two answers are the same.

24. (a) Complete the following factorization of $a^4 - b^4$:

$$(a-b) \cdot \underline{\ ?\ } = a^4 - b^4.$$

   (b) Imitate the method used in Section 3.2 for $\sqrt{x}$ and $\sqrt[3]{x}$ to find $\frac{d}{dx}(\sqrt[4]{x})$.

   (c) Generalize: use this method to find $\frac{d}{dx}(\sqrt[n]{x})$ for any integer $n \geq 2$.

25. Verify the entries in Table 3.1 for the derivatives of $\sec x$, $\cot x$, and $\csc x$.

26. Suppose that f is differentiable on $(-\infty, \infty)$.

   (a) Show that $\frac{d}{dx}((f(x))^2) = 2f(x)f'(x)$.

   (b) Find $\frac{d}{dx}((f(x))^3)$.

   (c) Make a conjecture about $\frac{d}{dx}((f(x))^n)$, for any positive integer n, and prove your conjecture by mathematical induction.

27. Find a formula for $\frac{d}{dx}(f(x)g(x)h(x))$ in terms of $f(x)$, $g(x)$, $h(x)$, $f'(x)$, $g'(x)$, and $h'(x)$.

3.4 The Chain Rule

We turn now to the differentiation of compositions of functions. The theorem that tells us how to do this is called the chain rule.

Theorem 3.4.1 (Chain Rule). *If g is differentiable at x and f is differentiable at $g(x)$, then $f \circ g$ is differentiable at x, and*

$$(f \circ g)'(x) = f'(g(x)) \cdot g'(x).$$

In other words,

$$\frac{d}{dx}(f(g(x))) = f'(g(x)) \cdot \frac{d}{dx}(g(x)).$$

Before proving the theorem, we try it out in a few examples.

Example 3.4.2. Find the derivatives of the following functions.

$$h(x) = (x^2 - 5x + 2)^7, \qquad j(x) = \sqrt{\sec x}.$$

Solution. We have $h = f \circ g$, where $f(x) = x^7$ and $g(x) = x^2 - 5x + 2$. By results from the last section, $f'(x) = 7x^6$ and $g'(x) = 2x - 5$. Therefore

$$h'(x) = f'(g(x)) \cdot g'(x) = f'(x^2 - 5x + 2) \cdot (2x - 5) = 7(x^2 - 5x + 2)^6(2x - 5).$$

For the function j, we write $j = f \circ g$, where $f(x) = \sqrt{x}$ and $g(x) = \sec x$. Again, the last sections tells us how to differentiate f and g:

$$f'(x) = \frac{d}{dx}(x^{1/2}) = \frac{1}{2}x^{-1/2} = \frac{1}{2\sqrt{x}}, \qquad g'(x) = \sec x \tan x.$$

Thus,

$$j'(x) = f'(g(x)) \cdot g'(x) = \frac{1}{2\sqrt{\sec x}} \cdot \sec x \tan x = \frac{\sqrt{\sec x} \tan x}{2}. \qquad \square$$

We can streamline our use of the chain rule by noticing the relationship between the following two equations:

$$\frac{d}{dx}(f(x)) = f'(x),$$

$$\frac{d}{dx}(f(g(x))) = f'(g(x)) \cdot \frac{d}{dx}(g(x)).$$

The first equation says that the derivative of $f(x)$ is $f'(x)$. Replacing x with the expression $g(x)$ on both sides, you might expect that the derivative of $f(g(x))$ would be $f'(g(x))$. The chain rule says that this is part of the answer, but it's not the whole thing. To get the correct answer you need to multiply by the derivative of $g(x)$.

This suggests a quick way to find the derivative of an expression of the form $f(g(x))$. First, identify the "outer" function f, and find its derivative:

$$\frac{d}{dx}(f(x)) = f'(x).$$

If you now replace x with $g(x)$ on both sides, and multiply by the derivative of $g(x)$ on the right, then the chain rule says that you will have the correct answer for the derivative of $f(g(x))$:

$$\frac{d}{dx}(f(g(x))) = f'(g(x)) \cdot \frac{d}{dx}(g(x)).$$

For example, consider again the function $h(x) = (x^2 - 5x + 2)^7$ from Example 3.4.2. The "outer" function in this case is the "raise to the 7th power" function, whose derivative is given by the equation

$$\frac{d}{dx}(x^7) = 7x^6.$$

Now we replace x with $x^2 - 5x + 2$ on both sides, and multiply by the derivative of $x^2 - 5x + 2$ on the right:

$$\frac{d}{dx}((x^2 - 5x + 2)^7) = 7(x^2 - 5x + 2)^6 \cdot \frac{d}{dx}(x^2 - 5x + 2)$$

$$= 7(x^2 - 5x + 2)^6 (2x - 5).$$

Notice that this agrees with the answer we got in Example 3.4.2.

With practice, you will be able to do much of this reasoning in your head. For example, suppose you want to differentiate the function $h(x) = \sin(1/x)$. You can just *think*

$$\frac{d}{dx}(\sin x) = \cos x,$$

and then *write*

$$\frac{d}{dx}(\sin(1/x)) = \cos(1/x) \cdot \frac{d}{dx}(1/x)$$

$$= \cos(1/x) \cdot \frac{d}{dx}(x^{-1})$$

$$= \cos(1/x) \cdot (-1)(x^{-2})$$

$$= -\frac{\cos(1/x)}{x^2}.$$

Example 3.4.3. Find the derivatives of the following functions:

$$f(x) = \sqrt{(x^2 + 1) \tan x}, \qquad g(x) = \sqrt{x^2 + 1} \tan x, \qquad h(x) = \sqrt{\tan(x^2 + 1)}.$$

Solution. How do you decide which derivative rule to use first when you are differentiating a complicated function? One approach is to imagine that you are computing the value of the function at a particular value of x, and think about what the last step in that computation would be. In the case of the function f, to evaluate $f(x)$ you would first compute $(x^2 + 1)$ and $\tan x$, then multiply them together, and finally take the square root of the product. Since the last step is taking the square root, the first step in differentiating f is to apply the chain rule, with the square root function as the outer function. We have

$$\frac{d}{dx}(\sqrt{x}) = \frac{d}{dx}(x^{1/2}) = \frac{1}{2}x^{-1/2},$$

so by the chain rule,

$$f'(x) = \frac{d}{dx}([(x^2 + 1) \tan x]^{1/2}) = \frac{1}{2}[(x^2 + 1) \tan x]^{-1/2} \cdot \frac{d}{dx}((x^2 + 1) \tan x)$$

$$= \frac{1}{2\sqrt{(x^2 + 1) \tan x}} \cdot \frac{d}{dx}((x^2 + 1) \tan x).$$

To evaluate the remaining derivative, notice that the last step in computing $(x^2+1)\tan x$ is to multiply x^2+1 by $\tan x$. This tells us that we should use the product rule:

$$f'(x) = \frac{1}{2\sqrt{(x^2+1)\tan x}} \cdot \left[(x^2+1)\cdot\frac{d}{dx}(\tan x) + \tan x \cdot \frac{d}{dx}(x^2+1)\right]$$

$$= \frac{(x^2+1)\sec^2 x + 2x\tan x}{2\sqrt{(x^2+1)\tan x}}.$$

In the case of the function g, to compute $g(x)$ we would first compute $\sqrt{x^2+1}$ and $\tan x$ and then multiply them together. Since the last step is multiplication, we start with the product rule:

$$g'(x) = \frac{d}{dx}(\sqrt{x^2+1}\tan x) = \sqrt{x^2+1}\cdot\frac{d}{dx}(\tan x) + \tan x \cdot \frac{d}{dx}(\sqrt{x^2+1})$$

$$= \sqrt{x^2+1}\sec^2 x + \tan x \cdot \frac{d}{dx}((x^2+1)^{1/2}).$$

To evaluate the last derivative, we apply the chain rule, with "raising to the power 1/2" as the outer function:

$$g'(x) = \sqrt{x^2+1}\sec^2 x + \tan x \cdot \frac{1}{2}(x^2+1)^{-1/2}\cdot\frac{d}{dx}(x^2+1)$$

$$= \sqrt{x^2+1}\sec^2 x + \frac{\tan x}{2\sqrt{x^2+1}}\cdot 2x$$

$$= \sqrt{x^2+1}\sec^2 x + \frac{x\tan x}{\sqrt{x^2+1}}.$$

Finally, the last step in evaluating $h(x)$ is taking the square root, so we use this as the outer function in the chain rule:

$$h'(x) = \frac{d}{dx}((\tan(x^2+1))^{1/2}) = \frac{1}{2}(\tan(x^2+1))^{-1/2}\cdot\frac{d}{dx}(\tan(x^2+1))$$

$$= \frac{1}{2\sqrt{\tan(x^2+1)}}\cdot\frac{d}{dx}(\tan(x^2+1)).$$

For the last derivative, we use the chain rule *again*, with the tangent function as the outer function. Since $\frac{d}{dx}(\tan x) = \sec^2 x$,

$$h'(x) = \frac{1}{2\sqrt{\tan(x^2+1)}}\cdot\sec^2(x^2+1)\cdot\frac{d}{dx}(x^2+1)$$

$$= \frac{\sec^2(x^2+1)}{2\sqrt{\tan(x^2+1)}}\cdot 2x = \frac{x\sec^2(x^2+1)}{\sqrt{\tan(x^2+1)}}. \qquad \square$$

There is another way of writing the chain rule that provides some enlightenment about why the rule is true, and points the way toward a proof. Suppose $y = f(g(x))$.

If we let $u = g(x)$, then we can write

$$y = (f \circ g)(x), \qquad y = f(u), \qquad u = g(x).$$

Therefore

$$\frac{dy}{dx} = (f \circ g)'(x), \qquad \frac{dy}{du} = f'(u), \qquad \frac{du}{dx} = g'(x). \tag{3.2}$$

Now, the chain rule says that

$$(f \circ g)'(x) = f'(g(x)) \cdot g'(x) = f'(u) \cdot g'(x).$$

Rewriting all of these derivatives in the Leibniz notation of equation (3.2), we see that the chain rule can be written

$$\frac{dy}{dx} = \frac{dy}{du} \cdot \frac{du}{dx}. \tag{3.3}$$

For example, consider again the function $h(x) = (x^2 - 5x + 2)^7$ from Example 3.4.2, and let $y = h(x)$. Then we can say that $y = u^7$, where $u = x^2 - 5x + 2$. Thus,

$$\frac{dy}{du} = 7u^6, \qquad \frac{du}{dx} = 2x - 5,$$

and therefore by the chain rule,

$$\frac{dy}{dx} = \frac{dy}{du} \cdot \frac{du}{dx} = 7u^6 \cdot (2x - 5) = 7(x^2 - 5x + 2)^6 \cdot (2x - 5).$$

Thus, we have yet another way of finding the derivative of h.

When written in the form of equation (3.3), the chain rule might seem obvious. But remember that the derivatives in equation (3.3) are not fractions, so we can't prove the chain rule by simply canceling du in the product $dy/du \cdot du/dx$. The chain rule could be thought of as saying that, although these derivatives are not fractions, they cancel as if they were.

It is not surprising that derivatives act in some ways like fractions, because they are close to fractions. If we let Δx stand for a small change in x, then this change will cause a change Δu in u because of the equation $u = g(x)$, and then since $y = f(u)$, this change in u will cause a change Δy in y. This "chain reaction" of changes explains why the term "chain rule" is an appropriate name for Theorem 3.4.1. The derivatives in equation (3.3) are close to the fractions $\Delta y/\Delta x$, $\Delta y/\Delta u$, and $\Delta u/\Delta x$, and since these really are fractions, we are justified in saying that

$$\frac{\Delta y}{\Delta x} = \frac{\Delta y}{\Delta u} \cdot \frac{\Delta u}{\Delta x}. \tag{3.4}$$

A natural approach to proving the chain rule would be to take the limit of both sides of equation (3.4) as Δx approaches 0 in the hope that this would lead to equation (3.3).

Unfortunately, there is a small flaw in this approach: even if $\Delta x \neq 0$, it is possible that $\Delta u = 0$. In that case, the fraction $\Delta y/\Delta u$ in equation (3.4) would be undefined. Thus, to turn our idea into a correct proof we must modify equation (3.4) to avoid dividing by Δu. In the proof we will give below, we will define a function r, and we will replace the fraction $\Delta y / \Delta u$ in equation (3.4) with $r(\Delta u)$. We will define r so that if $\Delta u \neq 0$ then $r(\Delta u) = \Delta y / \Delta u$. But unlike the expression $\Delta y / \Delta u$, $r(\Delta u)$ will be defined even if $\Delta u = 0$.

Proof of Theorem 3.4.1. Suppose that g is differentiable at some number x, and f is differentiable at $g(x)$. Let $u = g(x)$. Then f is differentiable at u, so $f'(u)$ defined. We now define a function r as follows:

$$r(h) = \begin{cases} \dfrac{f(u+h) - f(u)}{h}, & \text{if } h \neq 0, \\[2mm] f'(u), & \text{if } h = 0. \end{cases}$$

By the definition of derivative, we have

$$\lim_{h \to 0} r(h) = \lim_{h \to 0} \frac{f(u+h) - f(u)}{h} = f'(u) = r(0).$$

In other words, r is continuous at 0.

If $h \neq 0$ then $r(h) = (f(u+h) - f(u))/h$, and therefore

$$f(u+h) - f(u) = r(h) \cdot h. \tag{3.5}$$

But notice that even if $h = 0$, equation (3.5) is true, since in that case both sides of the equation are 0. Therefore equation (3.5) holds for all values of h.

Now let $\Delta x \neq 0$ represent a small change in x. As described before, this leads to corresponding changes in u and y:

$$\Delta u = g(x + \Delta x) - g(x),$$
$$\Delta y = f(u + \Delta u) - f(u) = f(g(x + \Delta x)) - f(g(x)).$$

We now rewrite Δy by plugging in Δu for h in equation (3.5):

$$\Delta y = f(u + \Delta u) - f(u) = r(\Delta u) \cdot \Delta u.$$

Dividing by Δx, we get

$$\frac{\Delta y}{\Delta x} = r(\Delta u) \cdot \frac{\Delta u}{\Delta x}.$$

Notice that this is exactly the same as equation (3.4), except that $\Delta y / \Delta u$ has been replaced with $r(\Delta u)$, as suggested earlier.

We want to see what happens now as Δx approaches 0. Since g is differentiable at x, it is continuous at x, and therefore

$$\lim_{\Delta x \to 0} \Delta u = \lim_{\Delta x \to 0} (g(x + \Delta x) - g(x)) = g(x) - g(x) = 0.$$

But then since r is continuous at 0, it follows that

$$\lim_{\Delta x \to 0} r(\Delta u) = r(0) = f'(u) = \frac{dy}{du}.$$

Thus,

$$\frac{dy}{dx} = \lim_{\Delta x \to 0} \frac{\Delta y}{\Delta x} = \lim_{\Delta x \to 0} \left(r(\Delta u) \cdot \frac{\Delta u}{\Delta x} \right) = \frac{dy}{du} \cdot \frac{du}{dx}.$$

In other words,

$$(f \circ g)'(x) = f'(g(x)) \cdot g'(x). \qquad \Box$$

One important lesson of the chain rule, when it is written in the form of equation (3.3), is that dy/dx is not the same as dy/du. The derivative dy/dx gives the rate of change of y with respect to x, which tells us how much y changes per unit of change of x. On the other hand, dy/du is equal to the rate of change of y with respect to u, which tells us how much y changes per unit of change of u. These are not the same thing; according to the chain rule, they are related by the equation $dy/dx = dy/du \cdot du/dx$. Thus, it would be ambiguous to talk about "the derivative of y"; we always need to specify: derivative with respect to which other variable?

In the last section, we saw that the power rule works if the exponent is a positive integer, a negative integer, or a fraction of the form $1/n$ for an integer $n \geq 2$. Using the chain rule, we can see that it works for any nonzero rational exponent.

Theorem 3.4.4. *Suppose $f(x) = x^r$, where r is a nonzero rational number. Then $f'(x) = rx^{r-1}$.*

Since r is rational and $r \neq 0$, we can write $r = m/n$, where n is a positive integer, m is a nonzero integer, and the fraction m/n is reduced to lowest terms. If n is even, then the domain of f' is $(0, \infty)$. If n is odd and $r < 1$, then the domain of f' is $(-\infty, 0) \cup (0, \infty)$. And if n is odd and $r \geq 1$ then the domain of f' is $(-\infty, \infty)$.

Proof. Since we have already proven the power rule for positive and negative integer exponents, we may as well assume $n \geq 2$. By definition,

$$f(x) = x^r = x^{m/n} = (\sqrt[n]{x})^m.$$

We have already shown that if n is even then $\sqrt[n]{x}$ is differentiable when $x > 0$, and if n is odd then $\sqrt[n]{x}$ is differentiable for all $x \neq 0$. For these values of x, we can therefore use the chain rule to find $f'(x)$:

$$f'(x) = m(\sqrt[n]{x})^{m-1} \cdot \frac{d}{dx}(\sqrt[n]{x}) = mx^{(m-1)/n} \cdot \frac{d}{dx}(x^{1/n})$$

$$= mx^{m/n-1/n} \cdot \frac{1}{n}x^{1/n-1} = \frac{m}{n}x^{m/n-1/n+1/n-1} = \frac{m}{n}x^{m/n-1} = rx^{r-1}.$$

We leave it to you to check (see Exercise 24) that if n is odd and $r > 1$ then $f'(0)$ is also defined, and

$$f'(0) = 0 = r \cdot 0^{r-1}. \qquad \Box$$

You might wonder at this point whether the power rule works for irrational exponents as well. For example, if $f(x) = x^\pi$, is it the case that $f'(x) = \pi x^{\pi-1}$? We answer this question with another question: What does x^π mean? We know that for a rational number m/n, $x^{m/n}$ means $(\sqrt[n]{x})^m$. But π is an irrational number, so this definition doesn't apply to x^π. In fact, one has to use methods of calculus to even say what x^π means. Later in this book (see Section 7.4) we will define x^r for irrational numbers r, and then we will be able to find the derivatives of these functions. (And if you can't wait to find out the answer: yes, it will turn out that even if r is irrational, $\frac{d}{dx}(x^r) = rx^{r-1}$.)

Example 3.4.5. Find the derivatives of the following functions:

$$f(x) = \frac{(3x+2)^{5/3}}{7}, \qquad g(x) = \frac{7}{(3x+2)^{5/3}}.$$

Solution. You might be tempted to use the quotient rule to compute these derivatives. While that wouldn't be wrong, there is an easier approach. For the function f, it is easier to rewrite "division by 7" as "multiplication by 1/7":

$$f'(x) = \frac{d}{dx}\left(\frac{1}{7} \cdot (3x+2)^{5/3}\right) = \frac{1}{7} \cdot \frac{d}{dx}((3x+2)^{5/3})$$

$$= \frac{1}{7} \cdot \frac{5}{3}(3x+2)^{2/3} \cdot \frac{d}{dx}(3x+2) = \frac{5(3x+2)^{2/3}}{7 \cdot 3} \cdot 3 = \frac{5(3x+2)^{2/3}}{7}.$$

We can also eliminate division from the formula for $g(x)$, by using a negative exponent:

$$g'(x) = \frac{d}{dx}(7 \cdot (3x+2)^{-5/3}) = 7 \cdot (-5/3)(3x+2)^{-8/3} \cdot \frac{d}{dx}(3x+2)$$

$$= -\frac{7 \cdot 5}{3(3x+2)^{8/3}} \cdot 3 = -\frac{35}{(3x+2)^{8/3}}. \qquad \Box$$

Notice that if f is a function, then its derivative f' is also a function. It therefore makes sense to differentiate f'. As usual, we put a prime after f' to denote its derivative. In other words, the derivative of f' is written f''; it is called the *second derivative* of f. Similarly, the derivative of f'' is called the *third derivative* of f, and it is denoted f'''. It is convenient to have a notation that doesn't require counting a long string of primes, so another notation for the third derivative is $f^{(3)}$. Similarly, if we differentiate f n times, then we get the *nth derivative* of f, which is denoted $f^{(n)}$. These are called *higher-order derivatives* of f. For example, if $f(x) = x^3$ then

$$f'(x) = 3x^2, \quad f''(x) = 6x, \quad f'''(x) = f^{(3)}(x) = 6, \quad f^{(4)}(x) = 0, \quad f^{(5)}(x) = 0, \quad \dots.$$

There is also a version of Leibniz notation for higher-order derivatives. If $y = f(x)$, then $dy/dx = f'(x)$. Therefore

$$f''(x) = \frac{d}{dx}(f'(x)) = \frac{d}{dx}\left(\frac{dy}{dx}\right).$$

Reading across the "numerator" and "denominator" in the last expression, we see two d's and a y in the numerator, and dx twice in the denominator. This suggests the following notation for the second derivative:

$$f''(x) = \frac{d}{dx}\left(\frac{dy}{dx}\right) = \frac{d^2 y}{dx^2}.$$

Similarly, we write $d^3 y/dx^3$ for the third derivative of y with respect to x, and, in general, $d^n y/dx^n$ for the nth derivative. As usual, we can also replace y with $f(x)$ in Leibniz notation and write

$$f^{(n)}(x) = \frac{d^n y}{dx^n} = \frac{d^n}{dx^n}(f(x)).$$

Rewriting our last example in Leibniz notation, we have

$$\frac{d}{dx}(x^3) = 3x^2, \quad \frac{d^2}{dx^2}(x^3) = 6x, \quad \frac{d^3}{dx^3}(x^3) = 6, \quad \frac{d^4}{dx^4}(x^3) = 0, \quad \frac{d^5}{dx^5}(x^3) = 0, \quad \dots.$$

Since derivatives give instantaneous rates of change, $d^2 y/dx^2$ can be interpreted as the rate of change of dy/dx with respect to x. For example, consider again our discussion in Section 2.1 of a ball thrown up in the air. Recall that the height of the ball, in feet, after t seconds was given by the formula

$$y = h(t) = 160t - 16t^2.$$

The derivative of y with respect to t gives the velocity of the ball, in ft/sec:

$$v = \frac{dy}{dt} = h'(t) = 160 - 32t.$$

For example, we computed before that after 3 seconds, the velocity is

$$\left.\frac{dy}{dt}\right|_{t=3} = h'(3) = 64 \text{ ft/sec}.$$

The second derivative of y with respect to t is

$$\frac{dv}{dt} = \frac{d^2 y}{dt^2} = h''(t) = -32.$$

This is the rate of change of velocity with respect to time, which is called the *acceleration* of the ball. It tells us how fast the velocity of the ball is changing. Since v is measured in ft/sec, dv/dt is measured in (ft/sec)/sec, which can also be written ft/sec^2. The acceleration of the ball is a constant -32 ft/sec^2, which means that the velocity of the ball is decreasing by 32 ft/sec every second. Of course, this is the result of gravity pulling the ball down. We will see other uses of higher-order derivatives later in this book.

We now know how to differentiate any function built up from constants, rational powers of x, and trigonometric functions by addition, subtraction, multiplication, division, and composition. This is a very wide range of functions, including many functions that come up in applications. We work one more example to illustrate how the rules we have learned allow us to compute complicated derivatives.

Example 3.4.6. Let $f(x) = \sqrt{\sin x}$. Find $f'''(x)$.

Solution. It is easiest to use fractional exponents throughout the calculation:

$$f'(x) = \frac{d}{dx}((\sin x)^{1/2}) = \frac{1}{2}(\sin x)^{-1/2} \cdot \cos x,$$

$$f''(x) = \frac{1}{2}(\sin x)^{-1/2} \cdot \frac{d}{dx}(\cos x) + \cos x \cdot \frac{d}{dx}\left(\frac{1}{2}(\sin x)^{-1/2}\right)$$

$$= \frac{1}{2}(\sin x)^{-1/2} \cdot (-\sin x) + \cos x \cdot \left(-\frac{1}{4}(\sin x)^{-3/2} \cdot \cos x\right)$$

$$= -\frac{1}{2}(\sin x)^{1/2} - \frac{1}{4}(\sin x)^{-3/2} \cos^2 x,$$

$$f'''(x) = -\frac{1}{4}(\sin x)^{-1/2} \cos x$$

$$- \left[\frac{1}{4}(\sin x)^{-3/2} \cdot \frac{d}{dx}(\cos^2 x) + \cos^2 x \cdot \frac{d}{dx}\left(\frac{1}{4}(\sin x)^{-3/2}\right)\right]$$

$$= -\frac{1}{4}(\sin x)^{-1/2} \cos x$$

$$- \left[\frac{1}{4}(\sin x)^{-3/2} \cdot 2\cos x \cdot (-\sin x) + \cos^2 x \cdot \left(-\frac{3}{8}(\sin x)^{-5/2} \cdot \cos x\right)\right]$$

$$= -\frac{1}{4}(\sin x)^{-1/2} \cos x - \left[-\frac{1}{2}(\sin x)^{-1/2} \cos x - \frac{3}{8}(\sin x)^{-5/2} \cos^3 x\right]$$

$$= \frac{1}{4}(\sin x)^{-1/2} \cos x + \frac{3}{8}(\sin x)^{-5/2} \cos^3 x$$

$$= \frac{\cos x}{4\sqrt{\sin x}} + \frac{3\cos^3 x}{8(\sin x)^{5/2}}. \qquad \square$$

Exercises 3.4

1–16: Find the derivative of the function.

1. $f(x) = (5x - 8)^{12}$.

2. $g(x) = \sin(x^2 + 7x - 9)$.

3. $h(x) = \sqrt{x^2 + 4}$.

4. $f(x) = x^3 \tan(x^2 + 3x)$.

5. $h(x) = \dfrac{2}{\sqrt[3]{3x+4}}$.

6. $g(x) = \sec^3(x^2 - 5x)$. (Hint: Recall that $\sec^3(x^2 - 5x) = (\sec(x^2 - 5x))^3$.)

7. $f(x) = \cos^2(3x)$.

8. $g(x) = \sqrt{\cos^2(3x) + 1}$.

9. $h(x) = x\sqrt[4]{2x+3}$.

10. $f(x) = x^2\sqrt[3]{3x - 5}$.

11. $g(x) = \sin(\cos(\tan(3x)))$.

12. $h(x) = \sqrt{5 + 4\sqrt{3 + 2\sqrt{1+x}}}$.

13. $f(x) = \left(\dfrac{2x+5}{2-5x}\right)^7$.

14. $g(x) = \dfrac{\cos(5x)}{\sin(3x)}$.

15. $f(x) = \sqrt{\tan x}$.

16. $g(x) = \sqrt{\dfrac{x^2 - 1}{x^2 + 1}}$.

17. Suppose $g(x) = f(x^2 - 3x + 1)$. If $f'(1) = 3$ and $f''(1) = -1$, find $g'(3)$ and $g''(3)$.

18. Suppose the line $y = 5x - 3$ is tangent to the curve $y = f(x)$ at the point $(3, 12)$. Let g be defined as follows:

$$g(x) = \frac{f(x^2 - 1)}{x}.$$

Find the equation of the line tangent to the curve $y = g(x)$ at the point $(2, g(2))$. (Hint: First compute $g(2)$ and $g'(2)$.)

19. Let $f(x) = (3x - 2)^7$. Find $f'''(x)$.

20. Let $g(x) = \sec(3x + 1)$. Find $g'''(x)$.

21. Let $h(x) = \sin(2x)$. Find $h^{(100)}(x)$. (Hint: Look for a pattern in the higher-order derivatives.)

22. An object moves along a number line (marked off in meters) in such a way that its position at time t (measured in seconds) is $y = \sin(\pi t^2)$. Find the velocity and acceleration of the object at time $t = 1$.

23. According to the double-angle formula for the sine function, for all x, $\sin(2x) = 2 \sin x \cos x$. By differentiating both sides of this equation, derive the double-angle formula for the cosine function.

24. Suppose $f(x) = x^{m/n}$, where m and n are positive integers, $m > n \geq 3$, n is odd, and the fraction m/n is reduced to lowest terms. Show that $f'(0) = 0$. (This completes the proof of Theorem 3.4.4.)

3.5   Implicit Differentiation

Suppose you remember that $\sqrt[3]{x}$ is differentiable for all $x \neq 0$, but you don't remember what the derivative is. Here's a clever way you could work out the derivative. We start with the fact that, by definition, $\sqrt[3]{x}$ is a number whose cube is x:

$$(\sqrt[3]{x})^3 = x.$$

Now take the derivative with respect to x of both sides of this equation, using the chain rule on the left:

$$\frac{d}{dx}((\sqrt[3]{x})^3) = \frac{d}{dx}(x),$$

$$3(\sqrt[3]{x})^2 \cdot \frac{d}{dx}(\sqrt[3]{x}) = 1.$$

(Notice that we need to know that $\sqrt[3]{x}$ is differentiable to justify the use of the chain rule here. The equation is therefore correct for all $x \neq 0$, but not for $x = 0$.) We can now solve this equation for $\frac{d}{dx}(\sqrt[3]{x})$:

$$\frac{d}{dx}(\sqrt[3]{x}) = \frac{1}{3(\sqrt[3]{x})^2}.$$

Of course, this agrees with our previous answer for this derivative.

Let's see what this calculation looks like if we let $y = \sqrt[3]{x}$. On the left below we have reproduced the steps in the calculation above, and on the right we have rewritten each step, substituting y for $\sqrt[3]{x}$:

$$(\sqrt[3]{x})^3 = x, \qquad\qquad\qquad y^3 = x,$$

$$\frac{d}{dx}((\sqrt[3]{x})^3) = \frac{d}{dx}(x), \qquad\qquad \frac{d}{dx}(y^3) = \frac{d}{dx}(x),$$

$$3(\sqrt[3]{x})^2 \cdot \frac{d}{dx}(\sqrt[3]{x}) = 1, \qquad\qquad 3y^2 \cdot \frac{dy}{dx} = 1,$$

$$\frac{d}{dx}(\sqrt[3]{x}) = \frac{1}{3(\sqrt[3]{x})^2}. \qquad\qquad \frac{dy}{dx} = \frac{1}{3y^2} = \frac{1}{3(\sqrt[3]{x})^2}.$$

One step on the right may look strange to you: Why did $\frac{d}{dx}(y^3)$ get rewritten as $3y^2 \cdot \frac{dy}{dx}$, rather than just $3y^2$? The reason is that y stands for the expression $\sqrt[3]{x}$, and, as the corresponding step on the left shows, to find the derivative of $(\sqrt[3]{x})^3$ you need to use the chain rule.

The lesson here is that in general there is a difference between $\frac{d}{dx}(f(x))$ and $\frac{d}{dx}(f(y))$. Of course,

$$\frac{d}{dx}(f(x)) = f'(x).$$

But $\frac{d}{dx}(f(y))$ only makes sense in a context in which y stands for some expression $g(x)$, and in that case you must use the chain rule to compute the derivative:

$$\frac{d}{dx}(f(g(x))) = f'(g(x)) \cdot \frac{d}{dx}(g(x)), \quad \text{so}$$

$$\frac{d}{dx}(f(y)) = f'(y) \cdot \frac{dy}{dx}.$$

Here's another example of this method of finding derivatives. The unit circle $x^2 + y^2 = 1$ passes through the point $P = (3/5, -4/5)$. What is the equation of the line tangent to the circle at the point P? The unit circle and the tangent line at P are shown in Figure 3.11.

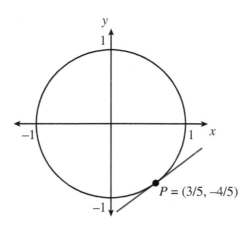

Figure 3.11: The unit circle (black) and the line tangent to the circle at $P = (3/5, -4/5)$ (red).

One way to answer this question is to solve for y in the equation of the circle:

$$y^2 = 1 - x^2,$$

$$y = \pm\sqrt{1 - x^2}.$$

Thus, the unit circle can be thought of as comprising the graphs of two functions,

$$f_t(x) = \sqrt{1 - x^2}, \qquad f_b(x) = -\sqrt{1 - x^2}. \tag{3.6}$$

The graph of f_t is the top half of the circle, and the graph of f_b is the bottom half. Both functions have domain $[-1, 1]$, and they are differentiable on the interval $(-1, 1)$.

The point P is on the bottom half of the circle, so we can find the slope of the tangent line by computing $f_b'(3/5)$, using the formula for $f_b(x)$ in (3.6). You are asked to do this calculation in Exercise 1.

But there is an easier way. If we let $y = f_b(x)$, then there is a way to find $dy/dx = f_b'(x)$ without having to use the formula for $f_b(x)$. For every $x \in (-1, 1)$, the point (x, y) is on the bottom half of the unit circle, and therefore it satisfies the equation $x^2 + y^2 = 1$. We can now find dy/dx by differentiating both sides of this equation:

$$x^2 + y^2 = 1,$$

$$\frac{d}{dx}(x^2 + y^2) = \frac{d}{dx}(1),$$

$$2x + 2y \cdot \frac{dy}{dx} = 0,$$

$$\frac{dy}{dx} = -\frac{2x}{2y} = -\frac{x}{y}.$$

In other words, $f_b'(x) = -x/f_b(x)$. Thus, the slope of the line tangent to the circle at P is

$$f_b'(3/5) = \left. \frac{dy}{dx} \right|_{(x,y)=(3/5,-4/5)} = -\frac{3/5}{-4/5} = \frac{3}{4}. \tag{3.7}$$

The equation of the tangent line is therefore $y + 4/5 = (3/4)(x - 3/5)$, or equivalently

$$y = \frac{3}{4}x - \frac{5}{4}.$$

The functions f_t and f_b are defined explicitly in (3.6). We say that these functions are defined *implicitly* by the equation $x^2 + y^2 = 1$. The method we used to find $dy/dx = f_b'(x)$, in which we differentiated both sides of this equation rather than using the explicit formula for $f_b(x)$, is therefore called *implicit differentiation*.

Let's give one more example of implicit differentiation. Consider the graph of the equation

$$y^3 - 4xy = x^3 - 1, \tag{3.8}$$

which is the curve shown in Figure 3.12 (part of it is black and part is blue). It is easy to verify that this curve passes through the point $(1, 2)$. We will find the equation of the line tangent to the curve at this point (shown in red in Figure 3.12).

Unfortunately, our curve is not the graph of a function, because it fails the vertical line test. For example, the vertical line $x = 1$ crosses the curve three times, at the points $(1, 2)$, $(1, 0)$, and $(1, -2)$. However, as with the unit circle, it is possible to break the curve into pieces, each of which is the graph of a function. For example, the part of the curve that is colored blue in Figure 3.12 passes the vertical line test, so it is the graph of some function f. And since the point $(1, 2)$ lies on this part of the curve, $f(1) = 2$. It can be shown, by methods that go beyond the scope of this book, that f is differentiable at 1. We won't worry about this detail here; we'll simply assume that f is differentiable at 1 and find the value of $f'(1)$. This will tell us the slope of the tangent line at $(1, 2)$.

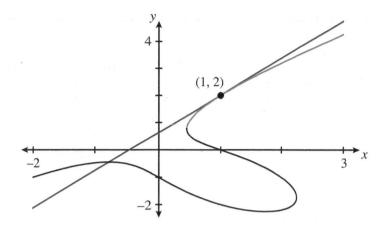

Figure 3.12: The graph of $y^3 - 4xy = x^3 - 1$ (black and blue) and the line tangent to this curve at $(1, 2)$ (red).

One way to proceed would be to solve for y in equation (3.8) in order to find a formula for $f(x)$. But this equation is very hard to solve, and the formula for $f(x)$ would be very complicated. A better idea is to use the fact that if $y = f(x)$, then x and y satisfy equation (3.8). We can therefore use implicit differentiation to find $f'(x) = dy/dx$:[4]

$$y^3 - 4xy = x^3 - 1,$$

$$\frac{d}{dx}(y^3 - 4xy) = \frac{d}{dx}(x^3 - 1),$$

$$3y^2 \cdot \frac{dy}{dx} - \left(4x \cdot \frac{dy}{dx} + y \cdot \frac{d}{dx}(4x)\right) = 3x^2,$$

$$(3y^2 - 4x)\frac{dy}{dx} - 4y = 3x^2,$$

$$\frac{dy}{dx} = \frac{3x^2 + 4y}{3y^2 - 4x}.$$

The slope of the tangent line is therefore

$$f'(1) = \frac{dy}{dx}\bigg|_{(x,y)=(1,2)} = \frac{3(1^2) + 4(2)}{3(2^2) - 4(1)} = \frac{11}{8},$$

and the equation of the tangent line is $y - 2 = (11/8)(x - 1)$, or

$$y = \frac{11}{8}x + \frac{5}{8}.$$

---

[4]In the last step of our computation of dy/dx, we divide by $3y^2 - 4x$. This step would not be correct at a point on the curve where $3y^2 - 4x = 0$, and our formula for dy/dx would be undefined at such a point. However, as long as we use our formula only at points where $3y^2 - 4x \neq 0$, the formula will give the correct value for dy/dx. Similar restrictions apply to our other derivative calculations in this section.

The key step in all of our examples of implicit differentiation was to take the derivative of both sides of an equation. It is worthwhile to pause, therefore, to discuss when it is correct to do this.

In Figure 3.13a, the blue and red curves are the graphs of two functions, f and g. The curves cross at $x = a$, so $f(a) = g(a)$. But the tangent lines to the curves at the crossing point have different slopes, so $f'(a) \neq g'(a)$. This illustrates that it is possible to have $f(a) = g(a)$, but $f'(a) \neq g'(a)$. Does that mean that if you take the derivative of both sides of a true equation, you can end up with a false equation?

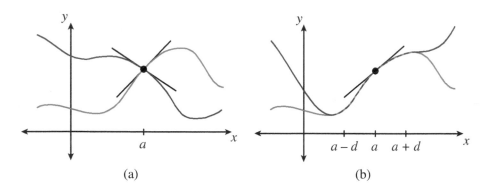

(a)   (b)

Figure 3.13: In both figures, the blue and red curves are the graphs of $y = f(x)$ and $y = g(x)$, and tangent lines are drawn in black. In (a), the tangent lines to the two curves at $x = a$ have different slopes, so $f'(a) \neq g'(a)$. In (b), the curves have the same tangent line at $x = a$, so $f'(a) = g'(a)$.

Let's look more closely at how it can happen that $f(a) = g(a)$ but $f'(a) \neq g'(a)$. According to the definition of derivative,

$$f'(a) = \lim_{h \to 0} \frac{f(a+h) - f(a)}{h}.$$

Thus, the value of $f'(a)$ depends not only on $f(a)$, but also on $f(a+h)$ for h close to 0. In Figure 3.13a, we have $f(a) = g(a)$, but it is not the case that $f(a+h) = g(a+h)$ for h close to 0. This explains why $f'(a) \neq g'(a)$.

Now consider the functions f and g whose graphs are shown in Figure 3.13b. In this case, not only do $f(x)$ and $g(x)$ agree at $x = a$, they agree at all values of x in the interval $(a - d, a + d)$, for some number $d > 0$. It follows that $f(a+h) = g(a+h)$ for h close to 0 (more precisely, for $|h| < d$), and therefore $f'(a) = g'(a)$. You can see this in the figure, because the tangent lines to the two curves at $x = a$ are the same. This motivates our next theorem.

Theorem 3.5.1. *Suppose that I is an open interval, and for every number x in the interval I, $f(x) = g(x)$. Then for every number x in I, $f'(x) = g'(x)$, where we interpret this equation to mean that either both derivatives are defined and they are equal, or both are undefined.*

Proof. Consider any number x in the interval I. Since I is an open interval, it does not include its endpoints, so x is not an endpoint of the interval. Therefore for h close to 0, $x + h$ is also in I. It follows that we have not only $f(x) = g(x)$, but also $f(x + h) = g(x + h)$ for all h sufficiently close to 0. Therefore

$$f'(x) = \lim_{h \to 0} \frac{f(x+h) - f(x)}{h} = \lim_{h \to 0} \frac{g(x+h) - g(x)}{h} = g'(x). \qquad \square$$

Thus, if you have an equation between two expressions involving an independent variable x, then it may be correct to differentiate both sides of the equation with respect to x, but you have to be careful about it. If the equation holds at just a single value of x, then you can't take the derivative of both sides of the equation and expect the resulting equation to be true. But if the equation holds for all values of x in some open interval, then you can differentiate both sides and the resulting equation will be true on the same interval.

To put it more informally, if an equation says that two expressions involving an independent variable x define the same function on some open interval, then you can take the derivative of both sides. But if it merely says that the two expressions have the same numerical value at a particular value of x, then you can't. This is a reflection of the fact that differentiation is an operation on *functions*, not *numbers*.

We'd better look back at the examples in which we differentiated both sides of an equation and make sure we did it right! Consider, for example, the unit circle $x^2 + y^2 = 1$, and the function f_b whose graph is the bottom half of this circle. With $y = f_b(x)$, the equation $x^2 + y^2 = 1$ is true for all x in the open interval $(-1, 1)$. It is therefore correct to differentiate both sides of the equation, and the resulting equation will also be true for all x in the interval $(-1, 1)$. We leave it to you to check that in our other examples, when we differentiated both sides of an equation, the equation was true throughout some open interval.

It is also possible to find higher-order derivatives of functions that have been defined implicitly. We give one example illustrating this.

Example 3.5.2. It can be shown that the curve $3y - \cos y = 2 \sin x$, which is shown in Figure 3.14, is the graph of a differentiable function f. Show that $f(7\pi/6) = 0$, and find $f''(7\pi/6)$.

Solution. If $x = 7\pi/6$ and $y = 0$ then

$$3y - \cos y = -1 = 2 \sin x,$$

so the curve passes through the point $(7\pi/6, 0)$. Since the curve is the graph of f, this means that $f(7\pi/6) = 0$.

If we let $y = f(x)$, then for every number x, $3y - \cos y = 2 \sin x$. We can therefore use implicit differentiation to find $f'(x) = dy/dx$:

$$3y - \cos y = 2 \sin x,$$

$$\frac{d}{dx}(3y - \cos y) = \frac{d}{dx}(2 \sin x),$$

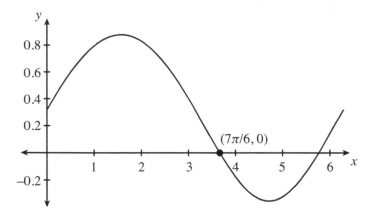

Figure 3.14: The graph of the equation $3y - \cos y = 2 \sin x$.

$$3 \cdot \frac{dy}{dx} + \sin y \cdot \frac{dy}{dx} = 2 \cos x,$$

$$\frac{dy}{dx} = \frac{2 \cos x}{3 + \sin y}.$$

To find $f''(x) = d^2y/dx^2$, we just differentiate dy/dx. In the course of this calculation, the term dy/dx shows up. When it does, we simply fill in the formula for it that we have already computed:

$$\frac{d^2y}{dx^2} = \frac{d}{dx}\left(\frac{dy}{dx}\right) = \frac{d}{dx}\left(\frac{2 \cos x}{3 + \sin y}\right)$$

$$= \frac{(3 + \sin y) \cdot \frac{d}{dx}(2 \cos x) - 2 \cos x \cdot \frac{d}{dx}(3 + \sin y)}{(3 + \sin y)^2}$$

$$= \frac{(3 + \sin y)(-2 \sin x) - 2 \cos x(\cos y \cdot \frac{dy}{dx})}{(3 + \sin y)^2}$$

$$= \frac{-2 \sin x(3 + \sin y) - 2 \cos x \cos y \cdot \frac{2 \cos x}{3 + \sin y}}{(3 + \sin y)^2} \cdot \frac{3 + \sin y}{3 + \sin y}$$

$$= \frac{-2 \sin x(3 + \sin y)^2 - 4 \cos^2 x \cos y}{(3 + \sin y)^3}.$$

Finally, we evaluate at the requested point:

$$f''(7\pi/6) = \left.\frac{d^2y}{dx^2}\right|_{(x,y)=(7\pi/6,0)} = \frac{-2(-1/2)(3+0)^2 - 4(-\sqrt{3}/2)^2(1)}{(3+0)^3} = \frac{2}{9}. \quad \square$$

Exercises 3.5

1. Use the formula for $f_b(x)$ given in (3.6) to find $f'_b(3/5)$, and verify that the answer is the same as the one we found in equation (3.7).

2–11: Find dy/dx. You may assume that y is a differentiable function of x.

2. $x^3 + y^3 = 1$.

3. $x^{1/3} + y^{1/3} = 1$.

4. $x^3 y - y^3 x = 3$.

5. $xy^2 + \sqrt{x+y} = 5$.

6. $x = \sin(xy)$.

7. $y = \sin(xy)$.

8. $y = \cos(x/y)$.

9. $xy^2 = \sin(3x - y)$.

10. $y \tan(x - y) = x$.

11. $y\sqrt{xy + 3} = 2x$.

12–16: Find $d^2 y/dx^2$. You may assume that y is a differentiable function of x.

12. $y = x^2 + \sin y$.

13. $y^2 = x - y$.

14. $y = \sqrt{xy + 1}$.

15. $x^{2/3} + y^{2/3} = 1$.

16. $xy^2 = x + y$.

17. Suppose that $y = f(x)$, for some differentiable function f, and
$$\frac{dy}{dx} = \frac{y^2 + 3}{2x}.$$
Find $d^2 y/dx^2$. Simplify your answer as much as possible.

18. Find the equation of the line tangent to the curve $\sin(xy) = \cos(x + y)$ at the point $(\pi/2, 0)$. (You may assume that this point lies on a part of the curve that is the graph of a differentiable function.)

19. Suppose that the curve $\cos(x^2 y) + 3y = c$ (where c is a constant) passes through the point $(2, \pi/8)$.

    (a) Find c.

    (b) Find the equation of the line tangent to the curve at the point $(2, \pi/8)$. (You may assume that this point lies on a part of the curve that is the graph of a differentiable function.)

20. Let f be the function in Example 3.5.2. The equation that implicitly defines f can be rewritten
$$y = \frac{2 \sin x + \cos y}{3}.$$

In this problem we will compute $f(\pi/6)$. The same method could be used to compute $f(x)$ for any value of x.

Let
$$a_1 = \frac{2 \sin(\pi/6)}{3} = \frac{1}{3},$$

and for every positive integer n let
$$a_{n+1} = \frac{2 \sin(\pi/6) + \cos a_n}{3} = \frac{1 + \cos a_n}{3}.$$

In other words,
$$a_1 = \frac{1}{3},$$
$$a_2 = \frac{1 + \cos a_1}{3} = \frac{1 + \cos(1/3)}{3},$$
$$a_3 = \frac{1 + \cos a_2}{3} = \frac{1 + \cos\left(\frac{1 + \cos(1/3)}{3}\right)}{3},$$
$$\vdots$$

(a) Prove that for every positive integer n, $|a_n - f(\pi/6)| \le 1/3^n$. (Hint: Use mathematical induction, and in the induction step, use Lemma 2.7.9.)

(b) Prove that $\lim_{n \to \infty} a_n = f(\pi/6)$.

(c) For what value of n can we be sure that a_n differs from $f(\pi/6)$ by at most 0.005? Use a calculator or computer to find a_n for that value of n.

Chapter 4

Applications of Differentiation

4.1 Related Rates

Imagine that a leaking oil well is creating a circular oil slick on the surface of the ocean. As the oil continues to leak, the area of the oil slick is increasing at a steady rate of 1000 square meters per hour. How fast is the radius of the slick increasing when the radius is 500 meters?

We will be solving word problems in this section, so it might be helpful to begin with some general advice about how to approach word problems. The most important thing to understand about word problems is that there are two phases to the solution of any word problem. In the first phase, you must translate the English of the problem into mathematical language. This turns the word problem into a mathematical problem. In the second phase, you solve this mathematical problem. The most common mistake students make when solving word problems is to try to begin phase two of the problem before they have completed phase one. When you are starting to work on a word problem, you shouldn't even be *thinking* about *solving* the problem. Your first task is simply to understand what the problem is asking, and to do this you must translate the English into mathematical language. It is only when you have translated the entire problem into mathematical language that you should begin to think about solving the problem.

Turning now to the problem in the first paragraph of this section, let's translate the problem into mathematical language. We begin by identifying the quantities relevant to the problem and introducing notation for those quantities. Clearly the problem concerns the area and radius of the oil slick, so we will need letters to stand for these quantities. A natural choice is to let A stand for the area, and r the radius. But our notation must be precise, so we need to specify the units in which these quantities are measured. Looking back at the problem, we see that it will be most convenient to measure the radius in meters and the area in square meters, so we define:

$$r = \text{radius of oil slick, in m,}$$

$$A = \text{area of oil slick, in m}^2.$$

But there is a third quantity relevant to this problem. The problem says that r and A are changing over time, and it talks about the rates at which they are changing as time passes. So time is another relevant variable:

$$t = \text{time since beginning of oil spill, in hr.}$$

Since r and A are changing over time, we can think of each of them as being given by some function of t:

$$r = f(t), \qquad A = g(t).$$

This means that r and A are dependent variables in this problem, and t is the independent variable. In a real-world problem like this it is reasonable to assume that f and g are differentiable at all values of t. Their derivatives give the rates at which r and A are changing.

Now that we have our notation fixed, we are ready to begin translating the problem from English into math. We will go through the problem sentence by sentence (or even word by word, if necessary) and translate it into mathematical language.

The first sentence says, "Imagine that a leaking oil well is creating a circular oil slick on the surface of the ocean." The crucial word in this sentence is "circular." Since the problem concerns the area and radius of the oil spill, surely the formula for the area of a circle will be relevant. Let's write it down:

$$A = \pi r^2.$$

We can now move on to the next sentence: "As the oil continues to leak, the area of the oil slick is increasing at a steady rate of 1000 square meters per hour." This sentence tells us how fast the area A is changing; in other words, it is telling us dA/dt, which is measured in m²/hr:

$$\frac{dA}{dt} = 1000.$$

But there is one word in this sentence that we haven't included in our translation: "steady." What does it mean for A to increase "at a steady rate"? It means that the rate of increase is the same at all times. So a more careful statement of the information in this sentence is:

$$\text{For all } t, \quad \frac{dA}{dt} = 1000.$$

Well, maybe not *all* t—we hope the oil leak will be stopped eventually! But the problem is concerned with a certain period of time—a certain open interval of values of t—during which the oil is leaking. When we say "for all t," we mean for all values of t in the interval of time under discussion in the problem.[1]

Now we can move on to the last sentence of the problem: "How fast is the radius of the slick increasing"—translation: What is dr/dt?—"when the radius is 500 meters"—translation: when $r = 500$.

Let's summarize our translation of the problem into mathematical language. Note that some sentences in the problem, such as the description of the steady rate of increase of the area of the oil slick, are about what is true at all times, whereas the last sentence is about a particular time, namely the moment in time when $r = 500$. This distinction will be important in our solution of the problem, so we include it in our summary.

For all values of t:	For a particular value of t:
$A = \pi r^2$	$r = 500$
$\dfrac{dA}{dt} = 1000$	$\dfrac{dr}{dt} = ?$

We have put the equation $A = \pi r^2$ in the first column because as the oil slick grows, it remains circular, so this formula for the area of the slick holds at all times.

Now that we have translated the problem into mathematical language, we are ready to begin thinking about how to solve it. We know dA/dt, and we need to find dr/dt. Can we find a relationship between these rates of change? Well, we have a relationship between A and r, namely $A = \pi r^2$. Taking the derivative of both sides of this equation with respect to t will give us a relationship between dA/dt and dr/dt:

$$A = \pi r^2, \tag{4.1}$$

$$\frac{d}{dt}(A) = \frac{d}{dt}(\pi r^2),$$

$$\frac{dA}{dt} = 2\pi r \frac{dr}{dt}. \tag{4.2}$$

Recall that, as we saw in Section 3.5, if an equation is true for all values of t in some open interval, then you can take the derivative with respect to t of both sides of the equation and the resulting equation will be true for the same range of values of t. Since equation (4.1) is true for all values of t (in the time period under consideration in the problem), equation (4.2) is also true for all values of t.

We can now fill in the given values of dA/dt and r, and solve for dr/dt. Once we fill in $r = 500$, we will no longer have an equation true at all times; the equation will be true only at the moment when $r = 500$:

$$\text{For all } t, \quad \frac{dA}{dt} = 2\pi r \frac{dr}{dt}.$$

---

[1] Why an *open* interval? If the oil leak lasts for 100 hours, then in the equations $r = f(t)$, $A = g(t)$ we might want to think of the domains of the functions f and g as the closed interval $[0, 100]$. But in that case the functions would only be differentiable on the open interval $(0, 100)$, and therefore the discussion of the rates of change of A and r would only make sense on this open interval.

$$\text{When } r = 500, \quad 1000 = 2\pi(500)\frac{dr}{dt},$$

$$\frac{dr}{dt} = \frac{1000}{1000\pi} = \frac{1}{\pi}.$$

Thus, when $r = 500$, the radius is increasing at a rate of $1/\pi \approx 0.32$ m/hr.

The key idea in solving this problem was to find a relationship between the rates of change of A and r. This is therefore called a *related rates* problem. We found this relationship between rates of change by taking the derivative of both sides of the equation $A = \pi r^2$, which gives a relationship between A and r. And to justify this step it was important that the equation $A = \pi r^2$ was true for all values of t. This means that it was important *not* to use the value $r = 500$ until after the differentiation step. You might have been tempted to plug in $r = 500$ in the equation $A = \pi r^2$ right away, to get $A = \pi(500^2) = 250000\pi$. This isn't wrong, but it is only correct at a single moment, not for all t. It would therefore not be correct to differentiate both sides of this equation.

In the rest of this section, we'll give several more examples of related rates problems. In each case, we will take the derivative of both sides of an equation to get a relationship between rates of change. And in each case, it will be crucial to make sure this equation is true for all values of t in some open interval. We will therefore need to make sure we don't use any information that is true at only one time before we do the differentiation step.

Example 4.1.1. A lighthouse is 2 miles west of a shoreline that runs north to south. The beam of the lighthouse rotates counterclockwise at a steady speed of 4 revolutions per minute. How fast is the point where the beam hits the shore moving when it is 3 miles north of the light?

Solution. It is often helpful in word problems to draw a diagram. The situation in this example is illustrated in Figure 4.1. The lighthouse is at the point L, 2 miles due west of the point Q on the shoreline. The beam of the lighthouse hits the shore at the point P. Ultimately, we will be interested in the moment when the distance PQ is 3 miles, but this distance is changing over time, and we will need to work out equations that are true at all times, not just at this particular moment. So in the figure we label this distance x, not 3. The problem says that the beam rotates counterclockwise at 4 revolutions per minute. This sounds like a rate of change, but the rate of change of what quantity?

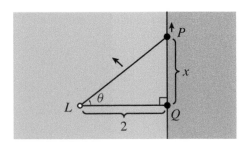

Figure 4.1: Lighthouse and shoreline in Example 4.1.1.

The most natural choice is the angle θ, measured in radians. Since the beam is rotating counterclockwise, θ is increasing, and since one revolution is 2π radians, it is increasing at a steady rate of 8π radians per minute. Thus, if t represents time, measured in minutes, then

$$\text{for all } t, \quad \frac{d\theta}{dt} = 8\pi.$$

The problem asks how fast the point P is moving when $x = 3$. If we think of the shoreline as a vertical number line marked off in miles, with the origin at Q, then x represents the position of P on this number line, and the velocity of P, in miles per minute, is dx/dt. Thus, the question we must answer is:

$$\text{When } x = 3, \quad \frac{dx}{dt} = ?$$

We have now translated the whole problem into mathematical language, so we are ready to start solving it. We know $d\theta/dt$, and we're looking for dx/dt. Imitating the approach we used in the oil slick problem, if we can find an equation relating θ and x and then differentiate both sides of this equation, then we will get a relationship between $d\theta/dt$ and dx/dt. Looking at Figure 4.1, the labeling on the right triangle LPQ suggests the relationship

$$\tan \theta = \frac{x}{2}.$$

Is this relationship true at all times? Well, there are times when the beam of the lighthouse is pointing out to sea and doesn't hit the shore at all, and at such times x isn't even defined. But the relationship does hold not only at the crucial moment when $x = 3$, but also for an interval of time that extends both before and after this crucial moment. Thus, there is an open interval of values of t when this equation is true, and this interval includes the moment we are interested in. We can therefore restrict attention to this interval of values of t and take the derivative of both sides of the equation:

$$\frac{d}{dt}(\tan \theta) = \frac{d}{dt}\left(\frac{x}{2}\right),$$
$$\sec^2 \theta \cdot \frac{d\theta}{dt} = \frac{1}{2} \cdot \frac{dx}{dt}. \tag{4.3}$$

When $x = 3$, the important points L, P, and Q are related as shown in Figure 4.2, where the length of LP comes from the Pythagorean theorem. We can see in this diagram that $\cos \theta = 2/\sqrt{13}$, and therefore $\sec \theta = \sqrt{13}/2$. Filling in all the values we know in equation (4.3), we have:

$$\text{When } x = 3, \quad \left(\frac{\sqrt{13}}{2}\right)^2 \cdot 8\pi = \frac{1}{2} \cdot \frac{dx}{dt},$$
$$\frac{dx}{dt} = 52\pi.$$

Thus, the point P is moving north at $52\pi \approx 163$ miles/min. $\qquad\square$

Figure 4.2: Triangle LPQ when $x = 3$.

Example 4.1.2. Water pours out of a conical funnel at a constant rate of 5 cubic inches per minute. The funnel is 12 inches high and the radius at the top is 4 inches. How fast is the depth of the water decreasing when it is 3 inches deep?

Solution. Figure 4.3a shows the funnel with water pouring out of the bottom. The first sentence of the problem says that the volume of water in the funnel is decreasing at a constant rate of 5 in^3/min. Thus, if we let V stand for the volume of water in the funnel, in cubic inches, and t for time, in minutes, then

$$\text{for all } t, \quad \frac{dV}{dt} = -5.$$

Notice that dV/dt is negative, because the volume is decreasing. The last sentence of the problem asks us to find dh/dt when $h = 3$, where h is the depth of the water, in inches, as shown in the figure. Thus, we need to find an equation relating V to h.

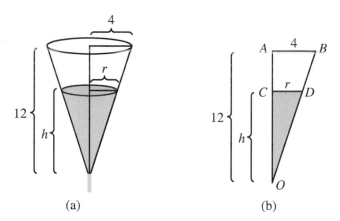

(a)   (b)

Figure 4.3: The funnel of Example 4.1.2 is shown in (a). Figure (b) shows the right half of a vertical cross section.

The water is in the shape of a cone, so we should probably use the formula for the volume of a cone. If we let r be the radius of the circular top of the water, in inches, then

$$\text{for all } t, \quad V = \frac{1}{3}\pi r^2 h. \tag{4.4}$$

This equation relates V to h, but unfortunately it brings in a third variable, r, whose value is also changing over time. If we take the derivative of both sides of equation (4.4), we'll get an equation involving dV/dt, dh/dt, and dr/dt. Since we know neither dh/dt nor dr/dt, we'll have two unknown quantities in this equation, so it seems unlikely that we'll be able to solve for dh/dt. It appears that the inclusion of the extra variable r is a problem.

Fortunately, there is a way to eliminate r from the problem. Figure 4.3b shows the right half of a vertical cross section through the funnel. In this figure, triangles ABO and CDO are similar, and therefore

$$\text{for all } t, \quad \frac{r}{h} = \frac{4}{12} = \frac{1}{3}.$$

Thus $r = h/3$, and substituting this into equation (4.4), we get

$$\text{for all } t, \quad V = \frac{1}{3}\pi \left(\frac{h}{3}\right)^2 h = \frac{\pi h^3}{27}.$$

Now we are ready to differentiate and then plug in all of the given values:

$$\text{For all } t, \quad \frac{d}{dt}(V) = \frac{d}{dt}\left(\frac{\pi h^3}{27}\right),$$

$$\frac{dV}{dt} = \frac{1}{27} \cdot 3\pi h^2 \cdot \frac{dh}{dt} = \frac{\pi h^2}{9} \cdot \frac{dh}{dt}.$$

$$\text{When } h = 3, \quad -5 = \frac{\pi(3^2)}{9} \cdot \frac{dh}{dt},$$

$$\frac{dh}{dt} = -\frac{5}{\pi}.$$

Thus, when $h = 3$, the depth of the water is decreasing at a rate of $5/\pi \approx 1.6$ in/min. □

Example 4.1.3. A garage door is made out of a rigid piece of wood 10 feet high. The top of the door moves along horizontal tracks attached to the ceiling of the garage, which is 10 feet above the floor, and the bottom moves along vertical tracks attached to the sides of the door opening. When the door is being closed, a motor pushes the top of the door along its tracks at a constant speed of 2 ft/sec. How fast is the bottom of the door moving when it is 2 feet above the floor? What is the acceleration of the bottom of the door at that moment?

Solution. Figure 4.4 shows a side view of the garage door while it is being closed. We are told that

$$\text{for all } t, \quad \frac{dx}{dt} = -2,$$

where x is measured in feet and t is time, measured in seconds. Notice that x is decreasing, so dx/dt is negative. To determine how fast the bottom of the door is moving when it is 2 feet above the floor we must find dy/dt when $y = 8$.

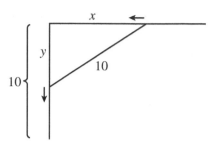

Figure 4.4: Side view of garage door in Example 4.1.3.

We see in Figure 4.4 that x and y are the legs of a right triangle with hypotenuse 10, so

$$\text{for all } t, \quad x^2 + y^2 = 10,$$

and therefore

$$\text{for all } t, \quad \frac{d}{dt}(x^2 + y^2) = \frac{d}{dt}(10),$$

$$2x \cdot \frac{dx}{dt} + 2y \cdot \frac{dy}{dt} = 0,$$

$$x \cdot \frac{dx}{dt} + y \cdot \frac{dy}{dt} = 0. \tag{4.5}$$

When $y = 8$, we have $x^2 + 8^2 = 10^2$, and therefore $x = 6$. Plugging in all the values we know, we find that

$$\text{when } y = 8, \quad (6)(-2) + (8) \cdot \frac{dy}{dt} = 0,$$

$$\frac{dy}{dt} = \frac{3}{2}.$$

Thus, when $y = 8$, y is increasing at a rate of 3/2 ft/sec. In other words, when the bottom of the door is 2 feet above the floor, it is moving down at a rate of $3/2 = 1.5$ ft/sec.

To find the acceleration of the bottom of the door, we must find d^2y/dt^2. The equation $dy/dt = 3/2$ is true at only one instant, so we can't differentiate that equation to find d^2y/dt^2. But equation (4.5) is true at all times. Furthermore, we know that for all t, $dx/dt = -2$, so we can plug in that value before differentiating:

$$\text{for all } t, \quad x(-2) + y \cdot \frac{dy}{dt} = 0,$$

$$\frac{d}{dt}\left(-2x + y \cdot \frac{dy}{dt}\right) = \frac{d}{dt}(0),$$

$$-2 \cdot \frac{dx}{dt} + y \cdot \frac{d^2y}{dt^2} + \frac{dy}{dt} \cdot \frac{dy}{dt} = 0.$$

Finally, plugging in known values again when $y = 8$, we have

$$\text{when } y = 8, \quad (-2)(-2) + 8 \cdot \frac{d^2y}{dt^2} + \left(\frac{3}{2}\right)^2 = 0,$$

$$\frac{d^2y}{dt^2} = -\frac{25}{32}.$$

Thus, dy/dt is decreasing, so the bottom of the door is slowing down. Its speed is decreasing at a rate of $25/32 \approx 0.78$ ft/sec^2. ☐

Example 4.1.4. A heavy weight is to be lifted using a block and tackle. A pulley is attached to the ceiling, 15 feet above the floor, and another pulley is attached to the weight. A 50-foot rope is attached to the weight, and it then runs up to and around the pulley attached to the ceiling, down to and around the pulley attached to the weight, back up to and around the ceiling pulley, and then down to a man. The man holds the end of the rope 3 feet above the floor and walks away from the weight at a constant speed. When he is 9 feet away from the point directly below the pulley and weight, the weight is rising at a rate of 1/2 ft/sec. How fast is it rising when he is 16 feet away? (You may assume that the weight and the pulleys are small enough that their size can be neglected.)

Solution. The situation in this problem is illustrated in Figure 4.5a. We measure all distances in feet, and time in seconds. Since the rope traverses the vertical distance $15 - x$ three times, the fact that the length of the rope is 50 feet means that

$$d + 3(15 - x) = 50,$$

and therefore

$$d = 5 + 3x.$$

We are told that dy/dt is a constant, but we are not told what this constant is. We also know that when $y = 9$, $dx/dt = 1/2$, and we are asked to find dx/dt when $y = 16$.

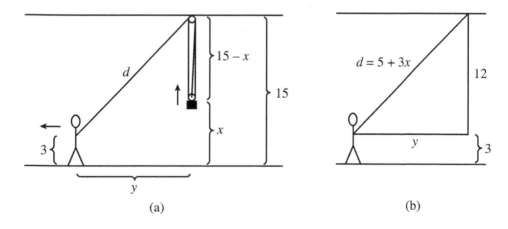

(a)   (b)

Figure 4.5: The block and tackle setup in Example 4.1.4.

We can get a relationship between x and y by noting that the diagonal length of rope is the hypotenuse of the right triangle shown in Figure 4.5b. This gives us the equation $y^2 + 12^2 = (5 + 3x)^2$. Notice that this problem is concerned with two different times, so we can summarize the problem as follows:

For all values of t: For one value of t: For another value of t:

$$y^2 + 144 = (5 + 3x)^2 \qquad\qquad y = 9 \qquad\qquad\qquad y = 16$$

$$\frac{dy}{dt} = c, \text{ a constant} \qquad\qquad \frac{dx}{dt} = \frac{1}{2} \qquad\qquad\qquad \frac{dx}{dt} = ?$$

As usual, we begin by differentiating the equation relating x and y:

$$\text{For all } t, \quad \frac{d}{dt}(y^2 + 144) = \frac{d}{dt}((5 + 3x)^2),$$

$$2y \cdot \frac{dy}{dt} = 2(5 + 3x) \cdot 3 \cdot \frac{dx}{dt},$$

$$y \cdot \frac{dy}{dt} = (15 + 9x) \cdot \frac{dx}{dt}. \tag{4.6}$$

Next, we look at the time when $y = 9$. We know that at this time, $dx/dt = 1/2$, and solving for x in the equation $y^2 + 144 = (5 + 3x)^2$, we find that $x = 10/3$. Thus, by equation (4.6) we have:

$$\text{When } y = 9, \quad 9 \cdot \frac{dy}{dt} = \left(15 + 9 \cdot \frac{10}{3}\right) \cdot \frac{1}{2},$$

$$\frac{dy}{dt} = \frac{5}{2}.$$

But dy/dt is a constant, so this is the value of dy/dt when $y = 16$ as well. The equation $y^2 + 144 = (5 + 3x)^2$ tells us that at that time, $x = 5$, so plugging into equation (4.6) again we get:

$$\text{When } y = 16, \quad 16 \cdot \frac{5}{2} = (15 + 9 \cdot 5) \cdot \frac{dx}{dt},$$

$$\frac{dx}{dt} = \frac{2}{3}.$$

Thus, when $y = 16$, the weight is rising at a rate of $2/3 \approx 0.67$ ft/sec. $\qquad\qquad\square$

Exercises 4.1

1. In Section 3.1, we discussed an electric heater whose power P (measured in watts) is related to the current I flowing through the heater (measured in amps) by the equation $P = 25I^2$. Suppose that at a certain moment the current flowing through the heater is 4 amps, and it is increasing at a rate of 2 amps/min. How fast is the power of the heater changing at that moment? Is it increasing or decreasing?

2. In Section 3.1, we discussed a gas-filled cylinder whose volume V (measured in cubic inches) is related to the pressure P of the gas (measured in kilopascals) by the equation $P = 3000/V$. Suppose that at a certain moment the volume of the cylinder is 20 cubic inches, and it is increasing at a rate of 3 in^3/sec. How fast is the pressure of the gas changing at that moment? Is it increasing or decreasing?

3. A spherical balloon is inflated in such a way that its volume increases at a steady rate of 5 cubic inches per second. How fast is the radius of the balloon increasing when the radius is 3 inches?

4. Alice is 5 feet tall, and she is walking away from a street light that is 9 feet tall. If she walks at a constant rate of 2 ft/sec, show that the length of her shadow changes at a constant rate. How fast is the length of her shadow changing, and is it getting longer or shorter?

5. Alice is standing 8 feet away from a street light that is 9 feet tall. She is getting shorter at a constant rate of 2 ft/min. How fast is the length of her shadow changing when she is 3 feet tall? Is the length of her shadow increasing or decreasing?

6. A road running west to east passes 50 feet north of a building. A car is driving east along the road at a speed of 30 ft/sec. How fast is the distance between the building and the car increasing when the car is 100 feet east of the point on the road closest to the building?

7. A 15-foot ladder is leaning against a wall. The bottom of the ladder is sliding away from the wall at a steady rate of 2 ft/sec, which causes the top of the ladder to slide down the wall. How fast is the top of the ladder sliding down when it is 12 feet above the ground?

8. An extendable ladder is leaning against a wall, with the bottom of the ladder held fixed at a point 3 feet from the bottom of the wall. The ladder is lengthened at a constant rate of 2 ft/sec, which causes the top of the ladder to slide up the wall. How fast is the top of the ladder sliding up the wall when the ladder is 5 feet long?

9. A camera is 2 km from the launch pad of a rocket. As the rocket rises from the launch pad, the camera will have to be tilted up to stay focused on the rocket. When the rocket is 1 km high, the camera is being tilted up at a rate of 1/10 rad/sec. How fast is the rocket rising at that time?

10. A police officer is 1/10 of a mile south of a road that runs west to east. He is using a telescope mounted on a tripod to watch a car that is driving east on the road. As the car drives by, he must rotate his telescope to keep it pointed at the car, and the angle θ in Figure 4.6 increases. The tripod measures both θ and the rate of change of θ. He discovers that when $\theta = \pi/3$, it is increasing at a rate of 1/20 rad/sec. The speed limit on the road is 65 mph. Should he give the driver a speeding ticket?

11. Sand pours into a conical pile at a steady rate of 2 ft^3/sec. If the height of the pile is always the same as the radius of the base, how fast is the height of the pile increasing when the pile is 5 feet high?

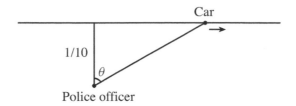

Figure 4.6: Police officer watching a car drive by.

12. There is a traffic light at the intersection of a north-south road and an east-west road. A police car is traveling toward the traffic light from the south, chasing a stolen van that is traveling east away from the light. The speed of the police car is 80 mph at the moment it is 2 miles from the light. At the same time, the van is 3 miles from the light going 60 mph. At this moment, how fast is the distance between the car and van changing? Is it increasing or decreasing?

13. A baseball diamond is a square 90 feet on each side. A runner leaves first base and runs toward second base at a speed of 20 ft/sec.

   (a) When he is 30 feet from first base, how fast is his distance from home plate changing?

   (b) At that same moment, how fast is his distance to the pitcher's mound changing? (The pitcher's mound is located in the center of the baseball diamond.)

14. The London Eye is a Ferris wheel on the south bank of the River Thames in London. It has a diameter of 120 meters, and it takes 30 minutes to complete one revolution. A passenger gets on and is lifted up into the air. When he reaches a height of 90 meters above the base of the wheel, how fast is his height increasing?

15. Elm Street and Oak Street cross at right angles. There is a statue in the middle of Oak St., 100 feet from the intersection of Elm and Oak. A bus is traveling along Elm St. toward the intersection at a speed of 40 ft/sec. A passenger on the bus is pointing his video camera at the statue. When the bus is still 50 feet from the intersection, how fast must he turn the camera (in rad/sec) to keep it pointed at the statue?

16. A man is standing at the end of a dock and pulling a boat toward the dock using a rope tied to the bow of the boat. If the man's hands are 5 feet higher than the point where the rope is attached to the boat, and he pulls the rope in at a rate of 4 ft/sec, how fast is the boat approaching the dock when it is 12 feet from the dock?

17. A light bulb is suspended 20 inches above a tabletop, and a horizontal board is positioned between the light bulb and the tabletop. The board has a circular hole, with radius 3 inches, centered below the light bulb, and light passing through this hole creates a circular lighted region on the tabletop (see Figure 4.7). If the board is raised at a rate of 5 in/sec, how fast is the area of the lighted region changing when the board is 10 inches below the light bulb? Is the area increasing or decreasing?

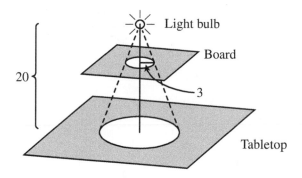

Figure 4.7: Light shines through a circular hole onto a tabletop.

18. One morning it started raining and it rained at a steady rate. A trough with triangular ends was sitting out in the rain and it began to fill with water. The trough was 2 feet wide at the top, 3 feet high, and 5 feet long (see Figure 4.8). At noon, the water in the trough was 1.5 feet deep, and the water level was rising at a rate of 0.1 ft/hr.

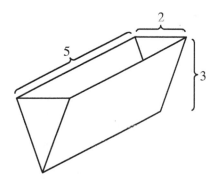

Figure 4.8: Trough that fills with rain water.

(a) How fast was the volume of the water in the trough increasing at that moment?

(b) What time did it start raining?

19. The Great Pyramid of Giza is 455 feet high, and its base is a square that is 756 feet on each side. Suppose that the pyramid has been hollowed out, and it is being filled with Egyptian musk oil at a rate of 10 ft^3/min. How fast is the depth of the oil increasing when it is 400 feet deep?

20. A man is looking at a painting hanging on a wall. The painting is 3 feet high, and the bottom of the painting is 1 foot above the man's eye level. The man walks toward the wall at a steady rate of 3 ft/sec.

(a) The angle θ in Figure 4.9 is called the *angle subtended by the painting at the man's eye*. How fast is θ changing when he is 3 feet from the wall?

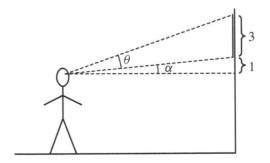

Figure 4.9: Looking at a painting hanging on a wall.

Is it increasing or decreasing? (Hint: First figure out how fast the angle α in Figure 4.9 is changing.)

(b) How fast is θ changing when he is 1 foot from the wall? Is it increasing or decreasing?

4.2 The Mean Value Theorem

The theoretical basis for all of the other applications of derivatives in this chapter will be an important theorem called the mean value theorem. Here's what it says:

Theorem 4.2.1 (Mean Value Theorem). *Suppose f is continuous on the closed interval $[a, b]$ and differentiable on (a, b). Then there is a number c such that $a < c < b$ and*

$$\frac{f(b) - f(a)}{b - a} = f'(c).$$

The mean value theorem is illustrated in Figure 4.10. The black curve is the graph of f, and the blue line is the secant line for this graph that passes through the points

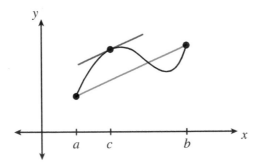

Figure 4.10: The mean value theorem. The red tangent line is parallel to the blue secant line, and therefore $(f(b) - f(a))/(b - a) = f'(c)$.

$(a, f(a))$ and $(b, f(b))$. The slope of this secant line is $(f(b) - f(a))/(b - a)$. The mean value theorem says that for some number c between a and b, the slope of this secant line is the same as the slope of the tangent line to the graph at $(c, f(c))$. In other words, the tangent line is parallel to the secant line. Such a number c is shown in the figure, where the tangent line is drawn in red. Another way to express the conclusion of the mean value theorem is to say that the average rate of change of $f(x)$ over the interval $[a, b]$ is equal to the instantaneous rate of change at some number c between a and b.

For example, consider the case of the function $f(x) = 1/x$, on the interval $[2, 8]$. We already know that f is continuous and differentiable at all numbers except 0, so it is certainly continuous on the interval $[2, 8]$ and differentiable on $(2, 8)$. The derivative is given by the formula $f'(x) = -1/x^2$. Thus, according to the mean value theorem, there should be a number c between 2 and 8 such that

$$\frac{f(8) - f(2)}{8 - 2} = f'(c) = -\frac{1}{c^2}. \tag{4.7}$$

Let's see if we can find such a number c. Filling in the formula for $f(x)$ on the left side of equation (4.7), we find that

$$\frac{f(8) - f(2)}{8 - 2} = \frac{1/8 - 1/2}{6} = -\frac{1}{16}.$$

Thus, equation (4.7) says that

$$-\frac{1}{16} = -\frac{1}{c^2}.$$

Of course, there is a number c between 2 and 8 that satisfies this equation, namely $c = 4$.

It may not be immediately clear why the mean value theorem is important. But again and again in this chapter we will find that the mean value theorem is just what we need to solve some problem. In particular, if you are studying a function f and you want to know something about the relationship between $f(a)$ and $f(b)$, for two numbers a and b with $a < b$, it is worthwhile to see what the mean value theorem says about the function f on the interval $[a, b]$. It is surprising how often this leads to useful information.

For example, what is the value of $\sqrt{20}$? Well, $20 > 16$, so $\sqrt{20}$ is bigger than $\sqrt{16} = 4$. But how much bigger? We are asking about the relationship between $f(20)$ and $f(16)$, where $f(x) = \sqrt{x}$. So let's see what the mean value theorem has to say about this function f on the interval $[16, 20]$.

Since f is continuous and differentiable at all positive numbers, it is continuous on $[16, 20]$ and differentiable on $(16, 20)$. The mean value theorem therefore tells us that there is some number c such that $16 < c < 20$ and

$$\frac{\sqrt{20} - \sqrt{16}}{20 - 16} = f'(c).$$

Since $f'(x) = 1/(2\sqrt{x})$, this means that

$$\frac{\sqrt{20} - 4}{4} = \frac{1}{2\sqrt{c}},$$

or in other words

$$\sqrt{20} = 4 + \frac{2}{\sqrt{c}}. \tag{4.8}$$

Thus, the mean value theorem has given us a formula for $\sqrt{20}$. Of course, we don't know the value of the number c in this formula. We could solve for c, as we did in our previous example. But this time we're going to show that, even without knowing the value of c, we can learn a lot about $\sqrt{20}$ from equation (4.8).

Although we don't know the exact value of c, we do know that $16 < c < 20$. Since $c > 16$, we have $\sqrt{c} > 4$, and therefore $2/\sqrt{c} < 2/4 = 1/2$. Thus, equation (4.8) tells us that

$$\sqrt{20} = 4 + \frac{2}{\sqrt{c}} < 4 + \frac{1}{2} = \frac{9}{2} = 4.5.$$

Combining this with the fact that $c < 20$, we conclude that $\sqrt{c} < \sqrt{20} < 9/2$, and therefore

$$\sqrt{20} = 4 + \frac{2}{\sqrt{c}} > 4 + \frac{2}{9/2} = \frac{40}{9} \approx 4.44.$$

Thus, we have $4.44 < \sqrt{20} < 4.5$. Of course, this isn't a terribly important discovery—your calculator will tell you almost instantly that $\sqrt{20} \approx 4.472$. But who would have guessed that the mean value theorem could tell us the approximate value of $\sqrt{20}$? When we plug other functions and intervals into the mean value theorem, what other interesting facts will pop out?

To prove the mean value theorem, we will use the nested interval theorem, Theorem 2.8.7.[2] Thus, our plan will be to find a nested sequence of closed intervals $[u_1, v_1] = [a, b], [u_2, v_2], [u_3, v_3], \ldots$ such that $\lim_{n \to \infty}(v_n - u_n) = 0$. Recall that to say that the intervals are nested means that $u_1 \le u_2 \le u_3 \le \cdots$ and $v_1 \ge v_2 \ge v_3 \ge \cdots$. The nested interval theorem will then tell us that there is a unique number c that is in all of the intervals in the nested sequence. If we choose the intervals $[u_n, v_n]$ carefully, then the number c will be the number we need for the mean value theorem.

One of the key steps in our proof will be the following fact:

Lemma 4.2.2. *Suppose that f is continuous on $[u, v]$. Then there are numbers u' and v' such that $u < u' < v' < v$, $v' - u' = (v - u)/3$, and*

$$\frac{f(v') - f(u')}{v' - u'} = \frac{f(v) - f(u)}{v - u}.$$

Proof. The lemma is illustrated in Figure 4.11. We begin by dividing the interval $[u, v]$ in thirds. Let $d = (v - u)/3$, one third of the length of the interval, and let

$$p = u + d, \qquad q = u + 2d.$$

Thus, $u < p < q < v$, and $p - u = q - p = v - q = d = (v - u)/3$, as shown in Figure 4.12.

---

[2]Our proof of the mean value theorem is based on F. Acker, The missing link, *Mathematical Intelligencer* **18**, no. 3 (1996), pp. 4–9. However, the idea behind the proof can be traced back to a paper by André-Marie Ampère (1775–1836) published in 1806; see T. M. Flett, Ampère and the horizontal chord theorem, *Bulletin of the Institute of Mathematics and its Applications* **11**, no. 1–2 (1975), p. 34.

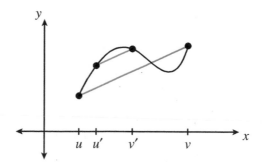

Figure 4.11: Lemma 4.2.2. The two blue secant lines are parallel, and $v' - u' = (v-u)/3$.

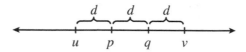

Figure 4.12: The points p and q divide the interval $[u, v]$ into three equal pieces.

Let

$$m = \frac{f(v) - f(u)}{v - u}, \quad m_1 = \frac{f(p) - f(u)}{p - u}, \quad m_2 = \frac{f(q) - f(p)}{q - p}, \quad m_3 = \frac{f(v) - f(q)}{v - q}.$$

In other words, m is the slope of the secant line for the interval $[u, v]$, and m_1, m_2, and m_3 are the slopes of the secant lines for the three pieces we have cut this interval into. Notice that

$$
\begin{aligned}
m &= \frac{f(v) - f(u)}{v - u} = \frac{f(v) - f(q) + f(q) - f(p) + f(p) - f(u)}{v - u} \\
&= \frac{f(v) - f(q)}{3(v - q)} + \frac{f(q) - f(p)}{3(q - p)} + \frac{f(p) - f(u)}{3(p - u)} \\
&= \frac{1}{3}\left[\frac{f(v) - f(q)}{v - q} + \frac{f(q) - f(p)}{q - p} + \frac{f(p) - f(u)}{p - u} \right] \\
&= \frac{m_3 + m_2 + m_1}{3}.
\end{aligned}
$$

Thus, the slope of the secant line for the interval $[u, v]$ is the average of the slopes of the secant lines for the intervals $[u, p]$, $[p, q]$, and $[q, v]$.

If we are lucky, $m_2 = m$. In that case, we can just let $u' = p$ and $v' = q$, and we will have $v' - u' = (v - u)/3$ and

$$\frac{f(v') - f(u')}{v' - u'} = \frac{f(q) - f(p)}{q - p} = m_2 = m = \frac{f(v) - f(u)}{v - u},$$

as required. If we are not so lucky, then either $m_2 < m$ or $m_2 > m$. We will only consider the case in which $m_2 < m$; the proof if $m_2 > m$ is similar.

So suppose now that $m_2 < m$. If $m_1 \leq m$ and $m_3 \leq m$, then we have

$$m = \frac{m_3 + m_2 + m_1}{3} < \frac{m + m + m}{3} = m,$$

which is impossible. So either $m_1 > m$ or $m_3 > m$. Once again, the proofs for the two cases are similar, so we will only do one of them. Let's assume that $m_3 > m$.

We now have $m_2 < m < m_3$. Recall that we are looking for an interval $[u', v']$ of length d for which the slope of the secant line is m. We can't let $[u', v'] = [p, q]$, because the slope of the secant line for the interval $[p, q]$ is too small, and similarly the slope for the interval $[q, v]$ is too big. The idea for the rest of the proof is to imagine sliding the interval $[p, q]$ to the right until it coincides with $[q, v]$. At the beginning of this process the slope of the secant line is too small, and at the end it is too big. So somewhere along the way, it should be just right.

To make this idea precise, we use the intermediate value theorem (Theorem 2.7.6). Define a function g as follows:

$$g(x) = \frac{f(x + d) - f(x)}{d}.$$

Then

$$g(p) = \frac{f(p + d) - f(p)}{d} = \frac{f(q) - f(p)}{q - p} = m_2 < m$$

and

$$g(q) = \frac{f(q + d) - f(q)}{d} = \frac{f(v) - f(q)}{v - q} = m_3 > m.$$

Also, since f is continuous on $[u, v]$, g is continuous on the interval $[p, q]$ (see Exercise 15). Therefore, by the intermediate value theorem, there is some number c such that $p < c < q$ and $g(c) = m$. According to the definition of g, this means that

$$\frac{f(c + d) - f(c)}{d} = m = \frac{f(v) - f(u)}{v - u}.$$

Thus, we can complete the proof of the lemma by letting $u' = c$ and $v' = c + d$. □

In our proof of the mean value theorem we will compute $f'(c)$ using a limit that is slightly different from the ones we have used in the past. It will be helpful to work out this derivative calculation in advance:

Lemma 4.2.3. *Suppose that f is differentiable at c, and $\{u_n\}_{n=1}^{\infty}$ and $\{v_n\}_{n=1}^{\infty}$ are sequences such that as $n \to \infty$, $u_n \to c^-$ and $v_n \to c^+$. Then*

$$\lim_{n \to \infty} \frac{f(v_n) - f(u_n)}{v_n - u_n} = f'(c).$$

Proof. For every n, we have

$$
\begin{aligned}
\frac{f(v_n) - f(u_n)}{v_n - u_n} &= \frac{f(v_n) - f(c) + f(c) - f(u_n)}{v_n - u_n} \\
&= \frac{f(v_n) - f(c)}{v_n - u_n} + \frac{f(c) - f(u_n)}{v_n - u_n} \\
&= \frac{v_n - c}{v_n - u_n} \cdot \frac{f(v_n) - f(c)}{v_n - c} + \frac{c - u_n}{v_n - u_n} \cdot \frac{f(c) - f(u_n)}{c - u_n}.
\end{aligned} \tag{4.9}
$$

To simplify our notation, we let

$$
s_n = \frac{f(c) - f(u_n)}{c - u_n}, \qquad t_n = \frac{f(v_n) - f(c)}{v_n - c}, \qquad p_n = \frac{v_n - c}{v_n - u_n}.
$$

Thus, s_n and t_n are the slopes of the secant lines for the intervals $[u_n, c]$ and $[c, v_n]$. Also, notice that since $u_n < c < v_n$, we have $0 < p_n < 1$, and

$$
1 - p_n = \frac{v_n - u_n}{v_n - u_n} - \frac{v_n - c}{v_n - u_n} = \frac{c - u_n}{v_n - u_n}.
$$

Substituting into equation (4.9), we get

$$
\frac{f(v_n) - f(u_n)}{v_n - u_n} = p_n \cdot t_n + (1 - p_n) \cdot s_n. \tag{4.10}
$$

We could say that the slope of the secant line for the interval $[u_n, v_n]$ is a *weighted average* of the slopes for the intervals $[u_n, c]$ and $[c, v_n]$. To compute the limit in the lemma, it will be helpful to rearrange equation (4.10) like this:

$$
\frac{f(v_n) - f(u_n)}{v_n - u_n} = s_n + p_n \cdot (t_n - s_n). \tag{4.11}
$$

Recall that in Theorem 3.2.3 we showed that

$$
\lim_{x \to c} \frac{f(x) - f(c)}{x - c} = f'(c).
$$

Since we know that as $n \to \infty$, $u_n \to c^-$ and $v_n \to c^+$, it follows that

$$
\begin{aligned}
\lim_{n \to \infty} t_n &= \lim_{n \to \infty} \frac{f(v_n) - f(c)}{v_n - c} = f'(c), \\
\lim_{n \to \infty} s_n &= \lim_{n \to \infty} \frac{f(c) - f(u_n)}{c - u_n} = \lim_{n \to \infty} \frac{f(u_n) - f(c)}{u_n - c} = f'(c).
\end{aligned} \tag{4.12}
$$

Thus, the first term on the right-hand side of equation (4.11) approaches $f'(c)$ as $n \to \infty$. Our plan now is to use the squeeze theorem to show that the rest of the right-hand side of that equation approaches 0.

Since $0 < p_n < 1$, we have

$$0 \le |p_n \cdot (t_n - s_n)| = p_n \cdot |t_n - s_n| \le |t_n - s_n|.$$

Using equations (4.12), we see that

$$\lim_{n \to \infty} |t_n - s_n| = |f'(c) - f'(c)| = 0,$$

and therefore, by the squeeze theorem,

$$\lim_{n \to \infty} |p_n \cdot (t_n - s_n)| = 0.$$

It follows (by the sequence version of Theorem 2.4.11) that

$$\lim_{n \to \infty} p_n \cdot (t_n - s_n) = 0. \tag{4.13}$$

We can now complete the proof of the lemma by combining equations (4.11), (4.12), and (4.13):

$$\lim_{n \to \infty} \frac{f(v_n) - f(u_n)}{v_n - u_n} = \lim_{n \to \infty} [s_n + p_n \cdot (t_n - s_n)] = f'(c) + 0 = f'(c). \qquad \square$$

We are now ready to prove the mean value theorem:

Proof of Theorem 4.2.1. We use Lemma 4.2.2 to define a nested sequence of intervals. Let $u_1 = a$ and $v_1 = b$. Since f is continuous on $[a, b] = [u_1, v_1]$, by the lemma there are numbers u_2 and v_2 such that $u_1 < u_2 < v_2 < v_1$, $v_2 - u_2 = (v_1 - u_1)/3 = (b - a)/3$, and

$$\frac{f(v_2) - f(u_2)}{v_2 - u_2} = \frac{f(v_1) - f(u_1)}{v_1 - u_1} = \frac{f(b) - f(a)}{b - a}.$$

Now apply the lemma again to find u_3 and v_3 so that $u_2 < u_3 < v_3 < v_2$, $v_3 - u_3 = (v_2 - u_2)/3 = (b - a)/9$, and

$$\frac{f(v_3) - f(u_3)}{v_3 - u_3} = \frac{f(v_2) - f(u_2)}{v_2 - u_2} = \frac{f(b) - f(a)}{b - a}.$$

Continuing in this way, we define a sequence of intervals $[u_n, v_n]$ such that for every n, $v_n - u_n = (b - a)/3^{n-1}$,

$$\frac{f(v_n) - f(u_n)}{v_n - u_n} = \frac{f(b) - f(a)}{b - a},$$

and $u_n < u_{n+1} < v_{n+1} < v_n$.

Since

$$\lim_{n \to \infty} (v_n - u_n) = \lim_{n \to \infty} \frac{b - a}{3^{n-1}} = 0,$$

the nested interval theorem tells us that there is a unique number c such that for all n, $u_n \le c \le v_n$. In fact, since $u_n < u_{n+1} \le c \le v_{n+1} < v_n$, we have $u_n < c < v_n$.

In particular $a = u_1 < c < v_1 = b$, and since f is differentiable on (a, b), it follows that f is differentiable at c. The nested interval theorem also asserts that $\lim_{n \to \infty} u_n = \lim_{n \to \infty} v_n = c$, so we can say that as $n \to \infty$, $u_n \to c^-$ and $v_n \to c^+$.

We can now complete the proof by applying Lemma 4.2.3:

$$f'(c) = \lim_{n \to \infty} \frac{f(v_n) - f(u_n)}{v_n - u_n} = \lim_{n \to \infty} \frac{f(b) - f(a)}{b - a} = \frac{f(b) - f(a)}{b - a}. \qquad \square$$

We close this section by considering one more example. Let $f(x) = |x|$ and $[a, b] = [-2, 2]$. Then

$$\frac{f(b) - f(a)}{b - a} = \frac{|2| - |-2|}{2 - (-2)} = 0.$$

But in Section 3.2 we showed that

$$f'(x) = \begin{cases} -1, & \text{if } x < 0, \\ \text{undefined}, & \text{if } x = 0, \\ 1, & \text{if } x > 0, \end{cases}$$

so there is no number c between 2 and -2 such that $f'(c) = 0$. Does this contradict the mean value theorem? No, the statement of the mean value theorem says that it can only be applied if the function f is continuous on $[a, b]$ and differentiable on (a, b). But in this case f is not differentiable on $(-2, 2)$, because it is not differentiable at 0. This example illustrates that it is important to check that f is continuous on $[a, b]$ and differentiable on (a, b) before applying the mean value theorem; applying the theorem without checking these conditions can lead to incorrect conclusions.

Exercises 4.2

1–5: Find a number c such that $a < c < b$ and $(f(b) - f(a))/(b - a) = f'(c)$.

1. $f(x) = x^2 - 3x$, $a = 1$, $b = 4$.

2. $f(x) = x^3$, $a = -2$, $b = 4$.

3. $f(x) = \sqrt{x}$, $a = 1$, $b = 9$.

4. $f(x) = \sin x$, $a = 0$, $b = \pi$.

5. $f(x) = 1/x^2$, $a = -4$, $b = -1$.

6. Let $f(x) = x^{2/3}$. Show that there is no number c such that $-1 < c < 8$ and $f'(c) = (f(8) - f(-1))/(8 - (-1))$. Why doesn't this contradict the mean value theorem?

7. (a) Show that $\sqrt{10} < 19/6 \approx 3.167$. (Hint: Apply the mean value theorem to the function $f(x) = \sqrt{x}$ on the interval $[9, 10]$.)

    (b) Show that $\sqrt{10} > 60/19 \approx 3.158$. (Hint: Apply the mean value theorem again and use part (a).)

8. (a) Show that $\sqrt[3]{10} < 13/6 \approx 2.167$. (Hint: Apply the mean value theorem to the function $f(x) = \sqrt[3]{x}$ on the interval $[8, 10]$.)

   (b) Show that $\sqrt[3]{10} > 362/169 \approx 2.142$. (Hint: Apply the mean value theorem again and use part (a).)

9. Suppose that for all $x > 0$, $f'(x) = 1/x$, and $f(1) = 0$. (We will study a function with these properties in Section 7.3.)

   (a) Show that $1/2 < f(2) < 1$. (Hint: Apply the mean value theorem on the interval $[1, 2]$.)

   (b) Show that $1/3 < f(3/2) < 1/2$.

   (c) Show that $7/12 < f(2) < 5/6$. (Hint: Apply the mean value theorem on the interval $[3/2, 2]$, and use part (b).)

10. Show that for all $b > 0$, $\sqrt[3]{b+1} < b/3 + 1$.

11. Suppose that for all $x > 0$, $f'(x) = \sqrt{x^2 + 4}$, and $f(0) = 5$. Show that for all $x > 0$, $f(x) > 2x + 5$.

12. (a) Show that if $x > 0$ then $\sin x < x$. (Hint: Apply the mean value theorem to the function $f(x) = \sin x$ on the interval $[0, x]$.)

   (b) Show that if $0 < x \leq \pi/2$ then $\sin x > x/\sqrt{x^2 + 1}$. (Hint: Apply the mean value theorem to the function $f(x) = \sin x$ on the interval $[0, x]$, and then use the fact that for $0 < c < x$, $\cos c > \cos x = \sqrt{1 - \sin^2 x}$.)

   (c) Show that $1/\sqrt{5} < \sin(1/2) < 1/2$.

13. Show that if $0 < x < \pi/2$ then $\tan x > x$.

14. Suppose that $0 < \alpha < \pi/2$ and $\sin \alpha = 3/5$. (The angle α is called the *inverse sine*, or *arcsine*, of 3/5. We will study the calculus of the inverse sine function in Section 7.5.) Show that $3/5 < \alpha < 3/4$. (Hint: Apply the mean value theorem to the function $f(x) = \sin x$ on the interval $[0, \alpha]$, and use the fact that if $0 < c < \alpha$ then $\cos \alpha < \cos c < 1$.)

15. Verify that the function g in the proof of Lemma 4.2.2 is continuous on $[p, q]$.

4.3 Increasing and Decreasing Functions

We now want to see what the derivative of a function can tell us about the shape of the graph of the function. In this section, we will see how to determine whether the graph goes up or down as we move from left to right. We begin by fixing our terminology. Imitating the terminology we introduced for monotone sequences in Section 2.9, we make the following definition.

Definition 4.3.1. Let f be a function and I an interval contained in the domain of f.

1. We say that f is *strictly increasing* on I if for all numbers $x_1, x_2 \in I$, if $x_1 < x_2$ then $f(x_1) < f(x_2)$.

2. We say that f is *weakly increasing* on I if for all numbers $x_1, x_2 \in I$, if $x_1 < x_2$ then $f(x_1) \le f(x_2)$.

3. We say that f is *strictly decreasing* on I if for all numbers $x_1, x_2 \in I$, if $x_1 < x_2$ then $f(x_1) > f(x_2)$.

4. We say that f is *weakly decreasing* on I if for all numbers $x_1, x_2 \in I$, if $x_1 < x_2$ then $f(x_1) \ge f(x_2)$.

5. We say that f is *constant* on I if for all numbers $x_1, x_2 \in I$, $f(x_1) = f(x_2)$.

For example, consider the function f whose graph is shown in Figure 4.13. The domain of f is $[p, t)$. For any two numbers x_1 and x_2 in the interval $[p, q]$, if $x_1 < x_2$ then $f(x_1) < f(x_2)$; an example of such a pair of numbers is shown in the figure. Therefore f is strictly increasing on the interval $[p, q]$. For any numbers x_1 and x_2 with $x_1 < x_2$, notice that $f(x_1) < f(x_2)$ if and only if the slope of the secant line through $(x_1, f(x_1))$ and $(x_2, f(x_2))$ is positive. Thus, another way to see that f is strictly increasing on $[p, q]$ is to note that if you pick two points on the graph whose x-coordinates are in the interval $[p, q]$, then the secant line through those two points will have positive slope. The blue line in the figure is an example of such a secant line. To put it more informally, to say that f is strictly increasing on $[p, q]$ means that if a point moves along the graph from $x = p$ to $x = q$, then as the point moves from left to right, it also moves up.

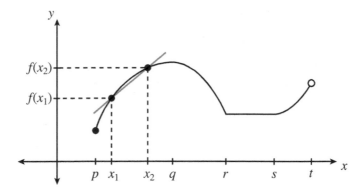

Figure 4.13: The graph of a function f that is strictly increasing on $[p, q]$, strictly decreasing on $[q, r]$, constant on $[r, s]$, and strictly increasing on $[s, t)$.

Similarly, f is strictly decreasing on the interval $[q, r]$, constant on $[r, s]$, and strictly increasing again on $[s, t)$. A point moving along the graph from left to right will go down for $q \le x \le r$, then stay at the same level for $r \le x \le s$, and then go up again for $s \le x < t$. All secant lines through two points whose x-coordinates are in the interval $[q, r]$ will have negative slope, secant lines on the interval $[r, s]$ will have slope 0, and secant lines on the interval $[s, t)$ will have positive slope. We could also say that f is weakly decreasing on the interval $[q, s]$, and weakly increasing on $[r, t)$.

This example illustrates that you can see, in the graph of a function, the intervals on which the function is strictly or weakly increasing, strictly or weakly decreasing, or constant. But what if you know a formula for $f(x)$, but don't have the graph of f? Is there a way to use the formula for $f(x)$ to determine intervals on which f is increasing, decreasing, or constant? Yes, in many cases the derivative of f will tell you the answer:

Theorem 4.3.2. *Suppose f is continuous on an interval I and differentiable on the interior of I.*

1. *If $f'(x) > 0$ for all x in the interior of I, then f is strictly increasing on I.*

2. *If $f'(x) \geq 0$ for all x in the interior of I, then f is weakly increasing on I.*

3. *If $f'(x) < 0$ for all x in the interior of I, then f is strictly decreasing on I.*

4. *If $f'(x) \leq 0$ for all x in the interior of I, then f is weakly decreasing on I.*

5. *If $f'(x) = 0$ for all x in the interior of I, then f is constant on I.*

Proof. The proofs are all very similar, so we will just prove part 1. (We ask you to prove two other parts in Exercises 23 and 24.) We could summarize the idea behind the proof of part 1 of the theorem like this: The assumption that $f'(x)$ is positive on the interior of I means that every tangent line in this interval has positive slope. According to the mean value theorem, every secant line on the interval I is parallel to one of these tangent lines. Therefore all of these secant lines have positive slope, so f is strictly increasing on I.

Now let's see if we can spell out the details more carefully. Suppose that $f'(x) > 0$ for all x in the interior of I. To prove that f is strictly increasing on I, we assume that x_1 and x_2 are elements of I and $x_1 < x_2$. According to Definition 4.3.1, we must prove that $f(x_1) < f(x_2)$.

We need to compare $f(x_1)$ to $f(x_2)$, so let's try applying the mean value theorem to f on the interval $[x_1, x_2]$. Notice that $[x_1, x_2] \subseteq I$, so since f is continuous on I and differentiable on the interior of I, it is continuous on $[x_1, x_2]$ and differentiable on (x_1, x_2). Therefore, according to the mean value theorem, there is some number c such that $x_1 < c < x_2$ and

$$\frac{f(x_2) - f(x_1)}{x_2 - x_1} = f'(c).$$

But we have assumed that $f'(x) > 0$ for all x in the interior of I, so $f'(c) > 0$. Since $x_1 < x_2$, it follows that

$$f(x_2) - f(x_1) = f'(c)(x_2 - x_1) > 0,$$

and therefore $f(x_1) < f(x_2)$. $\square$

For example, consider the function $f(x) = x^2 - 6x + 5$. Since f is a polynomial, it is continuous at all numbers, so the graph of f is a curve with no breaks in it.

We have $f'(x) = 2x - 6 = 2(x - 3)$, so $f'(3) = 0$, $f'(x) > 0$ for $x > 3$, and $f'(x) < 0$ for $x < 3$. It follows, by Theorem 4.3.2, that f is strictly decreasing on the interval $(-\infty, 3]$ and strictly increasing on $[3, \infty)$. Since $f'(3) = 0$, the tangent line at the point $(3, f(3)) = (3, -4)$ is horizontal. Thus, the curve "levels off" at $x = 3$, where the function switches from decreasing to increasing. The graph of f is shown in Figure 4.14.

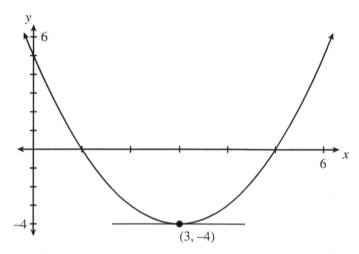

Figure 4.14: The graph of the function $f(x) = x^2 - 6x + 5$. The horizontal tangent line at the point $(3, -4)$ is shown in red.

Notice that for every number x, $f(3) \leq f(x)$. We can verify this by reasoning by cases:

Case 1. $x < 3$. Since f is strictly decreasing on $(-\infty, 3]$, we have $f(x) > f(3)$.

Case 2. $x = 3$. Of course, in this case $f(x) = f(3)$.

Case 3. $x > 3$. Since f is strictly increasing on $[3, \infty)$, $f(3) < f(x)$.

Thus, in every case, $f(3) \leq f(x)$, and therefore the point $(3, -4)$ is the lowest point on the graph. We say that -4 is the *minimum value* of the function f, and f *attains its minimum value* at 3.

Definition 4.3.3. Suppose that f is a function and a is a number in the domain of f.

1. Suppose that for every x in the domain of f, $f(a) \leq f(x)$. Then we say that $f(a)$ is the *minimum value of f*. We also say that f *attains its minimum value* at a.

2. Suppose that for every x in the domain of f, $f(a) \geq f(x)$. Then we say that $f(a)$ is the *maximum value of f*, and f *attains its maximum value* at a.

Not every function has maximum and minimum values. For example, the function $f(x) = x^2 - 6x + 5$, whose graph is shown in Figure 4.14, has a minimum value but no maximum value.

Now consider the function $g(x) = x|x - 4|$. In order to apply Theorem 4.3.2 we would like to compute $g'(x)$, but to do this we will need to consider $x < 4$, $x > 4$, and

$x = 4$ separately. For $x < 4$ we have $|x - 4| = 4 - x$, so $g(x) = x(4 - x) = 4x - x^2$. According to Theorem 3.5.1, if two functions agree on an open interval then their derivatives also agree on that interval, so we can conclude that for all x in the open interval $(-\infty, 4)$,

$$g'(x) = \frac{d}{dx}(4x - x^2) = 4 - 2x = 2(2 - x).$$

Similarly, for $x > 4$ we have $g(x) = x(x - 4) = x^2 - 4x$ and therefore $g'(x) = 2x - 4 = 2(x - 2)$. Note that Theorem 3.5.1 applies only to *open* intervals, so we cannot include $x = 4$ in either of these calculations. To compute $g'(4)$, we apply the definition of derivative:

$$g'(4) = \lim_{h \to 0} \frac{g(4 + h) - g(4)}{h} = \lim_{h \to 0} \frac{(4 + h)|4 + h - 4| - 0}{h} = \lim_{h \to 0} \frac{(4 + h)|h|}{h}.$$

Computing the two one-sided limits, we find that

$$\lim_{h \to 0^+} \frac{(4 + h)|h|}{h} = \lim_{h \to 0^+} \frac{(4 + h) \cdot h}{h} = \lim_{h \to 0^+} (4 + h) = 4,$$

$$\lim_{h \to 0^-} \frac{(4 + h)|h|}{h} = \lim_{h \to 0^-} \frac{(4 + h) \cdot (-h)}{h} = \lim_{h \to 0^-} -(4 + h) = -4.$$

Since the one-sided limits are different, $g'(4)$ is undefined. Putting all of our calculations together, we have

$$g'(x) = \begin{cases} 2(2 - x), & \text{if } x < 4, \\ \text{undefined}, & \text{if } x = 4, \\ 2(x - 2), & \text{if } x > 4. \end{cases}$$

Now that we have found $g'(x)$, we are ready to determine where $g'(x)$ is positive and negative. For $x > 4$, we see that $g'(x) = 2(x - 2) > 0$. For $x < 4$ we have $g'(x) = 2(2 - x)$, so $g'(2) = 0$, $g'(x) > 0$ for $x < 2$, and $g'(x) < 0$ for $2 < x < 4$. We can summarize our conclusions with the following number line:

$$
\begin{array}{c c c c c c c}
g'(x) & + & 0 & - & \text{u} & + & \\
\hline
x & & 2 & & 4 & &
\end{array}
$$

We have marked the important x values 2 and 4 below the number line and information about $g'(x)$ above the number line, using the letter "u" to indicate that $g'(x)$ is undefined when $x = 4$.

Although g is not differentiable at 4, it is continuous everywhere. We can therefore use Theorem 4.3.2 to conclude that g is strictly increasing on $(-\infty, 2]$, strictly decreasing on $[2, 4]$, and strictly increasing on $[4, \infty)$. The graph of g is shown in Figure 4.15. You can see in the graph that there is no tangent line at the point $(4, g(4)) = (4, 0)$, which explains why $g'(4)$ is undefined. Since $g'(2) = 0$, the tangent line at the point $(2, g(2)) = (2, 4)$ (shown in red in the figure) is horizontal.

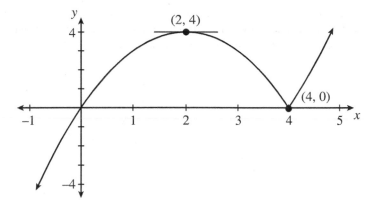

Figure 4.15: The graph of the function $g(x) = x|x - 4|$.

Notice that g does *not* attain its maximum value at 2. You can show this by checking that, for example, $g(5) = 5 > 4 = g(2)$. But it is more informative to note that

$$\lim_{x \to \infty} g(x) = \lim_{x \to \infty} x|x - 4| = \lim_{x \to \infty} x(x - 4) = \infty.$$

This limit tells us that the right side of the graph of g continues to rise higher and higher, eventually crossing any horizontal line $y = M$ that we might draw. In particular, it must cross the line $y = 4$, so $(2, 4)$ is not the highest point on the graph. Similarly,

$$\lim_{x \to -\infty} g(x) = \lim_{x \to -\infty} x|x - 4| = \lim_{x \to -\infty} x(4 - x) = -\infty,$$

so on the left side of the graph, $g(x)$ goes lower than $g(4) = 0$. Therefore g does not attain its minimum value at 4. However, the point $(2, 4)$ is higher than all other nearby points on the graph, and $(4, 0)$ is lower than all nearby points. We say that g has a *local maximum* at 2, and a *local minimum* at 4. Let us give careful definitions of these terms. First we define a restricted version of maximum and minimum values.

Definition 4.3.4. Suppose f is a function, S is a subset of the domain of f, and $a \in S$.

1. Suppose that for every $x \in S$, $f(a) \leq f(x)$. Then we say that $f(a)$ is the *minimum value of f on S*. We also say that f *attains its minimum value on the set S at a*.

2. Suppose that for every $x \in S$, $f(a) \geq f(x)$. Then we say that $f(a)$ is the *maximum value of f on S*, and that f *attains its maximum value on the set S at a*.

Notice that if S is the entire domain of f, then the minimum and maximum values of f on S are just the minimum and maximum values of f, as defined earlier.

Definition 4.3.5. Suppose f is a function and a is a number in the domain of f.

1. We say that f has a *local minimum* at a if there is some number $d > 0$ such that the interval $(a - d, a + d)$ is contained in the domain of f, and f attains its minimum value on $(a - d, a + d)$ at a.

2. We say that f has a *local maximum* at a if there is some $d > 0$ such that the interval $(a - d, a + d)$ is contained in the domain of f, and f attains its maximum value on $(a - d, a + d)$ at a.

Looking back at the graph of g in Figure 4.15, we see that g is strictly increasing on $(0, 2]$ and strictly decreasing on $[2, 4)$. Therefore g attains its maximum value on the interval $(0, 4)$ at 2, so g has a local maximum at 2. Similarly, g attains its minimum value on the interval $(2, 6)$ at 4, and therefore g has a local minimum at 4.

The local maximum of g occurs at a point where the derivative is 0, and the local minimum occurs at a point where the derivative is undefined. It turns out that these are instances of a general pattern:

Theorem 4.3.6. *Suppose f has a local maximum or minimum at a. Then either f is not differentiable at a or $f'(a) = 0$.*

Proof. We assume f has a local maximum at a; the proof for local minima is similar. We will assume that f is differentiable at a, and prove that $f'(a) = 0$; this will establish that either f is not differentiable at a or $f'(a) = 0$. Since f is differentiable at a, we know that

$$\lim_{x \to a} \frac{f(x) - f(a)}{x - a} = f'(a),$$

and therefore

$$\lim_{x \to a^-} \frac{f(x) - f(a)}{x - a} = \lim_{x \to a^+} \frac{f(x) - f(a)}{x - a} = f'(a).$$

We consider these two one-sided limits separately.

By the definition of local maximum, there is some number $d > 0$ such that the interval $(a - d, a + d)$ is contained in the domain of f, and for all x, if $a - d < x < a + d$ then $f(x) \le f(a)$. Thus, for $a - d < x < a$ we have $f(x) - f(a) \le 0$ and $x - a < 0$, and therefore $(f(x) - f(a))/(x - a) \ge 0$. Since weak inequalities are preserved by limits (Corollary 2.8.9), it follows that

$$f'(a) = \lim_{x \to a^-} \frac{f(x) - f(a)}{x - a} \ge 0.$$

Similarly, for $a < x < a + d$ we have $(f(x) - f(a))/(x - a) \le 0$, and therefore

$$f'(a) = \lim_{x \to a^+} \frac{f(x) - f(a)}{x - a} \le 0.$$

Combining these two inequalities, we conclude that $f'(a) = 0$.   □

The numbers identified in Theorem 4.3.6 are important enough to have a name:

Definition 4.3.7. A number a in the domain of f such that either $f'(a)$ is undefined or $f'(a) = 0$ is called a *critical number* of f.

Thus, our last theorem says that a local maximum or minimum of a function can occur only at a critical number of the function. How can you tell if a critical number is the location of either a local maximum or a local minimum? The following theorem gives a convenient test.

Theorem 4.3.8 (First Derivative Test). *Suppose that f is continuous at a, and for some number $d > 0$, f is differentiable on the intervals $(a - d, a)$ and $(a, a + d)$.*

1. *Suppose that for every number x, if $a - d < x < a$ then $f'(x) < 0$, and if $a < x < a + d$ then $f'(x) > 0$. Then f has a local minimum at a.*

2. *Suppose that for every number x, if $a - d < x < a$ then $f'(x) > 0$, and if $a < x < a + d$ then $f'(x) < 0$. Then f has a local maximum at a.*

Proof. The proofs of the two parts are similar, so we prove only part 1. Since f is differentiable on the intervals $(a - d, a)$ and $(a, a + d)$, it is also continuous on these intervals. It is also continuous at a, so it is continuous on the interval $(a - d, a + d)$. Since $f'(x)$ is negative on the interval $(a - d, a)$ and positive on $(a, a + d)$, f is strictly decreasing on $(a - d, a]$ and strictly increasing on $[a, a + d)$. It follows that f attains its minimum value on the interval $(a - d, a + d)$ at a, and therefore f has a local minimum at a. $\square$

Example 4.3.9. Let $f(x) = 3x^4 + 4x^3 - 6x^2 - 12x + 7$. Find all critical numbers of f, and determine whether f has a local maximum or minimum at each of these numbers. Determine intervals on which f is increasing or decreasing. Determine the maximum and minimum values of f, if they exist. Sketch the graph of f.

Solution. To find the critical numbers, we first compute

$$f'(x) = 12x^3 + 12x^2 - 12x - 12 = 12(x^3 + x^2 - x - 1)$$
$$= 12(x^2(x + 1) - (x + 1)) = 12(x + 1)(x^2 - 1) = 12(x + 1)^2(x - 1).$$

There are no points where f is not differentiable, but there are two points where the derivative is 0:

$$f'(x) = 12(x + 1)^2(x - 1) = 0,$$
$$x = -1, 1.$$

Thus, the only critical numbers are ± 1.

Notice that f' is continuous at all numbers. It follows, by the intermediate value theorem, that if $f'(x)$ is negative at one value of x and positive at another, then it must be 0 somewhere in between. In other words, $f'(x)$ can change sign only at points where it is equal to 0—that is, only at the critical numbers ± 1. Thus, $f'(x)$ must have one sign throughout each of the intervals $(-\infty, -1)$, $(-1, 1)$, and $(1, \infty)$. We can determine the sign of $f'(x)$ on each of these intervals by simply checking one number in each interval:

$$f'(-2) = -36 < 0, \qquad f'(0) = -12 < 0, \qquad f'(2) = 108 > 0.$$

Thus, $f'(x)$ is negative on the intervals $(-\infty, -1)$ and $(-1, 1)$, and positive on $(1, \infty)$. We summarize our findings as follows:

$$f'(x) \quad \underline{\quad - \quad\quad 0 \quad\quad - \quad\quad 0 \quad\quad + \quad\quad}$$
$$x \quad\quad\quad\quad\quad\quad -1 \quad\quad\quad\quad 1$$

To sketch the graph of f, we first plot the points at $x = \pm 1$:

$$f(-1) = 12, \qquad f(1) = -4.$$

The graph has horizontal tangent lines at the points $(-1, 12)$ and $(1, -4)$. Also, f is strictly decreasing on the intervals $(-\infty, -1]$ and $[-1, 1]$, and strictly increasing on $[1, \infty)$. Thus, the shape of the graph of f must be as shown in Figure 4.16.

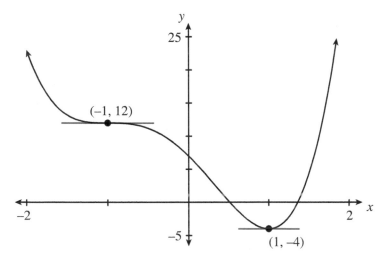

Figure 4.16: The graph of $f(x) = 3x^4 + 4x^3 - 6x^2 - 12x + 7$.

By the first derivative test, f has a local minimum at 1. But f is strictly decreasing on both of the intervals $(-\infty, -1]$ and $[-1, 1]$, so f has neither a local minimum nor a local maximum at -1. In fact, although the tangent line at $x = -1$ has slope 0, all *secant* lines in the interval $(-\infty, 1]$ have negative slope, and therefore f is strictly decreasing on $(-\infty, 1]$. From the graph, it is clear that the local minimum at $x = 1$ is actually the lowest point on the entire graph, so the minimum value of f is -4, and it is attained at 1. The function f has no maximum value. $\qquad \square$

Example 4.3.9 illustrates that a function need not have a local maximum or minimum at a critical number. In the example, the critical number $x = -1$ turned out to be neither a local maximum nor a local minimum, but merely the location of a "wiggle" in the graph of f. The only reason we detected the wiggle in this case is that the wiggle was large enough that the curve leveled off at this point, creating a critical number. A smaller wiggle might have gone undetected by the methods we have used in this section. And this raises the question of whether there might be other wiggles that we missed. In the next section, we will develop methods for detecting such wiggles.

But before discussing the detection of wiggles, we give one more example to illustrate that the methods of this section can also be useful in the study of sequences.

Example 4.3.10. Determine whether the sequence $(a_n)_{n=1}^{\infty}$ given by the formula

$$a_n = \frac{2n+3}{n+2}$$

is monotone.

Solution. The methods we have discussed in this section do not apply directly to the sequence $(a_n)_{n=1}^{\infty}$. The reason is that it would make no sense to compute the derivative of a_n with respect to n; since n stands for an integer, it cannot change by an amount Δn that approaches 0.

However, we can use derivatives to study the function

$$f(x) = \frac{2x+3}{x+2}$$

on the interval $[1, \infty)$. The function f is continuous on this interval, and its derivative is

$$f'(x) = \frac{(x+2) \cdot 2 - (2x+3) \cdot 1}{(x+2)^2} = \frac{1}{(x+2)^2} > 0.$$

Therefore f is strictly increasing on the interval $[1, \infty)$. This means that for any real numbers x_1 and x_2 with $1 \le x_1 < x_2$, $f(x_1) < f(x_2)$. In particular, for every positive integer n, $f(n) < f(n+1)$. But $f(n) = a_n$, so this means that for every positive integer n, $a_n < a_{n+1}$. In other words, $(a_n)_{n=1}^{\infty}$ is strictly increasing. $\square$

Exercises 4.3

1. Figure 4.17 shows the graphs of two functions. In each case, identify intervals on which the function is strictly increasing, weakly increasing, strictly decreasing,

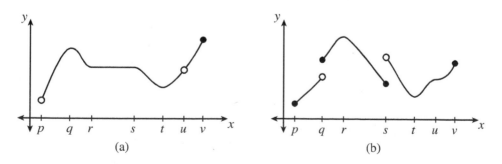

Figure 4.17: The graphs of two functions.

weakly decreasing, and constant. Also, determine whether or not the function has maximum and minimum values, and if so where they are attained, and find the locations of any local maxima and minima.

2–12: Find all critical numbers of the function. Determine intervals on which the function is strictly increasing and strictly decreasing, and find all local maxima and minima. Sketch the graph.

2. $f(x) = x^2 - 6x + 5$.

3. $g(x) = 3x - x^2$.

4. $f(x) = 12x - x^3$.

5. $g(x) = (x^2 - 1)^2$.

6. $f(x) = x^4/2 - 4x^3 + 9x^2 - 5$.

7. $g(x) = x^4/2 - 4x^3 + 5x^2 - 5$.

8. $f(x) = x - 4\sqrt{x}$.

9. $g(x) = x^2 - 4\sqrt{x}$.

10. $f(x) = \dfrac{1}{x^2 + 1}$. (Note: To make your graph more accurate, find the horizontal asymptote.)

11. $g(x) = \dfrac{x}{x^2 + 1}$.

12. $h(x) = \dfrac{x^2}{x^2 + 1}$.

13–20: Determine whether the sequence is strictly increasing, strictly decreasing, or neither.

13. $\left(\dfrac{n+1}{n^2+3}\right)_{n=1}^{\infty}$.

14. $\left(\dfrac{n-1}{n^2+8}\right)_{n=1}^{\infty}$.

15. $\left(1/n^2 - 1/n\right)_{n=1}^{\infty}$.

16. $\left(1/n^2 - 4/n\right)_{n=1}^{\infty}$.

17. $\left(\dfrac{2n^3}{3} - 5n^2 + 12n\right)_{n=1}^{\infty}$.

18. $\left(\dfrac{4n^3}{3} - 9n^2 + 20n\right)_{n=1}^{\infty}$.

19. $\left(\sqrt{n^2+2} - n\right)_{n=1}^{\infty}$.

20. $\left(\sqrt{n^2 + 2n} - n\right)_{n=1}^{\infty}$.

21. Suppose a and b are real numbers, and $f(x) = ax^4 + bx^3$. If the maximum value of f is 1, and f attains this maximum value at -1, what are a and b?

22. Suppose that the domain of f is an interval I, the domain of g is an interval J, and for all $x \in J$, $g(x) \in I$. It follows that $(f \circ g)(x)$ is defined for all $x \in J$.

    (a) Show that if f is strictly increasing on I and g is strictly increasing on J, then $f \circ g$ is strictly increasing on J.

    (b) Show that if f is strictly decreasing on I and g is strictly decreasing on J, then $f \circ g$ is strictly increasing on J.

    (c) Suppose that f is strictly increasing on I and g is strictly decreasing on J. Is $f \circ g$ strictly increasing or strictly decreasing on J?

23. Prove part 4 of Theorem 4.3.2.

24. Prove part 5 of Theorem 4.3.2.

4.4 Concavity

Consider the function f whose graph is shown in Figure 4.18. If this graph were a road and you were driving on this road from left to right, then you would have the steering wheel turned to the left from $x = a$ to $x = b$, and to the right from $x = b$ to $x = c$. We say that f is *concave up* on the interval $[a, b]$, and *concave down* on $[b, c]$.

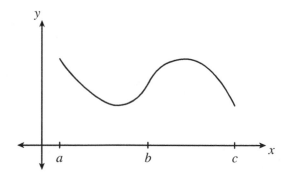

Figure 4.18: The graph of a function f that is concave up on $[a, b]$ and concave down on $[b, c]$.

While this description gives a good intuitive idea of the difference between concave up and concave down, it is not precise enough to be useful in mathematical reasoning. To formulate a precise definition, we consider triples of points in each interval. As you can see in Figure 4.19, if we choose a triple of points on the graph of f with x-coordinates in the interval $[a, b]$, then the intermediate point is below the secant line through the

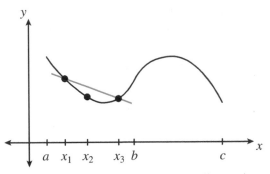

(a) In the interval $[a, b]$, the intermediate point is always below the secant line through the other two points.

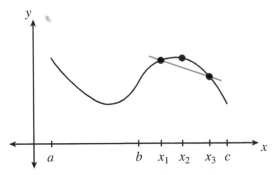

(b) In the interval $[b, c]$, the intermediate point is always above the secant line through the other two points.

Figure 4.19: Triples of points on the graph of f.

other two points, whereas in the interval $[b, c]$, the intermediate point is above the secant line through the other two points. We base our precise definition of concavity on this distinction.

Definition 4.4.1. Let f be a function and I an interval contained in the domain of f.

1. We say that f is *concave up* on I if for all numbers $x_1, x_2, x_3 \in I$, if $x_1 < x_2 < x_3$ then the point $(x_2, f(x_2))$ is below the secant line through the points $(x_1, f(x_1))$ and $(x_3, f(x_3))$.

2. We say that f is *concave down* on I if for all numbers $x_1, x_2, x_3 \in I$, if $x_1 < x_2 < x_3$ then the point $(x_2, f(x_2))$ is above the secant line through the points $(x_1, f(x_1))$ and $(x_3, f(x_3))$.

Let's work this out algebraically. Suppose $x_1 < x_2 < x_3$, and let ℓ be the secant line through the points $(x_1, f(x_1))$ and $(x_3, f(x_3))$. The slope of ℓ is

$$m = \frac{f(x_3) - f(x_1)}{x_3 - x_1}.$$

Since ℓ passes through the point $(x_1, f(x_1))$, we can write the equation of ℓ in the point-slope form $y - f(x_1) = m(x - x_1)$, or equivalently

$$y = f(x_1) + m(x - x_1).$$

Therefore ℓ passes through the point $(x_2, f(x_1) + m(x_2 - x_1))$.

Now suppose that the point $(x_2, f(x_2))$ is below ℓ. This means that $f(x_2) < f(x_1) + m(x_2 - x_1)$, and rearranging we get

$$\frac{f(x_2) - f(x_1)}{x_2 - x_1} < m. \tag{4.14}$$

We could have used the point $(x_3, f(x_3))$, instead of $(x_1, f(x_1))$, when writing the equation of ℓ in point-slope form. This would lead to the inequality $f(x_2) < f(x_3) + m(x_2 - x_3)$, which is equivalent to

$$\frac{f(x_3) - f(x_2)}{x_3 - x_2} > m. \tag{4.15}$$

Combining inequalities (4.14) and (4.15), we conclude that

$$\frac{f(x_2) - f(x_1)}{x_2 - x_1} < \frac{f(x_3) - f(x_2)}{x_3 - x_2}.$$

If we define

$$m_L = \frac{f(x_2) - f(x_1)}{x_2 - x_1}, \qquad m_R = \frac{f(x_3) - f(x_2)}{x_3 - x_2},$$

then m_L is the slope of the secant line through the leftmost pair of points, $(x_1, f(x_1))$ and $(x_2, f(x_2))$, and m_R is the slope of the secant line through the rightmost points, $(x_2, f(x_2))$ and $(x_3, f(x_3))$. What we have shown is that if $(x_2, f(x_2))$ is below ℓ, then $m_L < m_R$. This should be evident in Figure 4.19a.

Similar reasoning shows that if $(x_2, f(x_2))$ is above ℓ then $m_L > m_R$, and if $(x_2, f(x_2))$ lies on ℓ then $m_L = m_R$. Thus, we have found an algebraic test to determine the position of $(x_2, f(x_2))$ relative to ℓ. This gives us an alternative way to state the meaning of concavity.

Theorem 4.4.2. *Suppose that f is a function and I is an interval contained in the domain of f.*

1. *The function f is concave up on I if and only if for all numbers $x_1, x_2, x_3 \in I$, if $x_1 < x_2 < x_3$ then*

$$\frac{f(x_2) - f(x_1)}{x_2 - x_1} < \frac{f(x_3) - f(x_2)}{x_3 - x_2}.$$

2. *The function f is concave down on I if and only if for all numbers $x_1, x_2, x_3 \in I$, if $x_1 < x_2 < x_3$ then*

$$\frac{f(x_2) - f(x_1)}{x_2 - x_1} > \frac{f(x_3) - f(x_2)}{x_3 - x_2}.$$

Theorem 4.4.2 gives us a way to relate concavity to slopes of secant lines. Combining this with the mean value theorem, which relates slopes of secant lines to slopes of tangent lines, we should be able to relate concavity to derivatives. In fact, it turns out that the *second* derivative of f can be used to determine the concavity of f.

Theorem 4.4.3. *Suppose f is continuous on an interval I and twice differentiable on the interior of I; in other words, for all x in the interior of I, $f''(x)$ is defined.*

1. *If $f''(x) > 0$ for all x in the interior of I, then f is concave up on I.*

2. *If $f''(x) < 0$ for all x in the interior of I, then f is concave down on I.*

Proof. For part 1, suppose that $f''(x) > 0$ for all x in the interior of I. Then by the results of the last section, f' is strictly increasing on the interior of I. Now consider three numbers $x_1, x_2, x_3 \in I$ with $x_1 < x_2 < x_3$. Since $[x_1, x_2] \subseteq I$, we know that f is continuous on $[x_1, x_2]$ and differentiable on (x_1, x_2), so by the mean value theorem, there is a number c_1 such that $x_1 < c_1 < x_2$ and

$$\frac{f(x_2) - f(x_1)}{x_2 - x_1} = f'(c_1).$$

Similarly, by the mean value theorem there is a number c_2 such that $x_2 < c_2 < x_3$ and

$$\frac{f(x_3) - f(x_2)}{x_3 - x_2} = f'(c_2).$$

But now since f' is strictly increasing on the interior of I and $x_1 < c_1 < x_2 < c_2 < x_3$, we have $f'(c_1) < f'(c_2)$, and therefore

$$\frac{f(x_2) - f(x_1)}{x_2 - x_1} < \frac{f(x_3) - f(x_2)}{x_3 - x_2}.$$

It follows, by Theorem 4.4.2, that f is concave up on I. The proof of part 2 is similar. □

For example, consider again the function $f(x) = 3x^4 + 4x^3 - 6x^2 - 12x + 7$ from Example 4.3.9. We can gain a better understanding of the graph of f by determining intervals on which it is concave up and concave down. We begin by computing $f''(x)$ and determining when it is 0. Since we showed in Example 4.3.9 that $f'(x) = 12(x^3 + x^2 - x - 1)$, we have

$$f''(x) = 12(3x^2 + 2x - 1) = 12(x + 1)(3x - 1),$$

and therefore $f''(x) = 0$ when $x = -1$ or $x = 1/3$. Since f'' is continuous everywhere, these are the only points where $f''(x)$ can change sign. Thus, as before, we can determine the sign of $f''(x)$ on each of the intervals $(-\infty, -1)$, $(-1, 1/3)$, and $(1/3, \infty)$ by testing one sample point in each interval:

$$f''(-2) = 84 > 0, \qquad f''(0) = -12 < 0, \qquad f''(1) = 48 > 0.$$

The sign of $f''(x)$ can therefore be summarized as follows:

$$f''(x) \quad \underline{\quad + \quad 0 \quad - \quad 0 \quad + \quad}$$
$$ x \qquad\qquad -1 \qquad\quad 1/3$$

Thus, f is concave up on the interval $(-\infty, -1]$, concave down on $[-1, 1/3]$, and concave up on $[1/3, \infty)$.

At the end of the last section, we noted that the graph of f had a "wiggle" in it at the point $(-1, 12)$. We can now say more about the nature of this wiggle: this is a point where the concavity of f changes. Such a point is called an *inflection point*.

Definition 4.4.4. A point $(a, f(a))$ on the graph of a function f is called an *inflection point* if there is a number $d > 0$ such that either f is concave up on $(a - d, a]$ and concave down on $[a, a + d)$, or f is concave down on $(a - d, a]$ and concave up on $[a, a + d)$.

In fact, we have discovered that there is a second inflection point on the graph of f, at $x = 1/3$. Since

$$f(1/3) = 68/27 \approx 2.52,$$

the coordinates of the second inflection point are $(1/3, 68/27)$. The graph of f, with the inflection points and minimum point marked, is shown in Figure 4.20.

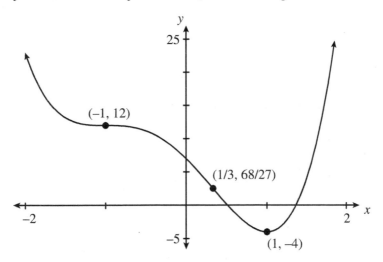

Figure 4.20: The graph of $y = 3x^4 + 4x^3 - 6x^2 - 12x + 7$.

Example 4.4.5. Sketch the graph of the function $f(x) = -x^3/3 - x^2 + 3x + 2$.

Solution. We begin by computing the first two derivatives of f:

$$f'(x) = -x^2 - 2x + 3 = -(x + 3)(x - 1),$$
$$f''(x) = -2x - 2 = -2(x + 1).$$

Notice that f, f', and f'' are all continuous everywhere. Clearly $f'(x) = 0$ when $x = -3, 1$, and $f''(x) = 0$ when $x = -1$. Next, we test sample points:

$$f'(-4) = -5 < 0, \qquad f'(0) = 3 > 0, \qquad f'(2) = -5 < 0,$$
$$f''(-2) = 2 > 0, \qquad f''(0) = -2 < 0.$$

We can summarize what we have learned on a number line:

$f''(x)$	$+$		$+$		$+$		0		$-$		$-$		$-$
$f'(x)$	$-$		0		$+$		$+$		$+$		0		$-$
x			-3				-1				1		

Clearly the most important points to plot are the points with $x = -3, -1, 1$, so we compute:

$$f(-3) = -7, \qquad f(-1) = -\frac{5}{3}, \qquad f(1) = \frac{11}{3}.$$

We can now draw the following conclusions: From the sign of $f'(x)$, we see that f is strictly decreasing on the interval $(-\infty, -3]$, strictly increasing on the interval $[-3, 1]$, and strictly decreasing on $[1, \infty)$, and f has a local minimum at -3 and a local maximum at 1. The tangent lines at the points $(-3, -7)$ and $(1, 11/3)$ are horizontal. From the sign of $f''(x)$, we see that f is concave up on $(-\infty, -1]$ and concave down on $[-1, \infty)$, so the point $(-1, -5/3)$ is an inflection point. For the purpose of getting the shape of each segment of the graph right when we draw the graph, it might be best to summarize our conclusions like this:

$(-\infty, -3]$: strictly decreasing, concave up.
$[-3, -1]$: strictly increasing, concave up.
$[-1, 1]$: strictly increasing, concave down.
$[1, \infty)$: strictly decreasing, concave down.

The graph of f is shown in Figure 4.21.   □

In addition to helping us determine concavity, the second derivative of a function can also help us classify critical numbers as local maxima or minima:

Theorem 4.4.6 (Second Derivative Test). *Suppose that $f'(a) = 0$.*

1. *If $f''(a) > 0$, then f has a local minimum at a.*

2. *If $f''(a) < 0$, then f has a local maximum at a.*

Proof. We will only prove part 1, since, as usual, the two parts are very similar. Suppose $f''(a) > 0$. Applying Theorem 3.2.3 and using the fact that $f'(a) = 0$, we have

$$f''(a) = \lim_{x \to a} \frac{f'(x) - f'(a)}{x - a} = \lim_{x \to a} \frac{f'(x)}{x - a}.$$

Since $f''(a) > 0$, by Theorem 2.8.8 there must be some number $\delta > 0$ such that if $0 < |x - a| < \delta$ then $f'(x)/(x - a) > 0$.

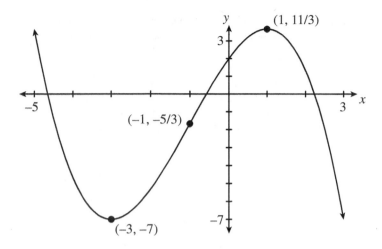

Figure 4.21: The graph of $y = -x^3/3 - x^2 + 3x + 2$.

Now, suppose x is any number such that $a - \delta < x < a$. Then $x - a < 0$, and multiplying the inequality $f'(x)/(x - a) > 0$ by the negative number $x - a$ we conclude that $f'(x) < 0$. Similarly, for $a < x < a + \delta$ we conclude that $f'(x) > 0$. But this means we can now apply the first derivative test (Theorem 4.3.8) to conclude that f has a local minimum at a. $\square$

The proof of Theorem 4.4.6 actually shows a little bit more than what the theorem says. If $f'(a) = 0$ and $f''(a) > 0$, then the proof shows that there is a positive number δ such that for every number x, if $a - \delta < x < a$ then $f'(x) < 0$, and if $a < x < a + \delta$ then $f'(x) > 0$. Therefore f is decreasing on $(a - \delta, a]$ and increasing on $[a, a + \delta)$. Similarly, if $f'(a) = 0$ and $f''(a) < 0$ then there is a positive number δ such that f is increasing on $(a - \delta, a]$ and decreasing on $[a, a + \delta)$.

Example 4.4.7. Sketch the graph of the function

$$f(x) = \frac{(x + 1)^2}{x^2 + 1}.$$

Solution. We use the quotient rule to compute $f'(x)$ and $f''(x)$. It is worthwhile to simplify both as much as possible:

$$\begin{aligned}
f'(x) &= \frac{(x^2 + 1) \cdot 2(x + 1) - (x + 1)^2 \cdot 2x}{(x^2 + 1)^2} \\
&= \frac{2(x + 1)[(x^2 + 1 - x(x + 1)]}{(x^2 + 1)^2} \\
&= \frac{2(x + 1)(1 - x)}{(x^2 + 1)^2} = \frac{2(1 - x^2)}{(x^2 + 1)^2},
\end{aligned}$$

$$f''(x) = \frac{(x^2+1)^2 \cdot (-4x) - 2(1-x^2) \cdot 2(x^2+1)(2x)}{(x^2+1)^4}$$

$$= \frac{4x(x^2+1)[-(x^2+1) - 2(1-x^2)]}{(x^2+1)^4} = \frac{4x(x^2-3)}{(x^2+1)^3}.$$

Since $x^2 + 1$ is never 0, all three of f, f', and f'' are defined and continuous everywhere. We have $f'(x) = 0$ at $x = \pm 1$, and $f''(x) = 0$ at $x = 0, \pm\sqrt{3}$. Testing sample points, we find that the signs of $f'(x)$ and $f''(x)$ can be summarized as follows:

$f''(x)$	$-$	0	$+$	$+$	$+$	0	$-$	$-$	$-$	0	$+$
$f'(x)$	$-$	$-$	$-$	0	$+$	$+$	$+$	0	$-$	$-$	$-$
x		$-\sqrt{3}$		-1		0		1		$\sqrt{3}$	

We have five important points to plot: inflection points at $x = -\sqrt{3}, 0, \sqrt{3}$, a local minimum at $x = -1$, and a local maximum at $x = 1$. We plug into the formula for $f(x)$ to find the y-coordinates of these points:

$$f(-\sqrt{3}) = 1 - \frac{\sqrt{3}}{2} \approx 0.13, \quad f(-1) = 0, \quad f(0) = 1,$$

$$f(1) = 2, \quad f(\sqrt{3}) = 1 + \frac{\sqrt{3}}{2} \approx 1.87.$$

Plotting these points, and filling in the shape of the graph using the signs of $f'(x)$ and $f''(x)$, we get the graph shown in Figure 4.22.

We know that f has a local minimum at -1, but is $f(-1) = 0$ the minimum value of f on its entire domain? We can't tell yet. We know that on the interval $[\sqrt{3}, \infty)$, f is strictly decreasing and concave up. If we extend the curve in Figure 4.22 further to the right, how low will it go? If it eventually goes below 0, then 0 will not be the minimum value of f. But if it never goes below 0, then 0 is the minimum value of f, attained at $x = -1$. Similarly, we can't tell whether $f(1) = 2$ is the maximum value of f on its entire domain until we investigate how high the curve goes if we extend it further to the left. The question marks in Figure 4.22 indicate that we still have more issues to resolve at the left and right ends of the graph.

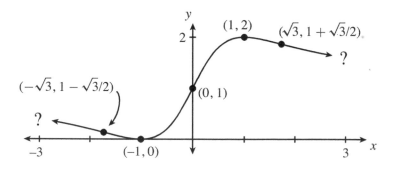

Figure 4.22: The middle part of the graph of $y = (x+1)^2/(x^2+1)$.

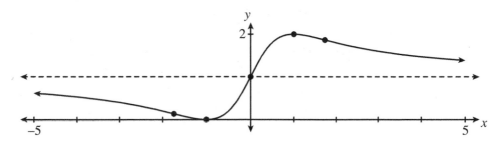

Figure 4.23: The graph of $y = (x+1)^2/(x^2+1)$. The dashed line is the asymptote $y = 1$.

To resolve these issues, we need to know what happens to $f(x)$ when x gets very large or very small, so we compute the limits of $f(x)$ as $x \to \pm\infty$. In fact, both computations are the same:

$$\lim_{x \to \pm\infty} f(x) = \lim_{x \to \pm\infty} \frac{x^2 + 2x + 1}{x^2 + 1} = \lim_{x \to \pm\infty} \frac{x^2 \cdot (1 + 2/x + 1/x^2)}{x^2 \cdot (1 + 1/x^2)}$$
$$= \lim_{x \to \pm\infty} \frac{1 + 2/x + 1/x^2}{1 + 1/x^2} = 1.$$

Thus, the line $y = 1$ is a horizontal asymptote, and the graph approaches this asymptote at both ends. In Figure 4.23 we have added this additional information to the graph. Since $f(x)$ is strictly decreasing on $[1, \infty)$ and approaches 1 as x approaches ∞, $f(x)$ must be larger than 1 for all positive x. Similarly, $f(x) < 1$ for $x < 0$. Therefore 0 is, indeed, the minimum value of f, and 2 is the maximum value. □

Exercises 4.4

1. Figure 4.24 shows the graphs of two functions. In each case, identify intervals on which the function is concave up and concave down. Also, find the x-coordinates of all points of inflection.

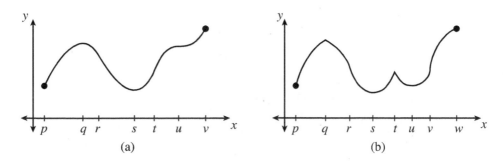

(a)   (b)

Figure 4.24: The graphs of two functions.

2–17: Determine intervals on which the function is strictly increasing, strictly decreasing, concave up, and concave down. Find all local maxima and minima and all points of inflection. Sketch the graph.

2. $f(x) = x^3 - 3x^2 + 3.$

3. $g(x) = x^3 + 6x^2 - 15x - 20.$

4. $f(x) = \dfrac{x^4}{4} - \dfrac{2x^3}{3} - \dfrac{15x^2}{2}.$

5. $f(x) = \dfrac{x^5}{5} - x^4 + 20.$

6. $g(x) = -x^3(x + 4).$

7. $f(x) = \sqrt[3]{x}.$

8. $g(x) = x^2 + \sqrt{x}.$

9. $f(x) = 3x^{5/3} - 5x.$

10. $f(x) = \dfrac{1}{x^2 + 3}.$

11. $f(x) = \dfrac{x}{x^2 + 3}.$

12. $f(x) = \dfrac{x^2}{x^2 + 3}.$

13. $g(x) = \dfrac{x^3}{x^2 + 3}.$

14. $f(x) = 16\sqrt{x^2 + 1} - x^2.$

15. $f(x) = x + 2\sin x, \ 0 \le x \le 2\pi.$

16. $g(x) = x^2 + 4\sin x, \ 0 \le x \le 2\pi.$ (Hint: First graph $g'(x)$, and determine the minimum and maximum values of g'.)

17. $f(x) = 3\tan x - 4x, \ -\pi/2 < x < \pi/2.$

18. Suppose that f is concave up on an interval I and $x_1, x_2, x_3,$ and x_4 are numbers in I with $x_1 < x_2 < x_3 < x_4.$

    (a) Show that the slope of the secant line through the points at $x = x_1$ and $x = x_2$ is less than the slope of the secant line through the points at $x = x_3$ and $x = x_4.$

    (b) Show that the slope of the secant line through the points at $x = x_1$ and $x = x_3$ is less than the slope of the secant line through the points at $x = x_2$ and $x = x_4.$

19. Suppose that the domain of f is an interval I, the domain of g is an interval J, and for all $x \in J$, $g(x) \in I$. It follows that $(f \circ g)(x)$ is defined for all $x \in J$.

(a) Show that if f is increasing and concave up on I and g is increasing and concave up on J, then $f \circ g$ is concave up on J. (Do not assume that the functions are differentiable.)

(b) If f is concave up on I and g is concave up on J, must it be the case that $f \circ g$ is concave up on J? Give either a proof or a counterexample.

4.5 Sophisticated Graphing

In this section we use the methods we have developed to graph some more complicated functions. In the previous section, the domain of every function we graphed was $\mathbb{R}$, the set of all real numbers. However, in this section some functions will be undefined at some numbers.

When sketching the graph of a function f, we will generally use the following procedure:

1. Determine the domain of f.

2. Find $f'(x)$, and use it to determine intervals on which f is increasing or decreasing, and local maximum and minimum points.

3. Find $f''(x)$, and use it to determine intervals on which f is concave up or down, and inflection points.

4. Use limits to find all vertical and horizontal asymptotes.

5. Plot all important points, and then sketch the curve.

Since the examples in this section will be more difficult, we add a couple of additional suggestions:

6. Use common sense. If there's something about the shape of the graph you're not sure about, do whatever additional calculations are necessary to resolve your doubts.

7. Check your work: after sketching the graph of a function, make sure your sketch is compatible with all of your calculations. If not, find your mistake and fix it.

As an example of item 6, if you find a number a such that $f(a)$ is defined but $f'(a)$ is not, you might investigate *why* the derivative is undefined. Is there a vertical tangent line at $x = a$? Does the graph have a sudden change of direction at $x = a$?

As an example of item 7, suppose you determine that f is strictly decreasing on a certain interval, and then you plot two points in that interval and find that the one on the right is higher. Something is wrong! If the function is strictly decreasing on the interval, then the point on the right should be lower. You must have made a mistake somewhere—go back and find it.

Example 4.5.1. Sketch the graph of the function

$$f(x) = \frac{x}{x^2 - 1}.$$

Solution. Notice that $f(x)$ is undefined when $x = \pm 1$. The domain of f is $(-\infty, -1) \cup (-1, 1) \cup (1, \infty)$, and f is continuous at all numbers in this domain. The first two derivatives are:

$$f'(x) = \frac{(x^2 - 1) \cdot 1 - x \cdot (2x)}{(x^2 - 1)^2} = -\frac{x^2 + 1}{(x^2 - 1)^2},$$

$$f''(x) = -\frac{(x^2 - 1)^2 \cdot (2x) - (x^2 + 1) \cdot 2(x^2 - 1)(2x)}{(x^2 - 1)^4}$$

$$= -\frac{2x(x^2 - 1)[(x^2 - 1) - (2x^2 + 2)]}{(x^2 - 1)^4} = \frac{2x(x^2 + 3)}{(x^2 - 1)^3}.$$

Looking at the numerator and denominator of $f'(x)$, it is easy to see that for all $x \neq \pm 1$, $x^2 + 1 > 0$ and $(x^2 - 1)^2 > 0$, and therefore $f'(x) < 0$. Clearly $f''(x) = 0$ when $x = 0$, but also $f''(x)$ is undefined when $x = \pm 1$. On each of the intervals $(-\infty, -1)$, $(-1, 0)$, $(0, 1)$, and $(1, \infty)$, f'' is continuous and nonzero, so it cannot change sign on any of these intervals. We can therefore determine its sign by checking a sample point in each interval. The results are as follows, where we have marked points where a function is undefined with the letter "u":

$f''(x)$	$-$	u	$+$	0	$-$	u	$+$
$f'(x)$	$-$	u	$-$	$-$	$-$	u	$-$
x		-1		0		1	

There are no local maximum or minimum points, but the concavity changes at $x = 0, \pm 1$. The numbers -1 and 1 are not in the domain of f, so we don't have inflection points at these x-coordinates. But $f(0) = 0$, so the point $(0, 0)$ is an inflection point.

Since $f(x)$ is undefined at $x = \pm 1$, we investigate what happens when x approaches ± 1 from either side. As $x \to 1^+$, $x^2 - 1 \to 0^+$, and therefore

$$\lim_{x \to 1^+} f(x) = \lim_{x \to 1^+} \frac{x}{x^2 - 1} = \infty.$$

Similarly, we have

$$\lim_{x \to 1^-} f(x) = -\infty, \qquad \lim_{x \to -1^+} f(x) = \infty, \qquad \lim_{x \to -1^-} f(x) = -\infty.$$

Thus, the vertical lines $x = -1$ and $x = 1$ are asymptotes.

To check for horizontal asymptotes, we compute

$$\lim_{x \to \pm\infty} f(x) = \lim_{x \to \pm\infty} \frac{x}{x^2 - 1} = \lim_{x \to \pm\infty} \frac{x}{x^2 \cdot (1 - 1/x^2)}$$

$$= \lim_{x \to \pm\infty} \frac{1}{x} \cdot \frac{1}{1 - 1/x^2} = 0 \cdot 1 = 0.$$

Thus, the graph will approach the horizontal asymptote $y = 0$ on both sides.

The graph of f comes in three pieces, one for each of the intervals $(-\infty, -1)$, $(-1, 1)$, and $(1, \infty)$ that make up the domain of f. We now have enough information about the shape of the graph on each of these intervals to draw a rough sketch. But if we want to make the sketch more accurate, there are a few additional calculations we could do.

On the interval $(-\infty, -1)$, we know that f is strictly decreasing and concave down, and the graph approaches the asymptotes $y = 0$ and $x = -1$. This tells us the shape of this part of the graph, but since we have not yet evaluated f at any number in this interval, we don't know exactly where to put it on the page. To give us a little more guidance, we could plot a point on this part of the graph:

$$f(-2) = -\frac{2}{3} \approx -0.67.$$

Similarly, to help with the part of the graph for $x > 1$ we compute

$$f(2) = \frac{2}{3} \approx 0.67.$$

On the interval $(-1, 1)$, we know there is an inflection point at $(0, 0)$. To make our graph more accurate, we will draw the tangent line at this point. To do this, we compute

$$f'(0) = -\frac{1}{1} = -1.$$

Thus, the tangent line is the line $y = -x$.

To sketch the graph of f, we plot the points we have computed, draw the tangent line at $(0, 0)$ and the asymptotes, and then use these as guides as we fill in the shape of the graph according to the signs of $f'(x)$ and $f''(x)$. The resulting graph is shown in Figure 4.25.

As a check on our work, let's run through everything we have computed and make sure it is reflected in our sketch. The signs of $f'(x)$ and $f''(x)$ tell us that on the interval

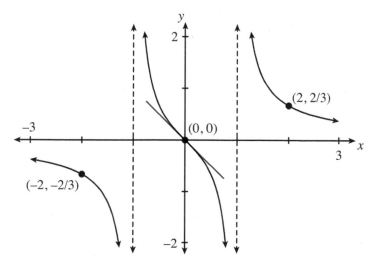

Figure 4.25: The graph of $y = x/(x^2 - 1)$. The red line is the line $y = -x$.

$(-\infty, -1)$, f is strictly decreasing and concave down, on $(-1, 0]$ it is strictly decreasing and concave up, on $[0, 1)$ it is strictly decreasing and concave down, and on $(1, \infty)$ it is strictly decreasing and concave up. All of this matches the curve we have drawn. Just to the right of $x = 1$ the curve goes up very high, which is in agreement with our calculation that $\lim_{x \to 1+} f(x) = \infty$. Similarly, just to the left of 1 the curve goes down very low, to the right of -1 it goes up, and to the left of -1 it goes down, just as all of our limits require. The curve also approaches the line $y = 0$ at both sides. The curve passes through all the points we computed, and it is tangent to the line $y = -x$ at the origin.   □

Example 4.5.2. Sketch the graph of the function

$$f(x) = \frac{2x^2 - 3x + 1}{x^2}.$$

Solution. Since $f(x)$ is undefined when $x = 0$, the domain of f is $(-\infty, 0) \cup (0, \infty)$. To compute derivatives, it is easiest to rewrite $f(x)$ in the form

$$f(x) = \frac{2x^2}{x^2} - \frac{3x}{x^2} + \frac{1}{x^2} = 2 - \frac{3}{x} + \frac{1}{x^2} = 2 - 3x^{-1} + x^{-2}.$$

We then have

$$f'(x) = 3x^{-2} - 2x^{-3} = \frac{3}{x^2} - \frac{2}{x^3} = \frac{3x - 2}{x^3},$$

$$f''(x) = -6x^{-3} + 6x^{-4} = -\frac{6}{x^3} + \frac{6}{x^4} = \frac{6(1 - x)}{x^4}.$$

Clearly $f'(2/3) = 0$, $f''(1) = 0$, and both $f'(0)$ and $f''(0)$ are undefined. On the intervals between these values the derivatives are continuous and nonzero, and by checking sample points we find that they have the following signs:

$f''(x)$	+	u	+	+	+	0	−
$f'(x)$	+	u	−	0	+	+	+
x		0		2/3		1	

We have a local minimum at $x = 2/3$ and an inflection point at $x = 1$, so we compute:

$$f(2/3) = -1/4, \qquad f(1) = 0.$$

Since $f(0)$ is undefined, we examine the behavior of $f(x)$ for x close to 0. Checking the numerator and denominator of $f(x)$ separately, we see that as $x \to 0^{\neq}$, $2x^2 - 3x + 1 \to 1$ and $x^2 \to 0^+$. Therefore

$$\lim_{x \to 0} f(x) = \lim_{x \to 0} \frac{2x^2 - 3x + 1}{x^2} = \infty,$$

so f has a vertical asymptote at $x = 0$. Also,

$$\lim_{x \to \pm\infty} f(x) = \lim_{x \to \pm\infty} \frac{2x^2 - 3x + 1}{x^2} = \lim_{x \to \pm\infty} \left(2 - \frac{3}{x} + \frac{1}{x^2} \right) = 2,$$

so the graph approaches the horizontal asymptote $y = 2$ on both sides.

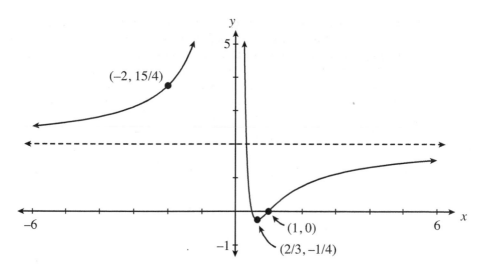

Figure 4.26: The graph of $y = (2x^2 - 3x + 1)/x^2$.

The graph of f comes in two pieces, one for $x < 0$ and one for $x > 0$, with a break at $x = 0$ where $f(x)$ is undefined. We have two points to plot on the right half, a local minimum at $(2/3, -1/4)$ and an inflection point at $(1, 0)$. We choose a point to plot on the left half as well:

$$f(-2) = \frac{15}{4} = 3.75.$$

Plotting all the points and asymptotes we have computed and then filling in the shape of the graph according to the signs of $f'(x)$ and $f''(x)$, we get the graph in Figure 4.26. □

Example 4.5.3. Sketch the graph of the function $g(x) = x\sqrt[3]{x+4}$.

Solution. The domain of g is $\mathbb{R}$, and g is continuous everywhere. To compute the derivatives, it is easiest to write $g(x) = x(x+4)^{1/3}$. To find $g'(x)$ we apply the product rule and simplify:

$$g'(x) = x \cdot \frac{1}{3}(x+4)^{-2/3} + (x+4)^{1/3} \cdot 1 = \frac{x}{3(x+4)^{2/3}} + (x+4)^{1/3} \cdot \frac{3(x+4)^{2/3}}{3(x+4)^{2/3}}$$

$$= \frac{x + 3(x+4)}{3(x+4)^{2/3}} = \frac{4(x+3)}{3(x+4)^{2/3}}.$$

For $g''(x)$, we use the quotient rule and simpify:

$$g''(x) = \frac{3(x+4)^{2/3} \cdot 4 - 4(x+3) \cdot 2(x+4)^{-1/3}}{9(x+4)^{4/3}} \cdot \frac{(x+4)^{1/3}}{(x+4)^{1/3}}$$

$$= \frac{12(x+4) - 8(x+3)}{9(x+4)^{5/3}} = \frac{4(x+6)}{9(x+4)^{5/3}}.$$

We have $g'(x) = 0$ when $x = -3$ and $g''(x) = 0$ when $x = -6$. Note also that $g'(x)$ and $g''(x)$ are undefined when $x = -4$. Checking sample points in the intervals between these points we get the following results:

$$
\begin{array}{c c c c c c c c}
g''(x) & + & 0 & - & u & + & + & + \\
g'(x) & - & - & - & u & - & 0 & + \\
\hline
x & & -6 & & -4 & & -3 &
\end{array}
$$

We have an inflection point at $x = -6$ and a local minimum at $x = -3$, so we will want to plot these points:

$$g(-6) = -6\sqrt[3]{-2} = 6\sqrt[3]{2} \approx 7.56, \qquad g(-3) = -3.$$

Notice that although $g'(-4)$ and $g''(-4)$ are undefined, $g(-4) = 0$, and in fact g is continuous at -4. Thus, g is concave down on $[-6, -4]$ and concave up on $[-4, \infty)$, and therefore $(-4, 0)$ is another inflection point.

Why is $g'(-4)$ undefined? Let's investigate:

$$g'(-4) = \lim_{h \to 0} \frac{g(-4+h) - g(-4)}{h} = \lim_{h \to 0} \frac{(-4+h)\sqrt[3]{h} - 0}{(\sqrt[3]{h})^3} = \lim_{h \to 0} \frac{h - 4}{(\sqrt[3]{h})^2}.$$

As $h \to 0^{\neq}$, $h - 4 \to -4$ and $(\sqrt[3]{h})^2 \to 0^+$, and therefore

$$\lim_{h \to 0} \frac{h - 4}{(\sqrt[3]{h})^2} = -\infty.$$

Thus, secant lines through $(-4, 0)$ and nearby points slope down very steeply, and the tangent line at $(-4, 0)$ is vertical. It is easy to check that $\lim_{x \to \pm\infty} g(x) = \infty$, so there are no horizontal asymptotes. The graph of g is shown in Figure 4.27.   □

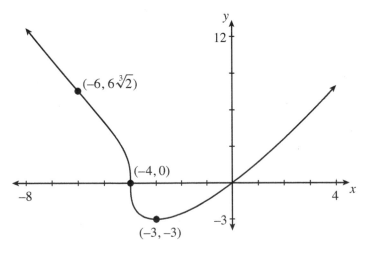

Figure 4.27: The graph of $y = x\sqrt[3]{x+4}$.

Example 4.5.4. Sketch the graph of the function

$$h(x) = \frac{x}{\sqrt{x^2 + x}}.$$

Solution. In order for $h(x)$ to be defined, we must have $x^2 + x > 0$. To solve this inequality, we first determine when $x^2 + x = 0$:

$$x^2 + x = x(x+1) = 0,$$
$$x = -1, 0.$$

Next, we check sample points in the intervals $(-\infty, -1)$, $(-1, 0)$, and $(0, \infty)$:

$$\text{When } x = -2, \ x^2 + x = 2,$$
$$\text{When } x = -1/2, \ x^2 + x = -1/4,$$
$$\text{When } x = 1, \ x^2 + x = 2.$$

The sign of $x^2 + x$ can therefore be summarized as follows:

$x^2 + x$	$+$	0	$-$	0	$+$
x		-1		0	

Thus, the domain of h is $(-\infty, -1) \cup (0, \infty)$.

Next, we compute derivatives:

$$h'(x) = \frac{(x^2 + x)^{1/2} \cdot 1 - x \cdot (1/2)(x^2 + x)^{-1/2}(2x + 1)}{x^2 + x} \cdot \frac{2(x^2 + x)^{1/2}}{2(x^2 + x)^{1/2}}$$

$$= \frac{(2x^2 + 2x) - (2x^2 + x)}{2(x^2 + x)^{3/2}} = \frac{x}{2(x^2 + x)^{3/2}},$$

$$h''(x) = \frac{2(x^2 + x)^{3/2} \cdot 1 - x \cdot 3(x^2 + x)^{1/2}(2x + 1)}{4(x^2 + x)^3}$$

$$= \frac{(x^2 + x)^{1/2}[(2x^2 + 2x) - (6x^2 + 3x)]}{4(x^2 + x)^3}$$

$$= \frac{-4x^2 - x}{4(x^2 + x)^{5/2}} = -\frac{x(4x + 1)}{4(x^2 + x)^{5/2}}.$$

These functions are continuous and nonzero on the intervals $(-\infty, -1)$ and $(0, \infty)$, and undefined on $[-1, 0]$. Their signs are very simple:

$h''(x)$	$-$	u	u	u	$-$
$h'(x)$	$-$	u	u	u	$+$
x		-1		0	

Thus, h is strictly decreasing and concave down on $(-\infty, -1)$, and strictly increasing and concave down on $(0, \infty)$. There are no local maxima, local minima, or inflection points.

To check for asymptotes, we need to compute several limits. For the limit as $x \to \infty$, we factor out the highest power of x inside the square root and then write $\sqrt{x^2}$ as x:

$$\lim_{x \to \infty} h(x) = \lim_{x \to \infty} \frac{x}{\sqrt{x^2 \cdot (1 + 1/x)}} = \lim_{x \to \infty} \frac{x}{\sqrt{x^2} \cdot \sqrt{1 + 1/x}}$$

$$= \lim_{x \to \infty} \frac{x}{x \cdot \sqrt{1 + 1/x}} = \lim_{x \to \infty} \frac{1}{\sqrt{1 + 1/x}} = \frac{1}{\sqrt{1}} = 1.$$

For the limit as x approaches $-\infty$, we need to be a little careful. For $x < 0$, $x = -\sqrt{x^2}$, so

$$\lim_{x \to -\infty} h(x) = \lim_{x \to -\infty} \frac{x}{\sqrt{x^2} \cdot \sqrt{1 + 1/x}}$$

$$= \lim_{x \to -\infty} \frac{x}{(-x) \cdot \sqrt{1 + 1/x}}$$

$$= \lim_{x \to -\infty} \frac{-1}{\sqrt{1 + 1/x}} = -1.$$

Thus, we have asymptotes $y = 1$ on the right and $y = -1$ on the left.

We should also see what happens as $x \to -1^-$ and as $x \to 0^+$. For the first, as $x \to -1^-$ we have $x \to -1$ and $\sqrt{x^2 + x} \to 0^+$, and therefore

$$\lim_{x \to -1^-} h(x) = \lim_{x \to -1^-} \frac{x}{\sqrt{x^2 + x}} = -\infty.$$

Thus, the line $x = -1$ is also an asymptote. The limit as $x \to 0^+$ is more complicated, because the numerator and denominator of $h(x)$ both approach 0. We need to see if we can cancel something:

$$\lim_{x \to 0^+} h(x) = \lim_{x \to 0^+} \frac{\sqrt{x^2}}{\sqrt{x^2 + x}} = \lim_{x \to 0^+} \sqrt{\frac{x^2}{x^2 + x}} = \lim_{x \to 0^+} \sqrt{\frac{x}{x + 1}} = \sqrt{0} = 0.$$

So the graph approaches the point $(0, 0)$ from the right. However, this point is not on the graph, since $h(0)$ is undefined.

The graph of h comes in two pieces, one for $x < -1$ and one for $x > 0$. It's probably a good idea to plot at least one point in each piece, so we compute:

$$h(-2) = \frac{-2}{\sqrt{2}} = -\sqrt{2} \approx -1.41, \qquad h(1) = \frac{1}{\sqrt{2}} = \frac{\sqrt{2}}{2} \approx 0.71.$$

We plot these points, an open circle at $(0, 0)$, and the asymptotes, and then use these as guides as we draw a curve that is strictly decreasing and concave down on $(-\infty, -1)$, and strictly increasing and concave down on $(0, \infty)$. The resulting sketch is shown in Figure 4.28.

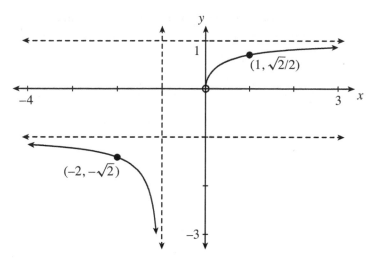

Figure 4.28: The graph of $y = x/\sqrt{x^2 + x}$. The dashed lines are the asymptotes $y = \pm 1$ and $x = -1$.

It appears in the graph that the curve is very steep just to the right of $x = 0$. Let's check:

$$\lim_{x \to 0^+} h'(x) = \lim_{x \to 0^+} \frac{x}{2(x^2 + x)^{3/2}} = \lim_{x \to 0^+} \frac{x}{2(x^2 + x)\sqrt{x^2 + x}}$$

$$= \lim_{x \to 0^+} \frac{1}{2(x + 1)\sqrt{x^2 + x}} = \infty.$$

This shows that as we approach 0 from the right along the curve, the curve slopes up more and more steeply, approaching vertical. □

We end this section by making an observation about the functions g and h that we graphed in the last two examples. Notice that g and h are both strictly increasing on the rightmost parts of their domains—on the interval $[-3, \infty)$ in the case of g, and on the interval $(0, \infty)$ in the case of h. Also, notice that $\lim_{x \to \infty} g(x) = \infty$ and $\lim_{x \to \infty} h(x) = 1$. This may remind you of our study of monotone sequences in Section 2.9. Recall that in that section we showed that if a sequence is weakly increasing, then either it converges or its terms approach ∞, and we established an analogous result for weakly decreasing sequences. The behavior of the functions g and h might lead you to ask if something similar is true for functions. The following theorem shows that the answer is yes.

Theorem 4.5.5. *Suppose f is a function whose domain contains an interval (a, ∞), and let A be the set of all values of f at numbers in this interval; that is, $A = \{f(x) : x > a\} = \{y : \text{for some } x > a, \, y = f(x)\}$.*

1. *If f is weakly increasing on (a, ∞) and A has an upper bound, then $\lim_{x \to \infty} f(x)$ is defined and is equal to the least upper bound of A.*

2. If f is weakly increasing on (a, ∞) and A does not have an upper bound, then $\lim_{x \to \infty} f(x) = \infty$.

3. If f is weakly decreasing on (a, ∞) and A has a lower bound, then $\lim_{x \to \infty} f(x)$ is defined and is equal to the greatest lower bound of A.

4. If f is weakly decreasing on (a, ∞) and A does not have a lower bound, then $\lim_{x \to \infty} f(x) = -\infty$.

Proof. The proof is very similar to the proof of Theorem 2.9.5. For part 1, suppose that f is weakly increasing and A has an upper bound. By the completeness of the real numbers (Theorem 2.9.4), we can let L be the least upper bound of A. We now verify that $\lim_{x \to \infty} f(x) = L$ by using the definition of limits.

Suppose $\epsilon > 0$. Since L is the least upper bound of A, $L - \epsilon$ is not an upper bound, and therefore there must be some number $N > a$ such that $f(N) > L - \epsilon$. Now consider any $x > N$. Since f is weakly increasing on (a, ∞), $f(x) \geq f(N) > L - \epsilon$, and since L is an upper bound for A, $f(x) \leq L$. Thus, $L - \epsilon < f(x) \leq L < L + \epsilon$, so $|f(x) - L| < \epsilon$.

To prove part 2, suppose that f is weakly increasing on (a, ∞) and A does not have an upper bound. Once again, we use the definition of limits to prove that $\lim_{x \to \infty} f(x) = \infty$. Let M be any number. Since A does not have an upper bound, M is not an upper bound for A, so there must be some $N > a$ such that $f(N) > M$. But then since f is weakly increasing on (a, ∞), for every $x > N$ we have $f(x) \geq f(N) > M$. This proves that $\lim_{x \to \infty} f(x) = \infty$.

The proofs of parts 3 and 4 are similar.   □

For a function that is weakly increasing or decreasing on an interval (a, b), there is a similar theorem about $\lim_{x \to b^-} f(x)$. And analogous theorems can be proven for limits as $x \to -\infty$ or $x \to a^+$. We ask you to formulate and prove one such theorem in Exercise 16.

Exercises 4.5

1–15: Sketch the graph of the function.

1. $f(x) = \dfrac{2x - 1}{x^2}$.

2. $g(x) = \dfrac{2}{x^2} - \dfrac{1}{x}$.

3. $f(x) = \dfrac{x}{4 - x^2}$.

4. $g(x) = \dfrac{x^2}{4 - x^2}$.

5. $f(x) = \dfrac{x^3 + 2}{x}$.

6. $g(x) = x + \dfrac{4}{x}$.

7. $f(x) = \dfrac{x^2 + x + 2}{x - 1}$.

8. $h(x) = \dfrac{\sqrt{x}}{x + 3}$.

9. $f(x) = \dfrac{x + 1}{\sqrt{1 - x^2}}$.

10. $g(x) = \dfrac{x + 1}{\sqrt{x^2 - 1}}$.

11. $f(x) = \dfrac{(x - 5)\sqrt{x}}{4}$.

12. $g(x) = \dfrac{(x - 5)^2 \sqrt{x}}{4}$.

13. $f(x) = x - 3\sqrt[3]{x}$.

14. $f(x) = \sqrt[3]{x}(\sqrt[3]{x} - 2)$.

15. $g(x) = 3x^{2/3} - 2x$.

16. Suppose that f is a function whose domain contains the interval $(-\infty, b)$, and let $A = \{f(x) : x < b\}$. Formulate and prove a theorem, analogous to Theorem 4.5.5, about $\lim_{x \to -\infty} f(x)$.

4.6 Optimization Problems

We now know how to use derivatives to determine the shape of the graph of a function. But why is this important? The answer is that often knowing the shape of the graph of a function can help us solve problems involving that function. For example, in some situations it is important to know how large or small some quantity can be. If this problem can be phrased as asking for the maximum or minimum value of some function, then we can use the methods we have developed to find the answer.

Example 4.6.1. A square sheet of metal is 12 inches on each side. A box with an open top is to be made by cutting equal-sized squares from the corners of the sheet and then folding up the sides, as shown in Figure 4.29. What is the largest possible volume of the box?

Solution. We begin by recalling our advice in Section 4.1 about word problems: before we even think about solving the problem, we need to translate the problem into mathematical language. We begin by introducing notation for the important quantities in the problem. We let x stand for the length of the sides of each of the squares that are to be cut out, in inches, and we let V be the volume of the final box, in cubic inches. As indicated in

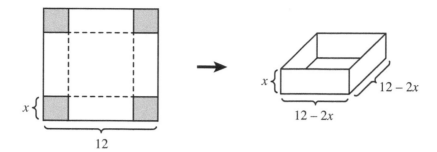

Figure 4.29: Making the box in Example 4.6.1. The shaded squares are removed, and then the sides are folded up along the dashed lines.

the figure, the dimensions of the box are $(12 - 2x) \times (12 - 2x) \times x$, so the volume of the box is given by the formula $V = x(12 - 2x)^2 = 4x(x - 6)^2$. We could write this as $V = f(x)$, where $f(x)$ is given by the formula

$$f(x) = 4x(x - 6)^2.$$

What is the domain of f? Since x is the length of the sides of the cut-out squares, it must be positive. And in order to have some metal left to form the box we must have $x < 6$. So the domain of f is the open interval $(0, 6)$. We can now rephrase the problem as follows: Find the maximum value of f on the interval $(0, 6)$. Now that we have translated the problem from English into mathematics, we are ready to solve it.

We can solve this problem using the methods we have developed in the previous sections. We begin by finding the derivative and determining where it is positive, negative, and zero. We have

$$f'(x) = 4x \cdot \frac{d}{dx}((x - 6)^2) + (x - 6)^2 \cdot \frac{d}{dx}(4x) = 4x \cdot 2(x - 6) + (x - 6)^2 \cdot 4$$
$$= 4(x - 6)(3x - 6) = 12(x - 6)(x - 2).$$

Clearly $f'(x) = 0$ when $x = 2$ (we don't include $x = 6$ as another critical number, since 6 is not in the domain of f), and checking the sign of $f'(x)$ on either side of 2 we get the following results:

$$f'(x) \qquad + \qquad 0 \qquad - $$
$$x \quad 0 \qquad\qquad 2 \qquad\qquad\qquad 6$$

Thus, f is strictly increasing on the interval $(0, 2]$ and strictly decreasing on $[2, 6)$. It follows that f attains its maximum value at $x = 2$, and the maximum value is $f(2) = 128$. Thus, the largest possible volume of the box is 128 in^3. The graph of the function f is shown in Figure 4.30.   □

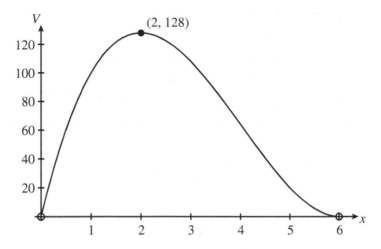

Figure 4.30: The graph of $V = f(x) = 4x(x-6)^2, 0 < x < 6$.

Problems that ask that some quantity be made as large or as small as possible arc called *optimization problems*. Our strategy for solving optimization problems is to identify the quantity that is to be maximized or minimized, and then find a formula for this quantity in terms of some other quantity that can vary over some interval. In the case of the last example, we had to maximize V, the volume of the box, and we expressed it in terms of x, the length of the sides of the cut-out squares, which could vary over the interval $(0, 6)$. This allows us to rephrase the problem as asking for the maximum or minimum value of some function on some interval, and we can then use our curve-sketching methods to solve this problem.

Example 4.6.2. A farmer wants to fence in a rectangular field with area 200 square meters. One side of the fence is to be made of bricks and will cost $30 per meter. The other three will be made of wood and will cost $10 per meter. What should the dimensions of the field be to minimize the cost?

Solution. The phrase "minimize the cost" identifies this problem as an optimization problem, and tells us that the quantity to be minimized is C, the cost of the fence (in dollars). We want to find the dimensions of the field that will minimize the cost, so we begin by finding a formula for C in terms of the dimensions of the field. Let x and y be the length and width of the field, in meters, and suppose that one of the sides of length x is the one along which the fence will be made of bricks, as shown in Figure 4.31. The cost of this side of the fence will be $30x$, and the cost of the other three sides will be $10(x+2y)$, so the formula for C is

$$C = 30x + 10(x+2y) = 40x + 20y.$$

There is a complication in this example that was not present in our previous example: C has been expressed in terms of *two* variables, x and y. But there is also one sentence of the problem that we have not yet translated into mathematics.

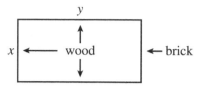

Figure 4.31: The rectangular field of Example 4.6.2.

The first sentence says that the field must have area 200 square meters, which means that $xy = 200$. Thus, we can eliminate one of these variables. We have

$$y = \frac{200}{x},$$

and therefore

$$C = 40x + 20 \cdot \frac{200}{x} = 40x + \frac{4000}{x}.$$

What is the range of possible values of x? Since x is the length of one side of the field, it must be positive. But there is no upper limit on the possible values of x; if x is very large then $y = 200/x$ will be very small, but nothing in the problem rules out a very long, narrow field. Thus, x can take any value in the interval $(0, \infty)$. So we can solve our problem by determining where the function

$$f(x) = 40x + \frac{4000}{x} = 40x + 4000x^{-1}$$

attains its minimum value on the interval $(0, \infty)$.
    We have

$$f'(x) = 40 - 4000x^{-2} = 40 - \frac{4000}{x^2} = \frac{40(x^2 - 100)}{x^2}.$$

Clearly $f'(10) = 0$, $f'(x) < 0$ when $0 < x < 10$, and $f'(x) > 0$ for $x > 10$. Thus, f is strictly decreasing on $(0, 10]$ and strictly increasing on $[10, \infty)$, so the minimum value of $C = f(x)$ is attained at $x = 10$.
    It is always a good idea to look over your solution to a word problem to make sure you have answered the question that was asked. In this case, the problem asks for "the dimensions of the field," so we should find y as well. We have

$$y = \frac{200}{x} = \frac{200}{10} = 20,$$

so the field should be 10 meters by 20 meters, with one of the 10 meter sides of the fence being made of brick.   □

Example 4.6.3. Find the dimensions of the cylinder of greatest volume that can be inscribed in a sphere of radius R.

Solution. We must maximize the volume of the cylinder, which we denote by V. If the radius and height of the cylinder are r and h, as in Figure 4.32a, then the formula for the volume is

$$V = \pi r^2 h.$$

Once again we have an extra variable: V has been expressed in terms of two variables, r and h. So we look for a relationship between r and h that we can use to eliminate one of them. The two variables are related because the cylinder must be inscribed in a sphere of radius R. Figure 4.32b shows a vertical cross section through the center of the sphere, and the thick lines in the figure form a right triangle. By the Pythagorean theorem, $r^2 + (h/2)^2 = R^2$, and therefore $r^2 = R^2 - h^2/4$. Substituting into the formula for V, we get

$$V = \pi \left(R^2 - \frac{h^2}{4} \right) h = \pi \left(R^2 h - \frac{h^3}{4} \right).$$

We have now expressed V in terms of one variable, h. (Note that R is a constant; its value has not been specified, but it is fixed throughout the problem.) Thus, we can write $V = f(h)$, where

$$f(h) = \pi \left(R^2 h - \frac{h^3}{4} \right).$$

What is the domain of f? It is clear in Figure 4.32a that $0 < h < 2R$, and h can take on any value in this range, so the domain of f is the open interval $(0, 2R)$. Thus, to solve the problem we must determine where f attains its maximum value on the interval $(0, 2R)$.

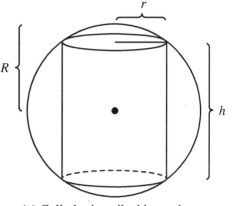

(a) Cylinder inscribed in a sphere.

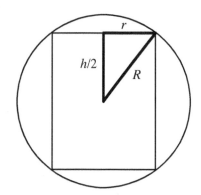

(b) Vertical cross section.

Figure 4.32

We begin by determining where the derivative is 0:

$$f'(h) = \pi \left(R^2 - \frac{3h^2}{4} \right) = 0,$$

$$h = \frac{2R}{\sqrt{3}}.$$

Checking the sign of $f'(h)$ on either side of this critical number we get the following results:

$$f'(h) \quad \overset{+}{\underset{\underset{\displaystyle h \quad 0}{|}}{\rule[0.5ex]{0pt}{0pt}}} \quad \overset{0}{\underset{\underset{\displaystyle 2R/\sqrt{3}}{|}}{\rule[0.5ex]{0pt}{0pt}}} \quad \overset{-}{\underset{\underset{\displaystyle 2R}{|}}{\rule[0.5ex]{0pt}{0pt}}}$$

Thus, f is strictly increasing on the interval $(0, 2R/\sqrt{3}]$ and strictly decreasing on $[2R/\sqrt{3}, 2R)$, and therefore V has its largest value when $h = 2R/\sqrt{3}$. Since the problem asks for "the dimensions of the cylinder," we also compute the radius. From the equation $r^2 = R^2 - h^2/4$ that we derived earlier, we get

$$r = \sqrt{R^2 - \frac{h^2}{4}} = \sqrt{R^2 - \frac{1}{4} \cdot \frac{4R^2}{3}} = \sqrt{\frac{2R^2}{3}} = \frac{\sqrt{2}R}{\sqrt{3}}.$$

Thus, the cylinder of largest volume that can be inscribed in a sphere of radius R has height $2R/\sqrt{3}$ and radius $\sqrt{2}R/\sqrt{3}$.   □

Example 4.6.4. Two pulleys are attached to the ceiling, 2 feet apart. A rope is 10 feet long, and it has a ring attached to one end and a handle at the other end. The rope is passed over the two pulleys so that the handle hangs down on one side and the ring on the other. Then the handle is passed through the ring and pulled down. As the handle is pulled down, the ring slides up the rope, staying centered between the two pulleys, as shown in Figure 4.33a. How far down below the ceiling can the handle be pulled? You may assume that the pulleys, ring, and handle are small enough that their sizes can be ignored.

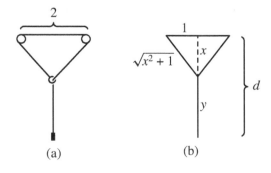

Figure 4.33: The rope and pulleys of Example 4.6.4.

Solution. Ignoring the sizes of the pulleys, ring, and handle, we idealize the situation as in Figure 4.33b, where all distances are measured in feet. We are asked to maximize d, the distance from the ceiling to the handle.

Clearly $d = x + y$, where x is the distance from the ceiling to the ring and y is the length of the part of the rope below the ring. As usual, we look for a relationship between x and y. In this case, the relationship comes from the fact that the total length of the rope is 10 feet, a fact that we have not yet used. The rope comes in four segments: two slanted segments of length $\sqrt{x^2 + 1}$ each, the horizontal segment at the top of length 2, and the vertical segment at the bottom of length y. Thus, $2\sqrt{x^2 + 1} + 2 + y = 10$, and therefore

$$y = 8 - 2\sqrt{x^2 + 1}. \tag{4.16}$$

So the quantity to be maximized is

$$d = x + y = x + 8 - 2\sqrt{x^2 + 1}.$$

The smallest possible value of x is $x = 0$, which would occur if the ring were pushed all the way up to the ceiling. The largest value occurs when the handle is first put through the ring, at which point $y = 0$. Setting $y = 0$ in equation (4.16) we get $8 - 2\sqrt{x^2 + 1} = 0$, and therefore $x = \sqrt{15} \approx 3.87$. Thus, we must find the maximum value of the function

$$f(x) = x + 8 - 2\sqrt{x^2 + 1}$$

on the interval $[0, \sqrt{15}]$.

We begin by differentiating:

$$f'(x) = 1 - (x^2 + 1)^{-1/2} \cdot 2x = 1 - \frac{2x}{\sqrt{x^2 + 1}} = \frac{\sqrt{x^2 + 1} - 2x}{\sqrt{x^2 + 1}}.$$

The derivative turns out to be easiest to understand if we rationalize the numerator:

$$f'(x) = \frac{\sqrt{x^2 + 1} - 2x}{\sqrt{x^2 + 1}} \cdot \frac{\sqrt{x^2 + 1} + 2x}{\sqrt{x^2 + 1} + 2x}$$

$$= \frac{x^2 + 1 - 4x^2}{\sqrt{x^2 + 1}(\sqrt{x^2 + 1} + 2x)} = \frac{1 - 3x^2}{x^2 + 1 + 2x\sqrt{x^2 + 1}}.$$

Clearly the denominator is positive for all x in the interval $(0, \sqrt{15})$. Looking at the numerator, we see that $f'(x) = 0$ when $x = 1/\sqrt{3} \approx 0.58$, $f'(x) > 0$ for $0 < x < 1/\sqrt{3}$, and $f'(x) < 0$ for $1/\sqrt{3} < x < \sqrt{15}$. Thus, f is strictly increasing on $[0, 1/\sqrt{3}]$ and strictly decreasing on $[1/\sqrt{3}, \sqrt{15}]$, and the maximum value of f is

$$f\left(\frac{1}{\sqrt{3}}\right) = \frac{1}{\sqrt{3}} + 8 - 2\sqrt{\frac{1}{3} + 1} = 8 - \frac{3}{\sqrt{3}} = 8 - \sqrt{3} \approx 6.27.$$

In other words, the handle can be pulled down to a point approximately 6.27 feet below the ceiling. □

Exercises 4.6

1. The bottom two vertices of a rectangle are on the x-axis, and the top two vertices are on the curve $y = 75 - x^2$. What is the largest possible area for such a rectangle?

2. A rectangular box with a square base and open top is to have volume 500 cubic inches. What should the dimensions be to minimize the amount of material needed?

3. A rectangular box with a square base and open top is to be made from 75 square inches of cardboard. What is the largest possible volume for such a box?

4. A rectangular piece of cardboard is 24 inches by 15 inches. A box with an open top is to be made by cutting equal-sized squares from the corners of the cardboard and then folding up the sides. What is the largest possible volume of the box?

5. Three little pigs are building a rectangular house. The walls of the house will be 10 feet high, and it will have a flat roof. The front wall of the house will be made out of bricks, which cost \$4 per square foot, the sides and back will be made out of sticks, which cost \$2 per square foot, and the roof will be made of straw, which costs \$1 per square foot. If they have \$3000 to spend, what should the dimensions of the house be to maximize the floor space?

6. A cardboard poster is to have a rectangular printed area of 150 square inches, surrounded by margins of 1 inch at the sides and top and 2 inches at the bottom. What should the dimensions of the poster be to minimize the amount of cardboard needed?

7. A sign is to be made out of a rectangular piece of wood. A strip of metal trim will go along the top of the sign (see Figure 4.34). The wood for the sign costs \$3 per square foot, and the metal for the trim costs \$2 per foot. If the sign is to have 48 square feet of printed area, surrounded by 1-foot margins at the top, bottom, and sides, what should the dimensions of the sign be to minimize the total cost of the materials?

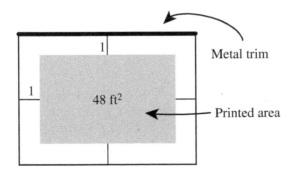

Figure 4.34: A wooden sign with metal trim along the top.

8. A closed rectangular box with square ends is to be constructed. There will be a square partition inside the box, parallel to the square ends (see Figure 4.35).

The box is to be constructed from material that costs $2 per square foot, except that the material for the partition will cost only $1 per square foot. Find the dimensions of the box of greatest volume that can be constructed for $60.

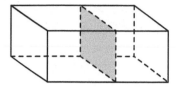

Figure 4.35: A box with square ends and a partition.

9. A cylindrical coffee mug is to be filled with hot coffee. In 5 minutes, the temperature of the coffee will decrease 4°F for each square inch of the surface of the coffee that is exposed to the air. In addition, the temperature will decrease another 2°F for each square inch of coffee touching the bottom and sides of the mug. If the mug must hold 3π cubic inches of coffee, what should the dimensions be in order to keep the coffee as hot as possible?

10. Find the dimensions of the cylinder of greatest volume that can be inscribed in a cone of height 18 whose base has radius 6 (see Figure 4.36).

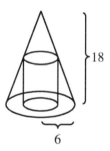

Figure 4.36: A cylinder inscribed in a cone.

11. You are baking a two-layer round cake. The total volume of the two layers of cake is to be 81π cubic inches. There will be a filling between the layers, and frosting on top and around the sides. The frosting costs 10 cents per square inch, but the filling is 20 cents per square inch. What should the dimensions of the cake be to minimize the cost of the filling and frosting?

12. Find the points on the curve $y = 4/x$ that are closest to the origin. (Hint: It is simplest to minimize the *square* of the distance from the origin to the point.)

13. A wall is 1 meter high and 8 meters from the side of a building. What is the shortest board that can go from the ground over the wall to the side of the building? (See Figure 4.37.)

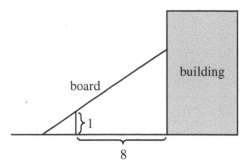

Figure 4.37: Board going over wall to side of building.

14. A cylindrical metal can is to have volume V. What should the dimensions be to minimize the amount of metal needed to make the can? Show that the height and diameter of the optimal can are equal.

15. A cylindrical can is to be wrapped with wrapping paper, as shown in Figure 4.38. The wrapping paper must be wide enough to wrap all the way around the can, and it must be tall enough that the top and the bottom can be folded in (with some crumpling) to cover the top and bottom of the can. If the volume of the can must be 32π cubic inches, what should the dimensions of the can be to minimize the area of the wrapping paper needed?

Figure 4.38: Wrapping a cylinder.

16. The two legs of a right triangle lie along the positive x- and y-axes. The hypotenuse is tangent to the ellipse $2x^2 + y^2 = 1$. What is the smallest possible area for such a triangle?

4.7 Maxima and Minima on Finite Closed Intervals

Often in optimization problems, the domain of the function whose maximum or minimum value must be found is a finite closed interval. For example, in Example 4.6.4 the domain of f was the interval $[0, \sqrt{15}]$. Sometimes the domain is a finite interval that is not closed, but we can make it closed by adding in the endpoints. For example, in Example 4.6.1 we had to find the maximum possible volume of a box with dimensions $(12 - 2x) \times (12 - 2x) \times x$, for $0 < x < 6$. We did this by finding the maximum value of the function $f(x) = x(12 - 2x)^2 = 4x(x - 6)^2$ on the interval $(0, 6)$. But the formula for $f(x)$ makes sense at $x = 0$ and $x = 6$, and these values of x could be thought of as corresponding to "degenerate boxes" in which at least one dimension is 0, and therefore the volume is also 0. If we had included these endpoints, then the domain of f would have been the finite closed interval $[0, 6]$. In the graph of f in Figure 4.30, the points $(0, 0)$ and $(6, 0)$ would have been black dots instead of open circles, but the rest of the solution would have been the same. Similarly, in Example 4.6.3 we had to maximize the volume of a cylinder, where the volume was given by the function $f(h) = \pi(R^2 h - h^3/4)$. We took the domain of f to be the interval $(0, 2R)$, but we could have used the interval $[0, 2R]$ instead, with the endpoints representing "degenerate cylinders" with height 0 or radius 0.

Why would we want to include the endpoints in the domains of these functions? The reason is given by the following theorem. We will give a proof of the theorem at the end of this section.

Theorem 4.7.1 (Extreme Value Theorem). *Suppose f is continuous on the finite closed interval $[a, b]$. Then f has a maximum value and a minimum value on $[a, b]$.*

We have seen that some functions don't have maximum or minimum values. The extreme value theorem says that this never happens in the case of a continuous function f on a finite closed interval $[a, b]$. Such a function always has both a maximum value and a minimum value. Where will these values be attained? If the maximum value is not attained at one of the endpoints, then it must occur at a number c in the interior of the interval, and f will then have a local maximum at c. It follows, by Theorem 4.3.6, that c must be a critical number of f; in other words, either $f'(c)$ is undefined or $f'(c) = 0$. Similarly, the minimum value of f must be attained at either an endpoint of the interval or a critical number.

This gives us another method that can be used to find the maximum and minimum values of a continuous function on a finite closed interval. Simply compute the value of the function at the endpoints and at all critical numbers in the interior of the interval. Whichever of these values is largest must be the maximum value of the function on the interval, and whichever value is smallest is the minimum value.

For example, consider again Example 4.6.4, in which we had to find the maximum value of the function

$$f(x) = x + 8 - 2\sqrt{x^2 + 1}$$

on the interval $[0, \sqrt{15}]$. We showed in the last section that the only critical number for this function in the interval $(0, \sqrt{15})$ is $1/\sqrt{3}$. Instead of checking the sign of $f'(x)$ to the left and right of this number, we could just check the value of the function at this critical number and the endpoints of the interval:

$$f(1/\sqrt{3}) = 8 - \sqrt{3} \approx 6.27,$$
$$f(0) = 6,$$
$$f(\sqrt{15}) = \sqrt{15} \approx 3.87.$$

We can now say that the maximum value of $f(x)$ on $[0, \sqrt{15}]$ is $8 - \sqrt{3}$, attained at $x = 1/\sqrt{3}$, and the minimum value is $\sqrt{15}$, attained at $x = \sqrt{15}$.

We can also use this method to solve the problem in Example 4.6.1. As we observed at the beginning of this section, we could solve that problem by finding the maximum value of the function

$$f(x) = 4x(x - 6)^2$$

on the interval $[0, 6]$. We showed in the last section that the only critical number for f in the interior of this interval is at $x = 2$, where the value of the function is $f(2) = 128$. Since the value at the endpoints is $f(0) = f(6) = 0$, the maximum value is 128. Similarly, this method could be used to solve the problem in Example 4.6.3; we leave the details to you. Let's try using it in one more example:

Example 4.7.2. A dog is standing at a point D on the shore of a lake. His owner is standing at a point A on the shore 20 meters away, and he throws the dog's favorite ball into the water. The ball lands at a point B, 14 meters from the shore (see Figure 4.39). If the dog can run 6.40 meters per second and swim 0.91 meters per second, what route should the dog take to get to the ball as fast as possible?

Solution. The dog could simply jump into the water and swim in a straight line to the ball. This is the shortest route to the ball, but perhaps not the fastest route, since the dog would be swimming the whole way, and he swims much more slowly than he runs. Another possibility would be to run along the shore to the point A and then swim from there to the ball. This route is longer, but since the dog would be running most of the way, it might be faster. And then there are intermediate possibilities: the dog could run to a point C between D and A and then swim from there. Our task is to determine which of these routes is fastest. (It is clear in Figure 4.39 that it would not be fastest for the dog to jump into the water either to the left of D or to the right of A.)

Suppose the dog runs to a point C that is x meters from A and then swims. If $x = 20$ then $C = D$, and we have the first route, in which the dog swims directly from D to B. If $x = 0$ then $C = A$ and we have the second route, which involves the least swimming.

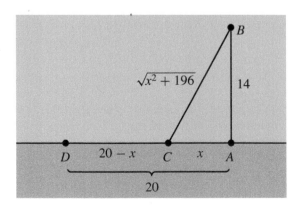

Figure 4.39: The dog in Example 4.7.2 is at point D, and he wants to get to the ball, which is at point B.

Since any intermediate value of x is also possible, we must consider all values of x in the interval $[0, 20]$.

How long will it take the dog to reach the ball? He will run $20 - x$ meters at a speed of 6.40 m/sec, which will take him $(20 - x)/6.40$ seconds. By the Pythagorean theorem, the distance he has to swim is $\sqrt{x^2 + 14^2} = \sqrt{x^2 + 196}$, and at 0.91 m/sec this will take $\sqrt{x^2 + 196}/0.91$ seconds. Thus, the total time will be

$$f(x) = \frac{20 - x}{6.40} + \frac{\sqrt{x^2 + 196}}{0.91} = \frac{1}{6.40}(20 - x) + \frac{1}{0.91}(x^2 + 196)^{1/2}$$

seconds. We must find the minimum value of this function on the interval $[0, 20]$.

We begin by solving for critical numbers:

$$f'(x) = \frac{1}{6.40} \cdot (-1) + \frac{1}{0.91} \cdot \frac{1}{2}(x^2 + 196)^{-1/2} \cdot 2x = -\frac{1}{6.40} + \frac{x}{0.91\sqrt{x^2 + 196}} = 0,$$

$$6.40x = 0.91\sqrt{x^2 + 196},$$

$$40.96x^2 = 0.8281(x^2 + 196),$$

$$x \approx 2.01.$$

Since f is continuous on the finite closed interval $[0, 20]$, it must have a minimum value, and this minimum value must occur at either 0, 20, or 2.01. We check the values of the function at these points:

$$f(0) \approx 18.51,$$
$$f(2.01) \approx 18.35,$$
$$f(20) \approx 26.83.$$

Since $f(2.01)$ is the smallest, this is the minimum value of the function. Thus, the dog should run to a point just a bit more than 2 meters short of A, and then jump into the water and swim from there.　　　　□

By the way, Example 4.7.2 is based on a true story. The dog, whose name is Elvis, actually ran to a point 2.6 meters from A before jumping into the water. Not bad for a dog who didn't have access to a calculator! For more details, see the article "Do Dogs Know Calculus?" by Elvis's owner Timothy Pennings (*College Mathematics Journal* **34** (2003), pp. 178–182).

We now turn to the proof of the extreme value theorem. Suppose f is a function that is continuous on the interval $[a, b]$. We must prove that f has both a maximum value and a minimum value. Since the proofs for the maximum and minimum values are similar, we will just prove that f has a maximum value.

If $a \leq u < v \leq b$, then we will say that the interval $[u, v]$ is *nonmaximizing* if there is some number $d \in [a, b]$ such that for every $x \in [u, v]$, $f(x) < f(d)$. In other words, the interval $[u, v]$ is nonmaximizing if there is a point $(d, f(d))$ on the graph of f that is higher than any point on the part of the graph of f with $u \leq x \leq v$, as shown in Figure 4.40. If $[u, v]$ is not nonmaximizing, then we say that it is *maximizing*.

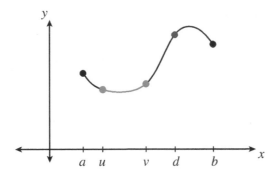

Figure 4.40: The interval $[u, v]$ is nonmaximizing, because the point $(d, f(d))$ (red) is higher than any point on the part of the graph of f with $u \leq x \leq v$ (blue).

Notice that the number d in the definition of nonmaximizing cannot be in the interval $[u, v]$, so a nonmaximizing interval cannot be all of $[a, b]$. In other words, the interval $[a, b]$ is maximizing. If we are looking for a number where f attains its maximum value on the interval $[a, b]$, there is no point in looking in a nonmaximizing interval: clearly the maximum value cannot be attained at any number in a nonmaximizing interval.

The following fact about maximizing intervals is the key to our proof:

Lemma 4.7.3. *Suppose that $[u, v]$ is a maximizing interval. Then there are numbers u' and v' such that $u \leq u' < v' \leq v$, $v' - u' = (v - u)/2$, and $[u', v']$ is maximizing.*

Proof. Let $w = (u + v)/2$, the midpoint of the interval $[u, v]$. Our plan is to show that either $[u, w]$ or $[w, v]$ is maximizing. If we can prove this, then we can let $[u', v']$ be whichever of these two intervals is maximizing.

Suppose that $[u, w]$ and $[w, v]$ are both nonmaximizing. Since $[u, w]$ is non-maximizing, there is some number d_1 such that for every $x \in [u, w]$, $f(x) < f(d_1)$. Similarly, since $[w, v]$ is nonmaximizing, there is some d_2 such that for every $x \in [w, v]$, $f(x) < f(d_1)$.

Either $f(d_1) < f(d_2)$ or $f(d_1) \geq f(d_2)$. Suppose first that $f(d_1) < f(d_2)$. Then for every $x \in [u, w]$, $f(x) < f(d_1) < f(d_2)$. Combining this with the fact that for all $x \in [w, v]$, $f(x) < f(d_2)$, we conclude that for all $x \in [u, v]$, $f(x) < f(d_2)$. But this means that $[u, v]$ is nonmaximizing, which is impossible, since we are told in the lemma that it is maximizing. Similarly, if $f(d_1) \geq f(d_2)$ then we can show that for all $x \in [u, v]$, $f(x) < f(d_1)$, again contradicting the fact that $[u, v]$ is maximizing. Thus, it is impossible for both $[u, w]$ and $[w, v]$ to be nonmaximizing, so at least one of them is maximizing. This proves the lemma. $\qquad \square$

We hope that by now you have guessed how we are going to prove the extreme value theorem. By applying the lemma repeatedly, we will create a nested sequence of maximizing intervals, and then we will use the nested interval theorem.

Proof of Theorem 4.7.1. Let $u_1 = a$ and $v_1 = b$. As we have already mentioned, $[u_1, v_1] = [a, b]$ is maximizing, so by the lemma we can find u_2 and v_2 such that

$u_1 \leq u_2 < v_2 \leq v_1$, $v_2 - u_2 = (v_1 - u_1)/2 = (b-a)/2$, and $[u_2, v_2]$ is maximizing. Applying the lemma again, we see that there are u_3 and v_3 such that $u_2 \leq u_3 < v_3 \leq v_2$, $v_3 - u_3 = (v_2 - u_2)/2 = (b-a)/4$, and $[u_3, v_3]$ is maximizing. Continuing in this way we get a nested sequence of maximizing intervals $[u_n, v_n]$ such that $v_n - u_n = (b-a)/2^{n-1}$.

Since $\lim_{n\to\infty}(v_n - u_n) = \lim_{n\to\infty}(b-a)/2^{n-1} = 0$, the nested interval theorem tells us that there is number c such that for every n, $u_n \leq c \leq v_n$, and $\lim_{n\to\infty} u_n = \lim_{n\to\infty} v_n = c$. Of course, our plan is to show that f attains its maximum value on the interval $[a, b]$ at c.

If f does not attain its maximum value at c, then there is some number $d \in [a, b]$ such that $f(c) < f(d)$. We now apply the fact that f is continuous at c. (To simplify the proof, we will assume that c is in the interior of the interval $[a, b]$. If c is one of the endpoints, then minor adjustments must be made in the following argument to use one-sided continuity.) Since f is continuous at c, $\lim_{x\to c} f(x) = f(c) < f(d)$. Therefore, by Theorem 2.8.8, there must be a number $\delta > 0$ such that if $c - \delta < x < c + \delta$ then $f(x) < f(d)$. Next we use the fact that $\lim_{n\to\infty} u_n = \lim_{n\to\infty} v_n = c$ to conclude that if we choose n large enough, then $c - \delta < u_n < v_n < c + \delta$. It follows that for all x, if $x \in [u_n, v_n]$ then $c - \delta < x < c + \delta$, and therefore $f(x) < f(d)$. But this means that $[u_n, v_n]$ is nonmaximizing, contradicting the fact that we chose our sequence of nested intervals to be maximizing. Thus, f must attain its maximum value at c. ⊔

Exercises 4.7

1. Internet service is to be provided to three houses located at the points $(3, 0)$, $(-3, 0)$, and $(0, 3)$ in the xy-plane. A switching center will be built at some point on the y-axis between $(0, 0)$ and $(0, 3)$, and cables will run in straight lines from this switching center to each of the houses. Where should the switching center be built to minimize the total amount of cable needed?

2. It costs the Acme Syrup Company $30 to produce a gallon of maple syrup, so when they sell their syrup they must charge at least $30 per gallon to keep from losing money. They have found that for $30 \leq x \leq 100$, if they charge $x per gallon, then they will be able to sell $1000 - 10x$ gallons.

   (a) How should they set their price to maximize their revenue (the amount of money they take in)?

   (b) How should they set their price to maximize their profit (the amount of money they take in, minus the cost of producing the syrup they sell)?

3. A rectangular poster is to have a total area of 600 square inches. It will have a rectangular printed region surrounded by margins of 1 inch at the sides and top and 2 inches a the bottom. What should the dimensions of the poster be to maximize the area of the printed region?

4. Find the dimensions of the cone of greatest volume that can be inscribed in a sphere of radius R.

5. A Norman window is a window in the shape of a rectangle with a semicircle on top (see Figure 4.41). If the perimeter of the window is P, what should the dimensions of the window be to maximize its area? Show that for the optimum dimensions, the width and height of the window are equal.

Figure 4.41: A Norman window.

6. A conical drinking cup is to be made by cutting a sector from a circular piece of paper with radius 3 inches and then gluing the two cut edges together, as shown in Figure 4.42. What is the largest possible volume of the cup?

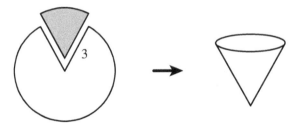

Figure 4.42: Making a drinking cup from a circular piece of paper.

7. Little Red Riding Hood is on the south shore of a river that is 60 feet wide. Her grandmother's house is on the north shore of the river, 100 feet west of the point directly across the river from her. To get to her grandmother's house, she must row across the river and then walk through the woods along the north shore of the river. If she rows 3 feet per second and walks 5 feet per second, what route should she take to get to her grandmother's house as fast as possible?

8. A farmer is going to fence in a rectangular field along the shore of a river with 200 feet of fence. The river will form one side of the field, so he will only need to use fence on three sides, as shown in Figure 4.43a. What should the dimensions of the field be to maximize its area?

9. A farmer is going to fence in a rectangular field along the back of his barn with 200 feet of fence. The field must be at least as wide as the barn, but it could be wider, with the barn forming part of one side of the field (see Figure 4.43b).

   (a) If the barn is 40 feet wide, what should the dimensions of the field be to maximize its area?

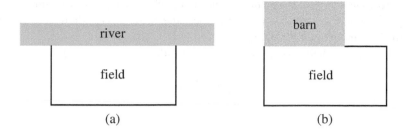

Figure 4.43: Fencing in a field along a river (a) or along the back of a barn (b).

(b) If the barn is 70 feet wide, what should the dimensions of the field be to maximize its area?

10. A farmer is going to fence in a rectangular field with 200 feet of fence. He plans to rent out the field to another farmer, charging a monthly rent of \$3 per square foot. One side of the field will be next to a busy highway, and he plans to rent the fence on that side to an advertising company, charging a monthly rent of \$30 per foot. What should the dimensions of the field be to maximize his monthly income?

4.8 L'Hôpital's Rule

In this section, we discuss a method for using derivatives to help us evaluate limits. The method is named for Guillaume de l'Hôpital (1661–1704), who published it in 1696 in his book *Analyse des Infiniment Petits pour l'Intelligence des Lignes Courbes*, the first textbook on differential calculus. However, it is believed that the method was actually discovered by Johann Bernoulli (1667–1748); l'Hôpital and Bernoulli had entered into a contract that gave l'Hôpital the right to publish Bernoulli's discoveries. The method is based on the following theorem:

Theorem 4.8.1 (L'Hôpital's Rule). *Suppose that* $\lim_{x \to a} f(x) = \lim_{x \to a} g(x) = 0$. *If*

$$\lim_{x \to a} \frac{f'(x)}{g'(x)} = L,$$

then

$$\lim_{x \to a} \frac{f(x)}{g(x)} = L.$$

L'Hôpital's rule is useful if you are trying to compute a limit of the form $\lim_{x \to a}(f(x)/g(x))$, and the limits of the numerator and denominator are both 0. A limit of this form is sometimes called an *indeterminate form of type* 0/0 because one cannot determine the value of the limit by simply computing the limits of numerator and denominator and then dividing. The rule says that in this situation you can try computing $\lim_{x \to a}(f'(x)/g'(x))$ instead. If you get some number L as the answer for this limit, then the answer for the original limit is also L.

We will discuss the proof of l'Hôpital's rule later in this section. As we will see, the proof is based on our old friend the mean value theorem. But before turning to the proof let's try the rule out in a few examples.

Consider the limit

$$\lim_{x \to 2} \frac{x^2 - x - 2}{x^2 - 3x + 2}.$$

Since $\lim_{x \to 2}(x^2 - x - 2) = \lim_{x \to 2}(x^2 - 3x + 2) = 0$, this limit fits the pattern of l'Hôpital's rule, with $f(x) = x^2 - x - 2$ and $g(x) = x^2 - 3x + 2$. We have

$$\lim_{x \to 2} \frac{f'(x)}{g'(x)} = \lim_{x \to 2} \frac{2x - 1}{2x - 3} = 3,$$

so by l'Hôpital's rule, the value of the original limit is also 3.

Of course, we could also have evaluated this limit by the calculation

$$\lim_{x \to 2} \frac{x^2 - x - 2}{x^2 - 3x + 2} = \lim_{x \to 2} \frac{(x - 2)(x + 1)}{(x - 2)(x - 1)} = \lim_{x \to 2} \frac{x + 1}{x - 1} = 3,$$

which confirms that l'Hôpital's rule has given us the right answer in this case. Thus, we didn't need l'Hôpital's rule to evaluate this limit. But there are other examples in which there is no easy alternative to the use of l'Hôpital's rule.

Example 4.8.2. Evaluate the limit

$$\lim_{x \to \pi} \frac{\sin(3x)}{x - \pi}.$$

Solution. Since $\lim_{x \to \pi} \sin(3x) = \sin(3\pi) = 0$ and $\lim_{x \to \pi}(x - \pi) = 0$, we have an indeterminate form of type 0/0. There is no obvious cancelation to be done, so we try l'Hôpital's rule. The rule tells us to evaluate

$$\lim_{x \to \pi} \frac{\cos(3x) \cdot 3}{1} = 3\cos(3\pi) = -3.$$

We conclude that the value of the original limit is also -3.   □

Example 4.8.3. Evaluate the limit

$$\lim_{x \to 0} \frac{1 - \cos(5x)}{x^2}.$$

Solution. Once again we have an indeterminate form of type 0/0. L'Hôpital's rule tells us to try to evaluate

$$\lim_{x \to 0} \frac{\sin(5x) \cdot 5}{2x}. \tag{4.17}$$

The limit in (4.17) is also an indeterminate form of type 0/0, so we use l'Hôpital's rule again:

$$\lim_{x \to 0} \frac{25\cos(5x)}{2} = \frac{25\cos(0)}{2} = \frac{25}{2}.$$

L'Hôpital's rule now tells us that the value of the limit in (4.17) is 25/2, and it follows by another application of l'Hôpital's rule that this is also the value of the original limit. □

L'Hôpital's rule can also be used for one-sided limits and limits as x approaches ∞ or $-\infty$. And it can also be used if we find that the limit of $f'(x)/g'(x)$ is ∞ or $-\infty$; in this case, once again, the same conclusion applies to the limit of $f(x)/g(x)$.

Example 4.8.4. Evaluate the limit

$$\lim_{t \to 1^+} \frac{\sqrt{t-1}}{\sqrt{t}-1}.$$

Solution. Of course, the fact that the variable in the limit is t rather than x makes no difference in the use of l'Hôpital's rule. We have a one-sided limit that is an indeterminate form of type $0/0$, so we compute the derivatives of the numerator and denominator:

$$\frac{d}{dt}((t-1)^{1/2}) = \frac{1}{2}(t-1)^{-1/2} = \frac{1}{2\sqrt{t-1}}, \qquad \frac{d}{dt}(t^{1/2}-1) = \frac{1}{2}t^{-1/2} = \frac{1}{2\sqrt{t}}.$$

Next we compute

$$\lim_{t \to 1^+} \frac{1/(2\sqrt{t-1})}{1/(2\sqrt{t})} = \lim_{t \to 1^+} \frac{\sqrt{t}}{\sqrt{t-1}} = \infty.$$

By l'Hôpital's rule, we conclude that

$$\lim_{t \to 1^+} \frac{\sqrt{t-1}}{\sqrt{t}-1} = \infty. \qquad\qquad □$$

Sometimes a difficult limit is not in the $0/0$ form, but it can be rearranged to put it in that form:

Example 4.8.5. Evaluate the limit

$$\lim_{x \to \infty} \left[x \sin\left(\frac{1}{2x+3} \right) \right].$$

Solution. We are asked to compute the limit of a product, not a quotient, so l'Hôpital's rule doesn't apply directly. We have

$$\lim_{x \to \infty} x = \infty, \qquad \lim_{x \to \infty} \sin\left(\frac{1}{2x+3} \right) = \sin(0) = 0,$$

and as we pointed out in our discussion after Theorem 2.5.14, if one factor in a product approaches ∞ and the other approaches 0, then the two factors are pulling the product in opposite directions, and there is no easy way to determine the limit of the product.

We might say that the limit in this example is an indeterminate form of type $\infty \cdot 0$. But we can rearrange this limit to turn it into an indeterminate form of type 0/0:

$$\lim_{x \to \infty} \left[x \sin\left(\frac{1}{2x+3} \right) \right] = \lim_{x \to \infty} \frac{\sin(1/(2x+3))}{1/x}.$$

Now we can apply l'Hôpital's rule. We begin by computing the derivatives of the numerator and denominator:

$$\frac{d}{dx}(\sin((2x+3)^{-1})) = \cos((2x+3)^{-1}) \cdot (-1)(2x+3)^{-2} \cdot 2$$

$$= -\frac{2}{(2x+3)^2} \cdot \cos\left(\frac{1}{2x+3} \right),$$

$$\frac{d}{dx}(x^{-1}) = -x^{-2} = -\frac{1}{x^2}.$$

Next, we compute

$$\lim_{x \to \infty} \frac{(-2/(2x+3)^2) \cdot \cos(1/(2x+3))}{-1/x^2} = \lim_{x \to \infty} \left[2 \cdot \left(\frac{x}{2x+3} \right)^2 \cdot \cos\left(\frac{1}{2x+3} \right) \right]$$

$$= \lim_{x \to \infty} \left[2 \cdot \left(\frac{1}{2+3/x} \right)^2 \cdot \cos\left(\frac{1}{2x+3} \right) \right]$$

$$= 2 \cdot \left(\frac{1}{2} \right)^2 \cdot \cos(0) = \frac{1}{2}.$$

Finally, by l'Hôpital's rule we conclude that the value of the original limit is 1/2.   □

Although l'Hôpital's rule does not apply directly to limits of sequences, it can still play a role in the evaluation of sequence limits. For example, consider the sequence $(a_n)_{n=1}^{\infty}$, where a_n is given by the formula

$$a_n = \frac{\sin(\pi/n)}{\sin(1/n+\pi)}.$$

The limit of this sequence is

$$\lim_{n \to \infty} a_n = \lim_{n \to \infty} \frac{\sin(\pi/n)}{\sin(1/n+\pi)},$$

which is an indeterminate form of type 0/0. L'Hôpital's rule does not apply to this limit, since the variable n here stands for an integer. (As we observed in Example 4.3.10, the idea of differentiating with respect to an integer n makes no sense.)

However, we can apply l'Hôpital's rule to the limit

$$\lim_{x \to \infty} \frac{\sin(\pi/x)}{\sin(1/x + \pi)}. \tag{4.18}$$

Taking the derivatives of the numerator and denominator, we compute

$$\lim_{x \to \infty} \frac{\cos(\pi/x) \cdot (-\pi/x^2)}{\cos(1/x + \pi) \cdot (-1/x^2)} = \lim_{x \to \infty} \frac{\pi \cos(\pi/x)}{\cos(1/x + \pi)} = \frac{\pi}{-1} = -\pi.$$

We conclude, by l'Hôpital's rule, that the value of limit (4.18) is also $-\pi$, and therefore, by Theorem 2.8.2, $\lim_{n \to \infty} a_n = -\pi$.

L'Hôpital's rule can also be applied to limits of fractions $f(x)/g(x)$ in which numerator and denominator both approach $\pm\infty$. In other words, we have the following theorem:

Theorem 4.8.6 (L'Hôpital's Rule, Part 2). *Suppose that either* $\lim_{x \to a} f(x) = \infty$ *or* $\lim_{x \to a} f(x) = -\infty$, *and also either* $\lim_{x \to a} g(x) = \infty$ *or* $\lim_{x \to a} g(x) = -\infty$. *If*

$$\lim_{x \to a} \frac{f'(x)}{g'(x)} = L,$$

then

$$\lim_{x \to a} \frac{f(x)}{g(x)} = L.$$

As before, the same rule applies to one-sided limits and limits as x approaches ∞ *or* $-\infty$, *and it also applies if L is replaced by either* ∞ *or* $-\infty$.

Limits of the kind considered in Theorem 4.8.6 are called *indeterminate forms of type* ∞/∞.

Example 4.8.7. Evaluate the limit

$$\lim_{x \to \infty} \frac{1 - 3x^2}{x^2 + 5x - 3}.$$

Solution. We already know an algebraic trick that can be used to evaluate this limit:

$$\lim_{x \to \infty} \frac{1 - 3x^2}{x^2 + 5x - 3} = \lim_{x \to \infty} \left[\frac{x^2 \cdot (1/x^2 - 3)}{x^2 \cdot (1 + 5/x - 3/x^2)} \right]$$

$$= \lim_{x \to \infty} \frac{1/x^2 - 3}{1 + 5/x - 3/x^2}$$

$$= \frac{-3}{1} = -3.$$

But let's check that l'Hôpital's rule would give the same result.

Since $\lim_{x \to \infty} (1 - 3x^2) = -\infty$ and $\lim_{x \to \infty} (x^2 + 5x - 3) = \infty$, we have an indeterminate form of type ∞/∞. L'Hôpital's rule suggests that we compute

$$\lim_{x \to \infty} \frac{-6x}{2x + 5}. \tag{4.19}$$

This is still an indeterminate form of type ∞/∞, so we use l'Hôpital's rule again:

$$\lim_{x \to \infty} \frac{-6}{2} = -3.$$

By l'Hôpital's rule, this is the answer to the limit in (4.19), and also the answer to the original limit. $\square$

As we mentioned earlier, the proof of l'Hôpital's rule is based on the mean value theorem. Actually, we will need the following more general form of the theorem.

Theorem 4.8.8 (Cauchy's Mean Value Theorem). *Suppose that f and g are continuous on $[a, b]$ and differentiable on (a, b), and for all $x \in (a, b)$, $g'(x) \neq 0$. Then there is a number c such that $a < c < b$ and*

$$\frac{f(b) - f(a)}{g(b) - g(a)} = \frac{f'(c)}{g'(c)}. \tag{4.20}$$

Notice that if $g(x) = x$ then $g'(x) = 1$, and equation (4.20) becomes

$$\frac{f(b) - f(a)}{b - a} = f'(c),$$

which is the equation in the conclusion of the original mean value theorem. Thus, Cauchy's mean value theorem is a generalization of the original mean value theorem: when we apply Cauchy's mean value theorem with $g(x) = x$, we get the same conclusion that we would get from the original mean value theorem, but Cauchy's mean value theorem can be applied with other choices for g.

Proof of Cauchy's Mean Value Theorem. First we would like to verify that we are not dividing by 0 in equation (4.20). We are told in the statement of the theorem that $g'(x) \neq 0$ for all x in the interval (a, b), so we know that no matter what number c we choose in this interval we will have $g'(c) \neq 0$, and therefore the denominator on the right-hand side of (4.20) will not be 0. But what about the denominator on the left?

Applying the original mean value theorem to g, we know that there is some number d such that $a < d < b$ and

$$\frac{g(b) - g(a)}{b - a} = g'(d).$$

Since $g'(d) \neq 0$, we conclude that $g(b) - g(a) \neq 0$, and therefore the denominator on the left-hand side of (4.20) is not 0.

Let
$$r = \frac{f(b) - f(a)}{g(b) - g(a)}.$$

Our goal is to find a number c between a and b such that $f'(c)/g'(c) = r$. To motivate the proof, we rewrite this last equation as $f'(c) = rg'(c)$, or equivalently $f'(c) - rg'(c) = 0$. This suggests that we study the function

$$h(x) = f(x) - rg(x).$$

Since f and g are both continuous on $[a, b]$ and differentiable on (a, b), the same is true of h, and
$$h'(x) = f'(x) - rg'(x).$$

Thus, we can apply the original mean value theorem to h to conclude that there is some number c such that $a < c < b$ and

$$\frac{h(b) - h(a)}{b - a} = h'(c) = f'(c) - rg'(c). \tag{4.21}$$

To complete the proof we must evaluate the left-hand side of equation (4.21). By the definition of h, we have

$$h(b) - h(a) = [f(b) - rg(b)] - [f(a) - rg(a)] = [f(b) - f(a)] - r[g(b) - g(a)]$$
$$= [f(b) - f(a)] - \frac{f(b) - f(a)}{g(b) - g(a)} \cdot [g(b) - g(a)]$$
$$= [f(b) - f(a)] - [f(b) - f(a)] = 0.$$

Therefore equation (4.21) tells us that

$$f'(c) - rg'(c) = \frac{h(b) - h(a)}{b - a} = 0.$$

Thus, $f'(c) = rg'(c)$, and since we know $g'(c) \neq 0$ we can divide by $g'(c)$ to conclude that

$$\frac{f'(c)}{g'(c)} = r = \frac{f(b) - f(a)}{g(b) - g(a)}.$$

This is the same as equation (4.20), so the proof is complete. □

Proof of L'Hôpital's Rule. Since l'Hôpital's rule applies in many different situations, there are many cases to consider. We will give the proof for just one case involving a one-sided limit that is an indeterminate form of type $0/0$. At the end of the proof we will make some brief comments about other cases.

Suppose that $\lim_{x \to a^+} f(x) = \lim_{x \to a^+} g(x) = 0$ and

$$\lim_{x \to a^+} \frac{f'(x)}{g'(x)} = L.$$

Notice that in order for this last limit statement to be true, the fraction $f'(x)/g'(x)$ must be defined for all x close to a but larger than a. More precisely, there must be some number $b > a$ such that f and g are differentiable on the interval (a, b), and for all x in that interval, $g'(x) \neq 0$. We must prove that $\lim_{x \to a^+}(f(x)/g(x)) = L$.

We begin by defining functions F and G on the domain $[a, b)$ as follows:

$$F(x) = \begin{cases} f(x), & \text{if } a < x < b, \\ 0, & \text{if } x = a; \end{cases} \qquad G(x) = \begin{cases} g(x), & \text{if } a < x < b, \\ 0, & \text{if } x = a. \end{cases}$$

In other words, $F(x)$ and $G(x)$ are the same as $f(x)$ and $g(x)$ for x in the interval (a, b), but we have defined $F(a)$ and $G(a)$ to be 0. It follows that F and G are differentiable, and therefore also continuous, on the interval (a, b), and for all x in that interval, $F'(x) = f'(x)$ and $G'(x) = g'(x) \neq 0$. Also notice that since $\lim_{x \to a^+} F(x) = \lim_{x \to a^+} f(x) = 0 = F(a)$, F is continuous from the right at a, and similarly G is continuous from the right at a.

For any number x such that $a < x < b$, we have

$$\frac{f(x)}{g(x)} = \frac{F(x)}{G(x)} = \frac{F(x) - F(a)}{G(x) - G(a)}.$$

We now apply Cauchy's mean value theorem to F and G on the interval $[a, x]$ to conclude that there is some number c_x such that $a < c_x < x$ and

$$\frac{F(x) - F(a)}{G(x) - G(a)} = \frac{F'(c_x)}{G'(c_x)} = \frac{f'(c_x)}{g'(c_x)}.$$

We have used the notation c_x, rather than simply c, to stress that c_x depends on x; for different values of x, the value of c_x will be different.

Since $a < c_x < x$, the squeeze theorem tells us that as $x \to a^+$, $c_x \to a^+$. And since $\lim_{x \to a^+}(f'(x)/g'(x)) = L$, we can say that as $c_x \to a^+$, $f'(c_x)/g'(c_x) \to L$. Combining these facts, we conclude that as $x \to a^+$,

$$\frac{f(x)}{g(x)} = \frac{f'(c_x)}{g'(c_x)} \to L,$$

and therefore

$$\lim_{x \to a^+} \frac{f(x)}{g(x)} = L.$$

Similar reasoning can be used to prove l'Hôpital's rule for limits in which $x \to a^-$, and the two one-sided versions of l'Hôpital's rule can be combined to prove the rule for two-sided limits. One can also replace L in the reasoning above with ∞ or $-\infty$ to prove that l'Hôpital's rule applies when $f'(x)/g'(x) \to \pm\infty$. For a proof of l'Hôpital's rule for limits in which $x \to \infty$, see Exercise 17. L'Hôpital's rule for indeterminate forms of type ∞/∞ can also be proven using Cauchy's mean value theorem, although the details are a bit trickier (see Exercise 18). $\qquad\qquad\square$

We end this section with some examples illustrating the limitations of l'Hôpital's rule. Students sometimes become enamored of l'Hôpital's rule and want to apply it to every limit they come across. Of course, it is never wrong to try using l'Hôpital's rule on any indeterminate form of type $0/0$ or ∞/∞, but the rule isn't always helpful.

For example, consider the limit

$$\lim_{x \to -\infty} \frac{\sqrt{x^2+1}}{x}.$$

This is an indeterminate form of type ∞/∞, so we compute

$$\frac{d}{dx}(\sqrt{x^2+1}) = \frac{1}{2}(x^2+1)^{-1/2} \cdot 2x = \frac{x}{\sqrt{x^2+1}}, \qquad \frac{d}{dx}(x) = 1.$$

L'Hôpital's rule now suggests that we compute the limit

$$\lim_{x \to -\infty} \frac{x/\sqrt{x^2+1}}{1} = \lim_{x \to -\infty} \frac{x}{\sqrt{x^2+1}}.$$

This is another indeterminate form of type ∞/∞, and another application of l'Hôpital's rule leads to the limit

$$\lim_{x \to -\infty} \frac{\sqrt{x^2+1}}{x}.$$

But this is the same as our original limit! L'Hôpital's rule is just leading us in circles; it is not going to solve this problem. For a solution, see Exercise 19.

Another interesting example is the limit

$$\lim_{x \to 0} \frac{x^2 \cos(1/x)}{\sin x}. \tag{4.22}$$

Clearly the denominator approaches 0, but the numerator is more complicated. We have

$$-1 \le \cos(1/x) \le 1,$$

and therefore

$$-x^2 \le x^2 \cos(1/x) \le x^2.$$

Since $\lim_{x \to 0} x^2 = \lim_{x \to 0}(-x^2) = 0$, the squeeze theorem tells us that $\lim_{x \to 0}(x^2 \cos(1/x)) = 0$. Thus, we have an indeterminate form of type $0/0$, so we can try l'Hôpital's rule.

We begin by computing

$$\frac{d}{dx}\left[x^2 \cos\left(\frac{1}{x}\right)\right] = x^2 \cdot \left[-\sin\left(\frac{1}{x}\right)\right] \cdot \left[-\frac{1}{x^2}\right] + \cos\left(\frac{1}{x}\right) \cdot 2x$$

$$= \sin\left(\frac{1}{x}\right) + 2x \cos\left(\frac{1}{x}\right).$$

Thus, l'Hôpital's rule suggests that we evaluate

$$\lim_{x \to 0} \frac{\sin(1/x) + 2x \cos(1/x)}{\cos x}. \tag{4.23}$$

As $x \to 0$, the denominator approaches 1, and we leave it to you to verify, using the squeeze theorem, that $2x \cos(1/x) \to 0$. But the term $\sin(1/x)$, like the function $h_1(x)$

from Section 2.2, oscillates back and forth between -1 and 1 infinitely many times as $x \to 0$. It appears, then, that the fraction in limit (4.23) is sometimes close to -1 and sometimes close to 1 when x is near 0, and therefore the limit is undefined. (You are asked to verify this conclusion more carefully in Exercise 20.) Does it follow that limit (4.22) is undefined? No, this would be a misuse of l'Hôpital's rule. The rule says that if limit (4.23) is equal to some number, or equal to $\pm\infty$, then the same is true for limit (4.22), but that is not the situation we are in.

In fact, limit (4.22) is defined, as the following calculation shows:

$$\lim_{x \to 0} \frac{x^2 \cos(1/x)}{\sin x} = \lim_{x \to 0} \frac{x \cos(1/x)}{(\sin x)/x} = \frac{0}{1} = 0.$$

Thus, this example illustrates that when using l'Hôpital's rule you must be careful to apply it only in cases that fit the rule exactly as it is written.

Finally, we consider the limit

$$\lim_{x \to \infty} \frac{x - \sin x \cos x}{2x - x \cos x + \sin x - 2 \sin x \cos x}. \tag{4.24}$$

This is an indeterminate form of type ∞/∞ (see Exercise 21), so let's try l'Hôpital's rule. We compute:

$$\frac{d}{dx}(x - \sin x \cos x) = 1 + \sin^2 x - \cos^2 x = 2 \sin^2 x$$

and

$$\frac{d}{dx}(2x - x \cos x + \sin x - 2 \sin x \cos x)$$
$$= 2 + x \sin x - \cos x + \cos x + 2 \sin^2 x - 2 \cos^2 x$$
$$= x \sin x + 2(1 - \cos^2 x) + 2 \sin^2 x$$
$$= x \sin x + 4 \sin^2 x.$$

Thus, l'Hôpital's rule leads us to compute the limit

$$\lim_{x \to \infty} \frac{2 \sin^2 x}{x \sin x + 4 \sin^2 x}. \tag{4.25}$$

After canceling a factor of $\sin x$ from numerator and denominator in this limit we get

$$\lim_{x \to \infty} \frac{2 \sin x}{x + 4 \sin x} = 0, \tag{4.26}$$

since the denominator in (4.26) approaches ∞ and the numerator is always between -2 and 2. You might think that we could use l'Hôpital's rule to conclude that the value of limit (4.24) is also 0, but this is incorrect. The mistake this time is very subtle. Although we have computed limit (4.26) correctly, limit (4.25) is actually undefined, because numerator and denominator are both 0 whenever x is an integer multiple of π. Thus, l'Hôpital's rule doesn't apply.

To get the correct answer for limit (4.24), we divide numerator and denominator by x:

$$\lim_{x\to\infty} \frac{x - \sin x \cos x}{2x - x\cos x + \sin x - 2\sin x \cos x} \cdot \frac{1/x}{1/x} = \lim_{x\to\infty} \frac{1 - \frac{\sin x \cos x}{x}}{2 - \cos x + \frac{\sin x - 2\sin x \cos x}{x}}.$$

$$(4.27)$$

As $x \to \infty$, the fractions $(\sin x \cos x)/x$ and $(\sin x - 2\sin x \cos x)/x$ both approach 0, and therefore for large x, the numerator on the right side of (4.27) is close to 1 and the denominator is close to $2 - \cos x$. Since $2 - \cos x$ oscillates between 1 and 3, for some large values of x the function in limit (4.27) is close to 1, and for some it is close to 1/3. Thus, the limit is undefined.

Exercises 4.8

1–16: Evaluate the limit.

1. $\lim_{x\to1} \dfrac{x^4 - 1}{x^5 - 1}$.

2. $\lim_{x\to1} \dfrac{\sqrt[4]{x} - 1}{\sqrt[5]{x} - 1}$.

3. $\lim_{u\to4} \dfrac{u^{3/2} - 8}{\sqrt{u} - 2}$.

4. $\lim_{x\to1} \dfrac{x^8 + 3x^6 - 5x^2 + 1}{x^7 + 4x^5 - 3x^3 - 2}$.

5. $\lim_{x\to1} \dfrac{x^8 + x^6 - 7x^2 + 5}{x^8 + 2x^5 - 9x^2 + 6}$.

6. $\lim_{x\to\pi/2} \dfrac{\cos(x + \pi)}{2x - \pi}$.

7. $\lim_{x\to2} \dfrac{\cos(\pi/x)}{x - 2}$.

8. $\lim_{t\to1} \dfrac{\sqrt{t + 3} - 2}{\sqrt{t + 8} - 3}$.

9. $\lim_{\theta\to0} (\tan(3\theta)\cot(4\theta))$.

10. $\lim_{t\to0} \dfrac{\sec t - 1}{t^2}$.

11. $\lim_{x\to0} \dfrac{x - \sin x}{x^3}$.

12. $\displaystyle\lim_{x\to 0}\frac{\tan x - x}{x^3}$.

13. $\displaystyle\lim_{x\to \pi/2^-}\frac{\tan x}{\cot(2x)}$.

14. $\displaystyle\lim_{x\to -\infty}\frac{\csc(1/x)}{x^2}$.

15. $\displaystyle\lim_{x\to 0}(x\tan(x+\pi/2))$.

16. $\displaystyle\lim_{x\to \infty}\left[x\cos\left(\frac{\pi x + 2}{2x+\pi}\right)\right]$.

17. In this problem you will prove l'Hôpital's rule for limits in which $x \to \infty$. Suppose $\lim_{x\to\infty} f(x) = \lim_{x\to\infty} g(x) = 0$ and $\lim_{x\to\infty}(f'(x)/g'(x)) = L$. Apply l'Hôpital's rule to the limit

$$\lim_{t\to 0^+}\frac{f(1/t)}{g(1/t)}$$

to prove that $\lim_{x\to\infty}(f(x)/g(x)) = L$.

18. In this problem you will prove one case of l'Hôpital's rule for indeterminate forms of type ∞/∞. Suppose $\lim_{x\to a^+} f(x) = \lim_{x\to a^+} g(x) = \infty$ and $\lim_{x\to a^+}(f'(x)/g'(x)) = L$. We will use the ϵ-δ method to prove that $\lim_{x\to a^+}(f(x)/g(x)) = L$. Suppose $\epsilon > 0$.

(a) Show that there is a number $b > a$ such that if $a < x < b$ then $g'(x) \neq 0$ and

$$\left|\frac{f'(x)}{g'(x)} - L\right| < \frac{\epsilon}{2}.$$

We will use this number b in the rest of the proof.

(b) Show that if $a < x < b$ then $g(x) \neq g(b)$ and

$$\left|\frac{f(x) - f(b)}{g(x) - g(b)} - L\right| < \frac{\epsilon}{2}.$$

(Hint: Use the Cauchy mean value theorem.)

(c) Show that there is some $\delta > 0$ such that $a + \delta \leq b$ and if $a < x < a + \delta$ then $g(x) \neq 0$ and

$$\left|\frac{f(x)}{g(x)} - \frac{f(x) - f(b)}{g(x) - g(b)}\right| < \frac{\epsilon}{2}.$$

(Hint: First verify the equation

$$\frac{f(x)}{g(x)} - \frac{f(x) - f(b)}{g(x) - g(b)} = \frac{f(b)}{g(x)} - \frac{g(b)}{g(x)}\cdot\frac{f(x) - f(b)}{g(x) - g(b)}.$$

Then use part (b) and the fact that $\lim_{x\to a^+} g(x) = \infty$.)

(d) Combine parts (b) and (c) to complete the proof.

19. Use the fact that for $x < 0$, $x = -\sqrt{x^2}$ to evaluate $\lim_{x \to -\infty}(\sqrt{x^2+1}/x)$.

20. Verify that limit (4.23) is undefined. (Hint: Suppose the limit is equal to L. First show that $\lim_{x \to 0} \sin(1/x) = L$. Then let $\epsilon = 1$ in the definition of limits and derive a contradiction.)

21. Verify that limit (4.24) is an indeterminate form of type ∞/∞.

22. Let

$$f(x) = \begin{cases} \dfrac{\sin x}{x}, & \text{if } x \neq 0, \\ 1, & \text{if } x = 0. \end{cases}$$

(a) Compute $f'(0)$. (Hint: You will need to use the definition of derivatives.)

(b) Is f' continuous at 0?

4.9 Antiderivatives

In all of our work with derivatives so far, we have started with a function and then computed its derivative. But in some applications of calculus it is useful to work backwards: given the derivative of a function, find the function. We investigate this process of "antidifferentiation" in this section.

For example, consider the function $f(x) = 6x^2$. Can you find a function $F(x)$ such that $F'(x) = f(x)$? In this case it is not hard to guess an answer. If we let $F(x) = 2x^3$, then $F'(x) = 6x^2 = f(x)$. But this isn't the only possible answer. For example, if $F(x) = 2x^3 + 1$ then it is also true that $F'(x) = f(x)$. In fact, it is clear that for any number C, if $F(x) = 2x^3 + C$ then $F'(x) = f(x)$.

Are there any other possibilities we haven't thought of? No, these are the only ones. To see why, suppose that F is a function such that for all x, $F'(x) = 6x^2$. Then

$$\frac{d}{dx}(F(x) - 2x^3) = 6x^2 - 6x^2 = 0.$$

Now recall from part 5 of Theorem 4.3.2 that if a function is continuous on an interval and its derivative is 0 on the interior of the interval, then the function is constant on the interval. It follows that $F(x) - 2x^3$ must be constant on the interval $(-\infty, \infty)$. In other words, there is some number C such that for all x, $F(x) - 2x^3 = C$, and therefore $F(x) = 2x^3 + C$. The graphs of some functions of this form are shown in Figure 4.44.

In this example we were able to find a family of functions F such that the equation $F'(x) = f(x)$ holds for all values of x; in other words, F' agrees with f throughout the interval $(-\infty, \infty)$. We say that F is an *antiderivative* of f on the interval $(-\infty, \infty)$. We will sometimes want to work with antiderivatives on smaller intervals as well. For open intervals, the definition is easy:

Definition 4.9.1. Suppose f is a function and I is an open interval contained in the domain of f. We say that a function F is an *antiderivative* of f on I if F is differentiable on I, and for all $x \in I$, $F'(x) = f(x)$.

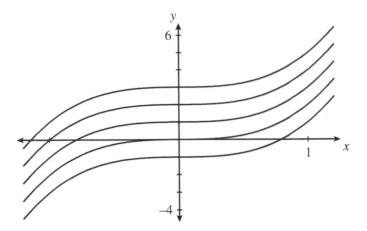

Figure 4.44: Graphs of the curves $y = 2x^3 - 1$, $y = 2x^3$, $y = 2x^3 + 1$, $y = 2x^3 + 2$, and $y = 2x^3 + 3$.

For example, the function $F(x) = 2\sqrt{x}$ is an antiderivative of $f(x) = 1/\sqrt{x}$ on the interval $(0, \infty)$, because for all $x > 0$,

$$F'(x) = \frac{d}{dx}(2x^{1/2}) = 2 \cdot \frac{1}{2}x^{-1/2} = \frac{1}{\sqrt{x}} = f(x).$$

We will also want to work with antiderivatives on closed intervals, but this concept is slightly more complicated. Just as continuity on closed intervals required us to use a one-sided version of continuity at the endpoints, differentiability on closed intervals will require a one-sided version of derivatives at the endpoints.

Definition 4.9.2. The *right-hand derivative* of a function f at a number a is the number $f'_+(a)$ defined by the formula

$$f'_+(a) = \lim_{h \to 0^+} \frac{f(a+h) - f(a)}{h}.$$

Similarly, the *left-hand derivative* of f at a is

$$f'_-(a) = \lim_{h \to 0^-} \frac{f(a+h) - f(a)}{h}.$$

Notice that our definitions of $f'_+(a)$ and $f'_-(a)$ are based on one-sided versions of the limit that defines $f'(a)$. Thus, the ordinary two-sided derivative $f'(a)$ is defined if and only if the two one-sided derivatives are defined and equal, and in that case $f'_+(a) = f'_-(a) = f'(a)$. But sometimes the one-sided derivatives are defined when the two-sided derivative is not. For example, we have already seen that if $f(x) = |x|$ then

$f'(0)$ is undefined. But the one-sided derivatives are

$$f'_+(0) = \lim_{h \to 0^+} \frac{|0+h| - |0|}{h} = \lim_{h \to 0^+} \frac{h}{h} = 1,$$

$$f'_-(0) = \lim_{h \to 0^-} \frac{|0+h| - |0|}{h} = \lim_{h \to 0^-} \frac{-h}{h} = -1.$$

All of our familiar rules for computing derivatives have versions that apply to one-sided derivatives as well. For example, if $f'_+(a)$ and $g'_+(a)$ are defined, then $(f + g)'_+(a)$ and $(f \cdot g)'_+(a)$ are also defined, and we have

$$(f + g)'_+(a) = f'_+(a) + g'_+(a), \qquad (f \cdot g)'_+(a) = f(a) \cdot g'_+(a) + g(a) \cdot f'_+(a).$$

As in our proof of Theorem 3.2.3, by letting $x = a + h$ in the definitions of the one-sided derivatives we can derive the alternative formulas

$$f'_+(a) = \lim_{x \to a^+} \frac{f(x) - f(a)}{x - a}$$

and

$$f'_-(a) = \lim_{x \to a^-} \frac{f(x) - f(a)}{x - a}.$$

Thus, $f'_+(a)$ is determined by the values of $f(a)$ and $f(x)$ for $x > a$, whereas $f'_-(a)$ depends on $f(a)$ and $f(x)$ for $x < a$. Also, by imitating the proof of Theorem 3.2.4, we can show that if $f'_+(a)$ is defined then f is continuous from the right at a, and if $f'_-(a)$ is defined then f is continuous from the left at a.

Consider the function $g(x) = x^{3/2} = x\sqrt{x}$. The domain of this function is $[0, \infty)$, and $g'(x) = (3/2)x^{1/2} = 3\sqrt{x}/2$ for $x > 0$. We have already observed that $g'(0)$ is undefined, because $g(x)$ is not defined for x close to 0 on both sides of 0. But $g(x)$ is defined for all $x > 0$, so it makes sense to compute the right-hand derivative of g at 0:

$$g'_+(0) = \lim_{h \to 0^+} \frac{(0+h)^{3/2} - 0^{3/2}}{h} = \lim_{h \to 0^+} h^{1/2} = 0.$$

This motivates our next definition.

Definition 4.9.3. We say that a function f is *differentiable on a closed interval* $[a, b]$ if f is differentiable on (a, b), and the one-sided derivatives $f'_+(a)$ and $f'_-(b)$ are defined.

Similarly, we say that f is differentiable on $[a, b)$ if it is differentiable on (a, b) and $f'_+(a)$ is defined, and it is differentiable on $(a, b]$ if it is differentiable on (a, b) and $f'_-(b)$ is defined. In general, f is differentiable on an interval I if it is differentiable on the interior of I, and the appropriate one-sided derivatives are defined at any endpoints that are included in I. For example, we can now say that the function $g(x) = x^{3/2}$ is differentiable on the interval $[0, \infty)$. The function $f(x) = |x|$ is differentiable on both of the intervals $(-\infty, 0]$ and $[0, \infty)$, although it is not differentiable on $(-\infty, \infty)$.

We are now ready to define antiderivatives on intervals that are not open.

Definition 4.9.4. Suppose f is a function and the closed interval $[a, b]$ is contained in the domain of f. We say that a function F is an *antiderivative* of f on $[a, b]$ if F is differentiable on $[a, b]$, $F'_+(a) = f(a)$, $F'_-(b) = f(b)$, and for all $x \in (a, b)$, $F'(x) = f(x)$. More generally, if I is any interval contained in the domain of f, then to say that F is an antiderivative of f on I means that F is differentiable on I, if I includes its left endpoint a then $F'_+(a) = f(a)$, if I includes its right endpoint b then $F'_-(b) = f(b)$, and for every x in the interior of I, $F'(x) = f(x)$.

When talking about derivatives of a function we may sometimes leave the subscript $+$ or $-$ out of our derivative notation if it is clear from context what kind of derivative we are talking about. For example, if F is an antiderivative of f on an interval I, then we may simply say that for all $x \in I$, $F'(x) = f(x)$, leaving it for you to fill in that in this equation $F'(x)$ denotes $F'_+(x)$ if x is the left endpoint of I, $F'_-(x)$ if x is the right endpoint of I, and $F'(x)$ if x is in the interior of I. Another example of this convention is the following version of the chain rule, which can be proven by making minor adjustments in our previous proof:

Theorem 4.9.5 (Chain Rule on Intervals). *Suppose that f is differentiable on an interval I, g is differentiable on an interval J, and for every $x \in J$, $g(x) \in I$. Then $f \circ g$ is differentiable on J, and for every $x \in J$,*

$$(f \circ g)'(x) = f'(g(x)) \cdot g'(x),$$

where each of the derivatives in this equation is interpreted as a one-sided derivative whenever it is evaluated at an endpoint.

As in our first example in this section, antiderivatives always come in families of functions that differ from each other by a constant. In other words, we have the following theorem.

Theorem 4.9.6. *Suppose that F is an antiderivative of f on an interval I. Then for every number C, the function $G(x) = F(x) + C$ is also an antiderivative of f on I. Furthermore, these are the only antiderivatives. In other words, if G is any antiderivative of f on I, then there is some number C such that for all $x \in I$, $G(x) = F(x) + C$.*

Proof. Since F is an antiderivative of f on I, for every $x \in I$, $F'(x) = f(x)$ (with appropriate one-sided derivatives being used at endpoints). Since the derivative of a constant function is 0, it follows that if $G(x) = F(x) + C$ then $G'(x) = F'(x) + 0 = f(x)$, so G is also an antiderivative of f on I.

To see that these are the only antiderivates of f on I, suppose that G is any antiderivative of f on I. Since F and G are differentiable on I, they are also continuous on I. We also know that for all $x \in I$, $G'(x) = f(x)$, but we will actually use this equation only for x in the interior of I. Let $H = G - F$. Then H is continuous on I, and for all x in the interior of I,

$$H'(x) = G'(x) - F'(x) = f(x) - f(x) = 0.$$

By part 5 of Theorem 4.3.2, it follows that H is constant on the interval I. In other words, there is some number C such that for all $x \in I$, $H(x) = C$. Therefore $G(x) - F(x) = C$, so $G(x) = F(x) + C$. $\square$

It is perhaps worth mentioning that the second paragraph of this proof actually proves a little more than what is stated in the theorem. The proof shows that if F is an antiderivative of f on I, G is continuous on I, and G is an antiderivative of f on the interior of I, then there is some number C such that for all $x \in I$, $G(x) = F(x) + C$, and therefore G is actually an antiderivative of f on all of I. For more on this, see Exercise 19.

According to Theorem 4.9.6, to find all antiderivatives of a function f on an interval I, we simply have to find one antiderivative F and then add "$+ C$" to the formula for $F(x)$. Finding one antiderivative can sometimes be done by reversing our rules for computing derivatives. For example, reversing the power rule for rational exponents (Theorem 3.4.4) gives us the following theorem.

Theorem 4.9.7. *Suppose r is a rational number and $r \neq -1$, and let $f(x) = x^r$. Then the function*

$$F(x) = \frac{x^{r+1}}{r+1}$$

is an antiderivative of f on every interval contained in the domain of f.

To list these intervals explicitly, we first write the rational number r in lowest terms as $r = m/n$. If n is odd and $r \geq 0$, then $f(x) = x^r$ is defined for all x, and F is an antiderivative of f on the interval $(-\infty, \infty)$. If n is odd and $r < 0$, then $f(x)$ is defined for all values of x except $x = 0$, and F is an antiderivative of f on each of the intervals $(-\infty, 0)$ and $(0, \infty)$. If n is even, then the domain of f is $[0, \infty)$ if $r > 0$, and $(0, \infty)$ if $r < 0$. In both of these cases, F is an antiderivative of f on the entire domain of f.

Proof. Notice that since $r \neq -1$, the denominator in the formula for $F(x)$ is not 0. According to the power rule for derivatives,

$$F'(x) = \frac{1}{r+1} \cdot (r+1)x^r = x^r = f(x).$$

The correctness of this equation on the intervals listed in the theorem follows, in almost all cases, from Theorem 3.4.4. The one exception is that if n is even and $r > 0$ then, since the domain of f includes the left endpoint 0, we must verify that $F'_+(0) = f(0)$:

$$F'_+(0) = \lim_{h \to 0^+} \frac{F(0+h) - F(0)}{h} = \lim_{h \to 0^+} \frac{h^{r+1}/(r+1) - 0}{h} = \lim_{h \to 0^+} \frac{h^r}{r+1} = 0 = f(0).$$

$\square$

Another derivative theorem that can easily be reversed is Theorem 3.3.2:

Theorem 4.9.8. *Suppose that F and G are antiderivatives of f and g, respectively, on an interval I. Then:*

1. *$F + G$ is an antiderivative of $f + g$ on I.*

2. *For any number c, cF is an antiderivative of cf on I.*

3. *$F - G$ is an antiderivative of $f - g$ on I.*

Proof. Since F and G are antiderivatives of f and g on I, we know that for all $x \in I$, $F'(x) = f(x)$ and $G'(x) = g(x)$. All three parts of the theorem now follow easily from the corresponding rules for derivatives. For example, $F + G$ is an antiderivative of $f + g$ because
$$(F + G)'(x) = F'(x) + G'(x) = f(x) + g(x) = (f + g)(x). \qquad \square$$

Combining our theorems, we can now easily find antiderivatives of all polynomials. For example, if $f(x) = 2x^3 + x^2 - 4x + 5$, then the antiderivatives of f on $(-\infty, \infty)$ are precisely the functions of the form
$$F(x) = 2 \cdot \frac{x^4}{4} + \frac{x^3}{3} - 4 \cdot \frac{x^2}{2} + 5 \cdot \frac{x^1}{1} + C = \frac{x^4}{2} + \frac{x^3}{3} - 2x^2 + 5x + C.$$

Notice in particular that, although the derivative of a constant function is 0, this is not true of the antiderivatives of a constant function. Rather, the antiderivatives of a constant function $f(x) = a$ are functions of the form $F(x) = ax + C$.

Reversing our trigonometric derivative rules gives us the trigonometric antiderivatives in Table 4.1. This table illustrates that antidifferentiation is often more difficult than differentiation. For example, the table gives an antiderivative for $\sec^2 x$, but not $\sec x$. In fact, $\sec x$ does have an antideriviative, but it involves a function that we haven't studied yet. We'll learn about it later in this book (see Example 8.3.3). In contrast, finding the derivative of any power of $\sec x$ is easy. For another example of the difficulty of antidifferentiation, notice that although we know the derivative of the function $f(x) = 1/x$, we don't yet know an antiderivative of f, since Theorem 4.9.7 doesn't apply when $r = -1$. Again, this is an antiderivative that we will have to return to later (see Section 7.3).

Function	Antiderivative	Function	Antiderivative
$\sin x$	$-\cos x$	$\cos x$	$\sin x$
$\sec^2 x$	$\tan x$	$\csc^2 x$	$-\cot x$
$\sec x \tan x$	$\sec x$	$\csc x \cot x$	$-\csc x$

Table 4.1: Some trigonometric antiderivatives. In each case, the antiderivative works on any interval contained in the domain of the function.

Example 4.9.9. Find a function F such that for all x, $F'(x) = \sqrt[3]{x} + \sin x$, and $F(0) = 3$.

Solution. By Theorem 4.9.7, one antiderivative of $\sqrt[3]{x} = x^{1/3}$ is $(3/4)x^{4/3}$, and by Table 4.1, one antiderivative of $\sin x$ is $-\cos x$. Thus, we must have
$$F(x) = \frac{3}{4}x^{4/3} - \cos x + C,$$

for some constant C. We can use the fact that $F(0) = 3$ to solve for C:
$$F(0) = \frac{3}{4}0^{4/3} - \cos 0 + C = 0 - 1 + C = 3,$$

and therefore $C = 4$. So the unique function F that meets the stated requirements is

$$F(x) = \frac{3}{4}x^{4/3} - \cos x + 4. \qquad \square$$

This example illustrates a common pattern: When we compute an antiderivative, there will always be an undetermined constant C in our answer. But we may be able to determine the value of C if we have some additional information. In particular, if we know the value of the antiderivative at a single number, then we can use this to solve for C.

Example 4.9.10. A ball is thrown straight up from ground level with an initial velocity of 160 feet per second. Find the height of the ball t seconds later. (You may ignore the effects of air resistance.)

Solution. If we let h denote the height of the ball, in feet, then $h = f(t)$, for some unknown function f. We wish to find a formula for $f(t)$. As we have discussed before, the velocity of the ball at any time will be $v = dh/dt = f'(t)$, and the acceleration will be $a = dv/dt = f''(t)$. In solving this problem we will make use of the fact, determined by experiment, that the earth's gravity will cause the ball (or, in fact, any object in free fall near the surface of the earth) to experience a constant downward acceleration of approximately 32 ft/sec^2.

The velocity $v = dh/dt$ will be positive if the height of the ball is increasing, and therefore the ball is moving up, and it will be negative if the ball is moving down. Since the acceleration caused by gravity is downward, the velocity of the ball will be decreasing at a constant rate of 32 ft/sec^2, and therefore for all t we have

$$a = f''(t) = -32.$$

We can now use antidifferentiation to find first $f'(t)$ and then $f(t)$.

Since $f'(t)$ must be an antiderivative of $f''(t)$, we have

$$v = f'(t) = -32t + C,$$

for some number C. We are also told that the initial velocity of the ball is 160 ft/sec, so

$$f'(0) = -32(0) + C = 160.$$

Therefore $C = 160$, and $v = f'(t) = -32t + 160$. Antidifferentiating again, we get

$$h = f(t) = -16t^2 + 160t + D,$$

for some D. Once again, we have a given value that we can use to solve for D. We are told that the ball is thrown "from ground level," so

$$f(0) = -16(0^2) + 160(0) + D = 0.$$

Therefore $D = 0$, so $h = f(t) = -16t^2 + 160t$. Notice that this is the formula we used for the height of the ball in Section 2.1. We have finally learned enough calculus to be able to see where that formula came from.

We have not specified the intervals on which our antiderivatives were computed. As we observed in Section 2.1, after 10 seconds the ball hits the ground. It is then no longer in free fall, and the formula $a = -32$ no longer applies. Thus, our formulas are correct only for the time interval [0, 10].   □

Example 4.9.11. A rock is dropped from the top of a tall building. It hits the ground 4 seconds later. How tall is the building?

Solution. Once again we let h be the height of the ball above the ground, in feet, and we have $h = f(t)$, where t is the number of seconds since the rock was dropped. As before, while the rock is falling—that is, for t in the interval [0, 4]—the acceleration is $a = f''(t) = -32$, and therefore

$$v = f'(t) = -32t + C.$$

This time we are told that the rock is *dropped*, not thrown, so the initial velocity is

$$f'(0) = -32(0) + C = 0.$$

This tells us that $C = 0$, and therefore $v = f'(t) = -32t$. Antidifferentiating again on the interval [0, 4], we get

$$f(t) = -16t^2 + D.$$

We are told that the rock hits the ground after 4 seconds, so

$$f(4) = -16(4^2) + D = 0,$$

and therefore $D = 256$. So the height of the rock after t seconds is $h = f(t) = -16t^2 + 256$. In particular, the rock starts at a height of $h = f(0) = 256$ feet. Thus, the building is 256 feet tall.   □

Example 4.9.12. You drop a rock into a well, and 4 seconds later you hear the splash when it hits the water. How deep is the well?

Solution. This problem is slightly different from the last one, because it takes some time for the sound to travel from the bottom of the well to the top. Thus, the rock hits the water in a little less than 4 seconds.

   Just for variety, in this case we will measure the rock's position *downward* from ground level. So we'll let d be the depth of the rock in the well, measured in feet, and we have $d = f(t)$, where t is the time, in seconds, since the rock was dropped. Once again the velocity is $v = dd/dt = f'(t)$, but now a positive velocity means that the depth d is increasing, which means that the rock is moving *down*. Now the downward acceleration of gravity will cause the velocity to *increase* at a steady rate of 32 ft/sec^2, so we have $a = dv/dt = f''(t) = 32$.

As before, we antidifferentiate twice and use the fact that the initial depth and velocity are both 0 to deduce that

$$v = f'(t) = 32t,$$
$$d = f(t) = 16t^2.$$

The rock hits the water at some time $t = t_0 < 4$, at which time the rock has fallen to a depth of $d = 16t_0^2$ feet. The sound must now travel the $16t_0^2$ feet back to the surface in $4 - t_0$ seconds. Since the speed of sound is approximately 1126 ft/sec, this means that

$$16t_0^2 = 1126(4 - t_0).$$

This is a quadratic equation, and we can use the quadratic formula to solve for t_0:

$$16t_0^2 + 1126t_0 - 4504 = 0,$$
$$t_0 = \frac{-1126 \pm \sqrt{1126^2 + 4(16)(4504)}}{32} \approx -74.2, 3.8.$$

Of course, the negative solution doesn't make sense in this problem—the rock can't hit the water 74 seconds before it was dropped! So the rock hits the water about 3.8 seconds after it was dropped, and the depth of the well is approximately $f(3.8) = 16(3.8^2) \approx 231$ ft. As a check, we can compute that the time it will take the sound to travel the 231 feet back to the surface is $231/1126 \approx 0.2$ sec, so the sound will arrive after $3.8 + 0.2 = 4$ seconds, as required. □

Exercises 4.9

1–9: Find an antiderivative of the function. On what interval or intervals does your antiderivative apply?

1. $f(x) = 2x^3 - 9x^2 + 3x - 1$.

2. $g(x) = \sqrt{x} + \sqrt[3]{x}$.

3. $f(x) = \sin x + \cos x$.

4. $g(x) = \sec x(\sec x + \tan x)$.

5. $f(x) = x^2 + \dfrac{1}{x^2}$.

6. $g(x) = \dfrac{x^3 + 3}{x^2}$.

7. $f(x) = (x^3 + 3)^2$.

8. $g(x) = \left(\dfrac{x^3 + 3}{x}\right)^2$.

9. $f(x) = \sqrt{2x + 3}$. (Hint: First guess an antiderivative. Then differentiate your guess and adjust if necessary.)

10–13: Find the function.

10. Domain of f is $(-\infty, \infty)$, $f'(x) = 4x - 3$, $f(1) = 4$.

11. Domain of g is $[0, \infty)$, $g'(x) = \sqrt{x} + \sin x$, $g(0) = 1$.

12. Domain of f is $(0, \pi/2)$, $f'(x) = \dfrac{1}{\sqrt{x}} + \sec x \tan x$, $\lim_{x \to 0^+} f(x) = 2$.

13. Domain of g is $(-\infty, \infty)$, $g'(x) = x^2 \sqrt[3]{x}$, $g(1) = 2$.

14. A rock is dropped from the top of a cliff. One second later, another rock is thrown downward from the top of the cliff with an initial velocity of 40 ft/sec. The two rocks reach the ground at the same time. How tall is the cliff?

15. A driver applied the brakes in his car, causing a constant deceleration of 10 ft/sec². The car traveled 125 feet before coming to a stop. How fast was the car moving when the driver applied the brakes?

16. A driver sees a police car by the side of the road 100 feet ahead and slams on his brakes, hoping to avoid a speeding ticket. The brakes produce a constant deceleration of 10 ft/sec². If he passes the police car 2 seconds later, how fast was he going when he applied the brakes?

17. Harry Potter is flying forward at a speed of 10 m/sec when he notices the snitch hovering at a point 100 meters directly ahead of him. If his broom accelerates at 4 m/sec², how long will it take him to reach the snitch?

18. A 500-gallon tank was full of water. At noon, water started leaking out of the tank, and the leak gradually got worse over time. At t minutes after noon, the water was leaking out of the tank at a rate of $5 + 2t$ gallons per minute, until the tank was empty. When did the tank become empty?

19. Suppose that f and F are continuous on $[a, b]$ and F is an antiderivaitve of f on (a, b). Show that in fact F is an antiderivative of f on $[a, b]$. (Hint: Use the mean value theorem.)

20. Suppose that g is differentiable at a, $g(a) = b$, and as $x \to a^{\neq}$, $g(x) \to b^{\geq}$. Show that if f is differentiable from the right at b, then $f \circ g$ is differentiable at a, and $(f \circ g)'(a) = f'_+(b) \cdot g'(a)$. (Do not use Theorem 4.9.5. This exercise asks you to verify one of the cases that must be checked in order to prove Theorem 4.9.5.)

21. Suppose that g is differentiable at a, $g(a) = b$, and as $x \to a^{\neq}$, $g(x) \to b^{\geq}$. Show that $g'(a) = 0$.

Chapter 5

Integrals

5.1 Summations

In this chapter we will often work with sums of lists of numbers. It will therefore be convenient to introduce notation for such sums. We will write $\sum_{i=1}^{n} a_i$ to denote the sum of the numbers $a_1, a_2, \ldots, a_n$. In other words,

$$\sum_{i=1}^{n} a_i = a_1 + a_2 + \cdots + a_n. \tag{5.1}$$

As in our notation for sequences, the subscript "$i = 1$" and superscript "n" indicate that in the expression a_i, we should plug in values for i starting with 1 and running up to n. The symbol "Σ" (the Greek uppercase letter sigma) indicates that the resulting numbers are to be added. For example,

$$\sum_{i=1}^{4} \frac{1}{i} = 1 + \frac{1}{2} + \frac{1}{3} + \frac{1}{4} = \frac{25}{12}.$$

It will be helpful to develop rules that will sometimes allow us to evaluate $\sum_{i=1}^{n} a_i$ when a_i is given by a formula. As with limits and derivatives, we proceed in a systematic way, working out the simplest summations, and then establishing rules that will allow us to combine simple summations to find more complicated ones.

Recall that the simplest functions are the constant functions and the identity function. Similarly, the simplest summations of the form (5.1) are those in which either a_i is equal to a constant or $a_i = i$. Consider first the case in which there is some constant c such that for every i, $a_i = c$. Then

$$\sum_{i=1}^{n} a_i = a_1 + a_2 + \cdots + a_n = \underbrace{c + c + \cdots + c}_{n \text{ times}} = nc.$$

Filling in $a_i = c$ in the summation on the left-hand side of this equation, we have

$$\sum_{i=1}^{n} c = nc. \tag{5.2}$$

For example, $\sum_{i=1}^{100} 6 = 100 \cdot 6 = 600$.

If a_i is given by the formula $a_i = i$, then the summation in (5.1) becomes

$$\sum_{i=1}^{n} i = 1 + 2 + \cdots + n.$$

Of course, it is straightforward to compute this sum for a particular value of n. But there is a clever trick that allows us to find a formula that gives the value of this sum for any value of n. First we write the summation twice, with the terms written in reverse order the second time, and then we add the two equations, term-by-term:

$$\sum_{i=1}^{n} i = \quad 1 \quad + \quad 2 \quad + \quad 3 \quad + \cdots + (n-1) + \quad n$$

$$+ \sum_{i=1}^{n} i = \quad n \quad + (n-1) + (n-2) + \cdots + \quad 2 \quad + \quad 1$$

$$2\sum_{i=1}^{n} i = (n+1) + (n+1) + (n+1) + \cdots + (n+1) + (n+1). \tag{5.3}$$

On the right-hand side of equation (5.3), the term $(n+1)$ appears n times, so we have

$$2\sum_{i=1}^{n} i = n(n+1),$$

and therefore

$$\sum_{i=1}^{n} i = \frac{n(n+1)}{2} = \frac{n^2 + n}{2}. \tag{5.4}$$

For example,

$$1 + 2 + \cdots + 10 = \sum_{i=1}^{10} i = \frac{10^2 + 10}{2} = 55,$$

as you can confirm by adding up the numbers from 1 to 10.

We can combine these simple summations to evaluate some more complicated ones by using the following theorem.

Theorem 5.1.1. *For any numbers c, a_1, a_2, ..., a_n, b_1, b_2, ..., b_n, the following equations are true:*

1. $\displaystyle\sum_{i=1}^{n}(ca_i) = c\sum_{i=1}^{n} a_i.$

2. $\displaystyle\sum_{i=1}^{n}(a_i + b_i) = \sum_{i=1}^{n} a_i + \sum_{i=1}^{n} b_i.$

3. $\displaystyle\sum_{i=1}^{n}(a_i - b_i) = \sum_{i=1}^{n} a_i - \sum_{i=1}^{n} b_i.$

Proof. We can prove this theorem by simply writing out the meaning of the summation notation and then applying basic rules of algebra. (For an alternative approach to the proof of this theorem, see Exercise 8.) For part 1, we have

$$\sum_{i=1}^{n}(ca_i) = ca_1 + ca_2 + \cdots + ca_n = c(a_1 + a_2 + \cdots + a_n) = c\sum_{i=1}^{n} a_i.$$

Similarly, to prove part 2 we compute

$$\sum_{i=1}^{n}(a_i + b_i) = (a_1 + b_1) + (a_2 + b_2) + \cdots + (a_n + b_n)$$

$$= (a_1 + a_2 + \cdots + a_n) + (b_1 + b_2 + \cdots + b_n) = \sum_{i=1}^{n} a_i + \sum_{i=1}^{n} b_i.$$

Part 3 follows from parts 1 and 2:

$$\sum_{i=1}^{n}(a_i - b_i) = \sum_{i=1}^{n}(a_i + (-1)b_i) = \sum_{i=1}^{n} a_i + \sum_{i=1}^{n}((-1)b_i) \qquad \text{(by part 2)}$$

$$= \sum_{i=1}^{n} a_i + (-1)\sum_{i=1}^{n} b_i \qquad \text{(by part 1)}$$

$$= \sum_{i=1}^{n} a_i - \sum_{i=1}^{n} b_i. \qquad \qquad \square$$

Example 5.1.2. Evaluate $\sum_{i=1}^{n}(2i - 1)$.

Solution. We begin by applying parts 1 and 3 of Theorem 5.1.1:

$$\sum_{i=1}^{n}(2i - 1) = \sum_{i=1}^{n}(2i) - \sum_{i=1}^{n} 1 = 2\sum_{i=1}^{n} i - \sum_{i=1}^{n} 1.$$

We can now get the solution by filling in the formulas from equations (5.2) and (5.4):

$$\sum_{i=1}^{n}(2i-1) = 2\sum_{i=1}^{n}i - \sum_{i=1}^{n}1 = 2\cdot\frac{n^2+n}{2} - n\cdot 1 = n^2+n-n = n^2. \qquad \square$$

Theorem 5.1.1 tells us what happens to a summation when all of its terms are multiplied by a constant, and it tells us what happens when two summations are combined by adding or subtracting their terms. Based on our experience with limits and derivatives, you might think that our next step should be to investigate what happens when two summations are combined by multiplying or dividing their terms. Are there simple formulas that express $\sum_{i=1}^{n}(a_i\cdot b_i)$ and $\sum_{i=1}^{n}(a_i/b_i)$ in terms of $\sum_{i=1}^{n}a_i$ and $\sum_{i=1}^{n}b_i$? Unfortunately, there are no such simple formulas.

With no formula for $\sum_{i=1}^{n}(a_i\cdot b_i)$, how are we to evaluate a sum like $\sum_{i=1}^{n}i^2$? Unfortunately, the method we used to compute $\sum_{i=1}^{n}i$ doesn't work for $\sum_{i=1}^{n}i^2$. But surprisingly, evaluating the summation $\sum_{i=1}^{n}[(i+1)^3 - i^3]$ in two different ways will lead us to the answer. Writing out the meaning of this summation, we see that most of the terms cancel out:

$$\sum_{i=1}^{n}[(i+1)^3 - i^3] = [2^3 - 1^3] + [3^3 - 2^3] + [4^3 - 3^3] + \cdots$$

$$+ [n^3 - (n-1)^3] + [(n+1)^3 - n^3]. \qquad (5.5)$$

Notice that the 2^3 in the first term cancels with the -2^3 in the second term, the 3^3 in the second term cancels with the -3^3 in the third term, and so on. After all of these cancelations, all that is left on the right-hand side of equation (5.5) is the -1^3 from the first term and the $(n+1)^3$ from the last term. Thus, we have

$$\sum_{i=1}^{n}[(i+1)^3 - i^3] = (n+1)^3 - 1 = (n^3 + 3n^2 + 3n + 1) - 1 = n^3 + 3n^2 + 3n. \quad (5.6)$$

Sums like (5.5), in which all but a few terms cancel out, are called *telescoping sums*.

But there is a second way to evaluate this summation. Since

$$(i+1)^3 - i^3 = (i^3 + 3i^2 + 3i + 1) - i^3 = 3i^2 + 3i + 1,$$

we have

$$\sum_{i=1}^{n}[(i+1)^3 - i^3] = \sum_{i=1}^{n}[3i^2 + 3i + 1] = 3\sum_{i=1}^{n}i^2 + 3\sum_{i=1}^{n}i + \sum_{i=1}^{n}1$$

$$= 3\sum_{i=1}^{n}i^2 + 3\cdot\frac{n^2+n}{2} + n\cdot 1 = 3\sum_{i=1}^{n}i^2 + \frac{3n^2+5n}{2}. \qquad (5.7)$$

Equations (5.6) and (5.7) give two different formulas for the same sum. Equating these formulas, we get

$$3\sum_{i=1}^{n}i^2 + \frac{3n^2+5n}{2} = n^3 + 3n^2 + 3n.$$

Rearranging, we find that

$$3\sum_{i=1}^{n} i^2 = n^3 + 3n^2 + 3n - \frac{3n^2 + 5n}{2} = \frac{2n^3 + 3n^2 + n}{2},$$

and therefore

$$\sum_{i=1}^{n} i^2 = \frac{2n^3 + 3n^2 + n}{6} = \frac{n(n+1)(2n+1)}{6}. \tag{5.8}$$

Similar methods can be used to find a formula for $\sum_{i=1}^{n} i^p$, for any positive integer p. For example, in Exercise 9 you are asked to derive the equation

$$\sum_{i=1}^{n} i^3 = \frac{n^4 + 2n^3 + n^2}{4} = \frac{n^2(n+1)^2}{4}. \tag{5.9}$$

Using such formulas, together with Theorem 5.1.1, we can evaluate $\sum_{i=1}^{n} f(i)$ for any polynomial f.

Example 5.1.3. Find

$$\sum_{i=1}^{n} (i^3 - 3i^2 + i - 5).$$

Solution.

$$\sum_{i=1}^{n} (i^3 - 3i^2 + i - 5) = \sum_{i=1}^{n} i^3 - 3\sum_{i=1}^{n} i^2 + \sum_{i=1}^{n} i - \sum_{i=1}^{n} 5$$

$$= \frac{n^4 + 2n^3 + n^2}{4} - 3 \cdot \frac{2n^3 + 3n^2 + n}{6} + \frac{n^2 + n}{2} - 5n$$

$$= \frac{n^4 - 2n^3 - 3n^2 - 20n}{4}. \qquad \square$$

So far we have written all of our summations with the variable i running through the values from 1 to n. But we can use other variables when writing summations, and we can let the variable run through a different range of values. Our next example illustrates this.

Example 5.1.4. Find

$$\sum_{k=1}^{m} (k+3)^2.$$

Solution. We give two solutions, to illustrate different methods that are sometimes useful for evaluating sums.

Solution 1. We begin by expanding $(k + 3)^2$ and applying Theorem 5.1.1:

$$\sum_{k=1}^{m}(k + 3)^2 = \sum_{k=1}^{m}(k^2 + 6k + 9) = \sum_{k=1}^{m}k^2 + 6\sum_{k=1}^{m}k + \sum_{k=1}^{m}9.$$

Next we apply equations (5.2), (5.4), and (5.8), with m substituted for n, to evaluate the three summations on the right. Notice that the use of the letter k instead of i in our summations does not affect the values of the sums. For example, we have

$$\sum_{k=1}^{m}k^2 = 1^2 + 2^2 + \cdots + m^2 = \sum_{i=1}^{m}i^2 = \frac{2m^3 + 3m^2 + m}{6}. \tag{5.10}$$

The variables k and i in the summations in equation (5.10) serve as placeholders, to be replaced by the numbers from 1 to m when the sum is evaluated. Once these replacements are made, the letters k and i disappear from the expression, and we see that the value of the sum does not depend on k or i (although it does depend on m). We say that k and i are *dummy variables* (or *bound variables*). In general, a dummy variable in an expression can be changed to a different variable without affecting the value of the expression.

Evaluating the other two sums in a similar way, we arrive at the answer:

$$\sum_{k=1}^{m}(k + 3)^2 = \sum_{k=1}^{m}k^2 + 6\sum_{k=1}^{m}k + \sum_{k=1}^{m}9$$
$$= \frac{2m^3 + 3m^2 + m}{6} + 6 \cdot \frac{m^2 + m}{2} + 9m = \frac{2m^3 + 21m^2 + 73m}{6}.$$

Solution 2. Writing out the meaning of the original summation we see that it can be written more simply in a different way:

$$\sum_{k=1}^{m}(k + 3)^2 = 4^2 + 5^2 + \cdots + (m + 3)^2 = \sum_{i=4}^{m+3}i^2.$$

Notice that in the final sum, the variable i starts at the value 4, not 1. Substituting $m + 3$ for n in equation (5.8), we see that

$$\sum_{i=1}^{m+3}i^2 = \frac{2(m + 3)^3 + 3(m + 3)^2 + (m + 3)}{6} = \frac{2m^3 + 21m^2 + 73m + 84}{6}.$$

But we need to evaluate the sum with i running through the values from 4 to $m + 3$, not 1 to $m + 3$. Writing out the meaning of the summation notation makes it clear what adjustment needs to be made. We have

$$\sum_{i=1}^{m+3} i^2 = 1^2 + 2^2 + 3^2 + 4^2 + 5^2 + \cdots + (m+3)^2 = \sum_{i=1}^{3} i^2 + \sum_{i=4}^{m+3} i^2,$$

and therefore

$$\sum_{i=4}^{m+3} i^2 = \sum_{i=1}^{m+3} i^2 - \sum_{i=1}^{3} i^2 = \frac{2m^3 + 21m^2 + 73m + 84}{6} - 14 = \frac{2m^3 + 21m^2 + 73m}{6}.$$

Of course, this is the same as the answer we got in Solution 1. □

Exercises 5.1

1. (a) Find a formula for $\sum_{i=1}^{n} (i^2 - 2i + 3)$.

   (b) Evaluate $\sum_{i=1}^{5} (i^2 - 2i + 3)$ in two ways: by using your formula from part (a), and by writing out the five terms in the sum and adding them.

2–6: Find a formula for the sum.

2. $\displaystyle\sum_{i=1}^{n} (i^3 - 4i)$.

3. $\displaystyle\sum_{j=1}^{n} (2j^3 - j)$.

4. $\displaystyle\sum_{i=1}^{t} (3i^2 - i + 2)$.

5. $\displaystyle\sum_{i=1}^{n} (i + 1)^3$.

6. $\displaystyle\sum_{i=5}^{n+3} i^2$.

7. (a) Find a formula for $\sum_{i=1}^{n} (i^2 + 5i)$.

   (b) Find a formula for $\sum_{j=6}^{n+5} (j^2 - 5j)$.

   (c) Show that your answers to part (a) and (b) are equal. Explain why. (Hint: Writing out a few terms of each sum may give you an idea.)

8. Our proof of Theorem 5.1.1 is not completely explicit, because of the use of "$\cdots$" in some formulas. Use mathematical induction to give proofs of parts 1 and 2 of Theorem 5.1.1 that do not rely on the use of "$\cdots$".

9. Derive equation (5.9). Begin by evaluating $\sum_{i=1}^{n}[(i+1)^4 - i^4]$ in two different ways.

10. Give an alternative derivation of equation (5.4) by evaluating $\sum_{i=1}^{n}[(i+1)^2 - i^2]$ in two different ways.

11. Use mathematical induction to prove equation (5.4).

12. (a) Compute $\sum_{i=1}^{n} \frac{1}{i(i+1)}$ for several values of n. Based on these calculations, guess a formula for this sum, and use mathematical induction to confirm that your guess is correct.

   (b) Verify that
   $$\frac{1}{i(i+1)} = \frac{1}{i} - \frac{1}{i+1}.$$
   Use this fact to rewrite the sum in part (a) as a telescoping sum, and use this to find a formula for the sum. Verify that your answer agrees with your answer to part (a).

13. Suppose r is a real number and $r \neq 1$.

   (a) Evaluate $\sum_{i=1}^{n}[r^{i+1} - r^i]$ by recognizing it as a telescoping sum.

   (b) Rewrite the sum in part (a) as $\sum_{i=1}^{n}[(r-1)r^i] = (r-1)\sum_{i=1}^{n} r^i$. Use this to find a formula for $\sum_{i=1}^{n} r^i$.

   (c) Find $\sum_{i=1}^{n}(1/2)^i$.

14. Use mathematical induction to show that for every positive integer n,
$$\sum_{i=1}^{n} i2^i = 2 + (n-1)2^{n+1}.$$

15. Assume $\sin(\theta/2) \neq 0$. Use mathematical induction to show that for every positive integer n,
$$\sum_{i=1}^{n} \cos(i\theta) = \frac{\cos\left(\frac{(n+1)\theta}{2}\right)\sin\left(\frac{n\theta}{2}\right)}{\sin\left(\frac{\theta}{2}\right)}.$$
(Hint: In the induction step, use the identity $\cos u \sin v = (1/2)[\sin(u+v) - \sin(u-v)]$.)

5.2   Accumulation and Area

A scientist is studying a stream that flows into a lake. She wants to know how much water flows from the stream into the lake, so she asks her research assistant to set up equipment to measure the flow rate of the stream. Since the flow rate might vary, she asks her assistant to report back to her about the flow rate on five consecutive days. The assistant reports measurements of the rate at which water is flowing from the stream

Monday 9 a.m.	Tuesday 9 a.m.	Wednesday 9 a.m.	Thursday 9 a.m.	Friday 9 a.m.
603	527	511	556	661

Table 5.1: Flow rate of stream on five consecutive days, in m^3/hr.

into the lake, in cubic meters per hour (abbreviated m^3/hr), at 9 a.m. on each of five consecutive days, as shown in Table 5.1.

The scientist suspects that the flow rate won't vary much over the course of a single day, so she thinks it will be approximately correct to assume that the stream flowed at a constant rate of 603 m^3/hr all day on Monday, 527 m^3/hr on Tuesday, and so on. Thus, she computes that during the 24 hours of Monday, a total of about $603 \cdot 24 = 14472$ cubic meters of water flowed from the stream into the lake. Similarly, the total volume of water flowing into the lake on Tuesday was about $527 \cdot 24 = 12648$ m^3, and so on. Adding up the contributions for all five days, she computes that the total volume of water that flowed from the stream into the lake over the five-day period was approximately

$$603 \cdot 24 + 527 \cdot 24 + 511 \cdot 24 + 556 \cdot 24 + 661 \cdot 24 = 68592 \text{ m}^3. \qquad (5.11)$$

This is only an approximation, because the flow rate was probably not constant on each day, but if the flow rate didn't vary much over the course of a single day, it should be fairly accurate.

The next day, the scientist shows her calculations to her research assistant. The assistant says that the equipment he used actually recorded the flow rate four times every day, at 3 a.m., 9 a.m., 3 p.m., and 9 p.m., and he noticed that there was a fair amount of variability in the flow rate over the course of each day. Thus, the scientist's calculations might not be very accurate. The scientist asks for this more detailed data, and the research assistant supplies the figures shown in Table 5.2.

Using this data, the scientist is able to do a more accurate calculation. She takes each measurement as applying to a 6-hour period. Thus, she assumes that from midnight to 6 a.m. on Monday, the flow rate was 531 m^3/hr, from 6 a.m. to noon it was 603 m^3/hr, and so on. Now she computes that from midnight to 6 a.m. a total of about $531 \cdot 6 = 3186$ m^3 of water flowed into the lake, from 6 a.m. to noon the total was about $603 \cdot 6 = 3618$ m^3,

Monday				Tuesday			
3 a.m.	9 a.m.	3 p.m.	9 p.m.	3 a.m.	9 a.m.	3 p.m.	9 p.m.
531	603	625	527	475	527	548	491

Wednesday				Thursday				Friday			
3	9	3	9	3	9	3	9	3	9	3	9
466	511	539	510	502	556	599	584	586	661	728	712

Table 5.2: Flow rate of stream at four times on each of five days, in m^3/hr.

and similarly for all of the other 6-hour periods. Adding up the contributions of all of the 6-hour periods over all five days, she estimates that the total volume of water that flowed into the lake was

$$531 \cdot 6 + 603 \cdot 6 + 625 \cdot 6 + \cdots + 712 \cdot 6 = 67686 \text{ m}^3. \qquad (5.12)$$

This is still an approximation, but it is probably more accurate than the first calculation, since it takes into account some of the variability in the flow rate over the course of each day.

Recall that when we studied rates of change, we found it helpful to interpret them geometrically as slopes of secant and tangent lines. There is also a helpful geometric interpretation of the calculations our scientist has done. To explain this geometric interpretation, we first introduce some notation. Let t denote time since midnight Sunday night, measured in hours. Since $24 \cdot 5 = 120$, the five-day period under consideration in this example corresponds to values of t in the interval $[0, 120]$. For each t in this interval, let $f(t)$ be the rate at which water was flowing from the stream into the lake, in m^3/hr, at time t. The graph of the equation $y = f(t)$ is shown in Figure 5.1. The five measurements in Table 5.1 correspond to the five black dots on the curve. For example, since the flow rate was 603 m^3/hr at 9 a.m. on Monday, we have $f(9) = 603$, and therefore the point $(9, 603)$ is on the graph. The measurement of 527 m^3/hr at 9 a.m. on Tuesday tells us that the curve passes through the point $(33, 527)$, since 9 a.m. on Tuesday was 33 hours after midnight on Sunday night. Similarly, the other three measurements in Table 5.1 correspond to the other three labeled points in Figure 5.1.

We have also drawn five blue rectangles in Figure 5.1, one for each of the five days of the experiment. The top of each rectangle passes through one of the labeled points on the graph. Thus, the first rectangle contains all points (t, y) with $0 \leq t \leq 24$

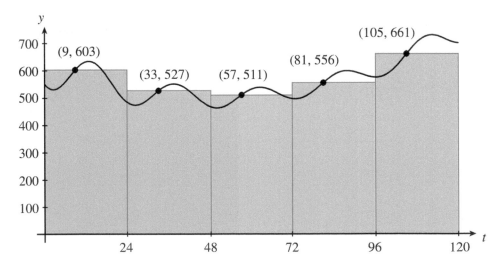

Figure 5.1: The graph of $y = f(t)$. The areas of the five blue rectangles are equal to the five terms in the sum in equation (5.11).

and $0 \leq y \leq 603$. This rectangle has height 603 and width 24, and therefore its area is $603 \cdot 24 = 14472$—exactly the same as the scientist's initial estimate of the volume of water that flowed into the lake on Monday. Similarly, the scientist's initial estimates of the volume of water flowing into the lake on the other four days are equal to the areas of the other four rectangles. Therefore, the sum in equation (5.11) is equal to the total area of the five rectangles in Figure 5.1.

The scientist's improved estimate in equation (5.12) can also be interpreted as the total area of a collection of rectangles. This time there are 20 terms in the sum, each giving an approximation of the volume of water flowing into the lake in a 6-hour time period, so we draw 20 rectangles, each of width 6 on the t-axis; see Figure 5.2. The tops of the rectangles pass through the points on the graph corresponding to the 20 measurements in Table 5.2. With more sample points on the graph, the tops of the rectangles follow the shape of the curve fairly closely in Figure 5.2. As a result, the total area of the rectangles, which is given by the sum in equation (5.12), is approximately equal to the area of the region under the curve $y = f(t)$ between $t = 0$ and $t = 120$—that is, the area of the region containing all points (t, y) with $0 \leq t \leq 120$ and $0 \leq y \leq f(t)$.

You should recognize that we now have the ingredients for applying the methods of calculus to the problem of computing the volume of water flowing into the lake over the five days: we have a way of computing an approximate answer to the problem, and a way of making the approximation more accurate. The scientist could have made her approximation even more accurate by dividing the time period $0 \leq t \leq 120$ into an even larger number of narrower intervals, taking a measurement of the flow rate in each interval, and adding up estimates of the volume of water that flowed into the lake in each of these intervals. The terms of this sum would again be equal to the areas of a sequence of rectangles. For example, if she had measured the flow rate once every hour, then she

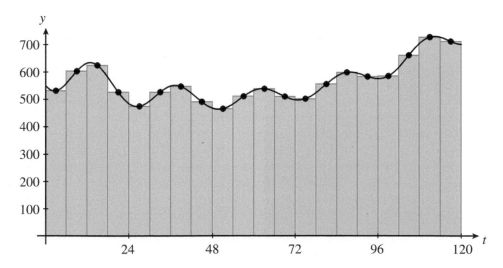

Figure 5.2: The areas of the blue rectangles are equal to the terms in the sum in equation (5.12).

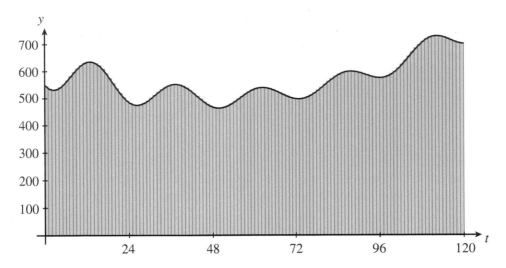

Figure 5.3: The region inside 120 rectangles, each of width 1, is almost indistinguishable from the region under the curve.

would have ended up adding up the areas of 120 rectangles, as shown in Figure 5.3. The region inside these rectangles is almost indistinguishable from the region under the curve $y = f(t)$ between $t = 0$ and $t = 120$. It appears that if we take the limit of these approximations as the number of rectangles approaches ∞, we will get the exact area under the curve, and this will be equal to the exact number of cubic meters of water that flowed into the lake.

Let's see if we can spell this idea out more precisely. Suppose we divide the interval $[0, 120]$ into n smaller intervals, each of width $\Delta t = 120/n$. These smaller intervals will be $[0, \Delta t]$, $[\Delta t, 2\Delta t]$, $[2\Delta t, 3\Delta t]$, and so on. It will be convenient to have notation for the endpoints of these intervals, so we let $t_0 = 0$, $t_1 = \Delta t$, $t_2 = 2\Delta t$, and, in general, $t_i = i\Delta t$ for $0 \le i \le n$. Notice that $t_n = n\Delta t = n(120/n) = 120$. The first interval is $[t_0, t_1]$, the second is $[t_1, t_2]$, and for $1 \le i \le n$, the ith interval is $[t_{i-1}, t_i]$. To estimate the volume of water flowing into the lake in each of these time intervals, we will need a measurement of the flow rate at some point in each interval. Thus, for $1 \le i \le n$ we choose a number $t_i^* \in [t_{i-1}, t_i]$, and measure the flow rate at time t_i^*; of course, this flow rate is $f(t_i^*)$. To estimate the volume of water flowing into the lake during the time interval $[t_{i-1}, t_i]$, we assume that the flow rate was a constant $f(t_i^*)$ m³/hr throughout this interval. Since the interval is Δt hours long, we compute that the total volume of water that flowed into the lake over this interval is approximately $f(t_i^*)\Delta t$ m³. Adding up the contributions of all the time intervals, we conclude that the total volume of water that flowed into the lake from $t = 0$ to $t = 120$, in cubic meters, is about

$$f(t_1^*)\Delta t + f(t_2^*)\Delta t + \cdots + f(t_n^*)\Delta t = \sum_{i=1}^{n} f(t_i^*)\Delta t. \qquad (5.13)$$

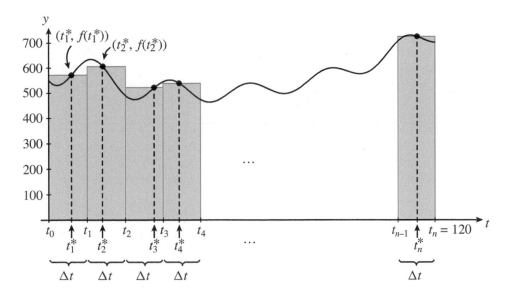

Figure 5.4: The areas of these rectangles are equal to the terms in the sum in equation (5.13).

The sum in equation (5.13) is called a *Riemann sum* for the function f on the interval [0, 120], after the mathematician Georg Friedrich Bernhard Riemann (1826–1866). Each term of the sum in equation (5.13) can be thought of as representing the area of a rectangle, as illustrated in Figure 5.4. If we want to specify that there are n terms in the sum, we may refer to it as a *Riemann n-sum*. Thus, the sum in equation (5.11) is an example of a Riemann 5-sum for the function f on the interval [0, 120], and the sum in equation (5.12) is an example of a Riemann 20-sum. Notice that in forming a Riemann n-sum, we can choose the sample points $t_1^*, t_2^*, \ldots, t_n^*$ however we please. Thus, for a single value of n there will be many possible Riemann n-sums. For example, we could get another Riemann 5-sum by using, say, the 3 p.m. measurements on Monday, Wednesday, and Friday, and the 9 p.m. measurements on Tuesday and Thursday.

Suppose that, for each n, R_n is a Riemann n-sum for f on the interval [0, 120]. The examples we have considered suggest that as n increases, R_n should get closer and closer to both the area under the graph of f for $0 \le t \le 120$ and also the total volume of water that flowed into the lake over the five days, in cubic meters. To get the exact value, we therefore take the limit as $n \to \infty$. In other words,

$$\lim_{n \to \infty} R_n = \text{area under graph of } f \text{ for } 0 \le t \le 120$$

$$= \text{volume of water that flowed into lake over five days, in m}^3.$$

Later in this chapter we will study methods for evaluating such limits. For the moment, we will simply be satisfied with having found a mathematical formula for the exact answer to our problem.

Limits of Riemann sums can be used to solve a wide variety of problems. Here's another example. Consider a piece of wire that is 10 centimeters long. The *density* of the wire measures the mass of the wire per unit of length. If mass is measured in grams (abbreviated g) and length in centimeters (abbreviated cm), then density is measured in grams per centimeter (g/cm). For example, if the density of the wire is 3 g/cm, then since the length of the wire is 10 cm, the total mass of the wire is $3 \cdot 10 = 30$ g.

But now suppose the density of the wire varies from one point to another. How can we compute the total mass of the wire? We will imagine that the wire is marked off in centimeters as a number line, with one end labeled 0 and the other labeled 10. Each point on the wire corresponds to some number in the interval [0, 10]. For each x in this interval, let $h(x)$ be the density of the wire, in g/cm, at the point corresponding to x.

For each positive integer n, we can now compute an approximation to the mass of the wire as follows. We divide the wire into n pieces, each of width $\Delta x = 10/n$. For $0 \leq i \leq n$, let $x_i = i\Delta x$. Then for $1 \leq i \leq n$, the ith piece of the wire corresponds to the interval $[x_{i-1}, x_i]$. Next, we choose a sample point $x_i^* \in [x_{i-1}, x_i]$. The density at this sample point is $h(x_i^*)$ g/cm, but if n is large, and therefore Δx is small, then the density probably doesn't vary a great deal over the interval $[x_{i-1}, x_i]$, so we will not be making a large error if we assume that the density is $h(x_i^*)$ g/cm over this entire interval. Under this assumption, since the length of the ith piece of the wire is Δx cm, the mass of this piece is $h(x_i^*)\Delta x$ g. Adding up the contributions of all the pieces, we conclude that the mass of the wire, in grams, is approximately

$$R_n = \sum_{i=1}^{n} h(x_i^*)\Delta x.$$

Of course, this is a Riemann n-sum for h on the interval [0, 10].

We now have a sequence $(R_n)_{n=1}^{\infty}$ of approximations to the mass of the wire. Each of these approximations could also be thought of as an approximation to the area under the graph of the curve $y = h(x)$, for $0 \leq x \leq 10$. As n increases, we expect these approximations to improve. Thus, to get the exact answer we use a limit:

$$\lim_{n \to \infty} R_n = \text{area under graph of } h \text{ for } 0 \leq x \leq 10$$

$$= \text{total mass of the wire, in grams.}$$

We will consider one more example of the use of Riemann sums. Your house is probably connected to an electrical grid from which it draws electrical power. For example, if you turn on ten 100-watt light bulbs, then these bulbs will draw $10 \cdot 100 = 1000$ watts of electrical power, or 1 kilowatt (kW), from the electrical grid. If you leave the bulbs on for an hour, then they will use 1 kilowatt hour (kWh) of electrical energy. More generally, if the electrical devices in your house draw a constant P kilowatts of electrical power for d hours, then over this time they will use a total of Pd kWh of electrical energy. The electric meter on your house measures the total number of kilowatt hours of electrical energy you have taken from the grid, and the electric company charges you for each kilowatt hour.

Computing your electricity usage is easy if the devices in your house draw a constant level of power. But this is unlikely to happen. As you turn various electrical devices in your house on and off, the total electrical power drawn by these devices will vary over time. How, then, can we compute the total number of kilowatt hours of electrical energy used during some period of time?

To make our question precise, we introduce some notation. Let t denote time, measured in hours, and let $g(t)$ be the number of kilowatts of power being drawn from the electrical grid by devices in your house at time t. For any numbers $a < b$, we would like to know how to compute the number of kilowatt hours of electrical energy you took from the grid during the time interval from $t = a$ to $t = b$.

Before answering this question, we add one additional complication. If you install solar panels on the roof of your house, you can decrease the amount of electrical energy you need to buy from the electric company. In fact, if at some time your solar panels generate more electricity than you are using, then your electric company may allow you to supply electrical energy back to the electrical grid, and any electrical energy you supply will be subtracted from your total usage (and the cost of that electricity will be subtracted from your bill!). This means that for some values of t, $g(t)$ may be negative. If $g(t) = -P$ kW for a period of d hours, then during that time your electric meter will run backwards, and Pd kWh will be subtracted from your total electrical energy usage.

Now we are ready to compute the total number of kilowatt hours of electrical energy you took from the grid for $a \leq t \leq b$. By now, the procedure should look familiar. For each positive integer n, we divide the interval $[a, b]$ into n smaller intervals, each of width $\Delta t = (b - a)/n$. The intervals will be $[a, a + \Delta t]$, $[a + \Delta t, a + 2\Delta t]$, and so on. Thus, if we let $t_i = a + i\Delta t$, then the ith interval is $[t_{i-1}, t_i]$. Next, we choose sample points $t_i^* \in [t_{i-1}, t_i]$, and we assume that during each time period $[t_{i-1}, t_i]$, you drew a constant $g(t_i^*)$ kW of power from the grid. The electrical energy you took from the grid during this time period was therefore approximately $g(t_i^*)\Delta t$ kWh (with a negative value indicating electrical energy that you supplied to the grid). Adding up the contributions of all of the time periods, we conclude that the total number of kilowatt hours of electrical energy that you took from the grid is about

$$R_n = \sum_{i=1}^{n} g(t_i^*)\Delta t. \tag{5.14}$$

Finally, to get an exact answer we let $n \to \infty$:

$$\lim_{n \to \infty} R_n = \text{total electrical energy taken from grid, in kWh.}$$

Can we also give a geometric interpretation of this limit in terms of area? Yes, but we will have to make some adjustments to take into account the fact that $g(t)$ may be negative for some values of t. For example, consider the Riemann sum illustrated in Figure 5.5. The value of $g(t_3^*)$ in this case is negative. If we write $g(t_3^*) = -P$, where P is positive, then the first red rectangle in the figure extends *down* P units below the t axis. Since the rectangle has width Δt, its area is $P\Delta t$. But the third term in the Riemann sum in equation (5.14) is $g(t_3^*)\Delta t = -P\Delta t$. Thus, in the Riemann sum,

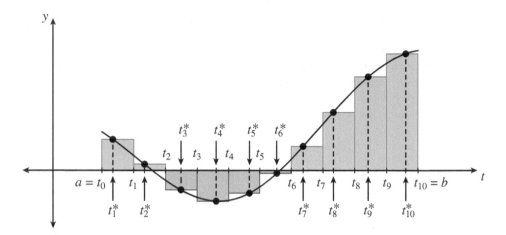

Figure 5.5: A Riemann 10-sum for the function g on the interval $[a, b]$.

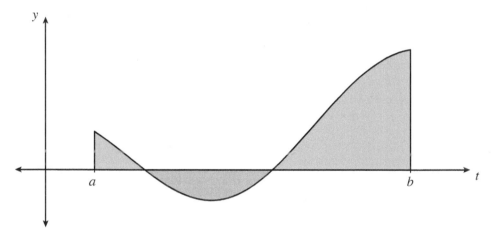

Figure 5.6: The limit of the Riemann sums is equal to the blue area minus the red area.

this area is *subtracted off*, rather than being added in. Similarly, the areas of the other red rectangles are also subtracted. Thus, the Riemann sum gives the sum of the areas of the blue rectangles minus the areas of the red rectangles. As the number of rectangles increases, the Riemann sums will approach the area of the blue region in Figure 5.6 minus the area of the red region.

All of our examples in this section have something in common. In every case, we were measuring something that could be thought as *accumulating*. In the case of the stream, the water from the stream accumulated in the lake over a period of five days. If you imagine an ant walking along the wire from one end to the other, we could say that the mass of the wire accumulates behind the ant as it walks. And as you use electricity, your electrical usage accumulates in the read-out of your electric meter. In each case, we had information about the rate at which the quantity was accumulating: the flow rate of the stream describes the rate at which water accumulates in the lake, the density of the wire gives the rate at which mass accumulates behind the ant as it walks, and the electrical power you draw from the grid measures the rate at which you are using electrical energy. There are many other situations in which we have information about the rate at which something is accumulating, and in these situations we can always use a limit of Riemann sums to compute the total accumulation.

Exercises 5.2

1. Consider the example in this section of a stream flowing into a lake. Using the data in Table 5.2, estimate the amount of water that flowed from the stream into the lake over the 5-day period as follows:

   (a) Compute a Riemann 5-sum, using the 3 p.m. measurement each day.

   (b) Compute a Riemann 10-sum, using the 9 a.m. and 9 p.m. measurements each day.

2. Consider the example in this section of a 10-centimeter wire. Recall that the density, in g/cm, at the point x centimeters from one end of the wire is given by the function $h(x)$. Suppose the density is measured at 1-centimeter intervals, giving the values in Table 5.3. Estimate the mass of the wire as follows:

   (a) Compute a Riemann 5-sum, using the midpoint of each interval as the sample point. (In other words, let $x_i^* = (x_{i-1} + x_i)/2$.)

   (b) Compute a Riemann 10-sum, using the left endpoint of each interval as the sample point. (In other words, let $x_i^* = x_{i-1}$.)

   (c) Compute a Riemann 10-sum, using the right endpoint of each interval as the sample point. (In other words, let $x_i^* = x_i$.)

x	0	1	2	3	4	5	6	7	8	9	10
$h(x)$ = density at x	2	4	3	4	4	5	4	3	3	2	1

Table 5.3: Density of wire, in g/cm, at 1-centimeter intervals.

3. Consider the example in this section involving electricity usage. Recall that the number of kilowatts of power being drawn from the electrical grid at time t (measured in hours) is given by the function $g(t)$. Suppose $g(t)$ is given by the formula $g(t) = 2t - 30$. Estimate the number of kilowatt hours of electrical energy taken from the grid during the time interval from $t = 12$ to $t = 24$ as follows:

   (a) Compute a Riemann 6-sum, using the midpoint of each interval as the sample point.

   (b) Compute a Riemann 12-sum, using the right endpoint of each interval as the sample point.

4. Let $f(x) = 6 - 2x$.

   (a) Sketch the graph of f for $0 \le x \le 3$.

   (b) Compute a Riemann 6-sum to estimate the area under the graph of f for $0 \le x \le 3$. Use the right endpoint of each interval as the sample point.

   (c) Use geometric reasoning to find the exact area under the graph of f for $0 \le x \le 3$.

5. Let $f(x) = 1/x$.

   (a) Sketch the graph of f for $1 \le x \le 3$.

   (b) Compute a Riemann 10-sum to estimate the area under the graph of f for $1 \le x \le 3$. Use the midpoint of each interval as the sample point.

6. In this exercise you will use a Riemann sum to estimate the value of π.

   (a) Find a function f such that the graph of f is the top half of the circle $x^2 + y^2 = 1$.

   (b) Use a Riemann 10-sum for f on the interval $[-1, 1]$ to estimate the area under the graph of f. Use the midpoint of each interval as the sample point.

   (c) Use geometric reasoning to show that the area under the graph of f is $\pi/2$. Multiply your answer to part (b) by 2 to get an estimate of π.

7. At t minutes after noon, water was leaking out of a water tank at a rate of $5 + 2t$ gallons per minute. Use a Riemann 10-sum to estimate the amount of water that leaked out of the tank from noon until 20 minutes past noon. (You may want to compare your answer to this exercise to your answer to Exercise 18 from Section 4.9.)

8. For $6 \le t \le 18$, at t hours past midnight a solar panel generates $100 \sin((t - 6) \pi/12)$ watts of electricity. Use a Riemann 12-sum to estimate the amount of electrical energy generated between 6 a.m. and 6 p.m.

9. Engineers testing the performance of a car measured its velocity once every 5 seconds for one minute. The velocity of the car, in ft/sec, t seconds after the beginning of the test is given by a function $v(t)$, and the measured values are shown in Table 5.4. Use a Riemann 12-sum to estimate the total distance traveled by the car during the test. Use the right endpoint of each interval as the sample point.

t	0	5	10	15	20	25	30	35	40	45	50	55	60
$v(t) =$ velocity after t seconds	0	40	75	90	95	90	80	75	80	85	60	25	0

Table 5.4: Velocity of car, in ft/sec, at 5-second intervals.

5.3 Definite Integrals

We now set aside the particular applications considered in the last section—water flowing into a lake, the mass of a wire, and electricity usage—and focus on the mathematical calculation that arose in all of these applications: limits of Riemann sums. We begin with a careful definition of the concept we will be studying.

Definition 5.3.1. Suppose that $a < b$, f is a function that is defined on the closed interval $[a, b]$, and n is a positive integer. Let $\Delta x = (b - a)/n$, and for $0 \le i \le n$ let $x_i = a + i \Delta x$. These numbers divide the interval $[a, b]$ into n smaller intervals $[x_0, x_1]$, $[x_1, x_2]$, ..., $[x_{n-1}, x_n]$, each of width Δx. Also, suppose that for $1 \le i \le n$, $x_i^* \in [x_{i-1}, x_i]$. Then the sum

$$f(x_1^*)\Delta x + f(x_2^*)\Delta x + \cdots + f(x_n^*)\Delta x = \sum_{i=1}^{n} f(x_i^*)\Delta x$$

is called a *Riemann sum* for f on the interval $[a, b]$. We may also call it a *Riemann n-sum* if we want to specify the number of terms in the sum.

   Suppose that, for each positive integer n, R_n is a Riemann n-sum for f on $[a, b]$. Then we will say that $(R_n)_{n=1}^{\infty}$ is a *Riemann sum sequence* for f on $[a, b]$.

   Recall that different choices of the sample points $x_i^* \in [x_{i-1}, x_i]$ may lead to different Riemann sums, and therefore there may be many different Riemann sum sequences for a given function f and interval $[a, b]$. If $(R_n)_{n=1}^{\infty}$ is a Riemann sum sequence for a function f on an interval $[a, b]$, then we will usually be interested in evaluating $\lim_{n \to \infty} R_n$. There are a couple of worries we might have about this limit. Might the limit be undefined in some cases? Even if it is defined, might different Riemann sum sequences for the same function have different limits? It turns out that if we stick to continuous functions then these awkward situations will not arise:

Theorem 5.3.2. *Suppose that $a < b$ and f is a function that is continuous on the closed interval $[a, b]$. Then every Riemann sum sequence for f on $[a, b]$ converges, and all such sequences converge to the same number. In other words, there is some number L such that if $(R_n)_{n=1}^{\infty}$ is any Riemann sum sequence for f on $[a, b]$, then $\lim_{n \to \infty} R_n = L$.*

   The proof of Theorem 5.3.2 is difficult; we will present it at the end of this chapter. For the moment we will simply accept the theorem and see where it leads us. Clearly the number L in the theorem is important and deserves a name.

Definition 5.3.3. Suppose that $a < b$ and f is continuous on $[a, b]$. Then the number L that is the limit of all Riemann sum sequences for f on $[a, b]$ is called the *definite integral of f from a to b*, and it is denoted $\int_a^b f(x)\, dx$. In other words, for any Riemann sum sequence $(R_n)_{n=1}^{\infty}$,

$$\int_a^b f(x)\, dx = \lim_{n \to \infty} R_n.$$

The number a is called the *lower limit*[1] *of integration*, b is called the *upper limit of integration*, and $f(x)$ is called the *integrand*.

More informally, in the notation of Definition 5.3.1 we can say that for large n,

$$\int_a^b f(x)\,dx \approx \sum_{i=1}^n f(x_i^*)\Delta x.$$

The expression $f(x)\,dx$ on the left is supposed to remind you of the formula $f(x_i^*)\Delta x$ that appears in the Riemann sum on the right. The symbol $\int$ is called an *integral sign*. It is an elongated S, which is intended to remind you that the expression on the right is a *sum*. The dx at the end of the integral notation, like the dx in the notation dy/dx, serves to identify the independent variable. It is an important part of the notation for integrals, and must not be left out. In particular, we will see in Section 5.5 that it plays an important role in the method of integration by substitution.

More advanced books use a somewhat more complicated definition of definite integrals. However, Definition 5.3.3 will be adequate for our purposes in this book.

In thinking about definite integrals, it may help to keep in mind the geometric interpretation discussed in the last section. If $f(x)$ is positive for all values of x, then $\int_a^b f(x)\,dx$ is equal to the area of the region under the graph of f for $a \le x \le b$. If $f(x)$ is not always positive, then we divide the region between the graph of f and the x-axis for $a \le x \le b$ into two parts, the part above the x-axis and the part below. The value of $\int_a^b f(x)\,dx$ is equal to the area of the part above the x-axis minus the area of the part below.

Example 5.3.4. Compute

$$\int_0^6 \left(2 - \frac{x}{2}\right) dx.$$

Solution. Suppose n is a positive integer. We begin by computing all of the quantities mentioned in Definition 5.3.1. We have $f(x) = 2 - x/2$, $a = 0$, and $b = 6$, so $\Delta x = (b - a)/n = (6 - 0)/n = 6/n$. Next, the points that divide the interval $[0, 6]$ into n smaller intervals are given by the formula

$$x_i = a + i\,\Delta x = 0 + i \cdot \frac{6}{n} = \frac{6i}{n}.$$

Finally, we must choose the sample points $x_i^* \in [x_{i-1}, x_i]$. According to Theorem 5.3.2, it doesn't matter what choice we make; all choices will lead to the same value when we take the limit of the Riemann n-sum as $n \to \infty$. We can therefore make the choice that seems simplest. We will let $x_i^* = x_i = 6i/n$. (To find out what happens with an alternative choice, see Exercise 7.)

---

[1] The use of the word "limit" here is somewhat unfortunate, since it has nothing to do with the mathematical concept of limits. However, it is standard terminology.

We can now work out a formula for the Riemann sum:

$$R_n = \sum_{i=1}^n f(x_i^*)\Delta x = \sum_{i=1}^n f\left(\frac{6i}{n}\right) \cdot \frac{6}{n} = \sum_{i=1}^n \left(2 - \frac{1}{2} \cdot \frac{6i}{n}\right) \cdot \frac{6}{n} = \sum_{i=1}^n \left(\frac{12}{n} - \frac{18i}{n^2}\right).$$

Notice that in this summation, n is a constant; as i runs through all the values from 1 to n, n doesn't change. We can therefore use Theorem 5.1.1 and equations (5.2) and (5.4) to evaluate the sum:

$$R_n = \sum_{i=1}^n \left(\frac{12}{n} - \frac{18i}{n^2}\right) = \frac{12}{n}\sum_{i=1}^n 1 - \frac{18}{n^2}\sum_{i=1}^n i = \frac{12}{n}\cdot n - \frac{18}{n^2}\cdot \frac{n^2+n}{2} = 3 - \frac{9}{n}.$$

Finally, to evaluate the definite integral we take the limit as $n \to \infty$:

$$\int_0^6 \left(2 - \frac{x}{2}\right) dx = \lim_{n\to\infty} R_n = \lim_{n\to\infty}\left(3 - \frac{9}{n}\right) = 3. \qquad \Box$$

The graph of the equation $y = 2 - x/2$ is shown in Figure 5.7, and according to the geometric interpretation we have given for definite integrals, the value of $\int_0^6 (2 - x/2)\,dx$ should be equal to the area of the blue triangle in this figure minus the area of the red triangle. The blue triangle has base 4 and height 2, so its area is $\frac{1}{2}\cdot 4\cdot 2 = 4$. Similarly, the area of the red triangle is $\frac{1}{2}\cdot 2\cdot 1 = 1$, so

$$(\text{area of blue triangle}) - (\text{area of red triangle}) = 4 - 1 = 3 = \int_0^6 \left(2 - \frac{x}{2}\right) dx,$$

exactly as expected.

Example 5.3.5. Evaluate

$$\int_{-1}^2 (x^2 - 2x)\,dx.$$

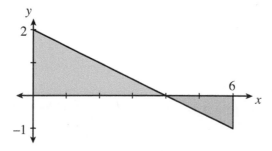

Figure 5.7: The graph of $y = 2 - x/2$ for $0 \le x \le 6$. The value of $\int_0^6 (2 - x/2)\,dx$ is equal to the area of the blue triangle minus the area of the red triangle.

Solution. In this example we have $f(x) = x^2 - 2x$, $a = -1$, and $b = 2$. As before, for any positive integer n we let $\Delta x = (b-a)/n = 3/n$, and for $0 \le i \le n$ we let

$$x_i = a + i\Delta x = -1 + \frac{3i}{n}.$$

We once again choose $x_i^* = x_i = -1 + 3i/n$. This choice leads to the Riemann sum

$$R_n = \sum_{i=1}^{n} f(x_i^*)\Delta x = \sum_{i=1}^{n} f\left(-1+\frac{3i}{n}\right)\cdot\frac{3}{n} = \sum_{i=1}^{n}\left[\left(-1+\frac{3i}{n}\right)^2 - 2\left(-1+\frac{3i}{n}\right)\right]\cdot\frac{3}{n}$$

$$= \sum_{i=1}^{n}\left[\left(1-\frac{6i}{n}+\frac{9i^2}{n^2}\right)-\left(-2+\frac{6i}{n}\right)\right]\cdot\frac{3}{n}$$

$$= \sum_{i=1}^{n}\left[3-\frac{12i}{n}+\frac{9i^2}{n^2}\right]\cdot\frac{3}{n}$$

$$= \sum_{i=1}^{n}\left[\frac{9}{n}-\frac{36i}{n^2}+\frac{27i^2}{n^3}\right] = \frac{9}{n}\sum_{i=1}^{n}1-\frac{36}{n^2}\sum_{i=1}^{n}i+\frac{27}{n^3}\sum_{i=1}^{n}i^2$$

$$= \frac{9}{n}\cdot n - \frac{36}{n^2}\cdot\frac{n^2+n}{2}+\frac{27}{n^3}\cdot\frac{2n^3+3n^2+n}{6}$$

$$= -\frac{9}{2n}+\frac{9}{2n^2}.$$

Finally, we take the limit as $n \to \infty$ to get the value of the integral:

$$\int_{-1}^{2}(x^2-2x)\,dx = \lim_{n\to\infty} R_n = \lim_{n\to\infty}\left(-\frac{9}{2n}+\frac{9}{2n^2}\right) = 0. \qquad \square$$

All that work for nothing! What does it mean that the value of the integral is 0? Since the integral is the blue area minus the red area in Figure 5.8, the blue and red areas must be exactly equal.

The examples we have done illustrate that computing a definite integral of even a fairly simple function by using the definition can be quite difficult. As with limits and derivatives, we need to look for theorems that will make the computation of definite integrals easier. Since definite integrals are so closely related to summations, we begin with an analogue of our main theorem about summations, Theorem 5.1.1:

Theorem 5.3.6. *Suppose that $a < b$, f and g are continuous on $[a, b]$, and c is any real number. Then:*

1. $\displaystyle\int_a^b cf(x)\,dx = c\int_a^b f(x)\,dx.$

2. $\displaystyle\int_a^b (f(x)+g(x))\,dx = \int_a^b f(x)\,dx + \int_a^b g(x)\,dx.$

3. $\displaystyle\int_a^b (f(x)-g(x))\,dx = \int_a^b f(x)\,dx - \int_a^b g(x)\,dx.$

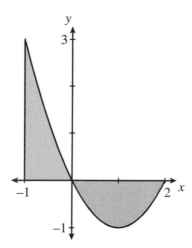

Figure 5.8: The graph of $y = x^2 - 2x$ for $-1 \le x \le 2$. Since $\int_{-1}^{2}(x^2 - 2x)\,dx = 0$, the blue area and the red area are exactly equal.

Proof. Each of the three parts can be proven by applying the corresponding part of Theorem 5.1.1 to the Riemann sums involved. We illustrate this by proving part 2.

For any positive integer n, as usual we let $\Delta x = (b-a)/n$ and $x_i = a + i\,\Delta x$, and we choose sample points $x_i^* \in [x_{i-1}, x_i]$. Using these sample points, we get the following Riemann sums for f and g on $[a, b]$:

$$R_n = \sum_{i=1}^{n} f(x_i^*)\Delta x, \qquad S_n = \sum_{i=1}^{n} g(x_i^*)\Delta x.$$

We can also use these sample points to compute a Riemann sum for $f + g$:

$$\sum_{i=1}^{n}(f(x_i^*) + g(x_i^*))\Delta x = \sum_{i=1}^{n}(f(x_i^*)\Delta x + g(x_i^*)\Delta x)$$

$$= \sum_{i=1}^{n} f(x_i^*)\Delta x + \sum_{i=1}^{n} g(x_i^*)\Delta x = R_n + S_n,$$

where we have used part 2 of Theorem 5.1.1 in the second step. Finally, by the definition of definite integrals, we have

$$\int_a^b (f(x) + g(x))\,dx = \lim_{n\to\infty}(R_n + S_n) = \lim_{n\to\infty} R_n + \lim_{n\to\infty} S_n$$

$$= \int_a^b f(x)\,dx + \int_a^b g(x)\,dx. \qquad \square$$

Unfortunately, as with summations, we don't have similar theorems for dealing with products and quotients. In other words, there are no simple formulas relating

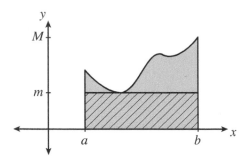

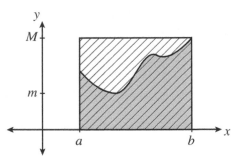

(a) The striped rectangle, whose area is $m(b - a)$, is contained in the blue region.

(b) The striped rectangle, whose area is $M(b - a)$, contains the blue region.

Figure 5.9: The graph of the function f in Theorem 5.3.7. The area of the region that is shaded blue is $\int_a^b f(x)\,dx$.

$\int_a^b (f(x) \cdot g(x))\,dx$ and $\int_a^b (f(x)/g(x))\,dx$ to $\int_a^b f(x)\,dx$ and $\int_a^b g(x)\,dx$. However, there are a few other basic properties of definite integrals that we will find useful. For the first of these, recall that, according to the extreme value theorem, if f is continuous on the interval $[a, b]$ then it has a maximum value and a minimum value on that interval. Our next theorem, which is illustrated in Figure 5.9, says that these values can be used to get some idea of the size of the definite integral.

Theorem 5.3.7. *Suppose that $a < b$ and f is continuous on $[a, b]$. Let m be the minimum value of f on $[a, b]$, and let M be the maximum value. Then*

$$m(b - a) \le \int_a^b f(x)\,dx \le M(b - a).$$

Proof. For any positive integer n, let

$$R_n = \sum_{i=1}^n f(x_i^*)\Delta x$$

be a Riemann n-sum for f, where all notation is as in Definition 5.3.1. Since M is the maximum value of f on $[a, b]$, for every i we have $f(x_i^*) \le M$, and therefore

$$R_n = \sum_{i=1}^n f(x_i^*)\Delta x \le \sum_{i=1}^n M\Delta x = M\Delta x \cdot n = M\frac{b-a}{n} \cdot n = M(b - a).$$

Taking the limit as $n \to \infty$ (and applying the preservation of weak inequalities by limits), we conclude that

$$\int_a^b f(x)\,dx = \lim_{n \to \infty} R_n \le M(b - a).$$

Similar reasoning establishes the other inequality in the theorem. $\square$

A useful consequence of this theorem is the following corollary.

Corollary 5.3.8. *Suppose that $a < b$, f and g are continuous on $[a, b]$, and for all $x \in [a, b]$, $f(x) \leq g(x)$. Then $\int_a^b f(x)\,dx \leq \int_a^b g(x)\,dx$.*

Proof. We will apply Theorem 5.3.7 to the function $g - f$. For all $x \in [a, b]$ we have $f(x) \leq g(x)$, and therefore $g(x) - f(x) \geq 0$. It follows that if we let m be the minimum value of $g - f$ on $[a, b]$, then $m \geq 0$. Applying Theorem 5.3.7, we conclude that

$$0 \leq m(b - a) \leq \int_a^b (g(x) - f(x))\,dx = \int_a^b g(x)\,dx - \int_a^b f(x)\,dx,$$

and therefore $\int_a^b f(x)\,dx \leq \int_a^b g(x)\,dx$. $\qquad\square$

There is one more theorem about definite integrals that we will find useful in the next section. Consider the function whose graph is shown in Figure 5.10. The region under the graph between $x = a$ and $x = c$, which is shaded blue, comes in two parts: the part to the left of $x = b$, which has stripes with positive slope, and the part to the right, which has stripes with negative slope. If we add the areas of the two striped parts, we should get the area of the entire blue region. This suggests that the following theorem should be true:

Theorem 5.3.9. *Suppose that $a < b < c$, and f is continuous on $[a, c]$. Then*

$$\int_a^b f(x)\,dx + \int_b^c f(x)\,dx = \int_a^c f(x)\,dx.$$

Although the interpretation of definite integrals in terms of areas is a useful guide to intuition, it is important to remember that it is not how we defined definite integrals. By definition, a definite integral is a limit of Riemann sums. Thus, although Figure 5.10 makes Theorem 5.3.9 very plausible, a rigorous proof of the theorem must show that the appropriate limits of Riemann sums satisfy the equation in the theorem. The details of this proof are a little tricky, so we will put it off until the end of this chapter.

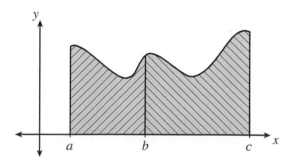

Figure 5.10: The graph of the function f in Theorem 5.3.9. The area of the blue region is $\int_a^c f(x)\,dx$. The part of this region where the stripes have positive slope is $\int_a^b f(x)\,dx$, and the part where the stripes have negative slope is $\int_b^c f(x)\,dx$.

So far we have defined $\int_a^b f(x)\,dx$ only when $a < b$, but it will turn out to be useful to extend the definition to the case $a \geq b$ as well. Suppose we try to carry out the calculations in Definition 5.3.1 even when $a \geq b$. For any positive integer n, we can still let $\Delta x = (b-a)/n$ and $x_i = a + i\Delta x$. We can still choose sample points x_i^* between x_{i-1} and x_i and form the sum

$$\sum_{i=1}^{n} f(x_i^*)\Delta x. \tag{5.15}$$

What will the result be?

If $a = b$, then for every n we will have $\Delta x = (b-a)/n = 0$, and therefore the sum (5.15) will be 0. It therefore seems reasonable that in this case $\int_a^b f(x)\,dx$ should be defined to be 0. We could think of it as representing the area of a region of width 0.

If $a > b$, then $\Delta x = (b-a)/n$ will be negative. The numbers x_i for $0 \leq i \leq n$ will still be evenly spaced between a and b, but they will run from right to left on the x-axis rather than left to right. As a result, we will have $x_i \leq x_i^* \leq x_{i-1}$ rather than $x_{i-1} \leq x_i^* \leq x_i$. The sum (5.15) will be just like a Riemann sum for the definite integral $\int_b^a f(x)\,dx$, except that Δx will be negative. Thus, instead of adding up the areas of rectangles above the x-axis and subtracting the areas of rectangles below, we will be adding up the areas of rectangles below the x-axis and subtracting the areas of rectangles above. As $n \to \infty$, it appears that the sum (5.15) will approach $-\int_b^a f(x)\,dx$.

The claims we have just made can be justified more carefully, but we won't bother, because we have presented this discussion for motivational purposes only. The simplest way to proceed is simply to take the values suggested by this discussion to be the *definition* of $\int_a^b f(x)\,dx$ when $a \geq b$.

Definition 5.3.10. For $a \geq b$, we define $\int_a^b f(x)\,dx$ as follows:

1. If $a = b$ then $\int_a^b f(x)\,dx = 0$.

2. If $a > b$ then $\int_a^b f(x)\,dx = -\int_b^a f(x)\,dx$.

Notice that according to this definition, whether $a < b$, $a > b$, or $a = b$, the equation $\int_a^b f(x)\,dx = -\int_b^a f(x)\,dx$ holds. As evidence of the merit of the definition, we note that Theorem 5.3.9 can now be extended.

Theorem 5.3.9, Extended. Suppose f is continuous on an interval I, and a, b, and c are numbers in I. Then

$$\int_a^b f(x)\,dx + \int_b^c f(x)\,dx = \int_a^c f(x)\,dx.$$

Proof. Of course, we already know that this theorem holds if $a < b < c$. We can prove the theorem by considering all other possible orderings of the numbers a, b, and c. Checking all of the possibilities is a bit tedious, so we will just check a couple of cases.

Suppose first that $b < a < c$. Then by the original version of Theorem 5.3.9 we know that

$$\int_b^c f(x)\,dx = \int_b^a f(x)\,dx + \int_a^c f(x)\,dx.$$

Combining this with Definition 5.3.10, we have

$$\int_a^b f(x)\,dx + \int_b^c f(x)\,dx = -\int_b^a f(x)\,dx + \left(\int_b^a f(x)\,dx + \int_a^c f(x)\,dx\right)$$

$$= \int_a^c f(x)\,dx,$$

as required.

Next, suppose that $a = b > c$. Then

$$\int_a^b f(x)\,dx + \int_b^c f(x)\,dx = 0 - \int_c^b f(x)\,dx = -\int_c^a f(x)\,dx = \int_a^c f(x)\,dx,$$

once again confirming the theorem. $\qquad\square$

We leave it to you to check that the requirement $a < b$ can also be eliminated from Theorem 5.3.6 (see Exercise 10):

Theorem 5.3.6, Extended. Suppose that f and g are continuous on an interval I, a and b are numbers in I, and c is any number. Then:

1. $\displaystyle\int_a^b cf(x)\,dx = c\int_a^b f(x)\,dx.$

2. $\displaystyle\int_a^b (f(x) + g(x))\,dx = \int_a^b f(x)\,dx + \int_a^b g(x)\,dx.$

3. $\displaystyle\int_a^b (f(x) - g(x))\,dx = \int_a^b f(x)\,dx - \int_a^b g(x)\,dx.$

Exercises 5.3

1–6: Use Definition 5.3.3 to evaluate the definite integral.

1. $\displaystyle\int_0^3 (2x + 1)\,dx.$

2. $\displaystyle\int_1^3 (2x + 1)\,dx.$

3. $\displaystyle\int_0^4 x^2\,dx.$

4. $\displaystyle\int_1^3 (4 - x^2)\,dx.$

5. $\displaystyle\int_{-1}^{1}(x^2-3x+1)\,dx.$

6. $\displaystyle\int_{1}^{4}x^3\,dx.$

7. Redo Example 5.3.4 with the sample points $x_i^* = x_{i-1} = 6(i-1)/n$, and verify that this leads to the same answer for the definite integral.

8. Use Theorem 5.3.7 to show that for any numbers a, b, and c with $a < b$, $\int_a^b c\,dx = c(b-a)$.

9. State and prove a version of Theorem 5.3.7 for $a > b$. In other words, suppose that $a > b$ and f is continuous on $[b, a]$, and let m and M be the minimum and maximum values of f on $[b, a]$. What inequalities hold among the quantities $m(b-a)$, $\int_a^b f(x)\,dx$, and $M(b-a)$?

10. Prove the extended version of Theorem 5.3.6.

5.4   The Fundamental Theorems of Calculus

In this section we will discover a remarkable shortcut for evaluating definite integrals. We begin by returning to the integral we evaluated in Example 5.3.5, $\int_{-1}^{2}(x^2-2x)\,dx$. We plan to use the variable x for another purpose in this section, so it will be convenient to eliminate x from this integral. Fortunately, the value of the integral will be unchanged if we replace the variable x with a different variable, say t. In other words,

$$\int_{-1}^{2}(x^2-2x)\,dx = \int_{-1}^{2}(t^2-2t)\,dt. \tag{5.16}$$

To see why this equation makes sense, recall the interpretation of $\int_{-1}^{2}(x^2-2x)\,dx$ in terms of the graph of $y = x^2 - 2x$ for $-1 \le x \le 2$. This graph is shown in Figure 5.8, and the value of the integral is the blue area in this figure minus the red area. The graph of $y = t^2 - 2t$ for $-1 \le t \le 2$ is exactly the same, except that the horizontal axis is labeled t rather than x. Of course, changing the label on the horizontal axis in Figure 5.8 has no effect on the areas of the blue and red regions, and therefore changing the variable in the integral from x to t has no effect on the value of the integral.

A more careful way to verify equation (5.16) is to work out the Riemann sums involved. Let $f(x) = x^2 - 2x$. Recall that in Example 5.3.5, we evaluated $\int_{-1}^{2}(x^2-2x)\,dx = \int_{-1}^{2}f(x)\,dx$ by letting $\Delta x = (2-(-1))/n = 3/n$ and $x_i^* = x_i = -1 + i\Delta x = -1 + 3i/n$, and then computing the Riemann n-sum

$$\sum_{i=1}^{n}f(x_i^*)\Delta x = \sum_{i=1}^{n}f\left(-1+\frac{3i}{n}\right)\cdot\frac{3}{n}. \tag{5.17}$$

To evaluate $\int_{-1}^{2}(t^2 - 2t)\,dt = \int_{-1}^{2} f(t)\,dt$ we would similarly let $\Delta t = 3/n$ and $t_i^* = t_i = -1 + i\,\Delta t = -1 + 3i/n$, and then we would compute the Riemann n-sum

$$\sum_{i=1}^{n} f(t_i^*)\Delta t = \sum_{i=1}^{n} f\left(-1 + \frac{3i}{n}\right)\cdot\frac{3}{n}. \tag{5.18}$$

Notice that the letters x and t have disappeared from these Riemann sums—these letters do not appear on the right-hand sides of equations (5.17) and (5.18)—and the two sums are identical. Therefore the two integrals in equation (5.16) are equal. The variables x and t in these integrals, like the variable i in the Riemann sums, are dummy variables.

Now, how can we find a shortcut for evaluating $\int_{-1}^{2}(t^2 - 2t)\,dt$? Recall the paradox of generalization: when we were studying derivatives, we found that it was easier to find the general formula for the rate of change of $f(x)$ at any value of x than to find the rate of change at a particular value of x. In other words, it was easier to find $f'(x)$ than to find, say, $f'(2)$. Let's see if the same thing is true for definite integrals: instead of computing $\int_{-1}^{2}(t^2 - 2t)\,dt$, let's try computing $\int_{-1}^{x}(t^2 - 2t)\,dt$ for any $x > -1$. It seems that this should make the problem harder—indeed, initially the computation will get harder. But don't let this brief setback discourage you; eventually we will discover that, paradoxically, generalizing makes the problem easier.

To compute $\int_{-1}^{x}(t^2 - 2t)\,dt$ for any $x > -1$, we let $\Delta t - (x \quad (-1))/n = (x+1)/n$ and $t_i^* = t_i = -1 + i\,\Delta t$. Next, we compute the Riemann n-sum R_n. To keep the calculations from getting too complicated, we won't fill in the formula for Δt right away.

$$R_n = \sum_{i=1}^{n} f(t_i^*)\Delta t = \sum_{i=1}^{n} f(-1 + i\,\Delta t)\Delta t = \sum_{i=1}^{n}[(-1 + i\,\Delta t)^2 - 2(-1 + i\,\Delta t)]\Delta t$$

$$= \sum_{i=1}^{n}[(1 - 2i\,\Delta t + i^2(\Delta t)^2) - (-2 + 2i\,\Delta t)]\Delta t$$

$$= \sum_{i=1}^{n}[3 - 4i\,\Delta t + i^2(\Delta t)^2]\Delta t = \sum_{i=1}^{n}[3\Delta t - 4i(\Delta t)^2 + i^2(\Delta t)^3]$$

$$= 3\Delta t\sum_{i=1}^{n}1 - 4(\Delta t)^2\sum_{i=1}^{n}i + (\Delta t)^3\sum_{i=1}^{n}i^2$$

$$= 3\Delta t\cdot n - 4(\Delta t)^2\cdot\frac{n^2 + n}{2} + (\Delta t)^3\cdot\frac{2n^3 + 3n^2 + n}{6}$$

$$= \Delta t\cdot 3n - (\Delta t)^2\cdot(2n^2 + 2n) + (\Delta t)^3\cdot\left(\frac{n^3}{3} + \frac{n^2}{2} + \frac{n}{6}\right).$$

To simplify further, we now need to fill in the formula for Δt:

$$R_n = \frac{x+1}{n}\cdot 3n - \frac{(x+1)^2}{n^2}\cdot(2n^2 + 2n) + \frac{(x+1)^3}{n^3}\cdot\left(\frac{n^3}{3} + \frac{n^2}{2} + \frac{n}{6}\right)$$

$$= (x+1)\cdot 3 - (x+1)^2\cdot\left(2 + \frac{2}{n}\right) + (x+1)^3\cdot\left(\frac{1}{3} + \frac{1}{2n} + \frac{1}{6n^2}\right).$$

At this point, we can compute the limit as $n \to \infty$ before simplifying further:

$$\lim_{n\to\infty} R_n = \lim_{n\to\infty} \left[(x+1) \cdot 3 - (x+1)^2 \cdot \left(2 + \frac{2}{n}\right) + (x+1)^3 \cdot \left(\frac{1}{3} + \frac{1}{2n} + \frac{1}{6n^2}\right) \right]$$

$$= (x+1) \cdot 3 - (x+1)^2 \cdot 2 + (x+1)^3 \cdot \frac{1}{3}$$

$$= \frac{x^3}{3} - x^2 + \frac{4}{3}.$$

Our final conclusion is that

$$\int_{-1}^{x} (t^2 - 2t) \, dt = \frac{x^3}{3} - x^2 + \frac{4}{3}.$$

Although we assumed $x > -1$ in our calculations, it can be shown that this formula is actually correct for all values of x.

There's something surprising about this answer: $x^3/3 - x^2 + 4/3$ happens to be an antiderivative of our original integrand $f(x) = x^2 - 2x$. Why would an antiderivative show up as the answer? The calculation we did seems to have absolutely nothing to do with derivatives. Is it just a coincidence that we ended up with an antiderivative? No, it is an example of a general pattern:

Theorem 5.4.1 (First Fundamental Theorem of Calculus). *Suppose f is continuous on an interval I and $a \in I$. For all $x \in I$ let*

$$F(x) = \int_{a}^{x} f(t) \, dt.$$

Then F is an antiderivative of f on I.

Proof. To prove the theorem we have to verify that for all $x \in I$, $F'(x) = f(x)$, where we use appropriate one-sided derivatives at any endpoints that are included in I. We compute $F'(x)$ by computing the two one-sided derivatives separately, using the definition of derivative. We begin by showing that if x is not the right endpoint of I, then

$$F'_+(x) = \lim_{h\to 0^+} \frac{F(x+h) - F(x)}{h} = f(x).$$

Since we are assuming that x is not the right endpoint of I, if $h > 0$ and h is close to 0 then $x + h \in I$. We can therefore use Theorem 5.3.9 to compute $F(x+h) - F(x)$:

$$F(x+h) - F(x) = \int_{a}^{x+h} f(t) \, dt - \int_{a}^{x} f(t) \, dt$$

$$= \left(\int_{a}^{x} f(t) \, dt + \int_{x}^{x+h} f(t) \, dt \right) - \int_{a}^{x} f(t) \, dt = \int_{x}^{x+h} f(t) \, dt.$$

Next, we use Theorem 5.3.7 to estimate the last integral above. Since f is continuous on I and $[x, x+h]$ is contained in I, f is continuous on $[x, x+h]$, and therefore it has a minimum value and a maximum value on that interval. Let m_h be the minimum value of f on $[x, x+h]$, attained at some number $c_h \in [x, x+h]$, and let M_h be the maximum value, attained at some $C_h \in [x, x+h]$. We have put a subscript h on all of these variables to remind you that they depend on h; as h varies, the numbers c_h, $m_h = f(c_h)$, C_h, and $M_h = f(C_h)$ will vary as well. By Theorem 5.3.7,

$$m_h(x + h - x) \leq \int_x^{x+h} f(t)\,dt \leq M_h(x + h - x).$$

In other words,

$$f(c_h)h \leq F(x+h) - F(x) \leq f(C_h)h.$$

This inequality is illustrated in Figure 5.11. Dividing through by the positive number h, we have

$$f(c_h) \leq \frac{F(x+h) - F(x)}{h} \leq f(C_h). \tag{5.19}$$

Inequality (5.19) suggests that we might be able to use the squeeze theorem to complete the proof. To do this we must compute the limits of $f(c_h)$ and $f(C_h)$ as $h \to 0^+$. Since $c_h \in [x, x+h]$, we have

$$x \leq c_h \leq x+h.$$

As $h \to 0^+$, $x+h \to x^>$, and therefore by the squeeze theorem $c_h \to x^\geq$. But f is continuous from the right at x, so by Theorem 2.7.12 it follows that $f(c_h) \to f(x)$. In other words,

$$\lim_{h \to 0^+} f(c_h) = f(x).$$

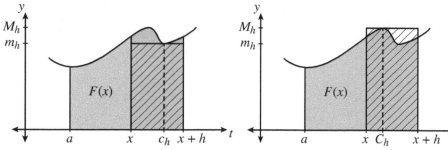

(a) The striped rectangle, whose area is $m_h h = f(c_h)h$, is contained in the blue region.

(b) The striped rectangle, whose area is $M_h h = f(C_h)h$, contains the blue region.

Figure 5.11: The graph of the function f in Theorem 5.4.1. The area of the region that is shaded gray is $F(x)$, and the area of the region that is shaded blue is $F(x+h) - F(x)$.

Similarly, as $h \to 0^+$, $C_h \to x^2$, and therefore

$$\lim_{h \to 0^+} f(C_h) = f(x).$$

Applying the squeeze theorem to inequality (5.19), we conclude that

$$\lim_{h \to 0^+} \frac{F(x+h) - F(x)}{h} = f(x).$$

Similar reasoning can be used to show that if x is not the left endpoint of I then

$$F'_-(x) = \lim_{h \to 0^-} \frac{F(x+h) - F(x)}{h} = f(x)$$

(see Exercise 15). Combining our results for the two one-sided derivatives, we conclude that F is an antiderivative of f on I. □

The first fundamental theorem of calculus gives us a way of producing an antiderivative for any continuous function on any interval. An important consequence of the theorem is that continuous functions always have antiderivatives. Continuous functions are not always differentiable, but they are always "antidifferentiable":

Corollary 5.4.2. *Suppose f is continuous on an interval I. Then f has an antiderivative on I.*

Proof. Let a be any number in I, and for $x \in I$ let $F(x) = \int_a^x f(t)\,dt$. Then by the first fundamental theorem of calculus, F is an antiderivative of f on I. □

For example, consider the function $f(x) = \sqrt{x^3 + 1}$. We will let you verify that the domain of f is the interval $I = [-1, \infty)$, and that f is continuous on this domain. Now for $x \geq -1$ let

$$F(x) = \int_0^x \sqrt{t^3 + 1}\,dt. \tag{5.20}$$

Then F is an antiderivative of f on $[-1, \infty)$. In other words, for all $x > -1$, $F'(x) = f(x) = \sqrt{x^3 + 1}$, and $F'_+(-1) = f(-1) = 0$. It would be very difficult to evaluate the integral in equation (5.20) by taking a limit of Riemann sums, and therefore we have no simpler formula for $F(x)$ than (5.20). But even without a simpler formula for $F(x)$, we know what $F'(x)$ is.

Now consider the function

$$G(x) = \int_{x^2}^1 \sqrt{t^3 + 1}\,dt.$$

Since x^2 is always nonnegative, $G(x)$ is defined for all values of x, and we have

$$G(x) = \int_{x^2}^1 \sqrt{t^3 + 1}\,dt = \int_{x^2}^0 \sqrt{t^3 + 1}\,dt + \int_0^1 \sqrt{t^3 + 1}\,dt$$

$$= -\int_0^{x^2} \sqrt{t^3 + 1}\,dt + \int_0^1 \sqrt{t^3 + 1}\,dt = -F(x^2) + F(1).$$

Again, we have no simple formula for $G(x)$, but we can find its derivative. By the chain rule,

$$G'(x) = -F'(x^2) \cdot \frac{d}{dx}(x^2) = -f(x^2) \cdot 2x = -2x\sqrt{x^6 + 1}.$$

Let's try one more example. For all x, let

$$H(x) = \int_1^x (t^3 - 4t)\, dt.$$

Then by the first fundamental theorem of calculus,

$$H'(x) = x^3 - 4x.$$

In this case, we can actually determine a formula for $H(x)$ by antidifferentiating. We know that one antiderivative of $x^3 - 4x$ on the interval $(-\infty, \infty)$ is $x^4/4 - 2x^2$, and all other antiderivatives can be found by adding a constant to this formula. Thus, there must be some constant C such that for all x,

$$H(x) = \frac{x^4}{4} - 2x^2 + C. \tag{5.21}$$

Is there any way to find the value of C? In the past, we have always done this by filling in a known value of the function. Is there any value of H that we know? Yes, one value is easy to compute:

$$H(1) = \int_1^1 (t^3 - 4t)\, dt = 0.$$

Filling in this value in equation (5.21), we have

$$0 = H(1) = \frac{1^4}{4} - 2(1^2) + C = -\frac{7}{4} + C,$$

and therefore $C = 7/4$. In other words, we have shown that

$$\int_1^x (t^3 - 4t)\, dt = H(x) = \frac{x^4}{4} - 2x^2 + \frac{7}{4}.$$

For example, taking $x = 2$ we have

$$\int_1^2 (t^3 - 4t)\, dt = \frac{2^4}{4} - 2(2^2) + \frac{7}{4} = -\frac{9}{4}. \tag{5.22}$$

Something truly amazing just happened, and it happened so quickly and effortlessly that you might not have noticed. We just found the value of a definite integral without having to compute any Riemann sums! You could compute the value of the integral in equation (5.22) by taking a limit of Riemann sums—you are asked to do this calculation in Exercise 13. But we have found the answer by a much easier method.

Let's see if we can turn the calculation we did in this example into a general method for evaluating definite integrals. Suppose f is continuous on an interval I that contains two numbers a and b, and we wish to compute $\int_a^b f(x)\,dx$. Since the variable x in this integral is a dummy variable, we can switch to a different variable without changing the value of the integral:

$$\int_a^b f(x)\,dx = \int_a^b f(t)\,dt.$$

Now we generalize the problem we are trying to solve. For all $x \in I$ we define

$$H(x) = \int_a^x f(t)\,dt.$$

We want to find $H(b)$, but it turns out it is easier to find the general formula for $H(x)$ than it is to find the particular value $H(b)$. (The paradox of generalization strikes again!) What makes it easier to find $H(x)$ than $H(b)$ is the first fundamental theorem of calculus: we know that H is an antiderivative of f on the interval I.

Suppose we happen to know an antiderivative F of f on I. Then by Theorem 4.9.6, there is some constant C such that for all $x \in I$,

$$H(x) = F(x) + C. \tag{5.23}$$

To find C, we evaluate H at the easiest number:

$$H(a) = \int_a^a f(t)\,dt = 0.$$

Substituting into equation (5.23) we get

$$0 = H(a) = F(a) + C,$$

and therefore $C = -F(a)$. Filling in this value for C in equation (5.23), we find that

$$H(x) = F(x) - F(a),$$

or in other words

$$\int_a^x f(t)\,dt = F(x) - F(a).$$

Now that we have solved the general problem, we can finally fill in the particular value we wanted:

$$\int_a^b f(x)\,dx = \int_a^b f(t)\,dt = F(b) - F(a).$$

We have just proven the following theorem:

Theorem 5.4.3 (Second Fundamental Theorem of Calculus). *Suppose f is a function that is continuous on an interval I and F is an antiderivative of f on I. Then for any numbers a and b in I,*

$$\int_a^b f(x)\,dx = F(b) - F(a).$$

Example 5.4.4. Evaluate

$$\int_{-1}^{2} (x^2 - 2x)\, dx.$$

Solution. We computed this integral in Example 5.3.5 by means of a somewhat messy Riemann sum calculation, and found that the answer was 0. Now we know a much easier method.

If we let $f(x) = x^2 - 2x$, then one antiderivative of f on $(-\infty, \infty)$ is $F(x) = x^3/3 - x^2$. Therefore, by the second fundamental theorem of calculus,

$$\int_{-1}^{2} (x^2 - 2x)\, dx = F(2) - F(-1) = \left(\frac{8}{3} - 4\right) - \left(-\frac{1}{3} - 1\right) = 0. \qquad \square$$

Since the difference $F(b) - F(a)$ will come up whenever we apply the second fundamental theorem of calculus, it will be convenient to have notation for it. This difference is often denoted either $F(x)|_a^b$ or $[F(x)]_a^b$. In other words,

$$[F(x)]_a^b = F(x)|_a^b = F(b) - F(a).$$

Using this notation, we could rewrite our solution to Example 5.4.4 like this:

$$\int_{-1}^{2} (x^2 - 2x)\, dx = \left[\frac{x^3}{3} - x^2\right]_{-1}^{2} = \left(\frac{8}{3} - 4\right) - \left(-\frac{1}{3} - 1\right) = 0.$$

Example 5.4.5. Evaluate the following definite integrals:

$$\text{(a) } \int_{1}^{3} \frac{1}{x^2}\, dx, \qquad \text{(b) } \int_{0}^{\pi/2} \cos\theta\, d\theta, \qquad \text{(c) } \int_{0}^{2} |x^2 - 1|\, dx.$$

Solution. For integral (a), we need an antiderivative of $1/x^2 = x^{-2}$. By Theorem 4.9.7, one antiderivative is given by the formula $x^{-1}/(-1) = -1/x$. This is an antiderivative on each of the intervals $(-\infty, 0)$ and $(0, \infty)$; for the evaluation of integral (a), we use $(0, \infty)$ as the interval I in the second fundamental theorem of calculus:

$$\int_{1}^{3} \frac{1}{x^2}\, dx = -\frac{1}{x}\bigg|_1^3 = -\frac{1}{3} - (-1) = \frac{2}{3}.$$

Notice that it was crucial in this calculation that 1 and 3 both belong to an interval I, namely $I = (0, \infty)$, for which the hypotheses of the second fundamental theorem hold: the integrand $1/x^2$ is continuous on this interval, and $-1/x$ is an antiderivative of $1/x^2$ on the interval. Thus, we could not have used the second fundamental theorem to evaluate, say, $\int_{-1}^{3} (1/x^2)\, dx$. Indeed, since $1/x^2$ is not continuous on the interval $[-1, 3]$, our Definition 5.3.3 does not even apply to this last integral. Later in this book we will extend our definition of the definite integral to include some discontinuous integrands, but for the moment we regard this integral as simply meaningless.

For integral (b), we apply the second fundamental theorem of calculus with the variable θ replacing x. Since $\frac{d}{d\theta}(\sin\theta) = \cos\theta$, the value of the integral is

$$\int_0^{\pi/2} \cos\theta \, d\theta = [\sin\theta]_0^{\pi/2} = \sin(\pi/2) - \sin 0 = 1.$$

What makes integral (c) difficult is the use of the absolute value function. As usual, we deal with this by considering the values of x in the interval $[0, 2]$ for which $x^2 - 1$ is positive and negative separately. We have

$$|x^2 - 1| = \begin{cases} 1 - x^2, & \text{if } 0 \le x \le 1, \\ x^2 - 1, & \text{if } 1 \le x \le 2. \end{cases}$$

Therefore, by Theorem 5.3.9,

$$\begin{aligned}
\int_0^2 |x^2 - 1| \, dx &= \int_0^1 |x^2 - 1| \, dx + \int_1^2 |x^2 - 1| \, dx \\
&= \int_0^1 (1 - x^2) \, dx + \int_1^2 (x^2 - 1) \, dx \\
&= \left[x - \frac{x^3}{3} \right]_0^1 + \left[\frac{x^3}{3} - x \right]_1^2 \\
&= \left[\left(1 - \frac{1}{3} \right) - 0 \right] + \left[\left(\frac{8}{3} - 2 \right) - \left(\frac{1}{3} - 1 \right) \right] = 2. \qquad \square
\end{aligned}$$

Example 5.4.6. An object moves along a number line (which is marked off in feet) in such a way that its velocity at time t (measured in seconds) is given by the formula

$$v(t) = \sqrt[3]{t} \text{ ft/sec.}$$

What is the displacement of the object from $t = 0$ to $t = 8$? (Recall that the *displacement* of the object is the change in its position.)

Solution. We will give two solutions to this problem.
Solution 1. Let $s(t)$ be the position of the object on the number line at time t. Then

$$s'(t) = v(t) = \sqrt[3]{t} = t^{1/3},$$

and by antidifferentiating we conclude that

$$s(t) = \frac{3}{4} t^{4/3} + C,$$

for some constant C. Therefore the displacement of the object is

$$s(8) - s(0) = \left(\frac{3}{4} \cdot 8^{4/3} + C \right) - \left(\frac{3}{4} \cdot 0^{4/3} + C \right) = 12 \text{ ft.}$$

Solution 2. If an object moves at a constant velocity of v ft/sec for d seconds, its displacement will be vd feet. The velocity of the object in this problem varies over time, so we use Riemann sums to approximate its displacement. For each positive integer n, we divide the time interval $[0, 8]$ into n smaller intervals, each of width $\Delta t = 8/n$. As usual, we let $t_i = i \Delta t$, and we choose sample points $t_i^* \in [t_{i-1}, t_i]$. If n is large, so that Δt is small, then the velocity of the object won't vary much over the time interval from $t = t_{i-1}$ to $t = t_i$, so we can approximate the displacement of the object over this time interval by assuming that the velocity is a constant $v(t_i^*)$ ft/sec throughout the interval. The displacement over the interval $[t_{i-1}, t_i]$ is therefore about $v(t_i^*) \Delta t$ ft. The displacement from $t = 0$ to $t = 8$ is the sum of the displacements over all of these smaller intervals, so it is approximately

$$R_n = \sum_{i=1}^{n} v(t_i^*) \Delta t.$$

Of course, this is a Riemann sum for v on the interval $[0, 8]$. The approximation gets better and better as n increases, so the exact displacement is

$$\lim_{n \to \infty} R_n = \int_0^8 v(t)\, dt.$$

We now evaluate this integral by the second fundamental theorem of calculus:

$$\int_0^8 v(t)\, dt = \int_0^8 t^{1/3}\, dt = \frac{3}{4} t^{4/3} \Big|_0^8 = \frac{3}{4} \cdot 8^{4/3} - \frac{3}{4} \cdot 0^{4/3} = 12 \text{ ft.} \qquad \square$$

When we proved the fundamental theorems of calculus, we suggested that these theorems were very surprising. But Example 5.4.6 suggests that perhaps they aren't so surprising after all. There are two ways of approaching the problem of finding the displacement of the moving object in the example. According to the first approach, the displacement is $s(8) - s(0)$, where s is an antiderivative of the velocity v. According to the second approach, the displacement is $\int_0^8 v(t)\, dt$. Of course, these two approaches should give the same answer, and this is exactly what the second fundamental theorem says.

Exercises 5.4

1–12: Use the second fundamental theorem of calculus to evaluate the integral.

1. $\int_1^3 (2x + 1)\, dx$. (Compare to Exercise 2 in Section 5.3.)

2. $\int_1^4 x^3\, dx$. (Compare to Exercise 6 in Section 5.3.)

3. $\int_{-1}^2 (2x^3 - 3x^2 + 4x - 5)\, dx$.

4. $\int_0^1 x^{100}\, dx$.

5. $\int_0^4 \sqrt{x} \, dx.$

6. $\int_{-1}^8 \sqrt[3]{x} \, dx.$

7. $\int_0^{2\pi} (1 + \sin x) \, dx.$

8. $\int_1^4 \dfrac{x^3 + 3}{x^2} \, dx.$

9. $\int_1^4 (\sqrt{x} + 1)^2 \, dx.$

10. $\int_1^4 \dfrac{(\sqrt{x} + 1)^2}{\sqrt{x}} \, dx.$

11. $\int_0^2 x^2 (x + 1)^3 \, dx.$

12. $\int_{\pi/6}^{\pi/3} \csc x \cot x \, dx.$

13. Compute the integral in (5.22) by taking a limit of Riemann sums.

14. Evaluate $\int_0^{\pi/2} \cos x \, dx$ by taking a limit of Riemann sums. You will find Exercise 15 in Section 5.1 helpful.

15. In the proof of the first fundamental theorem of calculus, verify that if x is not the left endpoint of I then

$$F'_-(x) = \lim_{h \to 0^-} \frac{F(x + h) - F(x)}{h} = f(x).$$

16–21: Find the derivative of the function.

16. $f(x) = \int_0^x \dfrac{1}{t^3 + 1} \, dt.$

17. $g(x) = \int_x^0 \dfrac{1}{t^3 + 1} \, dt.$

18. $h(x) = \int_0^{x^2} \dfrac{1}{t^3 + 1} \, dt.$

19. $j(x) = \int_x^{x^2} \dfrac{1}{t^3 + 1} \, dt.$

20. $f(x) = \int_1^x \sin(t^2) \, dt.$

21. $g(x) = \int_0^{\sqrt{x}} \sin(t^2)\, dt.$

22. Evaluate

$$\lim_{x \to 0} \frac{\int_0^x \sin(t^2)\, dt}{x^3}.$$

5.5 Integration by Substitution

According to the second fundamental theorem of calculus, if F is an antiderivative of a continuous function f on an interval I, then F can be thought of as giving a general formula for computing definite integrals of f; for any numbers a and b in I, we can compute $\int_a^b f(x)\, dx$ by applying F to b and a and subtracting. For this reason, antiderivatives are also often written using an integral sign. If F is an antiderivative of f on an interval I, then we write

$$\int f(x)\, dx = F(x) + C \text{ on } I. \tag{5.24}$$

The integral in equation (5.24) is called an *indefinite integral*—indefinite because no limits of integration a and b have been specified. It is customary to include the "$+ C$" in the formula on the right side of equation (5.24) so that it gives the general formula for all antiderivatives of f on I.

In practice, mathematicians rarely specify the interval on which an indefinite integral applies. The notation $\int f(x)\, dx = F(x) + C$ is usually used in situations in which the domain of f is a union of intervals, and F is an antiderivative of f on every interval contained in the domain of f. These intervals can be determined from the formula for $f(x)$, so there is no need to specify them. For example, we would write

$$\int \frac{1}{x^2}\, dx = -\frac{1}{x} + C \tag{5.25}$$

to indicate that $-1/x + C$ is the general formula for all antiderivatives of the function $f(x) = 1/x^2$ on each of the two intervals $(-\infty, 0)$ and $(0, \infty)$ that make up the domain of f.

In this book, we will generally use the notation $\int f(x)\, dx$ only when f is continuous on every interval contained in its domain. Corollary 5.4.2 guarantees that any such function will have an indefinite integral on each of these intervals. Our only problem, then, is finding it.

Unfortunately, finding indefinite integrals is significantly harder than finding derivatives. However, checking the correctness of an indefinite integral is usually easy: to check the correctness of the equation $\int f(x)\, dx = F(x) + C$, we just have to verify that $\frac{d}{dx}(F(x)) = f(x)$. For example, to check the correctness of equation (5.25),

$$\text{For every rational number } r \neq -1, \int x^r \, dx = \frac{x^{r+1}}{r+1} + C$$

$$\int \sin x \, dx = -\cos x + C \qquad\qquad \int \cos x \, dx = \sin x + C$$

$$\int \sec^2 x \, dx = \tan x + C \qquad\qquad \int \csc^2 x \, dx = -\cot x + C$$

$$\int \sec x \tan x \, dx = \sec x + C \qquad\qquad \int \csc x \cot x \, dx = -\csc x + C$$

Table 5.5: Some indefinite integrals.

we compute

$$\frac{d}{dx}\left(-\frac{1}{x}\right) = \frac{d}{dx}(-x^{-1}) = x^{-2} = \frac{1}{x^2}.$$

It might be helpful to summarize what we know about antiderivatives at this point, writing everything in our new indefinite integral notation. First of all, we know the indefinite integrals listed in Table 5.5. We can also summarize Theorem 4.9.8 as saying that if f and g have indefinite integrals on some interval I—for example, if they are continuous on I—and c is any number, then on I,

$$\int cf(x) \, dx = c \int f(x) \, dx, \tag{5.26a}$$

$$\int (f(x) + g(x)) \, dx = \int f(x) \, dx + \int g(x) \, dx, \tag{5.26b}$$

$$\int (f(x) - g(x)) \, dx = \int f(x) \, dx - \int g(x) \, dx. \tag{5.26c}$$

For example, equation (5.26b) tells us that on any interval, if $\int f(x) \, dx = F(x) + C$ and $\int g(x) \, dx = G(x) + C$ then $\int (f(x) + g(x)) \, dx = F(x) + G(x) + C$.

To find indefinite integrals, we will need to reverse our rules for computing derivatives. For example, equations (5.26a)–(5.26c) can be thought of as the reversal of Theorem 3.3.2. One of the most important techniques for computing indefinite integrals is the result of reversing the chain rule. Suppose F is differentiable on an interval I, g is differentiable on an interval J, and for every $x \in J$, $g(x) \in I$. Then according to the chain rule on intervals (Theorem 4.9.5), $F \circ g$ is differentiable on J, and for all $x \in J$, $\frac{d}{dx}(F(g(x)) = F'(g(x)) \cdot g'(x)$. In other words, the following statement is true:

$$\text{If } \frac{d}{du}(F(u)) = f(u) \text{ for all } u \in I \text{ then}$$

$$\frac{d}{dx}(F(g(x))) = f(g(x)) \cdot g'(x) \text{ for all } x \in J.$$

Reversing the two derivatives in this statement, we get the following statement about indefinite integrals:

If $\int f(u)\,du = F(u) + C$ on I then $\int f(g(x)) \cdot g'(x)\,dx = F(g(x)) + C$ on J.

$$(5.27)$$

For example, suppose we want to evaluate the indefinite integral

$$\int (x^2 + 1)^7 \cdot 2x\,dx.$$

If we let $f(u) = u^7$ and $g(x) = x^2 + 1$, then $g'(x) = 2x$, and therefore the integral can be written

$$\int (x^2 + 1)^7 \cdot 2x\,dx = \int f(g(x)) \cdot g'(x)\,dx.$$

Motivated by (5.27), we compute

$$\int f(u)\,du = \int u^7\,du = \frac{u^8}{8} + C.$$

Thus, if we let $F(u) = u^8/8$, then we have $\int f(u)\,du = F(u) + C$, and statement (5.27) (with $I = J = (-\infty, \infty)$) tells us that the answer to our original integral is

$$\int (x^2 + 1)^7 \cdot 2x\,dx = \int f(g(x)) \cdot g'(x)\,dx = F(g(x)) + C = \frac{(x^2 + 1)^8}{8} + C.$$

We can confirm this answer by differentiating:

$$\frac{d}{dx}\left(\frac{(x^2 + 1)^8}{8}\right) = \frac{1}{8} \cdot 8(x^2 + 1)^7 \cdot 2x = (x^2 + 1)^7 \cdot 2x.$$

Of course, the factor of $2x$ in this derivative came from the chain rule. Since we used the chain rule backwards to find the answer to this indefinite integral, it is not surprising that we used the chain rule forwards when checking it.

It will be convenient to develop a symbolic procedure for applying statement (5.27). If we are faced with an integral of the form

$$\int f(g(x)) \cdot g'(x)\,dx,$$

then the first step of our procedure will be to replace $g(x)$ with u and $g'(x)\,dx$ with du, which will convert our integral into

$$\int f(u)\,du.$$

Suppose we are able to evaluate this new integral, getting the solution

$$\int f(u)\,du = F(u) + C.$$

Then we just have to replace u in this solution with $g(x)$ to get $F(g(x)) + C$, and, according to (5.27), we will have the right answer for the original integral.

As a guide to the substitutions involved in this procedure, we will write the equations $u = g(x)$ and $du = g'(x)\,dx$. The first of these equations can be thought of as introducing the new variable u into the problem and specifying what it stands for. We will not try to give an interpretation for the second equation, since we have not assigned any meaning to the symbols du and dx by themselves. We think of it as just a symbolic guide, telling us that in our integral notation, du should be substituted for $g'(x)\,dx$.

For example, using this symbolic procedure, here is how we could rewrite our solution to the integral $\int (x^2 + 1)^7 \cdot 2x\,dx$. We begin by writing the equations $u = g(x) = x^2 + 1$ and $du = g'(x)\,dx = 2x\,dx$. Making the substitutions suggested by these equations in our integral, we can write our solution like this:

$$\int (x^2 + 1)^7 \cdot 2x\,dx = \int u^7\,du = \frac{u^8}{8} + C = \frac{(x^2 + 1)^8}{8} + C.$$

This method of evaluating integrals is called *integration by substitution*.

Let's try another example of integration by substitution. Suppose we want to find

$$\int 3x^2 (x^3 + 2)^4\,dx.$$

We let $u = x^3 + 2$, so that $du = 3x^2\,dx$. Then we have

$$\int 3x^2 (x^3 + 2)^4\,dx = \int (x^3 + 2)^4 \cdot 3x^2\,dx = \int u^4\,du = \frac{u^5}{5} + C = \frac{(x^3 + 2)^5}{5} + C.$$

Recall that, although the Leibniz derivative notation du/dx is not really a fraction, the chain rule says that in some ways it acts like one. Integration by substitution can be thought of as another instance of this phenomenon. If $u = g(x)$, then

$$\frac{du}{dx} = g'(x).$$

Now, if we pretend that du/dx is a fraction, then we can "multiply both sides by dx" to get

$$du = g'(x)\,dx.$$

Again, this is not correct, or even meaningful. But the method of integration by substitution says that if we pretend it is, then we can get correct answers for indefinite integrals.

The hardest part of the method of integration by substitution is deciding what formula to use for u. A good guideline is that u should be substituted for some expression $g(x)$ appearing in the integral whose derivative $g'(x)$ also appears in the integral. Sometimes if $g'(x)$ doesn't appear exactly, it is possible to make small adjustments to make the method work. For example, let's evaluate

$$\int x \sqrt{x^2 - 4}\,dx.$$

The quantity $x^2 - 4$ appears in the integral, and its derivative is $2x$, which *almost* appears as well—only the factor of 2 is missing. So we try the substitution $u = x^2 - 4$, $du = 2x\,dx$, and make adjustments for the missing factor of 2:

$$\int x\sqrt{x^2 - 4}\,dx = \int \frac{1}{2}\sqrt{x^2 - 4} \cdot 2x\,dx = \int \frac{1}{2}\sqrt{u}\,du = \frac{1}{2}\int u^{1/2}\,du \qquad \text{(by (5.26a))}$$

$$= \frac{1}{2} \cdot \frac{2}{3}u^{3/2} + C = \frac{1}{3}(x^2 - 4)^{3/2} + C.$$

Notice that in this substitution, $x\,dx$ got replaced by $(1/2) \cdot 2x\,dx$, which was then replaced by $(1/2)\,du$. Another way to get the same result would have been to rewrite the original substitution equation $du = 2x\,dx$ as $(1/2)\,du = x\,dx$.

In integral notation, mathematicians often treat the symbol "dx" as if it were a quantity that the integrand is being multiplied by. For example, when the integrand is a fraction, mathematicians sometimes include the "dx" in the numerator of the fraction, rather than putting it after the fraction. We use this convention in part (c) of the next example.

Example 5.5.1. Find the following indefinite integrals:

$$\text{(a)} \int x^2 \cos(1 - x^3)\,dx, \qquad \text{(b)} \int \frac{\sin x}{\cos^2 x}\,dx, \qquad \text{(c)} \int \frac{x\,dx}{\sqrt{2x + 1}}.$$

Solution. For integral (a), we use the substitution

$$u = 1 - x^3, \qquad du = -3x^2\,dx, \qquad -\frac{1}{3}\,du = x^2\,dx.$$

This leads to the solution

$$\int x^2 \cos(1 - x^3)\,dx = \int \cos u \cdot \left(-\frac{1}{3}\right) du$$

$$= -\frac{1}{3}\sin u + C = -\frac{1}{3}\sin(1 - x^3) + C.$$

The substitution for (b) is perhaps harder to see, but it turns out that the choice $u = \cos x$ works. This gives us $du = -\sin x\,dx$, or equivalently $(-1)\,du = \sin x\,dx$, so

$$\int \frac{\sin x}{\cos^2 x}\,dx = \int (\cos x)^{-2} \cdot \sin x\,dx$$

$$= \int u^{-2} \cdot (-1)\,du = u^{-1} + C = \frac{1}{\cos x} + C = \sec x + C.$$

Finally, for part (c), since $2x + 1$ appears inside the square root, we try

$$u = 2x + 1, \qquad du = 2\,dx, \qquad \frac{1}{2}\,du = dx.$$

So in integral (c) we can replace $\sqrt{2x+1}$ by $\sqrt{u}$ and dx by $(1/2)\,du$. But what about the x in the numerator? Can we also express it in terms of u? Yes: since $u = 2x + 1$, we have $x = (u-1)/2$. Therefore

$$\int \frac{x\,dx}{\sqrt{2x+1}} = \int \frac{(u-1)/2}{\sqrt{u}} \cdot \frac{1}{2}\,du = \int \frac{u-1}{4u^{1/2}}\,du = \frac{1}{4}\int (u^{1/2} - u^{-1/2})\,du$$

$$= \frac{1}{4}\left(\frac{2}{3}u^{3/2} - 2u^{1/2}\right) + C = \frac{1}{6}(2x+1)^{3/2} - \frac{1}{2}\sqrt{2x+1} + C. \qquad \square$$

Notice that in integration by substitution, the integral $\int f(g(x)) \cdot g'(x)\,dx$, which involves only the variable x, is converted into the integral $\int f(u)\,du$, which involves only the variable u. We never have an integral that contains both of the variables x and u. That's why it was important, in part (c) of Example 5.5.1, that we were able to rewrite the x in the numerator in terms of u. If you are trying to evaluate an integral by the method of substitution and you find that you are unable to convert the entire integral from one variable to another, then you need to try a different method of evaluating the integral.

For example, consider the integral

$$\int \frac{\sin^2 x}{\cos^2 x}\,dx.$$

Motivated by the similarity between this integral and the one in part (b) of Example 5.5.1, we might try the substitution $u = \cos x$, $du = -\sin x\,dx$, $(-1)\,du = \sin x\,dx$. If we rewrite the integral as

$$\int \frac{\sin^2 x}{\cos^2 x}\,dx = \int \sin x(\cos x)^{-2} \cdot \sin x\,dx,$$

then we can replace $(\cos x)^{-2}$ with u^{-2} and $\sin x\,dx$ with $(-1)\,du$. But there's another factor of $\sin x$, and there is no easy way to express this in terms of u. Thus, this substitution won't allow us to find the integral. However, it turns out that a completely different approach works for this integral; see Exercise 15.

In some cases you may find it necessary to do two substitutions in succession. For example, consider the integral

$$\int \frac{\sin(\sqrt{x})\cos(\sqrt{x})}{\sqrt{x}}\,dx.$$

A good first step is the substitution $u = \sqrt{x} = x^{1/2}$, $du = (1/2)x^{-1/2}\,dx = 1/(2\sqrt{x})\,dx$, $2\,du = (1/\sqrt{x})\,dx$:

$$\int \frac{\sin(\sqrt{x})\cos(\sqrt{x})}{\sqrt{x}}\,dx = \int \sin(\sqrt{x})\cos(\sqrt{x}) \cdot \frac{1}{\sqrt{x}}\,dx = \int 2\sin u \cos u\,du.$$

But now how can we evaluate the last integral? Another substitution will do it: $v = \sin u$, $dv = \cos u\,du$:

$$\int 2\sin u \cos u\,du = \int 2v\,dv = v^2 + C = \sin^2 u + C.$$

Returning to our original integral, we conclude that

$$\int \frac{\sin(\sqrt{x})\cos(\sqrt{x})}{\sqrt{x}}\,dx = \int 2\sin u \cos u \, du = \sin^2 u + C = \sin^2(\sqrt{x}) + C.$$

Integration by substitution can also help us to evaluate definite integrals:

Example 5.5.2. Evaluate

$$\int_0^3 x\sqrt[3]{x^2 - 1}\,dx.$$

Solution. We begin by finding the indefinite integral, using the substitution $u = x^2 - 1$, $du = 2x\,dx$, $(1/2)\,du = x\,dx$:

$$\int x\sqrt[3]{x^2 - 1}\,dx = \int \frac{1}{2}u^{1/3}\,du = \frac{1}{2}\cdot\frac{3}{4}u^{4/3} + C = \frac{3}{8}(x^2 - 1)^{4/3} + C.$$

Now we use this indefinite integral to evaluate the original definite integral:

$$\int_0^3 x\sqrt[3]{x^2 - 1}\,dx = \frac{3}{8}(x^2 - 1)^{4/3}\Big|_0^3 = \frac{3}{8}\cdot 8^{4/3} - \frac{3}{8}\cdot(-1)^{4/3} = \frac{45}{8}. \qquad \square$$

While this solution is correct, there is an easier way. The easier way is based on the following theorem:

Theorem 5.5.3. *Suppose f is continuous on an interval I, g is differentiable on an interval J, g' is continuous on J, and for every $x \in J$, $g(x) \in I$. Then for any a and b in J,*

$$\int_a^b f(g(x))\cdot g'(x)\,dx = \int_{g(a)}^{g(b)} f(u)\,du.$$

Proof. Notice that since a and b are in J and g maps elements of J to elements of I, $g(a)$ and $g(b)$ are in I. Also, the hypotheses of the theorem guarantee that $f(g(x))\cdot g'(x)$ is continuous on J and $f(u)$ is continuous on I. Therefore we can use the second fundamental theorem of calculus to evaluate both definite integrals in the theorem. We will show that this evaluation leads to the same answer for both integrals.

We begin with the second integral. Since f is continuous on I, it has an antiderivative F on I. Therefore by the second fundamental theorem,

$$\int_{g(a)}^{g(b)} f(u)\,du = F(u)\big|_{g(a)}^{g(b)} = F(g(b)) - F(g(a)).$$

For the first integral, we use the chain rule on intervals. Since F is differentiable on I and g is differentiable on J, the chain rule on intervals tells us that $F \circ g$ is differentiable on J, and $(F \circ g)'(x) = F'(g(x))\cdot g'(x) = f(g(x))\cdot g'(x)$. In other words, $F(g(x))$ is an antiderivative of $f(g(x))\cdot g'(x)$ on J, and therefore

$$\int_a^b f(g(x))\cdot g'(x)\,dx = F(g(x))\big|_a^b = F(g(b)) - F(g(a)).$$

Thus, the values of the two integrals are the same. $\qquad \square$

Let's evaluate $\int_0^3 x\sqrt[3]{x^2-1}\,dx$, the definite integral from Example 5.5.2, using Theorem 5.5.3. In the example, we used the substitution $u = g(x) = x^2 - 1, (1/2)\,du = x\,dx$. The limits of integration were $a = 0$ and $b = 3$, and Theorem 5.5.3 says that when we do the substitution, we should change these to

$$g(a) = g(0) = -1, \qquad g(b) = g(3) = 8.$$

Thus,

$$\int_0^3 x\sqrt[3]{x^2-1}\,dx = \int_{-1}^8 \frac{1}{2}u^{1/3}\,du = \frac{1}{2}\cdot\frac{3}{4}u^{4/3}\Big|_{-1}^8 = \frac{3}{8}\cdot 8^{4/3} - \frac{3}{8}\cdot(-1)^{4/3} = \frac{45}{8}.$$

Notice that after we found the antiderivative of $(1/2)u^{1/3}$, there was no need to put this antiderivative back in terms of x; we merely had to substitute the new limits of integration for u and subtract. Indeed, it would have been incorrect to rewrite the antiderivative in terms of x. The new limits of integration, -1 and 8, are the u-values corresponding to the original limits of integration 0 and 3, so they must be substituted for u, not x. Thus, the advantage of using Theorem 5.5.3 to evaluate a definite integral is that once the substitution has been made, you can forget about not only the original integral but also the original variable of integration, and simply evaluate the new integral in terms of its new variable.

The lesson of this example is that there are two ways to evaluate a definite integral by substitution: you can evaluate the corresponding indefinite integral by substitution and then use this answer to evaluate the definite integral, or you can do the substitution in the definite integral by applying Theorem 5.5.3. The one thing you must *not* do is to perform the substitution in the *definite* integral without converting the limits of integration as required by Theorem 5.5.3; that would result in an integral in which the integrand was written in terms of the new variable but the limits of integration were values of the original variable, a situation that is bound to cause confusion.

Example 5.5.4. Evaluate

$$\int_4^8 \frac{x^3\,dx}{\sqrt{x^2-15}}.$$

Solution. We give two solutions.

Solution 1. We first find the corresponding indefinite integral, using the substitution $u = x^2 - 15, du = 2x\,dx$. Notice that we have $x^2 = u + 15$ and $x\,dx = (1/2)\,du$.

$$\int \frac{x^3\,dx}{\sqrt{x^2-15}} = \int \frac{x^2}{(x^2-15)^{1/2}}\cdot x\,dx = \int \frac{u+15}{u^{1/2}}\cdot\frac{1}{2}\,du$$

$$= \int \left(\frac{1}{2}u^{1/2} + \frac{15}{2}u^{-1/2}\right)du$$

$$= \frac{1}{3}u^{3/2} + 15u^{1/2} + C = \frac{1}{3}(x^2-15)^{3/2} + 15\sqrt{x^2-15} + C.$$

Now we use this indefinite integral to evaluate the definite integral:

$$\int_4^8 \frac{x^3 \, dx}{\sqrt{x^2 - 15}} = \left[\frac{1}{3}(x^2 - 15)^{3/2} + 15\sqrt{x^2 - 15} \right]_4^8$$

$$= \left(\frac{1}{3}(49^{3/2}) + 15\sqrt{49} \right) - \left(\frac{1}{3}(1^{3/2}) + 15\sqrt{1} \right) = 204.$$

Solution 2. We use the substitution $u = x^2 - 15$ again, but this time we substitute in the definite integral, changing the limits of integration from the x-values 4 and 8 to the corresponding u-values $4^2 - 15 = 1$ and $8^2 - 15 = 49$:

$$\int_4^8 \frac{x^3 \, dx}{\sqrt{x^2 - 15}} = \int_1^{49} \left(\frac{1}{2} u^{1/2} + \frac{15}{2} u^{-1/2} \right) du = \left[\frac{1}{3} u^{3/2} + 15 u^{1/2} \right]_1^{49}$$

$$= \left(\frac{1}{3}(49^{3/2}) + 15\sqrt{49} \right) - \left(\frac{1}{3}(1^{3/2}) + 15\sqrt{1} \right) = 204. \qquad \square$$

Exercises 5.5

1–14: Evaluate the integral.

1. $\displaystyle\int \sin(x^2 + 3) \cdot 2x \, dx.$

2. $\displaystyle\int x^2 \sqrt[4]{x^3 - 2} \, dx.$

3. $\displaystyle\int x^2 \cos(x^3) \, dx.$

4. $\displaystyle\int \sin x \cos^3 x \, dx.$

5. $\displaystyle\int x^2 \sin(x^3) \cos^3(x^3) \, dx.$

6. $\displaystyle\int \frac{x \, dx}{(x+1)^3}.$

7. $\displaystyle\int \frac{dx}{\sqrt{x}(\sqrt{x} + 1)^4}.$

8. $\displaystyle\int \frac{\sqrt{x} \, dx}{(\sqrt{x} + 1)^4}.$

9. $\displaystyle\int \sec^2 x \tan^2 x \, dx.$

10. $\displaystyle\int \sqrt{1 + \sin x} \cos x \, dx.$

11. $\displaystyle\int_0^4 \frac{x\,dx}{\sqrt{x^2+9}}$.

12. $\displaystyle\int_0^1 (x+1)(x^2+2x-1)^4\,dx$.

13. $\displaystyle\int_0^{\pi/3} \sec^3 x \tan x\,dx$.

14. $\displaystyle\int_{\pi^2}^{4\pi^2} \frac{\cos(\sqrt{x}/4)}{\sqrt{x}}\,dx$.

15. Evaluate

$$\int \frac{\sin^2 x}{\cos^2 x}\,dx.$$

(Hint: First rewrite $\sin^2 x$ as $1 - \cos^2 x$.)

5.6   Proofs of Theorems

At this point, we owe you the proofs of two theorems: Theorems 5.3.2 and 5.3.9. We
have put off these proofs for two reasons: they require a subtle technical strengthening of
the concept of continuity, and they also require a generalization of the idea of Riemann
sums. Neither of these ideas will be needed anywhere else in this book. Readers who
are willing to accept these two theorems without proof don't need to know about these
ideas and can skip this section. But for those who want to see the proofs, we provide full
details here.

  We begin by reexamining the concept of continuity. To say that a function f is
continuous at a number t means that as $s \to t$, $f(s) \to f(t)$, which means that for every
$\epsilon > 0$ there is some $\delta > 0$ such that if $|s - t| < \delta$ then $|f(s) - f(t)| < \epsilon$. Recall that
we think of ϵ as a challenge, and δ as a response to this challenge. Thus, for f to be
continuous at t means that we can meet every challenge ϵ with an appropriate response δ.
To say that f is continuous on an interval I means that for every $t \in I$, f is continuous
at t (where we use one-sided continuity at endpoints of I). Thus, for f to be continuous
on I means that for every $t \in I$, we can meet every challenge ϵ for continuity at t with
an appropriate response δ. But for different values of t, the responses may be different.
For example, for one value of t we might respond to any $\epsilon > 0$ with $\delta = \epsilon/3$, and for
another we might respond with $\delta = \min(1, \epsilon/7)$. A slight strengthening of the definition
is to require that the δ responses be the same for all t. This leads to the following concept.

Definition 5.6.1. A function f is said to be *uniformly continuous* on an interval I if
for every $\epsilon > 0$ there is some $\delta > 0$ such that for all s and t in I, if $|s - t| < \delta$ then
$|f(s) - f(t)| < \epsilon$.

  It is not hard to show that if f is uniformly continuous on an interval I, then it is
continuous on I (see Exercise 1). The converse does not hold: it is possible for a function

to be continuous on an interval, but not uniformly continuous (see Exercise 5). However, this never happens on finite closed intervals:

Theorem 5.6.2. *If f is continuous on a finite closed interval $[a, b]$, then it is uniformly continuous on $[a, b]$.*

Proof. Suppose that f is continuous on $[a, b]$, and suppose that $\epsilon > 0$. We think of ϵ as a challenge, to be met with a response δ as specified in Definition 5.6.1. For any interval I contained in $[a, b]$, we will say that the challenge ϵ can be met on I if there is some $\delta > 0$ such that for all s and t in I, if $|s - t| < \delta$ then $|f(s) - f(t)| < \epsilon$. To complete the proof, we must show that the challenge ϵ can be met on $[a, b]$.

We will assume that the challenge ϵ cannot be met on $[a, b]$ and derive a contradiction. The key to our proof is the following fact.

Lemma 5.6.3. *Suppose that $a \le u < v \le b$, and the challenge ϵ cannot be met on $[u, v]$. Then there are numbers u' and v' such that $u \le u' < v' \le v$, $v' - u' = (v - u)/2$, and the challenge ϵ cannot be met on $[u', v']$.*

Proof. Let $w = (u + v)/2$. We will show that the challenge ϵ cannot be met on one of the two intervals $[u, w]$ and $[w, v]$. We can then let $[u', v']$ be either $[u, w]$ or $[w, v]$—whichever one is an interval on which the challenge cannot be met.

Suppose the challenge ϵ can be met on both of the intervals $[u, w]$ and $[w, v]$. Then there are numbers $\delta_1 > 0$ and $\delta_2 > 0$ such that

  (a) for all s and t in $[u, w]$, if $|s - t| < \delta_1$ then $|f(s) - f(t)| < \epsilon$;
  (b) for all s and t in $[w, v]$, if $|s - t| < \delta_2$ then $|f(s) - f(t)| < \epsilon$.

Since f is continuous at w, we can also choose a number $\delta_3 > 0$ such that

  (c) for all s, if $|s - w| < \delta_3$ then $|f(s) - f(w)| < \epsilon/2$.

Now let $\delta = \min(\delta_1, \delta_2, \delta_3)$. We will show that δ meets the challenge ϵ on $[u, v]$. To see why, suppose s and t belong to the interval $[u, v]$ and $|s - t| < \delta$. We consider four possibilities:

Case 1. $s \le w$, $t \le w$. Then s and t belong to $[u, w]$ and $|s - t| < \delta \le \delta_1$, so by statement (a), $|f(s) - f(t)| < \epsilon$.

Case 2. $s \ge w$, $t \ge w$. Then s and t belong to $[w, v]$ and $|s - t| < \delta \le \delta_2$, so by statement (b), $|f(s) - f(t)| < \epsilon$.

Case 3. $s < w < t$. Then $|s - w| = w - s < t - s = |s - t| < \delta \le \delta_3$, so by statement (c), $|f(s) - f(w)| < \epsilon/2$. Similarly, $|t - w| < \delta_3$, and therefore $|f(w) - f(t)| < \epsilon/2$. Therefore by the triangle inequality,

$$|f(s) - f(t)| = |(f(s) - f(w)) + (f(w) - f(t))|$$
$$\le |f(s) - f(w)| + |f(w) - f(t)| < \frac{\epsilon}{2} + \frac{\epsilon}{2} = \epsilon.$$

Case 4. $t < w < s$. As in case 3, we can show that $|f(s) - f(t)| < \epsilon$.

Thus we see that in every case $|f(s) - f(t)| < \epsilon$, so δ meets the challenge ϵ on $[u, v]$. But we assumed that the challenge could not be met on $[u, v]$, so this is impossible. Thus, the challenge ϵ cannot be met on either $[u, w]$ or $[w, v]$, which proves the lemma. $\square$

Continuing with the proof of Theorem 5.6.2, we assume that the challenge ϵ cannot be met on $[a, b]$, and then we use the lemma to construct a nested sequence of intervals. Let $u_1 = a$ and $v_1 = b$. Then the challenge ϵ cannot be met on $[u_1, v_1]$, so by the lemma we can choose u_2 and v_2 so that $u_1 \leq u_2 < v_2 \leq v_1$, $v_2 - u_2 = (v_1 - u_1)/2 = (b - a)/2$, and the challenge ϵ cannot be met on $[u_2, v_2]$. Applying the lemma again, we see that there must be u_3 and v_3 such that $u_2 \leq u_3 < v_3 \leq v_2$, $v_3 - u_3 = (v_2 - u_2)/2 = (b - a)/4$, and the challenge ϵ cannot be met on $[u_3, v_3]$. Continuing in this way, we construct a nested sequence of intervals $[u_n, v_n]$ such that for every n, $v_n - u_n = (b - a)/2^{n-1}$ and the challenge ϵ cannot be met on $[u_n, v_n]$.

By the nested interval theorem, there is a number c such that for every n, $u_n \leq c \leq v_n$, and $\lim_{n \to \infty} u_n = \lim_{n \to \infty} v_n = c$. In particular, $c \in [u_1, v_1] = [a, b]$, so f is continuous at c. (If either $c = a$ or $c = b$, then f is continuous at c from only one side. We leave it to you to make appropriate adjustments in the proof to deal with these cases.) Since f is continuous at c, as $s \to c$, $f(s) \to f(c)$, and therefore there must be some $\delta_1 > 0$ such that if $|s - c| < \delta_1$ then $|f(s) - f(c)| < \epsilon/2$. And since $\lim_{n \to \infty} u_n = \lim_{n \to \infty} v_n = c$, there must be some n large enough that

$$c - \delta_1 < u_n \leq c \leq v_n < c + \delta_1.$$

But now for any s and t in $[u_n, v_n]$ we have $|s - c| < \delta_1$ and $|t - c| < \delta_1$, and therefore $|f(s) - f(c)| < \epsilon/2$ and $|f(t) - f(c)| < \epsilon/2$. As in case 3 of the proof of the lemma, we can use the triangle inequality to conclude that $|f(s) - f(t)| < \epsilon$. It follows that the challenge ϵ can be met on the interval $[u_n, v_n]$ (indeed, *any* choice for δ would meet the challenge), contradicting the way that interval was chosen.

We conclude that our assumption that the challenge ϵ could not be met on $[a, b]$ was incorrect, and therefore f is uniformly continuous on $[a, b]$. $\square$

Next we turn to a generalization of the idea of Riemann sums. Suppose that f is a function that is continuous on an interval $[a, b]$—and therefore, by Theorem 5.6.2, uniformly continuous on $[a, b]$. In Section 5.3, to form a Riemann n-sum for f on $[a, b]$ we defined numbers $x_0, x_1, \ldots, x_n$ that were evenly spaced between a and b. We now extend this idea by allowing the numbers to be spaced unevenly.

If $x_0, x_1, \ldots, x_n$ are any numbers such that

$$a = x_0 < x_1 < \cdots < x_n = b,$$

then we will say that the set $P = \{x_0, x_1, \ldots, x_n\}$ is a *partition* of the interval $[a, b]$. As before, we think of the elements of P as dividing the interval $[a, b]$ into n smaller intervals $[x_0, x_1]$, $[x_1, x_2]$, $\ldots$, $[x_{n-1}, x_n]$. However, since the numbers need not be evenly spaced, the widths of these smaller intervals may vary. We let $\Delta x_1 = x_1 - x_0$, $\Delta x_2 = x_2 - x_1$, and, in general, $\Delta x_i = x_i - x_{i-1} =$ the width of the ith interval.

The *mesh* of the partition is the largest of the numbers $\Delta x_1, \Delta x_2, \ldots, \Delta x_n$; that is,

$$\text{mesh}(P) = \max(\Delta x_1, \Delta x_2, \ldots, \Delta x_n).$$

As usual, we choose sample points $x_i^* \in [x_{i-1}, x_i]$, and then we form the sum

$$R = f(x_1^*)\Delta x_1 + f(x_2^*)\Delta x_2 + \cdots + f(x_n^*)\Delta x_n = \sum_{i=1}^{n} f(x_i^*)\Delta x_i.$$

In this section, and this section only, we extend the definition of Riemann sums to allow sums of this form. We will say that R is a *Riemann sum for f based on the partition P*, or, more briefly, a *Riemann P-sum for f*. As before, R can be interpreted in terms of adding and subtracting areas of rectangles, but now the widths of the rectangles may vary. The Riemann sums we studied previously had this form, but they were based on partitions with $\Delta x_1 = \Delta x_2 = \cdots = \Delta x_n = (b-a)/n$. We will say that such partitions are *uniform*.

There are two ways of choosing the sample points for a Riemann sum that we will be particularly interested in. Since f is continuous on each interval $[x_{i-1}, x_i]$, it has a minimum value and a maximum value on each of these intervals. Suppose the minimum is attained at $c_i \in [x_{i-1}, x_i]$, and the maximum is attained at $C_i \in [x_{i-1}, x_i]$. Using these numbers as sample points, we can define two special Riemann sums based on P:

$$m_P = \sum_{i=1}^{n} f(c_i)\Delta x_i, \qquad M_P = \sum_{i=1}^{n} f(C_i)\Delta x_i.$$

For any choice of sample points $x_i^* \in [x_{i-1}, x_i]$ we will have $f(c_i) \leq f(x_i^*) \leq f(C_i)$, and therefore

$$m_P = \sum_{i=1}^{n} f(c_i)\Delta x_i \leq \sum_{i=1}^{n} f(x_i^*)\Delta x_i \leq \sum_{i=1}^{n} f(C_i)\Delta x_i = M_P.$$

In other words, if R is any Riemann P-sum, then $m_P \leq R \leq M_P$. Thus, we can refer to m_P and M_P as the minimum and maximum Riemann P-sums for f. If $m_P = M_P$, then all Riemann P-sums are equal. If not, then m_P and M_P determine a closed interval $[m_P, M_P]$, and all Riemann P-sums belong to this interval.

If Q is another partition of $[a, b]$ and $P \subseteq Q$, then we say that Q is a *refinement* of P. The partition Q includes all of the numbers in P, plus perhaps some more. Thus, each of the n pieces into which P divides the interval $[a, b]$ may be divided into still smaller pieces in the partition Q. We would like to investigate the relationship between Riemann sums based on P and those based on Q.

We begin with the simplest case, in which Q is the result of adding just one more number to P. The new number will belong to one of the intervals in the partition P. Say the new number is t, and it belongs to the ith interval of P. In other words, $x_{i-1} < t < x_i$, and $Q = P \cup \{t\} = \{x_0, x_1, \ldots, x_{i-1}, t, x_i, \ldots, x_n\}$. Riemann sums based on Q will be exactly the same as Riemann sums based on P, except that the term in the P-sum for the interval $[x_{i-1}, x_i]$ will be replaced in the Q-sum by two terms, one for the interval $[x_{i-1}, t]$ and one for $[t, x_i]$.

In particular, consider the minimum and maximum Riemann P-sums and Q-sums. As before, we assume that the minimum and maximum values of f on the interval $[x_{i-1}, x_i]$ are attained at c_i and C_i, respectively. Suppose that the minimum and maximum on the interval $[x_{i-1}, t]$ are attained at c_i' and C_i', and the minimum and maximum on $[t, x_i]$ are attained at c_i'' and C_i''. Then clearly

$$f(c_i) \le f(c_i') \le f(C_i') \le f(C_i), \qquad f(c_i) \le f(c_i'') \le f(C_i'') \le f(C_i).$$

Also, let $\Delta x_i' = t - x_{i-1}$ and $\Delta x_i'' = x_i - t$, and note that

$$\Delta x_i' + \Delta x_i'' = \Delta x_i.$$

Therefore

$$
\begin{aligned}
f(c_i)\Delta x_i = f(c_i)(\Delta x_i' + \Delta x_i'') &= f(c_i)\Delta x_i' + f(c_i)\Delta x_i'' \\
&\le f(c_i')\Delta x_i' + f(c_i'')\Delta x_i'' \\
&\le f(C_i')\Delta x_i' + f(C_i'')\Delta x_i'' \\
&\le f(C_i)\Delta x_i' + f(C_i)\Delta x_i'' \\
&= f(C_i)(\Delta x_i' + \Delta x_i'') = f(C_i)\Delta x_i.
\end{aligned}
$$

Since all remaining terms in the minimum and maximum Riemann sums are the same for the partition Q as for P, we conclude that

$$m_P \le m_Q \le M_Q \le M_P.$$

To summarize: when a number is added to a partition, the minimum Riemann sum can only increase, and the maximum Riemann sum can only decrease. If Q is a refinement of P, then we can add the extra elements of Q to P one at a time, and at each step the range of Riemann sums can only shrink. As a result, we have the following fact.

Lemma 5.6.4. *If Q is a refinement of P, then*

$$m_P \le m_Q \le M_Q \le M_P.$$

According to Lemma 5.6.4, when we refine a partition, the range of Riemann sums based on the partition *might* shrink. But will it actually shrink, and how small will it get? Our next lemma will help us to answer this question.

Lemma 5.6.5. *Suppose that for every positive integer n, P_n is a partition of $[a, b]$, and* $\lim_{n \to \infty} \text{mesh}(P_n) = 0$. *Then* $\lim_{n \to \infty}(M_{P_n} - m_{P_n}) = 0$.

Proof. Suppose $\epsilon > 0$. To prove that $\lim_{n \to \infty}(M_{P_n} - m_{P_n}) = 0$, we must find a number N such that if n is a positive integer and $n > N$ then $|(M_{P_n} - m_{P_n}) - 0| = M_{P_n} - m_{P_n} < \epsilon$.

Since f is uniformly continuous on $[a, b]$, there is some $\delta > 0$ such that

(a) if $s, t \in [a, b]$ and $|s - t| < \delta$ then $|f(s) - f(t)| < \dfrac{\epsilon}{b - a}$.

And since $\lim_{n \to \infty} \text{mesh}(P_n) = 0$, there is some N such that

(b) if n is a positive integer and $n > N$ then $\text{mesh}(P_n) < \delta$.

Now suppose that n is a positive integer and $n > N$. By statement (b), this means that $\text{mesh}(P_n) < \delta$. Suppose that $P_n = \{x_0, x_1, \ldots, x_k\}$, where $a = x_0 < x_1 < \cdots < x_k = b$. To find the minimum and maximum Riemann sums based on P_n, for $1 \le i \le k$ we let c_i and C_i be the numbers where f attains its minimum and maximum values on $[x_{i-1}, x_i]$. Notice that since c_i and C_i both belong to $[x_{i-1}, x_i]$ we have

$$|C_i - c_i| \le x_i - x_{i-1} = \Delta x_i \le \text{mesh}(P_n) < \delta,$$

and therefore, by statement (a), $f(C_i) - f(c_i) = |f(C_i) - f(c_i)| < \epsilon/(b - a)$. Thus,

$$M_{P_n} - m_{P_n} = \sum_{i=1}^{k} f(C_i)\Delta x_i - \sum_{i=1}^{k} f(c_i)\Delta x_i = \sum_{i=1}^{k}(f(C_i) - f(c_i))\Delta x_i$$

$$< \sum_{i=1}^{k} \frac{\epsilon}{b - a}\Delta x_i = \frac{\epsilon}{b - a}\sum_{i=1}^{k}\Delta x_i = \frac{\epsilon}{b - a}(b - a) = \epsilon.$$

This is the result we needed to verify that $\lim_{n \to \infty}(M_{P_n} - m_{P_n}) = 0$, so the proof is complete. $\square$

We now have all the background we need to prove Theorem 5.3.2.

Theorem 5.3.2 (restated). Suppose that $a < b$ and f is a function that is continuous on the closed interval $[a, b]$. Then every Riemann sum sequence for f on $[a, b]$ converges, and all such sequences converge to the same number. In other words, there is some number L such that if $(R_n)_{n=1}^{\infty}$ is any Riemann sum sequence for f on $[a, b]$, then $\lim_{n \to \infty} R_n = L$.

Proof. We begin by defining a sequence of partitions of $[a, b]$. Let Q_1 be the uniform partition of $[a, b]$ into 2 pieces, and let Q_2 be the uniform partition into 4 pieces. Then Q_2 is a refinement of Q_1, since it can be obtained by taking each of the 2 pieces of the partition Q_1 and cutting it in half. Similarly, if we cut each piece of Q_2 in half then we get the uniform partition into 8 pieces, which we will call Q_3. In general, we let Q_n be the uniform partition of $[a, b]$ into 2^n pieces. Then for every n, Q_{n+1} is a refinement of Q_n, obtained by cutting each piece of the partition Q_n in half. Thus, by Lemma 5.6.4,

$$m_{Q_1} \le m_{Q_2} \le m_{Q_3} \le \cdots \le M_{Q_3} \le M_{Q_2} \le M_{Q_1}.$$

Also, notice that the mesh of Q_n is $(b - a)/2^n$. Therefore $\lim_{n\to\infty} \text{mesh}(Q_n) = 0$, and by Lemma 5.6.5 it follows that

$$\lim_{n\to\infty} (M_{Q_n} - m_{Q_n}) = 0.$$

By the nested interval theorem, there is some number L such that for every n, $m_{Q_n} \leq L \leq M_{Q_n}$. We must prove now that if $(R_n)_{n=1}^{\infty}$ is any Riemann sum sequence for f then $\lim_{n\to\infty} R_n = L$. Recall that according to the definition of Riemann sum sequences in Section 5.3, this means that R_n is a Riemann sum for the uniform partition of $[a, b]$ into n pieces. We will actually prove a little more than what is stated in the theorem. We will show that even if the partitions are not uniform, as long as the meshes of the partitions approach 0, the Riemann sums will converge to L. In other words, we now prove the following lemma:

Lemma 5.6.6. *Suppose that for every positive integer n, P_n is a partition of $[a, b]$ and R_n is a Riemann sum for f based on the partition P_n. Suppose also that $\lim_{n\to\infty} \text{mesh}(P_n) = 0$. Then $\lim_{n\to\infty} R_n = L$.*

Proof. Suppose $\epsilon > 0$; we must find a number N such that if n is a positive integer and $n > N$ then $|R_n - L| < \epsilon$.

Since $\lim_{n\to\infty} \text{mesh}(P_n) = 0$, Lemma 5.6.5 tells us that $\lim_{n\to\infty} (M_{P_n} - m_{P_n}) = 0$, and therefore there is some N_1 such that

(a) if n is a positive integer and $n > N_1$ then $M_{P_n} - m_{P_n} < \dfrac{\epsilon}{2}$.

Similarly, since $\lim_{n\to\infty} (M_{Q_n} - m_{Q_n}) = 0$, there is some N_2 such that

(b) if n is a positive integer and $n > N_2$ then $M_{Q_n} - m_{Q_n} < \dfrac{\epsilon}{2}$.

Let $N = \max(N_1, N_2)$, and suppose that n is a positive integer and $n > N$. Then $n > N_1$ and $n > N_2$, so by statements (a) and (b), we have

$$M_{P_n} - m_{P_n} < \frac{\epsilon}{2}, \qquad M_{Q_n} - m_{Q_n} < \frac{\epsilon}{2}.$$

We also know that

$$m_{P_n} \leq R_n \leq M_{P_n}, \qquad m_{Q_n} \leq L \leq M_{Q_n}.$$

What we need now is some relationship between the minimum and maximum Riemann sums for P_n and those for Q_n. Unfortunately, it is possible that neither P_n nor Q_n is a refinement of the other, so Lemma 5.6.4 is not directly applicable. What can we do?

Our solution will be to introduce a third partition that is a refinement of both P_n and Q_n. It is easy to find such a partition: if we let $S = P_n \cup Q_n$, then S is a partition of $[a, b]$ that is a refinement of both P_n and Q_n. It follows, by Lemma 5.6.4, that

$$m_{P_n} \leq m_S \leq M_S \leq M_{P_n}, \qquad m_{Q_n} \leq m_S \leq M_S \leq M_{Q_n}.$$

Now we have $m_{P_n} \le m_S \le M_{P_n}$ and $m_{P_n} \le R_n \le M_{P_n}$, so

$$|R_n - m_S| \le M_{P_n} - m_{P_n} < \frac{\epsilon}{2}.$$

Similarly, since $m_{Q_n} \le m_S \le M_{Q_n}$ and $m_{Q_n} \le L \le M_{Q_n}$, we have

$$|m_S - L| \le M_{Q_n} - m_{Q_n} < \frac{\epsilon}{2}.$$

Thus, by the triangle inequality,

$$|R_n - L| = |(R_n - m_S) + (m_S - L)| \le |R_n - m_S| + |m_S - L| < \frac{\epsilon}{2} + \frac{\epsilon}{2} = \epsilon.$$

This completes the proof of the lemma. $\qquad\square$

To see that the lemma suffices to complete the proof of Theorem 5.3.2, suppose $(R_n)_{n=1}^{\infty}$ is a Riemann sum sequence for f. Let P_n be the uniform partition of $[a, b]$ into n pieces. Then R_n is a Riemann sum for f based on P_n and $\lim_{n\to\infty} \text{mesh}(P_n) = \lim_{n\to\infty} (b - a)/n = 0$. Therefore, by Lemma 5.6.6, $\lim_{n\to\infty} R_n = L$, as required. $\quad\square$

Of course, the number L in Theorem 5.3.2 is what we defined to be $\int_a^b f(x)\,dx$. Thus, our proof of Lemma 5.6.6 establishes the following fact:

Theorem 5.6.7. *Suppose f is continuous on $[a, b]$, and for every positive integer n, P_n is a partition of $[a, b]$ and R_n is a Riemann sum for f based on P_n. If $\lim_{n\to\infty} \text{mesh}(P_n) = 0$, then*

$$\lim_{n\to\infty} R_n = \int_a^b f(x)\,dx.$$

This turns out to be just what we need to prove Theorem 5.3.9.

Theorem 5.3.9 (restated). *Suppose that $a < b < c$, and f is continuous on $[a, c]$. Then*

$$\int_a^b f(x)\,dx + \int_b^c f(x)\,dx = \int_a^c f(x)\,dx.$$

Proof. For each positive integer n, let P_n be the uniform partition of $[a, b]$ into n pieces, and let P_n' be the uniform partition of $[b, c]$ into n pieces. Let R_n be a Riemann sum for f based on P_n, and let R_n' be a Riemann sum based on P_n'. Let $Q_n = P_n \cup P_n'$. Then Q_n is a partition of $[a, c]$, and $R_n + R_n'$ is a Riemann sum for f based on Q_n. Note that Q_n may not be uniform, but its mesh is

$$\text{mesh}(Q_n) = \max(\text{mesh}(P_n), \text{mesh}(P_n')) = \frac{\max(b - a, c - b)}{n}.$$

Therefore $\lim_{n\to\infty} \text{mesh}(P_n) = \lim_{n\to\infty} \text{mesh}(P_n') = \lim_{n\to\infty} \text{mesh}(Q_n) = 0$, so by Theorem 5.6.7,

$$\int_a^c f(x)\,dx = \lim_{n\to\infty} (R_n + R_n') = \lim_{n\to\infty} R_n + \lim_{n\to\infty} R_n' = \int_a^b f(x)\,dx + \int_b^c f(x)\,dx.$$

$\qquad\qquad\square$

Exercises 5.6

1. Prove that if f is uniformly continuous on an interval I, then f is continuous on I.

2. In this exercise you will use Theorem 5.6.7 to evaluate $\int_1^4 \sqrt{x}\, dx$.

   (a) Show that for every integer $n \geq 2$,
   $$1 < \sqrt[n]{4} < 1 + \frac{3}{n},$$
   and use this fact to show that as $n \to \infty$, $\sqrt[n]{4} \to 1^+$. (Hint: Apply the mean value theorem to the function $f(x) = \sqrt[n]{x}$ on the interval $[1, 4]$.)

   Suppose n is a positive integer. For $0 \leq i \leq n$ let
   $$x_i = 4^{i/n} = (\sqrt[n]{4})^i.$$

   Notice that $x_0 = 4^0 = 1$, $x_n = 4^1 = 4$, and $x_0 < x_1 < \cdots < x_n$. Let $P_n = \{x_0, x_1, \ldots, x_n\}$, which is a partition of $[1, 4]$. Note that

   $$\Delta x_i = x_i - x_{i-1} = (\sqrt[n]{4})^i - (\sqrt[n]{4})^{i-1} = (\sqrt[n]{4})^{i-1}(\sqrt[n]{4} - 1) = (\sqrt[n]{4})^i \cdot \frac{\sqrt[n]{4} - 1}{\sqrt[n]{4}}.$$

   Let $x_i^* = x_i$, and let R_n be the Riemann sum

   $$R_n = \sum_{i=1}^{n} \sqrt{x_i^*}\, \Delta x_i.$$

   (b) Show that
   $$R_n = 7 \cdot \frac{(\sqrt[n]{4})^{3/2} - (\sqrt[n]{4})^{1/2}}{(\sqrt[n]{4})^{3/2} - 1}.$$

   (Hint: Use Exercise 13 from Section 5.1. If you didn't do that exercise, you can find the answer in equation (10.3).)

   (c) Show that $\lim_{n \to \infty} \text{mesh}(P_n) = 0$. It follows, by Theorem 5.6.7, that $\lim_{n \to \infty} R_n = \int_1^4 \sqrt{x}\, dx$.

   (d) Compute $\lim_{n \to \infty} R_n$. (Hint: First compute $\lim_{x \to 1^+} (x^{3/2} - x^{1/2})/(x^{3/2} - 1)$.)

   (e) Use the second fundamental theorem of calculus to compute $\int_1^4 \sqrt{x}\, dx$, and verify that your answer matches your answer to the previous part.

3. Use Definition 5.6.1 (not Theorem 5.6.2) to show that the function $f(x) = x^2$ is uniformly continuous on the interval $[0, 1]$.

4. Let $f(x) = 1/x$. Show that f is uniformly continuous on the interval $[1, \infty)$.

5. Let $f(x) = 1/x$. We already know that f is continuous on $(0, \infty)$. Show that f is not uniformly continuous on $(0, \infty)$. (Hint: Let $\epsilon = 1$. Show that there is no acceptable response δ for this value of ϵ. To do this, show that for every $\delta > 0$ there are numbers s and t in $(0, \infty)$ such that $|s - t| < \delta$ but $|f(s) - f(t)| \geq 1$.)

Chapter 6

Applications of Integration

6.1 Area Between Curves

In Section 5.2, we saw several examples of quantities that could be computed as limits of Riemann sums. In each case, the quantity could be approximated by breaking it into n pieces for some positive integer n, approximating the contribution of each piece, and then adding up these approximate contributions to get a Riemann n-sum. And the approximation could be improved by breaking the quantity into a larger number of pieces—that is, by increasing n. We were then able to get the exact answer by finding the limit of the Riemann n-sums as $n \to \infty$. Such a limit is, by definition, a definite integral, and therefore it can be computed by the methods we developed in the last chapter.

In this chapter we will study a number of quantities that can be computed as definite integrals. It is probably easiest to understand this idea for geometric quantities, where we can think of physically cutting a geometric object into pieces. We therefore begin with several examples of geometric quantities that can be computed by integration.

We have already seen one such geometric example: the area under a curve $y = f(x)$ for $a \le x \le b$. We assume here that f is continuous on $[a, b]$ and $f(x) \ge 0$ for all $x \in [a, b]$. Let R denote the region under the curve, which consists of all points (x, y) such that $a \le x \le b$ and $0 \le y \le f(x)$, and let A denote its area.[1] Recall that to approximate A, we first choose a positive integer n and cut the interval $[a, b]$ into

---

[1] We will base our reasoning in this chapter on intuitive conceptions of the relevant quantities—in this case, area. In more advanced books, this reasoning would be justified using precise mathematical definitions.

n pieces, each of width $\Delta x = (b-a)/n$. The dividing points between these pieces are the numbers $x_i = a + i\Delta x$ for $i = 0, 1, 2, \ldots, n$, and for $1 \leq i \leq n$ the ith piece is the interval $[x_{i-1}, x_i]$. The lines $x = x_i$ cut the region R into n vertical strips. If we let A_i denote the area of the ith strip, which is the region defined by the inequalities $x_{i-1} \leq x \leq x_i$ and $0 \leq y \leq f(x)$, then $A = A_1 + A_2 + \cdots + A_n = \sum_{i=1}^{n} A_i$. This is illustrated in Figure 6.1a.

Next we choose a sample point x_i^* in each interval $[x_{i-1}, x_i]$. If the value of the function f doesn't vary much over the interval $[x_{i-1}, x_i]$, then the curve $y = f(x)$ will stay close to the horizontal line $y = f(x_i^*)$ on that interval, and therefore A_i will be approximately equal to the area of the rectangle $x_{i-1} \leq x \leq x_i$, $0 \leq y \leq f(x_i^*)$. This rectangle has height $f(x_i^*)$ and width Δx, and therefore area $f(x_i^*)\Delta x$. Adding these approximations for all of the pieces, we see that A is approximately equal to the Riemann n-sum

$$R_n = \sum_{i=1}^{n} f(x_i^*)\Delta x.$$

This Riemann sum represents the sum of the areas of the blue rectangles in Figure 6.1b, which is approximately equal to the area of the gray region in Figure 6.1a.

As n increases, the rectangles in Figure 6.1b get narrower, their tops follow the shape of the curve $y = f(x)$ more closely, and the area inside the rectangles approaches the area of R. We conclude that

$$A = \lim_{n \to \infty} R_n = \int_a^b f(x)\, dx.$$

For a more careful justification of this equation, see Exercise 14.

Now we would like to generalize this problem. Suppose that f and g are both continuous on the interval $[a, b]$, and for all $x \in [a, b]$, $g(x) \leq f(x)$. We want to compute the area A of the region R between the graphs of f and g for $x \in [a, b]$, that is, the area of the region defined by the inequalities $a \leq x \leq b$ and $g(x) \leq y \leq f(x)$. If g were the constant function $g(x) = 0$, then the bottom of R would be the x-axis and this would be the same as the problem we just solved. Imitating our previous solution, we divide the interval $[a, b]$ into n pieces of width $\Delta x = (b-a)/n$, with division points $x_i = a + i\Delta x$ for $i = 0, 1, 2, \ldots, n$. Once again, the lines $x = x_i$ will divide R into n vertical strips. An example of such a region is shown in Figure 6.2a.

In the previous problem, the top of each strip was a piece of the curve $y = f(x)$, and we approximated the area of the strip by flattening the top to get a rectangle. But in our new problem, both the top and the bottom of each strip are curved, so we will have to flatten both of them to get a rectangular approximation. As before, for $1 \leq i \leq n$ we choose a sample point $x_i^* \in [x_{i-1}, x_i]$. We now approximate the area of the ith strip with the area of the rectangle $x_{i-1} \leq x \leq x_i$, $g(x_i^*) \leq y \leq f(x_i^*)$, as shown in Figure 6.2b. This rectangle has height $f(x_i^*) - g(x_i^*)$ and width Δx, so its area is $(f(x_i^*) - g(x_i^*))\Delta x$. (Notice that to find a vertical distance—in this case, the height of the rectangle—we compute a difference of two y-coordinates: the top y-coordinate

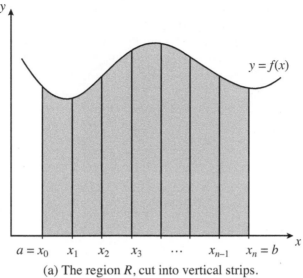

(a) The region R, cut into vertical strips.

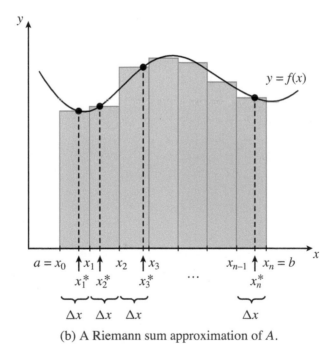

(b) A Riemann sum approximation of A.

Figure 6.1: Approximating the area under $y = f(x)$ for $a \le x \le b$ with a Riemann sum.

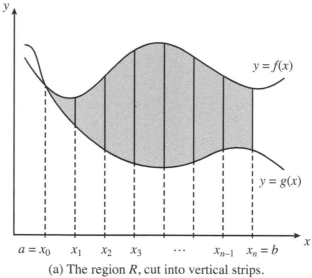

(a) The region R, cut into vertical strips.

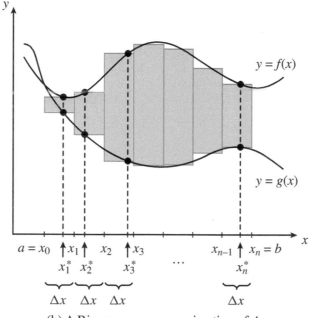

(b) A Riemann sum approximation of A.

Figure 6.2: Approximating the area between $y = f(x)$ and $y = g(x)$ for $a \leq x \leq b$ with a Riemann sum.

minus the bottom.) Adding up all of these approximations, we see that A is approximately equal to the sum

$$R_n = \sum_{i=1}^{n} (f(x_i^*) - g(x_i^*)) \Delta x.$$

Finally, as usual we get the exact answer by taking the limit as $n \to \infty$: $A = \lim_{n \to \infty} R_n$.

This calculation looks similar to the one we did in the previous problem, but is R_n a Riemann sum? If we let $h = f - g$, then we can write

$$R_n = \sum_{i=1}^{n} (f(x_i^*) - g(x_i^*)) \Delta x = \sum_{i=1}^{n} h(x_i^*) \Delta x,$$

and we see that R_n is a Riemann n-sum for the function h on the interval $[a, b]$. Thus,

$$A = \lim_{n \to \infty} R_n = \int_a^b h(x)\, dx = \int_a^b (f(x) - g(x))\, dx. \tag{6.1}$$

The lesson here is that if we have a sum of the form

$$\sum_{i=1}^{n} (\text{some expression involving } x_i^*) \cdot \Delta x,$$

then the sum is a Riemann sum for some function. In future examples, we won't bother to introduce a name for the function; we'll simply move from a limit of a sum to a definite integral in one step. For example, we could write the calculation we just completed like this:

$$\lim_{n \to \infty} \sum_{i=1}^{n} (f(x_i^*) - g(x_i^*)) \Delta x = \int_a^b (f(x) - g(x))\, dx.$$

Comparing the notation on the two sides of this equation, we see that when taking a limit of a Riemann sum, we can just change the summation sign to an integral sign, fill in the endpoints of the interval that was divided into pieces as the limits of integration, drop the subscript i and superscript $*$, and change the Δ to a d.

Another way to derive the formula for the area of the region R in Figure 6.2a would be to note that it is the area under the graph of f minus the area under the graph of g, so it can be computed by the formula

$$\int_a^b f(x)\, dx - \int_a^b g(x)\, dx = \int_a^b (f(x) - g(x))\, dx.$$

However, this derivation assumes that $g(x) \geq 0$ for all $x \in [a, b]$. Our original derivation, in contrast, works whether $g(x)$ is positive or negative, as long as $g(x) \leq f(x)$ for $x \in [a, b]$.

Example 6.1.1. Find the areas of the following regions:

(a) The region between the curves $y = 2x - 1$ and $y = x^2 + 1$ for $1 \leq x \leq 2$.
(b) The region bounded by the curves $y = x^2 - x$ and $y = 3x - x^2$.
(c) The region bounded by the curves $y = \sin x$ and $y = \sin 2x$, $0 \leq x \leq \pi$.
(d) The region bounded by the curves $x = y^2$ and $y = x - 2$.

Solution. The region in part (a) is shown in Figure 6.3. Since $(x^2 + 1) - (2x - 1) = x^2 - 2x + 2 = (x - 1)^2 + 1 \geq 1$, the curve $y = x^2 + 1$ is always above the curve $y = 2x - 1$, so the requested area is

$$\int_1^2 ((x^2 + 1) - (2x - 1))\, dx = \int_1^2 (x^2 - 2x + 2)\, dx = \left[\frac{x^3}{3} - x^2 + 2x\right]_1^2 = \frac{4}{3}.$$

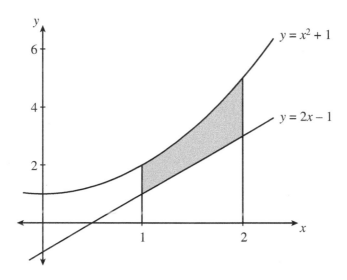

Figure 6.3: The region in part (a) of Example 6.1.1.

For part (b), setting $x^2 - x = 3x - x^2$ and solving, we find that the curves cross at $x = 0$ and $x = 2$. The phrase "bounded by" means "completely enclosed by," so the region is the one that is shaded in Figure 6.4. The fact that the region extends below the x-axis is irrelevant; since we have $x^2 - x \leq 3x - x^2$ for $0 \leq x \leq 2$, we can use equation (6.1) to compute the area of the region. The area is therefore

$$\int_0^2 ((3x - x^2) - (x^2 - x))\, dx = \int_0^2 (4x - 2x^2)\, dx = \left[2x^2 - \frac{2x^3}{3}\right]_0^2 = \frac{8}{3}.$$

The region in part (c) is shown in Figure 6.5. To find where the curves cross, we set $\sin 2x = \sin x$ and use the identity $\sin 2x = 2 \sin x \cos x$:

$$\sin 2x = \sin x,$$

$$2 \sin x \cos x - \sin x = 0,$$

$$\sin x (2 \cos x - 1) = 0,$$
$$\sin x = 0 \text{ or } \cos x = 1/2,$$
$$x = 0, \pi/3, \pi.$$

Since $\sin x \leq \sin 2x$ for $0 \leq x \leq \pi/3$ and $\sin 2x \leq \sin x$ for $\pi/3 \leq x \leq \pi$, we have to use two integrals to compute the area:

$$\text{Area} = \int_0^{\pi/3} (\sin 2x - \sin x)\, dx + \int_{\pi/3}^{\pi} (\sin x - \sin 2x)\, dx$$
$$= \left[-\frac{1}{2} \cos 2x + \cos x \right]_0^{\pi/3} + \left[-\cos x + \frac{1}{2} \cos 2x \right]_{\pi/3}^{\pi} = \frac{1}{4} + \frac{9}{4} = \frac{5}{2}.$$

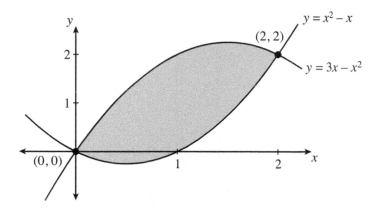

Figure 6.4: The region in part (b) of Example 6.1.1.

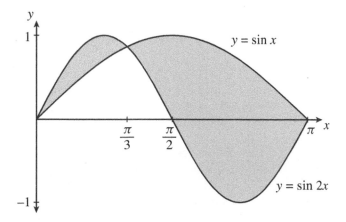

Figure 6.5: The region in part (c) of Example 6.1.1.

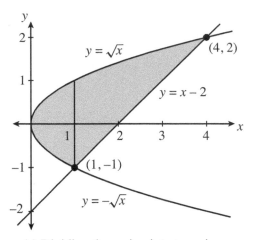

(a) Dividing the region into two pieces
with the line $x = 1$.

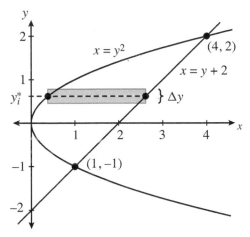

(b) Approximating a horizontal slice with
a rectangle.

Figure 6.6: The region in part (d) of Example 6.1.1.

Finally, the region in part (d) is shown in Figure 6.6a. We leave it to you to verify
that the curves intersect at the points $(1, -1)$ and $(4, 2)$, as shown in the figure. It is
possible to compute the area of this region using equation (6.1), but it requires two
integrations. If we rewrite the curve $x = y^2$ as $y = \pm\sqrt{x}$ and then divide the region into
two pieces with the line $x = 1$, as indicated in the figure, then we can compute the areas
of these two pieces separately and add them together to get the answer. You are asked to
work out the details of this calculation in Exercise 13.

However, it is possible to find the requested area with only one integral if we
proceed somewhat differently. Instead of dividing the region into thin vertical strips,

as we did before, we will divide it into horizontal strips. To do this, we reverse the roles of x and y in our earlier derivation. Rewriting the line $y = x - 2$ in the form $x = y + 2$, we see that the region whose area is to be computed can be defined by the inequalities $-1 \leq y \leq 2$, $y^2 \leq x \leq y + 2$. We now divide the interval $[-1, 2]$ on the y-axis into n pieces, each of size $\Delta y = 3/n$. The division points are $y_i = -1 + i\Delta y$ for $0 \leq i \leq n$. The lines $y = y_i$ divide the region into n horizontal strips, and we can approximate the area of each strip with the area of a rectangle. Rather than drawing all of these rectangles, we have drawn just one representative rectangle in Figure 6.6b; this is all we will need to work out the required formula. Since the left side of the rectangle passes through the indicated point on the curve $x = y^2$, the x-coordinate of the left side of the rectangle is $(y_i^*)^2$. Similarly, the x-coordinate of the right side of the rectangle is $y_i^* + 2$. Thus, the rectangle has width $y_i^* + 2 - (y_i^*)^2$ and height Δy, so its area is $(y_i^* + 2 - (y_i^*)^2)\Delta y$. (The width of the rectangle is a horizontal distance, which we compute as a difference of two x-coordinates: right minus left.) Adding up the areas of all the rectangles and then taking the limit as $n \to \infty$, we see that the requested area is

$$\lim_{n \to \infty} \sum_{i=1}^{n} (y_i^* + 2 - (y_i^*)^2)\Delta y = \int_{-1}^{2} (y + 2 - y^2)\, dy$$

$$= \left[\frac{y^2}{2} + 2y - \frac{y^3}{3} \right]_{-1}^{2} = \frac{9}{2}. \qquad \square$$

Exercises 6.1

1–12: Find the area of the region.

1. The region between the curves $y = x^3 + x^2$ and $y = x^3 - 1$ for $0 \leq x \leq 3$.

2. The region between the curves $y = 1/x^2$ and $y = 1/x^3$ for $1 \leq x \leq 2$.

3. The region bounded by the curves $y = x^2$ and $y = x^3$.

4. The region bounded by the curves $y = 1/2$ and $y = \sin x$, $\pi/6 \leq x \leq 5\pi/6$.

5. The region bounded by the curves $y = \sin x$ and $y = \cos x$, $\pi/4 \leq x \leq 5\pi/4$.

6. The region bounded by the curve $x = y^2 - 1$ and the y-axis.

7. The region bounded by the curves $y = \sqrt{x}$ and $y = x/2$.

8. The region bounded by the curves $y = x^3 - x$ and $y = 3x$.

9. The region bounded by the curves $x = y^2 + y$ and $x + y = 3$.

10. The region bounded by the curves $y = x$, $y = 2x$, and $y = 6 - x$.

11. The region bounded by the curves $y = x^2$ and $x = y^2$.

12. The region bounded by the curves $x = 2y - y^2$ and $y = 3x - 2x^2$. (Hint: First confirm that the curves cross at $(0, 0)$ and $(1, 1)$, and no other points. Then use the line $y = x$ to divide the region into two pieces, and compute the areas of the two pieces separately.)

13. Compute the area of the region in part (d) of Example 6.1.1 by using the line $x = 1$ to divide the region into two pieces as indicated in Figure 6.6a, finding the area of each piece with an integral with respect to x, and then adding the areas of the two pieces.

14. Suppose that f is continuous on $[a, b]$ and $f(x) \geq 0$ for all $x \in [a, b]$. Let R denote the region under the graph of f, and let A be the area of R. In this exercise you will give a more careful justification for the equation $A = \int_a^b f(x)\, dx$. As usual, we begin by dividing the interval $[a, b]$ into n pieces of width $(b-a)/n$, with division points $x_i = a + i\Delta x$.

   (a) Show that for each n, there is a choice of sample points $x_i^* \in [x_{i-1}, x_i]$ for which the rectangles in Figure 6.1b lie entirely inside the region R. It follows that if we let m_n denote the resulting Riemann n-sum, then $m_n \leq A$.

   (b) Show that for each n, there is a choice of sample points $x_i^* \in [x_{i-1}, x_i]$ for which the rectangles in Figure 6.1b completely cover the region R. It follows that if we let M_n denote the resulting Riemann n-sum, then $M_n \geq A$.

   (c) Use parts (a) and (b), together with the fact that $\lim_{n\to\infty} m_n = \lim_{n\to\infty} M_n = \int_a^b f(x)\, dx$, to conclude that $\int_a^b f(x)\, dx = A$.

6.2 Volume by Disks, Washers, and Slices

Consider again the region R under the graph of a continuous, nonnegative function f on an interval $[a, b]$. In this section we will be interested in computing, not the area of R, but rather the volume of the solid generated when the region R is rotated about the x-axis.

To visualize this solid, imagine that the x-axis is a wire and the region R is a piece of paper attached to the wire. Now imagine twirling the x-axis wire between your fingers. The piece of paper will spin around the x-axis, sweeping out a three-dimensional solid S, as illustrated in Figure 6.7. Let V denote the volume of S. We want to compute V.

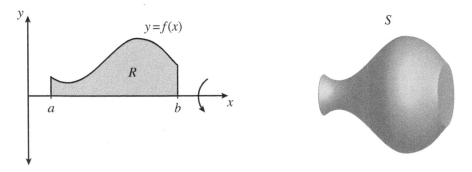

Figure 6.7: When the region R is rotated about the x-axis, it generates the solid S.

As before, we divide the interval $[a, b]$ into n pieces, each of width $\Delta x = (b - a)/n$, with division points $x_i = a + i \Delta x$. The lines $x = x_i$ once again divide R into thin vertical strips, but now we want to know how each strip contributes to the solid S. Figure 6.8 shows the ith strip, and the slice of the solid S that it generates when it is rotated about the x-axis. If we let V_i be the volume of this slice of S, then $V = \sum_{i=1}^{n} V_i$.

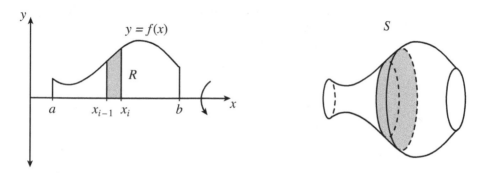

Figure 6.8: The ith vertical strip in R generates a slice of S.

To approximate V_i, we choose a sample point $x_i^* \in [x_{i-1}, x_i]$, and as before we flatten the top of the ith vertical strip of R to get the rectangle $x_{i-1} \le x \le x_i, 0 \le y \le f(x_i^*)$. When this rectangle is rotated about the x-axis, it generates a disk, as illustrated in Figure 6.9. We will use the volume of this disk as our approximation of V_i. The radius of the disk is the same as the height of the rectangle, which is $f(x_i^*)$, and its width is the same as the width of the rectangle, Δx. Thus, the volume of the disk is $\pi (f(x_i^*))^2 \Delta x$. Using this as our approximation of V_i, and adding up the contributions of all of the slices, we get

$$V = \sum_{i=1}^{n} V_i \approx \sum_{i=1}^{n} \pi (f(x_i^*))^2 \Delta x.$$

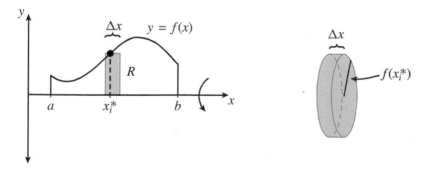

Figure 6.9: We approximate the ith vertical strip with a rectangle, which generates a disk.

Finally, to get the exact answer we let $n \to \infty$. The last sum above is a Riemann sum, so its limit is a definite integral:

$$V = \lim_{n \to \infty} \sum_{i=1}^{n} \pi(f(x_i^*))^2 \Delta x = \int_a^b \pi(f(x))^2 \, dx. \tag{6.2}$$

Example 6.2.1. Find the volumes of the solids generated when the regions below the following curves are rotated about the x-axis:

$$\text{(a) } y = 1/x, \ 1 \le x \le 2, \qquad \text{(b) } y = x^2, \ 0 \le x \le 1.$$

Solution. The region in part (a) is shown on the left in Figure 6.10, and the solid generated when this region is rotated about the x-axis is shown in the center. As usual, we slice the region into vertical strips and approximate each strip with a rectangle. The ith rectangle is shown in blue on the left in the figure, and it generates the disk on the right. The volume of this disk is $\pi(1/x_i^*)^2 \Delta x$. Adding up the volumes of these disks and letting $n \to \infty$, we find that the volume of the solid is

$$\lim_{n \to \infty} \sum_{i=1}^{n} \frac{\pi}{(x_i^*)^2} \Delta x = \int_1^2 \frac{\pi}{x^2} \, dx = -\frac{\pi}{x} \Big|_1^2 = \frac{\pi}{2}.$$

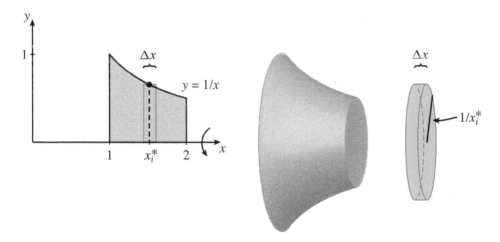

Figure 6.10: Computing the volume of the solid in part (a) of Example 6.2.1.

Figure 6.11 shows similar information for part (b). This time we are going to skip the Riemann sum and go straight to the definite integral. Since we know that the subscript i and superscript $*$ will be dropped when we set up the integral, we have not bothered to include them in Figure 6.11. Thus, the sample point for the rectangle on the left in the figure is simply called x, and the volume of the disk on the right is $\pi(x^2)^2 \Delta x = \pi x^4 \Delta x$.

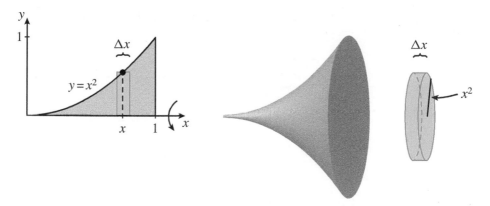

Figure 6.11: Computing the volume of the solid in part (b) of Example 6.2.1.

Therefore the volume of the solid is

$$\int_0^1 \pi x^4 \, dx = \left. \frac{\pi x^5}{5} \right|_0^1 = \frac{\pi}{5}. \qquad \square$$

In our solution to Example 6.2.1, rather than simply applying equation (6.2), we have chosen in each case to draw a picture from which the required definite integral could be worked out. One reason for doing this is that it makes it easier to remember the formula for the integral. But a more important reason is that this method makes it possible to compute a wide range of volumes that are similar to those we have computed, but that don't exactly fit the pattern of the derivation of equation (6.2). That equation doesn't apply directly to the volumes in our next example, but they can be computed by imitating the method we used in Example 6.2.1.

Example 6.2.2. Find the volumes of the solids generated when the following regions are rotated about the specified lines:

(a) The region bounded by $y = x^2$ and $y = 1$, rotated about the line $y = 1$.
(b) The region bounded by $y = x^2$ and $y = 1$, rotated about the x-axis.
(c) The region bounded by $y = x^3$, $y = 1$, and the y-axis, rotated about the y-axis.

Solution. Setting $x^2 = 1$ and solving, we find that the curves $y = x^2$ and $y = 1$ cross at $x = \pm 1$. Thus, the region in part (a) is the one shown in Figure 6.12. We are asked to rotate this region about the line $y = 1$, not the x-axis, so formula (6.2) doesn't apply, but we will be able to work out the required integral from the picture. As usual, we approximate a vertical strip of this region with a rectangle, and as in part (b) of Example 6.2.1, the label we have assigned to the sample point for this rectangle is simply x, rather than x_i^*. When this rectangle is rotated about the line $y = 1$, it generates the disk shown on the right in Figure 6.12. The radius of the disk is the height of the rectangle, which is $1 - x^2$, and the width is Δx, so its volume is $\pi(1 - x^2)^2 \Delta x$. Adding up the volumes of these

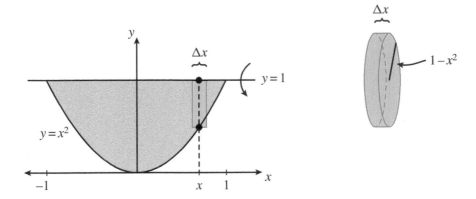

Figure 6.12: Computing the volume of the solid in part (a) of Example 6.2.2.

disks and then taking the limit as the number of disks approaches ∞, we conclude that the volume of the solid is

$$\int_{-1}^{1} \pi(1-x^2)^2 \, dx = \pi \int_{-1}^{1} (1-2x^2+x^4) \, dx = \pi \left[x - \frac{2x^3}{3} + \frac{x^5}{5} \right]_{-1}^{1} = \frac{16\pi}{15}.$$

Part (b) involves the same region, but this time it is rotated about the x-axis. You should be able to visualize that when a rectangular approximation of a vertical strip of this region is rotated about the x-axis, it generates a disk with a hole in it, as illustrated in Figure 6.13. This shape is usually called a *washer*, because it resembles the kind of washer that is used with a bolt and a nut. The volume of the washer can be computed

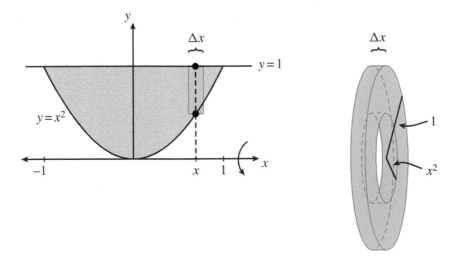

Figure 6.13: Computing the volume of the solid in part (b) of Example 6.2.2.

by determining what the volume would be if it were a disk with no hole in it, and then subtracting the volume of the hole:

$$\text{volume of washer} = \pi(1)^2 \Delta x - \pi(x^2)^2 \Delta x = \pi(1 - x^4)\Delta x.$$

As usual, we can now compute the volume of the entire solid by using a definite integral to add up the contributions of all of the slices:

$$\text{volume of solid} = \int_{-1}^{1} \pi(1 - x^4)\,dx = \pi\left[x - \frac{x^5}{5}\right]_{-1}^{1} = \frac{8\pi}{5}.$$

Finally, the region in part (c) must be rotated about the y-axis. To get disk-shaped slices, we must cut this region into horizontal strips, so we will need to treat y as the independent variable in this problem. Rewriting the curve $y = x^3$ in the form $x = \sqrt[3]{y}$, we see that the region can be defined by the inequalities $0 \le y \le 1$ and $0 \le x \le \sqrt[3]{y}$. A typical horizontal rectangle, and the disk it generates, are shown in Figure 6.14. The volume of the disk is $\pi(\sqrt[3]{y})^2\Delta y = \pi y^{2/3}\Delta y$, so the volume of the entire solid is

$$\int_{0}^{1} \pi y^{2/3}\,dy = \left.\frac{3\pi y^{5/3}}{5}\right|_{0}^{1} = \frac{3\pi}{5}. \qquad\qquad \square$$

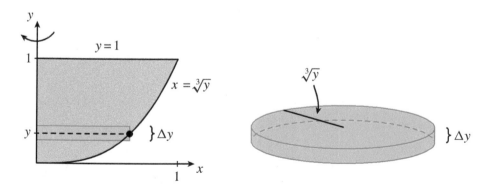

Figure 6.14: Computing the volume of the solid in part (c) of Example 6.2.2.

It may seem that the problems we have been solving in this section are rather contrived; when would one ever want to rotate a region around an axis and compute the resulting volume? But a number of important volumes can be computed by this method.

For example, you probably know that the volume of a sphere of radius r is $4\pi r^3/3$, but do you know where this formula comes from? We can generate a sphere of radius r by rotating the region below the semicircle $y = \sqrt{r^2 - x^2}$, $-r \le x \le r$ about the x-axis,

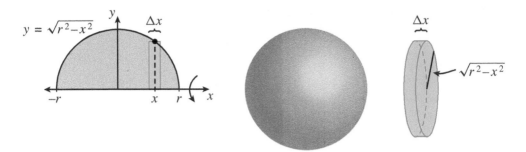

Figure 6.15: Computing the volume of a sphere of radius r.

as shown in Figure 6.15. The volume of the disk on the right in the figure is

$$\pi(\sqrt{r^2 - x^2})^2 \Delta x = \pi(r^2 - x^2)\Delta x,$$

and therefore the volume of the sphere is

$$\int_{-r}^{r} \pi(r^2 - x^2)\, dx = \pi\left[r^2 x - \frac{x^3}{3}\right]_{-r}^{r} = \frac{4\pi r^3}{3}.$$

Notice that in this integral, x is the independent variable and r is a constant.

Similarly, we can generate a cone of radius r and height h by drawing a line segment connecting the points $(0, h)$ and $(r, 0)$ in the plane and then rotating the region under this line segment about the y-axis. This is illustrated in Figure 6.16. The equation of the line through the points $(0, h)$ and $(r, 0)$ is $y = h - (h/r)x$, or equivalently $x = (r/h)(h - y)$, and therefore the volume of the disk shown on the right in the figure is

$$\pi\left(\frac{r}{h}(h - y)\right)^2 \Delta y.$$

Figure 6.16: Computing the volume of a cone of radius r and height h.

We conclude that the volume of the cone is

$$\int_0^h \frac{\pi r^2}{h^2}(h-y)^2\,dy = \frac{\pi r^2}{h^2}\int_0^h (h^2 - 2hy + y^2)\,dy$$

$$= \frac{\pi r^2}{h^2}\left[h^2 y - hy^2 + \frac{y^3}{3}\right]_0^h = \frac{\pi r^2 h}{3}.$$

In all of our examples in this section, we have found the volume of a solid by cutting it into slices, approximating the volumes of the slices, and then adding up these approximations. For the solids we have worked with so far, the slices were round, and therefore we were able to approximate the volume of each slice with the volume of a disk or a washer. But the method of slicing and adding can be applied even if the slices are not round. To demonstrate this, we close this section with one more example. We will find the volume V of a pyramid whose base is a square with sides of length s and whose height is h.

Figure 6.17 shows the pyramid whose volume is to be computed. To cut the pyramid into slices, we set up a vertical number line next to the pyramid. We will find it convenient to have the origin of the number line level with the top of the pyramid, with the numbers increasing as we move *down* the line, so that the point h on the number line is level with the bottom of the pyramid. Now we cut the interval $[0, h]$ on the number line into n pieces, each of size $\Delta x = h/n$, with division points $x_i = i\Delta x$ for $0 \le i \le n$. Horizontal planes passing through these division points will cut the pyramid into n slices. Figure 6.17 shows the ith slice. If we let V_i denote the volume of this slice, then $V = \sum_{i=1}^n V_i$.

The sides of slice i are slanted. To approximate V_i, we will flatten the sides of the slice to make them vertical. More precisely, we choose a sample point x_i^* in the interval $[x_{i-1}, x_i]$, and consider the horizontal plane passing though the point x_i^* on the number line. This plane intersects the pyramid in a square. Let w_i be the length of the sides of this square, as shown in Figure 6.18. We approximate V_i with the volume of the box on the right in the figure. This box has height Δx and a square base with sides of length w_i, so $V_i \approx (w_i)^2 \Delta x$, and therefore

$$V = \sum_{i=1}^n V_i \approx \sum_{i=1}^n (w_i)^2 \Delta x.$$

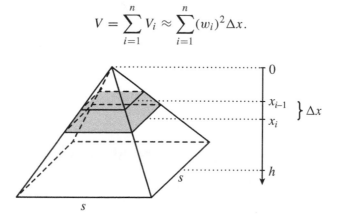

Figure 6.17: To find the volume of a pyramid, we cut it into horizontal slices.

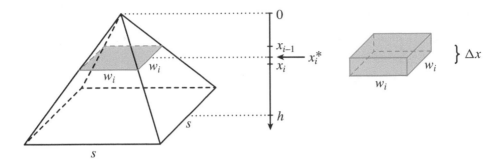

Figure 6.18: We approximate the ith slice with a box with vertical sides.

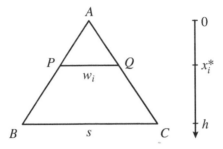

Figure 6.19: A vertical cross section through the center of the pyramid.

To turn this last sum into a Riemann sum, we will need a formula for w_i in terms of x_i^*. Consider the vertical cross section through the center of the pyramid shown in Figure 6.19. Triangles ABC and APQ are similar. Since $\triangle ABC$ has base s and height h, and $\triangle APQ$ has base w_i and height x_i^*, we have $s/h = w_i/x_i^*$, and therefore $w_i = sx_i^*/h$.

Substituting into our approximation for V, we find that

$$V \approx \sum_{i=1}^{n}(w_i)^2 \Delta x = \sum_{i=1}^{n}\frac{s^2(x_i^*)^2}{h^2}\Delta x.$$

Since the last sum has the form of a Riemann sum, when we take the limit as $n \to \infty$ to find the exact volume we get a definite integral:

$$V = \lim_{n\to\infty}\sum_{i=1}^{n}\frac{s^2(x_i^*)^2}{h^2}\Delta x = \int_{0}^{h}\frac{s^2 x^2}{h^2}\,dx = \frac{s^2 x^3}{3h^2}\Big|_{0}^{h} = \frac{s^2 h}{3}.$$

Exercises 6.2

1–11: Find the volume of the solid generated when the region is rotated about the specified line.

1. The region under the curve $y = \sqrt{x}$, $0 \le x \le 4$, rotated about the x-axis.

2. The region bounded by the line $y = 3 - x$ and the x- and y-axes, rotated about the x-axis.

3. The region bounded by the line $y = 3 - x$ and the x- and y-axes, rotated about the line $y = 3$.

4. The region under the curve $y = \tan x$, $0 \le x \le \pi/4$, rotated about the x-axis. (Hint: To evaluate the integral, use the identity $\tan^2 x = \sec^2 x - 1$.)

5. The region under the curve $y = 1/(3 - x)$, $0 \le x \le 2$, rotated about the x-axis.

6. The region under the curve $y = \sqrt{x^2 - 1}$, $1 \le x \le 4$, rotated about the x-axis.

7. The region bounded by the curves $x = y^2$ and $x = 3 - 2y$, rotated about the y-axis.

8. The region bounded by the curves $y = \sqrt{x}$ and $y = x/2$, rotated about the x-axis.

9. The region bounded by the curves $y = \sqrt{x}$ and $y = x/2$, rotated about the y-axis.

10. The region bounded by the curves $x = y^2$ and $y = x^3$, rotated about the x-axis.

11. The region bounded by the curves $x = y^2$ and $y = x^3$, rotated about the line $x = 2$.

12. Let a and b be positive constants. When the top half of the region inside the ellipse $x^2/a^2 + y^2/b^2 = 1$ is rotated about the x-axis, it generates a *solid ellipsoid*.

    (a) Find the volume of the solid ellipsoid.

    (b) When $a = b$, the ellipsoid is a sphere. Verify that in this case your answer to part (a) agrees with the formula for the volume of a sphere.

13. Let a and b be constants with $b > a > 0$. When the disk $x^2 + y^2 \le a^2$ is rotated about the line $x = b$, it generates a *solid torus*.

    (a) Set up an integral for the volume of the solid torus.

    (b) By giving the integral in part (a) a different geometric interpretation, find the volume of the solid torus.

    (c) Show that the volume of the solid torus is equal to the area of the disk times the distance traveled by the center of the disk when it rotates about the line $x = b$. This is an instance of *Pappus's second centroid theorem* (see Theorem 6.5.1).

14. The region under the curve $y = f(x)$, $1 \le x \le 4$ has area 3, and the volume of the solid generated when this region is rotated about the x-axis is 8. Find the volume of the solid generated when the region is rotated about the line $y = -2$. (Note: Although no formula for $f(x)$ has been given, there is enough information to solve the problem.)

15. Two cylinders of radius r meet at right angles. Find the volume of their intersection.

16. A board in the shape of an equilateral triangle, with each edge of length 1 ft, is wedged into the corner of a room at a slant so that one edge of the triangle rests on the floor and the other two edges are against the two walls. Find the volume of the region underneath the board.

17. You may have noticed some similarity between the formulas we derived in this section for the volumes of a cone and a pyramid: in both cases, the volume is equal to the area of the base of the solid times the height divided by 3. This is not a coincidence. Let R be a region in the plane with area A, and let P be a point h units above the plane. Let S be the solid consisting of all points that lie on a line segment connecting P to some point in R. Show that the volume of S is $Ah/3$.

6.3 Volume by Cylindrical Shells

In this section we will study one more method for computing volumes. Once again we consider the region R under the curve $y = f(x)$ for $a \le x \le b$; we assume that $f(x) \ge 0$ and $a \ge 0$, so that R is contained in the first quadrant. As usual, we divide the interval $[a, b]$ into n pieces of width $\Delta x = (b - a)/n$, using division points $x_i = a + i\Delta x$, and we cut R into vertical strips with the lines $x = x_i$. We then choose sample points $x_i^* \in [x_{i-1}, x_i]$, and approximate the ith vertical strip with the rectangle $x_{i-1} \le x \le x_i, 0 \le y \le f(x_i^*)$.

As in the last section, to find the volume of the solid generated when the region R is rotated about an axis, we add up approximations of the contributions that all of the vertical strips make to this volume. We approximate the contribution of the ith strip by computing the volume of the solid generated when we rotate the ith rectangle about the axis. If the region R is rotated about the x-axis, then the ith rectangle generates a disk, and we can use the methods of the last section to compute the volume. But what if R is rotated about the y-axis? In that case, the ith rectangle generates a thin cylindrical shell, as illustrated in Figure 6.20.

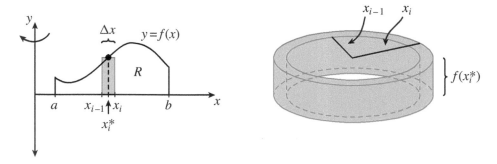

Figure 6.20: When a vertical rectangle is rotated about the y-axis, it generates a cylindrical shell.

When computing the volume of the cylindrical shell in Figure 6.20, we find it convenient to take the sample point x_i^* to be the midpoint of the interval $[x_{i-1}, x_i]$; in other words, we let $x_i^* = (x_{i-1} + x_i)/2$. The shell can be thought of as a cylinder with a slightly smaller cylindrical hole in it, so we compute its volume as the volume of the cylinder minus the volume of the hole:

$$\text{volume of cylindrical shell} = \pi x_i^2 f(x_i^*) - \pi x_{i-1}^2 f(x_i^*)$$
$$= \pi (x_i^2 - x_{i-1}^2) f(x_i^*)$$
$$= \pi (x_i + x_{i-1})(x_i - x_{i-1}) f(x_i^*)$$
$$= 2\pi \frac{x_{i-1} + x_i}{2} f(x_i^*)(x_i - x_{i-1})$$
$$= 2\pi x_i^* f(x_i^*) \Delta x.$$

As usual, we now add up these approximations of the contributions of all of the strips and then take the limit as $n \to \infty$. Thus, the volume of the solid generated when the region R in Figure 6.20 is rotated about the y-axis is

$$\lim_{n \to \infty} \sum_{i=1}^{n} 2\pi x_i^* f(x_i^*) \Delta x = \int_a^b 2\pi x f(x) \, dx. \tag{6.3}$$

When applying this formula in examples, it may be best to remember the formula for the volume of a cylindrical shell in words. As indicated on the left in Figure 6.21, x_i^* is the average radius of the shell—that is, the radius of a circle halfway between the inside edge and the outside edge of the shell, $f(x_i^*)$ is the height of the shell, and Δx is the thickness. Thus, we can say that

$$\text{volume of cylindrical shell} = 2\pi (\text{average radius})(\text{height})(\text{thickness}).$$

A good way to remember this formula is to imagine cutting the cylindrical shell open and flattening it out, as shown in Figure 6.21. The result would be a thin rectangular plate with height $f(x_i^*)$, width $2\pi x_i^*$, and thickness Δx. The volume of this plate is therefore $2\pi x_i^* f(x_i^*) \Delta x$, exactly the same as the volume of the cylindrical shell. Of course, flattening the shell would require some distortion, since the inner edge of the shell would be stretched when the shell was flattened, and the outer edge would be compressed.

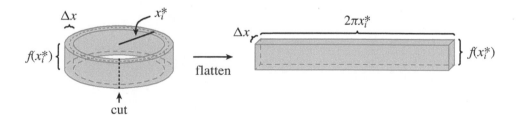

Figure 6.21: Cutting open and flattening a cylindrical shell.

Thus, this reasoning does not give a rigorous derivation of the formula for the volume of the shell. But it does provide an intuitive picture that may help you remember the formula.

Example 6.3.1. Find the volumes of the solids generated when the following regions are rotated about the specified lines:

(a) The region under the curve $y = 4 - x^2$, $0 \le x \le 2$, rotated about the y-axis.

(b) The region bounded by $y = 4 - x^2$ and $y = 4 - 2x$, rotated about the line $x = 2$.

(c) The region bounded by $x = (y - 1)^2$ and $x = 1$, rotated about the x-axis.

Solution. The region in part (a) is shown on the left in Figure 6.22. When a typical vertical rectangle is rotated about the y-axis, it generates a cylindrical shell, as shown on the right in the figure. The volume of this cylindrical shell is

$$2\pi(\text{average radius})(\text{height})(\text{thickness}) = 2\pi x(4 - x^2)\Delta x,$$

and therefore the volume of the solid is

$$\int_0^2 2\pi x(4 - x^2)\, dx = 2\pi \int_0^2 (4x - x^3)\, dx = 2\pi\left[2x^2 - \frac{x^4}{4}\right]_0^2 = 8\pi.$$

For part (b), we begin by setting $4 - x^2 = 4 - 2x$. Solving, we find that the curves cross at $x = 0$ and $x = 2$, as shown in Figure 6.23. For this volume, we are working with a region between two curves, not a region under a curve, and the region is being rotated about the line $x = 2$, not the y-axis, so equation (6.3) doesn't apply directly. Nevertheless, the shell method can be used to compute the required volume.

The rectangle in Figure 6.23 generates a cylindrical shell with average radius $2 - x$, height $(4 - x^2) - (4 - 2x) = 2x - x^2$, and thickness Δx. Therefore the volume of the

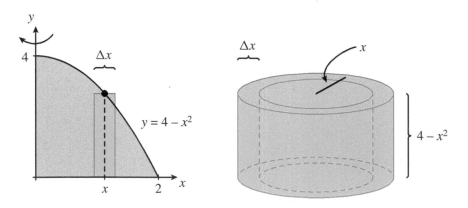

Figure 6.22: Computing the volume of the solid in part (a) of Example 6.3.1.

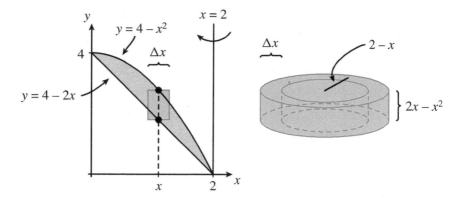

Figure 6.23: Computing the volume of the solid in part (b) of Example 6.3.1.

cylindrical shell is $2\pi(2-x)(2x-x^2)\Delta x$, and the volume of the entire solid is

$$\int_0^2 2\pi(2-x)(2x-x^2)\,dx = 2\pi\int_0^2(4x-4x^2+x^3)\,dx$$

$$= 2\pi\left[2x^2 - \frac{4x^3}{3} + \frac{x^4}{4}\right]_0^2 = \frac{8\pi}{3}.$$

Finally, as shown in Figure 6.24, the region in part (c) can be defined by the inequalities $0 \le y \le 2$, $(y-1)^2 \le x \le 1$. We therefore divide the interval $[0, 2]$ on the y-axis into pieces and cut the region into horizontal strips, approximating each with a rectangle. When the rectangle in Figure 6.24 is rotated about the x-axis, it generates a cylindrical shell with average radius y, height $1 - (y-1)^2 = 2y - y^2$, and thickness Δy.

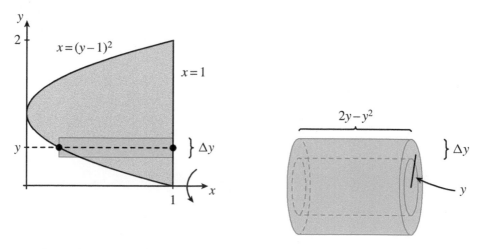

Figure 6.24: Computing the volume of the solid in part (c) of Example 6.3.1.

Therefore the volume of the solid is

$$\int_0^2 2\pi y(2y - y^2)\, dy = 2\pi \int_0^2 (2y^2 - y^3)\, dy = 2\pi \left[\frac{2y^3}{3} - \frac{y^4}{4}\right]_0^2 = \frac{8\pi}{3}. \qquad \square$$

Sometimes a volume can be computed by more than one method. For example, in part (c) of Example 6.2.2 we found the volume of the solid generated when the region bounded by $y = x^3$, $y = 1$, and the y-axis is rotated about the y-axis. In that example we cut the region into horizontal strips, but we can also find the volume by cutting the region into vertical strips. Figure 6.25 shows a rectangle approximating a vertical strip, and the cylindrical shell that it generates. The volume of this cylindrical shell is $2\pi x(1 - x^3)\Delta x$, and therefore the volume of the solid is

$$\int_0^1 2\pi x(1 - x^3)\, dx = 2\pi \int_0^1 (x - x^4)\, dx = 2\pi \left[\frac{x^2}{2} - \frac{x^5}{5}\right]_0^1 = \frac{3\pi}{5}.$$

Of course, this agrees with the answer we got in our original solution. For more examples of volumes that can be computed by two different methods, see Exercises 9–11.

If you are computing the volume of a solid generated when a region is rotated about an axis, how do you decide which method to use? It is generally best to decide first whether to use horizontal rectangles or vertical rectangles. If the region is bounded by curves of the form $y = f(x)$, then it is usually easiest to treat x as the independent variable. This means that you will divide an interval on the x-axis into pieces, cut the region into vertical strips, and approximate these strips with vertical rectangles. On the other hand, if the region is bounded by curves of the form $x = f(y)$, then you will probably want to cut an interval on the y-axis into pieces and then use horizontal rectangles to approximate horizontal strips of the region. Once you have decided whether to use horizontal or vertical rectangles, you can draw a picture to determine what shape these rectangles generate when they are rotated around the specified axis, and you can use this picture to help you set up the integral. Sometimes the curves bounding the region can be written in either the form $y = f(x)$ or $x = f(y)$. In that case, you can use

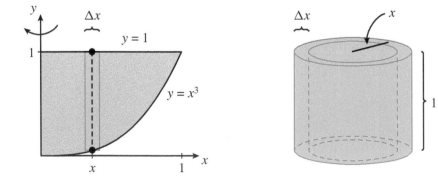

Figure 6.25: An alternative computation of the volume in part (c) of Example 6.2.2.

either method. You might even find it helpful to set up the integral both ways and then choose which integral seems easier to evaluate.

Example 6.3.2. Find the volumes of the solids generated when the following regions are rotated about the specified lines:

(a) The region under the curve $y = \sec x$, $0 \le x \le \pi/4$, rotated about the x-axis.

(b) The region bounded by the y-axis and the curve $x = \sin(y^2)$, $0 \le y \le \sqrt{\pi}$, rotated about the x-axis.

Solution. Since the upper boundary of the region in part (a) is the curve $y = \sec x$, it would be easiest to treat x as the independent variable and cut the region into vertical strips. When the vertical rectangle in Figure 6.26 is rotated about the x-axis, it generates a disk whose volume is $\pi \sec^2 x \Delta x$. Thus, the volume of the solid is

$$\int_0^{\pi/4} \pi \sec^2 x \, dx = \pi \tan x \big|_0^{\pi/4} = \pi.$$

For part (b), part of the boundary of the region is the curve $x = \sin(y^2)$, so it is easiest to treat y as the independent variable and cut the region into horizontal strips. As you can see in Figure 6.27, a typical horizontal rectangle generates a cylindrical shell whose volume is $2\pi y \sin(y^2)\Delta y$, so the volume of the solid is

$$\int_0^{\sqrt{\pi}} 2\pi y \sin(y^2) \, dy.$$

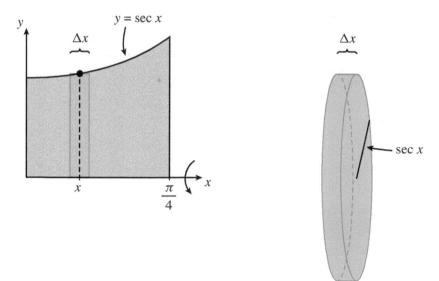

Figure 6.26: Computing the volume of the solid in part (a) of Example 6.3.2.

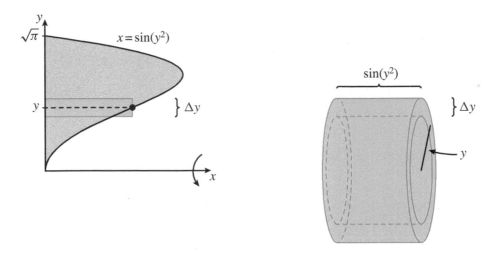

Figure 6.27: Computing the volume of the solid in part (b) of Example 6.3.2.

To evaluate this integral, we use the substitution $u = y^2$, $du = 2y\,dy$:

$$\int_0^{\sqrt{\pi}} 2\pi y \sin(y^2)\,dy = \int_0^{\pi} \pi \sin u\,du = -\pi \cos u\big|_0^{\pi} = 2\pi. \qquad \square$$

Exercises 6.3

1–8: Find the volume of the solid generated when the region is rotated about the specified line.

1. The region under the curve $y = 3/x$, $1 \le x \le 3$, rotated about the y-axis.

2. The region under the curve $y = 1/\sqrt{x^2 + 9}$, $0 \le x \le 4$, rotated about the y-axis.

3. The region bounded by the curve $y = \sqrt{x^2 + 9}$, the line $y = x$, the line $x = 4$, and the y-axis, rotated about the y-axis.

4. The region under the curve $y = \sqrt{x^2 - 1}$, $1 \le x \le 4$, rotated about the y-axis.

5. The region bounded by the curve $y = x^2 - 2x$ and the x-axis, rotated about the y-axis.

6. The region bounded by the curve $y = x^2 - 2x$ and the x-axis, rotated about the line $x = 3$.

7. The region bounded by the curves $x = y^2 + y$ and $x + y = 0$, rotated about the x-axis.

8. The region bounded by the curves $x = y^2$ and $x = y - y^2$, rotated about the line $y = 1$.

9. Redo part (c) of Example 6.3.1 using vertical slices rather than horizontal slices.

10. Use cylindrical shells to find the volume of a sphere of radius r.

11. Use cylindrical shells to find the volume of a cone of radius r and height h.

12. A vertical hole is drilled through the center of a ball of radius R. If the height of the remaining solid is h, show that the volume of this solid is $\pi h^3/6$. (Surprisingly, the answer doesn't depend on R!)

6.4 Work

Suppose that an object is at a point a on a number line. A constant force F is applied to the object, directed in the positive direction on the number line, and, under the influence of this force, the object moves to a point $b > a$, traveling through a distance $d = b - a$. In this situation, we say that the *work* done by the force is $W = Fd$. If d is measured in feet and F is measured in pounds, then W is measured in foot-pounds. If d is measured in meters and F in newtons, then the unit for W is called a *joule* (abbreviated J). In other words, 1 joule is the amount of work done by a constant force of 1 newton moving an object through a distance of 1 meter.

More complicated situations can be handled by applying the fact that work is additive: if a process can be decomposed as a sequence of subprocesses, then the work done in the course of the process is the sum of the work done in each of the subprocesses. For example, consider again the process of moving an object through a distance d while applying a constant force F. We can think of this process as a sequence of two subprocesses: moving the object through a distance $d_1 < d$, and then moving it through a further distance d_2, where $d_1 + d_2 = d$. The work done by the force in the first subprocess is Fd_1, while in the second it is Fd_2, and the sum of these is

$$Fd_1 + Fd_2 = F(d_1 + d_2) = Fd,$$

which is the work done by the force in the entire process.

What if the force is not a constant? Suppose once again that the object moves from a to b on the number line, but as the object moves the force varies. For $a \le x \le b$, let $F(x)$ be the force applied to the object when it is at the point x; we will assume that F is a continuous function. How much work does this varying force do?

We can use a Riemann sum to get an approximate answer. For some positive integer n, divide the interval $[a, b]$ into n pieces, each of width $\Delta x = (b - a)/n$. As usual, we let $x_i = a + i\Delta x$. If we let W_i be the work done in moving the object from x_{i-1} to x_i, then since work is additive, the total work done in moving the object from a to b is $W = \sum_{i=1}^{n} W_i$.

If n is large, so that Δx is small, then the force probably won't vary much as the object moves from x_{i-1} to x_i. We can therefore get a good approximation of W_i by choosing a sample point $x_i^* \in [x_{i-1}, x_i]$ and assuming that the force is a constant $F(x_i^*)$ throughout the interval $[x_{i-1}, x_i]$. Thus, $W_i \approx F(x_i^*)(x_i - x_{i-1}) = F(x_i^*)\Delta x$.

Adding up all of these approximations, we see that

$$W = \sum_{i=1}^{n} W_i \approx \sum_{i=1}^{n} F(x_i^*)\Delta x.$$

Of course, the last sum is a Riemann sum for F on the interval $[a, b]$.

As n increases, the terms in the Riemann sum track the varying force more and more closely, and the approximation improves. To get the exact value of the work, we take the limit as $n \to \infty$:

$$W = \lim_{n \to \infty} \sum_{i=1}^{n} F(x_i^*)\Delta x = \int_a^b F(x)\,dx.$$

For example, suppose that an object moves from 2 to 4 on a number line that is marked off in feet, and for $2 \le x \le 4$, the force on the object when it is at the point x is $F(x) = 3 + 2x$ pounds. Then the work done by this force is

$$\int_2^4 F(x)\,dx = \int_2^4 (3 + 2x)\,dx = \left[3x + x^2\right]_2^4 = 28 - 10 = 18 \text{ ft-lb}.$$

Example 6.4.1. An electron is held fixed at the origin of a number line that is marked off in meters, and a second electron is placed at the point 1. Since the electrons repel each other, the second electron will move away from the origin. How much work does the repulsive force do in pushing the second electron from 1 to 2 on the number line?

Solution. According to Coulomb's law, when the second electron is at the point x, the force pushing it away from the first electron, in newtons, is given by the formula

$$F(x) = \frac{C}{x^2},$$

where C is a constant whose value is approximately 2.3×10^{-28}. Thus, the work done by this force is

$$\int_1^2 \frac{C}{x^2}\,dx = \left[-\frac{C}{x}\right]_1^2 = -\frac{C}{2} + C = \frac{C}{2} \approx 1.15 \times 10^{-28} \text{ J}. \qquad \square$$

Example 6.4.2. A tank has the shape of the bottom half of a sphere whose radius is 3 feet. If the tank is full of water, how much work must be done to pump the water out over the edge of the tank?

Solution. Our strategy for solving this problem will be to divide the water into thin horizontal slices and compute the work required to lift each slice to the top of the tank. Adding these numbers will give the total amount of work required to pump all of the water out of the tank.

We can think of the water in the tank as the solid generated when the region R defined by the inequalities $-3 \le y \le 0$, $0 \le x \le \sqrt{9 - y^2}$ is rotated about the y-axis, as shown in Figure 6.28; we assume here that the x- and y-axes are marked off in feet.

Figure 6.28: The hemispherical tank in Example 6.4.2.

We now divide the interval $[-3, 0]$ on the y-axis into n pieces, each of width $\Delta y = 3/n$, with division points $y_i = -3 + i\Delta y$. The lines $y = y_i$ divide R into n horizontal strips, and when these strips are rotated about the y-axis they generate n horizontal slices of the water. For $i = 1, 2, \ldots, n$, let W_i be the work required to lift the ith slice of water up to the top of the tank. Then the work required to pump all of the water out of the tank is $W = \sum_{i=1}^{n} W_i$.

As usual, we now choose sample points $y_i^* \in [y_{i-1}, y_i]$ and approximate the ith horizontal strip of R with the rectangle $y_{i-1} \le y \le y_i$, $0 \le x \le \sqrt{9 - (y_i^*)^2}$. When this rectangle is rotated about the y-axis, it generates a disk whose volume is $\pi \left(\sqrt{9 - (y_i^*)^2}\right)^2 \Delta y = \pi(9 - (y_i^*)^2)\Delta y$ ft^3. One cubic foot of water weighs about 62.4 pounds, so the weight of the ith slice of water is about $62.4\pi(9 - (y_i^*)^2)\Delta y$ lb. This is the force that must be exerted to counteract the force of gravity in order to lift this slice of water.

The ith slice of water must be lifted a distance of about $-y_i^*$ feet to reach the top of the tank. (Notice that $y_i^* \le 0$, so $-y_i^* \ge 0$.) This is only an approximation; the water at the top of the slice must be lifted a slightly smaller distance than the water at the bottom. But if n is large, and therefore Δy is small, then this discrepancy will be negligible. The work required to lift the ith slice can now be approximated by multiplying force times distance:

$$W_i \approx (62.4\pi(9 - (y_i^*)^2)\Delta y)(-y_i^*) = 62.4\pi((y_i^*)^3 - 9y_i^*)\Delta y \text{ ft-lb.}$$

Adding the work for all the slices, we arrive at an approximation for the total amount of work required to pump out all of the water:

$$W = \sum_{i=1}^{n} W_i \approx \sum_{i=1}^{n} 62.4\pi((y_i^*)^3 - 9y_i^*)\Delta y \text{ ft-lb.}$$

Finally, to get the exact answer, we take the limit as $n \to \infty$, which will turn the last summation into an integral:

$$W = \lim_{n \to \infty} \sum_{i=1}^{n} 62.4\pi((y_i^*)^3 - 9y_i^*)\Delta y = \int_{-3}^{0} 62.4\pi(y^3 - 9y)\,dy$$

$$= 62.4\pi \left(\frac{y^4}{4} - \frac{9y^2}{2}\right)\Big|_{-3}^{0} = 62.4\pi \left[0 - \left(\frac{81}{4} - \frac{81}{2}\right)\right] \approx 3969.72 \text{ ft-lb.} \qquad \square$$

There is an interesting connection between the work done by a force on an object and the kinetic energy of the object. Consider an object moving to the right along the x-axis from $x = x_0$ to $x = x_1$ under the influence of a varying force $F(x)$ that is directed to the right. As we have already seen, the work done by this force is $W = \int_{x_0}^{x_1} F(x)\,dx$. But we can also think of the position x of the object as a function of time t—say, $x = s(t)$. If the motion of the object takes place between $t = t_0$ and $t = t_1$, then we have $x_0 = s(t_0)$ and $x_1 = s(t_1)$, and therefore we can write the work in the form

$$W = \int_{s(t_0)}^{s(t_1)} F(x)\,dx = \int_{t_0}^{t_1} F(s(t))s'(t)\,dt. \tag{6.4}$$

To see why the last two integrals are equal, use the substitution $x = s(t)$, $dx = s'(t)\,dt$ to convert the second integral into the first.

If we assume that $F(x)$ is the only force acting on the object, then we can apply Newton's second law of motion, which says that the force on the object is equal to the mass of the object times its acceleration. The acceleration is $d^2x/dt^2 = s''(t)$, so if the mass is m then we have $F(x) = F(s(t)) = ms''(t)$. Substituting into equation (6.4), we get

$$W = \int_{t_0}^{t_1} ms''(t)s'(t)\,dt.$$

Finally, to evaluate this integral we use the substitution $v = s'(t)$, $dv = s''(t)\,dt$; notice that v represents the velocity of the object. If we let $v_0 = s'(t_0)$ and $v_1 = s'(t_1)$ (the initial and final velocities of the object), then the substitution leads to the formula

$$W = \int_{t_0}^{t_1} ms'(t)s''(t)\,dt = \int_{v_0}^{v_1} mv\,dv = \frac{1}{2}mv^2 \Big|_{v_0}^{v_1} = \frac{1}{2}mv_1^2 - \frac{1}{2}mv_0^2.$$

If you have studied physics before, then you may recognize $(1/2)mv^2$ as the formula for the kinetic energy of the object. Thus, we have just shown that *the work done by the force on the object is equal to the change in its kinetic energy.*

Exercises 6.4

1. An object moves from 1 to 5 on a number line, and the force applied to it when it is at the point x is $F(x) = 2x^2 - 1$. Find the work done by this force.

2. An object moves from -5 to 8 on a number line, and the force applied to it when it is at the point x is $F(x) = \sqrt[3]{17 - 2x}$. Find the work done by this force.

3. An object moves on a number line so that its position at time t is $x = 2t + 1$, and the force applied to it at time t is $F(t) = t^2$. In this problem you will compute the work done by this force in the time interval $0 \le t \le 2$ in two ways.

   (a) What interval on the number line does the object traverse during the specified time interval? When the object is at the point x, what is the force on it?

   (b) Use your answers to part (a) to compute the work done by the force.

(c) Divide the interval $[0, 2]$ on the t-axis into n subintervals, approximate the work done in each subinterval, and add these approximations to get a Riemann sum approximation of the total work expressed in terms of the variable t.

(d) Take the limit of your answer to part (c) as $n \to \infty$ to get an integral with respect to t and evaluate this integral to find the work done by the force. Of course, your answer should agree with your answer to part (b).

4. A bucket full of water is at the bottom of a well that is 100 feet deep. A rope runs from the bucket to ground level and will be used to lift the bucket out of the well. If the bucket full of water weighs 25 lb and the rope weighs 1/10 lb/ft, how much work must be done to lift the bucket?

5. A 50-pound sandbag is lifted a distance of 100 feet at a steady speed of 2 ft/sec. As it is lifted, sand leaks out of the bag at a steady rate of 1/4 lb/sec. How much work is done?

6. A weather balloon is used to lift some scientific equipment to an altitude of 20 miles above the surface of the earth. The equipment weights 100 pounds on the surface of the earth. How much work does the weather balloon do? (The force of gravity on an object at a distance of x miles from the center of the earth is C/x^2, for some constant C. The radius of the earth is about 3959 miles.)

7. According to Hooke's law, when a spring is stretched, it resists further stretching with a force that is proportional to the distance it has been stretched from its natural length. Suppose that one end of a spring is held fixed, and the other end is pulled, increasing the length of the spring from its natural length by 5 inches. At that point, the spring is resisting further stretching with a force of 1 pound. How much work was done in stretching the spring?

8. A tank in the shape of an inverted cone is full of water. The tank is 4 feet deep, and the radius at the top is 2 feet. How much work must be done to pump all of the water out over the top of the tank?

9. In Exercise 8, suppose the tank is only full to a depth of 2 feet. Now how much work must be done to pump the water out?

10. In Exercise 8, suppose the tank is only half full—that is, the volume of water in the tank is only half the tank's capacity. Now how much work must be done to pump the water out?

6.5 Center of Mass

Alice and Bob are sitting on a seesaw. Alice weighs 120 pounds and Bob weighs 150, so the seesaw doesn't balance when they sit equally far from the center. But they discover that they can get it to balance if they shift their positions. In particular, after some experimentation they find that if Alice is 5 feet from the center of the seesaw and Bob is 4 feet from the center, as in Figure 6.29, then the seesaw balances.

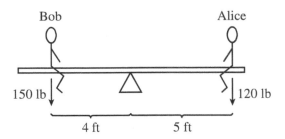

Figure 6.29: Alice and Bob create equal and opposite torques, so the seesaw balances.

To understand why the seesaw balances in this situation, we must consider the concept of *torque*. The seesaw is supported at a point in the center called the *fulcrum*. When Alice is sitting 5 feet to the right of the fulcrum, gravity causes her to push down on the seesaw at that point with a force of 120 lb, and if Bob were not present, this force would cause the seesaw to rotate clockwise about the fulcrum. The *torque* of this force about the fulcrum is a measure of the extent to which the force would cause the seesaw to rotate. It is computed as the magnitude of the force times the distance from the fulcrum. Thus, Alice's weight on the seesaw creates a torque of $120 \cdot 5 = 600$ ft-lb about the fulcrum. Similarly, Bob, sitting 4 feet to the left of the fulcrum and weighing 150 pounds, creates a torque of $150 \cdot 4 = 600$ ft-lb in the opposite direction, exactly balancing the torque created by Alice. We could indicate that Bob's torque is in the opposite direction from Alice's by saying that Bob creates a clockwise torque of -600 ft-lb. The net rotational effect of a collection of forces is found by adding their torques. In our situation, the total torque on the seesaw is $600 - 600 = 0$ ft-lb, and therefore the seesaw does not rotate at all; it balances.

Now let's generalize. Consider a horizontal beam that is marked off as a number line. There are objects whose masses are $m_1, m_2, \ldots, m_n$ resting on the beam at the points $x_1, x_2, \ldots, x_n$ (see Figure 6.30). We will assume that the mass of the beam itself is negligible. The weight of an object is proportional to its mass, so the ith object pushes down on the beam at the point x_i with a force $m_i g$, where g is a constant. If the beam rests on a fulcrum at the point c, then this force produces a clockwise torque of $(x_i - c)m_i g$ about the fulcrum. Notice that if $x_i < c$ then this torque is negative, indicating that the force would cause a counterclockwise rotation.

We would like to determine where we should position the fulcrum so that the beam will balance. In order for the beam to balance, the total torque must be 0:

$$\sum_{i=1}^{n} (x_i - c)m_i g = 0.$$

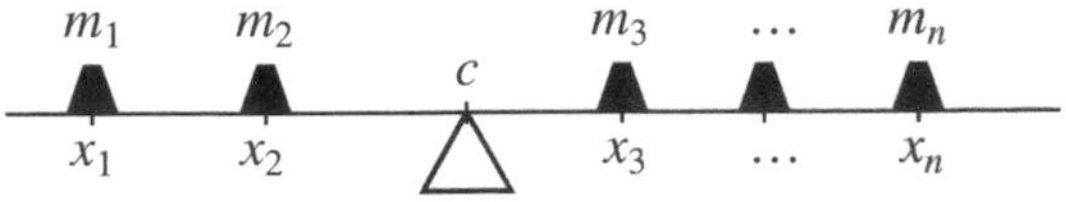

Figure 6.30: A beam with several masses on it.

We can now solve this equation for c to find the point where the beam will balance. We begin by expanding the summation, using Theorem 5.1.1:

$$0 = \sum_{i=1}^{n} (x_i - c)m_i g = \sum_{i=1}^{n} x_i m_i g - \sum_{i=1}^{n} cm_i g = g \sum_{i=1}^{n} x_i m_i - cg \sum_{i=1}^{n} m_i.$$

Solving for c, we find that the beam will balance if the fulcrum is at the point

$$c = \frac{g \sum_{i=1}^{n} x_i m_i}{g \sum_{i=1}^{n} m_i} = \frac{\sum_{i=1}^{n} x_i m_i}{\sum_{i=1}^{n} m_i}. \tag{6.5}$$

This point is called the *center of mass* of the collection of objects. Notice that the denominator of the fraction on the right in equation (6.5) is simply the total mass of the objects. The numerator is called the *first moment of mass* (about the origin) for the collection of objects.

Equation (6.5) tells us how to find the center of mass of a finite collection of objects arranged on a number line, if the mass of each object is concentrated at a point. But mass is generally spread out, not concentrated at a point. Thus, it would be more realistic to imagine mass spread continuously over some part of a number line. For example, consider a wire of varying density extending over the interval $[a, b]$ on a number line. We will assume that for $a \le x \le b$, the density of the wire at the point x is $d(x)$, where d is a function that is continuous on the interval $[a, b]$. We would like to find the center of mass of this wire.

We begin by showing how Riemann sums can be used to get an approximate answer. For a large positive integer n, we can imagine cutting the wire into n small pieces. If we treat these pieces as n point masses, then we can use equation (6.5) to find an approximate value for the center of mass of the wire. To cut the wire into pieces, we divide the interval $[a, b]$ into n smaller intervals, each of width $\Delta x = (b - a)/n$, with division points $x_i = a + i\Delta x$. For $1 \le i \le n$, the ith piece of the wire is the part that lies in the interval $[x_{i-1}, x_i]$. To approximate the mass of this piece of wire, we choose a sample point $x_i^* \in [x_{i-1}, x_i]$ and assume that the density is a constant $d(x_i^*)$ throughout the interval. Since density is mass per unit of length, the mass of the ith piece of wire is approximately $m_i = d(x_i^*)\Delta x$. By treating the ith piece of wire as a point mass located at the point x_i^* and applying equation (6.5), we find that the center of mass of the wire is approximately

$$\frac{\sum_{i=1}^{n} x_i^* m_i}{\sum_{i=1}^{n} m_i} = \frac{\sum_{i=1}^{n} x_i^* d(x_i^*)\Delta x}{\sum_{i=1}^{n} d(x_i^*)\Delta x}. \tag{6.6}$$

As n increases, the n pieces of wire become smaller, and this approximation, which is based on treating these pieces as points, becomes more accurate. The numerator and denominator of (6.6) are both Riemann sums, so as $n \to \infty$ they both approach integrals. Thus, the exact center of mass is

$$\lim_{n \to \infty} \frac{\sum_{i=1}^{n} x_i^* d(x_i^*)\Delta x}{\sum_{i=1}^{n} d(x_i^*)\Delta x} = \frac{\lim_{n \to \infty} \sum_{i=1}^{n} x_i^* d(x_i^*)\Delta x}{\lim_{n \to \infty} \sum_{i=1}^{n} d(x_i^*)\Delta x} = \frac{\int_a^b x d(x)\, dx}{\int_a^b d(x)\, dx}. \tag{6.7}$$

The denominator of the last fraction in equation (6.7) is the total mass of the wire, and the numerator is the first moment of mass about the origin.

For example, suppose that a piece of wire extends from 0 to 5 on a number line, and the density at the point x is $d(x) = 3x$. Then the mass of the wire is

$$\int_0^5 3x \, dx = \left. \frac{3x^2}{2} \right|_0^5 = \frac{75}{2},$$

and the first moment of mass is

$$\int_0^5 x \cdot 3x \, dx = \int_0^5 3x^2 \, dx = \left. x^3 \right|_0^5 = 125.$$

Therefore the center of mass is at

$$\frac{125}{75/2} = \frac{10}{3}.$$

In other words, the wire would balance on a fulcrum at the point 10/3.

It is interesting to work out what happens in the case of a wire of constant density. Suppose that the wire extends from a to b on the number line, and the density at every point in the interval $[a, b]$ is D. Then according to equation (6.7), the center of mass will be

$$\frac{\int_a^b x D \, dx}{\int_a^b D \, dx} = \frac{\left[Dx^2/2 \right]_a^b}{\left[Dx \right]_a^b} = \frac{Db^2/2 - Da^2/2}{Db - Da} = \frac{b^2 - a^2}{2(b - a)} = \frac{a + b}{2}.$$

In other words, the wire balances at its midpoint. Of course, in hindsight this answer is obvious, but it is nice to see that equation (6.7) leads to this obvious answer.

So far we have discussed balancing one-dimensional objects. Let's try moving up to a two-dimensional object. Suppose that a thin plate of uniform density covers the triangle with vertices $(0, 0)$, $(2, 0)$, and $(0, 4)$ in the plane. The hypotenuse of this triangle lies on the line $y = 4 - 2x$. For any number c_x between 0 and 2, we can draw the vertical line $x = c_x$ on the plate. We then lay the plate on top of a knife edge, with the line $x = c_x$ resting on the knife edge, as shown in Figure 6.31. For what value of c_x will the plate balance on the knife edge?

To find this balance point, we cut the triangular plate into n thin vertical strips and approximate these strips with rectangles as usual; the ith rectangle is shown in Figure 6.31. The rectangle has width Δx and height $4 - 2x_i^*$, so its area is $(4 - 2x_i^*)\Delta x$. Since the density of the plate is uniform, the weight of the ith strip of the plate is about $(4 - 2x_i^*)K \Delta x$, for some constant K. This means that when the plate is resting on the knife edge, the force of gravity on this strip of the plate exerts a torque of about $(x_i^* - c_x)(4 - 2x_i^*)K \Delta x$ about the knife edge. Adding up the torques for all of the strips

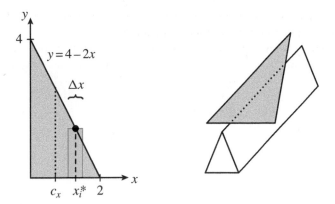

Figure 6.31: Balancing a triangular plate on a knife edge.

and taking the limit as the number of strips approaches ∞, we see that the total torque is

$$\lim_{n\to\infty}\sum_{i=1}^{n}(x_i^* - c_x)(4 - 2x_i^*)K\,\Delta x = \int_0^2 (x - c_x)(4 - 2x)K\,dx$$

$$= K\int_0^2 x(4 - 2x)\,dx - Kc_x\int_0^2 (4 - 2x)\,dx.$$

For the plate to balance on the knife edge, this total torque must be 0. Thus, to find the balance point, we set the torque equal to 0 and solve for c_x:

$$c_x = \frac{K\int_0^2 x(4 - 2x)\,dx}{K\int_0^2 (4 - 2x)\,dx} = \frac{\int_0^2 x(4 - 2x)\,dx}{\int_0^2 (4 - 2x)\,dx}.$$

Notice the similarity of this formula to the formula in equation (6.7): the denominator of the last fraction is the total area of the plate, and the numerator is the first moment of area about the line $x = 0$—that is, about the y-axis. Evaluating the integrals, we find that the balance point is

$$c_x = \frac{\int_0^2 (4x - 2x^2)\,dx}{\int_0^2 (4 - 2x)\,dx} = \frac{\left[2x^2 - 2x^3/3\right]_0^2}{\left[4x - x^2\right]_0^2} = \frac{8/3}{4} = \frac{2}{3}.$$

We can do a similar calculation to find a horizontal line on which the plate will balance. Rewriting the hypotenuse in the form $x = 2 - y/2$ and imitating the calculation above, we find that the first moment of area about the line $y = 0$ is

$$\int_0^4 y\left(2 - \frac{y}{2}\right)\,dy = \int_0^4 \left(2y - \frac{y^2}{2}\right)\,dy = \left[y^2 - \frac{y^3}{6}\right]_0^4 = \frac{16}{3}.$$

Since we have already computed that the area is 4, we conclude that the plate will balance on the line $y = c_y$, where

$$c_y = \frac{16/3}{4} = \frac{4}{3}.$$

Putting it all together, we have shown that the plate will balance on either a horizontal line or a vertical line through the point $(2/3, 4/3)$. In fact, it can be shown that the plate will balance on the point of a needle at this point. The point $(2/3, 4/3)$ is called the *centroid* of the triangle.

In general, if a region is defined by the inequalities $a \leq x \leq b$, $g(x) \leq y \leq f(x)$, where f and g are continuous on $[a, b]$ and for all $x \in [a, b]$, $g(x) \leq f(x)$, then the x-coordinate of the centroid is given by the formula

$$c_x = \frac{\int_a^b x(f(x) - g(x))\, dx}{A},$$

where A is the area of the region. A similar formula applies for the y-coordinate of the centroid of a region.

We close this chapter by noting a connection between centroids and volumes:

Theorem 6.5.1 (Pappus's Second Centroid Theorem[2]). *If a region in the plane is rotated about an axis that does not pass through the region, then the volume of the solid generated is equal to the area of the region times the distance traveled by the centroid of the region when it rotates about the axis.*

Proof. We will prove the theorem in the case of a region of the form $a \leq x \leq b$, $g(x) \leq y \leq f(x)$ that is rotated about the y-axis. We assume here that $0 < a < b$, f and g are continuous on $[a, b]$, and for all $x \in [a, b]$, $g(x) \leq f(x)$.

If the area of the region is A, then the x-coordinate of the centroid is

$$c_x = \frac{\int_a^b x(f(x) - g(x))\, dx}{A}.$$

When the centroid is rotated about the y-axis, it traces out a circle of radius c_x, so the distance it travels is the circumference of this circle, $2\pi c_x$.

Thus, the product of the area of the region and the distance traveled by the centroid is

$$A \cdot 2\pi c_x = A \cdot \frac{2\pi \int_a^b x(f(x) - g(x))\, dx}{A} = \int_a^b 2\pi x(f(x) - g(x))\, dx.$$

But this is precisely the volume of the solid generated when the region is rotated about the y-axis, computed by cylindrical shells.   □

Exercises 6.5

1. A wire extends from -2 to 1 on a number line, and the density at the point x is $d(x) = x^2 + 1$. Find the center of mass of the wire.

2. A wire extends from 1 to 4 on a number line, and the density at the point x is $d(x) = \sqrt{x}$. Find the center of mass of the wire.

---

[2]For Pappus's first centroid theorem, see Exercise 16 in Section 9.3.

3. A wire extends from 1 to 4 on a number line, and the density at the point x is $d(x) = \sqrt{3x + 13}$. Find the center of mass of the wire.

4. Find the centroid of the triangular region bounded by the lines $y = 1 - x$, $y = 1 + x$, and the x-axis.

5. Find the centroid of the region bounded by the curves $y = x^2$ and $y = 4$.

6. Find the centroid of the region bounded by the curves $y = x^2$ and $y = 2 - x$.

7. Find the centroid of the semicircular region $x^2 + y^2 \leq r^2$, $y \geq 0$. (Hint: You don't need to evaluate an integral to find the area of the semicircle.)

8. Use your answer to Exercise 7 and Pappus's second centroid theorem to find the volume of a sphere of radius r.

9. (a) Find the centroid of the triangular region with vertices $(0, 0)$, $(r, 0)$, and $(0, h)$.

   (b) Use your answer to part (a) and Pappus's second centroid theorem to find the volume of a cone of height h whose base has radius r.

10. Use Pappus's second centroid theorem to solve Exercise 14 from Section 6.2.

Chapter 7

Inverse Functions, the Natural Logarithm, and the Exponential Function

7.1 Inverse Functions

Consider the functions $f(x) = x^3$ and $g(x) = \sqrt[3]{x}$. Clearly these functions are closely related. But what, exactly, is the nature of the relationship?

We begin by examining the relationship between the graphs of these functions. The graph of g is the graph of the equation $y = \sqrt[3]{x}$. But by the definition of cube roots, to say that y is the cube root of x simply means that y is the number whose cube is x. In other words,

$$y = \sqrt[3]{x} \quad \text{if and only if} \quad x = y^3.$$

Thus, we could also describe the graph of g as the graph of the equation $x = y^3$. Of course, the graph of f is the graph of the equation $y = x^3$. These two graphs contain exactly the same points, but with the x- and y-coordinates reversed. You can see this reversal of x- and y-coordinates in Table 7.1, which lists some points on the graphs of f and g. Thus, we can construct the graph of g by determining the points on the graph of f and then reversing the coordinates of those points.

Geometrically, what is the effect of reversing the x- and y-coordinates of a point? The answer to this question is illustrated in Figure 7.1. Imagine that there is a wire embedded in the page along the line $y = x$, which is shown as a dashed line in the

x	$y = f(x) = x^3$
0	0
1	1
2	8
3	27
-1	-1
-2	-8
-3	-27

x	$y = g(x) = \sqrt[3]{x}$
0	0
1	1
8	2
27	3
-1	-1
-8	-2
-27	-3

Table 7.1: Some points on the graphs of f and g.

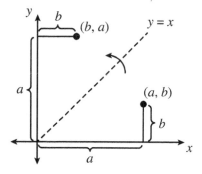

Figure 7.1: Reversing the coordinates of a point has the effect of flipping it over across the line $y = x$.

figure. If you were to spin the page 180 degrees, using this wire as an axis, then the point (a, b) in the figure would end up at the point (b, a). Thus, the geometric effect of reversing the coordinates of a point is to flip the point over across the line $y = x$. It follows that if we draw the graph of f and then flip this graph over across the line $y = x$, then we will get the graph of g. This is illustrated in Figure 7.2. The graph of f is shown in Figure 7.2a, and if you flip this graph over across the line $y = x$ then you get the graph of g, which is shown in Figure 7.2b.

The relationship between the graphs of f and g implies an algebraic relationship between these functions. For any real number x, the point $(x, f(x))$ is on the graph of f. It follows that $(f(x), x)$ is on the graph of g, and therefore $g(f(x)) = x$. Similar reasoning shows that for all x, $f(g(x)) = x$. We can verify these equations algebraically by filling in the formulas defining f and g:

$$g(f(x)) = \sqrt[3]{x^3} = x,$$
$$f(g(x)) = (\sqrt[3]{x})^3 = x.$$

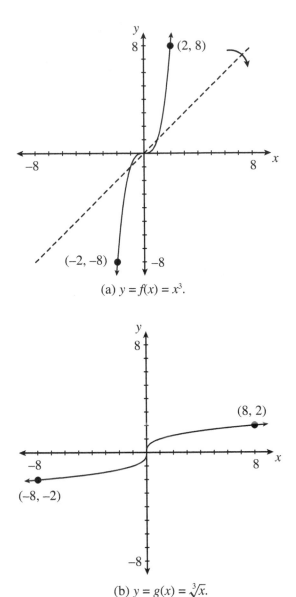

(a) $y = f(x) = x^3$.

(b) $y = g(x) = \sqrt[3]{x}$.

Figure 7.2: Flipping the graph of f over across the line $y = x$ produces the graph of g.

Thus, each of these functions "undoes" the other: If you apply f to a number x, then you can undo this operation and get back the original number x by applying g. Similarly, if you apply g to x, then you can undo this operation and get back x by applying f.

We say that g is the *inverse* of f, and we write $g = f^{-1}$. Thus, we can write $f^{-1}(x) = \sqrt[3]{x}$. Notice that $f^{-1}(x)$ is not the same thing as $(f(x))^{-1} = 1/f(x) = 1/x^3$.

Let's try another example. Let h be the function defined as follows:

$$h(x) = \frac{2x}{1+|x|} = \begin{cases} \dfrac{2x}{1+x}, & \text{if } x \geq 0, \\[2mm] \dfrac{2x}{1-x}, & \text{if } x < 0. \end{cases} \tag{7.1}$$

We will leave it to you to verify, using the methods of Chapter 4, that the graph of h is as shown in Figure 7.3a. In particular, note that h is strictly increasing on $(-\infty, \infty)$, and the lines $y = 2$ and $y = -2$ are asymptotes.

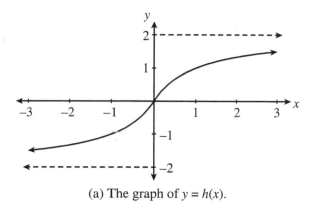

(a) The graph of $y = h(x)$.

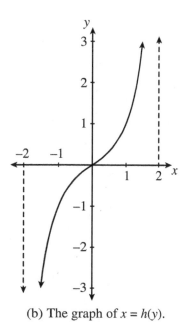

(b) The graph of $x = h(y)$.

Figure 7.3: To get the graph of h^{-1}, we flip over the graph of h.

Flipping this graph over across the line $y = x$ gives us the graph of the equation $x = h(y)$, which is shown in Figure 7.3b. This graph passes the vertical line test, and therefore it is the graph of a function. We call this function the inverse of h, and denote it by h^{-1}.

Notice that the domain of h is the interval $(-\infty, \infty)$, so every real number appears as the x-coordinate of a point on the graph of h. However, you can see in Figure 7.3a that the y-coordinates of the points on the graph are the numbers in the interval $(-2, 2)$. In other words,

$$\{h(x) : x \in \mathbb{R}\} = \{y : \text{for some real number } x, \ y = h(x)\} = (-2, 2).$$

Recall that this set is called the range of h. Because of the reversal of the roles of x and y in Figures 7.3a and 7.3b, this means that the set of x-coordinates of points on the graph of h^{-1} is $(-2, 2)$, and the set of y-coordinates is $(-\infty, \infty)$. In other words, the domain of h^{-1} is $(-2, 2)$—the same as the range of h—and the range of h^{-1} is $(-\infty, \infty)$—the same as the domain of h.

As before, the graphical relationship between h and h^{-1} implies an algebraic relationship. If x is any real number, then the point $(x, h(x))$ is on the graph of h. Therefore $(h(x), x)$ is on the graph of h^{-1}, so $h^{-1}(h(x)) = x$. Similarly, if $x \in (-2, 2)$, then $(x, h^{-1}(x))$ is on the graph of h^{-1}, so $(h^{-1}(x), x)$ is on the graph of h, which means that $h(h^{-1}(x)) = x$. In other words, h and h^{-1} undo each other.

We can verify this algebraically by finding a formula for $h^{-1}(x)$. Suppose $x \in (-2, 2)$, and let $y = h^{-1}(x)$. Then the point (x, y) is on the graph in Figure 7.3b, so it satisfies the equation

$$x = h(y) = \frac{2y}{1 + |y|}. \tag{7.2}$$

We can now find a formula for $h^{-1}(x)$ by solving for y in this equation. This is easiest to do if we treat the left and right halves of the graph separately. So suppose first that $0 \le x < 2$. Then it is clear in Figure 7.3b that $y \ge 0$, and therefore $|y| = y$. Thus, equation (7.2) becomes

$$x = \frac{2y}{1 + y},$$
$$x(1 + y) = 2y,$$
$$x = 2y - xy = y(2 - x),$$
$$y = \frac{x}{2 - x}.$$

Similarly, if $-2 < x < 0$ then $y < 0$, and therefore $|y| = -y$. Substituting this into equation (7.2) leads to the solution

$$y = \frac{x}{2 + x}.$$

Since $y = h^{-1}(x)$, we conclude that

$$h^{-1}(x) = \begin{cases} \dfrac{x}{2-x}, & \text{if } 0 \le x < 2, \\[2mm] \dfrac{x}{2+x}, & \text{if } -2 < x < 0. \end{cases}$$

In other words,

$$h^{-1}(x) = \frac{x}{2 - |x|}, \quad -2 < x < 2. \tag{7.3}$$

In Exercise 1, we ask you to use our formulas for $h(x)$ and $h^{-1}(x)$ to verify that h and h^{-1} undo each other.

We have now developed a graphical method for finding what we are calling the inverse of a function: graph the function and then flip the graph over across the line $y = x$, and you will have the graph of the inverse of the function. But there is something that could go wrong with this method: what if the flipped-over graph fails the vertical line test? In fact, this can happen. Consider, for example, the function $f(x) = x^2$. Figure 7.4a shows the graph of $y = x^2$, and flipping this graph over gives us the graph of $x = y^2$ in Figure 7.4b. This graph fails the vertical line test, as indicated by the red vertical line in Figure 7.4b. Tracing this problem back to its source, we see that the problem is caused by the fact that a horizontal line crosses the graph of f more than once. This horizontal line intersects the graph at two points that have different x-coordinates, labeled x_1 and x_2 in Figure 7.4a, but the same y-coordinate. In other words, $x_1 \ne x_2$, but $f(x_1) = f(x_2)$. This, then, is the situation we need to avoid if we want to construct the inverse of a function. There is a name for functions that avoid this situation:

Definition 7.1.1. A function f is called *one-to-one* if there do *not* exist numbers x_1 and x_2 in the domain of f such that $x_1 \ne x_2$ but $f(x_1) = f(x_2)$.

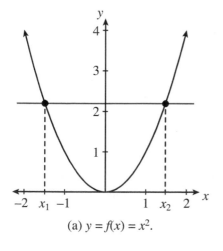

(a) $y = f(x) = x^2$.

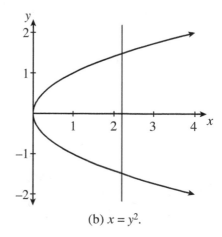

(b) $x = y^2$.

Figure 7.4

Example 7.1.2. Which of the following functions are one-to-one?

$$f(x) = x^3 + 2x, \qquad g(x) = x^3 - 2x, \qquad h(x) = 1/x.$$

Solution. We give a graphical solution. We leave it to you to verify that the graphs of the functions are as shown in Figure 7.5. The red lines in the figure illustrate that no horizontal line intersects the graph of f or h more than once, but there are horizontal lines intersecting the graph of g more than once. Therefore f and h are one-to-one, but g is not.   □

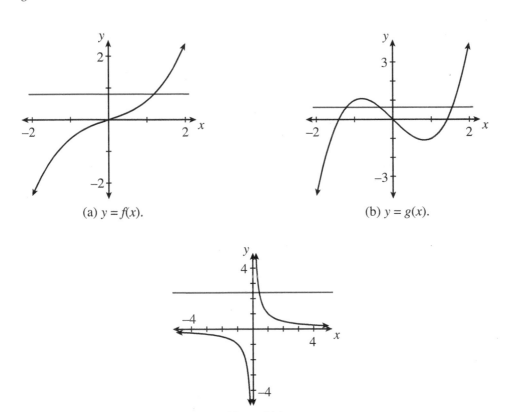

(a) $y = f(x)$.

(b) $y = g(x)$.

(c) $y = h(x)$.

Figure 7.5: The graphs of the functions in Example 7.1.2.

With this new terminology, we can now say more carefully what we mean by the inverse of a function. Suppose f is a one-to-one function. Then no horizontal line intersects the graph of the equation $y = f(x)$ more than once, and therefore no vertical line intersects the graph of $x = f(y)$ more than once. Thus, the graph of $x = f(y)$ passes the vertical line test, so it is the graph of a function. We define this function to be the *inverse* of f, and we denote it by f^{-1}. For any numbers x and y, we have $y = f^{-1}(x)$ if and only if $x = f(y)$, and as in the examples above we can use this to show that the functions f and f^{-1} undo each other. That is, for all x in the domain of f, $f^{-1}(f(x)) = x$, and for all x in the domain of f^{-1} (which is the same as the range of f), $f(f^{-1}(x)) = x$.

On the other hand, if f is not one-to-one then f does not have an inverse. For example, the reason we were unable to construct an inverse in the case of the function $f(x) = x^2$ is that this function is not one-to-one.

Although we cannot construct an inverse for a function that is not one-to-one, we can come close. The idea is to restrict the domain of the function in order to make it one-to-one. For example, although the function $f(x) = x^2$ is not one-to-one, we can turn it into a one-to-one function by restricting it to $x \geq 0$. More precisely, we define a function g with domain $[0, \infty)$ by saying that for all $x \in [0, \infty)$, $g(x) = x^2$. The function g is called the *restriction* of f to the interval $[0, \infty)$. The graph of g consists of all points (x, y) such that $y = x^2$ and $x \geq 0$. This is just the right half of the graph of f, as shown in Figure 7.6a, and it is clear from this graph that g is one-to-one, since no horizontal line intersects the curve more than once. Reversing the roles of x and y, we get the flipped graph in Figure 7.6b, which consists of all points (x, y) such that $x = y^2$ and $y \geq 0$. This curve passes the vertical line test, so it is the graph of the function g^{-1}, the inverse of g.

It is clear from Figure 7.6b that the domain of g^{-1} is $[0, \infty)$. Suppose $x \in [0, \infty)$, and let $y = g^{-1}(x)$. Then the point (x, y) is on the curve in Figure 7.6b, so

$$x = y^2, \quad y \geq 0. \tag{7.4}$$

As in the case of the function h^{-1}, we can now find a formula for $g^{-1}(x)$ by solving for y. The equation $x = y^2$ has two solutions for y: $y = \pm\sqrt{x}$. But the additional restriction $y \geq 0$ in (7.4) allows us to eliminate the negative solution and conclude that $y = \sqrt{x}$. Thus, the inverse of g is the function $g^{-1}(x) = \sqrt{x}$. Of course, these functions undo

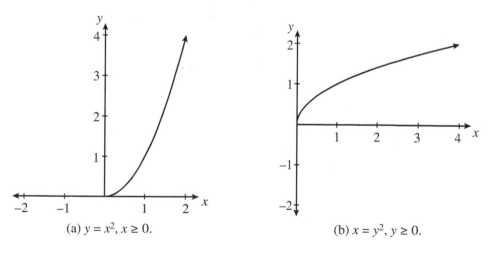

(a) $y = x^2$, $x \geq 0$.

(b) $x = y^2$, $y \geq 0$.

Figure 7.6: Graph (a) is the graph of g, and (b) is the graph of g^{-1}.

each other: both functions have domain $[0, \infty)$, and for all $x \geq 0$,

$$g(g^{-1}(x)) = (\sqrt{x})^2 = x,$$
$$g^{-1}(g(x)) = \sqrt{x^2} = x.$$

But notice that the restriction $x \geq 0$ is necessary here. For $x < 0$, we have $\sqrt{x^2} = |x| = -x$.

To state our conclusions more informally, we can say that although the cube root function is the inverse of the cubing function, the square root function is not the inverse of the squaring function. Rather, it is the inverse of the restriction of the squaring function to the interval $[0, \infty)$. In later sections of this chapter we will study the inverses of a number of functions, and in some cases we will have to use the trick of restricting the domain to make the function one-to-one before we can construct an inverse.

Exercises 7.1

1. Let h be the function defined in equation (7.1). We showed in this section that h is one-to-one, and that h^{-1} is given by equation (7.3). Use equations (7.1) and (7.3) to verify that for every $x \in (-\infty, \infty)$, $h^{-1}(h(x)) = x$, and for every $x \in (-2, 2)$, $h(h^{-1}(x)) = x$.

2–7: Determine whether or not the function is one-to-one. (Hint: You may find it helpful to graph the function.)

2. $f(x) = x^3 - 6x^2 + 9x$.

3. $g(x) = x^3 - 6x^2 + 12x$.

4. $f(x) = \dfrac{x+2}{x}$.

5. $g(x) = x + \dfrac{2}{x}$.

6. $f(x) = 2x + \sin x$.

7. $g(x) = x + 2\sin x$.

8. The graph of a function f is shown in Figure 7.7. What is the domain of f^{-1}? Find $f^{-1}(1)$, $f^{-1}(2)$, $f^{-1}(3)$, and $f^{-1}(5)$.

9–16: Find a formula for $f^{-1}(x)$. What are the domain and range of f^{-1}?

9. $f(x) = 5 - 2x$.

10. $f(x) = \dfrac{x+1}{x-2}$.

11. $f(x) = \sqrt[3]{x-8} + 2$.

12. $f(x) = \sqrt[3]{x^3 + 1}$.

13. $f(x) = \sqrt{1 + \sqrt{x}}$.

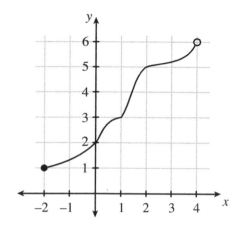

Figure 7.7: The graph of f.

14. $f(x) = \sqrt{1 - \sqrt{x}}$.

15. $f(x) = \dfrac{x}{\sqrt{x^2 + 1}}$.

16. $f(x) = \dfrac{1}{x^2 + 1}$, $x \geq 0$.

7.2 Calculus with Inverse Functions

We saw in the last section that if f is a one-to-one function, then it has an inverse function f^{-1}, and the domain of f^{-1} is the same as the range of f. Thus, if we want to be able to find the domains of inverse functions, then it will be helpful to develop a better understanding of the ranges of functions.

For example, suppose that the domain of f is a finite closed interval $[a, b]$, and f is continuous on this interval. Then as we saw in Section 4.7, f has a minimum value m, attained at some number $c \in [a, b]$, and it has a maximum value M, attained at some $C \in [a, b]$. If we also assume that f is not a constant function, then $m < M$. Clearly m and M are in the range of f, but no number smaller than m or larger than M can be in the range of f. Now consider any number r with $m < r < M$. Since f is continuous, according to the intermediate value theorem there must be some number d between c and C such that $f(d) = r$, and therefore r is in the range of f. Thus, the range of f includes m, M, and all numbers in between, but no other numbers; in other words, the range is the closed interval $[m, M]$.

It turns out something similar is true even if the domain of f is some other kind of interval:

Theorem 7.2.1. *Suppose the domain of f is an interval, f is continuous on that interval, and f is not a constant function. Then the range of f is an interval.*

Before we prove this theorem, let's consider a few examples. Let f, g, and h be the functions defined as follows:

$$f(x) = x^2, \qquad g(x) = \frac{1}{x^2 + 1}, \qquad h(x) = \frac{1}{\sqrt{1 - x^2}}.$$

The domains of f and g are $(-\infty, \infty)$, and the domain of h is $(-1, 1)$. If you sketch the graphs of these functions, using the methods of Chapter 4, you will find that the range of f is the interval $[0, \infty)$, the range of g is $(0, 1]$, and the range of h is $[1, \infty)$.

In our proof of Theorem 7.2.1, we will use the completeness of the real numbers (Theorem 2.9.4). To see why completeness is relevant to this proof, consider the interval $(0, 1]$. Clearly 1 is an upper bound for this interval, but no number smaller than 1 is an upper bound. Thus, 1 is the least upper bound of $(0, 1]$. Similarly, 0 is the greatest lower bound of $(0, 1]$. And similar reasoning applies to other intervals. In general, if an interval does not extend infinitely far to the right on the number line, then its right endpoint is its least upper bound, and if it does not extend infinitely far to the left, then its left endpoint is its greatest lower bound. Thus, to recognize a set as an interval we can try to locate its endpoints by looking for its least upper bound and greatest lower bound. In certain situations, the completeness of the real numbers guarantees the existence of these numbers.

Proof of Theorem 7.2.1. Suppose that the domain of f is an interval, f is continuous on that interval, and f is not a constant function. Let R be the range of f. We must prove that R is an interval.

There are several cases, depending on whether or not R has an upper bound or a lower bound. We will consider only one case, leaving the others as exercises for you (see Exercise 11 for another case). The case we will consider is the case in which R has an upper bound but no lower bound.

Since R has an upper bound, by the completeness of the real numbers, it has a least upper bound. Let b be the least upper bound of R. Since b is an upper bound, no number larger than b is in R. Our plan now is to show that every number smaller than b is in R.

Consider any number $r < b$. Since b is the *least* upper bound of R, r is not an upper bound, so there is some $y \in R$ such that $y > r$. And since we have assumed that R does not have a lower bound, r is not a lower bound for R, and therefore there is some $z \in R$ such that $z < r$.

By the definition of range, for y to belong to R, which is the range of f, means that there is some c in the domain of f such that $f(c) = y$. Similarly, since $z \in R$, there is some d in the domain of f such that $f(d) = z$. We now have $f(d) = z < r < y = f(c)$. Since f is continuous, we can apply the intermediate value theorem to conclude that there is some number a between c and d such that $f(a) = r$. Therefore r is in the range of f—that is, $r \in R$.

We have shown that no number larger than b is in R, and every number smaller than b is in R. Thus, if $b \in R$ then $R = (-\infty, b]$, and if not then $R = (-\infty, b)$. In either case, we can conclude that R is an interval.   □

We now turn to the main topic of this section, calculus with inverse functions. Suppose that f is a function whose domain is an interval I, and f is continuous on I. We will be most interested in the case in which f is either strictly increasing or strictly decreasing on I. If f is such a function, then it is not a constant function, so by Theorem 7.2.1, the range of f is some interval J. It should also be clear that no horizontal line can intersect the graph of f more than once, and therefore f is one-to-one. Thus, f^{-1} is defined, and its domain is the interval J. Must f^{-1} be continuous on J? If f is differentiable, must f^{-1} also be differentiable? We begin with an informal discussion of these questions. Once we have found tentative answers, we will state these answers in a theorem and prove the theorem.

Since f is continuous, the graph of f is a curve with no breaks in it. When this graph is flipped over across the line $y = x$, the resulting curve will also have no breaks in it. Thus, it seems that f^{-1} should also be continuous.

Now suppose that $a \in I$ and f is differentiable at a. Then the graph of f has a tangent line at the point where $x = a$. If we let $b = f(a)$ and $m = f'(a)$, then the equation of this tangent line, written in point-slope form, is $y - b = m(x - a)$. Now we flip over both the graph of f and also the tangent line across the line $y = x$, as shown in Figure 7.8. Reversing the roles of x and y, we see that the equation of the line tangent to the graph of f^{-1} at the point (b, a) is $x - b = m(y - a)$. If $m \neq 0$, then we can rewrite this equation in the form $y - a = (1/m)(x - b)$, and we see that the slope of the line is $1/m = 1/f'(a)$. Thus, it appears that if $f'(a) \neq 0$, then f^{-1} is differentiable at b and $(f^{-1})'(b) = 1/f'(a)$.

It turns out that our conclusions about the continuity and differentiability of f^{-1} are correct, but the reasoning we have used was based on intuition and pictures, rather than on the mathematical definitions of continuity and differentiability. To confirm the correctness of our conclusions, we formulate them as a precise theorem and give a careful proof.

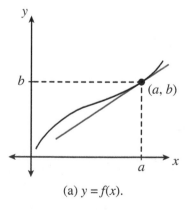

(a) $y = f(x)$.

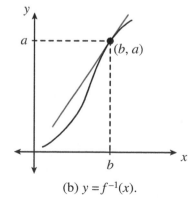

(b) $y = f^{-1}(x)$.

Figure 7.8: Flipping over the tangent line to the graph of f at (a, b) gives the tangent line to the graph of f^{-1} at (b, a).

Theorem 7.2.2. *Suppose the domain of f is an interval I, f is continuous on I, and f is either strictly increasing or strictly decreasing on I. Then f^{-1} is defined, and its domain is an interval J. Furthermore:*

1. *The function f^{-1} is continuous on J.*

2. *Suppose $a \in I$ and let $b = f(a)$. If f is differentiable at a and $f'(a) \neq 0$, then f^{-1} is differentiable at b, and*

$$(f^{-1})'(b) = \frac{1}{f'(a)}.$$

Proof. We will assume in our proof that f is strictly increasing; the proof for strictly decreasing functions is similar. We have already shown that f^{-1} is defined and that its domain is an interval J.

To prove part 1, suppose that b is in J. We must verify that f^{-1} is continuous at b, where as usual we use one-sided continuity at endpoints. We can accomplish this by showing that if b is not the right endpoint of J, then f^{-1} is continuous from the right at b, and if b is not the left endpoint, then f^{-1} is continuous from the left at b. The proofs of these two statements are very similar, so we will do only the first.

Suppose that b is not the right endpoint of J, and let $a = f^{-1}(b)$. Then $f(a) = b$ and, since f is strictly increasing, a is not the right endpoint of I. To show that f^{-1} is continuous from the right at b, we must prove that $\lim_{x \to b^+} f^{-1}(x) = f^{-1}(b) = a$. We use the ϵ-δ method.

Suppose $\epsilon > 0$. Since a is not the right endpoint of I, we can find a number a_1 in I such that $a < a_1 \leq a + \epsilon$. Let $b_1 = f(a_1)$, and note that since f is strictly increasing, $b < b_1$ (see Figure 7.9). Let $\delta = b_1 - b > 0$, so that $b + \delta = b_1$.

Now suppose that $b < x < b + \delta$; we must prove that $|f^{-1}(x) - a| < \epsilon$. We have

$$f(a) = b < x < b + \delta = b_1 = f(a_1),$$

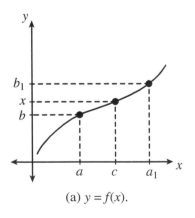

(a) $y = f(x)$.

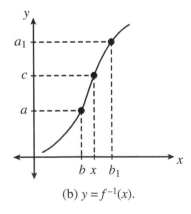

(b) $y = f^{-1}(x)$.

Figure 7.9: The graphs of f and f^{-1}.

so by the intermediate value theorem there must be some number c such that $a < c < a_1$ and $f(c) = x$. Therefore $f^{-1}(x) = c$, so

$$a < f^{-1}(x) < a_1 \leq a + \epsilon,$$

which implies that $|f^{-1}(x) - a| < \epsilon$, as required.

For part 2, suppose that f is differentiable at a and $f'(a) \neq 0$. We will assume initially that a is in the interior of I, so that b is in the interior of J, and we therefore use ordinary two-sided derivatives. To find $(f^{-1})'(b)$, we use the definition of derivatives:

$$(f^{-1})'(b) = \lim_{h \to 0} \frac{f^{-1}(b+h) - f^{-1}(b)}{h}.$$

To compute this limit, we let $x = f^{-1}(b+h)$. Then $f(x) = b+h$, so $h = f(x) - b = f(x) - f(a)$. It follows that if $h \neq 0$ then $f(x) \neq f(a)$, so $x \neq a$. Thus, we can write

$$\frac{f^{-1}(b+h) - f^{-1}(b)}{h} = \frac{x - a}{f(x) - f(a)} = \frac{1}{\frac{f(x) - f(a)}{x - a}}.$$

By part 1, f^{-1} is continuous at b, so as $h \to 0^{\neq}$, $x = f^{-1}(b+h) \to f^{-1}(b) = a$. But as we have just observed, if $h \neq 0$ then $x \neq a$, so we can say that as $h \to 0^{\neq}$, $x \to a^{\neq}$. And since f is differentiable at a, as $x \to a^{\neq}$,

$$\frac{f(x) - f(a)}{x - a} \to f'(a).$$

Putting these facts together, we see that as $h \to 0^{\neq}$,

$$\frac{f^{-1}(b+h) - f^{-1}(b)}{h} = \frac{1}{\frac{f(x) - f(a)}{x - a}} \to \frac{1}{f'(a)}.$$

Thus, $(f^{-1})'(b) = 1/f'(a)$. We leave it to you to make appropriate adjustments in the proof to verify that if a and b are endpoints of their intervals, then this equation holds for one-sided derivatives. $\square$

To illustrate Theorem 7.2.2, we return to our first example in the last section, the function $f(x) = x^3$, whose inverse is $f^{-1}(x) = \sqrt[3]{x} = x^{1/3}$. Since f is continuous and strictly increasing on the interval $(-\infty, \infty)$, we can apply the theorem. In particular, part 2 gives us a formula for the derivative of f^{-1}. For any real number $x \neq 0$, let $a = \sqrt[3]{x} \neq 0$. Then $f(a) = a^3 = x$ and $f'(a) = 3a^2 \neq 0$, so according to the theorem,

$$(f^{-1})'(x) = \frac{1}{f'(a)} = \frac{1}{3a^2} = \frac{1}{3(\sqrt[3]{x})^2} = \frac{1}{3} x^{-2/3},$$

in agreement with the answer given by the power rule.

The equation in part 2 of Theorem 7.2.2 takes an interesting form when written in Leibniz notation. If $y = f^{-1}(x)$ then $x = f(y)$, and according to part 2 of the theorem, if $f'(y) \neq 0$ then $(f^{-1})'(x) = 1/f'(y)$. But in Leibniz notation, $(f^{-1})'(x) = dy/dx$ and $f'(y) = dx/dy$, so this equation says

$$\frac{dy}{dx} = \frac{1}{dx/dy}.$$

Thus, this is another instance in which, although dy/dx is not a fraction, it acts like one.

Exercises 7.2

1–4: Find the range of the function.

1. $f(x) = 3x - 5$.

2. $g(x) = \sqrt{x - 2}$.

3. $h(x) = \sqrt{6x - x^2}$.

4. $f(x) = \cos\left(\dfrac{\pi}{x^2 + 3}\right)$.

5. Find the range of the function $f(x) = 1/x$. Why doesn't this contradict Theorem 7.2.1?

6. Let f be the function defined as follows:

$$f(x) = \begin{cases} |x|/x, & \text{if } x \neq 0, \\ 0, & \text{if } x = 0. \end{cases}$$

Find the range of f. Why doesn't this contradict Theorem 7.2.1?

7. Let $f(x) = (x + 1)/(x - 2)$. Use part 2 of Theorem 7.2.2 to compute $(f^{-1})'(2)$. Then use your answer to Exercise 10 of Section 7.1 to compute $(f^{-1})'(2)$ directly.

8. Let $f(x) = 1/(x^2 + 1)$, $x \geq 0$. Use part 2 of Theorem 7.2.2 to compute $(f^{-1})'(1/2)$. Then use your answer to Exercise 16 of Section 7.1 to compute $(f^{-1})'(1/2)$.

9. (a) Let $g(x) = \sqrt{x}$. Then $g = f^{-1}$, where $f(x) = x^2$, $x \geq 0$. Use part 2 of Theorem 7.2.2 to compute $g'(x)$, and verify that your answer agrees with the answer given by the power rule.

   (b) Generalize: Use part 2 of Theorem 7.2.2 to find $\frac{d}{dx}(\sqrt[n]{x})$ for any even positive integer n.

10. Let $g(x) = \sqrt[n]{x}$, where n is an odd positive integer. Use part 2 of Theorem 7.2.2 to find $g'(x)$, and verify that your answer agrees with the answer given by the power rule.

11. Prove Theorem 7.2.1 in the case that R has both a lower bound and an upper bound.

7.3 The Natural Logarithm

At this point, we know how to differentiate a wide range of functions, but our knowledge of antiderivatives is more limited. Perhaps the simplest function for which we don't yet know an antiderivative is the function $f(x) = x^{-1} = 1/x$. (Recall that the formula $\int x^r \, dx = x^{r+1}/(r+1) + C$ doesn't apply when $r = -1$, since it would lead to division by 0.) In this section we will learn about an antiderivative of f.

As we saw in Corollary 5.4.2, one consequence of the first fundamental theorem of calculus is that if a function is continuous on an interval then it has an antiderivative on that interval. Thus, f must have an antiderivative on each of the intervals $(-\infty, 0)$ and $(0, \infty)$. Indeed, the first fundamental theorem tells us how to write down a particular antiderivative. Motivated by that theorem, we make the following definition:

Definition 7.3.1. For $x > 0$, we define

$$\ln(x) = \int_1^x \frac{1}{t} \, dt.$$

We sometimes leave out the parentheses and write $\ln x$ rather than $\ln(x)$. (We will discuss the name of this function and the reason for the notation "ln" later.)

According to the first fundamental theorem of calculus, for $x > 0$ we have

$$\frac{d}{dx}(\ln x) = \frac{1}{x}.$$

It follows that

$$\int \frac{1}{x} \, dx = \ln x + C \text{ on } (0, \infty).$$

We have now solved the problem of finding an antiderivative of $1/x$, at least on the interval $(0, \infty)$. You may feel that, in some sense, it is cheating to call this a solution. Have we really found an antiderivative, or have we just made up a name for the antiderivative and then declared victory? To respond to this question, let us make several points about the significance of our solution.

First of all, we now know that $1/x$ *has* an antiderivative on the interval $(0, \infty)$. And in mathematics, once you know that something exists, it is often very useful to give it a name. Once you have a name for something, you can formulate questions about it and investigate its properties. We will eventually learn enough properties of the function $\ln x$ that you will begin to feel as comfortable with it as you do with other familiar functions, such as $\sqrt{x}$ or $\sin x$.

But Definition 7.3.1 does more than simply give a name to an antiderivative of $1/x$. The definition gives a formula for $\ln x$ that can be used to compute the numerical value of $\ln x$ for any positive number x. The easiest case is $x = 1$, since

$$\ln 1 = \int_1^1 \frac{1}{t} \, dt = 0.$$

But we can use Riemann sums to approximate other values of $\ln x$ to any desired degree of accuracy.

For example, consider the case $x = 2$:

$$\ln 2 = \int_1^2 \frac{1}{t}\, dt.$$

To estimate the value of ln 2, let's compute a Riemann 4-sum for the function $f(t) = 1/t$ on the interval $[1, 2]$. We divide the interval $[1, 2]$ into 4 smaller intervals, each of width $\Delta t = 1/4$, using the division points

$$t_0 = 1, \quad t_1 = \frac{5}{4}, \quad t_2 = \frac{3}{2}, \quad t_3 = \frac{7}{4}, \quad t_4 = 2.$$

Taking the right endpoint of each interval $[t_{i-1}, t_i]$ as the sample point—that is, letting $t_i^* = t_i$—leads to the Riemann sum

$$R_4 = \sum_{i=1}^4 f(t_i)\Delta t = \frac{4}{5}\cdot\frac{1}{4} + \frac{2}{3}\cdot\frac{1}{4} + \frac{4}{7}\cdot\frac{1}{4} + \frac{1}{2}\cdot\frac{1}{4} = \frac{533}{840} \approx 0.635.$$

In Figure 7.10a, the area of the blue shaded region is ln 2, and the Riemann sum R_4 is equal to the sum of the areas of the four striped rectangles. Since the rectangles lie entirely inside the blue region, it is clear that

$$\ln 2 \geq R_4 \approx 0.635. \tag{7.5}$$

(For a more careful verification of this inequality, see Exercise 20.) A similar calculation, using the left endpoint of each interval as the sample point, leads to a Riemann sum that is an overestimate of ln 2 (see Figure 7.10b):

$$\ln 2 \leq 1\cdot\frac{1}{4} + \frac{4}{5}\cdot\frac{1}{4} + \frac{2}{3}\cdot\frac{1}{4} + \frac{4}{7}\cdot\frac{1}{4} = \frac{319}{420} \approx 0.759.$$

Riemann n-sums for larger values of n require more arithmetic, but should give more accurate estimates of ln 2. And, indeed, similar calculations using Riemann 1000-sums rather than Riemann 4-sums lead to the conclusion that

$$0.6928 \leq \ln 2 \leq 0.6934.$$

Thus, we can say that $\ln 2 \approx 0.693$.

In Exercise 21, we invite you to use Riemann 4-sums to estimate the value of ln 3. Once again, using Riemann n-sums for larger values of n leads to more accurate estimates, allowing us to conclude that $\ln 3 \approx 1.099$.

We now know three points on the graph of $y = \ln x$: $(1, 0)$, $(2, \ln 2) \approx (2, 0.693)$, and $(3, \ln 3) \approx (3, 1.099)$. What does the rest of the graph look like? We can use the curve-sketching techniques we learned in Chapter 4 to answer this question. For $x > 0$

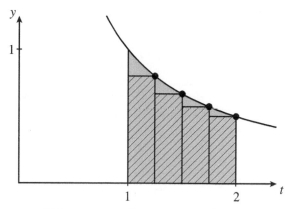

(a) Riemann sum is smaller than integral.

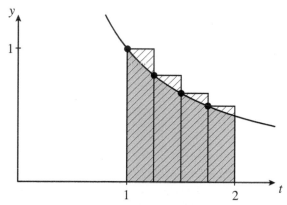

(b) Riemann sum is larger than integral.

Figure 7.10: The graph of $y = 1/t$.

we have

$$\frac{d}{dx}(\ln x) = \frac{1}{x} > 0,$$

$$\frac{d^2}{dx^2}(\ln x) = \frac{d}{dx}\left(\frac{1}{x}\right) = -\frac{1}{x^2} < 0.$$

It follows that the function $\ln x$ is continuous, strictly increasing, and concave down on its entire domain, the interval $(0, \infty)$. Thus, the graph of $y = \ln x$ must have the shape shown in Figure 7.11. We have put question marks at both ends of the graph to indicate that we don't know yet whether there are asymptotes. To determine this, we will need to evaluate $\lim_{x \to 0^+} \ln x$ and $\lim_{x \to \infty} \ln x$; we'll find the values of these limits later in this section.

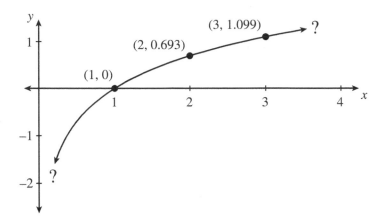

Figure 7.11: The graph of $y = \ln x$.

Example 7.3.2. Find the derivatives of the following functions:

$$f(x) = \sin(\ln x), \qquad g(x) = \sin x \ln(x^2 + 1), \qquad h(x) = \ln(5x).$$

Solution. We combine the equation $\frac{d}{dx}(\ln x) = 1/x$ with our previous formulas for derivatives:

$$f'(x) = \cos(\ln x) \cdot \frac{d}{dx}(\ln x) = \cos(\ln x) \cdot \frac{1}{x} = \frac{\cos(\ln x)}{x},$$

$$g'(x) = \sin x \cdot \frac{d}{dx}(\ln(x^2 + 1)) + \ln(x^2 + 1) \cdot \frac{d}{dx}(\sin x)$$

$$= \sin x \cdot \frac{1}{x^2 + 1} \cdot \frac{d}{dx}(x^2 + 1) + \ln(x^2 + 1) \cdot \cos x = \frac{2x \sin x}{x^2 + 1} + \ln(x^2 + 1) \cos x,$$

$$h'(x) = \frac{1}{5x} \cdot \frac{d}{dx}(5x) = \frac{1}{5x} \cdot 5 = \frac{1}{x}. \qquad \qquad \square$$

The derivative of the function h in Example 7.3.2 is a bit surprising: the factor of 5 ended up making no difference at all! Indeed, for any positive constant a we have

$$\frac{d}{dx}(\ln(ax)) = \frac{1}{ax} \cdot \frac{d}{dx}(ax) = \frac{1}{ax} \cdot a = \frac{1}{x}.$$

This means that $\ln(ax)$ is another antiderivative of $1/x$ on the interval $(0, \infty)$. But we already know that every such antiderivative must have the form $\ln x + C$. We conclude that there must be some number C such that for every $x > 0$,

$$\ln(ax) = \ln x + C.$$

Can we determine the value of C? We usually do this by plugging in a particular value for x, and in the case of the function $\ln x$, the easiest value to plug in is 1.

Setting $x = 1$ we find that

$$\ln a = \ln 1 + C = 0 + C = C.$$

Thus, $C = \ln a$, so we have shown that for all $x > 0$,

$$\ln(ax) = \ln x + \ln a \qquad (7.6)$$

This equation leads to our next theorem.

Theorem 7.3.3. *Let a and b be any positive numbers.*

1. $\ln(ab) = \ln a + \ln b$.

2. $\ln\left(\dfrac{1}{a}\right) = -\ln a$.

3. $\ln\left(\dfrac{a}{b}\right) = \ln a - \ln b$.

Proof. To prove part 1, we simply substitute b for x in equation (7.6). (Notice that this proof could be thought of as another example of the paradox of generalization: to understand $\ln(ab)$ for particular positive numbers a and b, we used a fact about the function $\ln(ax)$ for all $x > 0$.)

We use part 1 to prove part 2:

$$\ln a + \ln\left(\frac{1}{a}\right) = \ln\left(a \cdot \frac{1}{a}\right) \qquad \text{(by part 1)}$$
$$= \ln 1 = 0.$$

Rearranging, we get $\ln(1/a) = -\ln a$.

Finally, part 3 follows from parts 1 and 2:

$$\ln\left(\frac{a}{b}\right) = \ln\left(a \cdot \frac{1}{b}\right) = \ln a + \ln\left(\frac{1}{b}\right) = \ln a - \ln b. \qquad \square$$

We hope the equations in Theorem 7.3.3 look familiar to you: if you have studied logarithms before, then you will recognize that logarithms satisfy exactly the same equations. It appears, then, that the function $\ln x$ has something in common with logarithms. For this reason, $\ln x$ is called the *natural logarithm* of x. Although natural logarithms are not exactly the same as the logarithms that are usually studied in high school, the two are closely related; we will discuss the relationship later in this chapter.[1]

---

[1] Although the notation $\ln x$ is common, it is not clear how it originated. Logarithms were first developed by John Napier (1550–1617), so some suggest that "ln" stands for "logarithm of Napier," but there doesn't seem to be much evidence to support this interpretation. In 1668, in his book *Logarithmotechnia*, Nicolaus Mercator (1620–1687) (no relation to the Mercator of map fame) discussed natural logarithms using the Latin phrase "logarithmus naturalis." In his book *Lehrbuch der Mathematik*, published in 1875, Anton Steinhauser (1802–1890) suggested denoting the natural logarithm of a number a by "log. nat. a (spoken: logarithmus naturalis a) or ln. a" (p. 277). Steinhauser appears to have used the "log. nat." notation in the rest of his book, but the notation "ln x" then appeared without explanation in 1893 in the book *Uniplanar Algebra*, by Irving Stringham (1847–1909).

Our next theorem shows that natural logarithms satisfy another familiar property of logarithms:

Theorem 7.3.4. *For any positive number a and any rational number r,* $\ln(a^r) = r \ln a$.

Proof. Motivated by the paradox of generalization, to learn about $\ln(a^r)$ for a single positive number a we begin by studying the function $\ln(x^r)$ for all $x > 0$. We have

$$\frac{d}{dx}(\ln(x^r)) = \frac{1}{x^r} \cdot \frac{d}{dx}(x^r) = \frac{rx^{r-1}}{x^r} = \frac{r}{x}.$$

Thus, $\ln(x^r)$ is an antiderivative of r/x on the interval $(0, \infty)$. But we can find all such antiderivatives by computing the indefinite integral of r/x:

$$\int \frac{r}{x}\, dx = r \int \frac{1}{x}\, dx = r \ln x + C \text{ on } (0, \infty).$$

It follows that there must be some constant C such that for all $x > 0$,

$$\ln(x^r) = r \ln x + C.$$

Once again, to evaluate C we set $x = 1$, which gives us the equation

$$\ln 1 = r \ln 1 + C.$$

Since $\ln 1 = 0$, we conclude that $C = 0$, and therefore

$$\ln(x^r) = r \ln x.$$

Finally, substituting a for x proves the theorem. $\square$

As an example of Theorem 7.3.4, consider the case $a = 3$. Since $\ln 3 \approx 1.099 > 1$, for every positive integer n we have $\ln(3^n) = n \ln 3 > n$. Using this fact, we can now settle the question of whether the curve $y = \ln x$ has asymptotes:

Theorem 7.3.5.

1. $\lim\limits_{x \to \infty} \ln x = \infty$.

2. $\lim\limits_{x \to 0^+} \ln x = -\infty$.

Proof. By definition, the limit statement in part 1 means that for every number M, there is some number N such that if $x > N$ then $\ln x > M$. To prove this, consider any number M. Let n be an integer larger than M, and let $N = 3^n$. Then for any $x > N$ we have

$$\ln x > \ln N = \ln(3^n) = n \ln 3 > n > M,$$

as required.

For part 2, we use the fact that $\ln(1/x) = -\ln x$, and therefore $\ln x = -\ln(1/x)$. As $x \to 0^+$, we have $1/x \to \infty$, and therefore, by part 1,

$$\ln x = -\ln(1/x) \to -\infty.$$

Thus, $\lim_{x \to 0^+} \ln x = -\infty$. $\square$

Theorem 7.3.5 tells us that the graph of $y = \ln x$ has a vertical asymptote at $x = 0$, but no horizontal asymptote. Although the curve rises very slowly if we continue to follow it to the right in Figure 7.11, for any horizontal line $y = M$ that we might draw, the curve eventually crosses that line and continues to rise.

We have not yet found an antiderivative of $1/x$ on the interval $(-\infty, 0)$. For $x < 0$, $\ln x$ is undefined, but $\ln(-x)$ is defined, and we have

$$\frac{d}{dx}(\ln(-x)) = \frac{1}{-x} \cdot \frac{d}{dx}(-x) = -\frac{1}{x} \cdot (-1) = \frac{1}{x}.$$

Thus, $\ln(-x)$ is an antiderivative of $1/x$ on $(-\infty, 0)$. We can combine our antiderivatives on the intervals $(-\infty, 0)$ and $(0, \infty)$ into one function by observing that

$$\ln |x| = \begin{cases} \ln x, & \text{if } x > 0, \\ \ln(-x), & \text{if } x < 0. \end{cases}$$

Thus, we can write

$$\int \frac{1}{x}\, dx = \ln |x| + C,$$

and this answer applies on both of the intervals $(-\infty, 0)$ and $(0, \infty)$ that make up the domain of the function $1/x$. The graph of $y = \ln |x|$ is shown in Figure 7.12.

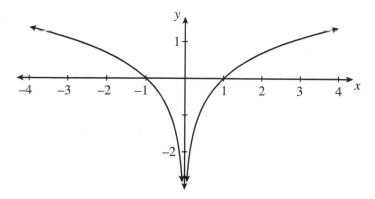

Figure 7.12: The graph of $y = \ln |x|$.

Example 7.3.6. Evaluate the following integrals:

(a) $\displaystyle\int \frac{x}{x^2 - 3}\, dx$, (b) $\displaystyle\int \frac{\ln x}{x}\, dx$, (c) $\displaystyle\int_1^2 \frac{dx}{1 - 2x}$.

Solution. To evaluate integral (a) we use the substitution $u = x^2 - 3$, $du = 2x\,dx$, $(1/2)\,du = x\,dx$:

$$\int \frac{x}{x^2 - 3}\,dx = \int \frac{1}{u} \cdot \frac{1}{2}\,du = \frac{1}{2} \ln |u| + C = \frac{\ln |x^2 - 3|}{2} + C.$$

For integral (b), we use the substitution $u = \ln x$, $du = (1/x)\,dx$:

$$\int \frac{\ln x}{x}\,dx = \int u\,du = \frac{u^2}{2} + C = \frac{(\ln x)^2}{2} + C.$$

Finally, for integral (c) we use the substitution $u = 1 - 2x$, $du = -2\,dx$, $(-1/2)\,du = dx$. Since this is a definite integral, we also change the limits of integration from the x-values 1 and 2 to the corresponding u-values -1 and -3:

$$\int_1^2 \frac{dx}{1 - 2x} = \int_{-1}^{-3} \frac{-1/2}{u}\,du = \left[-\frac{1}{2} \ln |u| \right]_{-1}^{-3} = -\frac{1}{2} \ln 3 + \frac{1}{2} \ln 1$$

$$= -\frac{\ln 3}{2} \approx -0.549. \qquad\qquad \square$$

Exercises 7.3

1–7: Find the derivative of the function.

1. $f(x) = \ln(2x - 3)$.

2. $f(x) = \ln(\tan x)$.

3. $g(x) = \tan(\ln x)$.

4. $h(x) = \ln(\tan(\ln x))$.

5. $f(x) = \ln(\sqrt{x^2 + 1})$.

6. $f(x) = \dfrac{\ln x}{x^2}$.

7. $g(x) = \sin x \cdot (\ln(\sin x) - 1)$.

8–15: Evaluate the integral.

8. $\displaystyle \int \frac{x^2}{x^3 + 5}\,dx$.

9. $\displaystyle \int \tan x\,dx$. (Hint: Start by filling in the definition of $\tan x$.)

10. $\displaystyle \int \frac{1}{x \ln x}\,dx$.

11. $\displaystyle \int \frac{x + 1}{x - 1}\,dx$.

12. $\displaystyle\int \frac{(x+2)^2}{x}\,dx.$

13. $\displaystyle\int \frac{1}{\sin x \cos x}\,dx.$ (Hint: First divide numerator and denominator by $\cos^2 x$.)

14. $\displaystyle\int_1^4 \frac{dx}{x+\sqrt{x}}.$ (Hint: First rewrite the denominator as $\sqrt{x}(\sqrt{x}+1)$.)

15. $\displaystyle\int_2^4 \frac{x\,dx}{1-x^2}.$

16–19: Evaluate the limit.

16. $\displaystyle\lim_{x\to 1}\frac{\ln(3-x)}{3-\ln x}.$

17. $\displaystyle\lim_{x\to 1^+}\frac{\ln(\sqrt{x})}{\sqrt{\ln x}}.$

18. $\displaystyle\lim_{x\to 1^+}[\ln(x^2-1)-\ln(x-1)].$

19. $\displaystyle\lim_{x\to 1}\frac{\ln x}{x-1}.$ (Hint: Use l'Hôpital's rule.)

20. Use Theorems 5.3.7 and 5.3.9 to verify inequality (7.5).

21. Use Riemann 4-sums to show that

$$\frac{19}{20} \le \ln 3 \le \frac{77}{60}.$$

22. Show that $2-\sqrt{2} < \ln 2 < 2\sqrt{2}-2$. (Start by applying the mean value theorem to the function $\ln x$ on the interval $[1, \sqrt{2}]$.)

23. Sketch the graph of the function $f(x) = \ln(x^2+1)$.

24. Show that for all $x > 0$, $\ln x < \sqrt{x}$. (Hint: Find the minimum value of $\sqrt{x}-\ln x$ on the interval $(0, \infty)$.)

25. Find the volume of the solid generated when the region under the curve $y = 1/\sqrt{x}$, $1 \le x \le 4$, is rotated about the x-axis.

26. Find the volume of the solid generated when the region under the curve $y = 1/(x^2+1)$, $0 \le x \le 1$, is rotated about the y-axis.

27. A hollow cylinder is 20 inches long, and it has a radius of 2 inches. It is closed at one end and open at the other. A piston (a disk that just fits inside the cylinder) is inserted into the open end and pushed in, and as a result the air inside the cylinder is compressed and the pressure inside the cylinder increases. According to Boyle's law, the pressure P of the air in the cylinder (measured in pounds per square inch, abbreviated psi) and the volume V of the air in the cylinder (measured in in^3) are related by the equation $P = K/V$, for some constant K.

(a) Assuming that the pressure inside the cylinder when the piston is first inserted is equal to atmospheric pressure (approximately 15 psi), find K.

(b) At any time, the air inside the cylinder pushes out on the piston with a force equal to the pressure inside the cylinder times the area of the piston. Similarly, the air outside the cylinder pushes in on the piston with a force equal to atmospheric pressure times the area of the piston. In order to push the piston further into the cylinder, one must exert a force equal to the difference between these forces. Show that when the piston has been pushed x inches into the cylinder, the force needed to push it further, in pounds, is given by the formula

$$F(x) = \frac{60\pi x}{20 - x}.$$

(c) Find the work done in pushing the piston 10 inches into the cylinder.

7.4 The Exponential Function

In the last section, we learned that the natural logarithm function is strictly increasing. It follows that the natural log is one-to-one, and therefore it has an inverse. In this section we study the inverse of the natural log. For reasons that will soon become clear, it is called the exponential function.

Definition 7.4.1. The inverse of the natural logarithm function is called the *exponential function*, and is denoted exp.

The graph of $y = \ln x$ is shown in Figure 7.13a. Flipping this graph over across the line $y = x$ produces the graph of $y = \exp(x)$, which is shown in Figure 7.13b.

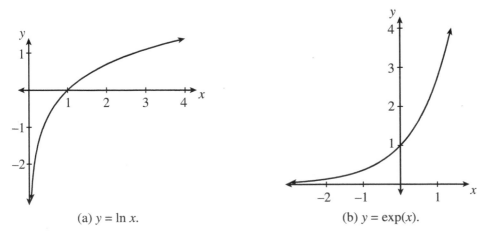

(a) $y = \ln x$. (b) $y = \exp(x)$.

Figure 7.13: The graphs of the natural logarithm and the exponential function.

And since flipping the graph over corresponds to reversing the roles of x and y, we have

$$y = \exp(x) \quad \text{if and only if} \quad x = \ln y.$$

Applying ideas from Sections 7.1 and 7.2, we can immediately make a few observations about the exponential function. The domain of the natural logarithm is $(0, \infty)$, and since $\lim_{x \to 0+} \ln x = -\infty$ and $\lim_{x \to \infty} \ln x = \infty$, the range is $(-\infty, \infty)$. It follows that the domain of the exponential function is $(-\infty, \infty)$, and the range is $(0, \infty)$. Thus, for all x, $\exp(x)$ is defined and $\exp(x) > 0$. We also know that the natural log and exponential function undo each other. In other words,

$$\text{for all } x, \ \ln(\exp(x)) = x,$$
$$\text{for all } x > 0, \ \exp(\ln x) = x.$$

By Theorem 7.2.2, the exponential function is continuous and differentiable at all real numbers. We will find its derivative later in this section.

Some further properties of the exponential function follow immediately from properties of the natural log:

Theorem 7.4.2. *Let a and b be any real numbers.*

1. $\exp(a) \cdot \exp(b) = \exp(a + b)$.

2. $\dfrac{\exp(a)}{\exp(b)} = \exp(a - b)$.

3. *If $a > 0$ and r is any rational number, then $a^r = \exp(r \ln a)$.*

Proof. We begin by observing that, by part 1 of Theorem 7.3.3,

$$\ln(\exp(a) \cdot \exp(b)) = \ln(\exp(a)) + \ln(\exp(b)) = a + b.$$

Therefore
$$\exp(a) \cdot \exp(b) = \exp(\ln(\exp(a) \cdot \exp(b))) = \exp(a + b),$$

which establishes part 1. The proof of part 2 is similar; see Exercise 1.

Finally, for part 3 we begin with the fact that, by Theorem 7.3.4, $\ln(a^r) = r \ln a$. Therefore

$$a^r = \exp(\ln(a^r)) = \exp(r \ln a). \qquad \square$$

As an example of part 3 of Theorem 7.4.2, consider the case $a = 2$. According to the theorem, for every rational number r,

$$2^r = \exp(r \ln 2) \approx \exp(0.693r).$$

What if r is an irrational number? Then, as we have observed before, we have not yet assigned a meaning to the expression 2^r. But notice that the domain of the exponential function includes all real numbers, and therefore $\exp(r \ln 2)$ is defined.

Another way to understand this point is to compare the graphs of the equations $y = 2^x$ and $y = \exp(x \ln 2)$. The graph of $y = 2^x$ includes only points whose

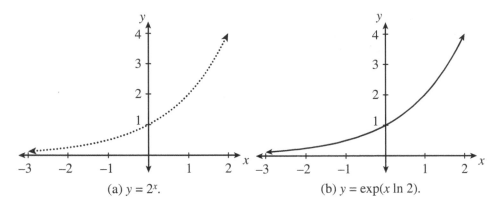

(a) $y = 2^x$.   (b) $y = \exp(x \ln 2)$.

Figure 7.14: The graphs of $y = 2^x$ and $y = \exp(x \ln 2)$.

x-coordinates are rational, so it looks something like the dotted curve in Figure 7.14a. It is a curve with holes in it at all irrational values of x. (Of course, it isn't possible to make the dots and holes in the figure small enough to accurately reflect the way the rational and irrational numbers are crowded together on the number line.) On the other hand, the graph of $y = \exp(x \ln 2)$ is a continuous curve with no holes in it, as shown in Figure 7.14b. What part 3 of Theorem 7.4.2 tells us is that this curve passes through all of the dots that make up the graph of $y = 2^x$. In other words, if you were to place Figure 7.14b on top of Figure 7.14a, the continuous curve $y = \exp(x \ln 2)$ would lie exactly on top of the dotted curve $y = 2^x$.

Up until now we have not tried to define 2^x for irrational values of x. But surely if we were to define it, we would want to do it in such a way that the graph of $y = 2^x$ would become a continuous curve. It wasn't clear before whether or not this was possible; that is, we didn't know whether there *was* a continuous curve passing through all points on the graph in Figure 7.14a. But now we have stumbled on such a curve: $y = \exp(x \ln 2)$. And in fact, it can be shown that this is the *only* continuous curve that passes through all of these points (see Exercise 22). This suggests that for irrational x, we should define 2^x to be $\exp(x \ln 2)$. More generally, we make the following definition:

Definition 7.4.3. If $a > 0$ and x is irrational, then we define a^x to be $\exp(x \ln a)$.

Notice that this definition can only be used when $a > 0$, since $\ln a$ is undefined for $a \le 0$ (see Exercise 23). Combining this definition with part 3 of Theorem 7.4.2, we can now say that for all $a > 0$ and all real numbers x, whether rational or irrational,

$$a^x = \exp(x \ln a), \tag{7.7}$$

and therefore

$$\ln a^x = x \ln a. \tag{7.8}$$

Using these facts, we can show that our new, extended definition of a^x satisfies all of the familiar properties of exponentiation:

Theorem 7.4.4. *Suppose $a > 0$, $b > 0$, and r and s are any real numbers.*

1. $a^r \cdot a^s = a^{r+s}$.

2. $\dfrac{a^r}{a^s} = a^{r-s}$.

3. $(a^r)^s = a^{rs}$.

4. $a^r \cdot b^r = (ab)^r$.

5. $\dfrac{a^r}{b^r} = \left(\dfrac{a}{b}\right)^r$.

Proof. For part 1, we apply equation (7.7) and part 1 of Theorem 7.4.2:

$$a^r \cdot a^s = \exp(r \ln a) \cdot \exp(s \ln a) = \exp(r \ln a + s \ln a) = \exp((r+s) \ln a) = a^{r+s}.$$

For part 3, we use equations (7.7) and (7.8) to compute

$$(a^r)^s = \exp(s \ln a^r) = \exp(s(r \ln a)) = \exp((rs) \ln a) = a^{rs}.$$

The proofs for parts 2, 4, and 5 are similar. We leave the details to you; see Exercise 2. □

We have now defined a^x for all $a > 0$ and all real numbers x, but there is a particular value of a that is especially convenient. Equation (7.7) is simplest if $\ln a = 1$, or equivalently $a = \exp(1)$. There is a standard notation in mathematics for this number.

Definition 7.4.5. The number $\exp(1)$ is denoted e.

We have $e = \exp(1)$, and since the exponential function is the inverse of the natural log, this means that $\ln e = 1$. We already know that $\ln 2 \approx 0.693 < 1$, $\ln 3 \approx 1.099 > 1$, and the natural log function is strictly increasing, so e must be between 2 and 3. Riemann sum calculations similar to the ones we did earlier can be used to show that $\ln 2.7 \approx 0.993 < 1$ and $\ln 2.8 \approx 1.030 > 1$, so $2.7 < e < 2.8$. And by taking these calculations even further it can be shown that $e \approx 2.71828$. It turns out that e, like π, is an irrational number whose decimal expansion goes on forever, with no obvious pattern in the digits. It has important uses throughout mathematics, and we will make extensive use of it for the rest of this book.

Taking $a = e$ in equation (7.7), we see that for all x,

$$e^x = \exp(x \ln e) = \exp(x \cdot 1) = \exp(x).$$

In other words, the exponential function $\exp(x)$ is the same as the function e^x! Of course, this is the reason it is called the exponential function. From now on we will use the notations $\exp(x)$ and e^x interchangeably. In particular, we can rewrite some of the most important facts we have learned about the exponential function as follows:

$$\text{for all } x, \ \ln e^x = x,$$

for all $x > 0$, $e^{\ln x} = x$,

for all $a > 0$ and all x, $a^x = e^{x \ln a}$.   (7.9)

A good way to remember equation (7.9) is to write $a = e^{\ln a}$ and then apply part 3 of Theorem 7.4.4:

$$a^x = (e^{\ln a})^x = e^{x \ln a}.$$

One reason that the exponential function and the natural log are important is that they can be useful for solving equations. We illustrate with a couple of examples.

Example 7.4.6. Solve for x:

$$\text{(a) } 3^x = 2^{x+1}, \qquad \text{(b) } \ln(2x + e) = 2 - \ln x.$$

Solution. The main difficulty in solving equation (a) is that the x's are in the exponents. We can get them out of the exponents by applying the natural logarithm function to both sides of the equation. Notice that both sides are positive for all values of x, so they are in the domain of the natural log.

$$\ln(3^x) = \ln(2^{x+1}),$$
$$x \ln 3 = (x + 1) \ln 2,$$
$$x(\ln 3 - \ln 2) = \ln 2,$$
$$x = \frac{\ln 2}{\ln 3 - \ln 2} \approx 1.71.$$

We leave it to you to verify (see Exercise 24) that this value is a solution to the original equation.

For equation (b), the difficulty is that the x's appear inside the natural log function, so we apply the exponential function to both sides:

$$e^{\ln(2x+e)} = e^{2-\ln x} = \frac{e^2}{e^{\ln x}},$$
$$2x + e = \frac{e^2}{x},$$
$$2x^2 + ex - e^2 = 0,$$
$$(2x - e)(x + e) = 0.$$

The final equation has two solutions, $x = e/2$ and $x = -e$. But $x = -e$ doesn't make sense in the original equation, since the natural log of a negative number is undefined. We ask you to check in Exercise 24 that $x = e/2 \approx 1.359$ is a solution to the original equation, so it is the only solution.   □

We pointed out earlier that, by Theorem 7.2.2, the exponential function is differentiable at all real numbers. What is its derivative? We could use the formula in part 2 of Theorem 7.2.2 to find this derivative, but it is perhaps just as easy to use implicit differentiation. If we let $y = e^x$, then $\ln y = x$, and therefore

$$\frac{d}{dx}(\ln y) = \frac{d}{dx}(x),$$
$$\frac{1}{y} \cdot \frac{dy}{dx} = 1,$$
$$\frac{dy}{dx} = y = e^x.$$

Thus,

$$\frac{d}{dx}(e^x) = e^x.$$

This is certainly a remarkable property of the exponential function; we have never before come across a function that is equal to its own derivative! There are many applications of the exponential function that are based on this property; see Exercises 27 and 28 for examples. In Exercise 25 you are asked to determine all functions that have this interesting property.

Notice that for all x, we have

$$\frac{d}{dx}(e^x) = e^x > 0,$$
$$\frac{d^2}{dx^2}(e^x) = \frac{d}{dx}(e^x) = e^x > 0.$$

It follows that the exponential function is strictly increasing and concave up on $(-\infty, \infty)$. This confirms the shape of the graph in Figure 7.13b. Since the exponential function is strictly increasing and its range is $(0, \infty)$, it is also not hard to verify (see Exercise 15) that

$$\lim_{x \to -\infty} e^x = 0, \qquad \lim_{x \to \infty} e^x = \infty.$$

These limits are also evident in Figure 7.13b. In particular, notice that the vertical line $x = 0$ is an asymptote of the graph of $y = \ln x$, and when the graph is flipped over it becomes the horizontal asymptote $y = 0$ of the graph of $y = e^x$.

Example 7.4.7. Find the derivatives of the following functions:

$$f(x) = e^{x^2 + 1}, \qquad g(x) = \ln(e^{2x} + 1).$$

Solution. We combine our new derivative formula $\frac{d}{dx}(e^x) = e^x$ with the chain rule:

$$f'(x) = e^{x^2 + 1} \cdot \frac{d}{dx}(x^2 + 1) = 2x e^{x^2 + 1},$$
$$g'(x) = \frac{1}{e^{2x} + 1} \cdot \frac{d}{dx}(e^{2x} + 1) = \frac{1}{e^{2x} + 1} \cdot e^{2x} \cdot \frac{d}{dx}(2x) = \frac{2e^{2x}}{e^{2x} + 1}. \qquad \square$$

When working with a function in which some number other than e is raised to a power, it is often easiest to use equation (7.9) to rewrite the function in terms of the exponential function. We illustrate this in our next example.

Example 7.4.8. Find the derivatives of the following functions:

$$f(x) = 2^x, \qquad g(x) = x^\pi, \qquad h(x) = x^x, \; x > 0.$$

Solution. To differentiate f, we use equation (7.9) to rewrite $f(x)$ in the form

$$f(x) = 2^x = e^{x \ln 2}.$$

Therefore

$$f'(x) = \frac{d}{dx}(e^{x \ln 2}) = e^{x \ln 2} \cdot \frac{d}{dx}(x \ln 2) = 2^x \cdot \ln 2.$$

Similar reasoning can be used to show that for any number $a > 0$,

$$\frac{d}{dx}(a^x) = a^x \ln a.$$

Notice that this equation is simplest when $a = e$, in which case $\ln a = \ln e = 1$ and the equation becomes $\frac{d}{dx}(e^x) = e^x$.

Now consider the function g. Since π is irrational, the meaning of x^π is given by equation (7.9). According to that definition, the domain of g is $(0, \infty)$, and for all $x > 0$, $g(x) = e^{\pi \ln x}$. Therefore

$$g'(x) = \frac{d}{dx}(e^{\pi \ln x}) = e^{\pi \ln x} \cdot \frac{d}{dx}(\pi \ln x) = x^\pi \cdot \frac{\pi}{x} = \pi x^{\pi - 1}.$$

Thus, the power rule works for the exponent π as well! In fact, similar reasoning can be used to show that it works for all irrational exponents.

Finally, turning to the function h, by equation (7.9) we have $h(x) = e^{x \ln x}$, and therefore

$$h'(x) = e^{x \ln x} \cdot \frac{d}{dx}(x \ln x) = x^x \cdot \left(x \cdot \frac{d}{dx}(\ln x) + \ln x \cdot \frac{d}{dx}(x) \right)$$

$$= x^x \cdot \left(x \cdot \frac{1}{x} + \ln x \cdot 1 \right) = x^x(1 + \ln x). \qquad \square$$

As another example of the value of equation (7.9), we establish the following facts, which will be useful later.

Theorem 7.4.9. *Let p be any real number.*

1. *If $p > 0$ then $\lim_{x \to \infty} x^p = \infty$.*

2. *If $p < 0$ then $\lim_{x \to \infty} x^p = 0$.*

3. *If $p > 1$ then $\lim_{x \to \infty} p^x = \infty$.*

4. *If $0 < p < 1$ then $\lim_{x \to \infty} p^x = 0$.*

For example, this theorem tells us that $\lim_{x \to \infty} x^{0.01} = \infty$, $\lim_{x \to \infty} x^{-0.01} = 0$, $\lim_{x \to \infty} 1.01^x = \infty$, and $\lim_{x \to \infty} 0.99^x = 0$.

Proof. By equation (7.9), $x^p = e^{p \ln x}$. To find the limit of this function as x approaches infinity, first note that as $x \to \infty$, $\ln x \to \infty$. Thus, if $p > 0$ then $p \ln x \to \infty$, and therefore $x^p = e^{p \ln x} \to \infty$. On the other hand, if $p < 0$ then $p \ln x \to -\infty$, so $x^p = e^{p \ln x} \to 0$. This proves statements 1 and 2.

The proofs for statements 3 and 4 are similar. Once again, we begin by applying equation (7.9) to write $p^x = e^{x \ln p}$. If $p > 1$ then $\ln p > 0$, so as $x \to \infty$, $x \ln p \to \infty$, and therefore $p^x = e^{x \ln p} \to \infty$. On the other hand, if $0 < p < 1$ then $\ln p < 0$, so as $x \to \infty$, $x \ln p \to -\infty$ and $p^x = e^{x \ln p} \to 0$. $\quad\square$

For every new derivative formula there is a corresponding integration formula. Since $\frac{d}{dx}(e^x) = e^x$, we also have $\int e^x \, dx = e^x + C$. Using this formula, we can solve a variety of other integrals involving the exponential function.

Example 7.4.10. Evaluate the following integrals:

$$\text{(a)} \int \frac{e^{\sqrt{x}}}{\sqrt{x}} \, dx, \quad \text{(b)} \int e^{x+3} \sin(e^x) \, dx, \quad \text{(c)} \int \frac{e^{2x}}{e^x + 1} \, dx.$$

Solution. For integral (a), we make the substitution

$$u = \sqrt{x} = x^{1/2}, \quad du = \frac{1}{2} x^{-1/2} \, dx = \frac{1}{2\sqrt{x}} \, dx, \quad 2 \, du = \frac{1}{\sqrt{x}} \, dx,$$

which leads to the solution

$$\int \frac{e^{\sqrt{x}}}{\sqrt{x}} \, dx = \int e^u \cdot 2 \, du = 2e^u + C = 2e^{\sqrt{x}} + C.$$

For integral (b), we write $e^{x+3} = e^x \cdot e^3$ and then substitute $u = e^x$, $du = e^x \, dx$:

$$\int e^{x+3} \sin(e^x) \, dx = \int e^3 \sin(e^x) \cdot e^x \, dx = e^3 \int \sin u \, du$$

$$= -e^3 \cos u + C = -e^3 \cos(e^x) + C.$$

Finally, for integral (c) we rewrite e^{2x} as $e^x \cdot e^x$ and then substitute $u = e^x + 1$, $e^x = u - 1$, $du = e^x \, dx$:

$$\int \frac{e^{2x}}{e^x + 1} \, dx = \int \frac{e^x}{e^x + 1} \cdot e^x \, dx = \int \frac{u - 1}{u} \, du = \int 1 - \frac{1}{u} \, du$$

$$= u - \ln |u| + C = e^x + 1 - \ln |e^x + 1| + C.$$

There are two simplifications we can make in this answer: we can drop the first "$+1$," since adding a constant to an antiderivative gives another antiderivative, and we can drop the absolute value signs, since $e^x + 1$ is always positive. This gives us the answer

$$\int \frac{e^{2x}}{e^x + 1} \, dx = e^x - \ln(e^x + 1) + C. \quad\square$$

We end this section by briefly considering the logarithms that you may have studied previously. If $a > 0$ and $a \neq 1$, then it can be shown that the function $f(x) = a^x$ is one-to-one, its domain is $(-\infty, \infty)$, and its range is $(0, \infty)$ (see Exercise 30). Therefore f has an inverse. This inverse is called the *logarithm to the base a*, and it is denoted $\log_a$. Thus, for all x and y,

$$y = \log_a x \quad \text{if and only if} \quad x = a^y.$$

The domain of $\log_a$ is $(0, \infty)$ and its range is $(-\infty, \infty)$.

To put it more informally, for any $x > 0$, $\log_a x$ is the number y such that $a^y = x$. For example, $\log_3 81 = 4$ since $3^4 = 81$, and $\log_2(1/4) = -2$ since $2^{-2} = 1/4$. You may have previously studied logarithms to the base 10, for which we have $\log_{10} 100 = 2$, $\log_{10} 1000 = 3$, and so on.

Using methods we developed in Section 7.1, we can find a formula for $\log_a x$. Suppose $x > 0$, and let $y = \log_a x$. Then $x = a^y$. We can now solve this equation for y by taking the natural log of both sides of the equation: $\ln x = \ln(a^y) = y \ln a$, and therefore $y = \ln x / \ln a$. But $y = \log_a x$, so we have shown that

$$\log_a x = \frac{\ln x}{\ln a}. \tag{7.10}$$

In the case $a = e$, we have

$$\log_e x = \frac{\ln x}{\ln e} = \frac{\ln x}{1} = \ln x.$$

Thus, we have finally found the relationship between the natural logarithm and other logarithms: the natural logarithm is simply the logarithm to the base e.

We have traveled a somewhat circuitous route to reach our conclusions in this section, so it may be worthwhile to summarize some of the conclusions we have reached about the functions a^x and $\log_a x$. When working with these functions, we have found that it is easiest to use the value $a = e$, for which these functions are $e^x = \exp(x)$ and $\log_e x = \ln x$. One of the reasons these particular functions are the easiest to use is that their derivatives take a very simple form:

$$\frac{d}{dx}(e^x) = e^x, \qquad \frac{d}{dx}(\ln x) = \frac{1}{x}.$$

For other values of a, the derivatives are slightly more complicated. But there is no need to spend much time studying a^x and $\log_a x$ for other values of a, because equations (7.9) and (7.10) tell us how to reexpress these functions in terms of e^x and $\ln x$:

$$a^x = e^{x \ln a}, \qquad \log_a x = \frac{\ln x}{\ln a}.$$

Thus, if we know how to work with the functions e^x and $\ln x$, then we can use these equations to solve problems involving a^x and $\log_a x$ for other values of a.

Exercises 7.4

1. Prove part 2 of Theorem 7.4.2.

2. Prove parts 2, 4, and 5 of Theorem 7.4.4.

3–7: Find the derivative of the function.

3. $f(x) = e^{3x-2}$.

4. $g(x) = x^2 e^{\sqrt{x}}$.

5. $f(x) = \sin(xe^{2x})$.

6. $g(x) = 5^{\cos x}$.

7. $h(x) = (x+1)^{x-1}$.

8–14: Evaluate the integral.

8. $\displaystyle \int e^{3x-2} \, dx$.

9. $\displaystyle \int \frac{e^{\tan x}}{\cos^2 x} \, dx$.

10. $\displaystyle \int e^x \cos(e^x) \, dx$.

11. $\displaystyle \int \frac{3}{e^{2x}} \, dx$.

12. $\displaystyle \int (e^x + 3)^2 \, dx$.

13. $\displaystyle \int e^{x+e^x} \, dx$.

14. $\displaystyle \int 3^{x+2} \, dx$.

15. Show that $\lim_{x \to -\infty} e^x = 0$ and $\lim_{x \to \infty} e^x = \infty$. (Do not use Theorem 7.4.9; the facts you are proving in this exercise were used in the proof of Theorem 7.4.9, so it would be circular to use the theorem to solve this exercise. Hint: Use the definition of limits, and the fact that the exponential function is positive and strictly increasing, and its range is $(0, \infty)$.)

16–19: Evaluate the limit.

16. $\displaystyle \lim_{x \to 3} e^{5-2x}$.

17. $\displaystyle \lim_{x \to \infty} \frac{e^x + e^{4x}}{e^{2x} + e^{3x}}$.

18. $\lim\limits_{x \to \infty} (e^{2x} - 2e^x)$.

19. $\lim\limits_{x \to 0} \dfrac{e^{3x} - 1}{x}$. (Hint: Use l'Hôpital's rule.)

20. Sketch the graph of the function $f(x) = e^{-2/x}$.

21. (a) Show that for all real numbers x, $e^x \geq x + 1$. (Hint: Find the minimum value of $e^x - x - 1$.)

    (b) Show that for all $x > 0$, $\ln x \leq x - 1$. (Hint: You can either apply part (a), or imitate the method used in part (a).)

22. (a) Suppose that f and g are continuous on an open interval I, and for every rational number $r \in I$, $f(r) = g(r)$. Prove that for every real number $x \in I$, $f(x) = g(x)$. (Hint: Suppose that for some real number $x \in I$, $f(x) \neq g(x)$. Apply the continuity of f and g at x, and use the fact that for every $\delta > 0$, there is a rational number r such that $|r - x| < \delta$.)

    (b) Show that $y = \exp(x \ln 2)$ is the *only* continuous curve that passes through all points on the dotted curve $y = 2^x$ shown in Figure 7.14a. In other words, show that if f is a continuous function and for every rational number r, $f(r) = 2^r$, then for every real number x, $f(x) = \exp(x \ln 2)$.

23. Sketch the graph of $y = (-2)^x$. (Hint: Notice that Definition 7.4.3 doesn't apply, since -2 is negative, so $(-2)^x$ is undefined if x is irrational. If x is rational, write $x = m/n$, reduced to lowest terms. Consider separately the cases n even, n odd and m even, n odd and m odd. In each case, what is the relationship between $(-2)^x$ and 2^x?)

24. Verify that the solutions we found in Exercise 7.4.6 satisfy the original equations.

25. (a) Show that if $f(x) = Ce^x$, for some constant C, then $f'(x) = f(x)$.

    (b) Show that if f is a function with the property that for all x, $f'(x) = f(x)$, then there is some constant C such that $f(x) = Ce^x$. (Hint: Assume that $f'(x) = f(x)$, and then compute the derivative of $f(x)/e^x$.)

26. Suppose k is a real number.

    (a) Show that if $f(x) = Ce^{kx}$, for some constant C, then $f'(x) = kf(x)$.

    (b) Show that if f is a function with the property that for all x, $f'(x) = kf(x)$, then there is some constant C such that $f(x) = Ce^{kx}$. (Hint: Imitate part (b) of Exercise 25.)

27. At noon, some bacteria are placed in a Petri dish. Let $f(t)$ be the mass of the bacteria in the dish, in milligrams (mg), t hours after noon. If there is nothing limiting the growth of the bacteria, such as a lack of food or space, then the bacteria culture will grow at a rate proportional to the size of the culture. In other words, there is some number k such that for all t, $f'(t) = kf(t)$. By Exercise 26, it follows that

$f(t) = Ce^{kx}$, for some constant C. Suppose that the mass of the culture at noon is 0.8 mg, and at 3:00 p.m. it is 1.2 mg.

(a) Find a formula for $f(t)$.

(b) What will the mass be at 6:00 p.m.?

28. Cobalt-60 is a radioactive metal with applications in medicine. It is produced by bombarding cobalt-59 with neutrons, and it decays over time into an isotope of nickel. Suppose a sample of cobalt-60 is produced, and let $f(t)$ be the number of grams of cobalt-60 remaining after t days. The rate at which the cobalt-60 decays is proportional to the amount present. In other words, there is some number k such that for all t, $f'(t) = kf(t)$. (Since the amount is decreasing, k is negative.) By Exercise 26, there is a constant C such that $f(t) = Ce^{kt}$.

(a) The *half-life* of cobalt-60 is approximately 1925 days. In other words, after 1925 days, half of the cobalt-60 will be left. Find k.

(b) Suppose that after 100 days there are 2 grams of cobalt-60 in the sample. How much was in the sample initially?

29. A light is hanging from the ceiling by a wire. Each piece of the wire must support the weight of the part of the wire that is below it, as well as the weight of the light. Thus, higher parts of the wire must support more weight than lower parts, and it makes sense that the wire might need to be thicker at the top than at the bottom. In this problem we will assume that the weight that a piece of wire can support is proportional to the cross-sectional area of the wire, and we will find the shape of a wire that is just barely thick enough to hold the light without breaking.

Suppose that the light weighs w pounds, the wire is L inches long, the wire is made of material that weighs d pounds per cubic inch, and a piece of the wire can support s pounds per square inch of cross-sectional area. We will assume that the wire is formed by rotating the region bounded by the curve $x = f(y)$, the y-axis, the x-axis, and the line $y = L$ about the y-axis, for some positive continuous function f, where the axes are marked off in inches. We also assume that at each point along its length, the wire is just barely wide enough to support the weight of the light and the part of the wire below that point.

(a) Show that for every number y in the interval $[0, L]$,

$$\pi s(f(y))^2 = w + d \int_0^y \pi (f(t))^2 \, dt.$$

(b) Show that for all $y \in [0, L]$,

$$f'(y) = \frac{d}{2s} f(y).$$

(Hint: Differentiate both sides of the equation in part (a).)

(c) Show that $f(0) = \sqrt{w/(\pi s)}$.

(d) Find $f(y)$. (Hint: Use Exercise 26.)

30. Show that if $a > 0$ and $a \neq 1$ then the function $f(x) = a^x$ is one-to-one, its domain is $(-\infty, \infty)$, and its range is $(0, \infty)$. (Hint: Consider the cases $0 < a < 1$ and $a > 1$ separately.)

31. Find the volume of the solid generated when the region under the curve $y = e^x$, $0 \le x \le 1$, is rotated about the x-axis.

32. Find the volume of the solid generated when the region bounded by the curves $y = 2^x$ and $y = x + 1$ is rotated about the x-axis.

7.5   The Inverse Trigonometric Functions

Now that we know how useful it can be to study inverses of functions, it is natural to ask if there are more functions whose inverses we should investigate. In this section we study inverses of some trigonometric functions.

For example, consider the sine function. Looking at the graph of $y = \sin x$ in Figure 7.15, we see that there is a problem with defining an inverse: the sine function is very far from being one-to-one. Indeed, for every number c between -1 and 1, the horizontal line $y = c$ intersects the curve $y = \sin x$ infinitely many times! But we already know how to deal with this problem: restrict the domain of the function to make it one-to-one.

Restricting the domain of the sine function to the interval $[-\pi/2, \pi/2]$ will give us the part of the graph shown in black in Figure 7.15, and no horizontal line intersects this part of the graph more than once. There are other intervals we could use. For example, restricting the domain of the sine function to the interval $[\pi/2, 3\pi/2]$ would also give us a one-to-one function. There is no particular reason to use one interval rather than another, but we must choose some interval, and the interval $[-\pi/2, \pi/2]$ is the standard choice that mathematicians use.

To state our idea more precisely, we define a function f with domain $[-\pi/2, \pi/2]$ by saying that for $-\pi/2 \le x \le \pi/2$, $f(x) = \sin x$. Thus, f is the restriction of the sine

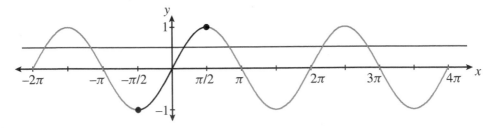

Figure 7.15: The function $\sin x$ is not one-to-one, but we can make it one-to-one by restricting it to $-\pi/2 \le x \le \pi/2$.

function to the interval $[-\pi/2, \pi/2]$, and it is a continuous, strictly increasing function with domain $[-\pi/2, \pi/2]$ and range $[-1, 1]$. The graph of f consists of all points (x, y) such that $y = \sin x$ and $-\pi/2 \le x \le \pi/2$. Flipping this graph over across the line $y = x$ gives us the points (x, y) such that $x = \sin y$ and $-\pi/2 \le y \le \pi/2$. This is the graph of the inverse of f, which is called the *inverse sine function*.

Definition 7.5.1. Let f be the restriction of the sine function to the interval $[-\pi/2, \pi/2]$. Then the inverse of f is called the *inverse sine function*, and it is denoted $\sin^{-1}$. Thus, for all numbers x and y,

$$y = \sin^{-1} x \quad \text{if and only if} \quad x = \sin y \text{ and } -\frac{\pi}{2} \le y \le \frac{\pi}{2}.$$

The graph of the inverse sine function is shown in Figure 7.16. Clearly the domain of the inverse sine is $[-1, 1]$, and its range is $[-\pi/2, \pi/2]$. Note that some books call this function the arcsine function, and write $\arcsin x$ rather than $\sin^{-1} x$. Also, remember that although $\sin^2 x = (\sin x)^2$, $\sin^{-1} x$ is not the same as $(\sin x)^{-1} = \csc x$.

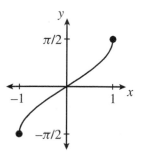

Figure 7.16: The graph of $y = \sin^{-1} x$.

Example 7.5.2. Compute:

$$\text{(a) } \sin^{-1}(\sqrt{3}/2), \quad \text{(b) } \sin^{-1}(-1/2), \quad \text{(c) } \sin^{-1}(1).$$

Solution. For part (a), let $\theta = \sin^{-1}(\sqrt{3}/2)$. According to Definition 7.5.1, this means that $\sin \theta = \sqrt{3}/2$ and $-\pi/2 \le \theta \le \pi/2$. Thus, θ is an angle in the first or fourth quadrant whose sine is $\sqrt{3}/2$. We know an angle that fits this description: $\sin(\pi/3) = \sqrt{3}/2$ and $-\pi/2 \le \pi/3 \le \pi/2$. Notice that there are other angles whose sine is $\sqrt{3}/2$. For example, $\sin(2\pi/3) = \sqrt{3}/2$. But $2\pi/3$ is not between $-\pi/2$ and $\pi/2$. In the interval $[-\pi/2, \pi/2]$, the *only* angle whose sine is $\sqrt{3}/2$ is $\pi/3$. Thus, $\sin^{-1}(\sqrt{3}/2) = \theta = \pi/3$.

When working with the inverse sine function, we will often use the strategy we just used in our solution to part (a): give a name to the particular value of the inverse sine function that is under discussion, and then apply Definition 7.5.1. Since values of the inverse sine function represent angles, we often use the letter θ for a value of the inverse sine function.

Applying this method in part (b), we let $\theta = \sin^{-1}(-1/2)$. This means $\sin \theta = -1/2$ and $-\pi/2 \le \theta \le \pi/2$. Since $\sin(-\pi/6) = -1/2$ and $-\pi/2 \le -\pi/6 \le \pi/2$, we conclude that $\sin^{-1}(-1/2) = \theta = -\pi/6$.

Finally, for part (c), if we let $\theta = \sin^{-1}(1)$, then $\sin \theta = 1$ and $-\pi/2 \le \theta \le \pi/2$, and the only angle that fits this description is $\theta = \pi/2$. Thus, $\sin^{-1}(1) = \pi/2$.   □

Since the inverse sine function is the inverse of a restricted version of the sine function, we know that these functions undo each other. In other words,

(a) for all $x \in [-\pi/2, \pi/2]$, $\sin^{-1}(\sin x) = x$;
(b) for all $x \in [-1, 1]$, $\sin(\sin^{-1} x) = x$.

The restrictions on the values of x in these statements are necessary. For example, consider $\sin^{-1}(\sin(8\pi/7))$. You might think that this must be equal to $8\pi/7$, but statement (a) doesn't apply, since $8\pi/7 > \pi/2$. Let's work it out carefully. We let $\theta = \sin^{-1}(\sin(8\pi/7))$, which means that $\sin \theta = \sin(8\pi/7)$ and $-\pi/2 \le \theta \le \pi/2$. We now see that $\theta = 8\pi/7$ can't be right, because $8\pi/7$ is not in the interval $[-\pi/2, \pi/2]$. To find θ, we must find the unique angle between $-\pi/2$ and $\pi/2$ whose sine is the same as the sine of $8\pi/7$. The easiest way to find the answer is to look at the unit circle. In Figure 7.17 you can see by symmetry that the points P and Q have the same y-coordinate, and therefore $\sin(-\pi/7) = \sin(8\pi/7)$. Since $-\pi/2 \le -\pi/7 \le \pi/2$, we conclude that $\sin^{-1}(\sin(8\pi/7)) = \theta = -\pi/7$.

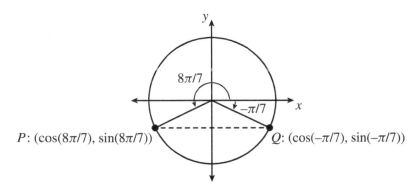

Figure 7.17

We can similarly define inverses for all of the trigonometric functions. In each case, we must restrict the domain of the trigonometric function in order to make it one-to-one. There is more than one restriction we could choose, but certain choices have become standard. For example, we define the inverse tangent function as follows:

Definition 7.5.3. Let f be the restriction of the tangent function to the interval $(-\pi/2, \pi/2)$. Then the inverse of f is called the *inverse tangent function*, and it is denoted $\tan^{-1}$. Thus, for all x and y,

$$y = \tan^{-1} x \quad \text{if and only if} \quad x = \tan y \text{ and } -\frac{\pi}{2} < y < \frac{\pi}{2}.$$

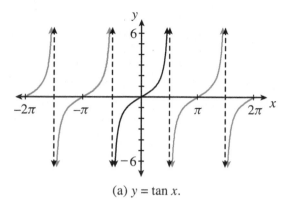

(a) $y = \tan x$.

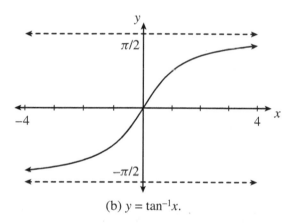

(b) $y = \tan^{-1}x$.

Figure 7.18: The graphs of the tangent and inverse tangent functions.

The graphs of the tangent and inverse tangent functions are shown in Figure 7.18. The restriction of the tangent function to the interval $(-\pi/2, \pi/2)$ gives us the part of the graph that is shown in black in Figure 7.18a, and flipping this part of the graph over gives us the graph of the inverse tangent in Figure 7.18b. Notice that the graph of the tangent function has vertical asymptotes at $x = \pm\pi/2$. When the graph is flipped over, these become horizontal asymptotes for the graph of the inverse tangent function at $y = \pm\pi/2$. Thus, $\lim_{x\to\infty} \tan^{-1} x = \pi/2$ and $\lim_{x\to-\infty} \tan^{-1} x = -\pi/2$.

The only other trigonometric function whose inverse we will make use of in this book is the secant function.

Definition 7.5.4. Let f be the restriction of the secant function to the set $[0, \pi/2) \cup (\pi/2, \pi]$. Then the inverse of f is called the *inverse secant function*, and it is denoted $\sec^{-1}$. Thus, for all x and y,

$$y = \sec^{-1}(x) \quad \text{if and only if} \quad x = \sec y \text{ and either } 0 \le y < \frac{\pi}{2} \text{ or } \frac{\pi}{2} < y \le \pi.$$

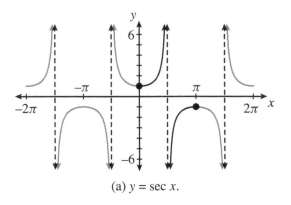

(a) $y = \sec x$.

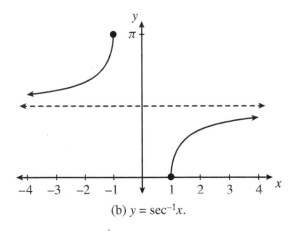

(b) $y = \sec^{-1}x$.

Figure 7.19: The graphs of the secant and inverse secant functions.

The graphs of secant and inverse secant are shown in Figure 7.19. The vertical asymptote of the graph of the secant function at $x = \pi/2$, when flipped over, turns into a horizontal asymptote for the graph of the inverse secant at $y = \pi/2$. Thus, $\lim_{x\to\infty} \sec^{-1} x = \lim_{x\to-\infty} \sec^{-1} x = \pi/2$.

It is clear in Figure 7.19b that the domain of the inverse secant function is $(-\infty, -1] \cup [1, \infty)$. Thus, the inverse secant function comes in two pieces, and it is sometimes helpful to work with these pieces separately. You can see in the figure that the graph of $y = \sec^{-1} x$ for $x \geq 1$ comes from flipping over the graph of $y = \sec x$ for $0 \leq x < \pi/2$, and $y = \sec^{-1} x$ for $x \leq -1$ comes from $y = \sec x$ for $\pi/2 < x \leq \pi$. Thus, for $x \geq 1$,

$$y = \sec^{-1} x \quad \text{if and only if} \quad x = \sec y \text{ and } 0 \leq y < \frac{\pi}{2},$$

while if $x \leq -1$ then

$$y = \sec^{-1} x \quad \text{if and only if} \quad x = \sec y \text{ and } \frac{\pi}{2} < y \leq \pi.$$

We should warn you that, while many mathematicians use the definition we have given for the inverse secant function, some use a slightly different definition. Everyone seems to agree that, for $x \geq 1$, $\sec^{-1} x$ should belong to the interval $[0, \pi/2)$. For $x \leq -1$, we have said that $\sec^{-1} x$ belongs to the interval $(\pi/2, \pi]$, but some mathematicians choose the interval $[\pi, 3\pi/2)$ instead. There are two lessons here. The first is that when reading another book, you should check to see which definition they use. And the second is that one can avoid potential confusions by using $\sec^{-1} x$ only for $x \geq 1$, if possible. We will see that it is sometimes possible to avoid the "left half" of the inverse secant function altogether.

Example 7.5.5. Compute:

$$\text{(a) } \tan^{-1}(-1), \qquad \text{(b) } \sec^{-1}(2).$$

Solution. For part (a), let $\theta = \tan^{-1}(-1)$. This means that $\tan \theta = -1$ and $-\pi/2 < \theta < \pi/2$. Since $\tan(-\pi/4) = -1$ and $-\pi/2 < -\pi/4 < \pi/2$, we see that $\tan^{-1}(-1) = \theta = -\pi/4$.

Similarly, for part (b) we let $\theta = \sec^{-1}(2)$, which means that $\sec \theta = 2$ and either $0 \leq \theta < \pi/2$ or $\pi/2 < \theta \leq \pi/2$. In fact, since $2 \geq 1$, we can restrict our attention to the "right half" of the inverse secant function and conclude that $0 \leq \theta < \pi/2$. Rewriting the equation $\sec \theta = 2$ as $\cos \theta = 1/2$ may help you see that $\sec^{-1}(2) = \theta = \pi/3$. □

Example 7.5.6. Compute:

$$\text{(a) } \cos(\sin^{-1}(-4/5)), \qquad \text{(b) } \tan(\sec^{-1}(-3)).$$

Solution. For part (a), we let $\theta = \sin^{-1}(-4/5)$. Then $\sin \theta = -4/5$ and $-\pi/2 \leq \theta \leq \pi/2$, and we are asked to find $\cos(\sin^{-1}(-4/5)) = \cos \theta$. Since we know the value of $\sin \theta$ and we have to find $\cos \theta$, a natural next step is to look for a relationship between $\sin \theta$ and $\cos \theta$. Of course, we know that $\cos^2 \theta + \sin^2 \theta = 1$, so

$$\cos^2 \theta = 1 - \sin^2 \theta = 1 - \left(-\frac{4}{5}\right)^2 = 1 - \frac{16}{25} = \frac{9}{25},$$

$$\cos \theta = \pm\sqrt{\frac{9}{25}} = \pm\frac{3}{5}.$$

But we're not done yet. There can be only one value for $\cos(\sin^{-1}(-4/5))$, so we need to decide which value is the right one, $3/5$ or $-3/5$. There is one piece of information about θ that we haven't used yet: $-\pi/2 \leq \theta \leq \pi/2$. This tells us that θ is in either the first or the fourth quadrant, and in those quadrants the cosine function is never negative. Therefore $\cos(\sin^{-1}(-4/5)) = \cos \theta = 3/5$.

We proceed in a similar way for part (b). Let $\theta = \sec^{-1}(-3)$, so that $\sec \theta = -3$ and either $0 \leq \theta < \pi/2$ or $\pi/2 < \theta \leq \pi$. We must find $\tan(\sec^{-1}(-3)) = \tan \theta$, so we use the identity $\tan^2 \theta + 1 = \sec^2 \theta$. (If you don't remember this identity, notice that you can derive it from $\cos^2\theta + \sin^2\theta = 1$ by dividing both sides by $\cos^2\theta$.) This gives us

$$\tan^2\theta = \sec^2\theta - 1 = (-3)^2 - 1 = 8,$$
$$\tan\theta = \pm\sqrt{8} = \pm 2\sqrt{2}.$$

Once again we look to the restriction on θ to decide which answer is right. We know that either $0 \le \theta < \pi/2$ or $\pi/2 < \theta \le \pi$, so θ is in either the first or second quadrant. Unfortunately, the tangent function is nonnegative in the first quadrant and nonpositive in the second, so this doesn't help us. This is a good example of why it is sometimes useful to treat the two halves of the inverse secant function separately. In this case we have $\theta = \sec^{-1}(-3)$, so we are using $\sec^{-1} x$ for $x \le -1$. For that part of the inverse secant function we have $\pi/2 < \theta \le \pi$, so θ is in the second quadrant and $\tan\theta \le 0$. Therefore $\tan(\sec^{-1}(-3)) = \tan\theta = -2\sqrt{2}$.   □

We turn now to the derivatives of the inverse trigonometric functions. The function $\sin x$ is differentiable at all numbers, and its derivative, $\cos x$, is nonzero whenever $-\pi/2 < x < \pi/2$. It follows, by Theorem 7.2.2, that the inverse sine function is differentiable on the interior of its domain, the interval $(-1, 1)$. We find it simplest to compute the derivative by implicit differentiation. Let $y = \sin^{-1} x$. Then $\sin y = x$ and $-\pi/2 \le y \le \pi/2$, so

$$\frac{d}{dx}(\sin y) = \frac{d}{dx}(x),$$
$$\cos y \cdot \frac{dy}{dx} = 1,$$
$$\frac{dy}{dx} = \frac{1}{\cos y} = \frac{1}{\cos(\sin^{-1} x)}.$$

Imitating the reasoning we used in Example 7.5.6, we can simplify the denominator by using the identity $\cos^2 y + \sin^2 y = 1$:

$$\cos^2 y = 1 - \sin^2 y = 1 - x^2,$$
$$\cos y = \pm\sqrt{1 - x^2}.$$

Since $-\pi/2 \le y \le \pi/2$, we have $\cos y \ge 0$, so we can eliminate the negative solution and conclude that $\cos y = \sqrt{1 - x^2}$. Thus,

$$\frac{d}{dx}(\sin^{-1} x) = \frac{dy}{dx} = \frac{1}{\cos y} = \frac{1}{\sqrt{1 - x^2}}.$$

Similar methods can be used to find the derivatives of the inverse tangent and secant. By Theorem 7.2.2, the inverse tangent is differentiable on its entire domain, the interval $(-\infty, \infty)$. And applying the theorem separately to the two pieces of the inverse secant function, we conclude that the inverse secant is differentiable on both of the intervals $(-\infty, -1)$ and $(1, \infty)$.

To find the derivative of the inverse tangent, we begin by letting $y = \tan^{-1} x$. This means that $\tan y = x$ and $-\pi/2 < y < \pi/$, so

$$\frac{d}{dx}(\tan y) = \frac{d}{dx}(x),$$

$$\sec^2 y \cdot \frac{dy}{dx} = 1,$$

$$\frac{dy}{dx} = \frac{1}{\sec^2 y} = \frac{1}{\tan^2 y + 1} = \frac{1}{x^2 + 1}.$$

Finally, for the inverse secant we let $y = \sec^{-1} x$. Then $\sec y = x$ and either $0 \le y < \pi/2$ or $\pi/2 < y \le \pi$, so

$$\frac{d}{dx}(\sec y) = \frac{d}{dx}(x),$$

$$\sec y \tan y \cdot \frac{dy}{dx} = 1,$$

$$\frac{dy}{dx} = \frac{1}{\sec y \tan y}.$$

To simplify the denominator, we start with the fact that $\sec y = x$, and therefore

$$\tan^2 y = \sec^2 y - 1 = x^2 - 1,$$

$$\tan y = \pm\sqrt{x^2 - 1}.$$

Unfortunately, the restriction on y tells us that y is in either the first or the second quadrant, and the sign of $\tan y$ is different in these two quadrants. Once again, we must treat the two halves of the inverse secant function separately. For $x \ge 1$ we have $0 \le y < \pi/2$, so $\tan y \ge 0$ and therefore $\tan y = \sqrt{x^2 - 1}$. Similar reasoning for $x \le -1$ leads to the conclusion that $\tan y = -\sqrt{x^2 - 1}$. Putting it all together, we conclude that

$$\frac{d}{dx}(\sec^{-1} x) = \frac{1}{\sec y \tan y} = \begin{cases} \dfrac{1}{x\sqrt{x^2 - 1}}, & \text{if } x > 1, \\[3mm] -\dfrac{1}{x\sqrt{x^2 - 1}}, & \text{if } x < -1. \end{cases}$$

We can combine these into one formula by writing

$$\frac{d}{dx}(\sec^{-1} x) = \frac{1}{|x|\sqrt{x^2 - 1}}.$$

The derivatives we have computed are summarized in Table 7.2. Notice that each derivative also gives us a new indefinite integral formula. These integration formulas illustrate one of the reasons why the inverse trigonometric functions are important in calculus: sometimes an integral that appears to have nothing to do with trigonometry can have an answer that contains an inverse trigonometric function.

$$\frac{d}{dx}(\sin^{-1} x) = \frac{1}{\sqrt{1-x^2}} \qquad \int \frac{1}{\sqrt{1-x^2}} \, dx = \sin^{-1} x + C$$

$$\frac{d}{dx}(\tan^{-1} x) = \frac{1}{x^2+1} \qquad \int \frac{1}{x^2+1} \, dx = \tan^{-1} x + C$$

$$\frac{d}{dx}(\sec^{-1} x) = \frac{1}{|x|\sqrt{x^2-1}} \qquad \int \frac{1}{|x|\sqrt{x^2-1}} \, dx = \sec^{-1} x + C$$

Table 7.2: Derivatives of inverse trigonometric functions, and the corresponding integration formulas.

In fact, the integrals in Table 7.2 are just the tip of the iceberg. We will eventually see that many integrals that don't appear to involve trigonometry will have inverse trigonometric functions in their answers. As a step in that direction, we evaluate some integrals related to those in Table 7.2.

Example 7.5.7. Evaluate the following integrals:

$$\text{(a)} \int \frac{1}{\sqrt{1-9x^2}} \, dx, \quad \text{(b)} \int \frac{1}{x^2+2x+5} \, dx, \quad \text{(c)} \int \frac{1}{x\sqrt{x^2-1}} \, dx.$$

Solution. For integral (a), motivated by the fact that $9x^2 = (3x)^2$, we make the substitution $u = 3x, du = 3\,dx, (1/3)\,du = dx$:

$$\int \frac{1}{\sqrt{1-9x^2}} \, dx = \int \frac{1}{\sqrt{1-u^2}} \cdot \frac{1}{3} \, du = \frac{1}{3}\sin^{-1} u + C = \frac{1}{3}\sin^{-1}(3x) + C.$$

To evaluate integral (b), it turns out it is helpful to complete the square in the denominator:

$$\int \frac{1}{x^2+2x+5} \, dx = \int \frac{1}{(x^2+2x+1)+4} \, dx$$

$$= \int \frac{1}{(x+1)^2+4} \, dx = \int \frac{1}{4} \cdot \frac{1}{(\frac{x+1}{2})^2+1} \, dx.$$

Now we can substitute $u = (x+1)/2, du = (1/2)\,dx, 2\,du = dx$:

$$\int \frac{1}{x^2+2x+5} \, dx = \int \frac{1}{4} \cdot \frac{2}{u^2+1} \, du = \frac{1}{2}\tan^{-1} u + C = \frac{1}{2}\tan^{-1}\left(\frac{x+1}{2}\right) + C.$$

Notice that the domain of the integrand in part (c) is $(-\infty, -1] \cup [1, \infty)$. For $x \geq 1$, the integral is the same as the third integral in Table 7.2, so the answer is $\sec^{-1} x + C$. For $x \le -1$, we'll use the substitution $u = -x$, $du = (-1)\,dx$; notice that $u \ge 1$:

$$\int \frac{1}{x\sqrt{x^2 - 1}}\,dx = \int \frac{-1}{(-u)\sqrt{u^2 - 1}}\,du = \int \frac{1}{u\sqrt{u^2 - 1}}\,du$$
$$= \sec^{-1} u + C = \sec^{-1}(-x) + C.$$

Combining our answers for the two halves of the domain, we can express our answer as follows:

$$\int \frac{1}{x\sqrt{x^2 - 1}}\,dx = \sec^{-1}|x| + C.$$

There are other ways of writing the answer to this integral, but one nice feature of the solution we have chosen is that it uses the inverse secant function only on the interval $[1, \infty)$. Thus, we have avoided the controversial left half of the inverse secant. □

We end this section with a few more examples of calculus problems we can solve with the formulas we have derived.

Example 7.5.8. Evaluate:

$$\text{(a) } \frac{d}{dx}(\sin^{-1}(2x - 1)), \qquad\qquad \text{(b) } \frac{d}{dx}(\sec^{-1}(e^x)),$$

$$\text{(c) } \int \frac{x + 1}{3x^2 + 1}\,dx, \qquad\qquad \text{(d) } \int \frac{dx}{x\sqrt{x^4 - 1}}.$$

Solution. For parts (a) and (b), we use the chain rule:

$$\frac{d}{dx}(\sin^{-1}(2x - 1)) = \frac{1}{\sqrt{1 - (2x - 1)^2}} \cdot \frac{d}{dx}(2x - 1) - \frac{2}{\sqrt{4x - 4x^2}} = \frac{1}{\sqrt{x - x^2}},$$

$$\frac{d}{dx}(\sec^{-1}(e^x)) = \frac{1}{|e^x|\sqrt{(e^x)^2 - 1}} \cdot \frac{d}{dx}(e^x) = \frac{e^x}{e^x\sqrt{e^{2x} - 1}} = \frac{1}{\sqrt{e^{2x} - 1}}.$$

For part (c), we begin by splitting the integral into two integrals:

$$\int \frac{x + 1}{3x^2 + 1}\,dx = \int \frac{x}{3x^2 + 1} + \frac{1}{3x^2 + 1}\,dx = \int \frac{x}{3x^2 + 1}\,dx + \int \frac{1}{3x^2 + 1}\,dx.$$

We can now solve the two integrals separately. For the first, we use the substitution $u = 3x^2 + 1$, $du = 6x\,dx$, $(1/6)\,du = x\,dx$:

$$\int \frac{x}{3x^2 + 1}\,dx = \int \frac{1}{3x^2 + 1} \cdot x\,dx = \int \frac{1}{u} \cdot \frac{1}{6}\,du = \frac{1}{6}\ln|u| + C = \frac{1}{6}\ln(3x^2 + 1) + C.$$

Notice that the absolute value signs can be dropped, since $3x^2 + 1 > 0$. For the second integral, we use the substitution $u = \sqrt{3}x$, $du = \sqrt{3}\,dx$, $(1/\sqrt{3})\,du = dx$:

$$\int \frac{1}{3x^2 + 1}\,dx = \int \frac{1}{u^2 + 1} \cdot \frac{1}{\sqrt{3}}\,du = \frac{1}{\sqrt{3}}\tan^{-1} u + C = \frac{1}{\sqrt{3}}\tan^{-1}(\sqrt{3}x) + C.$$

Combining these two integrals, we conclude that

$$\int \frac{x+1}{3x^2+1}\, dx = \frac{1}{6}\ln(3x^2+1) + \frac{1}{\sqrt{3}}\tan^{-1}(\sqrt{3}x) + C.$$

Finally, the substitution $u = x^2$, $du = 2x\, dx$ can be used to solve the integral in (d):

$$\int \frac{dx}{x\sqrt{x^4-1}} = \int \frac{2x\, dx}{2x^2\sqrt{(x^2)^2-1}} = \int \frac{1}{2}\cdot\frac{1}{u\sqrt{u^2-1}}\, du$$

$$= \frac{1}{2}\sec^{-1}|u| + C = \frac{1}{2}\sec^{-1}(x^2) + C.$$

Notice that we have used our answer to part (c) of the last example to get $\sec^{-1}|u|$, but once we replaced u with x^2 it became clear that the absolute value signs were unnecessary.   □

Exercises 7.5

1. Suppose $a > 0$. Verify the following integration formulas:

$$\int \frac{1}{\sqrt{a^2-x^2}}\, dx = \sin^{-1}\left(\frac{x}{a}\right) + C,$$

$$\int \frac{1}{x^2+a^2}\, dx = \frac{1}{a}\tan^{-1}\left(\frac{x}{a}\right) + C,$$

$$\int \frac{1}{x\sqrt{x^2-a^2}}\, dx = \frac{1}{a}\sec^{-1}\left|\frac{x}{a}\right| + C.$$

2–8: Find the value of the expression.

2. $\sin^{-1}(0)$.

3. $\tan^{-1}(\sqrt{3})$.

4. $\tan^{-1}(\tan(3\pi/5))$.

5. $\sec^{-1}(\sec(-\pi/7))$.

6. $\sec(\sin^{-1}(2/3))$.

7. $\tan(\sin^{-1}(-3/5))$.

8. $\tan(\sec^{-1}(4))$.

9–13: Find the derivative of the function.

9. $f(x) = \sin^{-1}(x^{3/2})$.

10. $g(x) = \sec^{-1}(1/x^2)$.

11. $f(x) = \tan^{-1}(\ln x)$.

12. $g(x) = \ln(\tan^{-1} x)$.

13. $h(x) = e^{\sin^{-1}(2x)}$.

14–20: Evaluate the integral.

14. $\displaystyle\int \frac{1}{\sqrt{4 - 9x^2}}\, dx$.

15. $\displaystyle\int \frac{2x}{x^2 - 2x + 2}\, dx$.

16. $\displaystyle\int \frac{e^{2x}}{\sqrt{1 - e^{2x}}}\, dx$.

17. $\displaystyle\int \frac{e^x}{\sqrt{1 - e^{2x}}}\, dx$.

18. $\displaystyle\int \frac{dx}{(x - 1)\sqrt{x^2 - 2x}}$.

19. $\displaystyle\int \frac{x - 1}{\sqrt{x^2 - 2x}}\, dx$.

20. $\displaystyle\int \frac{x\, dx}{x^4 + 1}$.

21–23: Evaluate the limit.

21. $\displaystyle\lim_{x \to 0^-} \tan^{-1}(1/x)$.

22. $\displaystyle\lim_{x \to \infty} \sec^{-1}\left(\frac{x + 1}{x - 1}\right)$.

23. $\displaystyle\lim_{x \to 1/2} \frac{6\sin^{-1} x - \pi}{2x - 1}$. (Hint: Use l'Hôpital's rule.)

24. Show that for all x,
$$\tan^{-1} x = \sin^{-1}\left(\frac{x}{\sqrt{x^2 + 1}}\right).$$

25. Show that for all $x \le -1$, $\sec^{-1} x = \pi - \sec^{-1}(-x)$.

26. The *inverse cosine* function is defined as follows:
$$y = \cos^{-1} x \quad \text{if and only if} \quad x = \cos y \text{ and } 0 \le y \le \pi.$$

(a) Show that if $-1 \le x \le 1$ then $\cos^{-1} x = \pi/2 - \sin^{-1} x$.

(b) Find $\frac{d}{dx}(\cos^{-1} x)$. (Note: One method is to imitate the computation of $\frac{d}{dx}(\sin^{-1} x)$ in the text. Another is to use part (a).)

(c) Show that if $|x| \ge 1$ then $\sec^{-1} x = \cos^{-1}(1/x)$.

27. A wire extends from 0 to 1 on a number line, and its density at the point x is $d(x) = 1/(x^2 + 1)$. Find the center of mass of the wire.

28. Find the volume of the solid generated when the region under the curve $y = 1/(x^4 + 1)$, $0 \le x \le 1$, is rotated about the y-axis.

29–35: These exercises are concerned with the functions cosh and sinh, which are defined as follows:

$$\cosh x = \frac{e^x + e^{-x}}{2}, \qquad \sinh x = \frac{e^x - e^{-x}}{2}.$$

29. Verify that for all real numbers t,

$$(\cosh t)^2 - (\sinh t)^2 = 1.$$

This implies that the point $(\cosh t, \sinh t)$ lies on the curve $x^2 - y^2 = 1$, which is a hyperbola. For this reason, the functions cosh and sinh are called the *hyperbolic cosine* and *hyperbolic sine*. (Of course, this should remind you of the identity $\cos^2 t + \sin^2 t = 1$, which implies that the point $(\cos t, \sin t)$ lies on the circle $x^2 + y^2 = 1$.)

30. Find the derivatives of $\cosh x$ and $\sinh x$.

31. Sketch the graphs of $y = \cosh x$ and $y = \sinh x$. You may find it helpful to first verify the following facts:

    (a) For all x, $\cosh x > 0$.

    (b) The function $\sinh x$ is strictly increasing on $(-\infty, \infty)$.

    (c) If $x < 0$ then $\sinh x < 0$, if $x > 0$ then $\sinh x > 0$, and $\sinh 0 = 0$.

32. Since $\sinh x$ is strictly increasing, it is one-to-one, so it has an inverse. For any real number x, let $y = \sinh^{-1} x$. By the definition of inverse, this means

$$x = \sinh y = \frac{e^y - e^{-y}}{2}.$$

Now solve this equation for y to show that $\sinh^{-1} x = \ln(x + \sqrt{x^2 + 1})$. (Hint: First derive the equation

$$(e^y)^2 - 2xe^y - 1 = 0.$$

Then use the quadratic formula to get a formula for e^y in terms of x. Use the fact that $e^y > 0$ to eliminate one of the two solutions given by the quadratic formula.)

33. The function $\cosh x$ is not one-to-one, but we can make it one-to-one by restricting it to $x \ge 0$. We define $\cosh^{-1}$ to be the inverse of this restricted version of the hyperbolic cosine function. In other words,

$$y = \cosh^{-1} x \quad \text{if and only if} \quad x = \cosh y \text{ and } y \ge 0.$$

Show that

$$\cosh^{-1} x = \ln(x + \sqrt{x^2 - 1}).$$

34. Find the derivatives of $\cosh^{-1} x$ and $\sinh^{-1} x$. Be sure to simplify your answers.

35. The hyperbolic tangent function tanh is defined by the formula

$$\tanh x = \frac{\sinh x}{\cosh x}.$$

(a) Sketch the graph of $y = \tanh x$. Be sure to check for asymptotes.

(b) Find a formula for $\tanh^{-1} x$.

(c) Find the derivative of $\tanh^{-1} x$.

7.6 L'Hôpital's Rule Again

With all of the new functions we have learned about in this chapter, we can now do a lot more with l'Hôpital's rule. We begin with a few examples.

Example 7.6.1. Evaluate the following limits:

(a) $\displaystyle\lim_{x \to 1^+} \frac{\sec^{-1} x}{\sqrt{x - 1}}$,

(b) $\displaystyle\lim_{x \to \infty} x \left(\tan^{-1} x - \frac{\pi}{2} \right)$,

(c) $\displaystyle\lim_{x \to 0} \frac{\ln(\cos x)}{x^2}$,

(d) $\displaystyle\lim_{x \to -\infty} x e^x$.

Solution. Limit (a) is an indeterminate form of type 0/0, so it is natural to try l'Hôpital's rule. We have

$$\frac{d}{dx}(\sec^{-1} x) = \frac{1}{|x|\sqrt{x^2 - 1}}, \qquad \frac{d}{dx}(\sqrt{x - 1}) = \frac{1}{2\sqrt{x - 1}},$$

so l'Hôpital's rule leads us to compute

$$\lim_{x \to 1^+} \frac{1/(|x|\sqrt{x^2 - 1})}{1/(2\sqrt{x - 1})} = \lim_{x \to 1^+} \frac{2\sqrt{x - 1}}{x\sqrt{x - 1}\sqrt{x + 1}} = \lim_{x \to 1^+} \frac{2}{x\sqrt{x + 1}} = \frac{2}{\sqrt{2}} = \sqrt{2}.$$

Thus, the value of limit (a) is $\sqrt{2}$.

In limit (b), we have $x \to \infty$ and $(\tan^{-1} x - \pi/2) \to (\pi/2 - \pi/2) = 0$, so the limit has the indeterminate form $\infty \cdot 0$. To apply l'Hôpital's rule, we first rewrite the limit as an indeterminate form of type 0/0:

$$\lim_{x \to \infty} x \left(\tan^{-1} x - \frac{\pi}{2} \right) = \lim_{x \to \infty} \frac{\tan^{-1} x - \pi/2}{1/x}.$$

Now l'Hôpital's rule leads us to evaluate the limit

$$\lim_{x \to \infty} \frac{1/(x^2 + 1)}{-1/x^2} = \lim_{x \to \infty} -\frac{x^2}{x^2 + 1} = \lim_{x \to \infty} -\frac{x^2}{x^2(1 + 1/x^2)} = \lim_{x \to \infty} -\frac{1}{1 + 1/x^2} = -1.$$

We conclude that the value of limit (b) is -1.

Limit (c) is also an indeterminate form of type 0/0. Differentiating numerator and denominator leads to the limit

$$\lim_{x \to 0} \frac{(1/\cos x)(-\sin x)}{2x} = \lim_{x \to 0} \frac{-\tan x}{2x}.$$

This is another indeterminate form of type 0/0, so we use l'Hôpital's rule again:

$$\lim_{x \to 0} \frac{-\sec^2 x}{2} = -\frac{1}{2}.$$

Thus, by two applications of l'Hôpital's rule, the value of limit (c) is $-1/2$.

Finally, in limit (d), as $x \to -\infty$, $e^x \to 0$, so the limit has the indeterminate form $\infty \cdot 0$. One way to rewrite it as a limit of a fraction would be

$$\lim_{x \to -\infty} x e^x = \lim_{x \to -\infty} \frac{e^x}{1/x}.$$

This is an indeterminate form of type 0/0, so we try differentiating numerator and denominator:

$$\lim_{x \to -\infty} \frac{e^x}{-1/x^2}.$$

This is another indeterminate form of type 0/0, but it looks more complicated than the original limit. We could apply l'Hôpital's rule again, but it is not hard to see that this will just make matters even worse. We'd better go back and try a different approach.

There is another way to rewrite the original limit as a limit of a fraction:

$$\lim_{x \to -\infty} x e^x = \lim_{x \to -\infty} \frac{x}{1/e^x} = \lim_{x \to -\infty} \frac{x}{e^{-x}}.$$

Now as $x \to -\infty$, in the denominator we have $-x \to \infty$, and therefore $e^{-x} \to \infty$. Thus, the limit is an indeterminate form of type ∞/∞, and l'Hôpital's rule prompts us to compute the limit

$$\lim_{x \to -\infty} \frac{1}{-e^{-x}} = 0.$$

We conclude that limit (d) is equal to 0.   □

Next we look at what l'Hôpital's rule can tell us about the growth rates of the exponential function and the natural logarithm. We know that $\lim_{x \to \infty} e^x = \lim_{x \to \infty} \ln x = \infty$, but looking at the graphs of the exponential function and the natural log, we can see that as $x \to \infty$, e^x grows very quickly, whereas $\ln x$ grows much more slowly. We make this statement precise in our next theorem by comparing the growth rates of e^x and $\ln x$ to the growth rates of some other familiar functions.

Theorem 7.6.2. *For every positive integer n, we have the following limits:*

1. $\displaystyle\lim_{x\to\infty} \frac{e^x}{x^n} = \infty.$

2. $\displaystyle\lim_{x\to\infty} \frac{\ln x}{\sqrt[n]{x}} = 0.$

Part 1 of Theorem 7.6.2 compares e^x to x^n. For every positive integer n, the function x^n is strictly increasing on the interval $[0, \infty)$, and $\lim_{x\to\infty} x^n = \infty$. The larger n is, the more quickly x^n increases as x increases. For example, if $f(x) = x^{100}$, then $f(0) = 0$ and $f(10)$ is a number that is 101 digits long, so f increases very quickly. But part 1 of the theorem says that no matter how large n is, e^x grows faster than x^n: when x gets very large, the ratio e^x/x^n gets very large, which means that e^x is much larger than x^n.

Similarly, part 2 of the theorem compares $\ln x$ to $\sqrt[n]{x}$. If n is large, then $\sqrt[n]{x}$ grows very slowly. But no matter how large n is, $\ln x$ grows more slowly than $\sqrt[n]{x}$: when x gets very large, $\ln x / \sqrt[n]{x}$ is close to 0, and therefore $\ln x$ is much smaller than $\sqrt[n]{x}$.

Proof of Theorem 7.6.2. To prove part 1, we use mathematical induction.
Base case. For the case $n = 1$, we must evaluate the limit

$$\lim_{x\to\infty} \frac{e^x}{x},$$

which is an indeterminate form of type ∞/∞. Differentiating numerator and denominator gives us the limit

$$\lim_{x\to\infty} \frac{e^x}{1} = \infty.$$

Thus, by l'Hôpital's rule, the original limit is also infinite.
Induction step. Suppose that n is a positive integer and $\lim_{x\to\infty} e^x/x^n = \infty$. To compute

$$\lim_{x\to\infty} \frac{e^x}{x^{n+1}}$$

we apply l'Hôpital's rule, which leads us to compute the limit

$$\lim_{x\to\infty} \frac{e^x}{(n+1)x^n} = \lim_{x\to\infty} \left(\frac{1}{n+1} \cdot \frac{e^x}{x^n}\right) = \infty,$$

by the inductive hypothesis. Thus, $\lim_{x\to\infty} e^x/x^{n+1} = \infty$.

To prove part 2, note that for any positive integer n, the limit

$$\lim_{x\to\infty} \frac{\ln x}{\sqrt[n]{x}} = \lim_{x\to\infty} \frac{\ln x}{x^{1/n}}$$

is an indeterminate form of type ∞/∞. To apply l'Hôpital's rule, we compute

$$\lim_{x\to\infty} \frac{1/x}{(1/n)x^{1/n-1}} = \lim_{x\to\infty} \frac{n}{x^{1/n}} = 0.$$

Thus, $\lim_{x\to\infty} \ln x / \sqrt[n]{x} = 0$.
For an alternative approach to this proof, see Exercise 30.   □

Although we have used the functions e^x and $\ln x$ in the statement of Theorem 7.6.2, it is worth mentioning that similar conclusions hold for the functions a^x and $\log_a x$, for any number $a > 1$. To see why, we can use Theorem 7.6.2 to justify the following calculations:

$$\lim_{x \to \infty} \frac{a^x}{x^n} = \lim_{x \to \infty} \frac{e^{x \ln a}}{(x \ln a)^n} \cdot (\ln a)^n = \infty,$$

$$\lim_{x \to \infty} \frac{\log_a x}{\sqrt[n]{x}} = \lim_{x \to \infty} \frac{\ln x}{\sqrt[n]{x}} \cdot \frac{1}{\ln a} = 0.$$

We can also use ideas from this chapter, together with l'Hôpital's rule, to evaluate limits of functions that have the form $f(x)^{g(x)}$. As we saw in Section 7.4, when solving calculus problems involving functions of this form, it is often helpful to use equation (7.9) to rewrite the exponentiation using the exponential function:

$$f(x)^{g(x)} = e^{g(x) \ln f(x)}.$$

We use this strategy in our next example. For more on limits of this form, see Exercises 26–28.

Example 7.6.3. Evaluate the following limits:

(a) $\displaystyle\lim_{x \to 0^+} x^x$,

(b) $\displaystyle\lim_{x \to \infty} x^x$,

(c) $\displaystyle\lim_{x \to \infty} \left(1 + \frac{1}{x}\right)^x$,

(d) $\displaystyle\lim_{x \to 0^+} (\csc x)^{1/\ln x}$.

Solution. Applying equation (7.9) to limit (a), we see that

$$\lim_{x \to 0^+} x^x = \lim_{x \to 0^+} e^{x \ln x}.$$

Now we work out the limit of the exponent:

$$\lim_{x \to 0^+} x \ln x = \lim_{x \to 0^+} \frac{\ln x}{1/x}.$$

This limit is an indeterminate form of type ∞/∞, so we apply l'Hôpital's rule:

$$\lim_{x \to 0^+} \frac{1/x}{-1/x^2} = \lim_{x \to 0^+} -x = 0.$$

By l'Hôpital's rule, we conclude that as $x \to 0^+$, $x \ln x \to 0$, and since the exponential function is continuous it follows that $x^x = e^{x \ln x} \to e^0 = 1$. Thus, the value of limit (a) is 1.

For limit (b), we again begin by writing

$$\lim_{x \to \infty} x^x = \lim_{x \to \infty} e^{x \ln x}.$$

This time there is no need to use l'Hôpital's rule to find the limit of the exponent. As $x \to \infty$, $\ln x \to \infty$, so $x \ln x \to \infty$ and therefore $x^x = e^{x \ln x} \to \infty$. Thus, $\lim_{x \to \infty} x^x = \infty$.

Applying the same strategy to limit (c), we begin with the equation

$$\lim_{x \to \infty} \left(1 + \frac{1}{x}\right)^x = \lim_{x \to \infty} \exp\left(x \ln\left(1 + \frac{1}{x}\right)\right).$$

Next we compute

$$\lim_{x \to \infty} x \ln\left(1 + \frac{1}{x}\right) = \lim_{x \to \infty} \frac{\ln(1 + 1/x)}{1/x},$$

which is an indeterminate form of type 0/0. L'Hôpital's rule leads us to compute

$$\lim_{x \to \infty} \frac{\frac{1}{1+1/x} \cdot -\frac{1}{x^2}}{-1/x^2} = \lim_{x \to \infty} \frac{1}{1 + 1/x} = 1.$$

Thus, as $x \to \infty$,

$$x \ln\left(1 + \frac{1}{x}\right) \to 1,$$

and therefore

$$\left(1 + \frac{1}{x}\right)^x = \exp\left(x \ln\left(1 + \frac{1}{x}\right)\right) \to \exp(1) = e.$$

In other words, the value of limit (c) is e.

Finally, to find limit (d) we rewrite the desired limit as

$$\lim_{x \to 0^+} (\csc x)^{1/\ln x} = \lim_{x \to 0^+} \exp\left(\frac{1}{\ln x} \cdot \ln(\csc x)\right) = \lim_{x \to 0^+} \exp\left(\frac{\ln(\csc x)}{\ln x}\right),$$

and then turn our attention to

$$\lim_{x \to 0^+} \frac{\ln(\csc x)}{\ln x} = \lim_{x \to 0^+} \frac{\ln(1/\sin x)}{\ln x} = \lim_{x \to 0^+} \frac{-\ln(\sin x)}{\ln x}.$$

As $x \to 0^+$, we have $\sin x \to 0^+$, so $\ln(\sin x) \to -\infty$ and therefore $-\ln(\sin x) \to \infty$. Also, as $x \to 0^+$, $\ln x \to -\infty$. Thus, our limit is an indeterminate form of type ∞/∞, and we can use l'Hôpital's rule:

$$\lim_{x \to 0^+} \frac{-(1/\sin x) \cdot \cos x}{1/x} = \lim_{x \to 0^+} \frac{-x \cos x}{\sin x}.$$

Now we have an indeterminate form of type 0/0, and another application of l'Hôpital's rule leads to

$$\lim_{x \to 0^+} \frac{x \sin x - \cos x}{\cos x} = -1.$$

Thus, as $x \to 0^+$ we have $\ln(\csc x)/\ln x \to -1$, and therefore

$$(\csc x)^{1/\ln x} = \exp\left(\frac{\ln(\csc x)}{\ln x}\right) \to \exp(-1) = e^{-1} = \frac{1}{e}.$$

We conclude that the value of limit (d) is $1/e$. $\qquad\square$

According to limit (c) in the last example, $\lim_{x\to\infty}(1+1/x)^x = e$. We end this section by doing some calculations to illustrate this limit. If we let $f(x) = (1+1/x)^x$, then you can use a calculator to check that

$$f(100) = (1.01)^{100} \approx 2.70481,$$

$$f(1000) = (1.001)^{1000} \approx 2.71692,$$

$$f(10000) = (1.0001)^{10000} \approx 2.71815.$$

As expected, the values seem to be getting close to $e \approx 2.71828$.

Exercises 7.6

1–25: Evaluate the limit.

1. $\lim\limits_{x\to 0} \dfrac{\tan^{-1} x}{\ln(x+1)}$.

2. $\lim\limits_{x\to 5} \dfrac{e^x - e^5}{\ln x - \ln 5}$.

3. $\lim\limits_{x\to\infty} \dfrac{e^x}{\sqrt{x^2+1}}$.

4. $\lim\limits_{x\to 0} \dfrac{e^{\cos x - 1} - 1}{x^2}$.

5. $\lim\limits_{x\to 1} \dfrac{\cos(\pi x) + 1}{(x-1)^2}$.

6. $\lim\limits_{x\to 1^+} \dfrac{\sec^{-1} x}{\sqrt{\ln x}}$.

7. $\lim\limits_{x\to 1/2} \dfrac{\sin^{-1} x - \pi/6}{2x - 1}$.

8. $\lim\limits_{x\to 2} \dfrac{3\sec^{-1} x - \pi}{x - 2}$.

9. $\lim\limits_{x\to 1^-} \dfrac{4\tan^{-1} x - \pi}{2\sin^{-1} x - \pi}$.

10. $\lim\limits_{x\to\infty} \dfrac{\ln(x^2+1)}{\ln x}$.

11. $\lim\limits_{x\to 0} \left[\dfrac{1}{\ln(x+1)} - \dfrac{1}{x} \right]$.

12. $\lim\limits_{x\to 0} \left[\cot x - \dfrac{e^x - 1}{x \sin x} \right]$.

13. $\lim_{x \to 0^+} (\ln x \ln(x + 1))$.

14. $\lim_{x \to 0} \dfrac{\sin^{-1} x - x}{x^3}$.

15. $\lim_{x \to 0} (\cos x)^{1/x}$.

16. $\lim_{x \to 0} (1 + x)^{\csc x}$.

17. $\lim_{x \to 1} (x + \ln x)^{1/(x-1)}$.

18. $\lim_{x \to \infty} x^{1/\sqrt{x}}$.

19. $\lim_{n \to \infty} \sqrt[n]{n}$. (Hint: First compute $\lim_{x \to \infty} x^{1/x}$.)

20. $\lim_{x \to \infty} (e^x + 1)^{1/x}$.

21. $\lim_{x \to 0^+} (e^x + 1)^{1/x}$.

22. $\lim_{x \to 0^-} (e^x + 1)^{1/x}$.

23. $\lim_{x \to \pi/2} \left(\dfrac{2x}{\pi}\right)^{\sec x}$.

24. $\lim_{x \to 3/2} \left(x - \dfrac{1}{2}\right)^{\sec(\pi x)}$.

25. $\lim_{x \to 0^+} (\ln(1 + x))^x$.

26. (a) Suppose that $\lim_{x \to a} f(x) = L > 0$ and $\lim_{x \to a} g(x) = M$. Show that $\lim_{x \to a} f(x)^{g(x)} = L^M$.

    (b) Suppose that as $x \to a^{\ne}$, $f(x) \to 0^+$ and $g(x) \to L > 0$. Show that $\lim_{x \to a} f(x)^{g(x)} = 0$.

    (c) Suppose that as $x \to a^{\ne}$, $f(x) \to 0^+$ and $g(x) \to L < 0$. What is $\lim_{x \to a} f(x)^{g(x)}$?

    (d) Suppose $c > 0$. Let $f(x) = e^{-1/x^2}$ and $g(x) = -x^2 \ln c$. Show that as $x \to 0^{\ne}$, $f(x) \to 0^+$, $g(x) \to 0$, and $f(x)^{g(x)} \to c$. (Since c here can be any positive number, this shows that if we know that $f(x) \to 0^+$ and $g(x) \to 0$, then we can't tell without further analysis what the limit of $f(x)^{g(x)}$ is. In other words, in limits, 0^0 is an indeterminate form.)

27. (a) Suppose that $\lim_{x \to a} f(x) = \infty$ and $\lim_{x \to a} g(x) = L > 0$. Show that $\lim_{x \to a} f(x)^{g(x)} = \infty$.

    (b) Suppose that $\lim_{x \to a} f(x) = \infty$ and $\lim_{x \to a} g(x) = L < 0$. What is $\lim_{x \to a} f(x)^{g(x)}$?

(c) Suppose $c > 0$. Find functions $f(x)$ and $g(x)$ such that $\lim_{x \to 0} f(x) = \infty$, $\lim_{x \to 0} g(x) = 0$, and $\lim_{x \to 0} f(x)^{g(x)} = c$. (Hint: Imitate part (d) of Exercise 26. This shows that in limits, ∞^0 is an indeterminate form.)

28. (a) Suppose that $\lim_{x \to a} f(x) = L > 1$ and $\lim_{x \to a} g(x) = \infty$. Show that $\lim_{x \to a} f(x)^{g(x)} = \infty$.

(b) Suppose that $\lim_{x \to a} f(x) = L$, $0 < L < 1$, and $\lim_{x \to a} g(x) = \infty$. What is $\lim_{x \to a} f(x)^{g(x)}$?

(c) Suppose $c > 0$. Find functions $f(x)$ and $g(x)$ such that $\lim_{x \to 0} f(x) = 1$, $\lim_{x \to 0} g(x) = \infty$, and $\lim_{x \to 0} f(x)^{g(x)} = c$. (Hint: Imitate part (d) of Exercise 26. This shows that in limits, 1^{∞} is an indeterminate form.)

29. Show that for any $c > 0$,

$$\lim_{p \to -1} \int_1^c t^p \, dt = \ln c.$$

30. In this exercise, we give an alternative approach to the proof of Theorem 7.6.2. Of course, you should not use Theorem 7.6.2 in your solution to this exercise.

(a) Use l'Hôpital's rule to show that $\lim_{x \to \infty} \dfrac{\ln x}{x} = 0$.

(b) For any $r > 0$, use part (a) to complete the following limit computation:

$$\lim_{x \to \infty} \frac{e^x}{x^r} = \lim_{x \to \infty} \frac{e^x}{e^{r \ln x}} = \lim_{x \to \infty} e^{x - r \ln x} = \lim_{x \to \infty} e^{x(1 - r \ln x / x)} = \cdots.$$

Use your answer to prove part 1 of Theorem 7.6.2.

(c) For any $r > 0$, use part (a) to complete the following limit computation:

$$\lim_{x \to \infty} \frac{\ln x}{x^r} = \lim_{x \to \infty} \frac{1}{r} \cdot \frac{\ln x^r}{x^r} = \cdots.$$

Use your answer to prove part 2 of Theorem 7.6.2.

Chapter 8

Techniques of Integration

8.1 Partial Fractions

As we have already observed, integration is generally harder than differentiation. In this chapter we will consider a number of techniques for evaluating integrals. These techniques will not allow us to evaluate all integrals, but they will greatly increase the range of integrals we can deal with.

We begin by considering algebraic reexpressions that sometimes make an integral easier.

Example 8.1.1. Evaluate the following integrals:

$$\text{(a) } \int (x^2 + 2)^3 \, dx, \quad \text{(b) } \int \frac{2x^2 + 3}{x} \, dx.$$

Solution. For integral (a), you might be tempted to try the substitution $u = x^2 + 2$. But for that substitution we would have $du = 2x \, dx$, and without an extra factor of x in the integral it would be difficult to make this part of the substitution work. A better approach is to simply multiply out the integrand:

$$\int (x^2 + 2)^3 \, dx = \int x^6 + 6x^4 + 12x^2 + 8 \, dx = \frac{x^7}{7} + \frac{6x^5}{5} + 4x^3 + 8x + C.$$

To evaluate integral (b), we split the fraction into a sum of two fractions:

$$\int \frac{2x^2 + 3}{x}\, dx = \int \frac{2x^2}{x} + \frac{3}{x}\, dx = \int 2x + \frac{3}{x}\, dx = x^2 + 3\ln |x| + C. \qquad \square$$

The integrals in Example 8.1.1 illustrate that it is usually easier to integrate a sum than a product or a quotient. For this reason, to integrate a complicated function it is sometimes helpful to rewrite the function as a sum of simpler functions if possible. Often this idea is useful for integrating rational functions.

For example, observe that

$$\frac{2}{x+2} + \frac{1}{x-3} = \frac{2(x-3) + 1(x+2)}{(x+2)(x-3)} = \frac{3x-4}{x^2 - x - 6}. \tag{8.1}$$

It follows that

$$\int \frac{3x-4}{x^2 - x - 6}\, dx = \int \frac{2}{x+2} + \frac{1}{x-3}\, dx = 2\ln |x+2| + \ln |x-3| + C. \tag{8.2}$$

But what if we were faced with solving the first integral in equation (8.2), and we didn't know that the integrand had come from the fraction addition in equation (8.1)? Is there a way to figure out how to break up the integrand as a sum of simpler fractions?

For example, let's try to evaluate

$$\int \frac{2x+9}{x^2 - x - 6}\, dx. \tag{8.3}$$

The denominator here is the same as in equation (8.2): $x^2 - x - 6 = (x+2)(x-3)$. So we can guess that, as in equation (8.1), the integrand might be a sum of fractions with denominators $x+2$ and $x-3$. To find the numerators, we look for numbers A and B such that

$$\frac{2x+9}{x^2 - x - 6} = \frac{A}{x+2} + \frac{B}{x-3}. \tag{8.4}$$

Adding the fractions on the right, we have

$$\frac{2x+9}{x^2 - x - 6} = \frac{A(x-3) + B(x+2)}{(x+2)(x-3)} = \frac{(A+B)x + (-3A + 2B)}{x^2 - x - 6},$$

and equating the numerators we get

$$2x+9 = (A+B)x + (-3A + 2B). \tag{8.5}$$

Notice that we think of equation (8.4), and therefore also equation (8.5), as an equation between *functions*; in other words, we want these equations to hold *for all values of x*. The formula $2x+9$ on the left-hand side of equation (8.5) defines a function whose graph is a straight line with slope 2 and y-intercept 9. The formula on the right defines a function whose graph is also a straight line, with slope $A+B$ and y-intercept $-3A + 2B$. For some values of A and B these two lines would cross, which means that equation (8.5) would be true for *one* value of x. But the only way the equation can be true for *all* values of x is if the two lines are exactly the same. In other words, the lines must have the same slope and y-intercept, which means that

$$A + B = 2, \tag{8.6a}$$
$$-3A + 2B = 9. \tag{8.6b}$$

These equations say that the coefficients of x on both sides of equation (8.5) are equal, and also the constant terms are equal.

This is a system of two simultaneous equations with two unknowns, A and B. We hope you have seen methods for solving such systems of equations before. One method is to multiply the first equation by 3 and then add it to the second equation, so that the variable A will cancel out:

$$3A + 3B = 6$$
$$+ \quad -3A + 2B = 9$$
$$\overline{5B = 15.}$$

We conclude that $B = 15/5 = 3$. Substituting into equation (8.6a), we get $A + 3 = 2$, and therefore $A = -1$. Let's confirm that the values $A = -1$, $B = 3$ work in equation (8.4):

$$\frac{-1}{x+2} + \frac{3}{x-3} = \frac{-1(x-3) + 3(x+2)}{(x+2)(x-3)} = \frac{2x+9}{x^2-x-6}.$$

As in our previous example, we can now use this equation to solve the integral in equation (8.3):

$$\int \frac{2x+9}{x^2-x-6} \, dx = \int \frac{-1}{x+2} + \frac{3}{x-3} \, dx = -\ln|x+2| + 3\ln|x-3| + C.$$

The method we have used to split the integrand into a sum of simpler fractions is called *partial fractions*. In the rest of this section we will discuss how to apply this method to any rational function.

Example 8.1.2. Evaluate the following integrals:

(a) $\displaystyle\int \frac{2\,dx}{2x^2 - 5x + 3}$, (b) $\displaystyle\int \frac{2x^3 + x^2 - x}{2x^2 - 5x + 3} \, dx$, (c) $\displaystyle\int \frac{u^2 - 3}{u^3 - u^2 + u - 1} \, du$.

Solution. For integral (a), as in our previous example we factor the denominator and use the factors to set up a guess about how to split up the integrand:

$$\frac{2}{2x^2 - 5x + 3} = \frac{2}{(2x-3)(x-1)} = \frac{A}{2x-3} + \frac{B}{x-1} = \frac{A(x-1) + B(2x-3)}{(2x-3)(x-1)}.$$

The denominators on the two sides are the same, so we only need to match up the numerators:

$$2 = A(x-1) + B(2x-3),$$

or in other words

$$0 \cdot x + 2 = (A + 2B) \cdot x + (-A - 3B).$$

As before, to ensure that this equation is true for all values of x we equate the coefficients on both sides, which leads to a system of two equations with two unknowns:

$$A + 2B = 0,$$
$$-A - 3B = 2.$$

Adding these two equations we find that $-B = 2$, and therefore $B = -2$. Substituting this value into either equation leads to the conclusion that $A = 4$. Therefore

$$\int \frac{2\,dx}{2x^2 - 5x + 3} = \int \frac{A}{2x - 3} + \frac{B}{x - 1}\,dx = \int \frac{4}{2x - 3} - \frac{2}{x - 1}\,dx$$
$$= \int \frac{4\,dx}{2x - 3} - \int \frac{2\,dx}{x - 1}.$$

For the first integral on the right, we substitute $u = 2x - 3$, $du = 2\,dx$:

$$\int \frac{4\,dx}{2x - 3} = \int \frac{2\,du}{u} = 2\ln|u| + C = 2\ln|2x - 3| + C.$$

Evaluating the second integral by a similar method, we conclude that

$$\int \frac{2\,dx}{2x^2 - 5x + 3} = 2\ln|2x - 3| - 2\ln|x - 1| + C.$$

The integrand in integral (b) has the same denominator, so we could try to split it up in a similar way:

$$\frac{2x^3 + x^2 - x}{2x^2 - 5x + 3} = \frac{A}{2x - 3} + \frac{B}{x - 1} = \frac{A(x - 1) + B(2x - 3)}{(2x - 3)(x - 1)}$$
$$= \frac{(A + 2B)x + (-A - 3B)}{2x^2 - 5x + 3}.$$

But at this point it should be clear that this decomposition won't work: the numerator on the left-hand side is $2x^3 + x^2 - x$, and the numerator on the right-hand side will not have terms involving x^3 and x^2, no matter what values we choose for A and B. There is no way to make the numerators match up.

Fortunately, there is a way to get rid of the terms involving x^3 and x^2 on the left: divide the polynomials. The long division calculation in Figure 8.1 shows that when we divide $2x^3 + x^2 - x$ by $2x^2 - 5x + 3$ we get a quotient of $x + 3$ with a remainder of $11x - 9$. In other words,

$$\frac{2x^3 + x^2 - x}{2x^2 - 5x + 3} = x + 3 + \frac{11x - 9}{2x^2 - 5x + 3}. \tag{8.7}$$

(You can check that equation (8.7) is correct by getting a common denominator on the right-hand side.)

$$
\begin{array}{r}
x + 3 \\
2x^2 - 5x + 3 \overline{)\,2x^3 + x^2 - x} \\
2x^3 - 5x^2 + 3x \\
\hline
6x^2 - 4x \\
6x^2 - 15x + 9 \\
\hline
11x - 9
\end{array}
$$

Figure 8.1: Dividing $2x^3 + x^2 - x$ by $2x^2 - 5x + 3$.

We can now split up the last fraction in equation (8.7):

$$
\frac{11x - 9}{2x^2 - 5x + 3} = \frac{A}{2x - 3} + \frac{B}{x - 1} = \frac{A(x - 1) + B(2x - 3)}{(2x - 3)(x - 1)}
$$
$$
= \frac{(A + 2B)x + (-A - 3B)}{2x^2 - 5x + 3}.
$$

Matching up the numerators and equating the coefficients gives us the system of equations

$$
A + 2B = 11,
$$
$$
-A - 3B = -9.
$$

We will leave it to you to verify that the unique solution to this system of equations is $A = 15$, $B = -2$. Thus,

$$
\frac{2x^3 + x^2 - x}{2x^2 - 5x + 3} = x + 3 + \frac{11x - 9}{2x^2 - 5x + 3} = x + 3 + \frac{15}{2x - 3} - \frac{2}{x - 1}.
$$

We can now finally evaluate integral (b):

$$
\int \frac{2x^3 + x^2 - x}{2x^2 - 5x + 3}\,dx = \int x + 3 + \frac{15}{2x - 3} - \frac{2}{x - 1}\,dx
$$
$$
= \frac{x^2}{2} + 3x + \frac{15}{2}\ln|2x - 3| - 2\ln|x - 1| + C.
$$

Finally, for integral (c) we again begin by factoring the denominator:

$$
u^3 - u^2 + u - 1 = (u^2 + 1)(u - 1).
$$

Based on our previous examples, we might try to split up the integrand like this:

$$
\frac{u^2 - 3}{u^3 - u^2 + u - 1} = \frac{A}{u^2 + 1} + \frac{B}{u - 1}
$$
$$
= \frac{A(u - 1) + B(u^2 + 1)}{(u^2 + 1)(u - 1)} = \frac{Bu^2 + Au + (-A + B)}{u^3 - u^2 + u - 1}. \tag{8.8}
$$

To make the numerators match, we must have

$$1 \cdot u^2 + 0 \cdot u - 3 = B \cdot u^2 + A \cdot u + (-A + B),$$

and equating coefficients leads to the system of equations

$$B = 1,$$
$$A = 0,$$
$$-A + B = -3.$$

Unfortunately, this system of equations doesn't have a solution; the values of A and B given by the first two equations don't work in the third equation. So there is no decomposition of the integrand of the form given in equation (8.8).

However, there is a different way of splitting up the integrand that does work:

$$\frac{u^2 - 3}{u^3 - u^2 + u - 1} = \frac{Au + B}{u^2 + 1} + \frac{C}{u - 1} = \frac{(Au + B)(u - 1) + C(u^2 + 1)}{(u^2 + 1)(u - 1)}$$
$$= \frac{(A + C)u^2 + (-A + B)u + (-B + C)}{u^3 - u^2 + u - 1}. \qquad (8.9)$$

Now when we equate coefficients in the numerator we get the system of equations

$$A + C = 1,$$
$$-A + B = 0,$$
$$-B + C = -3.$$

This system of equations can be solved by the same method we used in our earlier examples. But for the purpose of illustrating another possibility we will use a different method. We can rewrite the third equation as $C = B - 3$, and then substituting $B - 3$ for C in the first equation we get $A + B - 3 = 1$, or in other words $A + B = 4$. With these changes, the system of equations reads

$$A + B = 4,$$
$$-A + B = 0,$$
$$C = B - 3.$$

If we can solve the first two equations for A and B, then the third equation will give us C. Now we rewrite the second equation as $B = A$, and substitute A for B in the first equation:

$$2A = 4,$$
$$B = A,$$
$$C = B - 3.$$

Now we can read off the solution: from the first equation we get $A = 2$, the second then gives us $B = 2$, and the third says $C = 2 - 3 = -1$. So the decomposition of the integrand is

$$\frac{u^2 - 3}{u^3 - u^2 + u - 1} = \frac{2u + 2}{u^2 + 1} - \frac{1}{u - 1}.$$

We can now evaluate integral (c); we leave the details to you and just give the answer:

$$\int \frac{u^2 - 3}{u^3 - u^2 + u - 1}\, du = \int \frac{2u\, du}{u^2 + 1} + \int \frac{2\, du}{u^2 + 1} - \int \frac{du}{u - 1}$$
$$= \ln(u^2 + 1) + 2\tan^{-1} u - \ln|u - 1| + C. \qquad \square$$

Example 8.1.2 illustrates the basic idea behind the method of partial fractions: to write a rational function as a sum of simpler rational functions, we factor the denominator, use this factorization to set up a guess about how the rational function might be decomposed, with unknown numbers in the numerators of the decomposition, and then solve for these unknown numbers. We saw in parts (b) and (c) of the example that some guesses work out and some don't. In other words, for some guesses it is possible to solve for the unknown numbers in the numerators, and for some guesses there is no solution. Is there a way to set up the guess so that there will always be a solution? The answer is yes. It can be shown that the following procedure always works.

To decompose a rational function $\frac{P(x)}{Q(x)}$ as a sum of simpler rational functions:

1. Make sure that the degree of $P(x)$ is less than the degree of $Q(x)$. If not, divide $P(x)$ by $Q(x)$, as in part (b) of Example 8.1.2.

2. Factor $Q(x)$ into linear and quadratic factors. It can be shown that such a factorization always exists, although it is not always easy to find it.

3. Set up a guess about how to decompose $P(x)/Q(x)$, based on the factorization of $Q(x)$. Table 8.1 shows what terms to include in this guess for each factor of $Q(x)$.

Factor of $Q(x)$	Term(s) in Guess
$ax + b$	$\dfrac{A}{ax + b}$
$ax^2 + bx + c$	$\dfrac{Ax + B}{ax^2 + bx + c}$
$(ax + b)^k$	$\dfrac{A_1}{ax + b} + \dfrac{A_2}{(ax + b)^2} + \cdots + \dfrac{A_k}{(ax + b)^k}$
$(ax^2 + bx + c)^k$	$\dfrac{A_1 x + B_1}{ax^2 + bx + c} + \dfrac{A_2 x + B_2}{(ax^2 + bx + c)^2} + \cdots + \dfrac{A_k x + B_k}{(ax^2 + bx + c)^k}$

Table 8.1: Setting up a partial fractions decomposition of $\frac{P(x)}{Q(x)}$.

4. Add the fractions in your guess. The denominator of the sum should be $Q(x)$. Set the coefficients of the numerator of the sum equal to the corresponding coefficients of $P(x)$.

5. Solve the resulting system of simultaneous equations.

Example 8.1.3. Evaluate the integral

$$\int \frac{-2x^4 + 5x^3 - 4x^2 + x - 4}{(x+1)^2(x^2 - 2x + 5)}\,dx.$$

Solution. The denominator is

$$(x+1)^2(x^2 - 2x + 5) = (x^2 + 2x + 1)(x^2 - 2x + 5) = x^4 + 2x^2 + 8x + 5.$$

Numerator and denominator both have degree 4, so we divide, as shown in Figure 8.2, getting a quotient of -2 and a remainder of $5x^3 + 17x + 6$. This tells us that

$$\frac{-2x^4 + 5x^3 - 4x^2 + x - 4}{(x+1)^2(x^2 - 2x + 5)} = -2 + \frac{5x^3 + 17x + 6}{(x+1)^2(x^2 - 2x + 5)}.$$

Next, we use Table 8.1 to set up a guess about how to split up the last fraction above. Notice that according to row three of the table, the factor $(x+1)^2$ leads to two terms in the guess.

$$\frac{5x^3 + 17x + 6}{(x+1)^2(x^2 - 2x + 5)}$$
$$= \frac{A}{x+1} + \frac{B}{(x+1)^2} + \frac{Cx + D}{x^2 - 2x + 5}$$
$$= \frac{A(x+1)(x^2 - 2x + 5) + B(x^2 - 2x + 5) + (Cx + D)(x+1)^2}{(x+1)^2(x^2 - 2x + 5)}$$
$$= \frac{(A+C)x^3 + (-A+B+2C+D)x^2 + (3A-2B+C+2D)x + (5A+5B+D)}{(x+1)^2(x^2 - 2x + 5)}.$$

Equating the coefficients of the numerators leads to the system of equations

$$A + C = 5,$$
$$-A + B + 2C + D = 0,$$

$$
\begin{array}{r}
-2 \\
x^4 + 2x^2 + 8x + 5 \overline{\smash{\big)}\, -2x^4 + 5x^3 - 4x^2 + x - 4} \\
\underline{-2x^4 - 4x^2 - 16x - 10} \\
5x^3 + 17x + 6
\end{array}
$$

Figure 8.2: Dividing $-2x^4 + 5x^3 - 4x^2 + x - 4$ by $x^4 + 2x^2 + 8x + 5$.

$$3A - 2B + C + 2D = 17,$$
$$5A + 5B + D = 6.$$

It is tedious, but not really difficult, to show that the unique solution to this system of equations is

$$A = 3, \quad B = -2, \quad C = 2, \quad D = 1.$$

Thus, the integral can be rewritten in the form

$$\int \frac{-2x^4 + 5x^3 - 4x^2 + x - 4}{(x+1)^2(x^2 - 2x + 5)} \, dx = \int -2 + \frac{3}{x+1} - \frac{2}{(x+1)^2} + \frac{2x+1}{x^2 - 2x + 5} \, dx$$

$$= -2x + 3 \ln|x+1| + \frac{2}{x+1} + \int \frac{2x+1}{x^2 - 2x + 5} \, dx.$$

To evaluate the last integral, we complete the square in the denominator and then make the substitution $u = x - 1$, $x = u + 1$, $du = dx$:

$$\int \frac{2x+1}{x^2 - 2x + 5} \, dx = \int \frac{2x+1}{(x-1)^2 + 4} \, dx = \int \frac{2u+3}{u^2 + 4} \, du = \int \frac{2u\,du}{u^2 + 4} + \int \frac{3\,du}{u^2 + 4}$$

$$= \ln(u^2 + 4) + \frac{3}{2} \tan^{-1}\left(\frac{u}{2}\right) + C$$

$$= \ln(x^2 - 2x + 5) + \frac{3}{2} \tan^{-1}\left(\frac{x-1}{2}\right) + C.$$

Note that we have used Exercise 1 from Section 7.5 to evaluate the last integral. Putting everything together, we see that the answer to the original integral is

$$\int \frac{-2x^4 + 5x^3 - 4x^2 + x - 4}{(x+1)^2(x^2 - 2x + 5)} \, dx$$

$$= -2x + 3 \ln|x+1| + \frac{2}{x+1} + \ln(x^2 - 2x + 5) + \frac{3}{2} \tan^{-1}\left(\frac{x-1}{2}\right) + C. \qquad \square$$

Exercises 8.1

1–18: Evaluate the integral.

1. $\displaystyle\int \frac{x+2}{x^2} \, dx.$

2. $\displaystyle\int \frac{x^2}{x+2} \, dx.$

3. $\displaystyle\int \frac{5x+1}{x^2 - x - 12} \, dx.$

4. $\displaystyle\int \frac{5x+2}{2x^2 + x - 1} \, dx.$

5. $\int \dfrac{x+2}{x^2-2x}\,dx.$

6. $\int \dfrac{x^2+2}{x^2-2x}\,dx.$

7. $\int \dfrac{x^2+2}{x^3-2x^2}\,dx.$

8. $\int \dfrac{x^2-5x+2}{2x^3-x^2}\,dx.$

9. $\int \dfrac{t-1}{4t^3+t}\,dt.$

10. $\int \dfrac{t-1}{4t^3-t}\,dt.$

11. $\int \dfrac{dx}{x^3+x^2-2x}.$

12. $\int \dfrac{x^2+x}{x^3-x^2+x-1}\,dx$

13. $\int \dfrac{u^4}{u^2-4}\,du.$

14. $\int \dfrac{5y^3+y-3}{y^4+y^2}\,dy.$

15. $\int \dfrac{x^3+4x^2}{x^4-16}\,dx.$

16. $\int \dfrac{x^6+x^5}{x^4-16}\,dx.$

17. $\int \dfrac{\cos\theta}{4-\sin^2\theta}\,d\theta.$

18. $\int \dfrac{dx}{e^{2x}-e^x}.$

19. Find the volume of the solid generated when the region under the curve $y = 1/(x^2-1)$, $2 \le x \le 3$ is rotated about the line $x = -2$.

20. Find the volume of the solid generated when the region under the curve $y = 1/(x^2-1)$, $2 \le x \le 3$ is rotated about the x-axis.

8.2   Integration by Parts

The integration technique we will study in this section is simply the product rule for derivatives turned around backwards. According to the product rule, for any two differentiable functions f and g,

$$\frac{d}{dx}(f(x)g(x)) = f(x)g'(x) + g(x)f'(x).$$

Reversing this derivative equation, we get the integration formula

$$\int (f(x)g'(x) + g(x)f'(x))\,dx = f(x)g(x) + C.$$

While this equation is correct, it isn't very useful in this form; you wouldn't very often come across an integral in which the integrand happened to have the form $f(x)g'(x) + g(x)f'(x)$. But rearranging the equation leads to a more useful equation:

$$\int f(x)g'(x)\,dx + \int g(x)f'(x)\,dx = f(x)g(x) + C,$$

$$\int f(x)g'(x)\,dx = f(x)g(x) - \int g(x)f'(x)\,dx. \qquad (8.10)$$

Note that in equation (8.10) we have dropped the "$+C$" because we think of the integral on the right-hand side of the equation as including a "$+C$" term.

Example 8.2.1. Evaluate the integral

$$\int \ln x\,dx.$$

Solution. Let $f(x) = \ln x$ and $g(x) = x$. (For the moment, don't worry about how these functions were chosen. We'll discuss that later.) Then

$$f'(x) = \frac{1}{x}, \qquad g'(x) = 1.$$

We can now write the required integral in terms of f and g and then apply equation (8.10):

$$\int \ln x\,dx = \int f(x)g'(x)\,dx = f(x)g(x) - \int g(x)f'(x)\,dx$$

$$= x\ln x - \int x\cdot\frac{1}{x}\,dx = x\ln x - \int 1\,dx = x\ln x - x + C. \qquad \square$$

We can check our answer to the last example by differentiating. Of course, since we used the product rule backwards in our solution, we use the product rule forwards in our check.

$$\frac{d}{dx}(x\ln x - x) = x\cdot\frac{d}{dx}(\ln x) + \ln x\cdot\frac{d}{dx}(x) - 1 = x\cdot\frac{1}{x} + (\ln x)\cdot 1 - 1 = \ln x.$$

The method we used to evaluate the integral in the last example is called *integration by parts*. Before doing more examples, we introduce a shorthand version of equation (8.10). If we let $u = f(x)$ and $v = g(x)$, then as in integration by substitution we can write

$$du = f'(x)\,dx, \qquad dv = g'(x)\,dx.$$

Using these abbreviations, we can rewrite equation (8.10) in the form

$$\int u\,dv = uv - \int v\,du. \tag{8.11}$$

Example 8.2.2. Evaluate the integral

$$\int x \cos x \, dx.$$

Solution. Let $u = x$. If our integral is to have the form of the left-hand side of equation (8.11), then we will need to have $dv = \cos x\,dx$, so we let $v = \sin x$. With this choice of u and v we have $du = dx$ and $dv = \cos x\,dx$, so by equation (8.11) the value of the integral is

$$\int x \cos x \, dx = \int u\,dv = uv - \int v\,du = x \sin x - \int \sin x\,dx = x \sin x + \cos x + C.$$

$\square$

The hardest part of the method of integration by parts is choosing the functions to be represented by u and v in equation (8.11). In order for the method to be applicable, we need to choose u and v so that the integral we are trying to evaluate has the form $\int u\,dv$. But even if the method is *applicable*, it may not be *helpful*. The method will only be helpful if the integral on the right-hand side of equation (8.11) is easier to evaluate than the integral on the left. That is, $\int v\,du$ must be easier to evaluate than $\int u\,dv$.

In most cases, we choose u to be some function appearing in the integral that gets simpler when you differentiate it. With such a choice for u, the du in the integral on the right-hand side of equation (8.11) will be simpler than the u in the integral on the left. Once we have decided what to use for u, we know what dv must be if the integral is to have the form $\int u\,dv$, and we use this to determine v. Some good choices for u are powers of x, logarithms, and inverse trigonometric functions.

Example 8.2.3. Evaluate the following integrals:

(a) $\displaystyle\int x \ln x \, dx,$ (b) $\displaystyle\int x^2 e^{2x} \, dx,$

(c) $\displaystyle\int \sin^{-1} x \, dx,$ (d) $\displaystyle\int \ln(t^2 + 1) \, dt.$

Solution. In integral (a), both x and $\ln x$ are on our list of good choices for u; both get simpler when differentiated. But it seems more important to simplify the $\ln x$, so we choose $u = \ln x$. We must therefore have $dv = x\,dx$, so we let $v = x^2/2$. Since $u = \ln x$, we have $du = (1/x)\, dx$, so

$$\int x \ln x \, dx = \int u \, dv = uv - \int v \, du = \frac{x^2}{2} \ln x - \int \frac{x^2}{2} \cdot \frac{1}{x} \, dx$$

$$= \frac{x^2}{2} \ln x - \int \frac{x}{2} \, dx = \frac{x^2}{2} \ln x - \frac{x^2}{4} + C.$$

For integral (b), the factor x^2 makes the integral more difficult to deal with, and this factor will get simpler if we differentiate it. So we choose

$$u = x^2, \qquad dv = e^{2x} \, dx.$$

To find du and v, we differentiate u and integrate dv;

$$du = 2x \, dx, \qquad v = \frac{e^{2x}}{2}.$$

Applying integration by parts, we get

$$\int x^2 e^{2x} \, dx = \int u \, dv = uv - \int v \, du$$

$$= \frac{x^2 e^{2x}}{2} - \int \frac{e^{2x}}{2} \cdot 2x \, dx = \frac{x^2 e^{2x}}{2} - \int x e^{2x} \, dx.$$

We must still evaluate $\int x e^{2x} \, dx$. Comparing this to the original integral, $\int x^2 e^{2x} \, dx$, we see that we have made progress. The troublesome factor x^2 has changed to x; it has gotten simpler, but it hasn't disappeared altogether. The natural next step is to use integration by parts again! This time we use

$$u = x, \qquad dv = e^{2x} \, dx,$$

$$du = dx, \qquad v = \frac{e^{2x}}{2}.$$

(If you are uncomfortable using the letters u and v twice in the same problem to stand for different things, you could use, say, u_1 and v_1 for this second application of integration by parts. But at this point we are done with our original values of u and v, so there should be no danger of confusion if we reuse these letters.) Integration by parts now gives us

$$\int x e^{2x} \, dx = \int u \, dv = uv - \int v \, du = \frac{x e^{2x}}{2} - \int \frac{e^{2x}}{2} \, dx = \frac{x e^{2x}}{2} - \frac{e^{2x}}{4} + C.$$

Putting it all together, we conclude that

$$\int x^2 e^{2x} \, dx = \frac{x^2 e^{2x}}{2} - \int x e^{2x} \, dx$$

$$= \frac{x^2 e^{2x}}{2} - \left(\frac{x e^{2x}}{2} - \frac{e^{2x}}{4} \right) + C = \frac{(2x^2 - 2x + 1) e^{2x}}{4} + C.$$

Integral (c) involves an inverse trigonometric function, which is a good choice for u. So we let

$$u = \sin^{-1} x, \qquad dv = dx,$$
$$du = \frac{1}{\sqrt{1-x^2}}\,dx, \qquad v = x.$$

Integration by parts gives us

$$\int \sin^{-1} x\,dx = x \sin^{-1} x - \int \frac{x}{\sqrt{1-x^2}}\,dx.$$

To evaluate the remaining integral, we use the substitution $u = 1 - x^2$, $du = -2x\,dx$, $(-1/2)\,du = x\,dx$:

$$\int \frac{x}{\sqrt{1-x^2}}\,dx = \int -\frac{1}{2}u^{-1/2}\,du = -u^{1/2} + C = -\sqrt{1-x^2} + C.$$

Thus, the value of the original integral is

$$\int \sin^{-1} x\,dx = x \sin^{-1} x - \int \frac{x}{\sqrt{1-x^2}}\,dx = x \sin^{-1} x + \sqrt{1-x^2} + C.$$

Finally, the natural logarithm in integral (d) is a good choice for u in integration by parts, so we let

$$u = \ln(t^2 + 1), \qquad dv = dt,$$
$$du = \frac{2t}{t^2 + 1}\,dt, \qquad v = t.$$

This gives us

$$\int \ln(t^2 + 1)\,dt = t \ln(t^2 + 1) - \int \frac{2t^2}{t^2 + 1}\,dt.$$

Next we divide $2t^2$ by $t^2 + 1$, getting a quotient of 2 and a remainder of -2. In other words,

$$\int \ln(t^2 + 1)\,dt = t \ln(t^2 + 1) - \int \left(2 - \frac{2}{t^2 + 1} \right) dt$$
$$= t \ln(t^2 + 1) - 2t + 2 \tan^{-1} t + C. \qquad \square$$

In all of the integrals in Example 8.2.3, we let u stand for a function that got simpler when it was differentiated. Sometimes it works to let u stand for a function that doesn't get simpler when it is differentiated, as long as it also doesn't get significantly more complicated. We illustrate this with our next example.

Example 8.2.4. Evaluate the integral

$$\int e^x \cos x \, dx.$$

Solution. If we let $u = e^x$, then we will have $du = e^x \, dx$. In order to be able to apply integration by parts we need to have $dv = \cos x \, dx$, so we let $v = \sin x$. Integration by parts now gives us

$$\int e^x \cos x \, dx = e^x \sin x - \int e^x \sin x \, dx.$$

It's not clear that we've made any progress, but surprisingly we can get the answer by using integration by parts again on the integral on the right-hand side. This time we use $u = e^x$, $dv = \sin x \, dx$, $du = e^x \, dx$, $v = -\cos x$:

$$\int e^x \cos x \, dx = e^x \sin x - \int e^x \sin x \, dx = e^x \sin x - \left(-e^x \cos x + \int e^x \cos x \, dx \right)$$

$$= e^x (\sin x + \cos x) - \int e^x \cos x \, dx. \tag{8.12}$$

The original integral now appears on the right-hand side of equation (8.12). It may seem that we are just going in circles, but we can actually use equation (8.12) to solve for the desired integral. Since $e^x \cos x$ is continuous at all numbers, we know that it has an antiderivative on the interval $(-\infty, \infty)$. Let $F(x)$ be such an antiderivative. Then according to equation (8.12),

$$\int e^x \cos x \, dx = e^x (\sin x + \cos x) - F(x) + C.$$

In other words, this is a general formula for all antiderivatives of $e^x \cos x$ on $(-\infty, \infty)$. But since $F(x)$ is one such antiderivative, there must be some number C_0 for which

$$F(x) = e^x (\sin x + \cos x) - F(x) + C_0.$$

We can now solve this equation for $F(x)$. We get $2F(x) = e^x (\sin x + \cos x) + C_0$, and therefore

$$F(x) = \frac{e^x (\sin x + \cos x)}{2} + \frac{C_0}{2}.$$

The term $C_0/2$ is a constant; if we rename it C, then we have the following formula for all antiderivatives of $e^x \cos x$:

$$\int e^x \cos x \, dx = \frac{e^x (\sin x + \cos x)}{2} + C. \qquad \square$$

The reasoning we used in the last example is usually written more briefly, but less carefully, by treating $\int e^x \cos x \, dx$ in equation (8.12) as an unknown and solving for it.

Adding $\int e^x \cos x \, dx$ to both sides of equation (8.12), we get

$$2 \int e^x \cos x \, dx = e^x (\sin x + \cos x),$$

and dividing through by 2 gives the answer—except that the " $+ C$ " is missing. The more careful reasoning we gave in our solution to Example 8.2.4 shows why the " $+ C$ " belongs in the answer.

Since our solution to Example 8.2.4 was a bit unusual, it might be worthwhile to check it by differentiating:

$$\frac{d}{dx} \left(\frac{e^x (\sin x + \cos x)}{2} \right) = \frac{1}{2} \left(e^x \cdot \frac{d}{dx} (\sin x + \cos x) + (\sin x + \cos x) \cdot \frac{d}{dx} (e^x) \right)$$

$$= \frac{1}{2} \left(e^x \cdot (\cos x - \sin x) + (\sin x + \cos x) \cdot e^x \right) = e^x \cos x,$$

as expected.

There is also a version of integration by parts for definite integrals. If f' and g' are continuous on an interval containing the numbers a and b, then by the product rule and the second fundamental theorem of calculus,

$$\int_a^b (f(x)g'(x) + g(x)f'(x)) \, dx = [f(x)g(x)]_a^b.$$

Splitting up the integral on the left and rearranging the equation, we get

$$\int_a^b f(x)g'(x) \, dx = [f(x)g(x)]_a^b - \int_a^b g(x)f'(x) \, dx.$$

In other words, in the shorthand notation we have been using in this section,

$$\int_a^b u \, dv = [uv]_a^b - \int_a^b v \, du.$$

To illustrate the use of this version of integration by parts, we evaluate

$$\int_0^1 \tan^{-1} x \, dx.$$

The appearance of an inverse trigonometric function in the integral suggests the use of integration by parts. We let $u = \tan^{-1} x$ and $dv = dx$, and therefore

$$du = \frac{dx}{x^2 + 1}, \qquad v = x.$$

Applying integration by parts, we get

$$\int_0^1 \tan^{-1} x \, dx = \left[x \tan^{-1} x \right]_0^1 - \int_0^1 \frac{x \, dx}{x^2 + 1}$$

$$= [\tan^{-1} 1 - 0] - \left[\frac{1}{2} \ln(x^2 + 1) \right]_0^1 = \frac{\pi}{4} - \frac{\ln 2}{2}.$$

Exercises 8.2

1–17: Evaluate the integral.

1. $\displaystyle\int x \sin(3x+1)\,dx.$

2. $\displaystyle\int \ln(x^2-1)\,dx.$

3. $\displaystyle\int \ln(x^3-8)\,dx.$

4. $\displaystyle\int t \ln(t+2)\,dt.$

5. $\displaystyle\int x e^{x-2}\,dx.$

6. $\displaystyle\int x e^{x^2-2}\,dx.$

7. $\displaystyle\int x^2 e^{x-2}\,dx.$

8. $\displaystyle\int x^3 e^{x^2-2}\,dx.$

9. $\displaystyle\int x^2 \sin^{-1} x\,dx.$

10. $\displaystyle\int x^2 \tan^{-1} x\,dx.$

11. $\displaystyle\int \frac{\tan^{-1} x}{x^2}\,dx.$

12. $\displaystyle\int_1^2 x \sec^{-1} x\,dx.$

13. $\displaystyle\int \theta \sec^2 \theta\,d\theta.$

14. $\displaystyle\int x^2 \cos(\pi x)\,dx.$

15. $\displaystyle\int \tan^{-1}(x^2)\,dx.$ (Hint: x^4+1 can be factored. Try to find a factorization of the form $x^4+1=(x^2+ax+b)(x^2+cx+d)$.)

16. $\displaystyle\int \sin(\ln x)\,dx.$ (Hint: This is similar to Example 8.2.4.)

17. $\displaystyle\int x \sin(\ln x)\,dx.$

18. Just for fun, try computing $\int x^2 \, dx$ using integration by parts, with $u = x^2$, $dv = dx$. You will have to use the method of Example 8.2.4.

19. Suppose we use integration by parts to evaluate $\int (1/x) \, dx$, with

$$u = x^{-1}, \qquad dv = dx, \qquad du = -x^{-2} \, dx, \qquad v = x.$$

This gives us

$$\int \frac{1}{x} \, dx = \int u \, dv = uv - \int v \, du = 1 + \int \frac{1}{x} \, dx.$$

Subtracting $\int (1/x) \, dx$ from both sides, we get $0 = 1$. Clearly, something is wrong! Where is the mistake?

20. A wire extends from 0 to 1 on a number line, and the density of the wire at the point x is $\tan^{-1} x$. Find the center of mass of the wire.

21. Find the volume of the solid generated when the region under the curve $y = \sin x$, $0 \leq x \leq \pi$ is rotated about the y-axis.

22. Find the volume of the solid generated when the region bounded by the curves $y = e^x$, $y = e$, and the y-axis is rotated about the y-axis.

23. Find the volume of the solid generated when the region under the curve $y = \ln x$, $1 \leq x \leq 2$ is rotated about the x-axis.

8.3   Trigonometric Integrals

In this section we discuss some techniques that can be used to evaluate integrals involving trigonometric functions. Since all trigonometric functions can be expressed in terms of the sine and cosine functions, we begin by considering integrals that involve just these two functions.

For example, consider

$$\int \sin^3 \theta \cos \theta \, d\theta.$$

We can evaluate this integral by using the substitution $u = \sin \theta$, $du = \cos \theta \, d\theta$:

$$\int \sin^3 \theta \cos \theta \, d\theta = \int u^3 \, du = \frac{u^4}{4} + C = \frac{\sin^4 \theta}{4} + C.$$

What made this substitution work is the fact that the integrand has the form of an expression involving $\sin \theta$ that is multiplied by $\cos \theta$. The factor $\cos \theta$, together with $d\theta$, got replaced with du, and the rest of the integrand could be expressed in terms of u. Similarly, if the integrand is an expression involving $\cos \theta$ that is multiplied by $\sin \theta$, then the substitution $u = \cos \theta$, $du = -\sin \theta \, d\theta$ should work.

Sometimes the identity $\cos^2\theta + \sin^2\theta = 1$ can be used to put an integral into a form in which one of these substitutions can be used. For example, consider $\int \sin^3\theta \cos^2\theta \, d\theta$. We have

$$\int \sin^3\theta \cos^2\theta \, d\theta = \int \sin^2\theta \cos^2\theta \sin\theta \, d\theta = \int (1 - \cos^2\theta) \cos^2\theta \sin\theta \, d\theta.$$

The integral on the right is now an expression involving $\cos\theta$ that is multiplied by $\sin\theta$, so we can use the substitution $u = \cos\theta$, $du = -\sin\theta \, d\theta$:

$$\int \sin^3\theta \cos^2\theta \, d\theta = \int (1 - \cos^2\theta) \cos^2\theta \sin\theta \, d\theta = \int -(1 - u^2)u^2 \, du$$

$$= \int u^4 - u^2 \, du = \frac{u^5}{5} - \frac{u^3}{3} + C = \frac{\cos^5\theta}{5} - \frac{\cos^3\theta}{3} + C.$$

A key step in this example was replacing $\sin^2\theta$ with $1 - \cos^2\theta$. We think of this as "trading in two factors of $\sin\theta$ for two factors of $\cos\theta$." Since we started with three factors of $\sin\theta$, after the trade-in we had one $\sin\theta$ left, and this is exactly what we needed to make the substitution $u = \cos\theta$ work. In general, we can trade in factors of either $\sin\theta$ or $\cos\theta$ two at a time, and we usually want to trade in all factors except one. This means that we should look for *odd powers of* $\sin\theta$ *or* $\cos\theta$.

Example 8.3.1. Evaluate the following integrals:

$$\text{(a)} \int \sin^5\theta \cos^4\theta \, d\theta, \qquad \text{(b)} \int (\sin x - 2) \tan x \, dx.$$

Solution. Since the power of $\sin\theta$ in integral (a) is odd, we can trade in all factors of $\sin\theta$ except one:

$$\int \sin^5\theta \cos^4\theta \, d\theta = \int (\sin^2\theta)^2 \cos^4\theta \sin\theta \, d\theta = \int (1 - \cos^2\theta)^2 \cos^4\theta \sin\theta \, d\theta.$$

Now we are ready for the substitution $u = \cos\theta$, $du = -\sin\theta \, d\theta$:

$$\int \sin^5\theta \cos^4\theta \, d\theta = \int (1 - \cos^2\theta)^2 \cos^4\theta \sin\theta \, d\theta = \int -(1 - u^2)^2 u^4 \, du$$

$$= \int -u^8 + 2u^6 - u^4 \, du = -\frac{u^9}{9} + \frac{2u^7}{7} - \frac{u^5}{5} + C$$

$$= -\frac{\cos^9\theta}{9} + \frac{2\cos^7\theta}{7} - \frac{\cos^5\theta}{5} + C.$$

Integral (b) involves $\tan x$, which we have not yet discussed. Later in this section we will study some techniques that are sometimes helpful for integrals involving the tangent function. But one method we can always use is to fill in the definition of $\tan x$ in terms of $\sin x$ and $\cos x$:

$$\int (\sin x - 2) \tan x \, dx = \int (\sin x - 2) \cdot \frac{\sin x}{\cos x} \, dx = \int \frac{\sin^2 x - 2\sin x}{\cos x} \, dx.$$

The last integral involves both even and odd powers of $\sin x$, but the $\cos x$ in the denominator can be thought of as $(\cos x)^{-1}$, which is an odd power of $\cos x$. So let's see if we can put the integrand into the form of an expression involving $\sin x$ that is multiplied by $\cos x$. We begin by multiplying and dividing by $\cos x$:

$$\int \frac{\sin^2 x - 2\sin x}{\cos x}\, dx = \int \frac{\sin^2 x - 2\sin x}{\cos^2 x}\cdot \cos x\, dx = \int \frac{\sin^2 x - 2\sin x}{1 - \sin^2 x}\cdot \cos x\, dx.$$

Now we can use the substitution $u = \sin x$, $du = \cos x\, dx$:

$$\int (\sin x - 2)\tan x\, dx = \int \frac{\sin^2 x - 2\sin x}{1 - \sin^2 x}\cdot \cos x\, dx = \int \frac{u^2 - 2u}{1 - u^2}\, du.$$

The integrand in the last integral is a rational function, so we use techniques from Section 8.1. We leave the details to you, and just record the main steps:

$$\int \frac{u^2 - 2u}{1 - u^2}\, du = \int -1 + \frac{2u - 1}{u^2 - 1}\, du = \int -1 + \frac{1/2}{u - 1} + \frac{3/2}{u + 1}\, du$$
$$= -u + \frac{1}{2}\ln|u - 1| + \frac{3}{2}\ln|u + 1| + C.$$

Finally, we fill in $u = \sin x$ to get the answer to the original integral. Notice that since we know $-1 \le \sin x \le 1$, we can eliminate the absolute value signs:

$$\int (\sin x - 2)\tan x\, dx = -\sin x + \frac{1}{2}\ln(1 - \sin x) + \frac{3}{2}\ln(1 + \sin x) + C. \qquad \square$$

The techniques we have been using so far can be applied when either the sine function or the cosine function is raised to an odd power. But what if both are raised to even powers? In that case, there are some trigonometric identities that can be helpful. To derive these identities, we combine the equation $1 = \cos^2\theta + \sin^2\theta$ with the double-angle formula $\cos(2\theta) = \cos^2\theta - \sin^2\theta$. First we add and subtract these two equations:

$$
\begin{array}{rr}
\quad\quad 1 = \cos^2\theta + \sin^2\theta & \quad\quad 1 = \cos^2\theta + \sin^2\theta \\
+\ \underline{\quad \cos(2\theta) = \cos^2\theta - \sin^2\theta\quad} & -\ \underline{\quad \cos(2\theta) = \cos^2\theta - \sin^2\theta\quad} \\
1 + \cos(2\theta) = 2\cos^2\theta & 1 - \cos(2\theta) = 2\sin^2\theta
\end{array}
$$

Dividing the resulting equations by 2, we get the identities

$$\cos^2\theta = \frac{1 + \cos(2\theta)}{2}, \qquad \sin^2\theta = \frac{1 - \cos(2\theta)}{2}. \tag{8.13}$$

Example 8.3.2. Evaluate the integral

$$\int \sin^4\theta\cos^2\theta\, d\theta.$$

Solution. Since both $\sin\theta$ and $\cos\theta$ are raised to even powers, we rewrite the integrand using the identities in equation (8.13):

$$\int \sin^4\theta \cos^2\theta \, d\theta = \int (\sin^2\theta)^2 \cos^2\theta \, d\theta = \int \left(\frac{1-\cos(2\theta)}{2}\right)^2 \cdot \frac{1+\cos(2\theta)}{2} \, d\theta$$

$$= \frac{1}{8}\int (1-\cos(2\theta))^2(1+\cos(2\theta)) \, d\theta$$

$$= \frac{1}{8}\int \cos^3(2\theta) - \cos^2(2\theta) - \cos(2\theta) + 1 \, d\theta$$

$$= \frac{1}{8}\left(\int \cos^3(2\theta) - \cos(2\theta) \, d\theta + \int 1 - \cos^2(2\theta) \, d\theta\right).$$

In the last step, we have separated out the odd powers of $\cos(2\theta)$, which can now be handled by the methods discussed earlier in this section. We evaluate the final two integrals separately. For the first, we use the substitution $u = \sin(2\theta)$, $du = 2\cos(2\theta) \, d\theta$:

$$\int \cos^3(2\theta) - \cos(2\theta) \, d\theta = \int (\cos^2(2\theta) - 1)\cos(2\theta) \, d\theta = \int -\sin^2(2\theta)\cos(2\theta) \, d\theta$$

$$= -\frac{1}{2}\int u^2 \, du = -\frac{u^3}{6} + C = -\frac{\sin^3(2\theta)}{6} + C.$$

For the second integral, we use (8.13) again:

$$\int 1 - \cos^2(2\theta) \, d\theta = \int \sin^2(2\theta) \, d\theta = \int \frac{1-\cos(4\theta)}{2} \, d\theta = \frac{1}{2}\int 1 - \cos(4\theta) \, d\theta$$

$$= \frac{1}{2}\left(\theta - \frac{\sin(4\theta)}{4}\right) + C = \frac{\theta}{2} - \frac{\sin(4\theta)}{8} + C.$$

Putting it all together, our final answer is

$$\int \sin^4\theta \cos^2\theta \, d\theta = \frac{1}{8}\left(\int \cos^3(2\theta) - \cos(2\theta) \, d\theta + \int 1 - \cos^2(2\theta) \, d\theta\right)$$

$$= \frac{1}{8}\left(-\frac{\sin^3(2\theta)}{6} + \frac{\theta}{2} - \frac{\sin(4\theta)}{8}\right) + C$$

$$= -\frac{\sin^3(2\theta)}{48} + \frac{\theta}{16} - \frac{\sin(4\theta)}{64} + C. \qquad \square$$

Integration by parts is also sometimes helpful for integrals involving powers of trigonometric functions. We illustrate this with the integral

$$\int \sin^n\theta \, d\theta,$$

where n is an integer and $n \geq 2$. To apply integration by parts, we use

$$u = \sin^{n-1}\theta, \qquad\qquad dv = \sin\theta \, d\theta,$$

$$du = (n-1)\sin^{n-2}\theta \cos\theta \, d\theta, \qquad v = -\cos\theta.$$

Integration by parts now yields

$$\int \sin^n \theta \, d\theta = -\cos \theta \sin^{n-1} \theta + \int (n-1) \sin^{n-2} \theta \cos^2 \theta \, d\theta$$

$$= -\cos \theta \sin^{n-1} \theta + (n-1) \int \sin^{n-2} \theta (1 - \sin^2 \theta) \, d\theta$$

$$= -\cos \theta \sin^{n-1} \theta + (n-1) \int \sin^{n-2} \theta \, d\theta - (n-1) \int \sin^n \theta \, d\theta.$$

$$(8.14)$$

The integral we are trying to evaluate now appears on the right-hand side of (8.14). Imitating the method we used in Example 8.2.4, we can use equation (8.14) to solve for the desired integral. Adding $(n-1) \int \sin^n \theta \, d\theta$ to both sides of equation (8.14), we get

$$n \int \sin^n \theta \, d\theta = -\cos \theta \sin^{n-1} \theta + (n-1) \int \sin^{n-2} \theta \, d\theta,$$

and therefore

$$\int \sin^n \theta \, d\theta = -\frac{\cos \theta \sin^{n-1} \theta}{n} + \frac{n-1}{n} \int \sin^{n-2} \theta \, d\theta. \qquad (8.15)$$

Notice that there is no need to include a " $+ C$" on the right-hand side of this equation, since the integral on the right includes an arbitrary constant term.

Equation (8.15) tells us how to rewrite an integral of a power of $\sin \theta$ in terms of an integral in which the power of $\sin \theta$ has been reduced by 2. It is therefore called a *reduction formula* for integrals of powers of $\sin \theta$. It is often helpful to use this formula repeatedly when evaluating an integral. We illustrate this by giving an alternative solution to Example 8.3.2:

$$\int \sin^4 \theta \cos^2 \theta \, d\theta$$

$$= \int \sin^4 \theta (1 - \sin^2 \theta) \, d\theta = \int \sin^4 \theta \, d\theta - \int \sin^6 \theta \, d\theta$$

$$= \int \sin^4 \theta \, d\theta - \left(-\frac{\cos \theta \sin^5 \theta}{6} + \frac{5}{6} \int \sin^4 \theta \, d\theta \right) \qquad \text{(by (8.15), } n = 6\text{)}$$

$$= \frac{\cos \theta \sin^5 \theta}{6} + \frac{1}{6} \int \sin^4 \theta \, d\theta$$

$$= \frac{\cos \theta \sin^5 \theta}{6} + \frac{1}{6} \left(-\frac{\cos \theta \sin^3 \theta}{4} + \frac{3}{4} \int \sin^2 \theta \, d\theta \right) \qquad \text{(by (8.15), } n = 4\text{)}$$

$$= \frac{\cos \theta \sin^5 \theta}{6} - \frac{\cos \theta \sin^3 \theta}{24} + \frac{1}{8} \int \sin^2 \theta \, d\theta$$

$$= \frac{\cos\theta \sin^5\theta}{6} - \frac{\cos\theta \sin^3\theta}{24}$$

$$+ \frac{1}{8}\left(-\frac{\cos\theta \sin\theta}{2} + \frac{1}{2}\int 1\, d\theta\right) \qquad\qquad \text{(by (8.15), } n = 2\text{)}$$

$$= \frac{\cos\theta \sin^5\theta}{6} - \frac{\cos\theta \sin^3\theta}{24} - \frac{\cos\theta \sin\theta}{16} + \frac{\theta}{16} + C.$$

You may be worried that this answer looks different from the answer we got in Example 8.3.2. In Exercise 23 we ask you to show that the two answers are equal.

We will see in a later section of this chapter that the tangent and secant functions often appear together in integrals. One way to deal with them is to rewrite them in terms of the sine and cosine functions. We begin by using this approach to find antiderivatives of tangent and secant.

Example 8.3.3. Evaluate the following integrals:

$$\text{(a)} \int \tan\theta\, d\theta, \qquad \text{(b)} \int \sec\theta\, d\theta.$$

Solution. In integral (a), once we write $\tan\theta$ in terms of $\sin\theta$ and $\cos\theta$, we can use the substitution $u = \cos\theta$, $du = -\sin\theta\, d\theta$:

$$\int \tan\theta\, d\theta = \int \frac{\sin\theta}{\cos\theta}\, d\theta = \int -\frac{1}{u}\, du = -\ln|u| + C = -\ln|\cos\theta| + C.$$

Some mathematicians prefer to reexpress this answer using the fact that

$$-\ln|\cos x| = \ln\left(\frac{1}{|\cos x|}\right) = \ln\left|\frac{1}{\cos x}\right| = \ln|\sec x|.$$

Thus, we can also say that

$$\int \tan x\, dx = \ln|\sec x| + C.$$

The integrand in integral (b) involves an odd power of $\cos\theta$, so we use the substitution $u = \sin\theta$, $du = \cos\theta\, d\theta$:

$$\int \sec\theta\, d\theta = \int \frac{d\theta}{\cos\theta} = \int \frac{\cos\theta\, d\theta}{\cos^2\theta} = \int \frac{\cos\theta\, d\theta}{1 - \sin^2\theta} = \int \frac{du}{1 - u^2}.$$

We can now use partial fractions to finish the integration:

$$\int \sec\theta\, d\theta = \int \frac{du}{1 - u^2} = \int \frac{1/2}{u+1} - \frac{1/2}{u-1}\, du$$

$$= \frac{1}{2}\ln|u+1| - \frac{1}{2}\ln|u-1| + C = \frac{1}{2}\ln(1+\sin\theta) - \frac{1}{2}\ln(1-\sin\theta) + C.$$

Once again, this answer can be expressed in a different form. In Exercise 24 we ask you to show that this answer can be rewritten in the form

$$\int \sec\theta\, d\theta = \ln|\sec\theta + \tan\theta| + C. \qquad\qquad \square$$

Many of the ideas we used for integrals involving sine and cosine can be adapted to work with tangent and secant. We will often find it useful to use one of the following substitutions:

$$u = \tan\theta, \qquad\qquad\qquad u = \sec\theta,$$
$$du = \sec^2\theta\, d\theta, \qquad\qquad du = \sec\theta\tan\theta\, d\theta.$$

We can "trade in" secants and tangents two at a time, using the identity $\tan^2\theta + 1 = \sec^2\theta$. Thus, if an integral involves an *even* power of $\sec\theta$, then we can trade in all factors of $\sec\theta$ except two and then use the substitution $u = \tan\theta$, $du = \sec^2\theta\, d\theta$. If an integral involves an *odd* power of $\tan\theta$, then we can trade in all factors of $\tan\theta$ except one and then use the substitution $u = \sec\theta$, $du = \sec\theta\tan\theta\, d\theta$.

Example 8.3.4. Evaluate the following integrals:

$$\text{(a)}\ \int \tan^4\theta\,\sec^4\theta\, d\theta, \qquad \text{(b)}\ \int \tan^5\theta\,\sec^5\theta\, d\theta.$$

Solution. Integral (a) contains an even power of $\sec\theta$, so we use the substitution $u = \tan\theta$, $du = \sec^2\theta\, d\theta$:

$$\int \tan^4\theta\,\sec^4\theta\, d\theta = \int \tan^4\theta\,\sec^2\theta\,\sec^2\theta\, d\theta = \int \tan^4\theta(\tan^2\theta + 1)\sec^2\theta\, d\theta$$
$$= \int u^4(u^2 + 1)\, du = \int u^6 + u^4\, du = \frac{u^7}{7} + \frac{u^5}{5} + C$$
$$= \frac{\tan^7\theta}{7} + \frac{\tan^5\theta}{5} + C.$$

In integral (b), the power of $\tan\theta$ is odd, so we use the substitution $u = \sec\theta$, $du = \sec\theta\tan\theta\, d\theta$:

$$\int \tan^5\theta\,\sec^5\theta\, d\theta = \int (\tan^2\theta)^2 \sec^4\theta\,\sec\theta\tan\theta\, d\theta$$
$$= \int (\sec^2\theta - 1)^2 \sec^4\theta\,\sec\theta\tan\theta\, d\theta$$
$$= \int (u^2 - 1)^2 u^4\, du = \int u^8 - 2u^6 + u^4\, du$$
$$= \frac{u^9}{9} - \frac{2u^7}{7} + \frac{u^5}{5} + C = \frac{\sec^9\theta}{9} - \frac{2\sec^7\theta}{7} + \frac{\sec^5\theta}{5} + C. \quad \square$$

We can also use similar methods to derive a reduction formula for integrals of powers of the tangent function. For any integer $n \geq 2$ we have

$$\int \tan^n\theta\, d\theta = \int \tan^{n-2}\theta(\sec^2\theta - 1)\, d\theta = \int \tan^{n-2}\theta\,\sec^2\theta\, d\theta - \int \tan^{n-2}\theta\, d\theta.$$
$$\tag{8.16}$$

We now use the substitution $u = \tan\theta$, $du = \sec^2\theta\, d\theta$ to compute

$$\int \tan^{n-2}\theta \sec^2\theta\, d\theta = \int u^{n-2}\, du = \frac{u^{n-1}}{n-1} + C = \frac{\tan^{n-1}\theta}{n-1} + C.$$

Substituting into equation (8.16), we get the reduction formula

$$\int \tan^n \theta\, d\theta = \frac{\tan^{n-1}\theta}{n-1} - \int \tan^{n-2}\theta\, d\theta. \tag{8.17}$$

As an example of the use of this reduction formula, we compute $\int \tan^4\theta\, d\theta$:

$$\int \tan^4\theta\, d\theta = \frac{\tan^3\theta}{3} - \int \tan^2\theta\, d\theta \qquad \text{(by (8.17), } n = 4)$$

$$= \frac{\tan^3\theta}{3} - \left(\tan\theta - \int 1\, d\theta\right) \qquad \text{(by (8.17), } n = 2)$$

$$= \frac{\tan^3\theta}{3} - \tan\theta + \theta + C.$$

There is also a reduction formula for $\int \sec^n \theta\, d\theta$. To derive it, we use integration by parts, with

$$u = \sec^{n-2}\theta, \qquad\qquad dv = \sec^2\theta\, d\theta,$$
$$du = (n-2)\sec^{n-2}\theta \tan\theta\, d\theta, \qquad v = \tan\theta.$$

This gives us

$$\int \sec^n\theta\, d\theta = \tan\theta \sec^{n-2}\theta - (n-2)\int \sec^{n-2}\theta \tan^2\theta\, d\theta$$

$$= \tan\theta \sec^{n-2}\theta - (n-2)\int \sec^{n-2}\theta(\sec^2\theta - 1)\, d\theta$$

$$= \tan\theta \sec^{n-2}\theta - (n-2)\int \sec^n\theta\, d\theta + (n-2)\int \sec^{n-2}\theta\, d\theta.$$

Adding $(n-2)\int \sec^n\theta\, d\theta$ to both sides of this equation we get

$$(n-1)\int \sec^n\theta\, d\theta = \tan\theta \sec^{n-2}\theta + (n-2)\int \sec^{n-2}\theta\, d\theta,$$

and therefore

$$\int \sec^n\theta\, d\theta = \frac{\tan\theta \sec^{n-2}\theta}{n-1} + \frac{n-2}{n-1}\int \sec^{n-2}\theta\, d\theta. \tag{8.18}$$

The reduction formula (8.18) can be used to handle the most difficult combination of powers of tangent and secant: an even power of tangent times an odd power of secant.

For example,

$$\int \tan^2 \theta \sec^3 \theta \, d\theta$$

$$= \int (\sec^2 \theta - 1) \sec^3 \theta \, d\theta = \int \sec^5 \theta \, d\theta - \int \sec^3 \theta \, d\theta$$

$$= \left(\frac{\tan \theta \sec^3 \theta}{4} + \frac{3}{4} \int \sec^3 \theta \, d\theta \right) - \int \sec^3 \theta \, d\theta \qquad \text{(by (8.18), } n = 5)$$

$$= \frac{\tan \theta \sec^3 \theta}{4} - \frac{1}{4} \int \sec^3 \theta \, d\theta$$

$$= \frac{\tan \theta \sec^3 \theta}{4} - \frac{1}{4} \left(\frac{\tan \theta \sec \theta}{2} + \frac{1}{2} \int \sec \theta \, d\theta \right) \qquad \text{(by (8.18), } n = 3)$$

$$= \frac{\tan \theta \sec^3 \theta}{4} - \frac{\tan \theta \sec \theta}{8} - \frac{1}{8} \ln |\sec \theta + \tan \theta| + C.$$

We have not yet said anything about integrals involving the cotangent and cosecant functions, but the methods for dealing with these functions are almost exactly the same as those for the tangent and secant functions. We illustrate this with one example.

Example 8.3.5. Evaluate

$$\int \cot^2 x \csc^4 x \, dx.$$

Solution. For an integral involving $\sec^4 x$, we would trade in two factors of $\sec x$ and then use the substitution $u = \tan x$, $du = \sec^2 x \, dx$. So in this example, we trade in two factors of $\csc x$ and then use the substitution $u = \cot x$, $du = -\csc^2 x \, dx$:

$$\int \cot^2 x \csc^4 x \, dx = \int \cot^2 x \csc^2 x \csc^2 x \, dx = \int \cot^2 x (\cot^2 x + 1) \csc^2 x \, dx$$

$$= \int -u^2 (u^2 + 1) \, du = \int -u^4 - u^2 \, du = -\frac{u^5}{5} - \frac{u^3}{3} + C$$

$$= -\frac{\cot^5 x}{5} - \frac{\cot^3 x}{3} + C. \qquad \square$$

Exercises 8.3

1–20: Evaluate the integral.

1. $\displaystyle\int \sin^3 \theta \, d\theta.$

2. $\displaystyle\int \sin^2 \theta \cos^3 \theta \, d\theta.$

3. $\displaystyle\int \frac{\cos^3 (2\theta)}{\sin^2 (2\theta)} \, d\theta.$

4. $\int \dfrac{\sin^2(2\theta)}{\cos(2\theta)}\, d\theta.$

5. $\int \dfrac{\sin^2(2\theta)}{\cos\theta}\, d\theta.$

6. $\int \sin^2\theta \cos^2\theta\, d\theta.$

7. $\int (2+\sin\theta)^2\, d\theta.$

8. $\int (2+\sin\theta)^3\, d\theta.$

9. $\int (\sin x - 2)\cot x\, dx.$

10. $\int \tan^2\theta \sec^4\theta\, d\theta.$

11. $\int (\tan\theta \sec\theta)^3\, d\theta.$

12. $\int \tan^2\theta \sec\theta\, d\theta.$

13. $\int \dfrac{\tan\theta}{\sec^3\theta}\, d\theta.$

14. $\int \dfrac{\sec^3\theta}{\tan\theta}\, d\theta.$

15. $\int \dfrac{\tan^2\theta}{\sec\theta}\, d\theta.$

16. $\int \dfrac{\sec^4\theta}{2+\tan\theta}\, d\theta.$

17. $\int \sec\theta(1+\tan^3\theta)\, d\theta.$

18. $\int \sec^2\theta \cot\theta\, d\theta.$

19. $\int \sec^2 x \ln(\cos x)\, dx.$

20. $\int \cot^3\theta \csc^4\theta\, d\theta.$

21. Derive a reduction formula for $\int \cos^n\theta\, d\theta$. (Hint: Imitate the derivation in this section of a reduction formula for $\int \sin^n\theta\, d\theta$.)

22. (a) Use the reduction formula derived in the last exercise to evaluate $\int \cos^4 \theta \, d\theta$.

    (b) Use the reduction formula for $\int \sin^n \theta \, d\theta$ that was derived in the text to complete the following calculation:

    $$\int \cos^4 \theta \, d\theta = \int (1 - \sin^2 \theta)^2 \, d\theta = \int 1 - 2\sin^2 \theta + \sin^4 \theta \, d\theta = \cdots .$$

    (c) Show that the answers in parts (a) and (b) are the same.

23. In this section, we found two different answers for the integral in Example 8.3.2. Show that these answers are equal. In other words, verify the identity

    $$-\frac{\sin^3(2\theta)}{48} + \frac{\theta}{16} - \frac{\sin(4\theta)}{64} = \frac{\cos\theta \sin^5 \theta}{6} - \frac{\cos\theta \sin^3 \theta}{24} - \frac{\cos\theta \sin\theta}{16} + \frac{\theta}{16}.$$

24. Show that, as claimed in Example 8.3.3(b),

    $$\frac{1}{2}\ln(1 + \sin\theta) - \frac{1}{2}\ln(1 - \sin\theta) = \ln|\sec\theta + \tan\theta|.$$

25. Find the volume of the solid generated when the region under the curve $y = \sin x$, $0 \le x \le \pi$ is rotated about the x-axis.

26. Find the volume of the solid generated when the region under the curve $y = \sin x$, $0 \le x \le \pi$ is rotated about the line $y = 1$.

27. Many integrals involving trigonometric functions can be evaluated using the substitution $u = \tan(\theta/2)$. This substitution is sometimes called the *Weierstrass substitution*, after Karl Theodor Wilhelm Weierstrass (1815–1897), although it was used much earlier by Leonhard Euler (1707–1783). In this problem you will use this substitution to compute

    $$\int_{-\pi/2}^{\pi/2} \frac{d\theta}{\sin\theta + 2\cos\theta + 3}.$$

    Let $u = \tan(\theta/2)$.

    (a) Show that

    $$\frac{2}{u^2 + 1} \, du = d\theta.$$

    (b) Show that

    $$\sin\theta = \frac{2u}{u^2 + 1}.$$

    (c) Find a formula for $\cos\theta$ in terms of u.

    (d) Evaluate $\displaystyle\int_{-\pi/2}^{\pi/2} \frac{d\theta}{\sin\theta + 2\cos\theta + 3}$.

28. Use the Weierstrass substitution $u = \tan(\theta/2)$ (see Exercise 27) to evaluate

    $$\int_0^{\pi/2} \frac{1 + \sin\theta}{2 + \cos\theta} \, d\theta.$$

8.4   Substitution with Inverse Functions

Consider the integral

$$\int_a^b \frac{1}{(\sqrt[3]{x})^2 + 1} \, dx. \tag{8.19}$$

We are going to evaluate this integral by using the substitution $u = \sqrt[3]{x} = x^{1/3}$. You might expect that our first step would be to compute

$$du = \frac{1}{3} x^{-2/3} \, dx = \frac{1}{3(\sqrt[3]{x})^2} \, dx.$$

However, there is a small problem with this equation: this formula for du is undefined when $x = 0$, but the domain of the integrand in (8.19) includes all real numbers. We will resolve this problem by doing the details of the substitution a bit differently from the way we have done previous substitutions. Initially we will simply proceed in a way that seems natural, without worrying about justifying our steps. Afterwards we'll come back and show that our reasoning can be justified.

Since the cube root function is the inverse of the cubing function, the equation $u = \sqrt[3]{x}$ is equivalent to $x = u^3$, and from that equation we compute $dx = 3u^2 \, du$. Replacing $\sqrt[3]{x}$ with u, dx with $3u^2 \, du$, and the limits of integration $x = a$ and $x = b$ with the corresponding u-values $u = \sqrt[3]{a}$ and $u = \sqrt[3]{b}$, we get

$$\int_a^b \frac{1}{(\sqrt[3]{x})^2 + 1} \, dx = \int_{\sqrt[3]{a}}^{\sqrt[3]{b}} \frac{3u^2}{u^2 + 1} \, du. \tag{8.20}$$

We can now justify the procedure we have followed by observing that it is just a backwards form of our usual substitution method. If we were starting with the integral on the right-hand side of equation (8.20), then our usual substitution method would allow us to use the substitution $x = u^3$, $dx = 3u^2 \, du$ to get the integral on the left. We would substitute $\sqrt[3]{x}$ for u in the denominator, and the limits of integration $u = \sqrt[3]{a}$ and $u = \sqrt[3]{b}$ in the integral on the right would be converted to the corresponding x-values $x = (\sqrt[3]{a})^3 = a$ and $x = (\sqrt[3]{b})^3 = b$, which are used in the integral on the left. Thus, by Theorem 5.5.3, the two integrals are equal, and therefore equation (8.20) is correct. However, in this example we are going in the opposite direction: we are starting with the integral on the left, and converting it into the integral on the right.

To evaluate the integral on the right-hand side of equation (8.20), we now divide $3u^2$ by $u^2 + 1$:

$$\int_{\sqrt[3]{a}}^{\sqrt[3]{b}} \frac{3u^2}{u^2 + 1} \, du = \int_{\sqrt[3]{a}}^{\sqrt[3]{b}} 3 - \frac{3}{u^2 + 1} \, du = \left[3u - 3 \tan^{-1} u \right]_{\sqrt[3]{a}}^{\sqrt[3]{b}}.$$

Thus, the answer to our original integral is

$$\int_a^b \frac{1}{(\sqrt[3]{x})^2 + 1} \, dx = (3\sqrt[3]{b} - 3 \tan^1 (\sqrt[3]{b})) - (3\sqrt[3]{a} - 3 \tan^{-1}(\sqrt[3]{a})). \tag{8.21}$$

The "backwards" substitution method we used in this example can also be used if the integral is indefinite. Once again, we work out the answer first by proceeding in the way that seems most natural, and justify our work afterwards. As before, we use the substitution $u = \sqrt[3]{x}$, $x = u^3$, $dx = 3u^2\,du$. In the case of an indefinite integral, there are no limits of integration, and our final step is to reexpress the answer in terms of the original variable x:

$$\int \frac{1}{(\sqrt[3]{x})^2 + 1}\,dx = \int \frac{3u^2}{u^2 + 1}\,du = \int 3 - \frac{3}{u^2 + 1}\,du$$

$$= 3u - 3\tan^{-1} u + C = 3\sqrt[3]{x} - 3\tan^{-1}(\sqrt[3]{x}) + C. \qquad (8.22)$$

In Exercise 19 we ask you to check that this answer is correct. But we would like to know not merely that the answer is correct, but also that the method we used to reach that answer can be justified. So let's see if we can use the calculations we have done to show that equation (8.22) is correct.

We begin by rewriting equation (8.21) with different letters:

$$\int_a^x \frac{1}{(\sqrt[3]{t})^2 + 1}\,dt = (3\sqrt[3]{x} - 3\tan^{-1}(\sqrt[3]{x})) - (3\sqrt[3]{a} - 3\tan^{-1}(\sqrt[3]{a})).$$

Next we take the derivative with respect to x of both sides of this equation:

$$\frac{d}{dx}\left(\int_a^x \frac{1}{(\sqrt[3]{t})^2 + 1}\,dt\right) = \frac{d}{dx}\left[(3\sqrt[3]{x} - 3\tan^{-1}(\sqrt[3]{x})) - (3\sqrt[3]{a} - 3\tan^{-1}(\sqrt[3]{a}))\right].$$

Finally, we apply the first fundamental theorem of calculus on the left, and we note that the terms on the right involving a are constants, so their derivatives are 0:

$$\frac{1}{(\sqrt[3]{x})^2 + 1} = \frac{d}{dx}(3\sqrt[3]{x} - 3\tan^{-1}(\sqrt[3]{x})).$$

This shows that the answer we found in equation (8.22) is correct.

Before doing more examples, let's see if we can formulate the procedure we used in this example as a general method. Suppose that some function f is continuous on an interval I, and we want to evaluate $\int_a^b f(x)\,dx$, where a and b are numbers in I. Our plan is to do a substitution of the form $u = g^{-1}(x)$. Here g is a one-to-one function whose domain is some interval J and whose range is I. It follows that the domain of g^{-1} is I and its range is J. Thus, the equation $u = g^{-1}(x)$ makes sense for all $x \in I$, and we have $u = g^{-1}(x) \in J$. In our example, g was given by the formula $g(x) = x^3$, leading to the substitution $u = g^{-1}(x) = \sqrt[3]{x}$, and the intervals I and J were both $(-\infty, \infty)$. To carry out the substitution, we rewrite the equation $u = g^{-1}(x)$ as $x = g(u)$. Assuming that g is differentiable on J, we can then write $dx = g'(u)\,du$. Performing the substitution suggested by these equations leads to the equation

$$\int_a^b f(x)\,dx = \int_{g^{-1}(a)}^{g^{-1}(b)} f(g(u))g'(u)\,du. \qquad (8.23)$$

Again, to justify this equation we can perform the substitution $x = g(u)$, $dx = g'(u)\,du$ in the integral on the right-hand side to get the integral on the left. When we do this substitution, the limits of integration $u = g^{-1}(a)$ and $u = g^{-1}(b)$ in the integral on the right get converted to the corresponding values $x = g(g^{-1}(a)) = a$ and $x = g(g^{-1}(b)) = b$ for the integral on the left. As long as g' is continuous on J, we can use Theorem 5.5.3 to conclude that equation (8.23) is correct.

We can also use this new substitution method in indefinite integrals. To see why, suppose we are able to find an antiderivative for $f(g(u))g'(u)$ on the interval J:

$$\int f(g(u))g'(u)\, du = H(u) + C.$$

Then we can use the second fundamental theorem of calculus to evaluate the integral on the right-hand side of equation (8.23):

$$\int_{g^{-1}(a)}^{g^{-1}(b)} f(g(u))g'(u)\, du = [H(u)]_{g^{-1}(a)}^{g^{-1}(b)} = H(g^{-1}(b)) - H(g^{-1}(a)).$$

Thus, this is also the value of the integral on the left-hand side of (8.23):

$$\int_a^b f(x)\, dx = H(g^{-1}(b)) - H(g^{-1}(a)).$$

Switching letters in this equation, we can say that for all x in the interval I,

$$\int_a^x f(t)\, dt = H(g^{-1}(x)) - H(g^{-1}(a)).$$

Taking the derivative of both sides of this equation, and applying the first fundamental theorem of calculus, we can say that for all $x \in I$,

$$f(x) = \frac{d}{dx}(H(g^{-1}(x))).$$

We will therefore get the correct answer to the indefinite integral if we perform the substitution $u = g^{-1}(x)$, $x = g(u)$, $dx = g'(u)\, du$ as follows:

$$\int f(x)\, dx = \int f(g(u))g'(u)\, du = H(u) + C = H(g^{-1}(x)) + C.$$

Example 8.4.1. Evaluate the following integrals:

(a) $\displaystyle\int \sec^2(\sqrt{x})\, dx,$ (b) $\displaystyle\int \sqrt{\sqrt[3]{x} + 1}\, dx,$

(c) $\displaystyle\int e^{\sin^{-1} t}\, dt,$ (d) $\displaystyle\int \frac{dx}{\sqrt{x} + \sqrt[3]{x}}.$

Solution. For integral (a), we begin with the substitution $u = \sqrt{x}$. According to the definition of the square root function, this means that $x = u^2$ and $u \geq 0$. From the equation $x = u^2$ we get $dx = 2u\,du$, and substituting into the integral we get

$$\int \sec^2(\sqrt{x})\,dx = \int \sec^2 u \cdot 2u\,du.$$

Next, we use integration by parts. Since our integral is now written in terms of u, we will use the variables u_1 and v_1 in our integration by parts. We use

$$u_1 = 2u, \qquad dv_1 = \sec^2 u\,du,$$
$$du_1 = 2\,du, \qquad v_1 = \tan u.$$

Integration by parts now gives us

$$\int \sec^2 u \cdot 2u\,du = 2u \tan u - \int 2 \tan u\,du = 2u \tan u - 2 \ln |\sec u| + C,$$

where we have used part (a) of Example 8.3.3 to evaluate the last integral. Finally, we get the answer for the original integral by substituting $\sqrt{x}$ for u:

$$\int \sec^2(\sqrt{x})\,dx = 2\sqrt{x} \tan(\sqrt{x}) - 2 \ln \left|\sec(\sqrt{x})\right| + C.$$

In integral (b), we find it simplest to use the substitution $u = \sqrt[3]{x} + 1$. We rewrite this as $x = (u-1)^3$, compute $dx = 3(u-1)^2\,du$, and then substitute:

$$\int \sqrt{\sqrt[3]{x} + 1}\,dx = \int \sqrt{u} \cdot 3(u-1)^2\,du = \int 3(u^2 - 2u + 1)u^{1/2}\,du$$

$$= 3 \int u^{5/2} - 2u^{3/2} + u^{1/2}\,du = 3\left(\frac{2u^{7/2}}{7} - \frac{4u^{5/2}}{5} + \frac{2u^{3/2}}{3}\right) + C$$

$$= \frac{6(\sqrt[3]{x} + 1)^{7/2}}{7} - \frac{12(\sqrt[3]{x} + 1)^{5/2}}{5} + 2(\sqrt[3]{x} + 1)^{3/2} + C.$$

Since integral (c) contains the inverse function $\sin^{-1} t$, we use the substitution $u = \sin^{-1} t$. By definition of the inverse sine function, this means

$$t = \sin u, \qquad -\frac{\pi}{2} \leq u \leq \frac{\pi}{2},$$

and from the equation $t = \sin u$ we get $dt = \cos u\,du$. Substitution now leads to the equation

$$\int e^{\sin^{-1} t}\,dt = \int e^u \cos u\,du.$$

Fortunately, we solved the integral on the right-hand side of this equation in Example 8.2.4, so we can just fill in the answer:

$$\int e^u \cos u\,du = \frac{e^u(\sin u + \cos u)}{2} + C.$$

Finally, we must rewrite the answer in terms of t. We have $u = \sin^{-1} t$ and $\sin u = t$. To express $\cos u$ in terms of t, we use the identity $\cos^2 u = 1 - \sin^2 u = 1 - t^2$, so $\cos u = \pm\sqrt{1-t^2}$. And since we also know $-\pi/2 \le u \le \pi/2$, we can rule out the negative value for $\cos u$, so $\cos u = \sqrt{1-t^2}$. Therefore

$$\int e^{\sin^{-1}t}\,dt = \int e^u \cos u \, du = \frac{e^u(\sin u + \cos u)}{2} = \frac{e^{\sin^{-1}t}(t+\sqrt{1-t^2})}{2} + C.$$

It seems that for integral (d) we should substitute either $u = \sqrt{x}$ or $u = \sqrt[3]{x}$, but which should we use? In fact, we will use $u = \sqrt[6]{x}$, because for all $x \ge 0$,

$$\sqrt{x} = (\sqrt[6]{x})^3, \qquad \sqrt[3]{x} = (\sqrt[6]{x})^2.$$

From $u = \sqrt[6]{x}$ we get $x = u^6$ and $dx = 6u^5\,du$, and therefore

$$\int \frac{dx}{\sqrt{x}+\sqrt[3]{x}} = \int \frac{6u^5\,du}{u^3+u^2} = \int \frac{6u^3\,du}{u+1} = \int 6u^2 - 6u + 6 - \frac{6}{u+1}\,du$$
$$= 2u^3 - 3u^2 + 6u - 6\ln|u+1| + C$$
$$= 2\sqrt{x} - 3\sqrt[3]{x} + 6\sqrt[6]{x} - 6\ln(\sqrt[6]{x}+1) + C. \qquad \square$$

We end this section by giving one more example of an inverse function substitution in a definite integral. We will compute the integral

$$\int_0^8 \frac{dx}{\sqrt{2x+1}}$$

by using the substitution $u = \sqrt{2x+1}$. Rewriting this as $x = (u-1)^2/2$, we get $dx = (u-1)\,du$. For the limits of integration $x = 0$ and $x = 8$, the corresponding u-values are $u = \sqrt{2\cdot0+1} = 1$ and $u = \sqrt{2\cdot8+1} = 5$. Thus,

$$\int_0^8 \frac{dx}{\sqrt{2x+1}} = \int_1^5 \frac{u-1}{u}\,du = \int_1^5 1 - \frac{1}{u}\,du$$
$$= [u - \ln|u|]_1^5 = (5 - \ln 5) - 1 = 4 - \ln 5.$$

Exercises 8.4

1–18: Evaluate the integral.

1. $\int e^{\sqrt{x}}\,dx$.

2. $\int e^{\sqrt[3]{x}}\,dx$.

3. $\int \frac{e^{\sqrt[3]{x}}}{\sqrt[3]{x}}\,dx$.

4. $\int e^{\sqrt{1+\sqrt{x}}}\,dx$.

5. $\displaystyle\int \frac{dx}{\sqrt[3]{x}+1}.$

6. $\displaystyle\int \frac{dx}{\sqrt{x-4}+4}.$

7. $\displaystyle\int \frac{dx}{x+\sqrt{x}}.$

8. $\displaystyle\int \frac{dx}{x+x^{3/2}}.$

9. $\displaystyle\int \frac{dx}{\sqrt{x}+\sqrt[4]{x}-2}.$

10. $\displaystyle\int \sqrt[3]{\sqrt{x}+1}\,dx.$

11. $\displaystyle\int \frac{dx}{\sqrt[3]{\sqrt{x}+1}}.$

12. $\displaystyle\int \ln(\sqrt{t}+1)\,dt.$

13. $\displaystyle\int \sqrt{1+e^{x}}\,dx.$

14. $\displaystyle\int \frac{\sec(3\sqrt{x}+2)}{\sqrt{x}}\,dx.$

15. $\displaystyle\int \sin(\sqrt[3]{x})\,dx.$

16. $\displaystyle\int \tan(\sin^{-1}x)\,dx.$

17. $\displaystyle\int \sin(\tan^{-1}x)\,dx.$

18. $\displaystyle\int (\sin^{-1}x)^{2}\,dx.$

19. In this problem you will verify that equation (8.22) is correct. Let $f(x) = 3\sqrt[3]{x} - 3\tan^{-1}(\sqrt[3]{x})$.

   (a) Show that for $x \neq 0$,
   $$f'(x) = \frac{1}{(\sqrt[3]{x})^{2}+1}.$$

   (b) Show that
   $$f'(0) = 1 = \frac{1}{(\sqrt[3]{0})^{2}+1}.$$

   (Hint: Use the definition of derivatives.)

20. Find the volume of the solid generated when the region under the curve $y = 1/(\sqrt{x}+1), 0 \le x \le 1$ is rotated about the y-axis.

21. Find the volume of the solid generated when the region under the curve $y = 1/(\sqrt{x}+1), 0 \le x \le 1$ is rotated about the x-axis.

8.5 Trigonometric Substitutions

Sometimes, an integral that appear to have nothing to do with trigonometry can be evaluated by making a substitution that introduces a trigonometric function into the integral. For example, consider this integral:

$$\int \frac{dx}{\sqrt{x^2+1}}.$$

We are going to evaluate this integral by using the substitution $\theta = \tan^{-1} x$. This substitution may seem completely mysterious at this point: the inverse tangent function doesn't appear anywhere in the integral, so why would we introduce it into the problem by making this substitution? But as you will see, this substitution will simplify the integral.

We begin with the fact that according to the definition of the inverse tangent function, $\theta = \tan^{-1} x$ means

$$x = \tan \theta, \qquad -\frac{\pi}{2} < \theta < \frac{\pi}{2}.$$

From the equation $x = \tan \theta$ we get $dx = \sec^2 \theta \, d\theta$, and then the substitution yields

$$\int \frac{dx}{\sqrt{x^2+1}} = \int \frac{\sec^2 \theta \, d\theta}{\sqrt{\tan^2 \theta + 1}}.$$

The reason this substitution is helpful is that it allows us to simplify the square root in the denominator:

$$\sqrt{\tan^2 \theta + 1} = \sqrt{\sec^2 \theta} = |\sec \theta| = \pm \sec \theta.$$

Which answer is correct, $\sec \theta$ or $-\sec \theta$? Well, since $-\pi/2 < \theta < \pi/2$, we have $\sec \theta > 0$, and therefore

$$\sqrt{\tan^2 \theta + 1} = \sqrt{\sec^2 \theta} = \sec \theta.$$

We can now evaluate the integral:

$$\int \frac{dx}{\sqrt{x^2+1}} = \int \frac{\sec^2 \theta \, d\theta}{\sqrt{\tan^2 \theta + 1}} = \int \frac{\sec^2 \theta \, d\theta}{\sec \theta} = \int \sec \theta \, d\theta = \ln |\sec \theta + \tan \theta| + C,$$

where we have used part (b) of Example 8.3.3 in the last step. To complete the evaluation we must rewrite the answer in terms of x. We already know that $\tan \theta = x$, and when we simplified the square root we discovered that

$$\sec \theta = \sqrt{\tan^2 \theta + 1} = \sqrt{x^2 + 1}.$$

Therefore the value of the integral is

$$\int \frac{dx}{\sqrt{x^2 + 1}} = \ln |\sqrt{x^2 + 1} + x| + C.$$

It is not hard to show that the expression inside the absolute value signs is always positive, so we can actually drop the absolute value signs. Readers who have solved Exercises 29–35 in Section 7.5 may recognize that our answer is equal to $\sinh^{-1} x$.

Note that not only does the integrand in this example not involve any trigonometric functions, but the answer also does not contain a trigonometric function. And yet, somehow a detour through the land of trigonometry allowed us to evaluate the integral! The reason is that the trigonometric identity $\tan^2 \theta + 1 = \sec^2 \theta$ allowed us to simplify $\sqrt{x^2 + 1}$, once we replaced x with $\tan \theta$. Table 8.2 lists several trigonometric identities, each of which makes it possible for us to simplify a certain square root by making a trigonometric substitution.

Trigonometric Identity	To Simplify	Use Substitution
$1 - \sin^2 \theta = \cos^2 \theta$	$\sqrt{1 - x^2}$	$x = \sin \theta$
$\tan^2 \theta + 1 = \sec^2 \theta$	$\sqrt{x^2 + 1}$	$x = \tan \theta$
$\sec^2 \theta - 1 = \tan^2 \theta$	$\sqrt{x^2 - 1}$	$x = \sec \theta$

Table 8.2: Trigonometric substitutions.

Example 8.5.1. Evaluate the following integrals:

$$\text{(a)} \int \frac{\sqrt{1 - x^2}}{x^2} \, dx, \quad \text{(b)} \int \frac{dx}{x\sqrt{4x^2 + 1}}, \quad \text{(c)} \int \frac{dx}{x^2\sqrt{x^2 - 4}}.$$

Solution. Integral (a) contains $\sqrt{1 - x^2}$, and Table 8.2 suggests that we can simplify this if we replace x with $\sin \theta$. We therefore use the substitution $\theta = \sin^{-1} x$, which means

$$x = \sin \theta, \quad -\frac{\pi}{2} \le \theta \le \frac{\pi}{2}.$$

Notice that the domain of the integrand in integral (a) is $[-1, 0) \cup (0, 1]$, and for values of x in this set, $\sin^{-1} x$ is defined, so this substitution makes sense.

Since $x = \sin\theta$, we have $dx = \cos\theta\,d\theta$. The purpose of the substitution is to simplify the square root, and it is often helpful to do this simplification first:

$$\sqrt{1 - x^2} = \sqrt{1 - \sin^2\theta} = \sqrt{\cos^2\theta} = \pm\cos\theta.$$

But $-\pi/2 \le \theta \le \pi/2$, so $\cos\theta \ge 0$, and therefore

$$\sqrt{1 - x^2} = \cos\theta. \tag{8.24}$$

We are now ready to evaluate the integral:

$$\int \frac{\sqrt{1 - x^2}}{x^2}\,dx = \int \frac{\cos\theta}{\sin^2\theta}\cos\theta\,d\theta = \int \frac{\cos^2\theta}{\sin^2\theta}\,d\theta = \int \cot^2\theta\,d\theta = \int \csc^2\theta - 1\,d\theta$$

$$= -\cot\theta - \theta + C = -\frac{\cos\theta}{\sin\theta} - \theta + C = -\frac{\sqrt{1 - x^2}}{x} - \sin^{-1}x + C.$$

Note that we used equation (8.24) not only in the first step, but also in the last step, when we reexpressed the answer in terms of x. This is one reason why it is usually helpful, when doing a trigonometric substitution in an integral, to write out the simplification of the square root before evaluating the integral.

The square root in integral (b) doesn't exactly fit the pattern of any of the entries in Table 8.2, but it is close enough that we can adjust for the differences. We will be able to simplify the square root if $4x^2$ gets replaced by $\tan^2\theta$, or in other words if $2x = \tan\theta$, so we use the substitution $\theta = \tan^{-1}(2x)$. According to the definition of the inverse tangent function, this means

$$2x = \tan\theta, \qquad -\frac{\pi}{2} < \theta < \frac{\pi}{2}.$$

Once again, before evaluating the integral we work out the square root:

$$\sqrt{4x^2 + 1} = \sqrt{\tan^2\theta + 1} = \sqrt{\sec^2\theta} = \sec\theta, \tag{8.25}$$

where we have used the fact that $-\pi/2 < \theta < \pi/2$ to determine that $\sec\theta$ is positive. We also need to compute

$$x = \frac{1}{2}\tan\theta, \qquad dx = \frac{1}{2}\sec^2\theta\,d\theta.$$

Now we are ready to perform the substitution in the integral:

$$\int \frac{dx}{x\sqrt{4x^2 + 1}} = \int \frac{(1/2)\sec^2\theta\,d\theta}{(1/2)\tan\theta\sec\theta} = \int \frac{\sec\theta\,d\theta}{\tan\theta}.$$

We have eliminated the square root, but now we have a slightly tricky trigonometric integral. There is more than one way to evaluate this integral. Motivated by the fact that the power of $\tan\theta$ is odd, we will use the substitution $u = \sec\theta$, $du = \sec\theta\tan\theta\,d\theta$. For a different approach see Exercise 19.

$$\int \frac{\sec\theta\,d\theta}{\tan\theta} = \int \frac{\sec\theta\tan\theta\,d\theta}{\tan^2\theta} = \int \frac{\sec\theta\tan\theta\,d\theta}{\sec^2\theta - 1} = \int \frac{du}{u^2 - 1}.$$

The denominator in the last integral factors, so we use partial fractions:

$$\int \frac{\sec\theta\,d\theta}{\tan\theta} = \int \frac{du}{u^2 - 1} = \int \frac{1/2}{u - 1} - \frac{1/2}{u + 1}\,du$$

$$= \frac{1}{2}\ln|u - 1| - \frac{1}{2}\ln|u + 1| + C = \frac{1}{2}\ln(\sec\theta - 1) - \frac{1}{2}\ln(\sec\theta + 1) + C.$$

We have eliminated the absolute value signs because for $-\pi/2 < \theta < \pi/2$, $\sec\theta \geq 1$. Finally, we get the answer to the original integral by applying equation (8.25):

$$\int \frac{dx}{x\sqrt{4x^2 + 1}} = \frac{1}{2}\ln(\sqrt{4x^2 + 1} - 1) - \frac{1}{2}\ln(\sqrt{4x^2 + 1} + 1) + C.$$

For integral (c), we want to simplify $\sqrt{x^2 - 4}$. If we multiply the identity $\sec^2\theta - 1 = \tan^2\theta$ by 4 we get $4\sec^2\theta - 4 = 4\tan^2\theta$, which suggests that we will be able to simplify the square root if we replace x with $2\sec\theta$. We therefore use the substitution $\theta = \sec^{-1}(x/2)$, which means

$$\frac{x}{2} = \sec\theta, \qquad 0 \leq \theta < \frac{\pi}{2} \text{ or } \frac{\pi}{2} < \theta \leq \pi.$$

Notice that the domain of the integrand is $(-\infty, -2) \cup (2, \infty)$, and $\sec^{-1}(x/2)$ is defined for all x in this domain.

From $x = 2\sec\theta$ we get $dx = 2\sec\theta\tan\theta\,d\theta$, and

$$\sqrt{x^2 - 4} = \sqrt{4\sec^2\theta - 4} = \sqrt{4\tan^2\theta} = |2\tan\theta|.$$

Unfortunately, if $0 \leq \theta < \pi/2$ then $\tan\theta \geq 0$, and if $\pi/2 < \theta \leq \pi$ then $\tan\theta \leq 0$. Thus, as with so many problems involving the inverse secant function, we need to consider the two parts of the domain separately.

For $x > 2$ we have $0 < \theta < \pi/2$, so $\tan\theta > 0$ and therefore

$$\sqrt{x^2 - 4} = 2\tan\theta.$$

We can now perform the substitution and evaluate the integral:

$$\int \frac{dx}{x^2\sqrt{x^2 - 4}} = \int \frac{2\sec\theta\tan\theta\,d\theta}{4\sec^2\theta \cdot 2\tan\theta} = \int \frac{d\theta}{4\sec\theta} = \int \frac{\cos\theta}{4}\,d\theta = \frac{\sin\theta}{4} + C.$$

To finish the problem, we must express $\sin\theta$ in terms of x. We know $\sec\theta = x/2$ and $\tan\theta = \sqrt{x^2 - 4}/2$, and often once you know the values of two trigonometric functions it is easy to find the rest of them. In this case, from $\sec\theta = x/2$ we get $\cos\theta = 2/x$, and therefore

$$\sin\theta = \frac{\sin\theta}{\cos\theta}\cdot\cos\theta = \tan\theta\cdot\cos\theta = \frac{\sqrt{x^2-4}}{2}\cdot\frac{2}{x} = \frac{\sqrt{x^2-4}}{x}.$$

Another way to determine the formula for $\sin\theta$ is to draw the right triangle shown in Figure 8.3. Thus, we have

$$\int \frac{dx}{x^2\sqrt{x^2-4}} = \frac{\sin\theta}{4} + C = \frac{\sqrt{x^2-4}}{4x} + C \text{ on } (2,\infty).$$

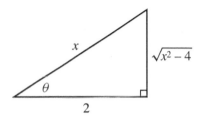

Figure 8.3: Since $0 < \theta < \pi/2$ and $\sec\theta = x/2$, θ is one of the angles of a right triangle in which the hypotenuse has length x and one leg has length 2. The Pythagorean theorem determines the length of the other leg, and all trigonometric functions can then be determined. In particular, we can see that $\sin\theta = \sqrt{x^2-4}/x$.

We still have to consider $x < -2$. For these values of x we have $\pi/2 < \theta < \pi$ and $\tan\theta < 0$, and therefore

$$\sqrt{x^2-4} = -2\tan\theta.$$

We leave it to you to carry out the substitution in this case and verify that the answer is exactly the same. Alternatively, you can differentiate the answer we computed in the case $x > 2$ and verify that it is an antiderivative of the integrand not only on $(2,\infty)$, but on $(-\infty,-2)$ as well. □

We can extend the method of trigonometric substitution to a wider range of integrals by combining it with algebraic reexpressions such as completing the square. We illustrate this in our next example.

Example 8.5.2. Evaluate the following integrals:

(a) $\displaystyle\int \frac{x\,dx}{\sqrt{4x-x^2}}$, (b) $\displaystyle\int (t^2-2t+4)^{3/2}\,dt$, (c) $\displaystyle\int \frac{dt}{(t^2-2t+4)^2}$.

Solution. The square root in integral (a) contains a quadratic expression, but it doesn't fit any of the patterns in Table 8.2. However, we can match it with one of those patterns if we first complete the square. To do this, it is usually easiest to begin by factoring out the coefficient of x^2, which in this case is -1:

$$4x - x^2 = -(x^2 - 4x) = -(x^2 - 4x + 4 - 4) = -((x-2)^2 - 4) = 4 - (x-2)^2.$$

Thus, the square root inside the integral can be rewritten in the form

$$\sqrt{4x - x^2} = \sqrt{4 - (x-2)^2}.$$

Now the trigonometric identity $4 - 4\sin^2\theta = 4\cos^2\theta$ suggests that we will be able to simplify the square root if we replace $x - 2$ with $2\sin\theta$. We therefore use the substitution $\theta = \sin^{-1}((x-2)/2)$, which means

$$\frac{x-2}{2} = \sin\theta, \qquad -\frac{\pi}{2} \le \theta \le \frac{\pi}{2}.$$

We have $x = 2\sin\theta + 2$ and $dx = 2\cos\theta\,d\theta$, and

$$\sqrt{4x - x^2} = \sqrt{4 - (x-2)^2} = \sqrt{4 - 4\sin^2\theta} = \sqrt{4\cos^2\theta} = 2\cos\theta,$$

where we have used the fact that $-\pi/2 \le \theta \le \pi/2$ to conclude that $\cos\theta \ge 0$.

Substitution now allows us to evaluate the integral:

$$\int \frac{x\,dx}{\sqrt{4x - x^2}} = \int \frac{(2\sin\theta + 2)\cdot 2\cos\theta\,d\theta}{2\cos\theta} = \int 2\sin\theta + 2\,d\theta$$

$$= -2\cos\theta + 2\theta + C = -\sqrt{4x - x^2} + 2\sin^{-1}\left(\frac{x-2}{2}\right) + C.$$

Integral (b) also contains a square root, since $(t^2 - 2t + 4)^{3/2} = (\sqrt{t^2 - 2t + 4})^3$. Again, we begin by completing the square:

$$t^2 - 2t + 4 = t^2 - 2t + 1 + 3 = (t-1)^2 + 3.$$

We will be able to use the trigonometric identity $3\tan^2\theta + 3 = 3\sec^2\theta$ if we replace $t - 1$ with $\sqrt{3}\tan\theta$, so we make the substitution

$$\theta = \tan^{-1}\left(\frac{t-1}{\sqrt{3}}\right), \qquad t = \sqrt{3}\tan\theta + 1, \qquad -\frac{\pi}{2} < \theta < \frac{\pi}{2}.$$

With this substitution, we will have $dt = \sqrt{3}\sec^2\theta\,d\theta$ and

$$\sqrt{t^2 - 2t + 4} = \sqrt{(t-1)^2 + 3} = \sqrt{3\tan^2\theta + 3} = \sqrt{3\sec^2\theta} = \sqrt{3}\sec\theta.$$

Thus,

$$\int (t^2 - 2t + 4)^{3/2}\,dt = \int (\sqrt{3}\sec\theta)^3 \cdot \sqrt{3}\sec^2\theta\,d\theta = \int 9\sec^5\theta\,d\theta.$$

We can now use the reduction formula (8.18) to complete the evaluation:

$$\int (t^2 - 2t + 4)^{3/2}\, dt = 9 \int \sec^5 \theta\, d\theta = 9 \left(\frac{\tan \theta \sec^3 \theta}{4} + \frac{3}{4} \int \sec^3 \theta\, d\theta \right)$$

$$= \frac{9 \tan \theta \sec^3 \theta}{4} + \frac{27}{4} \left(\frac{\tan \theta \sec \theta}{2} + \frac{1}{2} \int \sec \theta\, d\theta \right)$$

$$= \frac{9 \tan \theta \sec^3 \theta}{4} + \frac{27 \tan \theta \sec \theta}{8} + \frac{27}{8} \ln |\sec \theta + \tan \theta| + C$$

$$= \frac{(t - 1)(t^2 - 2t + 4)^{3/2}}{4} + \frac{9(t - 1)\sqrt{t^2 - 2t + 4}}{8}$$

$$+ \frac{27}{8} \ln \left| \frac{\sqrt{t^2 - 2t + 4} + t - 1}{\sqrt{3}} \right| + C.$$

Substitutions involving the inverse tangent function are sometimes useful even in integrals that don't contain square roots. In integral (c) we'll use the same substitution we used in integral (b):

$$\int \frac{dt}{(t^2 - 2t + 4)^2} = \int \frac{\sqrt{3} \sec^2 \theta\, d\theta}{(3 \sec^2 \theta)^2} = \frac{\sqrt{3}}{9} \int \cos^2 \theta\, d\theta$$

$$= \frac{\sqrt{3}}{9} \int \frac{1 + \cos(2\theta)}{2}\, d\theta = \frac{\sqrt{3}}{18} \left(\theta + \frac{\sin(2\theta)}{2} \right) + C$$

$$= \frac{\sqrt{3}}{18} (\theta + \sin \theta \cos \theta) + C.$$

To reexpress this answer in terms of t, we compute

$$\cos \theta = \frac{1}{\sec \theta} = \frac{\sqrt{3}}{\sqrt{t^2 - 2t + 4}},$$

$$\sin \theta = \tan \theta \cos \theta = \frac{t - 1}{\sqrt{3}} \cdot \frac{\sqrt{3}}{\sqrt{t^2 - 2t + 4}} = \frac{t - 1}{\sqrt{t^2 - 2t + 4}}.$$

Thus,

$$\int \frac{dt}{(t^2 - 2t + 4)^2} = \frac{\sqrt{3}}{18} (\theta + \sin \theta \cos \theta) + C$$

$$= \frac{\sqrt{3}}{18} \left(\tan^{-1} \left(\frac{t - 1}{\sqrt{3}} \right) + \frac{t - 1}{\sqrt{t^2 - 2t + 4}} \cdot \frac{\sqrt{3}}{\sqrt{t^2 - 2t + 4}} \right) + C$$

$$= \frac{\sqrt{3}}{18} \tan^{-1} \left(\frac{t - 1}{\sqrt{3}} \right) + \frac{t - 1}{6(t^2 - 2t + 4)} + C. \qquad \square$$

Exercises 8.5

1–18: Evaluate the integral.

1. $\displaystyle\int \frac{dx}{x^2\sqrt{1-x^2}}.$

2. $\displaystyle\int \frac{dx}{x^2\sqrt{1+x^2}}.$

3. $\displaystyle\int \frac{\sqrt{x^2+1}}{x}\,dx.$

4. $\displaystyle\int \sqrt{9-x^2}\,dx.$

5. $\displaystyle\int \sqrt{x^2-9}\,dx.$

6. $\displaystyle\int \sqrt{x^2+9}\,dx.$

7. $\displaystyle\int_0^1 \frac{dx}{(4-x^2)^{3/2}}.$

8. $\displaystyle\int_0^1 \frac{dx}{(4+x^2)^{3/2}}.$

9. $\displaystyle\int_{-2}^{-1} \frac{\sqrt{x^2-1}}{x^2}\,dx.$

10. $\displaystyle\int \frac{x^2}{\sqrt{4x-x^2}}\,dx.$

11. $\displaystyle\int \frac{x}{\sqrt{x^2-4x}}\,dx.$

12. $\displaystyle\int \frac{x}{\sqrt{x^2-10x+29}}\,dx.$

13. $\displaystyle\int \frac{x^2}{\sqrt{3+2x-x^2}}\,dx.$

14. $\displaystyle\int \frac{dx}{(x^2+1)^3}.$

15. $\displaystyle\int x\sin^{-1}x\,dx.$

16. $\displaystyle\int_1^2 \sec^{-1}x\,dx.$

17. $\int \dfrac{\sqrt{1-x}}{\sqrt{x}}\, dx.$

18. $\int \dfrac{x^2\, dx}{(x^2 + 2x + 2)^{3/2}}.$

19. In this exercise, we consider an alternative approach to Example 8.5.1(b).

   (a) As in the text, we use the substitution $\theta = \tan^{-1}(2x)$ to get

$$\int \frac{dx}{x\sqrt{4x^2 + 1}} = \int \frac{\sec\theta\, d\theta}{\tan\theta} = \int \frac{d\theta}{\sin\theta} = \int \csc\theta\, d\theta.$$

   We know (see Example 8.3.3(b)) that $\int \sec\theta\, d\theta = \ln|\sec\theta + \tan\theta| + C$. Based on this, see if you can guess a formula for $\int \csc\theta\, d\theta$. Check your guess by differentiating, and adjust your guess if necessary. Once you have an answer for $\int \csc\theta\, d\theta$, complete the evaluation of the integral in Example 8.5.1(b).

   (b) Show that your answer is equivalent to the answer for this integral given in the text.

20. Evaluate the integral

$$\int \frac{\sqrt{x^2 - 1}}{x}\, dx$$

   by using the substitution $\theta = \sec^{-1}|x|$. Notice that the domain of the integrand is $(-\infty, -1] \cup [1, \infty)$, and for all values of x in this domain, $|x| \geq 1$. This means that you only need to apply the inverse secant function to numbers in the interval $[1, \infty)$, and you will have $0 \leq \theta < \pi/2$. However, you will still need to consider the cases $x \leq -1$ and $x \geq 1$ separately.

21. Find the volume of the solid generated when the region under the curve $y = 1/(x^2 + 1), 0 \leq x \leq 1$ is rotated about the x-axis.

22. Find the centroid of the region under the curve $y = \sin x, 0 \leq x \leq \pi/2$.

23. This problem came up on the National Public Radio program Car Talk on Nov. 6, 2010. A truck has a gas tank in the shape of a cylinder whose axis is horizontal. The diameter of the cylinder is 20 inches. The gas gauge is broken, so the trucker checks his gas by inserting a dowel into the tank vertically and checking how deep the gas is. Of course, when the tank is full, the depth is 20 inches, and if the depth is 10 inches, then the tank is half full.

   (a) How deep will the gas be when the tank is one quarter full? (Note: You will only be able to find an approximate answer.)

   (b) A few months later, the question in part (a) was used as a puzzler on Car Talk. The hosts of the show, Tom and Ray Magliozzi, suggested the following solution: Cut out a piece of cardboard in the shape of a semicircle with radius 10 inches. Then find the centroid of this half disk by finding the point where the piece of cardboard will balance on the tip of a pencil. Draw a line from the center of the disk through the centroid to the circumference, and measure the distance along this line from the centroid to the circumference. This distance is the depth of the gas when the tank is one quarter full. Is this solution correct?

8.6   Numerical Integration

In the last few sections we have studied integration techniques that allow us to evaluate a variety of integrals. However, there are many integrals that cannot be evaluated by any of these techniques. For example, integrals of the form

$$\int_a^b e^{-x^2}\, dx$$

play an important role in probability theory, but they cannot be evaluated by the techniques we have studied.[1] What can we do if we need to know the value of such an integral?

Of course, a definite integral is, by definition, a limit of Riemann sums, so we can get an approximate answer for any definite integral by computing a Riemann sum. Recall that to compute a Riemann sum for $\int_a^b f(x)\, dx$, where $a < b$ and f is continuous on $[a, b]$, we first pick a positive integer n and divide the interval $[a, b]$ into n smaller intervals, each of width $\Delta x = (b - a)/n$. The endpoints of these smaller intervals are $x_i = a + i\Delta x$ for $i = 0, 1, 2, \ldots, n$. Next, for $1 \le i \le n$ we choose a sample point x_i^* between x_{i-1} and x_i. Our Riemann sum is then $\sum_{i=1}^n f(x_i^*)\Delta x$.

A natural way to choose the sample points is to let x_i^* be the midpoint of the interval $[x_{i-1}, x_i]$. With this choice, the Riemann sum is often called the *midpoint rule* approximation for the definite integral, and it is denoted M_n. In other words, if we let $x_i^* = (x_{i-1} + x_i)/2$, then

$$M_n = \sum_{i=1}^n f(x_i^*)\Delta x = \Delta x(f(x_1^*) + f(x_2^*) + \cdots + f(x_n^*)). \tag{8.26}$$

If $f(x)$ is positive for $a \le x \le b$, then the midpoint rule can be thought of as giving the total area of a collection of n rectangles, as illustrated in Figure 8.4. Recall that if $f(x)$ is negative for some values of x, then some rectangles will extend below the x-axis, and the areas of these rectangles will be subtracted rather than added.

For example, let's try computing the midpoint rule, with $n = 6$, for the integral

$$\int_{-1}^2 e^{-x^2}\, dx.$$

We have $f(x) = e^{-x^2}$ and $\Delta x = (2 - (-1))/6 = 1/2$, and this leads to the values shown in Table 8.3. (The numbers in the last column of the table are approximate values,

---

[1] In fact, although the function e^{-x^2} has an antiderivative, it can be proven that this antiderivative cannot be written using any combination of addition, subtraction, multiplication, division, roots, logarithms, exponential functions, trigonometric functions, and inverse trigonometric functions.

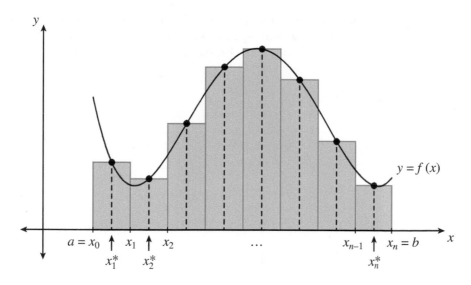

Figure 8.4: The midpoint rule approximation for $\int_a^b f(x)\,dx$ is the total area of the blue rectangles.

i	x_i	x_i^*	$f(x_i^*) = e^{-(x_i^*)^2}$
0	-1		
1	-0.5	-0.75	0.5698
2	0	-0.25	0.9394
3	0.5	0.25	0.9394
4	1	0.75	0.5698
5	1.5	1.25	0.2096
6	2	1.75	0.0468

Table 8.3: Computing the midpoint rule for $\int_{-1}^{2} e^{-x^2}\,dx$.

accurate to four decimal places.) Plugging these values into the formula in equation (8.26) gives us

$$M_6 \approx \frac{1}{2}(0.5698 + 0.9394 + 0.9394 + 0.5698 + 0.2096 + 0.0468) = 1.6374. \quad (8.27)$$

The region inside the rectangles in Figure 8.4 doesn't match the shape of the region under the graph of f very closely. This suggests that we might be able to get a better approximation if we compute the area of a region that matches the region under the graph more closely. In Figure 8.5 we have replaced the rectangles from Figure 8.4 with

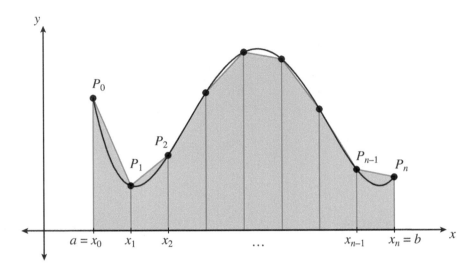

Figure 8.5: The trapezoid rule approximation for $\int_a^b f(x)\,dx$ is the total area of the blue trapezoids.

trapezoids whose upper vertices lie on the curve $y = f(x)$. For each i, let P_i be the point $(x_i, f(x_i))$. Then the top of the first trapezoid is the line segment from P_0 to P_1, the top of the next is the segment from P_1 to P_2, and so on. The sum of the areas of these trapezoids is called the *trapezoid rule* approximation for the definite integral, and it is denoted T_n.

To find a formula for the trapezoid rule, note that the left and right sides of the first trapezoid have height $f(x_0)$ and $f(x_1)$, and therefore its area is $\Delta x(f(x_0) + f(x_1))/2$. Computing the areas of the other trapezoids similarly and adding, we get the formula

$$
\begin{aligned}
T_n &= \Delta x\,\frac{f(x_0) + f(x_1)}{2} + \Delta x\,\frac{f(x_1) + f(x_2)}{2} + \cdots + \Delta x\,\frac{f(x_{n-1}) + f(x_n)}{2} \\
&= \frac{\Delta x}{2}\left(f(x_0) + 2f(x_1) + 2f(x_2) + \cdots + 2f(x_{n-1}) + f(x_n)\right).
\end{aligned}
\tag{8.28}
$$

Notice the pattern of the coefficients of $f(x_i)$ in this formula: 1, 2, 2, ..., 2, 1. As before, if $f(x)$ is negative for some values of x then this formula will result in some areas being subtracted rather than added.

Returning to the example $\int_{-1}^{2} e^{-x^2}\,dx$ with $n = 6$, we compute the values in Table 8.4. Filling in these values in formula (8.28), we get the approximation

$$
\begin{aligned}
T_6 &\approx \frac{1/2}{2}(0.3679 + 2\cdot 0.7788 + 2\cdot 1.0000 + 2\cdot 0.7788 + 2\cdot 0.3679 \\
&\quad + 2\cdot 0.1054 + 0.0183) \\
&= 1.6120.
\end{aligned}
\tag{8.29}
$$

i	x_i	$f(x_i) = e^{-x_i^2}$
0	-1	0.3679
1	-0.5	0.7788
2	0	1.0000
3	0.5	0.7788
4	1	0.3679
5	1.5	0.1054
6	2	0.0183

Table 8.4: Computing the trapezoid rule approximation for $\int_{-1}^{2} e^{-x^2}\, dx$.

Is the trapezoid rule more accurate than the midpoint rule? The difference $\int_a^b f(x)\, dx - M_n$ tells us how much the true value of the integral differs from the midpoint rule approximation; we call it the *error* in the midpoint rule. Similarly, $\int_a^b f(x)\, dx - T_n$ is the error in the trapezoid rule. The following theorem gives us some information about the errors in the midpoint and trapezoid rules. (For a proof, see Exercises 13 and 14.)

Theorem 8.6.1. *Suppose f is continuous on $[a, b]$ and $f''(x)$ is defined for all x in (a, b). Suppose also that K is a number such that for all x in (a, b), $|f''(x)| \le K$. Then for any positive integer n:*

1. $$\left| \int_a^b f(x)\, dx - M_n \right| \le \frac{K(b-a)^3}{24n^2}.$$

2. $$\left| \int_a^b f(x)\, dx - T_n \right| \le \frac{K(b-a)^3}{12n^2}.$$

Theorem 8.6.1 does not tell us the exact value of the error in either the midpoint rule or the trapezoid rule. Rather, for each rule it gives an *upper bound* on the absolute value of the error. This upper bound tells us how much confidence we can have in the accuracy of the rules. Notice that the upper bound for the error in the trapezoid rule is twice as large as the upper bound for the midpoint rule! This doesn't mean that the error in the trapezoid rule is always twice as large as the error in the midpoint rule, but in practice the trapezoid rule is often less accurate than the midpoint rule.

In order to apply Theorem 8.6.1, we have to find a number that can play the role of K in the theorem. For example, let's try applying the theorem to our estimates of $\int_{-1}^{2} e^{-x^2}\, dx$. We have $f(x) = e^{-x^2}$, so $f'(x) = -2xe^{-x^2}$ and $f''(x) = (4x^2 - 2)e^{-x^2}$. To determine a suitable value for K we will use methods from Chapter 4 find the minimum and maximum values of f'' on the interval $[-1, 2]$. Since f'' is continuous on $[-1, 2]$, it must have minimum and maximum values, and each of these values must be attained at either an endpoint of the interval or at a critical number of f''. To find the critical numbers, we compute the derivative of f''—that is, f''':

$$f'''(x) = (-8x^3 + 12x)e^{-x^2} = -4x(2x^2 - 3)e^{-x^2}.$$

Setting $f'''(x)$ equal to 0, we see that the critical numbers are 0 and $\pm\sqrt{3/2} \approx \pm 1.22$. Thus, minimum and maximum values of $f''(x)$ on $[-1, 2]$ can occur only at $x = -1, 0, \sqrt{3/2}$, or 2. Next we compute $f''(x)$ for these values of x:

$$f''(-1) = 2e^{-1} \approx 0.74, \qquad\qquad f''(0) = -2,$$
$$f''(\sqrt{3/2}) = 4e^{-3/2} \approx 0.89, \qquad\qquad f''(2) = 14e^{-4} \approx 0.26.$$

We conclude that the minimum value of $f''(x)$ for $-1 \leq x \leq 2$ is -2, and the maximum value is $4e^{-3/2} \approx 0.89$.

We now know that for all x in $(-1, 2)$, $-2 \leq f''(x) \leq 4e^{-3/2} \approx 0.89$, and therefore $|f''(x)| \leq 2$. Thus, we can use $K = 2$ in Theorem 8.6.1. Filling in $a = -1, b = 2$, and $n = 6$, we conclude that

$$\left| \int_{-1}^{2} e^{-x^2}\, dx - M_6 \right| \leq \frac{2(2 - (-1))^3}{24(6^2)} = \frac{1}{16} = 0.0625,$$

$$\left| \int_{-1}^{2} e^{-x^2}\, dx - T_6 \right| \leq \frac{2(2 - (-1))^3}{12(6^2)} = \frac{1}{8} = 0.125.$$

In other words, the midpoint rule approximation $M_6 \approx 1.6374$ differs from the true value of $\int_{-1}^{2} e^{-x^2}\, dx$ by at most 0.0625, and the trapezoid rule approximation $T_6 \approx 1.6120$ differs from the true value by at most 0.125.

We can improve the accuracy of these approximations by increasing n. For example, for $n = 500$ we have

$$\left| \int_{-1}^{2} e^{-x^2}\, dx - M_{500} \right| \leq \frac{2(2 - (-1))^3}{24(500^2)} = 9 \times 10^{-6} = 0.000009.$$

Using a computer, we find that $M_{500} \approx 1.6289$. With this very accurate estimate of the value of the integral, we can now see how large the errors really are when $n = 6$:

$$\int_{-1}^{2} e^{-x^2}\, dx - M_6 \approx 1.6289 - 1.6374 = -0.0085,$$

$$\int_{-1}^{2} e^{-x^2}\, dx - T_6 \approx 1.6289 - 1.6120 = 0.0169.$$

The error in the midpoint rule is negative in this case, indicating that the midpoint rule gives a value slightly larger than the true value of the integral. The absolute values of both errors are smaller than the bounds we computed from Theorem 8.6.1, as expected, and the midpoint rule is about twice as accurate as the trapezoid rule.

This example illustrates our earlier observation that, despite the appearance of Figures 8.4 and 8.5, the trapezoid rule is often less accurate than the midpoint rule. One weakness of the trapezoid rule is that it ignores the concavity of f. This is evident in Figure 8.5, where the tops of the trapezoids are straight line segments, even though the graph of f is curved. But this weakness is also an opportunity: by replacing the straight tops of the trapezoids in Figure 8.5 with curves that match the shape of the graph of f more closely, we will be able to get a more accurate approximation.

Here's another way of describing the trapezoid rule. Let g_1 be the linear function whose graph is the straight line through the points P_0 and P_1 in Figure 8.5. Then the area of the first trapezoid in that figure is $\int_{x_0}^{x_1} g_1(x)\,dx$, and in the trapezoid rule we use this as an approximation for $\int_{x_0}^{x_1} f(x)\,dx$. Similarly, if g_2 is the function whose graph is the line through P_1 and P_2, then the area of the second trapezoid is $\int_{x_1}^{x_2} g_2(x)\,dx$, and we use this as our approximation for $\int_{x_1}^{x_2} f(x)\,dx$. Adding up similar approximations for the integral of f on each interval $[x_{i-1}, x_i]$ leads to the trapezoid rule.

To improve on the trapezoid rule, we will replace g_1 and g_2 with a single function g whose graph passes through all three of the points P_0, P_1, and P_2. We will use a function g of the form

$$g(x) = A(x - x_1)^2 + B(x - x_1) + C,$$

whose graph is a parabola, as shown in Figure 8.6. We leave it to you to verify (see Exercise 15) that the graph of g passes through the points P_0, P_1, and P_2 if and only if the coefficients are given by the formulas

$$A = \frac{f(x_0) - 2f(x_1) + f(x_2)}{2\Delta x^2}, \qquad B = \frac{f(x_2) - f(x_0)}{2\Delta x}, \qquad C = f(x_1). \qquad (8.30)$$

Our hope is that with this choice of coefficients, not only will the graph of g pass through P_0, P_1, and P_2, but it will also curve in the same direction as f between these points, thus staying closer to the graph of f than g_1 and g_2 do. We can then use $\int_{x_0}^{x_2} g(x)\,dx$ as an approximation for $\int_{x_0}^{x_2} f(x)\,dx$.

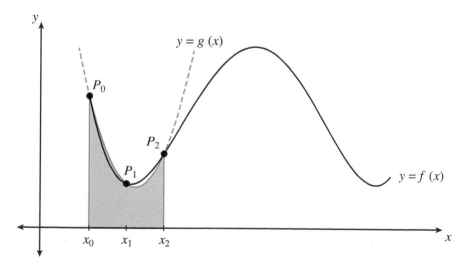

Figure 8.6: The graph of g is a parabola passing through the points P_0, P_1, and P_2. The area of the blue region is $\int_{x_0}^{x_2} g(x)\,dx$, which is approximately equal to $\int_{x_0}^{x_2} f(x)\,dx$.

To carry out this plan, we evaluate $\int_{x_0}^{x_2} g(x)\,dx$, using the substitution $u = x - x_1$, and then we fill in the formulas from (8.30):

$$\int_{x_0}^{x_2} g(x)\,dx = \int_{x_0}^{x_2} A(x - x_1)^2 + B(x - x_1) + C\,dx = \int_{-\Delta x}^{\Delta x} Au^2 + Bu + C\,du$$

$$= \left[A\frac{u^3}{3} + B\frac{u^2}{2} + Cu \right]_{-\Delta x}^{\Delta x} = 2A\frac{\Delta x^3}{3} + 2C\Delta x$$

$$= \frac{f(x_0) - 2f(x_1) + f(x_2)}{3}\Delta x + 2f(x_1)\Delta x$$

$$= \frac{\Delta x}{3}(f(x_0) + 4f(x_1) + f(x_2)).$$

Thus, it appears that

$$\int_{x_0}^{x_2} f(x)\,dx \approx \frac{\Delta x}{3}(f(x_0) + 4f(x_1) + f(x_2)). \tag{8.31}$$

Similar reasoning, using a parabola passing through the points P_2, P_3, and P_4, leads to the approximation

$$\int_{x_2}^{x_4} f(x)\,dx \approx \frac{\Delta x}{3}(f(x_2) + 4f(x_3) + f(x_4)). \tag{8.32}$$

And we can continue in this way to estimate the integral of f on the intervals $[x_4, x_6]$, $[x_6, x_8]$, and so on. Notice that since each approximation is on an interval of width $2\Delta x$, we must require n to be even so that these intervals will fill up the interval $[a, b]$ exactly.

Adding these approximations leads to the *Simpson's rule* approximation for $\int_a^b f(x)\,dx$, which is denoted S_n. It is named after Thomas Simpson (1710–1761), who published it in 1743, although it was known already a century before that. For any even positive integer n, using (8.31), (8.32), and similar formulas for the other intervals, we get the formula

$$S_n = \frac{\Delta x}{3}(f(x_0) + 4f(x_1) + f(x_2)) + \frac{\Delta x}{3}(f(x_2) + 4f(x_3) + f(x_4))$$

$$+ \cdots + \frac{\Delta x}{3}(f(x_{n-2}) + 4f(x_{n-1}) + f(x_n))$$

$$= \frac{\Delta x}{3}(f(x_0) + 4f(x_1) + 2f(x_2) + 4f(x_3) + \cdots + 2f(x_{n-2}) + 4f(x_{n-1}) + f(x_n)). \tag{8.33}$$

As with the trapezoid rule, it is important to notice the pattern of the coefficients of $f(x_i)$ in this formula: 1, 4, 2, 4, 2, ..., 4, 2, 4, 1. The Simpson's rule approximation is equal to the area of the blue shaded region in Figure 8.7.

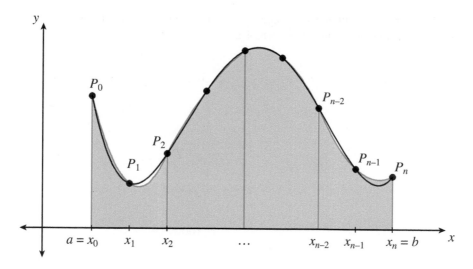

Figure 8.7: The Simpson's rule approximation for $\int_a^b f(x)\,dx$ is equal to the area of the blue region.

To try out Simpson's rule on the integral $\int_{-1}^2 e^{-x^2}\,dx$ with $n = 6$, we can use the values from Table 8.4. Plugging into formula (8.33), we get

$$
\begin{aligned}
S_6 \approx {}& \frac{1/2}{3}(0.3679 + 4 \cdot 0.7788 + 2 \cdot 1.0000 + 4 \cdot 0.7788 + 2 \cdot 0.3679 \\
& + 4 \cdot 0.1054 + 0.0183) \\
={}& 1.6290.
\end{aligned}
$$

This is remarkably close to the value $\int_{-1}^2 e^{-x^2}\,dx \approx M_{500} \approx 1.6289$ that we found earlier.

As with the midpoint and trapezoid rules, there is a theorem that gives a bound on the size of the error in Simpson's rule:

Theorem 8.6.2. *Suppose f is continuous on $[a, b]$ and $f^{(4)}(x)$ is defined for all x in (a, b), where $f^{(4)}$ is the fourth derivative of f. Suppose also that K is a number such that for all x in (a, b), $|f^{(4)}(x)| \le K$. Then for any even positive integer n,*

$$
\left| \int_a^b f(x)\,dx - S_n \right| \le \frac{K(b-a)^5}{180n^4}.
$$

Example 8.6.3. Find the decimal value of $\ln 3$ with error at most 0.001.

Solution. According to the definition of the natural logarithm function,

$$
\ln 3 = \int_1^3 \frac{dx}{x}.
$$

We now estimate the value of this integral by using Simpson's rule. We begin by applying Theorem 8.6.2 to see how to achieve the required accuracy. We have $f(x) = 1/x = x^{-1}$, and therefore $f^{(4)}(x) = 24x^{-5} = 24/x^5$. For $1 < x < 3$ we have

$$|f^{(4)}(x)| = \frac{24}{x^5} < \frac{24}{1^5} = 24,$$

so we can use $K = 24$ in the theorem. Thus,

$$\left| \int_1^3 \frac{dx}{x} - S_n \right| \le \frac{24(3-1)^5}{180n^4} = \frac{64}{15n^4}.$$

We can therefore achieve the desired accuracy by ensuring that

$$\frac{64}{15n^4} \le 0.001 = \frac{1}{1000},$$

or in other words $n \ge \sqrt[4]{64000/15} \approx 8.08$. Since n must be an even integer, we will use $n = 10$.

Since $a = 1$ and $b = 3$, we have $\Delta x = 1/5$, which gives us the values in Table 8.5. Therefore

$$\ln 3 = \int_1^3 \frac{dx}{x} \approx S_{10} = \frac{1/5}{3}\left(1 + 4 \cdot \frac{5}{6} + 2 \cdot \frac{5}{7} + 4 \cdot \frac{5}{8} + \cdots + 2 \cdot \frac{5}{13} + 4 \cdot \frac{5}{14} + \frac{1}{3}\right)$$

$$\approx 1.09866. \qquad \qquad \square$$

i	x_i	$f(x_i)$	i	x_i	$f(x_i)$
0	1	1	6	11/5	5/11
1	6/5	5/6	7	12/5	5/12
2	7/5	5/7	8	13/5	5/13
3	8/5	5/8	9	14/5	5/14
4	9/5	5/9	10	3	1/3
5	2	1/2			

Table 8.5: Computing Simpson's rule, with $n = 10$, for $\int_1^3 1/x\,dx$.

Exercises 8.6

1–5: Use the midpoint rule, with the specified value of n, to estimate the value of the integral. Use Theorem 8.6.1 to find an upper bound for the absolute value of the error in your estimate. Then use the second fundamental theorem of calculus to compute the exact value of the integral, and the exact value of the error.

1. $\displaystyle\int_0^2 x^2\,dx, n = 4.$

2. $\displaystyle\int_{-1}^3 x^3\,dx, n = 4.$

3. $\displaystyle\int_1^6 2x - 3\,dx, n = 5.$

4. $\displaystyle\int_0^\pi \sin x\,dx, n = 6.$

5. $\displaystyle\int_0^\pi \sin^3 x\,dx, n = 6.$

6–10: Use Simpson's rule, with the specified value of n, to estimate the value of the integral. Use Theorem 8.6.2 to find an upper bound for the absolute value of the error of your estimate. Then use the second fundamental theorem of calculus to compute the exact value of the integral, and the exact value of the error.

6. $\displaystyle\int_0^2 x^4\,dx, n = 4.$

7. $\displaystyle\int_{-1}^3 x^3\,dx, n = 4.$

8. $\displaystyle\int_1^4 \frac{dx}{x}, n = 6.$

9. $\displaystyle\int_0^\pi \sin x\,dx, n = 6.$

10. $\displaystyle\int_0^\pi \sin^3 x\,dx, n = 6.$

11–12: Use the midpoint rule to estimate the value of the integral. Choose n large enough to ensure that the absolute value of the error is no more than the specified bound.

11. $\displaystyle\int_1^4 \frac{dx}{x}$, absolute value of error at most 0.1.

12. $\displaystyle\int_1^3 e^{1/x}\,dx$, absolute value of error at most 0.05.

13. In this problem you will prove part 1 of Theorem 8.6.1. Suppose f is continuous on $[a, b]$, and $f''(x)$ is defined and $|f''(x)| \le K$ for all x in (a, b). Let n be a positive integer; we will study the error in the midpoint rule approximation M_n. Let x_i and x_i^* be defined as in the midpoint rule. We focus first on a single interval $[x_{i-1}, x_i]$. Define the function g by the formula:

$$g(x) = f(x_i^*) + f'(x_i^*)(x - x_i^*).$$

(a) Show that $g(x_i^*) = f(x_i^*)$ and for all x, $g'(x) = f'(x_i^*)$. It follows that the graph of g is the line tangent to the graph of f at $(x_i^*, f(x_i^*))$.

(b) Show that $\int_{x_{i-1}}^{x_i} g(x)\,dx = \Delta x f(x_i^*)$.

Now let $w(x) = (x - x_i^*)^2$. Our next step is to show that for every number x in the interval $[x_{i-1}, x_i]$,

$$|f(x) - g(x)| \le \frac{K}{2} w(x). \qquad (*)$$

It is easy to see that this holds when $x = x_i^*$, since both sides of $(*)$ are 0. In the next part you will prove that $(*)$ holds if $x < x_i^*$; similar reasoning can be used for $x > x_i^*$.

(c) Consider a fixed number x such that $x_{i-1} \le x < x_i^*$. Show that there is some number c such that $x < c < x_i^*$ and

$$f(x) - g(x) = \frac{f''(c)}{2} w(x),$$

and use this to conclude that $(*)$ is true. (Hint: Define a function h by the formula

$$h(t) = w(t)(f(x) - g(x)) - w(x)(f(t) - g(t)).$$

Apply the mean value theorem to h on the interval $[x, x_i^*]$ to show that there is some number d such that $x < d < x_i^*$ and $h'(d) = 0$. Then apply the mean value theorem to h' on the interval $[d, x_i^*]$.)

(d) Use $(*)$ to show that

$$-\frac{K \Delta x^3}{24} \le \int_{x_{i-1}}^{x_i} f(x)\,dx - \Delta x f(x_i^*) \le \frac{K \Delta x^3}{24}.$$

(e) Add the inequalities in the last part, for all i from 1 to n, to prove part 1 of Theorem 8.6.1.

14. Prove part 2 of Theorem 8.6.1. (Hint: Imitate Exercise 13, using the functions

$$g(x) = f(x_{i-1}) + \frac{f(x_i) - f(x_{i-1})}{\Delta x}(x - x_{i-1}), \qquad w(x) = (x - x_{i-1})(x_i - x).$$

To prove $(*)$ for $x_{i-1} < x < x_i$, apply the mean value theorem to h on the intervals $[x_{i-1}, x]$ and $[x, x_i]$ to find two numbers d_1 and d_2, and then apply the mean value theorem to h' on $[d_1, d_2]$.)

15. Verify that, as stated in the text, the values for A, B, and C given in (8.30) are the unique values for which the graph of the function $g(x) = A(x - x_1)^2 + B(x - x_1) + C$ passes through the points P_0, P_1, and P_2.

16. (a) Show that

$$\int_0^1 \frac{4}{x^2 + 1}\,dx = \pi.$$

(b) Use Simpson's rule, with $n = 8$, to estimate the value of the integral in part (a).

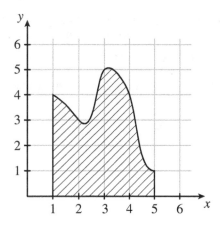

Figure 8.8: The region in Exercise 18.

17. Let

$$f(x) = \begin{cases} \dfrac{\sin x}{x}, & \text{if } x \neq 0, \\ 1, & \text{if } x = 0. \end{cases}$$

Use Simpson's rule with $n = 6$ to estimate the volume of the solid generated when the region under the curve $y = f(x), 0 \leq x \leq \pi$ is rotated about the x-axis. (Hint: First write down an integral for the required volume, and then use Simpson's rule to estimate the value of this integral.)

18. Use Simpson's rule with $n = 4$ to estimate the volume of the solid generated when the striped region in Figure 8.8 is rotated about the x-axis. (Hint: First give a name to the function whose graph is shown in the figure and write down an integral in terms of this function for the required volume. Then use Simpson's rule to estimate the value of this integral.)

8.7 Improper Integrals

In this section we will study integrals like

$$\int_1^\infty \frac{dx}{x^2}.$$

The definition of definite integrals in Section 5.3 does not apply to this integral, because ∞ is not a number. Thus, we will need to extend the definition of integration to cover this new kind of integral. Since a variable can approach ∞, but it cannot be equal to ∞, a natural approach is to interpret the integral as follows:

$$\int_1^\infty \frac{dx}{x^2} = \lim_{t \to \infty} \int_1^t \frac{dx}{x^2}.$$

Integrals of this kind are called *improper integrals*. Here is the general definition.

Definition 8.7.1. If f is continuous on $[a, \infty)$, then we define the improper integral $\int_a^\infty f(x)\,dx$ by the equation

$$\int_a^\infty f(x)\,dx = \lim_{t \to \infty} \int_a^t f(x)\,dx.$$

If this limit is defined, then we say that the integral *converges*; otherwise, it *diverges*.

For example, for any $t > 1$ we have

$$\int_1^t \frac{dx}{x^2} = -\frac{1}{x}\Big|_1^t = -\frac{1}{t} - (-1) = 1 - \frac{1}{t}.$$

Therefore, according to Definition 8.7.1,

$$\int_1^\infty \frac{dx}{x^2} = \lim_{t \to \infty} \int_1^t \frac{dx}{x^2} = \lim_{t \to \infty} \left(1 - \frac{1}{t}\right) = 1.$$

This is an example of an improper integral that converges.

Geometrically, these calculations tell us that the area of the region under the curve $y = 1/x^2$ from $x = 1$ to $x = t$ is $1 - 1/t$. As t increases, the region extends further and further to the right and the area increases, approaching 1 as $t \to \infty$. It seems reasonable, then, to say that the area of the region under the curve $y = 1/x^2$ that begins at $x = 1$ and extends infinitely far to the right is 1. This is the striped region in Figure 8.9.

Making a small change in this integral has a large effect on the answer:

$$\int_1^\infty \frac{dx}{x} = \lim_{t \to \infty} \int_1^t \frac{dx}{x} = \lim_{t \to \infty} \ln x\big|_1^t = \lim_{t \to \infty} \ln t = \infty.$$

Thus, the area under the curve $y = 1/x$ from $x = 1$ to $x = t$ gets larger and larger as t increases, eventually exceeding any number we might name. We therefore say that the area of the entire region under the curve $y = 1/x$ to the right of $x = 1$ is infinite. This is the blue region in Figure 8.9. We also say that $\int_1^\infty (1/x)\,dx$ diverges.

The two improper integrals we have worked out are part of a more general pattern. For future reference, it will be useful to record this pattern.

Theorem 8.7.2. *Let p be any real number.*

1. *If $p > 1$, then* $\displaystyle\int_1^\infty \frac{dx}{x^p}$ *converges.*

2. *If $p \le 1$, then* $\displaystyle\int_1^\infty \frac{dx}{x^p} = \infty.$

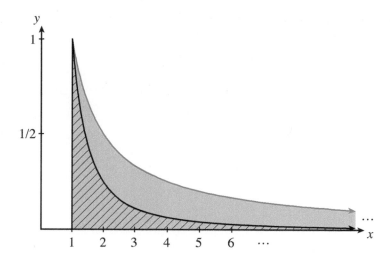

Figure 8.9: The black curve is the graph of $y = 1/x^2$, and the blue curve is $y = 1/x$. The area of the striped region is 1, and the area of the blue shaded region is infinite. Both regions extend infinitely far to the right.

Proof. We have already worked out the cases $p = 1$ and $p = 2$. The calculation for any $p > 1$ is similar to the case $p = 2$:

$$\int_1^\infty \frac{dx}{x^p} = \lim_{t \to \infty} \int_1^t x^{-p}\, dx$$

$$= \lim_{t \to \infty} \frac{x^{-p+1}}{-p+1}\bigg|_1^t = \lim_{t \to \infty}\left(-\frac{1}{(p-1)t^{p-1}} + \frac{1}{p-1}\right) = \frac{1}{p-1}.$$

Notice that we have used the fact that $p > 1$ to conclude that as $t \to \infty$, $t^{p-1} \to \infty$ (see Theorem 7.4.9). Thus, the integral converges.

For $p < 1$, we compute

$$\int_1^\infty \frac{dx}{x^p} = \lim_{t \to \infty} \int_1^t x^{-p}\, dx = \lim_{t \to \infty} \frac{x^{-p+1}}{-p+1}\bigg|_1^t = \lim_{t \to \infty}\left(\frac{t^{1-p}}{1-p} - \frac{1}{1-p}\right) = \infty. \quad \square$$

We can use a similar idea to define improper integrals whose lower limit is $-\infty$.

Definition 8.7.3. If f is continuous on $(-\infty, b]$, then

$$\int_{-\infty}^b f(x)\, dx = \lim_{t \to -\infty} \int_t^b f(x)\, dx.$$

For example,

$$\int_{-\infty}^{0} \frac{dx}{x^2+1} = \lim_{t\to-\infty} \int_{t}^{0} \frac{dx}{x^2+1} = \lim_{t\to-\infty} \tan^{-1} x \Big|_{t}^{0} = \lim_{t\to-\infty} [0 - \tan^{-1} t] = \frac{\pi}{2},$$

where in the final step we used our observation, from Section 7.5, that $\lim_{t\to-\infty} \tan^{-1} t = -\pi/2$. This integral represents the area of the blue region in Figure 8.10, which extends infinitely far to the left.

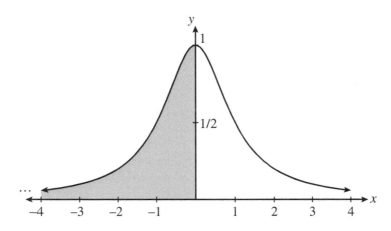

Figure 8.10: The area of the blue region is $\int_{-\infty}^{0} 1/(x^2+1)\,dx = \pi/2$.

The symmetry of the figure suggests that the area under $y = 1/(x^2+1)$ for $x \geq 0$ is also $\pi/2$, and this is confirmed by the following calculation:

$$\int_{0}^{\infty} \frac{dx}{x^2+1} = \lim_{t\to\infty} \int_{0}^{t} \frac{dx}{x^2+1} = \lim_{t\to\infty} \tan^{-1} x \Big|_{0}^{t} = \lim_{t\to\infty} \tan^{-1} t = \frac{\pi}{2}.$$

And this might lead us to say that the total area under the curve $y = 1/(x^2+1)$, for all values of x, both positive and negative, is $\pi/2 + \pi/2 = \pi$. To express this using an integral requires a new definition.

Definition 8.7.4. Suppose that f is continuous on $(-\infty, \infty)$. If $\int_{-\infty}^{0} f(x)\,dx$ and $\int_{0}^{\infty} f(x)\,dx$ both converge, then we say that $\int_{-\infty}^{\infty} f(x)\,dx$ converges and

$$\int_{-\infty}^{\infty} f(x)\,dx = \int_{-\infty}^{0} f(x)\,dx + \int_{0}^{\infty} f(x)\,dx.$$

If either $\int_{-\infty}^{0} f(x)\,dx$ or $\int_{0}^{\infty} f(x)\,dx$ diverges, then we say that $\int_{-\infty}^{\infty} f(x)\,dx$ diverges.

For example, according to this definition,

$$\int_{-\infty}^{\infty} \frac{dx}{x^2+1} = \int_{-\infty}^{0} \frac{dx}{x^2+1} + \int_{0}^{\infty} \frac{dx}{x^2+1} = \frac{\pi}{2} + \frac{\pi}{2} = \pi.$$

Although we have specified in Definition 8.7.4 that an integral from $-\infty$ to ∞ is to be split into two integrals at 0, in fact we will get the same answer if we split it at any number we choose. In other words, for any number a we can write

$$\int_{-\infty}^{\infty} f(x)\,dx = \int_{-\infty}^{a} f(x)\,dx + \int_{a}^{\infty} f(x)\,dx.$$

This is justified by part 3 of the following theorem.

Theorem 8.7.5. *Suppose $a < b$.*

1. *If f is continuous on $[a, \infty)$, then*

$$\int_{a}^{\infty} f(x)\,dx = \int_{a}^{b} f(x)\,dx + \int_{b}^{\infty} f(x)\,dx,$$

where we interpret this equation to mean that either both improper integrals converge and the two sides of the equation are equal, or both improper integrals diverge.

2. *If f is continuous on $(-\infty, b]$, then*

$$\int_{-\infty}^{b} f(x)\,dx = \int_{-\infty}^{a} f(x)\,dx + \int_{a}^{b} f(x)\,dx,$$

where again the equation means that either both sides are defined and equal, or both sides are undefined.

3. *If f is continuous on $(-\infty, \infty)$ then*

$$\int_{-\infty}^{a} f(x)\,dx + \int_{a}^{\infty} f(x)\,dx = \int_{-\infty}^{b} f(x)\,dx + \int_{b}^{\infty} f(x)\,dx,$$

where once again either both sides are defined and equal or both sides are undefined.

Proof. For part 1, we have

$$\int_{a}^{\infty} f(x)\,dx = \lim_{t\to\infty} \int_{a}^{t} f(x)\,dx = \lim_{t\to\infty} \left(\int_{a}^{b} f(x)\,dx + \int_{b}^{t} f(x)\,dx \right)$$

$$= \int_{a}^{b} f(x)\,dx + \lim_{t\to\infty} \int_{b}^{t} f(x)\,dx = \int_{a}^{b} f(x)\,dx + \int_{b}^{\infty} f(x)\,dx.$$

The proof of part 2 is similar.

Part 3 follows from parts 1 and 2:

$$\int_{-\infty}^{a} f(x)\,dx + \int_{a}^{\infty} f(x)\,dx$$

$$= \int_{-\infty}^{a} f(x)\,dx + \int_{a}^{b} f(x)\,dx + \int_{b}^{\infty} f(x)\,dx \qquad \text{(by part 1)}$$

$$= \int_{-\infty}^{b} f(x)\,dx + \int_{b}^{\infty} f(x)\,dx \qquad \text{(by part 2).} \qquad \square$$

Example 8.7.6. Evaluate the following expressions:

(a) $\displaystyle\int_{-\infty}^{\infty} \frac{2x}{x^2+1}\,dx,$ (b) $\displaystyle\lim_{t\to\infty} \int_{-t}^{t} \frac{2x}{x^2+1}\,dx,$ (c) $\displaystyle\lim_{t\to\infty} \int_{-t}^{2t} \frac{2x}{x^2+1}\,dx.$

Solution. For part (a), we split the integral into two integrals, as suggested by Definition 8.7.4:

$$\int_{-\infty}^{\infty} \frac{2x}{x^2+1}\,dx = \int_{-\infty}^{0} \frac{2x}{x^2+1}\,dx + \int_{0}^{\infty} \frac{2x}{x^2+1}\,dx.$$

The first integral on the right is

$$\int_{-\infty}^{0} \frac{2x}{x^2+1}\,dx = \lim_{t\to-\infty} \int_{t}^{0} \frac{2x}{x^2+1}\,dx = \lim_{t\to-\infty} \ln(x^2+1)\Big|_{t}^{0}$$

$$= \lim_{t\to-\infty} -\ln(t^2+1) = -\infty.$$

Since this integral diverges, there is no need to evaluate the second one; according to Definition 8.7.4, integral (a) diverges.

However, the limit in part (b) is defined, as the following calculation shows:

$$\lim_{t\to\infty} \int_{-t}^{t} \frac{2x}{x^2+1}\,dx = \lim_{t\to\infty} \ln(x^2+1)\Big|_{-t}^{t} = \lim_{t\to\infty} \left(\ln(t^2+1) - \ln(t^2+1)\right) = 0.$$

The limit in part (c) is also defined, but the value is different from the limit in part (b):

$$\lim_{t\to\infty} \int_{-t}^{2t} \frac{2x}{x^2+1}\,dx = \lim_{t\to\infty} \ln(x^2+1)\Big|_{-t}^{2t} = \lim_{t\to\infty} \left(\ln(4t^2+1) - \ln(t^2+1)\right)$$

$$= \lim_{t\to\infty} \ln\left(\frac{4t^2+1}{t^2+1}\right) = \lim_{t\to\infty} \ln\left(\frac{4+1/t^2}{1+1/t^2}\right) = \ln 4. \qquad \square$$

Example 8.7.6 illustrates how subtle improper integrals can be. In both parts (b) and (c), the upper limit of integration approaches ∞ and the lower limit approaches $-\infty$, but the values of the two limits are different! Perhaps this makes it seem reasonable that the improper integral in part (a) diverges; there would be no way to assign a value to it that is compatible with both parts (b) and (c). In Exercise 28 we ask you to show that this unusual situation never arises with improper integrals that converge.

So far we have discussed areas of regions that extend infinitely far to the left or right, but a region can also extend infinitely far up or down. For example, $\lim_{x \to 0^+} (1/\sqrt{x}) = \infty$, and therefore the region under the curve $y = 1/\sqrt{x}$ for $0 < x \le 1$ extends up infinitely far just to the right of the y-axis, as shown in Figure 8.11. The area of this region should be represented by the integral $\int_0^1 (1/\sqrt{x})\, dx$, but we cannot apply the definition of definite integrals in Section 5.3 to this integral, because $1/\sqrt{x}$ is not continuous on $[0, 1]$. Once again, we must extend our definition of integration to make sense of this integral, and as before we do this by taking a limit of an integral.

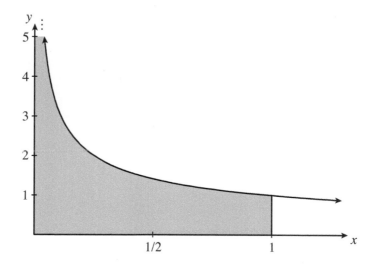

Figure 8.11: The region under the curve $y = 1/\sqrt{x}$ for $0 < x \le 1$.

Definition 8.7.7. Suppose that f is continuous on $(a, b]$, but f is not continuous from the right at a. Then we define the improper integral $\int_a^b f(x)\, dx$ by the equation

$$\int_a^b f(x)\, dx = \lim_{t \to a^+} \int_t^b f(x)\, dx.$$

Similarly, if f is continuous on $[a, b)$, but f is not continuous from the left at b, then

$$\int_a^b f(x)\, dx = \lim_{t \to b^-} \int_a^t f(x)\, dx.$$

For example, according to this definition,

$$\int_0^1 \frac{dx}{\sqrt{x}} = \lim_{t \to 0^+} \int_t^1 x^{-1/2}\, dx = \lim_{t \to 0^+} 2x^{1/2}\Big|_t^1 = \lim_{t \to 0^+} (2 - 2\sqrt{t}) = 2.$$

Thus, the area of the blue shaded region in Figure 8.11 is 2. Similarly, we have

$$\int_0^1 \frac{dx}{\sqrt[3]{x-1}} = \lim_{t \to 1^-} \int_0^t (x-1)^{-1/3}\,dx = \lim_{t \to 1^-} \frac{3}{2}(x-1)^{2/3}\Big|_0^t$$

$$= \lim_{t \to 1^-} \left(\frac{3}{2}(t-1)^{2/3} - \frac{3}{2}\right) = -\frac{3}{2}.$$

This means that the area of the red shaded region in Figure 8.12 is 3/2. Of course, the integral is negative because the region is below the x-axis.

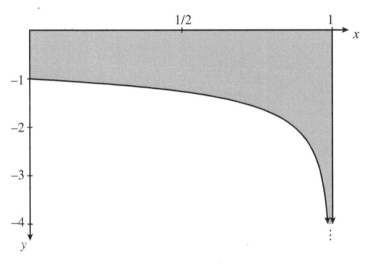

Figure 8.12: The region between the curve $y = 1/\sqrt[3]{x-1}$ and the x-axis for $0 \le x < 1$.

Improper integrals of the type considered in Definition 8.7.7 are harder to recognize than those in Definitions 8.7.1, 8.7.3, and 8.7.4, because the symbol ∞ does not appear in the integral notation. To see that an integral of this type is improper, you must examine the integrand and notice that it has a discontinuity.

There are other kinds of improper integrals, but rather than trying to list all the possibilities, we just state the general principles that we use to evaluate improper integrals:

1. An integral $\int_?^? f(x)\,dx$ is improper if either one of the limits of integration is $\pm\infty$ or f has a discontinuity.

2. If there is only one problem of the kind identified in 1, and the problem occurs at one of the limits of integration, then we evaluate the integral by taking a limit of a (proper) integral, as in Definitions 8.7.1, 8.7.3, and 8.7.7.

3. In other situations, we split up the integral as a sum of two or more integrals that can be evaluated as in 2. If any of these integrals diverge, then the original integral diverges.

Example 8.7.8. Evaluate the following integrals:

(a) $\displaystyle\int_0^\pi \tan x \, dx,$ (b) $\displaystyle\int_2^\infty \frac{dx}{x\sqrt{x^2-4}},$ (c) $\displaystyle\int_1^\infty \left(\frac{1}{x} - \frac{2}{2x+1}\right) dx.$

Solution. Integral (a) may not appear to be improper, and you might be tempted to evaluate it like this:

$$\text{WRONG!} \qquad \int_0^\pi \tan x \, dx = [\ln|\sec x|]_0^\pi = \ln 1 - \ln 1 = 0. \qquad \text{WRONG!}$$

The problem with this calculation is that although $\tan x$ is continuous on the intervals $[0, \pi/2)$ and $(\pi/2, \pi]$, $\tan(\pi/2)$ is undefined, so the second fundamental theorem of calculus does not apply. In fact, since $\lim_{x\to\pi/2^-} \tan x = \infty$ and $\lim_{x\to\pi/2^+} \tan x = -\infty$, the graph of the tangent function has a vertical asymptote at $x = \pi/2$. The integral is therefore improper, and we must split it up as suggested in principle 3 above:

$$\int_0^\pi \tan x \, dx = \int_0^{\pi/2} \tan x \, dx + \int_{\pi/2}^\pi \tan x \, dx.$$

We must now evaluate the two integrals on the right-hand side separately.

The first integral on the right is

$$\int_0^{\pi/2} \tan x \, dx = \lim_{t\to\pi/2^-} \int_0^t \tan x \, dx = \lim_{t\to\pi/2^-} [\ln|\sec x|]_0^t = \lim_{t\to\pi/2^-} \ln|\sec t| = \infty.$$

Thus, the first integral diverges, and therefore integral (a) diverges.

Integral (b) is clearly improper, because the upper limit of integration is ∞. But it is also improper at the lower limit of integration, since the graph of the integrand has a vertical asymptote at $x = 2$. We therefore split the integral into a sum of two improper integrals. As in part 3 of Theorem 8.7.5, we can split it wherever we choose; the answer will be the same no matter where we split it. We have arbitrarily chosen to split the integral at $x = 3$:

$$\int_2^\infty \frac{dx}{x\sqrt{x^2-4}} = \int_2^3 \frac{dx}{x\sqrt{x^2-4}} + \int_3^\infty \frac{dx}{x\sqrt{x^2-4}}.$$

To evaluate the two integrals on the right-hand side, it will be easiest to first evaluate the indefinite integral. We use the substitution

$$\theta = \sec^{-1}(x/2), \qquad x = 2\sec\theta, \qquad dx = 2\sec\theta\tan\theta \, d\theta,$$

which allows us to simplify the square root:

$$\sqrt{x^2-4} = \sqrt{4\sec^2\theta - 4} = \sqrt{4\tan^2\theta} = 2\tan\theta.$$

(Notice that since $x \geq 2$ and $\theta = \sec^{-1}(x/2)$, we have $0 \leq \theta < \pi/2$, and therefore $\tan\theta$ is positive.) The indefinite integral on the interval $(2, \infty)$ is therefore

$$\int \frac{dx}{x\sqrt{x^2-4}} = \int \frac{2\sec\theta\tan\theta \, d\theta}{(2\sec\theta)(2\tan\theta)} = \int \frac{1}{2} \, d\theta = \frac{1}{2}\theta + C = \frac{1}{2}\sec^{-1}\left(\frac{x}{2}\right) + C.$$

Now we can evaluate the improper integrals:

$$\int_2^3 \frac{dx}{x\sqrt{x^2-4}} = \lim_{t\to 2^+} \int_t^3 \frac{dx}{x\sqrt{x^2-4}} = \lim_{t\to 2^+} \frac{1}{2}\sec^{-1}\left(\frac{x}{2}\right)\Big|_t^3$$

$$= \lim_{t\to 2^+} \left(\frac{1}{2}\sec^{-1}\left(\frac{3}{2}\right) - \frac{1}{2}\sec^{-1}\left(\frac{t}{2}\right)\right) = \frac{1}{2}\sec^{-1}\left(\frac{3}{2}\right),$$

$$\int_3^\infty \frac{dx}{x\sqrt{x^2-4}} = \lim_{t\to\infty} \int_3^t \frac{dx}{x\sqrt{x^2-4}} = \lim_{t\to\infty} \frac{1}{2}\sec^{-1}\left(\frac{x}{2}\right)\Big|_3^t$$

$$= \lim_{t\to\infty} \left(\frac{1}{2}\sec^{-1}\left(\frac{t}{2}\right) - \frac{1}{2}\sec^{-1}\left(\frac{3}{2}\right)\right) = \frac{\pi}{4} - \frac{1}{2}\sec^{-1}\left(\frac{3}{2}\right),$$

and therefore

$$\int_2^\infty \frac{dx}{x\sqrt{x^2-4}} = \int_2^3 \frac{dx}{x\sqrt{x^2-4}} + \int_3^\infty \frac{dx}{x\sqrt{x^2-4}}$$

$$= \frac{1}{2}\sec^{-1}\left(\frac{3}{2}\right) + \left(\frac{\pi}{4} - \frac{1}{2}\sec^{-1}\left(\frac{3}{2}\right)\right) = \frac{\pi}{4}.$$

The integrand in integral (c) is continuous on $[1,\infty)$, so we can evaluate the integral with a single limit:

$$\int_1^\infty \left(\frac{1}{x} - \frac{2}{2x+1}\right) dx = \lim_{t\to\infty} \int_1^t \left(\frac{1}{x} - \frac{2}{2x+1}\right) dx = \lim_{t\to\infty} [\ln x - \ln(2x+1)]_1^t$$

$$= \lim_{t\to\infty} \ln\left(\frac{x}{2x+1}\right)\Big|_1^t = \lim_{t\to\infty} \left[\ln\left(\frac{t}{2t+1}\right) - \ln\left(\frac{1}{3}\right)\right]$$

$$= \lim_{t\to\infty} \left[\ln\left(\frac{1}{2+1/t}\right) + \ln 3\right] = \ln\left(\frac{1}{2}\right) + \ln 3 = \ln 3 - \ln 2.$$

Notice that $\int_1^\infty 1/x\, dx$ and $\int_1^\infty 2/(2x+1)\, dx$ both diverge, and therefore

$$\int_1^\infty \left(\frac{1}{x} - \frac{2}{2x+1}\right) dx \neq \int_1^\infty \frac{1}{x}\, dx - \int_1^\infty \frac{2}{2x+1}\, dx.$$

This does not contradict Theorem 5.3.6. According to that theorem, for any $t > 1$,

$$\int_1^t \left(\frac{1}{x} - \frac{2}{2x+1}\right) dx = \int_1^t \frac{1}{x}\, dx - \int_1^t \frac{2}{2x+1}\, dx,$$

and therefore

$$\int_1^\infty \left(\frac{1}{x} - \frac{2}{2x+1}\right) dx = \lim_{t\to\infty} \int_1^t \left(\frac{1}{x} - \frac{2}{2x+1}\right) dx$$

$$= \lim_{t\to\infty} \left[\int_1^t \frac{1}{x}\, dx - \int_1^t \frac{2}{2x+1}\, dx\right].$$

But the last limit above cannot be evaluated by finding the limits of the two integrals separately and then subtracting, since those limits are undefined. The lesson of this example is that facts about proper integrals do not necessarily apply to improper integrals. □

Exercises 8.7

1–26: Evaluate the integral.

1. $\displaystyle\int_2^\infty \frac{dx}{x^2}.$

2. $\displaystyle\int_2^\infty \frac{dx}{x^2+2x}.$

3. $\displaystyle\int_2^\infty \frac{dx}{x^2+2x+1}.$

4. $\displaystyle\int_2^\infty \frac{dx}{x^2+2x+2}.$

5. $\displaystyle\int_2^\infty \left(\frac{1}{x^2}+\frac{1}{2x}\right) dx.$

6. $\displaystyle\int_{-\infty}^0 e^{3x}\, dx.$

7. $\displaystyle\int_0^\infty \frac{x}{e^{2x}}\, dx.$

8. $\displaystyle\int_{-\infty}^\infty \frac{e^x}{e^{2x}+1}\, dx.$

9. $\displaystyle\int_0^2 \frac{x\, dx}{\sqrt{4-x^2}}.$

10. $\displaystyle\int_0^2 \frac{dx}{\sqrt{4-x^2}}.$

11. $\displaystyle\int_0^2 \frac{dx}{(4-x^2)^{3/2}}.$

12. $\displaystyle\int_1^2 \frac{dx}{\sqrt{x^2-1}}.$

13. $\displaystyle\int_1^\infty \frac{dx}{\sqrt{x^2-1}}.$

14. $\displaystyle\int_0^1 \ln x\, dx.$

15. $\displaystyle\int_{-\infty}^0 \left(\frac{2}{2x-1}-\frac{3}{3x-2}\right) dx.$

16. $\int_0^\infty \left(\dfrac{1}{\sqrt{x}} - \dfrac{1}{\sqrt{x+1}} \right) dx.$

17. $\int_{-\infty}^\infty \dfrac{dx}{x^4 + 5x^2 + 4}.$

18. $\int_\pi^{2\pi} \sec\theta \, d\theta.$

19. $\int_{-\infty}^0 \dfrac{dx}{(x+1)^{5/3}}.$

20. $\int_1^\infty \dfrac{dx}{x^{3/2}}.$

21. $\int_1^\infty \dfrac{dx}{x^{3/2} + x}.$

22. $\int_{-1}^1 \dfrac{dx}{x^{2/3}}.$

23. $\int_{-1}^1 \dfrac{dx}{2x^{2/3} + x}.$

24. $\int_{-1}^1 \dfrac{dx}{x^{2/3} + x}.$

25. $\int_0^\infty \dfrac{dx}{(x+4)\sqrt{x}}.$

26. $\int_0^\infty \left(\dfrac{\pi}{2} - \tan^{-1} x \right) dx.$

27. Evaluate each integral.

(a) $\int_1^\infty \dfrac{1}{x} \, dx.$

(b) $\int_1^\infty \dfrac{x}{x^2 + 1} \, dx.$

(c) $\int_1^\infty \left(\dfrac{1}{x} - \dfrac{x}{x^2 + 1} \right) dx.$

28. Suppose that $\int_{-\infty}^\infty f(x)\,dx$ converges, and its value is L. Show that if $g(t)$ and $h(t)$ are any two functions such that $\lim_{t \to \infty} g(t) = -\infty$ and $\lim_{t \to \infty} h(t) = \infty$, then

$$\lim_{t \to \infty} \int_{g(t)}^{h(t)} f(x)\,dx = L.$$

29. (a) Show that $\int_{-\infty}^{\infty} \dfrac{x}{\sqrt{x^2+1}}\, dx$ diverges.

(b) Find

$$\lim_{t\to\infty} \int_{-t}^{t} \frac{x}{\sqrt{x^2+1}}\, dx.$$

(c) Find

$$\lim_{t\to\infty} \int_{-t}^{t+1} \frac{x}{\sqrt{x^2+1}}\, dx.$$

30. In this section, we saw that the region $0 \le y \le 1/x$, $x \ge 1$ has infinite area. But surprisingly, the solid generated when this region is rotated about the x-axis has finite volume! Find the volume of this solid.

Chapter 9

Parametric Equations and Polar Coordinates

9.1 Parametric Equations

Consider a particle moving in the xy-plane. To describe how its position changes over time, we might define two functions as follows:

$$f(t) = \text{the } x\text{-coordinate of the particle at time } t,$$
$$g(t) = \text{the } y\text{-coordinate of the particle at time } t.$$

In other words, the coordinates (x, y) of the particle at any time t are given by the equations $x = f(t)$, $y = g(t)$.

For example, suppose a particle moves in the plane during the time interval $0 \leq t \leq 5$, and its coordinates at any time during this interval are given by the equations

$$x = 9 - 2t, \qquad y = t^2 - 4t, \qquad 0 \leq t \leq 5. \tag{9.1}$$

By substituting values of t between 0 and 5 into these equations, we can find the position of the particle at any time in this interval. In Figure 9.1 we have computed the position of the particle at several times and then plotted these points. The trajectory of the particle must be a curve passing through these points, as illustrated in the figure.

In this example, it is not hard to determine what this trajectory is. As t runs through all values from 0 to 5, $x = 9 - 2t$ runs through all values from 9 down to -1.

500

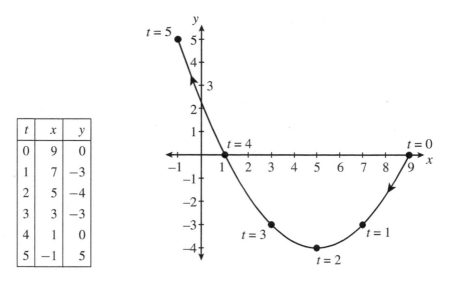

t	x	y
0	9	0
1	7	-3
2	5	-4
3	3	-3
4	1	0
5	-1	5

Figure 9.1: Graphing the trajectory of the particle in equations (9.1).

Solving for t in this equation, we see that each value of x in the interval $[-1, 9]$ arises from exactly one value of t, namely $t = (9 - x)/2$, and therefore there is exactly one point on the trajectory with this x-coordinate. The y-coordinate of this point is

$$y = t^2 - 4t = \left(\frac{9 - x}{2}\right)^2 - 4 \cdot \frac{9 - x}{2} = \frac{x^2 - 10x + 9}{4}.$$

Thus, the trajectory is the part of the parabola $y = (x^2 - 10x + 9)/4$ between $x = -1$ and $x = 9$, traversed from right to left.

The equations we started with in this example, equations (9.1), express the coordinates (x, y) of a varying point in terms of a third variable t. The variable t is called a *parameter*, and the equations are called *parametric equations*. In many examples the parameter represents time, but this need not be the case. Any equations that express the coordinates of a varying point in terms of another variable are called parametric equations. Such equations determine a curve in the plane; the curve is the trajectory of the varying point as the parameter runs through all values in its domain.

For example, according to the definitions of the trigonometric functions, if we start at the point P with coordinates $(1, 0)$ and travel a distance of θ units counterclockwise around the unit circle, we will arrive at the point Q with coordinates $(\cos \theta, \sin \theta)$ (see Figure 9.2a). Thus, the equations $x = \cos \theta$ and $y = \sin \theta$, for $0 \le \theta \le 2\pi$, are parametric equations for the unit circle. As θ runs through all values from 0 to 2π, the point (x, y) given by these equations travels counterclockwise once around the circle. The parameter in this case is θ, which does not represent time, but rather represents the distance along the circumference of the circle from P to Q, or, equivalently, the radian measure of the angle POQ in the figure. More generally, for any constant $r > 0$,

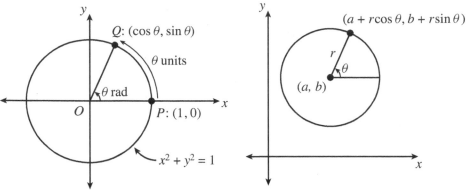

(a) The definitions of cos θ and sin θ give parametric equations for the unit circle.

(b) Parametric equations for a circle of radius r centered at (a, b).

Figure 9.2

we can rescale the circle by a factor of r to get a circle of radius r centered at the origin, which is represented by the parametric equations $x = r \cos \theta$, $y = r \sin \theta$. And for any numbers a and b, if we shift this circle a units horizontally and b units vertically, then we will get a circle of radius r centered at the point (a, b), with parametric equations $x = a + r \cos \theta$, $y = b + r \sin \theta$, for $0 \le \theta \le 2\pi$. This is illustrated in Figure 9.2b.

A well-known curve that is most easily described using parametric equations is the cycloid. Consider a wheel rolling along a straight road, and choose a distinguished point on the circumference of the wheel. The path traced by this distinguished point as the wheel rolls along the road is called a *cycloid*. For example, if we attach a light to the circumference of a bicycle wheel, then an observer watching a bicyclist ride past on a dark night will see the light trace out a cycloid.

To derive parametric equations for the cycloid, we take the road to be the x-axis, and we assume that a wheel of radius r starts out resting on the origin and then rolls to the right. Our distinguished point on the circumference of the wheel will be the point that is initially in contact with the road at the origin. For our parameter, we will use the angle θ through which the wheel has turned after it has rolled some distance to the right.

The initial position of the wheel is shown as a blue circle in Figure 9.3. After it has rolled through some angle θ, the wheel will be resting on the point Q in the figure, and the distinguished point will be at P. The part of the wheel that has been in contact with the road up to this point is the arc PQ, and the part of the road it has been in contact with is the segment OQ; these are shown in red in the figure. Since the wheel has rolled without sliding, arc PQ and segment OQ have the same length.

To find this common length, notice that angle PCQ is θ radians, so the fraction of the full circumference of the wheel that is taken up by arc PQ is $\theta/(2\pi)$. Since the circumference of the wheel is $2\pi r$, this means that the length of arc PQ is $(\theta/(2\pi))(2\pi r) = r\theta$. Therefore OQ has length $r\theta$, and it follows that the coordinates of C, the center of the wheel, are $(r\theta, r)$.

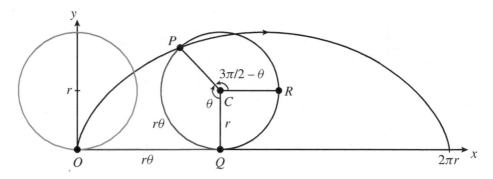

Figure 9.3: Finding parametric equations for a cycloid.

Thus, at this time the wheel is a circle with radius r centered at the point $(r\theta, r)$. Fortunately, we have already worked out parametric equations for such a circle. Since angle PCR is $3\pi/2 - \theta$, the coordinates of the point P are $(r\theta + r\cos(3\pi/2 - \theta),\ r + r\sin(3\pi/2 - \theta))$. We can simplify these coordinates by applying trigonometric identities:

$$\cos(3\pi/2 - \theta) = \cos(3\pi/2)\cos\theta + \sin(3\pi/2)\sin\theta$$
$$= 0 \cdot \cos\theta + (-1) \cdot \sin\theta = -\sin\theta,$$
$$\sin(3\pi/2 - \theta) = \sin(3\pi/2)\cos\theta - \cos(3\pi/2)\sin\theta$$
$$= (-1) \cdot \cos\theta + 0 \cdot \sin\theta = -\cos\theta.$$

Thus, the coordinates of P are given by the functions

$$x = f(\theta) = r\theta - r\sin\theta,$$
$$y = g(\theta) = r - r\cos\theta. \tag{9.2}$$

These are our parametric equations for the cycloid. The range of values $0 \le \theta \le 2\pi$ represents one full rotation of the wheel, during which the distinguished point traverses the entire curve shown in Figure 9.3, beginning at the origin and returning to contact with road at the point $(2\pi r, 0)$.

In Figure 9.3, it appears that the cycloid passes the vertical line test, so it is the graph of some function h. We can confirm this by observing that for all $\theta \in (0, 2\pi)$,

$$\frac{dx}{d\theta} = f'(\theta) = r(1 - \cos\theta) > 0, \tag{9.3}$$

and therefore f is strictly increasing on the interval $[0, 2\pi]$. It follows that f has an inverse function f^{-1}. The domain of f^{-1} is $[0, 2\pi r]$, and its range is $[0, 2\pi]$. This means that each value of x between 0 and $2\pi r$ corresponds to exactly one value of θ, namely $\theta = f^{-1}(x)$, so there is exactly one point on the cycloid with this x-coordinate. The y-coordinate of this point is $y = g(\theta) = g(f^{-1}(x))$. In other words, the cycloid is the graph of the function $h = g \circ f^{-1}$.

Recall that in Section 7.2, we worked out the calculus of inverse functions. In particular, in Theorem 7.2.2 we showed that if a function is strictly increasing and continuous, then its inverse is continuous as well. Thus, f^{-1} is continuous, and it follows that h is also continuous, since it is a composition of continuous functions. In fact, according to Theorem 7.2.2, since f is differentiable and $f'(\theta) \neq 0$ for $\theta \in (0, 2\pi)$, f^{-1} is differentiable on $(0, 2\pi r)$, and therefore $h = g \circ f^{-1}$ is as well. One way to find $h'(x)$ would be to use the formula for the derivative of f^{-1} given in Theorem 7.2.2 (see Exercise 21). But it is perhaps a little easier to proceed as follows. We begin with the equations

$$y = h(x), \qquad x = f(\theta).$$

By the chain rule, we have

$$\frac{dy}{d\theta} = \frac{dy}{dx} \cdot \frac{dx}{d\theta},$$

and therefore as long as $dx/d\theta \neq 0$,

$$\frac{dy}{dx} = \frac{dy/d\theta}{dx/d\theta}. \tag{9.4}$$

We have already computed $dx/d\theta$ in equation (9.3) above and verified that it is nonzero on the interval $(0, 2\pi)$, and

$$\frac{dy}{d\theta} = \frac{d}{d\theta}(r - r\cos\theta) = r\sin\theta.$$

Thus,

$$h'(x) = \frac{dy}{dx} = \frac{dy/d\theta}{dx/d\theta} = \frac{r\sin\theta}{r(1 - \cos\theta)} = \frac{\sin\theta}{1 - \cos\theta}.$$

For example, consider a cycloid generated by a wheel of radius $r = 6$. When $\theta = 4\pi/3$ we have

$$x = f\left(\frac{4\pi}{3}\right) = 6 \cdot \frac{4\pi}{3} - 6\sin\left(\frac{4\pi}{3}\right) = 8\pi + 3\sqrt{3},$$

$$y = g\left(\frac{4\pi}{3}\right) = 6 - 6\cos\left(\frac{4\pi}{3}\right) = 9.$$

Thus, the point $(8\pi + 3\sqrt{3}, 9) \approx (30.33, 9)$ is on the cycloid. At this point, the slope of the tangent line to the cycloid is

$$\left.\frac{dy}{dx}\right|_{\theta = 4\pi/3} = \frac{\sin(4\pi/3)}{1 - \cos(4\pi/3)} = \frac{-\sqrt{3}/2}{1 + 1/2} = -\frac{\sqrt{3}}{3} \approx -0.577.$$

This is illustrated in Figure 9.4.

The difficulty we faced in computing dy/dx was caused by the fact that our parametric equations (9.2) give us a formula for y in terms of θ, not x. By differentiating

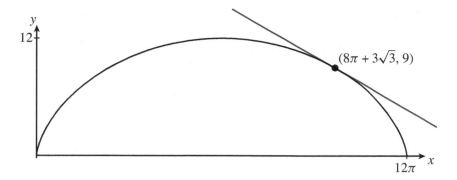

Figure 9.4: A cycloid with $r = 6$. The tangent line to the cycloid at the point $(8\pi + 3\sqrt{3}, 9)$ has slope $-\sqrt{3}/3 \approx -0.577$.

this formula we can compute $dy/d\theta$, but it is not so easy to see how to compute dy/dx. Equation (9.4), which is simply a rearranged form of the chain rule, shows how $dy/d\theta$ can be used to compute dy/dx. One way to remember equation (9.4) is to think of it as saying that the derivative dy/dx will be unchanged if we "divide numerator and denominator by $d\theta$." Of course, this doesn't really make sense; derivatives are not fractions, and $d\theta$ is not a number. But once again we have a form of the chain rule that tells us that derivatives act in certain ways as if they were fractions. This idea can be useful in other situations as well.

For example, let's compute d^2y/dx^2. It will make our calculations easier to follow if we give a name to dy/dx, so we introduce the notation

$$y' = \frac{dy}{dx} = \frac{\sin\theta}{1 - \cos\theta}. \tag{9.5}$$

We now want to compute $d^2y/dx^2 = dy'/dx$.

Notice that equation (9.5) expresses y' in terms of θ, not x. From this equation we can easily find $dy'/d\theta$, but we want dy'/dx. This is exactly the same as the difficulty we faced when computing dy/dx, and we can use the same idea as before to deal with the difficulty. Imitating our previous reasoning, we can show that we will get the right answer if we "divide numerator and denominator of dy'/dx by $d\theta$":

$$\frac{dy'}{dx} = \frac{dy'/d\theta}{dx/d\theta}.$$

Using equation (9.5), we find that

$$\frac{dy'}{d\theta} = \frac{(1 - \cos\theta)(\cos\theta) - (\sin\theta)(\sin\theta)}{(1 - \cos\theta)^2} = \frac{\cos\theta - 1}{(1 - \cos\theta)^2} = -\frac{1}{1 - \cos\theta},$$

so

$$\frac{d^2y}{dx^2} = \frac{dy'}{dx} = \frac{dy'/d\theta}{dx/d\theta} = \frac{-1/(1 - \cos\theta)}{r(1 - \cos\theta)} = -\frac{1}{r(1 - \cos\theta)^2}.$$

Notice that for $0 < \theta < 2\pi$, d^2y/dx^2 is negative, and therefore h is concave down. This confirms the shape of the curve in Figures 9.3 and 9.4.

Curves that are given by parametric equations do not always pass the vertical line test, so they are not always graphs of functions. But often it is possible to break the curve into pieces so that each piece is the graph of some function. For each piece, we can then imitate the reasoning we used with the cycloid to study the function and determine the shape of its graph.

Example 9.1.1. Sketch the curve C given by the parametric equations

$$x = f(t) = 3t^3 - 18t^2 + 27t + 10,$$
$$y = g(t) = t^3 - 6t^2 + 35.$$

Solution. We first compute

$$\frac{dx}{dt} = f'(t) = 9t^2 - 36t + 27 = 9(t^2 - 4t + 3) = 9(t-1)(t-3),$$
$$\frac{dy}{dt} = g'(t) = 3t^2 - 12t = 3(t^2 - 4t) = 3t(t-4).$$

Clearly $f'(t) = 0$ when t is 1 or 3 and $g'(t) = 0$ when t is 0 or 4, and checking sample points we find that the signs of $f'(t)$ and $g'(t)$ can be summarized as follows:

$dx/dt = f'(t)$	+	+	+	0	−	0	+	+	+
$dy/dt = g'(t)$	+	0	−	−	−	−	−	0	+

t		0		1		3		4	

Looking at this number line, we can see that f and g are both strictly increasing on the interval $(-\infty, 0]$. In other words, as t increases through this interval, x and y both increase, and the point (x, y) on the curve C moves to the right and up. On the interval $[0, 1]$, f is strictly increasing and g is strictly decreasing, so as t increases from 0 to 1, x increases and y decreases, and the point (x, y) moves to the right and down. Similarly, as t increases from 1 to 3, the point (x, y) moves to the left and down, as t increases from 3 to 4, (x, y) moves to the right and down, and as t increases beyond 4, (x, y) moves to the right and up. Plotting the points at $t = 0, 1, 3, 4$ and filling in the direction of motion between these points, we can see that the general shape of the curve must be as shown in Figure 9.5.

Figure 9.5 is not a very accurate picture of C; we will improve on it shortly. But it is already accurate enough to allow us to see that C is not the graph of a function. Indeed, any vertical line $x = c$ for c between 10 and 22 will cross C three times, violating the vertical line test. But we can break C into three pieces, indicated by the colors black, red, and blue in the figure, so that each piece is the graph of a function.

To spell this out more precisely, we let C_1 be the part of C generated by values of t in the interval $(-\infty, 1]$, C_2 the part for values of t in the interval $[1, 3]$, and C_3 the part for t in $[3, \infty)$. The overall shape of C_1 is indicated in black in Figure 9.5, C_2 in red, and C_3 in blue. Imitating the reasoning we used for the cycloid, we can show that

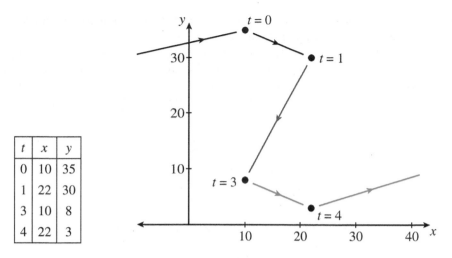

t	x	y
0	10	35
1	22	30
3	10	8
4	22	3

Figure 9.5: The general shape of the curve C in Example 9.1.1.

these three pieces of C are the graphs of three differentiable functions h_1, h_2, and h_3. To find the derivatives of these functions, we "divide the numerator and denominator of dy/dx by dt" to get the formula

$$\frac{dy}{dx} = \frac{dy/dt}{dx/dt} = \frac{3t^2 - 12t}{9t^2 - 36t + 27} = \frac{t^2 - 4t}{3t^2 - 12t + 9} = \frac{t(t-4)}{3(t-1)(t-3)}. \qquad (9.6)$$

Formula (9.6) gives the derivatives of all three of the functions h_1, h_2, and h_3. For $t < 1$ it is equal to $h_1'(x)$, for $1 < t < 3$ it is $h_2'(x)$, and for $t > 3$ it is $h_3'(x)$. The formula cannot be applied at $t = 1$ or 3, since at these points the denominator is 0.

Similarly, we can find a single formula that gives the second derivatives of all three functions. Letting $y' = dy/dx$, we compute

$$\frac{dy'}{dt} = \frac{(3t^2 - 12t + 9)(2t - 4) - (t^2 - 4t)(6t - 12)}{(3t^2 - 12t + 9)^2} = \frac{2(t-2)}{(t-1)^2(t-3)^2},$$

and therefore

$$\frac{d^2y}{dx^2} = \frac{dy'}{dx} = \frac{dy'/dt}{dx/dt} = \frac{2(t-2)}{9(t-1)^3(t-3)^3}. \qquad (9.7)$$

Using this information, let's try to draw C_1 more accurately. Since $f(1) = 22$ and

$$\lim_{t \to -\infty} f(t) = \lim_{t \to -\infty} (3t^3 - 18t^2 + 27t + 10)$$

$$= \lim_{t \to -\infty} t^3 \left(3 - \frac{18}{t} + \frac{27}{t^2} + \frac{10}{t^3} \right) = -\infty,$$

the set of x-coordinates of points on C_1 is the interval $(-\infty, 22]$. This interval is the domain of the function h_1. For $t < 1$, the only point where either dy/dx or d^2y/dx^2

changes sign occurs when $t = 0$ and $x = f(0) = 10$. Substituting appropriate sample points into equations (9.6) and (9.7), we find that

$$\left.\frac{dy}{dx}\right|_{t=-1} = \frac{5}{24}, \qquad \left.\frac{dy}{dx}\right|_{t=0} = 0, \qquad \left.\frac{dy}{dx}\right|_{t=1/2} = -\frac{7}{15}, \qquad \left.\frac{d^2y}{dx^2}\right|_{t=0} = -\frac{4}{243}.$$

We conclude that the signs of h_1' and h_1'' are as shown on the following number line:

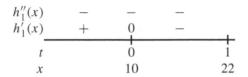

Thus, h_1 is strictly increasing on $(-\infty, 10]$ and strictly decreasing on $[10, 22]$, with a local maximum at the point where $t = 0$ and $x = 10$, and it is concave down on the entire interval $(-\infty, 22]$. The graph of h_1—that is, the curve C_1—is shown in black in Figure 9.6.

We use a similar method to sketch C_2 and C_3. We begin by making number lines indicating the signs of the derivatives of h_2 and h_3. Notice that since f is strictly decreasing on the interval $[1, 3]$, C_2 is traversed from right to left, as indicated in Figure 9.5. We have therefore chosen to draw our number line for h_2 with the values of t running from right to left, so that the x values will appear in their proper order.

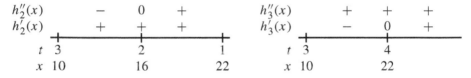

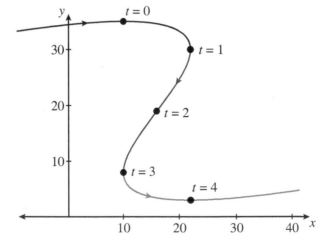

Figure 9.6: The curve C in Example 9.1.1. The curve C_1 is shown in black, C_2 in red, and C_3 in blue.

In the first of these number lines we see that $t = 2$ is another important point to plot. At $t = 2$, we have $x = f(2) = 16$ and $y = g(2) = 19$. We can read off from this number line that h_2 is strictly increasing on $[10, 22]$, concave down on $[10, 16]$, and concave up on $[16, 22]$, with an inflection point at $x = 16$. We see from the second number line that h_3 is strictly decreasing on $[10, 22]$, strictly increasing on $[22, \infty)$, and concave up on $[10, \infty)$; it has a local minimum at $x = 22$. The graphs are shown in Figure 9.6.

There is one more point that is perhaps worth commenting on. In Figure 9.6, it appears that C has vertical tangent lines at $t = 1$ and $t = 3$. The easiest way to confirm this is to reverse the roles of x and y in our calculations. The part of the curve between $t = 0$ and $t = 4$ is the graph of an equation $x = j(y)$, for some function j, and

$$j'(y) = \frac{dx}{dy} = \frac{dx/dt}{dy/dt} = \frac{3(t-1)(t-3)}{t(t-4)}.$$

At the points where $t = 1$ and $t = 3$ we have $dx/dy = 0$, and therefore the tangent lines at these points have equations of the form $x = 0 \cdot y + c = c$; that is, they are vertical. For an alternative way to see this, see Exercise 23.   □

The lesson of Example 9.1.1 is that if a curve is given by parametric equations expressing x and y as differentiable functions of a parameter t, then the curve can often be broken into pieces so that each piece is the graph of a differentiable function. The pieces will correspond to intervals of t values on which dx/dt has one sign, with each piece being traversed either left to right or right to left, depending on the sign of dx/dt. The first and second derivatives of these functions can be computed using the formulas

$$y' = \frac{dy}{dx} = \frac{dy/dt}{dx/dt},$$
$$\frac{d^2y}{dx^2} = \frac{dy'}{dx} = \frac{dy'/dt}{dx/dt}.$$

The signs of these derivatives can then be used in the usual way to determine the shape of the curve.

Exercises 9.1

1–9: Sketch the curve given by the parametric equations.

1. $x = 2t - 1$, $y = t^3$, $-2 \le t \le 2$.

2. $x = t^2$, $y = t^3$, $-2 \le t \le 2$.

3. $x = t^3 - 3t$, $y = t^2$, $-2 \le t \le 2$.

4. $x = t^3 - 3t$, $y = t - t^3$, $-2 \le t \le 2$.

5. $x = \cos\theta$, $y = 2\sin\theta$, $0 \le \theta \le 2\pi$.

6. $x = \cos\theta$, $y = \sin(2\theta)$, $0 \le \theta \le 2\pi$.

7. $x = \cos^3 \theta$, $y = \sin^3 \theta$, $0 \le \theta \le \pi$.

8. $x = \sec \theta$, $y = \tan \theta$, $-\pi/2 < \theta < \pi/2$.

9. $x = e^t - e^{-t}$, $y = e^t + e^{-t}$, $-\infty < t < \infty$.

10. Find the equation of the line tangent to the curve in Exercise 5 at the point where $\theta = 5\pi/6$.

11. Find the equation of the line tangent to the curve in Exercise 6 at the point where $\theta = \pi/6$.

12. Find the equation of the line tangent to the curve in Exercise 2 at the point $(1, -1)$.

13. Show that there are two lines that are tangent to the curve in Exercise 3 at the point $(0, 3)$. What are their equations?

14. Show that if (x, y) is any point on the curve in Exercise 5, then $x^2 + y^2/4 = 1$. What is the shape of the curve?

15. Show that if (x, y) is any point on the curve in Exercise 8, then $x^2 - y^2 = 1$. What is the shape of the curve?

16. Show that if (x, y) is any point on the curve in Exercise 9, then $y^2 - x^2 = 4$. What is the shape of the curve?

17. Show that the curve in Exercise 6 consists of the graphs of the two functions $f_1(x) = 2x\sqrt{1 - x^2}$ and $f_2(x) = -2x\sqrt{1 - x^2}$.

18. Two particles travel in the plane in such a way that their coordinates at time t are given by the following formulas:

$$\text{Particle 1: } x = 2t + 1, \ y = t^2 - 3,$$
$$\text{Particle 2: } x = t + 2, \ y = t^2 - 4t + 1.$$

(a) Let C_1 be the path traced by particle 1, and C_2 the path traced by particle 2. Find all intersection points of C_1 and C_2.

(b) Do the particles ever collide? If so, when and where?

19. Let P be a distinguished point attached to a wheel of radius r, and let a be the distance from the center of the wheel to P. The path traced by P as the wheel rolls along a road is called a *trochoid*. (If $a = r$, then the point P is on the circumference of the wheel, and the trochoid is a cycloid. But in this problem, we consider other possible values of a.)

(a) Find parametric equations for the trochoid.

(b) Sketch the graph of the trochoid in the case $r = 2$, $a = 1$. (For example, if a reflector is attached halfway along one of the spokes of a bicycle wheel of radius 2, then this trochoid would be the path of the reflector as a bicyclist rides by.)

(c) Sketch the graph of the trochoid in the case $r = 1, a = 2$. (In this case, $a > r$, so the distinguished point P is *beyond* the circumference of the wheel. Such a trochoid could be the path of a point on the flange of a train wheel as the train goes by.)

20. Figure 9.7 shows the design of an elliptical trainer. The axle of a wheel of radius 10 inches is attached at the origin O to a track lying along the positive x-axis. One end of a rigid rod is attached to the circumference of the wheel at P, and the other end rests on the track at Q. The rod can pivot around P, and the end of the rod at Q can slide along the track. The rod is 40 inches long, and a foot pad is attached to the center of the rod at M. When an exerciser uses the trainer, the wheel rotates, and the foot pad at M traces a closed curve C.

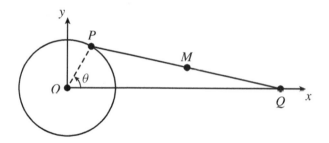

Figure 9.7: The design of an elliptical trainer.

(a) Find parametric equations for C by expressing the coordinates of M in terms of the angle θ in Figure 9.7.

(b) Show that C is *almost* an ellipse by finding an ellipse that is very close to C. (You should be able to find an ellipse that stays within a fraction of an inch of C.)

21. Let f and g be the functions given in equations (9.2). As shown in the text, the cycloid is the graph of the function $h = g \circ f^{-1}$. Use the chain rule and the formula for the derivative of f^{-1} given in Theorem 7.2.2 to find $h'(x)$, and verify that your answer agrees with the answer in the text.

22. Find the area under the cycloid in Figure 9.3. (Hint: The cycloid is the graph of the function $h = g \circ f^{-1}$, where f and g are the functions given in equations (9.2). Thus, the requested area is

$$\int_0^{2\pi r} h(x)\,dx = \int_0^{2\pi r} g(f^{-1}(x))\,dx.$$

To evaluate this integral, use the substitution $\theta = f^{-1}(x)$, and apply the methods of Section 8.4.)

23. For the functions f and g in Example 9.1.1, compute the following limits:

$$\lim_{t \to 1^-} \frac{g(t) - g(1)}{f(t) - f(1)}, \qquad \lim_{t \to 1^+} \frac{g(t) - g(1)}{f(t) - f(1)},$$

$$\lim_{t \to 3^-} \frac{g(t) - g(3)}{f(t) - f(3)}, \qquad \lim_{t \to 3^+} \frac{g(t) - g(3)}{f(t) - f(3)}.$$

What do these limits represent in Figure 9.6?

9.2   Arc Length

Consider a curve C given by parametric equations $x = f(t)$, $y = g(t)$, for $a \leq t \leq b$. In this section, we will address the question: What is the length of C?

As usual in calculus, we begin by trying to find a way to get an approximate answer. Our method will be to cut the curve into pieces, approximate the length of each piece, and then add up these approximations. If this leads to a Riemann sum, then we should be able to use an integral to compute the exact length of the curve.

To carry out this plan, we first divide the interval $[a, b]$ on the t-axis into n equal pieces, for some positive integer n. Each piece will have width $\Delta t = (b - a)/n$, and the division points will be $t_i = a + i\,\Delta t$ for $i = 0, 1, 2, \ldots, n$. The ith piece is the interval $[t_{i-1}, t_i]$, and when values of t in this interval are plugged into the parametric equations they generate a piece C_i of the curve C. If we let s_i be the length of C_i, then the total length of C is $s = \sum_{i=1}^{n} s_i$. We now want to approximate s_i.

For each i, let P_i be the point on C corresponding to $t = t_i$; in other words, P_i is the point with coordinates $(f(t_i), g(t_i))$. Then C_i is the piece of C that begins at P_{i-1} and ends at P_i. We will approximate its length, s_i, with the length of the line segment from P_{i-1} to P_i. In other words, if we let $\Delta x_i = f(t_i) - f(t_{i-1})$ and $\Delta y_i = g(t_i) - g(t_{i-1})$, then by the distance formula, our approximation is

$$s_i \approx \sqrt{(\Delta x_i)^2 + (\Delta y_i)^2}.$$

Adding these approximations, we conclude that the total length of C is approximately

$$Q_n = \sum_{i=1}^{n} \sqrt{(\Delta x_i)^2 + (\Delta y_i)^2}. \tag{9.8}$$

The line segments $\overline{P_{i-1}P_i}$, for $i = 1, 2, \ldots, n$, form a polygonal path that follows the general shape of the curve C, as illustrated in Figure 9.8. Since Q_n is the sum of the lengths of all of these line segments, it is the length of this polygonal path. Intuitively, it seems that as n increases, this polygonal path will follow the shape of C more and more closely, and Q_n will get closer and closer to the length of C. We therefore compute the exact length as a limit:

$$s = \lim_{n \to \infty} Q_n = \lim_{n \to \infty} \sum_{i=1}^{n} \sqrt{(\Delta x_i)^2 + (\Delta y_i)^2}.$$

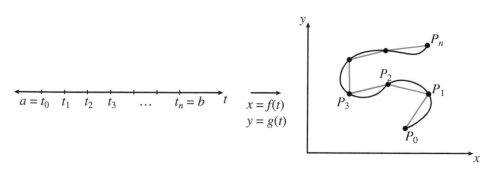

Figure 9.8: The curve C, shown in black, is approximated by the blue polygonal path.

Unfortunately, the sum that defines Q_n does not have the form of a Riemann sum. In a Riemann sum, every term should include a factor of Δt, but this factor does not appear in the sum in equation (9.8). But we can easily introduce such a factor by dividing by $(\Delta t)^2$ inside the square root and multiplying by Δt outside the square root:

$$Q_n = \sum_{i=1}^{n} \sqrt{\left(\frac{\Delta x_i}{\Delta t}\right)^2 + \left(\frac{\Delta y_i}{\Delta t}\right)^2} \,\Delta t. \tag{9.9}$$

The fractions inside the square root in this formula look like they are related to derivatives. Indeed, if we assume that f is differentiable on $[a, b]$, then we can use the mean value theorem to conclude that for each i, there is some number t_i^* between t_{i-1} and t_i such that

$$\frac{\Delta x_i}{\Delta t} = \frac{f(t_i) - f(t_{i-1})}{t_i - t_{i-1}} = f'(t_i^*).$$

Similarly, if g is also differentiable on $[a, b]$, then for each i, there is a number t_i^{**} between t_{i-1} and t_i such that

$$\frac{\Delta y_i}{\Delta t} = g'(t_i^{**}).$$

Substituting into equation (9.9), we find that

$$Q_n = \sum_{i=1}^{n} \sqrt{(f'(t_i^*))^2 + (g'(t_i^{**}))^2} \,\Delta t. \tag{9.10}$$

We are making progress; the sum in equation (9.9) is *almost* a Riemann sum. The only remaining problem is that each term in the sum involves *two* sample points in interval number i, t_i^* and t_i^{**}, whereas in a Riemann sum there should be only one sample point. But if n is large, so that Δt is small, then t_i^* and t_i^{**} will be very close together. If we assume that f' and g' are continuous on $[a, b]$, then the small discrepancy in the sample points shouldn't make much difference. In other words, the sum

$$R_n = \sum_{i=1}^{n} \sqrt{(f'(t_i^*))^2 + (g'(t_i^*))^2}\,\Delta t \qquad (9.11)$$

should be close to Q_n, and therefore also a good approximation of the length of C. In fact, it can be shown (see Exercise 15) that $\lim_{n\to\infty} Q_n = \lim_{n\to\infty} R_n$. Therefore the exact length of C is

$$s = \lim_{n\to\infty} Q_n = \lim_{n\to\infty} R_n = \lim_{n\to\infty} \sum_{i=1}^{n} \sqrt{(f'(t_i^*))^2 + (g(t_i^*))^2}\,\Delta t$$

$$= \int_a^b \sqrt{(f'(t))^2 + (g'(t))^2}\, dt.$$

Rewriting the derivatives in Leibniz notation, we have

$$s = \int_a^b \sqrt{\left(\frac{dx}{dt}\right)^2 + \left(\frac{dy}{dt}\right)^2}\, dt. \qquad (9.12)$$

Example 9.2.1. Find the lengths of the following curves:

(a) $x = \cos^3 t$, $\quad y = \sin^3 t$, $\quad 0 \le t \le \dfrac{\pi}{2}$,

(b) $x = \dfrac{t^2}{2} - t$, $\quad y = \dfrac{4}{3}(2-t)^{3/2}$, $\quad 0 \le t \le 2$,

(c) $y = x^2$, $\quad 0 \le x \le 1$.

Solution. For part (a), we have

$$\frac{dx}{dt} = -3\cos^2 t \sin t, \qquad \frac{dy}{dt} = 3\sin^2 t \cos t,$$

and therefore by equation (9.12) the length of the curve is

$$s = \int_0^{\pi/2} \sqrt{\left(\frac{dx}{dt}\right)^2 + \left(\frac{dy}{dt}\right)^2}\, dt = \int_0^{\pi/2} \sqrt{(-3\cos^2 t \sin t)^2 + (3\sin^2 t \cos t)^2}\, dt$$

$$= \int_0^{\pi/2} \sqrt{9\cos^4 t \sin^2 t + 9\sin^4 t \cos^2 t}\, dt$$

$$= \int_0^{\pi/2} \sqrt{9\cos^2 t \sin^2 t (\cos^2 t + \sin^2 t)}\, dt$$

$$= \int_0^{\pi/2} \sqrt{9\cos^2 t \sin^2 t}\, dt = \int_0^{\pi/2} 3\,|\cos t \sin t|\, dt.$$

Since $\cos t$ and $\sin t$ are both nonnegative for t between 0 and $\pi/2$, we can drop the absolute value signs and evaluate the integral using the substitution $u = \sin t$, $du = \cos t \, dt$:

$$s = \int_0^{\pi/2} 3 \cos t \sin t \, dt = \int_0^1 3u \, du = \frac{3}{2} u^2 \Big|_0^1 = \frac{3}{2}.$$

Our method for part (b) is similar:

$$\frac{dx}{dt} = t - 1, \qquad \frac{dy}{dt} = -2(2 - t)^{1/2} = -2\sqrt{2 - t},$$

and therefore

$$s = \int_0^2 \sqrt{(t - 1)^2 + (-2\sqrt{2 - t})^2} \, dt = \int_0^2 \sqrt{(t^2 - 2t + 1) + 4(2 - t)} \, dt$$

$$= \int_0^2 \sqrt{t^2 - 6t + 9} \, dt = \int_0^2 \sqrt{(t - 3)^2} \, dt = \int_0^2 |t - 3| \, dt.$$

For $0 \leq t \leq 2$, $t - 3$ is negative, so $|t - 3| = 3 - t$, and the length of the curve is

$$s = \int_0^2 3 - t \, dt = \left[3t - \frac{t^2}{2} \right]_0^2 = 4.$$

The curve in part (c) is not given by parametric equations, so equation (9.12) doesn't apply. But we can easily convert it to parametric equations by writing it like this:

$$x = t, \qquad y = t^2, \qquad 0 \leq t \leq 1.$$

Now we have

$$\frac{dx}{dt} = 1, \qquad \frac{dy}{dt} = 2t,$$

and therefore

$$s = \int_0^1 \sqrt{1 + 4t^2} \, dt.$$

Unfortunately, this is a somewhat challenging integral. To evaluate it, we use the substitution $\theta = \tan^{-1}(2t)$, which means

$$2t = \tan \theta, \qquad -\frac{\pi}{2} < \theta < \frac{\pi}{2}.$$

This substitution gives us

$$\sqrt{1 + 4t^2} = \sqrt{1 + \tan^2 \theta} = \sqrt{\sec^2 \theta} = \sec \theta, \qquad dt = \frac{1}{2} \sec^2 \theta \, d\theta.$$

We first evaluate the indefinite integral, using our reduction formula for powers of secant, equation (8.18):

$$\int \sqrt{1 + 4t^2}\, dt = \int \sec\theta \cdot \frac{1}{2}\sec^2\theta\, d\theta = \frac{1}{2}\int \sec^3\theta\, d\theta$$

$$= \frac{1}{2}\left[\frac{\tan\theta\,\sec\theta}{2} + \frac{1}{2}\int \sec\theta\, d\theta\right]$$

$$= \frac{\tan\theta\,\sec\theta}{4} + \frac{1}{4}\ln|\sec\theta + \tan\theta| + C$$

$$= \frac{t\sqrt{1 + 4t^2}}{2} + \frac{1}{4}\ln\left|\sqrt{1 + 4t^2} + 2t\right| + C.$$

Finally, we evaluate the definite integral to find the length of the curve:

$$s = \int_0^1 \sqrt{1 + 4t^2}\, dt = \left[\frac{t\sqrt{1 + 4t^2}}{2} + \frac{1}{4}\ln\left|2t + \sqrt{1 + 4t^2}\right|\right]_0^1 = \frac{\sqrt{5}}{2} + \frac{\ln(2 + \sqrt{5})}{4}.$$

$$\square$$

The method we used in part (c) of Example 9.2.1 can be used to find the length of any curve of the form $y = f(x)$, for $a \le x \le b$, as long as f' is continuous on $[a, b]$. We first rewrite the curve in parametric form using the equations

$$x = t, \qquad y = f(t), \qquad a \le t \le b,$$

and then equation (9.12) tells us that the length of the curve is

$$s = \int_a^b \sqrt{\left(\frac{dx}{dt}\right)^2 + \left(\frac{dy}{dt}\right)^2}\, dt = \int_a^b \sqrt{1 + (f'(t))^2}\, dt = \int_a^b \sqrt{1 + (f'(x))^2}\, dx.$$

Rewriting this answer in Leibniz notation, this means that

$$s = \int_a^b \sqrt{1 + \left(\frac{dy}{dx}\right)^2}\, dx. \tag{9.13}$$

Similarly, the length of a curve of the form $x = f(y)$, $a \le y \le b$ is

$$s = \int_a^b \sqrt{1 + \left(\frac{dx}{dy}\right)^2}\, dy. \tag{9.14}$$

Example 9.2.2. Find the length of the curve

$$x = \frac{y^2}{8} - \ln y, \qquad 1 \le y \le 2.$$

Solution. We first compute

$$\sqrt{1+\left(\frac{dx}{dy}\right)^2} = \sqrt{1+\left(\frac{y}{4}-\frac{1}{y}\right)^2} = \sqrt{1+\frac{y^2}{16}-\frac{1}{2}+\frac{1}{y^2}}$$

$$= \sqrt{\frac{y^2}{16}+\frac{1}{2}+\frac{1}{y^2}} = \sqrt{\left(\frac{y}{4}+\frac{1}{y}\right)^2} = \left|\frac{y}{4}+\frac{1}{y}\right|.$$

For y between 1 and 2 we can drop the absolute value signs, and the length of the curve is

$$s = \int_1^2 \sqrt{1+\left(\frac{dx}{dy}\right)^2}\,dy = \int_1^2 \frac{y}{4}+\frac{1}{y}\,dy = \left[\frac{y^2}{8}+\ln y\right]_1^2 = \frac{1}{2}+\ln 2 - \frac{1}{8} = \frac{3}{8}+\ln 2.$$

$\square$

We have now derived three formulas for the length of a curve, given in equations (9.12), (9.13), and (9.14). There is a relationship between these formulas that can be made explicit by introducing some new notation. In equation (9.12), if we pretend that dt is a number, and not part of our notation for derivatives and integrals, then we can move the final dt inside the square root as a dt^2 and then "cancel" with the "denominators" of the derivatives:

$$s = \int_a^b \sqrt{\left(\frac{dx}{dt}\right)^2+\left(\frac{dy}{dt}\right)^2}\,dt = \int_a^b \sqrt{dx^2+dy^2}.$$

Of course, this calculation doesn't really make sense, and the final integral above does not use standard integral notation. But it is striking that similar calculations with equations (9.13) and (9.14), moving either dx or dy inside the square root, lead to exactly the same result.

We therefore introduce the notation $ds = \sqrt{dx^2+dy^2}$, and say that the length of a curve is given by the formula

$$s = \int_a^b \sqrt{dx^2+dy^2} = \int_a^b ds. \tag{9.15}$$

Again, this formula doesn't really mean anything, but it is a useful mnemonic that can help you remember how to compute the length of a curve. By "factoring" either dx, dy, or dt out of the square root in the formula for ds, you can obtain the integrands in equations (9.12), (9.13), and (9.14):

$$ds = \sqrt{dx^2+dy^2} = \begin{cases} \sqrt{\left(\frac{dx}{dt}\right)^2+\left(\frac{dy}{dt}\right)^2}\,dt \\ \sqrt{1+\left(\frac{dy}{dx}\right)^2}\,dx \\ \sqrt{1+\left(\frac{dx}{dy}\right)^2}\,dy. \end{cases}$$

Thus, equation (9.15) can be thought of as a shorthand version of all of our length formulas. Equation (9.15) also has the advantage that the formula $\sqrt{dx^2 + dy^2}$ is reminiscent of the distance formula $\sqrt{(\Delta x)^2 + (\Delta y)^2}$, which was the basis of all of our length calculations.

Exercises 9.2

1–14: Find the length of the curve.

1. $x = t^3 - 3t, y = 3t^2, 0 \le t \le 2$.

2. $x = t^2, y = t^3, 0 \le t \le 2$.

3. $x = \dfrac{t}{t^2 + 1}, y = \dfrac{1}{t^2 + 1}, 0 \le t \le 1$.

4. $y = 2x^{3/2}, 0 \le x \le 7$.

5. $y = \ln(\sec x), 0 \le x \le \pi/3$.

6. $x = \dfrac{y^3}{3} + \dfrac{1}{4y}, 1 \le y \le 3$.

7. $x = e^{y/2} + e^{-y/2}, -2 \le y \le 2$.

8. $y = \ln(x + \sqrt{x^2 - 1}), 2 \le x \le 3$.

9. $y = \ln x, 1 \le x \le 2$.

10. $x = t^2 + t, y = \dfrac{(4t + 3)^{3/2}}{6}, 0 \le t \le 2$.

11. $x = 2(\sin t - t \cos t), y = 2(\cos t + t \sin t), 0 \le t \le \pi/2$.

12. $x = (4 - t^2)^{3/2}, y = t^3, -2 \le t \le 2$.

13. $x = 2 \tan^{-1} t, y = \ln(t^2 + 1), 0 \le t \le 1$.

14. $x = \tan t - t, y = \ln(\cos t), 0 \le t \le \pi/3$.

15. This exercise is for readers who have read Section 5.6.

(a) Show that for any numbers a, b, and k,

$$|\sqrt{k^2 + b^2} - \sqrt{k^2 + a^2}| \le |b - a|.$$

(Hint: If $a = b = 0$, then the inequality above is clearly true. If not, then start with

$$|\sqrt{k^2 + b^2} - \sqrt{k^2 + a^2}| = \left| \frac{(\sqrt{k^2 + b^2} - \sqrt{k^2 + a^2})(\sqrt{k^2 + b^2} + \sqrt{k^2 + a^2})}{\sqrt{k^2 + b^2} + \sqrt{k^2 + a^2}} \right|.$$

Another approach is to apply the mean value theorem to the function $f(x) = \sqrt{k^2 + x^2}$.)

Now let Q_n and R_n be the sums defined in equations (9.10) and (9.11). As in the text, we assume that f' and g' are continuous on $[a, b]$. In the next part, you will prove that $\lim_{n\to\infty} Q_n = \lim_{n\to\infty} R_n$.

(b) Suppose $\epsilon > 0$. Since g' is continuous on $[a, b]$, by Theorem 5.6.2 it is uniformly continuous on that interval. Thus, there is some $\delta > 0$ such that if $s, t \in [a, b]$ and $|s - t| < \delta$, then $|g'(s) - g'(t)| < \epsilon/(b-a)$. Let N be an integer larger than $(b - a)/\delta$. Now show that if $n > N$ then $|Q_n - R_n| < \epsilon$. Conclude that $\lim_{n\to\infty}(Q_n - R_n) = 0$, and therefore $\lim_{n\to\infty} Q_n = \lim_{n\to\infty} R_n$.

16. Suppose that f' and g' are continuous on $[a, b]$, and the curve $x = f(t)$, $y = g(t)$, $a \le t \le b$, has length 3. Find the length of the curve $x = 2f(t) + g(t)$, $y = f(t) - 2g(t)$, $a \le t \le b$.

9.3 Surface Area

In the last section, we found a formula for the length of a curve C given by parametric equations $x = f(t)$, $y = g(t)$, for $a \le t \le b$. In this section we assume that $y = g(t) \ge 0$, so that C does not go below the x-axis, and we compute the area of the surface generated when the curve C is rotated about the x-axis. This is illustrated in Figure 9.9. Once again, our method will be to approximate this area with a Riemann sum, and then to find the exact answer with an integral.

While the problem we have proposed no doubt reminds you of the volumes that we computed in Chapter 6, there are important differences. In Chapter 6, we rotated a region bounded by given curves about an axis to generate a solid. To find the volume of the solid, we cut the region into slices and approximated each slice with a rectangle. But in this section, it is not a region bounded by the curve C that is being rotated about an axis, but rather the curve itself. Thus, we will be cutting the curve C into pieces, not a

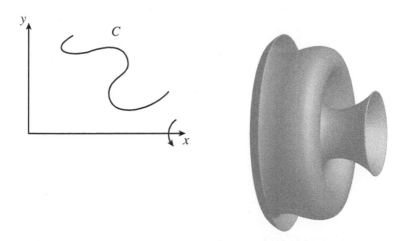

Figure 9.9: When the curve C is rotated about the x-axis, it generates a surface.

region bounded by C. We will approximate each piece of C with a line segment, as we did in our length calculations in the last section. Thus, our method in this section will combine ideas from the last section with ideas from Chapter 6.

We begin by cutting the curve C into n pieces, for some positive integer n, exactly as we did in the last sections. Thus, we again let $\Delta t = (b-a)/n$, and for $i = 0, 1, 2, \ldots, n$ we let $t_i = a + i\Delta t$ and we let P_i be the point with coordinates $(f(t_i), g(t_i))$. The numbers t_i divide the interval $[a, b]$ on the t-axis into n pieces, and the points P_i divide the curve C into corresponding pieces. The ith piece of $[a, b]$ is the interval $[t_{i-1}, t_i]$, and values of t in this interval correspond to the piece C_i of C that runs from P_{i-1} to P_i. As before, we will approximate C_i with the line segment from P_{i-1} to P_i. The length of this line segment is $\sqrt{(\Delta x_i)^2 + (\Delta y_i)^2}$, where $\Delta x_i = f(t_i) - f(t_{i-1})$ and $\Delta y_i = g(t_i) - g(t_{i-1})$.

Let S be the area of the surface generated when C is rotated about the x-axis; this is what we want to compute. If we let S_i be the area of the part of this surface that is generated by C_i, which is shown in gray in the center of Figure 9.10, then $S = \sum_{i=1}^{n} S_i$. We now approximate C_i with the line segment from P_{i-1} to P_i, and we approximate S_i with the area of the surface generated when this line segment is rotated about the x-axis. This surface is shown in blue on the right in Figure 9.10.

You may not be familiar with the formula for the area of the blue surface in Figure 9.10. We therefore temporarily set aside our computation of S to derive the formula for the area of this surface. The surface is called the *lateral surface of a frustum of a cone*, because we can think of it as being cut off of the end of a cone, as illustrated in the center of Figure 9.11. We will find a formula for the lateral surface area of a frustum in terms of the radii at either end of the frustum, labeled r_1 and r_2 in Figure 9.11, and the length l, which is called the *slant height* of the frustum. Let x be the distance along the surface of the cone from the frustum to the apex of the cone, as indicated in the figure.

If we cut the lateral surface of the cone along a line from the base to the apex and then flatten it out, we get a sector of a circle, as illustrated on the right in Figure 9.11.

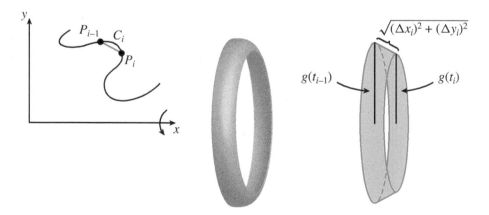

Figure 9.10: When C_i is rotated about the x-axis, it generates the gray surface in the center. We approximate C_i with the line segment from P_{i-1} to P_i, which generates the blue surface on the right.

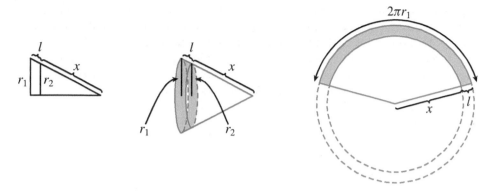

Figure 9.11: Computing the lateral surface area of a frustum of a cone.

The lateral surface of the frustum becomes the blue band along the outer edge of this sector. The length of the outer edge of the sector is the same as the circumference of the base of the cone, which is $2\pi r_1$, and the circumference of the entire outer circle on the right in the figure is $2\pi(l + x)$. Therefore the fraction of the circle taken up by the sector is $2\pi r_1/(2\pi(l + x)) = r_1/(l + x)$. The area of the outer circle is $\pi(l + x)^2$ and the area of the inner circle is πx^2, so the area of the ring between the two circles is

$$\pi(l + x)^2 - \pi x^2 = \pi[(l + x)^2 - x^2] = \pi[l^2 + 2lx] = \pi l(l + 2x).$$

The fraction of this ring taken up by the blue band is $r_1/(l + x)$, so the area of the blue band, which is the same as the lateral surface area of the frustum, is

$$\pi l(l + 2x) \cdot \frac{r_1}{l + x} = \pi l r_1 \left(\frac{l + 2x}{l + x}\right) = \pi l r_1 \left(1 + \frac{x}{l + x}\right). \tag{9.16}$$

We would now like to eliminate x from this formula. To do this, we use the triangle on the left in Figure 9.11, which is the top half of a vertical cross section through the cone. By similar triangles, we see that

$$\frac{x}{r_2} = \frac{l + x}{r_1},$$

and therefore

$$\frac{x}{l + x} = \frac{r_2}{r_1}.$$

Substituting into equation (9.16), we see that the lateral surface area of the frustum is

$$\pi l r_1 \left(1 + \frac{x}{l + x}\right) = \pi l r_1 \left(1 + \frac{r_2}{r_1}\right) = \pi l(r_1 + r_2).$$

It may be a little easier to understand and remember the formula for the lateral surface area of the frustum if we rewrite it in the form

$$2\pi \left(\frac{r_1 + r_2}{2}\right) l. \tag{9.17}$$

We can think of $(r_1 + r_2)/2$ as the average radius of the frustum, and therefore 2π $(r_1 + r_2)/2$ is the average circumference. Thus, equation (9.17) says that the lateral surface area of the frustum is the average circumference times the slant height. Note that in our calculations we assumed that $r_1 \neq r_2$. If $r_1 = r_2$, then the blue surface in Figure 9.11 becomes a cylinder, whose lateral surface area is the circumference times the height. Thus, equation (9.17) is correct even if $r_1 = r_2$.

We are ready now to return to the computation of S, the area of the surface generated when the curve C is rotated about the x-axis. Recall that our plan was to approximate S_i, the area of the part of the surface generated by the ith piece of C, with the area of the blue surface on the right in Figure 9.10. Using equation (9.17), we now have a formula for this approximation:

$$S_i \approx 2\pi \left(\frac{g(t_{i-1}) + g(t_i)}{2} \right) \sqrt{(\Delta x_i)^2 + (\Delta y_i)^2}.$$

Adding up these pieces, we conclude that

$$S = \sum_{i=1}^{n} S_i \approx \sum_{i=1}^{n} 2\pi \left(\frac{g(t_{i-1}) + g(t_i)}{2} \right) \sqrt{(\Delta x_i)^2 + (\Delta y_i)^2}.$$

As in the last section, to turn this into a Riemann sum we first factor out Δt and apply the mean value theorem to conclude that if f and g are differentiable on $[a, b]$, then

$$S \approx \sum_{i=1}^{n} 2\pi \left(\frac{g(t_{i-1}) + g(t_i)}{2} \right) \sqrt{\left(\frac{\Delta x_i}{\Delta t} \right)^2 + \left(\frac{\Delta y_i}{\Delta t} \right)^2} \Delta t$$

$$= \sum_{i=1}^{n} 2\pi \left(\frac{g(t_{i-1}) + g(t_i)}{2} \right) \sqrt{(f'(t_i^*))^2 + (g'(t_i^{**}))^2} \Delta t,$$

for some two numbers t_i^* and t_i^{**} between t_{i-1} and t_i.

Once again, the last sum above is still not a Riemann sum. This time, there are four values in the interval $[t_{i-1}, t_i]$ that appear in the ith term of the sum: the two endpoints, t_{i-1} and t_i, and two values in between, t_i^* and t_i^{**}. But as before, if we assume that g, f', and g' are continuous on $[a, b]$, then it shouldn't make much difference if we change all of these values of t to a single sample point in the ith interval. In other words, we have the approximation

$$S \approx \sum_{i=1}^{n} 2\pi \left(\frac{g(t_i^*) + g(t_i^*)}{2} \right) \sqrt{(f'(t_i^*))^2 + (g'(t_i^*))^2} \Delta t$$

$$= \sum_{i=1}^{n} 2\pi g(t_i^*) \sqrt{(f'(t_i^*))^2 + (g'(t_i^*))^2} \Delta t.$$

This approximation is a Riemann sum.

Finally, to get the exact answer, we take the limit as $n \to \infty$, which turns the Riemann sum into an integral:

$$S = \lim_{n \to \infty} \sum_{i=1}^{n} 2\pi g(t_i^*)\sqrt{(f'(t_i^*))^2 + (g'(t_i^*))^2}\,\Delta t = \int_a^b 2\pi g(t)\sqrt{(f'(t))^2 + (g'(t))^2}\,dt.$$

Alternatively, we could write the integral in the form

$$S = \int_a^b 2\pi y \sqrt{\left(\frac{dx}{dt}\right)^2 + \left(\frac{dy}{dt}\right)^2}\,dt. \tag{9.18}$$

But perhaps the most useful form of the integral comes from recognizing the geometric origin of each term in the integral. The y-coordinate of a point on C represents the radius of the circle traced by the point when it rotates about the x-axis, so that $2\pi y$ is the circumference. The rest of the integrand is what we called ds in the last section. So it may be best to remember the formula for surface area in the form

$$S = \int_a^b 2\pi (\text{radius})\,ds = \int_a^b (\text{circumference})\,ds. \tag{9.19}$$

This formula should make sense if you compare it to equation (9.17): the circumference factor in the integral corresponds to the average circumference of the frustum, and ds corresponds to the slant height.

Example 9.3.1. Find the areas of the following surfaces:

(a) The surface generated when the curve $x = t^3$, $y = t^2$, for $0 \le t \le 1$, is rotated about the x-axis.
(b) The surface generated when the curve $y = x^2$, for $0 \le x \le 1$, is rotated about the y-axis.
(c) The surface of a sphere of radius r.
(d) The surface generated when the circle $x^2 + y^2 = a^2$ is rotated about the line $x = b$, where $b > a > 0$.

Solution. For part (a), we can simply apply equation (9.18) to see that the surface area is

$$S = \int_0^1 2\pi t^2 \sqrt{(3t^2)^2 + (2t)^2}\,dt = \int_0^1 2\pi t^2 \sqrt{9t^4 + 4t^2}\,dt = \int_0^1 2\pi t^3 \sqrt{9t^2 + 4}\,dt.$$

To evaluate the integral, we use the substitution

$$u = 9t^2 + 4, \qquad t^2 = \frac{u - 4}{9}, \qquad du = 18t\,dt, \qquad \frac{1}{18}\,du = t\,dt.$$

This leads to the calculation

$$S = \int_0^1 2\pi t^3 \sqrt{9t^2 + 4}\,dt = \int_0^1 2\pi t^2 (9t^2 + 4)^{1/2} \cdot t\,dt = \int_4^{13} 2\pi \cdot \frac{u - 4}{9} \cdot u^{1/2} \cdot \frac{1}{18}\,du$$

$$= \frac{\pi}{81} \int_4^{13} u^{3/2} - 4u^{1/2}\,du = \frac{\pi}{81}\left[\frac{2}{5}u^{5/2} - \frac{8}{3}u^{3/2}\right]_4^{13} = \frac{(128 + 494\sqrt{13})\pi}{1215}.$$

The curve in part (b) is not given by parametric equations, and it is rotated about the y-axis, not the x-axis, so we cannot use equation (9.18) to find the requested surface area. However, we can use equation (9.19). The independent variable in the specification of the curve is x, so it is easiest to write ds in the form

$$ds = \sqrt{dx^2 + dy^2} = \sqrt{1 + \left(\frac{dy}{dx}\right)^2}\, dx = \sqrt{1 + (2x)^2}\, dx = \sqrt{1 + 4x^2}\, dx.$$

When a point on the curve is rotated about the y-axis, the radius of the circle it traces is x, so by equation (9.19) the surface area is

$$S = \int_0^1 2\pi x \sqrt{1 + 4x^2}\, dx.$$

The substitution $u = 1 + 4x^2$, $du = 8x\, dx$ now gives the solution:

$$S = \int_0^1 2\pi \sqrt{1 + 4x^2} \cdot x\, dx = \int_1^5 2\pi u^{1/2} \cdot \frac{1}{8}\, du = \frac{\pi}{6} u^{3/2} \Big|_1^5 = \frac{(5\sqrt{5} - 1)\pi}{6}.$$

To generate the surface of a sphere of radius r for part (c), we let C be the top half of the circle of radius r centered at the origin and then rotate C about the x-axis. We can represent C by the parametric equations

$$x = r\cos\theta, \qquad y = r\sin\theta, \qquad 0 \le \theta \le \pi.$$

Using θ as our independent variable, we have

$$ds = \sqrt{dx^2 + dy^2} = \sqrt{\left(\frac{dx}{d\theta}\right)^2 + \left(\frac{dy}{d\theta}\right)^2}\, d\theta = \sqrt{(-r\sin\theta)^2 + (r\cos\theta)^2}\, d\theta$$

$$= \sqrt{r^2(\cos^2\theta + \sin^2\theta)}\, d\theta = r\, d\theta.$$

The radius of the circle generated when a point on C is rotated about the x-axis is $y = r\sin\theta$, so by equation (9.19) the surface area of a sphere of radius r is

$$\int_0^\pi 2\pi r \sin\theta \cdot r\, d\theta = -2\pi r^2 \cos\theta \Big|_0^\pi = 2\pi r^2 - (-2\pi r^2) = 4\pi r^2.$$

You probably knew this formula already, but you may not have known where it came from.

Finally, the surface in part (d) is called a *torus*; it is the surface of a donut, as illustrated in Figure 9.12. We can represent the circle with the parametric equations $x = a\cos\theta$, $y = a\sin\theta$, for $0 \le \theta \le 2\pi$. When a point on this circle is rotated about

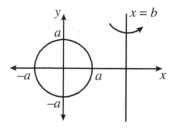

Figure 9.12: When a circle is rotated about a line not passing through the circle, it generates a torus.

the line $x = b$, it generates a circle of radius $b - x = b - a\cos\theta$, and we have

$$ds = \sqrt{\left(\frac{dx}{d\theta}\right)^2 + \left(\frac{dy}{d\theta}\right)^2}\, d\theta = \sqrt{(-a\sin\theta)^2 + (a\cos\theta)^2}\, d\theta = a\, d\theta.$$

Thus, the surface area of the torus is

$$\int_0^{2\pi} 2\pi(b - a\cos\theta) \cdot a\, d\theta = [2\pi a(b\theta - a\sin\theta)]_0^{2\pi} = 2\pi a \cdot 2\pi b = 4\pi^2 ab. \qquad \square$$

Exercises 9.3

1–13: Find the area of the surface generated when the curve is rotated about the specified line.

1. $x = t^3 - 3t$, $y = 3t^2$, $0 \le t \le 2$, rotated about the x-axis.

2. $x = t^2 - 2t$, $y = \frac{8}{3}t^{3/2}$, $1 \le t \le 2$, rotated about the y-axis.

3. $x = \frac{2(2 - 2t)^{3/2}}{3}$, $y = \frac{t^2}{2} + t$, $0 \le t \le 1$, rotated about the x-axis.

4. $x = \frac{t}{t^2 + 1}$, $y = \frac{1}{t^2 + 1}$, $0 \le t \le 1$, rotated about the y-axis.

5. $x = \frac{t}{t^2 + 1}$, $y = \frac{1}{t^2 + 1}$, $0 \le t \le 1$, rotated about the x-axis.

6. $x = 2(\sin t - t\cos t)$, $y = 2(\cos t + t\sin t)$, $0 \le t \le \pi/2$, rotated about the x-axis.

7. $y = x^3$, $0 \le x \le 1$, rotated about the x-axis.

8. $y = x^3$, $0 \le x \le 1$, rotated about the y-axis.

9. $y = 2\sqrt{x}$, $1 \le x \le 3$, rotated about the x-axis.

10. $x = 2y^{3/2}$, $0 \le y \le 1$, rotated about the x-axis.

11. $x = 2y^{3/2}$, $0 \le y \le 1$, rotated about the line $y = -1$.

12. $y = \ln x$, $1 \le x \le 2$, rotated about the y-axis.

13. $y = \sin x$, $0 \le x \le \pi$, rotated about the x-axis.

14. In Exercise 30 of Section 8.7, you used an improper integral to see that when the region under the curve $y = 1/x$, for $x \ge 1$, is rotated about the x-axis, it generates a solid whose volume is finite. In this problem we consider the area of the surface generated when the curve $y = 1/x$, $x \ge 1$, is rotated about the x-axis.

   (a) Suppose that $t > 1$. Write down an integral for the area of the surface generated when the curve $y = 1/x$, $1 \le x \le t$, is rotated about the x-axis.

   (b) Show that the value of the integral in part (a) is larger than $2\pi \ln t$, and use this to show that when the curve $y = 1/x$, $x \ge 1$, is rotated about the x-axis, it generates a surface whose area is infinite.

   (c) Find the exact value of the integral in part (a).

15. Let C be the curve $y = f(x)$, $0 \le x \le 3$, shown in Figure 9.13, and assume that f' is continuous on $[0, 3]$. Suppose that the length of C is 5 and the area of the surface generated when C is rotated about the x-axis is 30. Find the area of the surface generated when C is rotated about the line $y = 2$. (Hint: Begin by writing down integrals representing all of the lengths and areas mentioned in the problem.)

16. Consider a curve C given by parametric equations $x = f(t)$, $y = g(t)$, $a \le t \le b$, where f' and g' are continuous on $[a, b]$. Imitating the reasoning in Section 6.5, we say that the *centroid* of the curve C is the point (c_x, c_y) whose coordinates are given by the formulas

$$c_x = \frac{\int_a^b x \sqrt{\left(\frac{dx}{dt}\right)^2 + \left(\frac{dy}{dt}\right)^2}\, dt}{\int_a^b \sqrt{\left(\frac{dx}{dt}\right)^2 + \left(\frac{dy}{dt}\right)^2}\, dt}, \qquad c_y = \frac{\int_a^b y \sqrt{\left(\frac{dx}{dt}\right)^2 + \left(\frac{dy}{dt}\right)^2}\, dt}{\int_a^b \sqrt{\left(\frac{dx}{dt}\right)^2 + \left(\frac{dy}{dt}\right)^2}\, dt}.$$

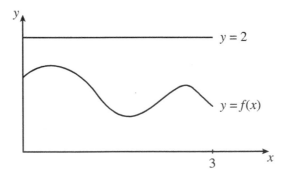

Figure 9.13: The curve in Exercise 15.

(a) Suppose that C lies entirely on one side of a line l. Pappus's first centroid theorem says that the area of the surface generated when C is rotated about l is equal to the length of C times the distance traveled by the centroid of C when it rotates about l. Prove Pappus's first centroid theorem for the surface generated when C is rotated about the x-axis, assuming C is entirely above the x-axis.

(b) Verify Pappus's theorem in the case of the torus in part (d) of Example 9.3.1.

9.4 Polar Coordinates

So far, whenever we have wanted to specify the location of a point in the plane, we have done so by giving the x- and y-coordinates of the point. This method of identifying points is called the *Cartesian coordinate system*. It is named for René Descartes (1596–1650), who proposed the use of such coordinates in 1637. However, for some problems it is convenient to use a different method of assigning coordinates to points. In this section we study one such alternative coordinate system, the *polar coordinate system*.

The Cartesian coordinate system requires two *axes*, a horizontal x-axis and a vertical y-axis, crossing at a distinguished point called the *origin*. The x- and y-coordinates of a point P could be thought of as giving us directions for getting from the origin to P; the x-coordinate tells us how far to move horizontally from the origin, and the y-coordinate tells us how far to move vertically.

In the polar coordinate system there is also an origin, but there is only one axis, called the *polar axis*. The polar axis is a ray—that is, a half-line—beginning at the origin O, and usually drawn extending to the right. Once again, the coordinates of a point P give directions for traveling from O to P, but now the directions specify how far to travel, and in which direction. The polar coordinates of P are (r, θ), where r is the distance from O to P, and θ is the counterclockwise angle from the polar axis to the ray from O through P, as shown in Figure 9.14.

Figure 9.14: The polar coordinate system.

Example 9.4.1. Plot the following points in polar coordinates:

$$P: \left(3, \frac{\pi}{4}\right), \qquad Q: \left(2, -\frac{\pi}{2}\right), \qquad R: \left(3, \frac{9\pi}{4}\right).$$

Solution. The solution is shown in Figure 9.15. To locate the point P, we draw a ray from the origin making an angle of $\pi/4$ in the counterclockwise direction from the polar axis, and then travel 3 units along this ray. The value $\theta = -\pi/2$ for the point Q means that we should make an angle of $\pi/2$ in the *clockwise* direction from the polar axis. Thus, Q is located 2 units down from the origin.

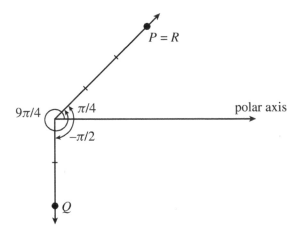

Figure 9.15: The solution to Example 9.4.1.

Finally, for the point R, note that $9\pi/4 = 2\pi + \pi/4$, and therefore the direction indicated by $\theta = 9\pi/4$ is the same as the direction for $\theta = \pi/4$. Thus, R is exactly the same point as P. □

Example 9.4.1 illustrates a feature of polar coordinates that does not occur in Cartesian coordinates: two different sets of coordinates can refer to the same point. For another example of this, note that the point Q in Figure 9.15 could also be represented by the polar coordinates $(2, 3\pi/2)$. In fact, each point in the plane has infinitely many sets of polar coordinates: the point (r, θ) is the same as $(r, \theta + 2\pi n)$ for any integer n. The origin can be represented by the coordinates $(0, \theta)$, for any value of θ.

It will be convenient to introduce one more convention. We have described Q as the point you reach by traveling 2 units down from the origin, but we could also say that it is the point you reach by traveling -2 units up from the origin. Since $\theta = \pi/2$ represents the up direction, we will also assign the polar coordinates $(-2, \pi/2)$ to Q. More generally, if r is positive, then we will say that the coordinates $(-r, \theta)$ represent the point obtained by determining the direction indicated by the angle θ, and then moving in the *opposite* direction a distance of r units, as illustrated in Figure 9.16. This means that the coordinates $(-r, \theta)$ and $(r, \theta + \pi)$ always represent the same point.

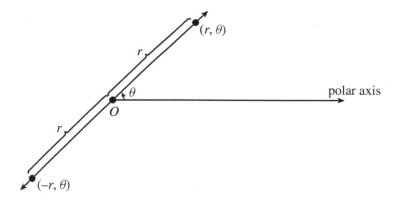

Figure 9.16: In polar coordinates, $(-r, \theta)$ represents the same point as $(r, \theta + \pi)$.

It is often useful to be able to convert between polar and Cartesian coordinates. When using both coordinate systems, we will set up our axes so that the origins of the two coordinate systems coincide and the polar axis lies along the positive x-axis. If $r > 0$, then the point P with polar coordinates (r, θ) will lie on the circle of radius r centered at the origin O, with the ray from O through P making an angle of θ measured counterclockwise from the positive x-axis. As we saw when we studied parametric equations for this circle in Section 9.1 (see Figure 9.2), the Cartesian coordinates of P will be

$$x = r \cos \theta,$$
$$y = r \sin \theta. \tag{9.20}$$

Notice that the point with polar coordinates $(-r, \theta)$ also has polar coordinates $(r, \theta + \pi)$, and therefore its Cartesian coordinates are

$$x = r \cos(\theta + \pi) = -r \cos \theta, \qquad y = r \sin(\theta + \pi) = -r \sin \theta.$$

It follows that equations (9.20) are correct when r is negative as well. And when $r = 0$, the point (r, θ) is the origin, and equations (9.20) give $x = y = 0$, which are the Cartesian coordinates of the origin. Thus, equations (9.20) can be used to convert any polar coordinates (r, θ) to Cartesian coordinates.

The conversion in the other direction is not quite so simple, since each point has multiple sets of polar coordinates. However, equations (9.20) imply that

$$x^2 + y^2 = r^2 \cos^2 \theta + r^2 \sin^2 \theta = r^2(\cos^2 \theta + \sin^2 \theta) = r^2, \tag{9.21}$$

and if $x \neq 0$ then

$$\frac{y}{x} = \frac{r \sin \theta}{r \cos \theta} = \tan \theta. \tag{9.22}$$

These equations can be useful for converting Cartesian coordinates to polar.

Example 9.4.2. Perform the following coordinate system conversions:

(a) The point P has polar coordinates $(4, \pi/3)$. What are its Cartesian coordinates?

(b) The point Q has Cartesian coordinates $(-3, 3)$. What are its polar coordinates?

Solution. We can solve part (a) by simply plugging the polar coordinates of P into equations (9.20):

$$x = r \cos\theta = 4\cos\left(\frac{\pi}{3}\right) = 4 \cdot \frac{1}{2} = 2,$$

$$y = r \sin\theta = 4\sin\left(\frac{\pi}{3}\right) = 4 \cdot \frac{\sqrt{3}}{2} = 2\sqrt{3}.$$

Thus, P has Cartesian coordinates $(2, 2\sqrt{3})$.

For part (b), we begin by applying equations (9.21) and (9.22) to see that

$$r^2 = x^2 + y^2 = 9 + 9 = 18, \qquad \tan\theta = \frac{y}{x} = \frac{3}{-3} = -1.$$

We conclude that

$$r = \pm\sqrt{18} = \pm 3\sqrt{2}, \qquad \theta = \dots, -\frac{\pi}{4}, \frac{3\pi}{4}, \frac{7\pi}{4}, \dots.$$

However, not every combination of these values of r and θ gives polar coordinates for Q. One way to see this is to convert each of these combinations back to Cartesian coordinates using equations (9.20), to see which ones represent Q. We leave it to you to verify that Q has the following polar coordinates:

$$\left(3\sqrt{2}, \frac{3\pi}{4} + 2\pi n\right), \qquad \left(-3\sqrt{2}, -\frac{\pi}{4} + 2\pi n\right), \qquad n \text{ any integer.}$$

The other combinations of these values of r and θ give polar coordinates for the point $(3, -3)$, and therefore don't represent Q. □

We often represent a curve in the plane as the graph of an equation involving x and y in Cartesian coordinates. But sometimes it is more convenient to represent a curve as the graph of an equation involving r and θ in polar coordinates. As usual, to draw the graph of such an equation, we plot all points whose coordinates make the equation true. But in polar coordinates, there is an extra complication: each point has multiple sets of polar coordinates. What do we do if one set of coordinates for a point makes the equation true, but another set of coordinates for the same point makes the equation false? We will follow the convention that such a point should be included in the graph; if *any* set of coordinates for a point makes an equation true, then that point is included in the graph of the equation.

For example, the graph of the equation $r = 3$ is the circle of radius 3 centered at the origin, as shown in Figure 9.17a. Any point on this circle could be written in polar coordinates as either $(3, \theta)$ or $(-3, \theta + \pi)$, for some value of θ. The first of these sets of coordinates satisfies the equation $r = 3$, and the second doesn't. But since there is at

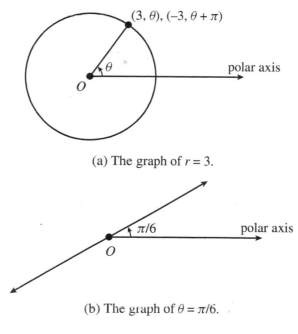

(a) The graph of $r = 3$.

(b) The graph of $\theta = \pi/6$.

Figure 9.17

least one set of coordinates that satisfies the equation, the point is included in the graph. Figure 9.17b shows the graph of the equation $\theta = \pi/6$, which is a straight line. The points on the line above and to the right of the origin have coordinates $(r, \pi/6)$ with $r > 0$, and those below and to the left of the origin have $r < 0$.

In Cartesian coordinates, we frequently graph equations of the form $y = f(x)$. Similarly, in polar coordinates we will often graph equations of the form $r = f(\theta)$. For example, consider the equation $r = 1 + \cos\theta$. As a first step in sketching the graph of this equation, we could plot some points. In Figure 9.18a, we have computed the values of r corresponding to several values of θ, and we have plotted the corresponding points in polar coordinates in Figure 9.18b. The graph of the equation is a curve passing through these points.

For each value of θ, the corresponding point (r, θ) on the graph lies on a ray from the origin making an angle of θ with the polar axis. You might think of this ray as representing a stick with one end attached to the origin, and the point (r, θ) as a bead on this stick that is r units from the origin. Several positions of the stick and bead are shown in Figure 9.18b.

To understand the shape of the curve between the points we have plotted, it might help to look at the graph of the equation $r = 1 + \cos\theta$ in *Cartesian* coordinates, but with the horizontal axis labeled θ and the vertical axis labeled r. This graph is shown in Figure 9.18c; it is simply the familiar graph of the cosine function, shifted up one unit. It is clear from this graph that as θ increases from 0 to π, r decreases from 2 down to 0. In our bead-and-stick metaphor, when θ increases from 0 to π, the stick starts out lying

θ	$r = 1 + \cos\theta$
0	2
$\pi/6$	$1 + \sqrt{3}/2 \approx 1.866$
$\pi/3$	1.5
$\pi/2$	1
$2\pi/3$	0.5
$5\pi/6$	$1 - \sqrt{3}/2 \approx 0.134$
π	0

(a) Table of values.

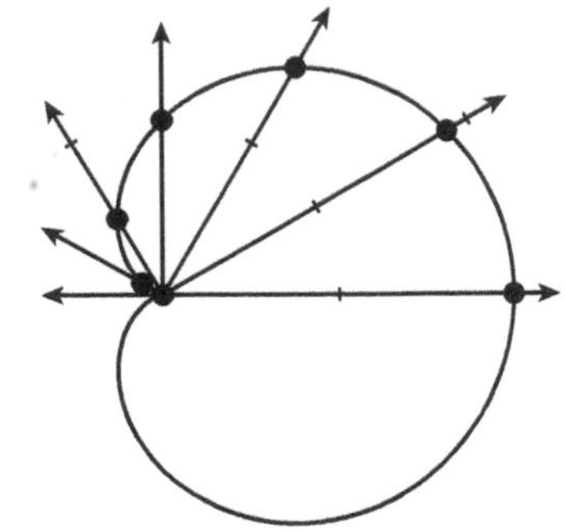

(b) Polar coordinates.

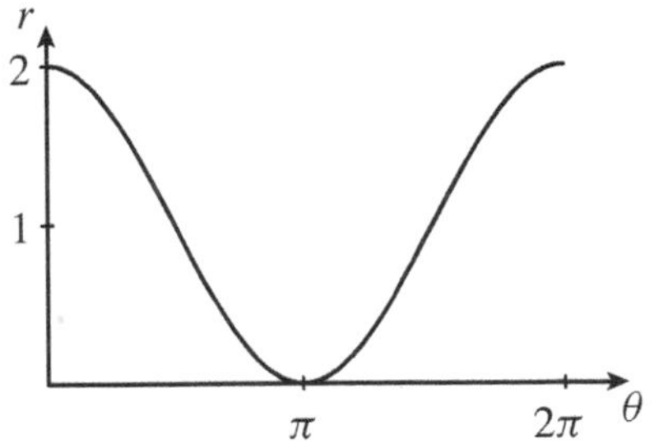

(c) Cartesian coordinates.

Figure 9.18: Graphing the equation $r = 1 + \cos\theta$.

along the polar axis and then rotates counterclockwise π radians. As the stick rotates, r decreases from 2 to 0, so the bead starts out 2 units from the origin and slides along the stick toward the origin. The top half of the graph is the path traced out by the bead, which spirals in toward the origin. As $\theta \to \pi^-$, $r \to 0^+$, and the curve approaches the origin from the left. As θ continues to increase from π to 2π, r increases from 0 back to 2, and the curve spirals out again, producing the bottom half of the graph. Notice that if $0 \le \theta \le \pi$ and $r = 1 + \cos\theta$, so that (r, θ) is a point on the top half of the graph, then since $\cos(2\pi - \theta) = \cos\theta$, the point $(r, 2\pi - \theta)$ is also on the graph. It follows that the bottom half of the graph is the mirror image of the top half, as seen in Figure 9.18b. This curve is called a *cardioid*, because it looks like a heart.

Example 9.4.3. Graph the equation $r = \sin(3\theta)$ in polar coordinates.

Solution. We begin by graphing $r = \sin(3\theta)$ in the Cartesian $r\theta$-plane; see Figure 9.19a. We see in this graph that as θ increases from 0 to $\pi/3$, r increases from 0 to 1 and then decreases back to 0. Thus, for $0 \le \theta \le \pi/3$ the graph in polar coordinates spirals out from the origin and then back in again. This generates the upper-right "leaf" in Figure 9.19b.

For $\pi/3 \le \theta \le 2\pi/3$, r decreases from 0 down to -1 and then increases again to 0. This generates another leaf, but since r is negative, this leaf points down rather than up. Finally, the upper-left leaf in the figure corresponds to $2\pi/3 \le \theta \le \pi$.

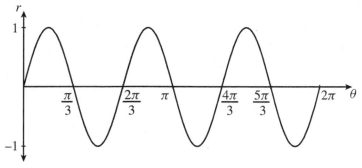

(a) Cartesian coordinates.

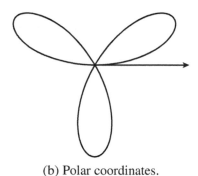

(b) Polar coordinates.

Figure 9.19: Graphing the equation $r = \sin(3\theta)$.

For values of θ in the range $\pi \leq \theta \leq 4\pi/3$, r is negative, and we trace the upper-right leaf again. In fact, all further values of θ repeat parts of the curve we have already graphed. To see why, let θ_0 be any angle and let $r_0 = \sin(3\theta_0)$, so that the point (r_0, θ_0) is on the graph. Now let (r_1, θ_1) be the point on the graph with $\theta_1 = \theta_0 + \pi$ and $r_1 = \sin(3\theta_1)$. Then

$$r_1 = \sin(3(\theta_0 + \pi)) = \sin(3\theta_0 + 3\pi) = -\sin(3\theta_0) = -r_0.$$

Thus, $(r_1, \theta_1) = (-r_0, \theta_0 + \pi)$, which represents the same point as (r_0, θ_0). The curve in this example is called a *three-leaf rose*. □

Sometimes it is possible to figure out the graph of an equation involving r and θ in polar coordinates by reexpressing the equation in terms of the Cartesian coordinates x and y, as illustrated in our next example.

Example 9.4.4. Graph the following equations in polar coordinates by finding equivalent equations in terms of the Cartesian coordinates x and y:

$$\text{(a) } r = \csc\theta, \qquad \text{(b) } r = \cos\theta.$$

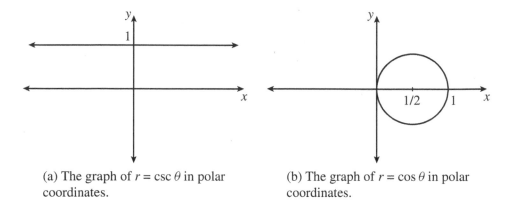

(a) The graph of $r = \csc\theta$ in polar coordinates.

(b) The graph of $r = \cos\theta$ in polar coordinates.

Figure 9.20

Solution. In part (a), the given equation means $r = 1/\sin\theta$, or in other words $r\sin\theta = 1$. But by the second equation in (9.20), this is equivalent to the Cartesian equation $y = 1$, whose graph is a horizontal line through 1 on the y-axis, as shown in Figure 9.20a.

Part (b) is harder to convert to Cartesian coordinates. We begin by multiplying both sides of the given equation by r to get the equation

$$r^2 = r\cos\theta. \tag{9.23}$$

We need to be careful here, because we have changed the meaning of the equation: there are solutions to equation (9.23) that are not solutions to the original equation in part (b). Specifically, if $r = 0$ then equation (9.23) will be true for all values of θ, but this is not the case for the equation in (b). But all of these solutions are sets of polar coordinates that represent the origin, and the origin is on the graph of the equation in (b), since that equation is true when $r = 0$ and $\theta = \pi/2$. Thus, although equation (9.23) has more *solutions* than equation (b), when we plot these solutions in polar coordinates, we don't get any more *points*.

It is easy now to reexpress equation (9.23) in terms of x and y. By equations (9.20) and (9.21), equation (9.23) is equivalent to

$$x^2 + y^2 = x.$$

Rewriting this equation in the form $(x - 1/2)^2 + y^2 = 1/4$, we see that its graph is the circle of radius $1/2$ centered at the point $(1/2, 0)$, which is shown in Figure 9.20b. □

We close this section by considering one more example, the equation $r = 1 + 2\cos^2\theta$. The graphs of this equation, in both Cartesian and polar coordinates, are shown in Figure 9.21. As θ increases from 0 to π, r decreases from 3 to 1 and then increases to 3 again, producing a dip at the top of the polar curve. It appears that there are inflection points on either side of this dip. Can we find them?

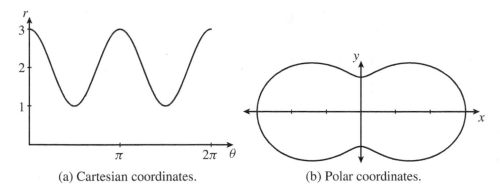

(a) Cartesian coordinates. (b) Polar coordinates.

Figure 9.21: Graphing the equation $r = 1 + 2\cos^2\theta$.

Equations (9.20) give us formulas for the x- and y-coordinates of any point on the polar graph in Figure 9.21b:

$$x = r\cos\theta = (1 + 2\cos^2\theta)\cos\theta = \cos\theta + 2\cos^3\theta,$$
$$y = r\sin\theta = (1 + 2(1 - \sin^2\theta))\sin\theta = (3 - 2\sin^2\theta)\sin\theta = 3\sin\theta - 2\sin^3\theta. \tag{9.24}$$

These are parametric equations: they express x and y in terms of the parameter θ. As θ runs from 0 to π, the top half of the curve is traversed from right to left, and as θ runs from π to 2π, the bottom half is traversed from left to right. We can now use the methods of Section 9.1 to give a detailed analysis of the shape of the curve. By symmetry, we may as well focus on just the top half of the curve.

We begin by computing

$$\frac{dx}{d\theta} = -\sin\theta - 6\cos^2\theta\sin\theta = -\sin\theta - 6(1 - \sin^2\theta)\sin\theta = -7\sin\theta + 6\sin^3\theta,$$
$$\frac{dy}{d\theta} = 3\cos\theta - 6\sin^2\theta\cos\theta = 3\cos\theta - 6(1 - \cos^2\theta)\cos\theta = -3\cos\theta + 6\cos^3\theta.$$

Thus,

$$\frac{dy}{dx} = \frac{dy/d\theta}{dx/d\theta} = \frac{-3\cos\theta + 6\cos^3\theta}{-7\sin\theta + 6\sin^3\theta} = \frac{3\cos\theta(1 - 2\cos^2\theta)}{\sin\theta(7 - 6\sin^2\theta)}.$$

Setting $dy/dx = 0$, we find that the top half of the curve has horizontal tangent lines when $\cos\theta = 0$ or $\cos\theta = \pm\sqrt{2}/2$. In the range $0 \le \theta \le \pi$, this happens when $\theta = \pi/4, \pi/2, 3\pi/4$, and according to the parametric equations (9.24), the x-coordinates of these points are $x = \sqrt{2}, 0, -\sqrt{2}$. The sign of dy/dx is shown in the number line below.

$\dfrac{dy}{dx}$		+	0	−	0	+	0	−
θ	π		$3\pi/4$		$\pi/2$		$\pi/4$	0
x	-3		$-\sqrt{2}$		0		$\sqrt{2}$	3

Thus we see that the top half of the curve has local maximum points at $x = \pm\sqrt{2}$ and a local minimum at $x = 0$. These points are easily visible in Figure 9.21b.

Next we compute d^2y/dx^2. Setting $y' = dy/dx$, we find (after some extensive algebraic calculations that we leave to you) that

$$\frac{d^2y}{dx^2} = \frac{dy'}{dx} = \frac{dy'/d\theta}{dx/d\theta} = \frac{3(4\cos^4\theta - 12\cos^2\theta + 1)}{\sin^3\theta(7 - 6\sin^2\theta)^3}.$$

The numerator here is a quadratic in $\cos^2\theta$, and the quadratic formula tells us that d^2y/dx^2 will be 0 when $\cos^2\theta = (3\pm2\sqrt{2})/2$. But $\cos^2\theta$ must be between 0 and 1, so the only possible solution is

$$\cos^2\theta = \frac{3 - 2\sqrt{2}}{2} = \left(\frac{2-\sqrt{2}}{2}\right)^2.$$

Thus, $d^2y/dx^2 = 0$ when $\cos\theta = \pm(2 - \sqrt{2})/2$, which happens when $\theta \approx 1.274$, 1.868. The corresponding values of x are $x = \pm(6 - 4\sqrt{2}) \approx \pm0.343$. These are the x-coordinates of the inflection points that we noticed in Figure 9.21b. The concavity can now be read off from the following number line:

$\dfrac{d^2y}{dx^2}$		$-$		0	$+$	0		$-$	
θ	π			1.868		1.274			0
x	-3			-0.343		0.343			3

Generalizing from our last example, we see that for any function f, the polar curve $r = f(\theta)$ can be converted to parametric form by using the equations

$$x = r\cos\theta = f(\theta)\cos\theta, \qquad y = r\sin\theta = f(\theta)\sin\theta.$$

We can then use the methods developed earlier in this chapter to study the curve. For more on this, see Exercises 21–23.

Exercises 9.4

1. Plot the following points in polar coordinates:

    (a) $(2, \pi)$.   (b) $(1, 3\pi/4)$.   (c) $(-1, 4\pi/3)$.

2. Convert the following polar coordinates to Cartesian coordinates:

    (a) $(3, \pi/2)$.   (b) $(-2, -\pi/3)$.   (c) $(4, \pi/4)$.

3. Convert the following Cartesian coordinates to polar coordinates:

    (a) $(-3, 0)$.   (b) $(-2, -2)$.   (c) $(\sqrt{3}, -1)$.

4. Find the distance between the points with polar coordinates $(4, \pi/6)$ and $(3, 3\pi/4)$.

5–9: Graph the equation in polar coordinates by first finding an equivalent equation in terms of the Cartesian coordinates x and y.

5. $r = 2 \sec \theta$.

6. $r = \sec(\theta + \pi/4)$.

7. $r = 4 \sin \theta$.

8. $r = \sin \theta - \cos \theta$.

9. $r = \sec \theta \tan \theta$.

10–17: Graph the equation in polar coordinates.

10. $r = \sin(2\theta), 0 \le \theta \le 2\pi$.

11. $r = 1 + 2 \sin \theta, 0 \le \theta \le 2\pi$.

12. $r = 2 + \sin \theta, 0 \le \theta \le 2\pi$.

13. $r = 1 + 2 \cos(3\theta), 0 \le \theta \le 2\pi$.

14. $r = \dfrac{1}{2 + \cos \theta}, 0 \le \theta \le 2\pi$.

15. $r = \dfrac{1}{1 + 2 \cos \theta}, -2\pi/3 < \theta < 2\pi/3$.

16. $r = \dfrac{1}{1 + \cos \theta}, -\pi < \theta < \pi$.

17. $r = 1 + \theta, 0 \le \theta \le 4\pi$.

18. Show that if P is any point on the curve in Exercise 14, then the Cartesian coordinates of P satisfy the equation

$$\frac{(x + 1/3)^2}{4/9} + \frac{y^2}{1/3} = 1.$$

What is the shape of the curve?

19. Show that if P is any point on the curve in Exercise 15, then the Cartesian coordinates of P satisfy the equation

$$\frac{(x - 2/3)^2}{1/9} - \frac{y^2}{1/3} = 1.$$

What is the shape of the curve?

20. Find an equation in Cartesian coordinates for the curve in Exercise 16. What is the shape of the curve? (Hint: Imitate Exercises 18 and 19.)

21. By rewriting the polar coordinates curve $r = f(\theta)$ in the parametric form $x = f(\theta)\cos\theta$, $y = f(\theta)\sin\theta$, derive the following formulas for dy/dx and d^2y/dx^2:

$$\frac{dy}{dx} = \frac{f'(\theta)\sin\theta + f(\theta)\cos\theta}{f'(\theta)\cos\theta - f(\theta)\sin\theta},$$

$$\frac{d^2y}{dx^2} = \frac{(f(\theta))^2 + 2(f'(\theta))^2 - f(\theta)f''(\theta)}{(f'(\theta)\cos\theta - f(\theta)\sin\theta)^3}.$$

22. Let C be the curve $r = 10 + \sec\theta$, $0 \le \theta < \pi/2$.

   (a) Find formulas for the x- and y-coordinates of any point on C in terms of θ.

   (b) Show that C is the graph of $y = h(x)$, for some function h. What is the domain of h? (You don't need to find a formula for $h(x)$.)

   (c) Show that h is strictly decreasing on its domain.

   (d) Show that C has an inflection point at $\theta = \pi/3$. What are the x- and y-coordinates of the inflection point? What is the concavity on either side of the inflection point?

   (e) Show that as $\theta \to \pi/2^-$, $x \to 1^+$ and $y \to \infty$. What does this tell you about C?

   (f) Sketch C.

23. (a) Let C be the polar coordinates curve $r = f(\theta)$, $\alpha \le \theta \le \beta$, where f' is continuous on $[\alpha, \beta]$. By rewriting the curve in the parametric form $x = f(\theta)\cos\theta$, $y = f(\theta)\sin\theta$, show that the length of C is

$$\int_\alpha^\beta \sqrt{(f(\theta))^2 + (f'(\theta))^2}\,d\theta = \int_\alpha^\beta \sqrt{r^2 + \left(\frac{dr}{d\theta}\right)^2}\,d\theta.$$

   (b) Find the length of the polar coordinates curve $r = \theta^2$, $0 \le \theta \le 2\pi$.

9.5   Areas in Polar Coordinates

One of the most important applications of integration is the calculation of areas. If f is continuous on an interval $[a, b]$ and $f(x) \ge 0$ for all $x \in [a, b]$, then as we saw in Chapter 5, $\int_a^b f(x)\,dx$ is the area of the region under the curve $y = f(x)$ for $a \le x \le b$. This region consists of all points (x, y) satisfying the inequalities $a \le x \le b$, $0 \le y \le f(x)$; it is shaded blue in Figure 9.22.

In this section, we study the analogous problem for polar coordinates. Consider the curve given in polar coordinates by the equation $r = f(\theta)$ for $\alpha \le \theta \le \beta$. We will assume that f is continuous on $[\alpha, \beta]$ and $f(\theta) \ge 0$ for all $\theta \in [\alpha, \beta]$, and we wish to compute the area of the region determined by the inequalities $\alpha \le \theta \le \beta$, $0 \le r \le f(\theta)$. We might describe this as the region inside the curve $r = f(\theta)$ for $\alpha \le \theta \le \beta$; it is shaded

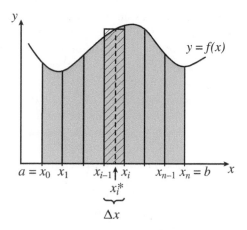

Figure 9.22: Computing areas in Cartesian coordinates: The lines $x = x_i$ divide the region under the curve $y = f(x)$ into vertical strips. We approximate the ith strip with the region $x_{i-1} \le x \le x_i, 0 \le y \le f(x_i^*)$, which is a rectangle. The area of this rectangle is $f(x_i^*)\Delta x$.

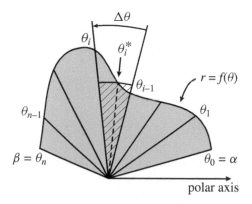

Figure 9.23: Computing areas in polar coordinates: The lines $\theta = \theta_i$ divide the region inside the curve $r = f(\theta)$ into wedges. We approximate the ith wedge with the region $\theta_{i-1} \le \theta \le \theta_i, 0 \le r \le f(\theta_i^*)$, which is a sector of a circle. The area of this sector is $(1/2)(f(\theta_i^*))^2 \Delta \theta$.

blue in Figure 9.23. We will assume that $\alpha < \beta \le \alpha + 2\pi$, so that the region does not wrap all the way around the origin and overlap with itself.

Recall that to compute the area in Cartesian coordinates under the curve $y = f(x)$, $a \le x \le b$, we begin by dividing the interval $[a, b]$ into n pieces, each of width $\Delta x = (b-a)/n$, with division points $x_0, x_1, \ldots, x_n$. The lines $x = x_i$ divide the region under the curve into vertical strips, and the area under the curve is the sum of the areas of these strips. For each i, we choose a sample point $x_i^* \in [x_{i-1}, x_i]$ and we approximate the ith strip with the rectangle $x_{i-1} \leq x \leq x_i$, $0 \leq y \leq f(x_i^*)$. You could think of this rectangle as what the ith strip would be if the function f were constant on the interval $[x_{i-1}, x_i]$; it is the striped rectangle in Figure 9.22. The area of this rectangle is $f(x_i^*)\Delta x$, so the area under the curve is approximated by the Riemann sum $\sum_{i=1}^{n} f(x_i^*)\Delta x$. The exact area is the limit of this Riemann sum as $n \to \infty$, which is $\int_a^b f(x)\, dx$.

To find the area in polar coordinates inside the curve $r = f(\theta)$, $\alpha \leq \theta \leq \beta$, we imitate this procedure. Thus, we begin by dividing the interval $[\alpha, \beta]$ into n pieces, each of width $\Delta\theta = (\beta - \alpha)/n$. The division points are $\theta_i = \alpha + i\Delta\theta$ for $0 \leq i \leq n$. The lines $\theta = \theta_i$ divide the region inside the curve into wedges, as illustrated in Figure 9.23. For each i, we choose a sample point $\theta_i^* \in [\theta_{i-1}, \theta_i]$, and we approximate the ith wedge with the region $\theta_{i-1} \leq \theta \leq \theta_i$, $0 \leq r \leq f(\theta_i^*)$. This region, which is striped in Figure 9.23, is a sector of a circle. This sector is what the ith wedge would be if f were constant on the interval $[\theta_{i-1}, \theta_i]$. To continue with our area calculation, we need to find the area of this sector.

We can think of the sector as being cut from a circle of radius $f(\theta_i^*)$. The area of this circle is $\pi(f(\theta_i^*))^2$. The angle of the sector is $\Delta\theta$, so the fraction of the circle represented by the sector is $\Delta\theta/(2\pi)$. Thus, the area of the sector is

$$\frac{\Delta\theta}{2\pi} \cdot \pi(f(\theta_i^*))^2 = \frac{1}{2}(f(\theta_i^*))^2 \Delta\theta.$$

This is our approximation of the area of the ith wedge.

Adding these approximations, we see that the area A inside the curve $r = f(\theta)$, $\alpha \leq \theta \leq \beta$, is approximately $\sum_{i=1}^{n}(1/2)(f(\theta_i^*))^2 \Delta\theta$. This approximation improves as n increases, so as usual we find the exact answer as a limit:

$$A = \lim_{n \to \infty} \sum_{i=1}^{n} \frac{1}{2}(f(\theta_i^*))^2 \Delta\theta = \int_\alpha^\beta \frac{1}{2}(f(\theta))^2\, d\theta.$$

Substituting r for $f(\theta)$, we can say that the area inside the curve $r = f(\theta)$, $\alpha \leq \theta \leq \beta$, is

$$A = \int_\alpha^\beta \frac{1}{2} r^2\, d\theta. \tag{9.25}$$

Example 9.5.1. Find the areas of the following regions:

    (a) The region inside the cardioid $r = 1 + \cos\theta$ (see Figure 9.18b).
    (b) The region inside one leaf of the three-leaf rose $r = \sin(3\theta)$ (see Figure 9.19b).
    (c) The region that is inside one leaf of the three-leaf rose $r = \sin(3\theta)$ and outside the circle $r = \sqrt{3}/2$.

Solution. To get the entire interior of the cardioid in part (a), we use the range of θ values $0 \leq \theta \leq 2\pi$. Thus, by equation (9.25), the area inside the cardioid is

$$A = \int_0^{2\pi} \frac{1}{2}(1 + \cos\theta)^2\, d\theta = \int_0^{2\pi} \left(\frac{1}{2} + \cos\theta + \frac{1}{2}\cos^2\theta\right) d\theta.$$

To evaluate this integral, we use the identity $\cos^2\theta = (1/2)(1+\cos(2\theta))$:

$$A = \int_0^{2\pi} \left(\frac{1}{2} + \cos\theta + \frac{1}{4}(1+\cos(2\theta)) \right) d\theta = \int_0^{2\pi} \left(\frac{3}{4} + \cos\theta + \frac{1}{4}\cos(2\theta) \right) d\theta$$

$$= \left[\frac{3}{4}\theta + \sin\theta + \frac{1}{8}\sin(2\theta) \right]_0^{2\pi} = \frac{3\pi}{2}.$$

For part (b), recall from Example 9.4.3 that the upper-right leaf of the three-leaf rose corresponds to values of θ in the range $0 \le \theta \le \pi/3$. Thus, the area of this leaf is

$$\int_0^{\pi/3} \frac{1}{2}\sin^2(3\theta)\, d\theta = \int_0^{\pi/3} \frac{1}{4}(1-\cos(6\theta))\, d\theta = \left[\frac{1}{4}\theta - \frac{1}{24}\sin(6\theta) \right]_0^{\pi/3} = \frac{\pi}{12}.$$

Finally, for part (c) we must first find where the circle $r = \sqrt{3}/2$ intersects the three-leaf rose $r = \sin(3\theta)$, so we set $\sin(3\theta) = \sqrt{3}/2$. To find the intersection points on the upper-right leaf we consider only $0 \le \theta \le \pi/3$, which means $0 \le 3\theta \le \pi$. For these values of θ, the solutions are $3\theta = \pi/3, 2\pi/3$, or in other words $\theta = \pi/9, 2\pi/9$, as shown in Figure 9.24.

The area we want to compute is the area of the region in Figure 9.24 that is shaded blue but not striped. To find this area, we will subtract the area of the striped region from the area of the blue region. Applying equation (9.25) to both $r = \sin(3\theta)$ and $r = \sqrt{3}/2$, we see that the desired area is

$$\int_{\pi/9}^{2\pi/9} \frac{1}{2}\sin^2(3\theta)\, d\theta - \int_{\pi/9}^{2\pi/9} \frac{1}{2}\left(\frac{\sqrt{3}}{2} \right)^2 d\theta$$

$$= \int_{\pi/9}^{2\pi/9} \left(\frac{1}{4}(1-\cos(6\theta)) - \frac{3}{8} \right) d\theta = \int_{\pi/9}^{2\pi/9} \left(-\frac{1}{8} - \frac{1}{4}\cos(6\theta) \right) d\theta$$

$$= \left[-\frac{1}{8}\theta - \frac{1}{24}\sin(6\theta) \right]_{\pi/9}^{2\pi/9} = \frac{3\sqrt{3} - \pi}{72}. \qquad \square$$

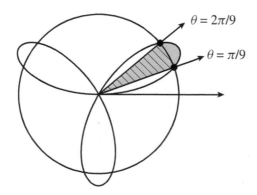

$\theta = 2\pi/9$

$\theta = \pi/9$

Figure 9.24: Part (c) of Example 9.5.1 asks for the area of the region that is shaded blue but not striped.

Exercises 9.5

1–12: Find the area of the region.

1. The region inside the curve $r = \theta(\pi - \theta)$, $0 \le \theta \le \pi$.

2. The region bounded by the curve $r = \theta + \sin\theta$, $0 \le \theta \le \pi$, and the line $\theta = \pi$.

3. The region bounded by the curve $r = \tan\theta$, $0 \le \theta \le \pi/4$, and the line $\theta = \pi/4$.

4. The region inside one leaf of the four-leaf rose $r = \sin(2\theta)$.

5. The region inside the curve $r = \sqrt{\sin(2\theta)}$, $0 \le \theta \le \pi/2$.

6. The region inside the curve $r = 2 + \cos(4\theta)$.

7. The region inside the curve $r = 1 + 2\cos^2\theta$ (see Figure 9.21b).

8. The region inside the cardioid $r = 1 + \cos\theta$ and outside the circle $r = 3/2$.

9. The region inside the cardioid $r = 1 + \sin\theta$ and outside the circle $r = 3\sin\theta$.

10. The region inside both of the circles $r = \cos\theta$ and $r = 1/2$.

11. The region inside both of the circles $r = \cos\theta$ and $r = \sqrt{3}\sin\theta$.

12. The region bounded by the lines $r = \csc\theta$, $\theta = \pi/4$, and $\theta = 2\pi/3$.

Chapter 10

Infinite Series and Power Series

10.1 Infinite Series

As we observed in Section 2.8, the meaning of decimal notation suggests that an infinite decimal expansion represents a sum of infinitely many numbers. For example, you are probably familiar with the fact that $1/3 = 0.333\ldots$, where the digit 3 is repeated forever. The first 3 after the decimal point stands for 3/10, the next represents 3/100, and so on. Thus, this decimal expansion says that

$$\frac{1}{3} = \frac{3}{10} + \frac{3}{100} + \frac{3}{1000} + \cdots .$$

But the right-hand side of this equation is a sum of infinitely many numbers, and we could never actually add up infinitely many numbers. So what does this sum mean?

In this chapter, we will study such infinite sums. To express these sums, we introduce the notation

$$\sum_{n=1}^{\infty} a_n = a_1 + a_2 + a_3 + \cdots .$$

A sum of this kind is called an *infinite series*. The symbol "∞" above the summation sign means that the terms in the series go on forever, and therefore there are infinitely many numbers to be added up. But we still need to explain what it means to write a sum of infinitely many numbers.

Our problem here is similar to the situation we faced when dealing with improper integrals of the form $\int_a^\infty f(x)\,dx$. Recall that when we studied improper integrals, we defined $\int_a^\infty f(x)\,dx$ to be $\lim_{t\to\infty}\int_a^t f(x)\,dx$. We use a similar approach for infinite series:

Definition 10.1.1. We define the value of an infinite series $\sum_{n=1}^\infty a_n$ as follows:

$$\sum_{n=1}^\infty a_n = \lim_{N\to\infty}\sum_{n=1}^N a_n = \lim_{N\to\infty}(a_1 + a_2 + \cdots + a_N).$$

If the limit above is defined, then we say that the infinite series *converges*, or that it is *convergent*. Otherwise, it *diverges*, or it is *divergent*.

Thus, to evaluate an infinite series we don't actually add up infinitely many numbers. Rather, we get approximate answers by adding up the first N terms of the series, for large integers N. As N increases, we add more and more terms, and the approximations improve. As usual in calculus, we define the exact answer to be a limit of approximations.

The quantity $\sum_{n=1}^N a_n$ is called the *Nth partial sum* of the infinite series $\sum_{n=1}^\infty a_n$. We will often denote the Nth partial sum by s_N; in other words, we write

$$s_N = \sum_{n=1}^N a_n = a_1 + a_2 + \cdots + a_N.$$

The sequence $(s_N)_{N=1}^\infty$ is called the *sequence of partial sums* for the infinite series. Definition 10.1.1 says that the value of an infinite series is the limit of its sequence of partial sums:

$$\sum_{n=1}^\infty a_n = \lim_{N\to\infty}\sum_{n=1}^N a_n = \lim_{N\to\infty} s_N.$$

For example, consider the infinite series

$$\sum_{n=1}^\infty \frac{1}{2^n} = \frac{1}{2} + \frac{1}{4} + \frac{1}{8} + \cdots. \tag{10.1}$$

The first few partial sums are

$$s_1 = \sum_{n=1}^1 \frac{1}{2^n} = \frac{1}{2},$$

$$s_2 = \sum_{n=1}^2 \frac{1}{2^n} = \frac{1}{2} + \frac{1}{4} = \frac{3}{4},$$

$$s_3 = \sum_{n=1}^3 \frac{1}{2^n} = \frac{1}{2} + \frac{1}{4} + \frac{1}{8} = \frac{7}{8}.$$

Based on these values, you might guess that in general, $s_N = (2^N - 1)/2^N = 1 - 1/2^N$. In fact, this formula is correct; we ask you to verify it by mathematical induction in Exercise 20. Using this formula for s_N, we can evaluate the infinite series:

$$\sum_{n=1}^{\infty} \frac{1}{2^n} = \lim_{N \to \infty} s_N = \lim_{N \to \infty} \left(1 - \frac{1}{2^N}\right) = 1.$$

Example 10.1.2. Evaluate the following infinite series:

$$\text{(a) } \sum_{n=1}^{\infty} (-1)^{n-1}, \quad \text{(b) } \sum_{n=1}^{\infty} \frac{1}{4n^2 - 1}.$$

Solution. For series (a), we begin by listing the first few terms of the series:

$$\sum_{n=1}^{\infty} (-1)^{n-1} = 1 + (-1) + 1 + (-1) + \cdots.$$

The first few partial sums are:

$$s_1 = 1,$$
$$s_2 = 1 + (-1) = 0,$$
$$s_3 = 1 + (-1) + 1 = 1,$$
$$s_4 = 1 + (-1) + 1 + (-1) = 0.$$

It should be clear at this point that the general pattern for all partial sums is

$$s_N = \begin{cases} 1, & \text{if } N \text{ is odd,} \\ 0, & \text{if } N \text{ is even.} \end{cases}$$

These partial sums do not approach a limiting value; rather, they continue to bounce back and forth between 1 and 0 forever. Thus, $\lim_{N \to \infty} s_N$ is undefined, and the series in (a) diverges.

Definition 10.1.1 was motivated by an analogy between series and integrals. To compute series (b), we continue this analogy by borrowing an idea from our study of integration: partial fractions. Factoring the denominator in series (b) leads us to set up the following partial fractions decomposition:

$$\frac{1}{4n^2 - 1} = \frac{1}{(2n - 1)(2n + 1)} = \frac{A}{2n - 1} + \frac{B}{2n + 1}.$$

We leave it to you to solve for A and B in order to derive the equation

$$\frac{1}{4n^2 - 1} = \frac{1/2}{2n - 1} - \frac{1/2}{2n + 1} = \frac{1}{4n - 2} - \frac{1}{4n + 2}.$$

Substituting into our series, we see that the Nth partial sum of the series is

$$s_N = \sum_{n=1}^{N} \frac{1}{4n^2 - 1} = \sum_{n=1}^{N} \left(\frac{1}{4n - 2} - \frac{1}{4n + 2} \right)$$
$$= \left(\frac{1}{2} - \frac{1}{6} \right) + \left(\frac{1}{6} - \frac{1}{10} \right) + \left(\frac{1}{10} - \frac{1}{14} \right) + \cdots + \left(\frac{1}{4N - 2} - \frac{1}{4N + 2} \right).$$

Notice that in the sum on the right-hand side above, the $-1/6$ in the first term cancels with the $1/6$ in the second term, the $-1/10$ in the second term cancels with the $1/10$ in the third term, and so on. The only numbers that *don't* cancel are the $1/2$ in the first term and the $-1/(4N + 2)$ in the last term. Thus, we have

$$s_N = \frac{1}{2} - \frac{1}{4N + 2}.$$

Sums that can be evaluated by recognizing patterns of cancellation like the one in this example are called *telescoping sums*. (You may recall that we saw an example of a telescoping sum in Section 5.1; see equation (5.5).)

Now that we have a formula for s_N, we can evaluate series (b):

$$\sum_{n=1}^{\infty} \frac{1}{4n^2 - 1} = \lim_{N \to \infty} s_N = \lim_{N \to \infty} \left(\frac{1}{2} - \frac{1}{4N + 2} \right) = \frac{1}{2}. \qquad \square$$

In the rest of this chapter, we will often find it useful to consider series of the form

$$\sum_{n=1}^{\infty} ar^{n-1} = a + ar + ar^2 + ar^3 + \cdots. \tag{10.2}$$

A series of this form is called a *geometric* series. In such a series, each term after the first is r times the term before it. The number a is the first term in the series.

In fact, we have already seen three examples of geometric series in this section:

1. If $a = 3/10$ and $r = 1/10$, then series (10.2) takes the form

$$\sum_{n=1}^{\infty} \frac{3}{10} \cdot \left(\frac{1}{10} \right)^{n-1} = \sum_{n=1}^{\infty} \frac{3}{10^n} = \frac{3}{10} + \frac{3}{100} + \frac{3}{1000} + \cdots.$$

This is the example with which we started this section.

2. If $a = r = 1/2$, then the geometric series becomes

$$\sum_{n=1}^{\infty} \frac{1}{2} \cdot \left(\frac{1}{2} \right)^{n-1} = \sum_{n=1}^{\infty} \frac{1}{2^n} = \frac{1}{2} + \frac{1}{4} + \frac{1}{8} + \cdots.$$

This is the next example we considered, the series in (10.1).

3. If $a = 1$ and $r = -1$, then we get

$$\sum_{n=1}^{\infty} 1 \cdot (-1)^{n-1} = \sum_{n=1}^{\infty} (-1)^{n-1} = 1 + (-1) + 1 + (-1) + \cdots.$$

This is series (a) from Example 10.1.2.

To find the value of the geometric series (10.2), we need a formula for the Nth partial sum. There is a clever trick that leads to such a formula. First, we write out the terms in the Nth partial sum, s_N. Next, we multiply this sum by $-r$ and write this new sum beneath the first, shifting the terms over so that corresponding terms in the two sums match up. Finally, we add the two equations and note that most terms cancel out:

$$
\begin{aligned}
s_N &= a + ar + ar^2 + \cdots + ar^{N-1} \\
+ \quad -rs_N &= \quad\quad - ar - ar^2 - \cdots - ar^{N-1} - ar^N \\
\hline
s_N - rs_N &= a \qquad\qquad\qquad\qquad\qquad\qquad\qquad - ar^N.
\end{aligned}
$$

We conclude that $s_N - rs_N = a - ar^N$, or in other words $s_N(1 - r) = a(1 - r^N)$. As long as $r \neq 1$, we can divide by $1 - r$ to get the formula

$$s_N = \frac{a(1 - r^N)}{1 - r}. \tag{10.3}$$

Using this formula, we can now find the value of any geometric series.

Theorem 10.1.3. *Suppose $a \neq 0$.*

1. *If $-1 < r < 1$ then the geometric series $\sum_{n=1}^{\infty} ar^{n-1}$ converges, and*

$$\sum_{n=1}^{\infty} ar^{n-1} = \frac{a}{1 - r}.$$

2. *If $r \leq -1$ or $r \geq 1$ then the geometric series $\sum_{n=1}^{\infty} ar^{n-1}$ diverges.*

Proof. For part 1, suppose that $-1 < r < 1$. We plan to use equation (10.3) to evaluate the geometric series, so we need to find $\lim_{N \to \infty} r^N$. If $r \neq 0$, then $0 < |r| < 1$, and by part 4 of Theorem 7.4.9, $\lim_{N \to \infty} |r^N| = \lim_{N \to \infty} |r|^N = 0$. It follows, by Theorem 2.4.11, that $\lim_{N \to \infty} r^N = 0$. Of course, if $r = 0$ then $\lim_{N \to \infty} r^N = 0$ as well. Thus, by equation (10.3),

$$\sum_{n=1}^{\infty} ar^{n-1} = \lim_{N \to \infty} s_N = \lim_{N \to \infty} \frac{a(1 - r^N)}{1 - r} = \frac{a}{1 - r}.$$

Now we turn to the proof of part 2. If $r > 1$, then by part 3 of Theorem 7.4.9, $\lim_{N \to \infty} r^N = \infty$. Plugging into equation (10.3), we conclude that

$$\lim_{N \to \infty} s_N = \lim_{N \to \infty} \frac{a(1 - r^N)}{1 - r} = \pm\infty,$$

with the sign depending on the sign of a. If $r = 1$, then equation (10.3) doesn't apply, but fortunately it is easy to find the value of s_N:

$$s_N = \sum_{n=1}^{N} a = \underbrace{a + a + \cdots + a}_{N \text{ times}} = Na.$$

We conclude that, once again, $\lim_{N \to \infty} s_N = \pm\infty$, with the sign depending on the sign of a.

Next we consider the case $r = -1$. In this case,

$$s_N = \frac{a(1 - (-1)^N)}{1 - (-1)} = \begin{cases} a, & \text{if } N \text{ is odd,} \\ 0, & \text{if } N \text{ is even.} \end{cases}$$

Since we have assumed $a \neq 0$, this means that s_N is bouncing back and forth between two different values, and as in part (a) of Example 10.1.2, $\lim_{N \to \infty} s_N$ is undefined and the series diverges.

Finally, suppose that $r < -1$. Then $r = -p$ for some number $p > 1$, and

$$s_N = \frac{a(1 - (-p)^N)}{1 - (-p)} = \frac{a(1 - (-1)^N p^N)}{1 + p}.$$

As before, we have $\lim_{N \to \infty} p^N = \infty$. This means that if N is a large integer, then the factor $1 - (-1)^N p^N$ will be a large positive number if N is odd, and a large magnitude negative number if N is even. We conclude that as $N \to \infty$, the partial sums s_N bounce back and forth between large magnitude positive and negative numbers, without approaching a limit. Thus, the geometric series diverges.   □

We can use Theorem 10.1.3 to evaluate the three geometric series that we saw earlier in this section. If $a = 3/10$ and $r = 1/10$, then by part 1 of the theorem,

$$\sum_{n=1}^{\infty} \frac{3}{10^n} = \frac{3}{10} + \frac{3}{100} + \frac{3}{1000} + \cdots = \frac{a}{1 - r} = \frac{3/10}{1 - 1/10} = \frac{1}{3}.$$

Thus, the value of the decimal expansion $0.333\ldots$ is $1/3$, as expected. In the case $a = r = 1/2$, we get

$$\sum_{n=1}^{\infty} \frac{1}{2^n} = \frac{1}{2} + \frac{1}{4} + \frac{1}{8} + \cdots = \frac{a}{1 - r} = \frac{1/2}{1 - 1/2} = 1,$$

in agreement with our evaluation of series (10.1). And when $a = 1$ and $r = -1$, part 2 of Theorem 10.1.3 tells us that the series $\sum_{n=1}^{\infty} (-1)^{n-1}$ diverges, as we found in part (a) of Example 10.1.2.

So far we have written all of our infinite series with the terms numbered by the variable n, and with the value of n starting at 1 and running through all the positive integers. But as with finite sums, the terms of an infinite series can be numbered by a different variable, and the value of the variable can start at a number other than 1. We extend Definition 10.1.1 to such series in the obvious way. For example, the geometric series (10.2) could also be written like this:

$$\sum_{i=0}^{\infty} ar^i = a + ar + ar^2 + ar^3 + \cdots . \tag{10.4}$$

Since the list of numbers added up in (10.4) is exactly the same as in (10.2), we have $\sum_{i=0}^{\infty} ar^i = \sum_{n=1}^{\infty} ar^{n-1}$. More precisely,

$$\sum_{i=0}^{\infty} ar^i = \lim_{N \to \infty} \sum_{i=0}^{N} ar^i = \lim_{N \to \infty} (a + ar + \cdots + ar^N)$$

$$= \lim_{N \to \infty} \sum_{n=1}^{N+1} ar^{n-1} = \sum_{n=1}^{\infty} ar^{n-1}.$$

Many facts about finite sums carry over to infinite series. For example, the first two parts of the following theorem are based on Theorem 5.1.1.

Theorem 10.1.4. *Let $\sum_{n=1}^{\infty} a_n$ and $\sum_{n=1}^{\infty} b_n$ be any two infinite series.*

1. *If $\sum_{n=1}^{\infty} a_n$ and $\sum_{n=1}^{\infty} b_n$ both converge, then so does $\sum_{n=1}^{\infty} (a_n + b_n)$, and*

$$\sum_{n=1}^{\infty} (a_n + b_n) = \sum_{n=1}^{\infty} a_n + \sum_{n=1}^{\infty} b_n.$$

2. *For any number $c \neq 0$, the series $\sum_{n=1}^{\infty} a_n$ and $\sum_{n=1}^{\infty} ca_n$ either both converge or both diverge. If they converge, then*

$$\sum_{n=1}^{\infty} ca_n = c \sum_{n=1}^{\infty} a_n.$$

3. *For any integer $k \geq 2$, the series $\sum_{n=1}^{\infty} a_n$ and $\sum_{n=k}^{\infty} a_n$ either both converge or both diverge. If they converge, then*

$$\sum_{n=1}^{\infty} a_n = \sum_{n=1}^{k-1} a_n + \sum_{n=k}^{\infty} a_n.$$

Proof. All three parts of the theorem can be proven by simply applying the definition of the value of an infinite series, Definition 10.1.1, together with facts about finite sums and limits. For example, to prove part 1, suppose that $\sum_{n=1}^{\infty} a_n$ and $\sum_{n=1}^{\infty} b_n$ both converge; say $\sum_{n=1}^{\infty} a_n = L$ and $\sum_{n=1}^{\infty} b_n = M$. This means that

$$\lim_{N \to \infty} \sum_{n=1}^{N} a_n = L, \qquad \lim_{N \to \infty} \sum_{n=1}^{N} b_n = M.$$

Therefore

$$\sum_{n=1}^{\infty}(a_n+b_n) = \lim_{N\to\infty}\sum_{n=1}^{N}(a_n+b_n) = \lim_{N\to\infty}\left(\sum_{n=1}^{N}a_n + \sum_{n=1}^{N}b_n\right)$$

$$= L + M = \sum_{n=1}^{\infty}a_n + \sum_{n=1}^{\infty}b_n.$$

We leave the proofs of the other two parts as exercises for you (see Exercise 21).   □

Example 10.1.5. Evaluate the following series:

$$\text{(a) } \sum_{n=1}^{\infty}\frac{4^n+1}{5^n}, \qquad \text{(b) } \sum_{k=0}^{\infty}(-2)^{3k+1}\cdot 3^{-2k+1}.$$

Solution. Like integrals, series are often easier to evaluate if they can be split up as sums, as in part 1 of Theorem 10.1.4. We therefore rewrite series (a) as follows:

$$\sum_{n=1}^{\infty}\frac{4^n+1}{5^n} = \sum_{n=1}^{\infty}\left(\frac{4^n}{5^n}+\frac{1}{5^n}\right).$$

Before we can apply Theorem 10.1.4 to write this as a sum of two simpler series, we must check that these simpler series converge. Fortunately, this is easy to verify, since they are both geometric series. To see this, and to identify the values of a and r for each series, we write them in the form of series (10.2).

$$\sum_{n=1}^{\infty}\frac{4^n}{5^n} = \sum_{n=1}^{\infty}\frac{4}{5}\cdot\left(\frac{4}{5}\right)^{n-1} = \frac{4/5}{1-4/5} = 4,$$

$$\sum_{n=1}^{\infty}\frac{1}{5^n} = \sum_{n=1}^{\infty}\frac{1}{5}\cdot\left(\frac{1}{5}\right)^{n-1} = \frac{1/5}{1-1/5} = \frac{1}{4}.$$

Thus, the value of series (a) is

$$\sum_{n=1}^{\infty}\frac{4^n+1}{5^n} = \sum_{n=1}^{\infty}\left(\frac{4^n}{5^n}+\frac{1}{5^n}\right) = \sum_{n=1}^{\infty}\frac{4^n}{5^n} + \sum_{n=1}^{\infty}\frac{1}{5^n} = 4 + \frac{1}{4} = \frac{17}{4}.$$

Series (b) is also a geometric series, but we must apply some exponent rules to see this. This time, we write it in the form of series (10.4).

$$\sum_{k=0}^{\infty}(-2)^{3k+1}\cdot 3^{-2k+1} = \sum_{k=0}^{\infty}(-2\cdot 3)\cdot\frac{((-2)^3)^k}{(3^2)^k}$$

$$= \sum_{k=0}^{\infty}(-6)\cdot\left(-\frac{8}{9}\right)^k = \frac{-6}{1-(-8/9)} = -\frac{54}{17}.$$ □

Exercises 10.1

1–16: Evaluate the series.

1. $\displaystyle\sum_{n=1}^{\infty} \frac{2^{n-1}}{3^n}$.

2. $\displaystyle\sum_{n=1}^{\infty} \frac{2^n - 1}{3^n}$.

3. $\displaystyle\sum_{n=1}^{\infty} \frac{3^n}{2^{n+2}}$.

4. $\displaystyle\sum_{n=1}^{\infty} \frac{(-3)^n}{2^{2n}}$.

5. $\displaystyle\sum_{n=1}^{\infty} \frac{3\cdot 2^{n-1} + 2\cdot 3^{n+1}}{5^n}$.

6. $\displaystyle\sum_{n=1}^{\infty} \frac{2 + (-1)^n}{3^n}$.

7. $\displaystyle\sum_{n=1}^{\infty} \frac{3^{1-n} + (-1)^{n+1}}{2^n}$.

8. $\displaystyle\sum_{n=1}^{\infty} (-1)^n \frac{3^{2n}}{2^{n+1} 5^{n-1}}$.

9. $\displaystyle\sum_{n=1}^{\infty} \left(\sin\left(\frac{\pi}{3n}\right) - \sin\left(\frac{\pi}{3n+3}\right) \right)$.

10. $\displaystyle\sum_{n=1}^{\infty} \left(\cos\left(\frac{\pi}{3n}\right) - \cos\left(\frac{\pi}{3n+3}\right) \right)$.

11. $\displaystyle\sum_{n=1}^{\infty} \left(\frac{n}{2n-1} - \frac{n+2}{2n+3} \right)$.

12. $\displaystyle\sum_{n=0}^{\infty} \left(\tan^{-1}(n+1) - \tan^{-1}(n) \right)$.

13. $\displaystyle\sum_{n=1}^{\infty} \frac{1}{n^2 + n}$. (Hint: Start with partial fractions.)

14. $\displaystyle\sum_{n=1}^{\infty} \frac{4}{4n^2 - 1}$.

15. $\displaystyle\sum_{n=1}^{\infty}\frac{2}{n^2+2n}$.

16. $\displaystyle\sum_{n=1}^{\infty}\frac{4}{4n^2+4n-3}$.

17. (a) Show that for every positive integer N,

$$\sum_{n=1}^{N}\frac{n}{2^n}=\frac{2^{N+1}-N-2}{2^N}.$$

(Hint: Use mathematical induction.)

(b) Find $\displaystyle\sum_{n=1}^{\infty}\frac{n}{2^n}$.

18. (a) Show that for every positive integer N,

$$\sum_{n=1}^{N}\frac{n}{3^n}=\frac{3^{N+1}-2N-3}{4\cdot 3^N}.$$

(b) Find $\displaystyle\sum_{n=1}^{\infty}\frac{n}{3^n}$.

19. For every integer $N \geq 2$ let

$$s_N=\sum_{n=2}^{N}\frac{2n-1}{n^2(n-1)^2}.$$

(a) Compute s_2, s_3, and s_4.

(b) Find a formula for s_N. (Hint: Make a guess based on part (a), and then prove your guess by mathematical induction. Alternatively, you could use partial fractions and then recognize the series as a telescoping series.)

(c) Find $\displaystyle\sum_{n=2}^{\infty}\frac{2n-1}{n^2(n-1)^2}$.

20. Use mathematical induction to prove that for every positive integer N, the Nth partial sum of the series (10.1) is given by the formula $s_N = (2^N - 1)/2^N$.

21. Prove parts 2 and 3 of Theorem 10.1.4.

22. Give an example of two series $\sum_{n=1}^{\infty} a_n$ and $\sum_{n=1}^{\infty} b_n$ such that both series diverge, but $\sum_{n=1}^{\infty}(a_n+b_n)$ converges.

23. Show that if $\sum_{n=1}^{\infty} a_n$ converges and $\sum_{n=1}^{\infty} b_n$ diverges, then $\sum_{n=1}^{\infty} (a_n + b_n)$ diverges. (Hint: Use proof by contradiction. Assume that $\sum_{n=1}^{\infty} (a_n + b_n)$ converges, and derive a contradiction.)

24. Show that $0.999\ldots = 1$. (Begin by interpreting $0.999\ldots$ as an infinite series. This demonstrates that decimal expansions are not unique: sometimes a single number has two different decimal expansions.)

10.2 Convergence Tests

Unfortunately, many infinite series are very hard to evaluate, because it is often difficult to find a formula for the Nth partial sum of a series. We therefore lower our sights, at least temporarily, and rather than trying to find the value of a given series, we will simply try to determine whether the series converges or diverges. To do this, we will develop a number of tests that can be used to establish convergence or divergence of a series.

There are several reasons for studying such tests. First of all, if we are able to establish that a series diverges, then there is no need to try to compute the value of the series; the series does not have a value. Secondly, if we determine that a series converges, then we know that the sequence $(s_N)_{N=1}^{\infty}$ of partial sums converges to the value of the series. Even if we are not able to find the exact value of the series, we can get an approximate answer by computing s_N for some large integer N. In many cases we will be able to determine the degree of accuracy of the approximation, and an approximate answer that achieves some desired degree of accuracy may be all that we need in some situations. Finally, sometimes showing that a series converges is the first step in finding its exact value. (We have already seen similar reasoning in some previous examples, such as Examples 2.9.6 and 2.9.7.) In several cases in this chapter we will show that a series converges, and then later, once we have developed more advanced techniques, we will use the fact that the series converges to find its exact value.

Our first convergence test is based on the following fact about convergent series.

Theorem 10.2.1. *If $\sum_{n=1}^{\infty} a_n$ converges, then $\lim_{n \to \infty} a_n = 0$.*

Proof. Suppose that $\sum_{n=1}^{\infty} a_n$ converges; say $\sum_{n=1}^{\infty} a_n = L$. Then $\lim_{N \to \infty} s_N = L$, where as usual, s_N is the Nth partial sum of the series:

$$s_N = \sum_{n=1}^{N} a_n = a_1 + a_2 + \cdots + a_N.$$

To complete the proof, we must compute $\lim_{n \to \infty} a_n$. To do this, we express a_n in terms of partial sums:

$$a_n = (a_1 + a_2 + \cdots + a_n) - (a_1 + a_2 + \cdots + a_{n-1}) = s_n - s_{n-1}.$$

Taking the limit as $n \to \infty$, we get

$$\lim_{n \to \infty} a_n = \lim_{n \to \infty} (s_n - s_{n-1}) = L - L = 0. \qquad \square$$

As an immediate consequence of this theorem, we get the following convergence test:

Corollary 10.2.2 (Test for Divergence). *If it is not the case that* $\lim_{n\to\infty} a_n = 0$, *then* $\sum_{n=1}^{\infty} a_n$ *diverges.*

Proof. Assume that $\lim_{n\to\infty} a_n$ is not equal to 0. If $\sum_{n=1}^{\infty} a_n$ converges then, by Theorem 10.2.1, $\lim_{n\to\infty} a_n = 0$, contrary to our assumption. Therefore it cannot be the case that $\sum_{n=1}^{\infty} a_n$ converges, so it must diverge. $\square$

For example, consider the series

$$\sum_{n=1}^{\infty} \frac{2n-1}{5n+3} = \frac{1}{8} + \frac{3}{13} + \frac{5}{18} + \cdots . \tag{10.5}$$

Since

$$\lim_{n\to\infty} \frac{2n-1}{5n+3} = \lim_{n\to\infty} \frac{2-1/n}{5+3/n} = \frac{2}{5} \neq 0,$$

the test for divergence tells us that series (10.5) diverges. Notice that we were able to reach this conclusion without having to find a formula for the Nth partial sum of the series.

It is important to recognize that the test for divergence does *not* say that if $\lim_{n\to\infty} a_n = 0$, then $\sum_{n=1}^{\infty} a_n$ converges. In fact, if $\lim_{n\to\infty} a_n = 0$ then $\sum_{n=1}^{\infty} a_n$ may converge and it may diverge. We will see many examples of both of these possibilities in the rest of this chapter. To distinguish between these two possibilities, we will need more convergence tests.

It turns out that it is much easier to tell whether or not a series converges if all of the terms of the series are nonnegative. The reason is that if all of the terms are nonnegative, then as we add more and more terms of the series, the sum can only increase. This means that the sequence of partial sums will be weakly increasing.

Let's work this out more carefully. Consider a series $\sum_{n=1}^{\infty} a_n$, and suppose that for every n, $a_n \geq 0$. Let $(s_N)_{N=1}^{\infty}$ be the sequence of partial sums of the series. Then for every positive integer N, we have

$$s_{N+1} = a_1 + a_2 + \cdots + a_N + a_{N+1} = s_N + a_{N+1} \geq s_N,$$

since $a_{N+1} \geq 0$. Therefore the sequence $(s_N)_{N=1}^{\infty}$ is weakly increasing. To test the series for convergence, we must determine whether or not this weakly increasing sequence converges.

Now, recall what we learned about weakly increasing sequences in Section 2.9. According to Theorem 2.9.5, if the set of partial sums has an upper bound, then the sequence $(s_N)_{N=1}^{\infty}$ converges, and if not then $\lim_{N\to\infty} s_N = \infty$. Thus, to determine whether or not the sequence of partial sums converges we can try to find an upper bound for the partial sums. We will do this by trying to *overestimate* s_N, which is often much easier than finding the exact value. If we are able to find an upper bound for the partial sums, then $\lim_{n\to\infty} s_N$ exists—in fact, it is equal to the least upper bound—and therefore the series $\sum_{n=1}^{\infty} a_n$ converges. If we can show that there is no upper bound, then we can conclude that $\lim_{N \to \infty} s_N = \infty$, and therefore $\sum_{n=1}^{\infty} a_n$ diverges.

One way to overestimate a partial sum s_N is with an integral. As we know, integrals are closely related to sums, but they are often easier to compute. This idea leads to our next convergence test.

Theorem 10.2.3 (Integral Test). *Consider a series $\sum_{n=1}^{\infty} a_n$ whose terms are given by a formula $a_n = f(n)$, for some function f. Suppose that f is continuous and weakly decreasing on $[1, \infty)$, and for all $x \geq 1$, $f(x) \geq 0$.*

1. *If $\int_1^{\infty} f(x)\, dx$ converges, then $\sum_{n=1}^{\infty} a_n$ converges.*

2. *If $\int_1^{\infty} f(x)\, dx = \infty$, then $\sum_{n=1}^{\infty} a_n = \infty$.*

Proof. The fact that $f(x) \geq 0$ for all $x \geq 1$ implies that for every positive integer n, $a_n \geq 0$. As we have observed, it follows that the sequence $(s_N)_{N=1}^{\infty}$ of partial sums of the series is weakly increasing, and therefore we can discover whether or not the series converges by determining whether or not the partial sums have an upper bound.

To prove part 1, suppose that $\int_1^{\infty} f(x)\, dx$ converges. This means that we can let $L = \int_1^{\infty} f(x)\, dx$. Figure 10.1a shows a Riemann sum approximation for $\int_1^N f(x)\, dx$, where N is some positive integer. All of the striped rectangles have width 1. The height of the first rectangle is $f(2) = a_2$, and therefore its area is $a_2 \cdot 1 = a_2$. Similarly, the next rectangle has height and area $f(3) = a_3$, the next has height and area a_4, and so on. Since f is weakly decreasing, if $1 \leq x \leq 2$ then $f(x) \geq f(2)$, which means that the top of the first rectangle stays below the curve $y = f(x)$. Similarly, the tops of all of the other rectangles stay below the curve $y = f(x)$, and therefore the sum of the areas of the rectangles, which is a Riemann sum for f on the interval $[1, N]$, is less than or equal to the integral. In other words,

$$a_2 + a_3 + \cdots + a_N \leq \int_1^N f(x)\, dx.$$

Adding a_1 to both sides of this inequality, we get

$$s_N = a_1 + a_2 + a_3 + \cdots + a_N \leq a_1 + \int_1^N f(x)\, dx. \tag{10.6}$$

Thus, using an integral we have found an overestimate of s_N.

But this overestimate is not yet an upper bound for the partial sums, because the right-hand side of (10.6) depends on N. To turn the overestimate into an upper bound, we need an upper bound for $\int_1^N f(x)\, dx$. This is where we make use of the fact that $\int_1^{\infty} f(x)\, dx$ converges. By Theorem 8.7.5 we have

$$\int_1^N f(x)\, dx + \int_N^{\infty} f(x)\, dx = \int_1^{\infty} f(x)\, dx = L.$$

Since $f(x) \geq 0$ for all $x \geq 1$, it is clear that $\int_N^{\infty} f(x)\, dx \geq 0$. Therefore

$$\int_1^N f(x)\, dx \leq \int_1^N f(x)\, dx + \int_N^{\infty} f(x)\, dx = L,$$

and combining this with (10.6), we see that for every positive integer N,

$$s_N \leq a_1 + \int_1^N f(x)\,dx \leq a_1 + L.$$

In other words, $a_1 + L$ is an upper bound for the set of partial sums of the series, and therefore the series converges.

Now we turn to the proof of part 2. Suppose that $\int_1^\infty f(x)\,dx = \infty$. This time we will use the Riemann sum approximation illustrated in Figure 10.1b. The area of the first striped rectangle in Figure 10.1b is a_1, the area of the next is a_2, and so on. Since the tops of the rectangle are now *above* the curve $y = f(x)$, we see that

$$\int_1^N f(x)\,dx \leq a_1 + a_2 + \cdots + a_{N-1} = s_{N-1} \leq s_N. \tag{10.7}$$

In other words, $\int_1^N f(x)\,dx$ is an *underestimate* of s_N. By our assumption,

$$\lim_{N\to\infty} \int_1^N f(x)\,dx = \int_1^\infty f(x)\,dx = \infty,$$

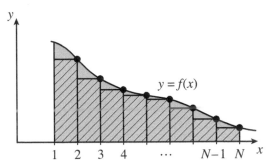

(a) The striped rectangles are contained in the blue region.

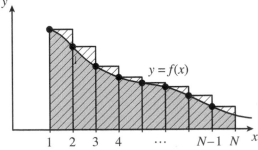

(b) The striped rectangles contain the blue region.

Figure 10.1: Using $\int_1^N f(x)\,dx$ to estimate partial sums.

so (10.7) implies that

$$\sum_{n=1}^{\infty} a_n = \lim_{N \to \infty} s_N = \infty.$$ □

Example 10.2.4. Determine whether the following series converge or diverge:

$$(a) \sum_{n=1}^{\infty} \frac{1}{n}, \quad (b) \sum_{n=1}^{\infty} \frac{1}{n^2}.$$

Solution. For series (a), let $f(x) = 1/x$. Clearly f is continuous and weakly decreasing on $[1, \infty)$, and $f(x) \geq 0$ for all $x \geq 1$. We can therefore apply the integral test. Since

$$\int_{1}^{\infty} f(x) \, dx = \lim_{t \to \infty} \int_{1}^{t} \frac{1}{x} \, dx = \lim_{t \to \infty} [\ln x]_{1}^{t} = \lim_{t \to \infty} \ln t = \infty,$$

part 2 of Theorem 10.2.3 tells us that series (a) diverges.

Similarly, for series (b) we apply the integral test with $f(x) = 1/x^2$. This time we find that

$$\int_{1}^{\infty} f(x) \, dx = \lim_{t \to \infty} \int_{1}^{t} \frac{1}{x^2} \, dx = \lim_{t \to \infty} \left[-\frac{1}{x} \right]_{1}^{t} = \lim_{t \to \infty} \left(1 - \frac{1}{t} \right) = 1,$$

so by part 1 of Theorem 10.2.3, series (b) converges. □

Notice that $\lim_{n \to \infty} 1/n = \lim_{n \to \infty} 1/n^2 = 0$. Thus, the terms of both of the series in Example 10.2.4 approach 0 as $n \to \infty$, but one series converges and the other diverges. This illustrates our earlier statement that if $\lim_{n \to \infty} a_n = 0$, then $\sum_{n=1}^{\infty} a_n$ may converge and it may diverge.

Let's see if we can generalize Example 10.2.4. Both of the series in the example have the form $\sum_{n=1}^{\infty} 1/n^p$; in series (a), we have $p = 1$, while $p = 2$ in series (b). Series of this form are sometimes called *p-series*. Recall that in Section 8.7 we studied improper integrals of the form $\int_{1}^{\infty} 1/x^p \, dx$, and we found that this integral converges if $p > 1$ and diverges if $p \leq 1$. Applying the integral test, we get an analogous result for *p*-series.

Theorem 10.2.5 (*p*-Series Test). *Let p be any real number.*

1. *If $p > 1$, then* $\displaystyle\sum_{n=1}^{\infty} \frac{1}{n^p}$ *converges.*

2. *If $p \leq 1$, then* $\displaystyle\sum_{n=1}^{\infty} \frac{1}{n^p} = \infty.$

Proof. For $x \geq 1$, let $f(x) = 1/x^p \geq 0$. As long as $p \geq 0$, $f'(x) = -px^{-p-1} \leq 0$, and therefore f is continuous and weakly decreasing on $[1, \infty)$. Thus, we can apply the integral test. According to Theorem 8.7.2, $\int_{1}^{\infty} 1/x^p \, dx$ converges if $p > 1$ and $\int_{1}^{\infty} 1/x^p \, dx = \infty$ if $p \leq 1$, and by the integral test we get the same conclusions for $\sum_{n=1}^{\infty} 1/n^p$.

If $p < 0$, then f is strictly increasing on $[1, \infty)$, so we cannot use the integral test. But in this case it is easy to see that for every positive integer n, $1/n^p \geq 1$, and therefore for every positive integer N,

$$s_N = \sum_{n=1}^{N} \frac{1}{n^p} \geq \sum_{n=1}^{N} 1 = N.$$

Thus,

$$\sum_{n=1}^{\infty} \frac{1}{n^p} = \lim_{N \to \infty} s_N = \infty. \qquad \square$$

In the p-series test, the case $p = 1$ is particularly important, because it is the borderline between convergent and divergent p-series; the series $\sum_{n=1}^{\infty} 1/n^p$ diverges if $p = 1$, but converges if, for example, $p = 1.01$, or $p = 1.001$. You might say that the p-series with $p = 1$ diverges, but just barely. This series is called the *harmonic series*. In other words, the harmonic series is the series

$$\sum_{n=1}^{\infty} \frac{1}{n} = 1 + \frac{1}{2} + \frac{1}{3} + \cdots .$$

The partial sums of the harmonic series are called the *harmonic numbers*, and are denoted H_N. Thus, for any positive integer N,

$$H_N = \sum_{n=1}^{N} \frac{1}{n} = 1 + \frac{1}{2} + \frac{1}{3} + \cdots + \frac{1}{N}.$$

The sequence of harmonic numbers approaches ∞, but very slowly. For example, $H_{100} \approx 5.187$, $H_{1000} \approx 7.485$, and $H_{10000} \approx 9.788$. For more on the harmonic numbers, see Exercise 24.

Example 10.2.6. Determine whether the following series converge or diverge:

$$\text{(a)} \sum_{n=1}^{\infty} \frac{1}{n\sqrt[3]{n}}, \qquad \text{(b)} \sum_{n=2}^{\infty} \frac{1}{n \ln n}, \qquad \text{(c)} \sum_{n=1}^{\infty} \frac{\ln n - 1}{n\sqrt[3]{n}}.$$

Solution. Series (a) is a p-series:

$$\sum_{n=1}^{\infty} \frac{1}{n\sqrt[3]{n}} = \sum_{n=1}^{\infty} \frac{1}{n^{4/3}}.$$

Since $p = 4/3 > 1$, the series converges by the p-series test.

For series (b), we would like to apply the integral test, using the function $f(x) = 1/(x \ln x)$. But notice that the values of n in the series start at $n = 2$ rather than $n = 1$—in fact, the series could not start at $n = 1$, since $f(1)$ is undefined. So we cannot apply the integral test exactly as written in Theorem 10.2.3. However, you can easily check that in the proof of the integral test, the number 1, which was both the beginning value for n in the series and also the lower limit of the improper integral, could be replaced by any integer. Thus, although we stated the integral test for series whose terms start at $n = 1$, the test actually works for series in which n starts at any integer. Indeed, all of our theorems about series can be generalized in this way.

It is easy to see that f is continuous and weakly decreasing on $[2, \infty)$, and for all $x \geq 2$, $f(x) \geq 0$. Therefore the integral test applies. To evaluate the improper integral, we use the substitution $u = \ln x$, $du = (1/x)\,dx$:

$$\int_2^\infty \frac{1}{x \ln x}\,dx = \lim_{t \to \infty} \int_2^t \frac{1}{\ln x} \cdot \frac{1}{x}\,dx = \lim_{t \to \infty} \int_{\ln 2}^{\ln t} \frac{1}{u}\,du$$

$$= \lim_{t \to \infty} [\ln u]_{\ln 2}^{\ln t} = \lim_{t \to \infty} (\ln \ln t - \ln \ln 2) = \infty.$$

Thus, by the integral test, series (b) diverges.

Finally, for series (c), we try the integral test with

$$f(x) = \frac{\ln x - 1}{x \sqrt[3]{x}} = \frac{\ln x - 1}{x^{4/3}}.$$

Unfortunately, the hypotheses of the integral test are not satisfied. Although f is continuous on $[1, \infty)$, we have $f(x) \geq 0$ only for $x \geq e \approx 2.718$. To see if f is increasing or decreasing, we compute

$$f'(x) = \frac{x^{4/3} \cdot (1/x) - (\ln x - 1) \cdot (4/3)x^{1/3}}{(x^{4/3})^2}$$

$$= \frac{x^{1/3}(1 - (4/3)(\ln x - 1))}{x^{8/3}}$$

$$= \frac{7 - 4\ln x}{3x^{7/3}}.$$

Thus, $f'(x) = 0$ when $\ln x = 7/4$, or $x = e^{7/4} \approx 5.755$. Checking the sign of $f'(x)$ on either side of this critical number, we find that f is strictly increasing on $[1, e^{7/4}]$ and strictly decreasing on $[e^{7/4}, \infty)$. Thus, we cannot apply the integral test directly to series (c). But f is weakly decreasing on $[6, \infty)$, and $f(x) \geq 0$ for $x \geq 6$, so we can apply the integral test to the series

$$\sum_{n=6}^\infty \frac{\ln n - 1}{n \sqrt[3]{n}}, \tag{10.8}$$

and according to part 3 of Theorem 10.1.4, this series converges if and only if series (c) converges.

To evaluate the required improper integral, we first evaluate the indefinite integral using integration by parts, with $u = \ln x - 1$, $dv = x^{-4/3}\,dx$, $du = x^{-1}\,dx$, and $v = -3x^{-1/3}$:

$$\int f(x)\,dx = \int (\ln x - 1)\cdot x^{-4/3}\,dx = (\ln x - 1)(-3x^{-1/3}) + \int 3x^{-4/3}\,dx$$

$$= -3x^{-1/3}\ln x + 3x^{-1/3} - 9x^{-1/3} + C = -\frac{3\ln x + 6}{\sqrt[3]{x}} + C.$$

Thus,

$$\int_6^\infty f(x)\,dx = \lim_{t\to\infty}\left[-\frac{3\ln x + 6}{\sqrt[3]{x}}\right]_6^t = \lim_{t\to\infty}\left(-\frac{3\ln t}{\sqrt[3]{t}} - \frac{6}{\sqrt[3]{t}} + \frac{3\ln 6 + 6}{\sqrt[3]{6}}\right)$$

$$= \frac{3\ln 6 + 6}{\sqrt[3]{6}}.$$

Note that we have used Theorem 7.6.2 to see that $\lim_{t\to\infty}(3\ln t/\sqrt[3]{t}) = 0$. Since $\int_6^\infty f(x)\,dx$ converges, we conclude by the integral test that series (10.8) converges, and by part 3 of Theorem 10.1.4, series (c) converges as well. □

Although the integral test can be useful for showing that a series converges, it does not tell us what the series converges to. In particular, although part 1 of Theorem 10.2.3 says that we can show that a series converges by computing an improper integral, it does not say that the value of the series is the same as the value of the improper integral. So how can we compute the value of the series?

We can get an approximate answer by computing a partial sum. Suppose that we have determined that a series $\sum_{n=1}^\infty a_n$ converges. If we let s denote the exact value of the series and s_N the Nth partial sum, then for large integers N, $s \approx s_N$. How good is this approximation? As in our study of numerical integration, we define the *error* of the approximation to be the difference between the exact value and the approximation. Since

$$s = \sum_{n=1}^\infty a_n = \sum_{n=1}^N a_n + \sum_{n=N+1}^\infty a_n = s_N + \sum_{n=N+1}^\infty a_n,$$

the error is

$$s - s_N = \sum_{n=N+1}^\infty a_n.$$

The sum on the right-hand side above is called the *Nth tail* of the series; it consists of all terms of the series that are missing from the Nth partial sum. We will denote the Nth tail by t_N:

$$t_N = \sum_{n=N+1}^\infty a_n.$$

Using this notation, what we have shown is that $s - s_N = t_N$, and this suggests that one way to try to determine the accuracy of the approximation $s \approx s_N$ is to study t_N.

For example, in part (b) of Example 10.2.4 we showed that $\sum_{n=1}^{\infty} 1/n^2$ converges. If we let s denote the value of this series, then we can approximate s by computing a partial sum. Using a computer to do the arithmetic, we find that the 100th partial sum is

$$s_{100} = \sum_{n=1}^{100} \frac{1}{n^2} = 1 + \frac{1}{4} + \frac{1}{9} + \cdots + \frac{1}{10000} \approx 1.63498.$$

This gives us an approximate value for s, and the error in the approximation is given by the 100th tail of the series:

$$s - s_{100} = t_{100} = \sum_{n=101}^{\infty} \frac{1}{n^2}.$$

To determine the accuracy of the approximation, we now examine t_{100}.

All the terms in the series are positive, so it is clear that $t_{100} \geq 0$. We can use the idea behind the proof of the integral test to overestimate t_{100}. In Figure 10.2, the area of the blue region is $\int_{100}^{\infty} (1/x^2)\, dx$, and the areas of the striped rectangles are $1/101^2$, $1/102^2$, and so on. Since the rectangles are all contained in the blue region, we see that

$$t_{100} = \sum_{n=101}^{\infty} \frac{1}{n^2} \leq \int_{100}^{\infty} \frac{1}{x^2}\, dx = \lim_{t \to \infty} \int_{100}^{t} x^{-2}\, dx$$

$$= \lim_{t \to \infty} \left[-x^{-1} \right]_{100}^{t} = \lim_{t \to \infty} \left(-\frac{1}{t} + \frac{1}{100} \right) = 0.01.$$

(For a more careful justification of the inequality in this calculation, see Exercise 20.) Since $s - s_{100} = t_{100}$, we conclude that $0 \leq s - s_{100} \leq 0.01$, and therefore

$$s_{100} \leq s \leq s_{100} + 0.01.$$

In other words, the exact value of the series is between about 1.63498 and 1.64498. Of course, we can get greater accuracy by computing s_N for larger N. (For more on using partial sums to approximate this series, see Exercise 25.)

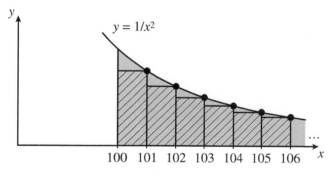

Figure 10.2: Estimating $t_{100} = \sum_{n=101}^{\infty} 1/n^2$.

Finding the exact value of this series is far more difficult. The problem of finding the exact value of $\sum_{n=1}^{\infty} 1/n^2$ was first proposed by Pietro Mengoli (1626–1686) in 1644, and the problem remained unsolved for almost a century, despite the efforts of many of the best mathematicians of the time. In 1689, Jacob Bernoulli (1654–1705) mentioned the problem in his *Tractatus de Seriebus Infinitis*, which was published in Basel, Switzerland, and as a result the problem came to be known as the Basel problem. The problem was finally solved in 1735, when Leonhard Euler (1707–1783) showed that the answer is $\pi^2/6$, whose decimal value is about 1.64493. For two proofs of this result, see Exercises 26 and 27.

Exercises 10.2

1–19: Determine whether the series converges or diverges.

1. $\displaystyle\sum_{n=1}^{\infty} \frac{1}{n+4}$.

2. $\displaystyle\sum_{n=1}^{\infty} \frac{n}{n+4}$.

3. $\displaystyle\sum_{n=1}^{\infty} \frac{1}{n^2+4}$.

4. $\displaystyle\sum_{n=1}^{\infty} \frac{n}{n^2+4}$.

5. $\displaystyle\sum_{n=1}^{\infty} \frac{1}{\sqrt{n^3}}$.

6. $\displaystyle\sum_{n=1}^{\infty} \frac{1}{\sqrt[3]{n^2}}$.

7. $\displaystyle\sum_{n=1}^{\infty} \frac{\sqrt{n}}{n^2}$.

8. $\displaystyle\sum_{n=1}^{\infty} \frac{\sqrt{n}+4}{n^2}$.

9. $\displaystyle\sum_{n=1}^{\infty} \frac{1}{4n^2+2n}$.

10. $\displaystyle\sum_{n=2}^{\infty} \frac{1}{n(3+\ln n)}$.

11. $\displaystyle\sum_{n=2}^{\infty} \frac{1}{n(\ln n)^3}.$

12. $\displaystyle\sum_{n=2}^{\infty} \frac{n}{\ln(n^3)}.$

13. $\displaystyle\sum_{n=2}^{\infty} \frac{\ln n}{n^3}.$

14. $\displaystyle\sum_{n=1}^{\infty} \frac{1}{(\tan^{-1} n)^2}.$

15. $\displaystyle\sum_{n=1}^{\infty} \frac{n}{e^n}.$

16. $\displaystyle\sum_{n=1}^{\infty} n e^{1/n}.$

17. $\displaystyle\sum_{n=1}^{\infty} \frac{1}{2^{\ln n}}.$

18. $\displaystyle\sum_{n=1}^{\infty} \frac{1}{3^{\ln n}}.$

19. $\displaystyle\sum_{n=1}^{\infty} \frac{1}{e^{\sqrt{n}}}.$

20. In this exercise you will give a careful justification for the claim in the text that $\sum_{n=101}^{\infty} 1/n^2 \leq \int_{100}^{\infty} (1/x^2)\, dx.$

    (a) Use Theorems 5.3.7 and 5.3.9 to show that for every integer $N \geq 101$,

    $$\int_{100}^{N} \frac{1}{x^2}\, dx - \sum_{n=101}^{N} \frac{1}{n^2} \geq 0.$$

    (b) Use part (a) together with the preservation of weak inequalities by limits (Corollary 2.8.9) to show that

    $$\sum_{n=101}^{\infty} \frac{1}{n^2} \leq \int_{100}^{\infty} \frac{1}{x^2}\, dx.$$

21. Let

$$s = \sum_{n=1}^{\infty} \frac{1}{n^3}.$$

(a) Compute the partial sum

$$s_{10} = \sum_{n=1}^{10} \frac{1}{n^3}.$$

(b) Show that the error in the approximation $s \approx s_{10}$ is at most 0.005. (Hint: Overestimate the tail t_{10} with an improper integral. Surprisingly, no simple formula for the exact value of this series is known.)

22. Let

$$s = \sum_{n=1}^{\infty} \frac{1}{n^4}.$$

(a) Approximate the value of s with error at most 0.005. (Hint: First find a positive integer N such that the partial sum s_N is within 0.005 of s.)

(b) Euler showed that the exact value of s is $\pi^4/90$. Verify that your approximation in part (a) is within 0.005 of $\pi^4/90$.

23. Let

$$s = \sum_{n=1}^{\infty} \frac{1}{e^n}.$$

(a) Compute

$$s_4 = \sum_{n=1}^{4} \frac{1}{e^n}.$$

(b) Show that the error in the approximation $s \approx s_4$ is less than 0.02 by overestimating a tail of the series with an improper integral.

(c) Find the exact value of s, and verify that $|s - s_4| < 0.02$.

24. Recall that the harmonic numbers are defined by the formula

$$H_N = \sum_{n=1}^{N} \frac{1}{n} = 1 + \frac{1}{2} + \cdots + \frac{1}{N}.$$

(a) Show that for every positive integer N,

$$\ln(N+1) \leq H_N \leq 1 + \ln N.$$

(Hint: Use Riemann sums, as in the proof of the integral test.)

(b) Show that

$$\lim_{N \to \infty} \frac{H_N}{\ln N} = 1.$$

(c) For each positive integer N, let $u_N = H_N - \ln(N+1)$ and $v_N = H_N - \ln N$. Clearly $u_N < v_N$. Show that $(u_N)_{N=1}^{\infty}$ is strictly increasing, $(v_N)_{N=1}^{\infty}$ is strictly decreasing, and $\lim_{N \to \infty}(v_N - u_N) = 0$. (Hint: Consider $u_{N+1} - u_N$ and $v_{N+1} - v_N$. Notice that it follows, by the nested interval theorem, that there is a unique number γ such that for all N, $u_N < \gamma < v_N$, and $\lim_{N \to \infty} u_N = \lim_{N \to \infty} v_N = \gamma$. The number $\gamma \approx 0.5772$ is known as the *Euler-Mascheroni constant*, after Leonhard Euler and Lorenzo Mascheroni (1750–1800). For large N, $v_N = H_N - \ln N \approx \gamma$, and therefore $H_N \approx \ln N + \gamma$.)

25. Let s denote the exact value of the series $\sum_{n=1}^{\infty} 1/n^2$, s_N the Nth partial sum, and t_N the Nth tail.

(a) By imitating the reasoning in the text, show that for every positive integer N, $t_N \leq 1/N$, and therefore

$$s_N \leq s \leq s_N + \frac{1}{N}.$$

(b) Show that for every positive integer N, $t_N \geq 1/(N+1)$. (Hint: Use a Riemann sum with rectangles above the curve, as in Figure 10.1b.) Combine this with part (a) to conclude that

$$s_N + \frac{1}{N+1} \leq s \leq s_N + \frac{1}{N}.$$

(c) Use part (b) in the case $N = 100$ to improve on the range of values for s given in the text. Verify that your range of values includes the exact answer $s = \pi^2/6$.

26. In this exercise, you will show that $\sum_{n=1}^{\infty} 1/n^2 = \pi^2/6$.[1] For every integer $n \geq 0$, let

$$I_n = \int_0^{\pi/2} \cos^{2n} x \, dx, \qquad J_n = \int_0^{\pi/2} x^2 \cos^{2n} x \, dx.$$

(a) Show that for every $n \geq 1$,

$$I_n = \frac{2n-1}{2n} \cdot I_{n-1}.$$

(Hint: Begin by applying integration by parts to I_n, with $u = \cos^{2n-1} x$ and $dv = \cos x \, dx$.)

(b) Show that for every $n \geq 1$,

$$I_n = n(2n-1)J_{n-1} - 2n^2 J_n.$$

(Hint: Apply integration by parts to I_n twice. The first time, use $dv = dx$, and the second time, use $dv = 2x \, dx$.)

---

[1] This exercise is based on Y. Matsuoka, An elementary proof of the formula $\sum_{k=1}^{\infty} 1/k^2 = \pi^2/6$, *Amer. Math. Monthly* **68** (1961), pp. 485–487.

(c) Combine parts (a) and (b) to show that for every $n \geq 1$,

$$\frac{1}{n^2} = 2\left(\frac{J_{n-1}}{I_{n-1}} - \frac{J_n}{I_n}\right).$$

(d) Use part (c) to show that for every positive integer N,

$$\sum_{n=1}^{N} \frac{1}{n^2} = \frac{\pi^2}{6} - 2 \cdot \frac{J_N}{I_N}.$$

(e) Show that for $0 \leq x \leq \pi/2$,

$$x \leq \frac{\pi}{2} \sin x.$$

(Hint: Apply curve sketching methods to the function $f(x) = (\pi/2)\sin x - x$ on the interval $[0, \pi/2]$.)

(f) Show that for every positive integer N,

$$0 \leq \frac{J_N}{I_N} \leq \frac{\pi^2}{8(N+1)}.$$

(Hint: Begin by applying part (e) to overestimate J_N as follows:

$$J_N = \int_0^{\pi/2} x^2 \cos^{2N} x\, dx \leq \int_0^{\pi/2} \left(\frac{\pi}{2}\sin x\right)^2 \cos^{2N} x\, dx.$$

Then show that the right-hand side above is equal to $(\pi^2/4)(I_N - I_{N+1})$ and apply part (a).)

(g) Combine parts (d) and (f) to conclude that

$$\sum_{n=1}^{\infty} \frac{1}{n^2} = \frac{\pi^2}{6}.$$

27. In this exercise you will give another proof that $\sum_{n=1}^{\infty} 1/n^2 = \pi^2/6$.[2]

(a) Show that the following series converges:

$$\sum_{n=1}^{\infty} \frac{1}{(2n-1)^2} = 1 + \frac{1}{3^2} + \frac{1}{5^2} + \cdots.$$

(b) Use the identities $\sin(2x) = 2\sin x \cos x$ and $\sin(\pi/2 - x) = \cos x$ to show that for $0 < x < \pi/4$,

$$\frac{1}{\sin^2 x} + \frac{1}{\sin^2(\pi/2 - x)} = \frac{4}{\sin^2(2x)}.$$

---

[2]This exercise is based on J. Hofbauer, A simple proof of $1 + \frac{1}{2^2} + \frac{1}{3^2} + \cdots = \frac{\pi^2}{6}$ and related identities, *Amer. Math. Monthly* **109** (2002), pp. 196–200.

(c) Use part (b) to show that

$$\frac{1}{\sin^2(\pi/4)} = 2,$$

$$\frac{1}{\sin^2(\pi/8)} + \frac{1}{\sin^2(3\pi/8)} = 2 \cdot 4,$$

$$\frac{1}{\sin^2(\pi/16)} + \frac{1}{\sin^2(3\pi/16)} + \frac{1}{\sin^2(5\pi/16)} + \frac{1}{\sin^2(7\pi/16)} = 2 \cdot 4^2,$$

and in general, for any $N \geq 0$,

$$\sum_{n=1}^{2^N} \frac{1}{\sin^2((2n-1)\pi/2^{N+2})} = 2 \cdot 4^N.$$

(d) For $0 < x < \pi/2$, use the fact that $0 \leq \cos x \leq (\sin x)/x \leq 1$ (see equation (2.18)) to show that

$$\frac{1}{\sin^2 x} - 1 \leq \frac{1}{x^2} \leq \frac{1}{\sin^2 x}.$$

(e) Combine parts (c) and (d) to show that for every integer $N \geq 0$,

$$\frac{\pi^2}{8}\left(1 - \frac{1}{2^{N+1}}\right) \leq \sum_{n=1}^{2^N} \frac{1}{(2n-1)^2} \leq \frac{\pi^2}{8}.$$

(f) Use parts (a) and (e) to show that

$$\sum_{n=1}^{\infty} \frac{1}{(2n-1)^2} = \frac{\pi^2}{8}.$$

(g) Show that for every integer $N \geq 1$,

$$\sum_{n=1}^{2N} \frac{1}{n^2} = \sum_{n=1}^{N} \frac{1}{(2n-1)^2} + \frac{1}{4}\sum_{n=1}^{N} \frac{1}{n^2}.$$

(h) Use parts (f) and (g) to show that

$$\sum_{n=1}^{\infty} \frac{1}{n^2} = \frac{\pi^2}{6}.$$

10.3 The Comparison and Limit Comparison Tests

In the last section, we found that we can sometimes determine whether or not a series converges by comparing the series to an improper integral. In this section, we show that we can get similar results by comparing a series to another series. When we first stated the integral test in the last section, we assumed that the numbering of the terms in our series started with 1, but later we realized that we could generalize to series in which the term numbers started at some other integer. This time we state our results in the more general form from the start. Our main result in this section is the following convergence test.

Theorem 10.3.1 (Comparison Test). *Suppose that $\sum_{n=k}^{\infty} a_n$ and $\sum_{n=k}^{\infty} b_n$ are two infinite series, and for all integers $n \geq k$,*

$$0 \leq a_n \leq b_n.$$

1. *If $\sum_{n=k}^{\infty} b_n$ converges, then $\sum_{n=k}^{\infty} a_n$ converges.*

2. *If $\sum_{n=k}^{\infty} a_n$ diverges, then $\sum_{n=k}^{\infty} b_n$ diverges.*

Proof. To prove part 1, suppose that $\sum_{n=k}^{\infty} b_n = L$. As we saw in the last section, since we have $a_n \geq 0$ for all $n \geq k$, the partial sums of the series $\sum_{n=k}^{\infty} a_n$ are weakly increasing, and therefore we can show that the series converges by finding an upper bound for the partial sums. But this is easy to do, because for every $N \geq k$,

$$\sum_{n=k}^{N} a_n \leq \sum_{n=k}^{N} b_n \qquad\qquad \text{(since } a_n \leq b_n)$$

$$\leq \sum_{n=k}^{N} b_n + \sum_{n=N+1}^{\infty} b_n \qquad\qquad \text{(since } b_n \geq 0)$$

$$= \sum_{n=k}^{\infty} b_n = L.$$

Thus, L is an upper bound for the partial sums, and therefore the series $\sum_{n=k}^{\infty} a_n$ converges.

Part 2 follows immediately from part 1: if $\sum_{n=k}^{\infty} a_n$ diverges, then $\sum_{n=k}^{\infty} b_n$ must diverge as well, because if it converged, then by part 1, $\sum_{n=k}^{\infty} a_n$ would converge, contrary to our assumption.   □

Notice that in the proof of part 1 we showed that for every $N \geq k$, $\sum_{n=k}^{N} a_n \leq \sum_{n=k}^{\infty} b_n$. By the preservation of weak inequalities by limits, it follows that

$$\sum_{n=k}^{\infty} a_n = \lim_{N \to \infty} \sum_{n=k}^{N} a_n \leq \sum_{n=k}^{\infty} b_n.$$

We can also say more about the series in part 2 of the theorem. The only way that a weakly increasing sequence of partial sums can fail to converge is if the partial sums approach ∞. It follows that in part 2 we must have $\sum_{n=k}^{\infty} a_n = \sum_{n=k}^{\infty} b_n = \infty$.

Intuitively, you could think of the comparison test as saying that a series with nonnegative terms will converge if the terms are "small enough," and diverge if the terms are "too large." In part 1, the terms of the series $\sum_{n=k}^{\infty} b_n$ are small enough for the series to converge, and the terms of the series $\sum_{n=k}^{\infty} a_n$ are even smaller, so they must also be small enough for the series to converge. In part 2, the terms of $\sum_{n=k}^{\infty} a_n$ are too large for the series to converge, and the terms of $\sum_{n=k}^{\infty} b_n$ are even larger, so they are also too large for convergence.

Example 10.3.2. Determine whether the following series converge or diverge:

(a) $\sum_{n=1}^{\infty} \dfrac{1}{n^2+1}$,

(b) $\sum_{n=1}^{\infty} \dfrac{1}{2n-1}$,

(c) $\sum_{n=1}^{\infty} \dfrac{1}{(3+\sin n)^n}$,

(d) $\sum_{n=1}^{\infty} \dfrac{1}{2^{(n^2)}}$.

Solution. You could test series (a) for convergence by using the integral test, but it is easier to recognize that it is similar to the p-series $\sum_{n=1}^{\infty} 1/n^2$, which we already know converges. Before we can use the comparison test, we must work out the necessary inequalities. Clearly for every positive integer n we have $n^2 + 1 \geq n^2 > 0$, and therefore

$$0 \leq \frac{1}{n^2+1} \leq \frac{1}{n^2}.$$

We can now apply part 1 of the comparison test, with $a_n = 1/(n^2+1)$ and $b_n = 1/n^2$, to conclude that series (a) converges.

Series (b) consists of the odd-numbered terms of the harmonic series:

$$\sum_{n=1}^{\infty} \frac{1}{2n-1} = 1 + \frac{1}{3} + \frac{1}{5} + \cdots.$$

This time we compare to the series

$$\sum_{n=1}^{\infty} \frac{1}{2n} = \sum_{n=1}^{\infty} \frac{1}{2} \cdot \frac{1}{n} = \frac{1}{2} + \frac{1}{4} + \frac{1}{6} + \cdots, \tag{10.9}$$

which is 1/2 times the familiar harmonic series. We know that the harmonic series diverges, and by part 2 of Theorem 10.1.4, multiplying the terms by 1/2 does not change this. Thus, series (10.9) diverges. For $n \geq 1$ it is clear that $2n \geq 2n - 1 > 0$, and therefore

$$0 \leq \frac{1}{2n} \leq \frac{1}{2n-1},$$

so by part 2 of the comparison test, series (b) diverges.

What makes series (c) hard to analyze is that the quantity $3 + \sin n$ fluctuates as n increases. But we know that $\sin n$ is always between -1 and 1, so for every positive integer n, $2 \leq 3 + \sin n \leq 4$ and therefore

$$0 \leq \frac{1}{4^n} \leq \frac{1}{(3+\sin n)^n} \leq \frac{1}{2^n}.$$

The series $\sum_{n=1}^{\infty} 1/4^n$ and $\sum_{n=1}^{\infty} 1/2^n$ are geometric series, with $r = 1/4$ in the first case and $r = 1/2$ in the second. In both cases we have $-1 < r < 1$, so both series converge.

But which one should we compare series (c) to? Looking back at the statement of the comparison test, we see that if we let $a_n = 1/(3 + \sin n)^n$ and $b_n = 1/2^n$ then we can use part 1 of the test to conclude that series (c) converges.

Finally, we also compare series (d) to the convergent geometric series $\sum_{n=1}^{\infty} 1/2^n$. For every $n \geq 1$ we have $n^2 \geq n$, so $2^{(n^2)} \geq 2^n > 0$ and

$$0 \leq \frac{1}{2^{(n^2)}} \leq \frac{1}{2^n}.$$

By part 1 of the comparison test, series (d) converges. $\square$

As with the integral test, we can use the idea behind the comparison test to estimate the tail of a convergent series. For example, let s be the exact value of series (d) in the last example:

$$s = \sum_{n=1}^{\infty} \frac{1}{2^{(n^2)}} = \frac{1}{2} + \frac{1}{16} + \frac{1}{512} + \frac{1}{65536} + \cdots.$$

The terms of this series shrink very quickly, and as a result one doesn't need to add very many terms to get a good approximation of s. For example, the third partial sum of the series is

$$s_3 = \sum_{n=1}^{3} \frac{1}{2^{(n^2)}} = \frac{1}{2} + \frac{1}{16} + \frac{1}{512} = \frac{289}{512} \approx 0.564453.$$

The error in the approximation $s \approx s_3$ is given by the third tail of the series:

$$s - s_3 = t_3 = \sum_{n=4}^{\infty} \frac{1}{2^{(n^2)}} = \frac{1}{2^{16}} + \frac{1}{2^{25}} + \cdots.$$

To overestimate t_3, we note that for $n \geq 4$, $n^2 \geq 4n$, which implies that $1/2^{(n^2)} \leq 1/2^{4n}$. We can therefore compare t_3 to the series $\sum_{n=4}^{\infty} 1/2^{4n}$. As we observed after the proof of the comparison test, the proof shows that

$$t_3 = \sum_{n=4}^{\infty} \frac{1}{2^{(n^2)}} \leq \sum_{n=4}^{\infty} \frac{1}{2^{4n}}.$$

But the last series above is a geometric series, so we can evaluate it easily:

$$\sum_{n=4}^{\infty} \frac{1}{2^{4n}} = \sum_{n=4}^{\infty} \left(\frac{1}{2^4}\right)^n = \left(\frac{1}{16}\right)^4 + \left(\frac{1}{16}\right)^5 + \cdots = \frac{(1/16)^4}{1 - 1/16} = \frac{1}{61440} \approx 0.000016.$$

Clearly $t_3 \geq 0$, so we have $0 \leq t_3 \leq 1/61440$. Since $s - s_3 = t_3$, this means that $0 \leq s - s_3 \leq 1/61440$, so

$$s_3 \leq s \leq s_3 + \frac{1}{61440}.$$

In other words, the value of the series is between about 0.564453 and 0.564469. We don't know the exact value of the series, but just by adding the first three terms we have found a very good approximation.

It is important to check the necessary inequalities before applying the comparison test; if the inequalities go in the wrong direction, then the test cannot be used. For example, consider the series

$$\sum_{n=1}^{\infty} \frac{1}{\sqrt{n}+3}. \tag{10.10}$$

A natural idea is to compare this series to the p-series

$$\sum_{n=1}^{\infty} \frac{1}{\sqrt{n}} = \sum_{n=1}^{\infty} \frac{1}{n^{1/2}}. \tag{10.11}$$

Since $p = 1/2 < 1$, series (10.11) diverges. But $\sqrt{n}+3 \geq \sqrt{n} > 0$, so

$$0 \leq \frac{1}{\sqrt{n}+3} \leq \frac{1}{\sqrt{n}},$$

and that means that the comparison test won't help us. The comparison test only applies if either the series with larger terms converges (part 1) or the series with smaller terms diverges (part 2). But in this case, the series with larger terms diverges, and the comparison test has nothing to say about that situation. A good way to think about it is that the terms of series (10.11) are too large for the series to converge, but the terms of series (10.10) are smaller. Maybe they are so much smaller that the series converges; without further analysis, we can't be sure.

It is possible to use the comparison test to determine whether or not the series (10.10) converges—see Exercise 29. But it is probably easier to use the following variant of the comparison test.

Theorem 10.3.3 (Limit Comparison Test). *Suppose that $\sum_{n=k}^{\infty} a_n$ and $\sum_{n=k}^{\infty} b_n$ are two infinite series, and for all integers $n \geq k$, $a_n > 0$ and $b_n > 0$. Suppose also that $\lim_{n \to \infty} a_n/b_n$ is defined, and $\lim_{n \to \infty} a_n/b_n > 0$. Then the series $\sum_{n=k}^{\infty} a_n$ and $\sum_{n=k}^{\infty} b_n$ either both converge or both diverge.*

The idea behind the proof of the limit comparison test is very simple. Suppose that $\lim_{n \to \infty} a_n/b_n = L > 0$. Then for large integers n, $a_n/b_n \approx L$, and therefore $a_n \approx L \cdot b_n$. According to part 2 of Theorem 10.1.4, multiplying the terms of a series by a nonzero constant does not affect the convergence or divergence of the series. Therefore it seems reasonable that $\sum_{n=k}^{\infty} a_n$ and $\sum_{n=k}^{\infty} b_n$ should either both converge or both diverge.

Turning this idea into a proof requires that we work out the approximation $a_n \approx L \cdot b_n$ more carefully. Here are the details.

Proof. Let $L = \lim_{n \to \infty} a_n/b_n > 0$. We now apply the definition of sequence limits. Let $\epsilon = L/2 > 0$. Then by the limit definition, there must be some integer $N \geq k$ such that if n is an integer and $n > N$, then $|a_n/b_n - L| < \epsilon$. In other words, for every integer $n > N$, a_n/b_n is between $L - \epsilon = L - L/2 = L/2$ and $L + \epsilon = L + L/2 = 3L/2$, and therefore

$$0 \le \frac{L}{2} \cdot b_n \le a_n \le \frac{3L}{2} \cdot b_n. \tag{10.12}$$

Suppose now that $\sum_{n=k}^{\infty} b_n$ converges. Then by Theorem 10.1.4, $\sum_{n=N+1}^{\infty} (3L/2)b_n$ converges. By inequality (10.12), we can now apply the comparison test to conclude that $\sum_{n=N+1}^{\infty} a_n$ converges, and therefore $\sum_{n=k}^{\infty} a_n$ converges.

Similarly, if $\sum_{n=k}^{\infty} b_n$ diverges, then $\sum_{n=N+1}^{\infty} (L/2)b_n$ diverges. By inequality (10.12) and the comparison test, $\sum_{n=N+1}^{\infty} a_n$ diverges, and therefore $\sum_{n=k}^{\infty} a_n$ diverges. $\square$

Let's return now to the comparison of the series $\sum_{n=1}^{\infty} 1/(\sqrt{n}+3)$ and $\sum_{n=1}^{\infty} 1/\sqrt{n}$. We were unable to apply the comparison test to these series, so let's try the limit comparison test. We begin by computing

$$\lim_{n\to\infty} \frac{1/(\sqrt{n}+3)}{1/\sqrt{n}} = \lim_{n\to\infty} \frac{\sqrt{n}}{\sqrt{n}+3} = \lim_{n\to\infty} \frac{1}{1+3/\sqrt{n}} = 1 > 0.$$

Since this limit is defined and positive, the limit comparison test tells us that the two series either both converge or both diverge. Since $\sum_{n=1}^{\infty} 1/\sqrt{n}$ is a divergent p-series, we conclude that $\sum_{n=1}^{\infty} 1/(\sqrt{n}+3)$ diverges as well.

Example 10.3.4. Determine whether the following series converge or diverge:

$$\text{(a)} \sum_{n=2}^{\infty} \frac{\sqrt{n}}{n^2-n}, \qquad \text{(b)} \sum_{n=1}^{\infty} \frac{1}{n^{1+1/n}}.$$

Solution. In series (a), the term $-n$ in the denominator seems less important than the term n^2, so we try comparing the series to the simpler series

$$\sum_{n=2}^{\infty} \frac{\sqrt{n}}{n^2} = \sum_{n=2}^{\infty} \frac{1}{n^{3/2}}.$$

This is a p-series with $p = 3/2 > 1$, so it converges. Unfortunately, the terms of series (a) are larger, so the comparison test doesn't apply. We try the limit comparison test instead. Since

$$\lim_{n\to\infty} \frac{\sqrt{n}/(n^2-n)}{1/n^{3/2}} = \lim_{n\to\infty} \frac{n^2}{n^2-n} = \lim_{n\to\infty} \frac{1}{1-1/n} = 1 > 0,$$

the limit comparison test tells us that series (a) converges.

In series (b), when n is large, the exponent $1 + 1/n$ in the denominator is just slightly larger than 1. This means that the terms of the series are slightly smaller than the terms of the harmonic series. The harmonic series diverges, but just barely. Are the terms of series (b) *enough* smaller to make the series converge? The limit comparison test gives us the answer. We have

$$\lim_{n\to\infty} \frac{1/n^{1+1/n}}{1/n} = \lim_{n\to\infty} \frac{n}{n^{1+1/n}} = \lim_{n\to\infty} \frac{1}{n^{1/n}} = \lim_{n\to\infty} \frac{1}{e^{(\ln n)/n}} = \frac{1}{e^0} = 1 > 0.$$

Since the harmonic series diverges, by the limit comparison test, series (b) diverges as well. □

Exercises 10.3

1–24: Determine whether the series converges or diverges.

1. $\displaystyle\sum_{n=1}^{\infty} \frac{1}{2^n + 1}$.

2. $\displaystyle\sum_{n=1}^{\infty} \frac{1}{2^n - n}$.

3. $\displaystyle\sum_{n=1}^{\infty} \frac{2^n - 1}{3^n + n}$.

4. $\displaystyle\sum_{n=1}^{\infty} \frac{2^n + 1}{3^n - n}$.

5. $\displaystyle\sum_{n=1}^{\infty} \frac{1}{\sqrt{n^3} + \sqrt[3]{n^2}}$.

6. $\displaystyle\sum_{n=1}^{\infty} \left(\frac{1}{\sqrt{n^3}} + \frac{1}{\sqrt[3]{n^2}} \right)$.

7. $\displaystyle\sum_{n=1}^{\infty} \frac{1}{n^{3+\sin n}}$.

8. $\displaystyle\sum_{n=1}^{\infty} \frac{n + 4}{n^3 - 3n + 5}$.

9. $\displaystyle\sum_{n=1}^{\infty} \frac{n^2 + 4}{n^3 - 3n + 5}$.

10. $\displaystyle\sum_{n=1}^{\infty} \frac{n^3 + 4}{n^3 - 3n + 5}$.

11. $\displaystyle\sum_{n=1}^{\infty} \frac{\sqrt{6^n + 1}}{3^n}$.

12. $\displaystyle\sum_{n=1}^{\infty} \frac{\sqrt{4^n+1}}{2^n}.$

13. $\displaystyle\sum_{n=1}^{\infty} \frac{n!+1}{(n+1)!}.$

14. $\displaystyle\sum_{n=1}^{\infty} \frac{n!+2}{(n+2)!}.$

15. $\displaystyle\sum_{n=1}^{\infty} \frac{1}{(3+(-1)^n)^n}.$

16. $\displaystyle\sum_{n=1}^{\infty} \frac{1}{(2+(-1)^n)^n}.$

17. $\displaystyle\sum_{n=1}^{\infty} \frac{1}{2^{\sqrt{n^2-1}}}.$

18. $\displaystyle\sum_{n=1}^{\infty} \frac{1}{n^{1+1/\sqrt{n}}}.$

19. $\displaystyle\sum_{n=1}^{\infty} \frac{1}{n^{1+1/\ln n}}.$

20. $\displaystyle\sum_{n=1}^{\infty} \frac{1}{n^{1+1/\sqrt{\ln n}}}.$

21. $\displaystyle\sum_{n=1}^{\infty} \frac{1}{n^{\ln n}}.$

22. $\displaystyle\sum_{n=2}^{\infty} \frac{1}{(\ln n)^{\ln n}}.$

23. $\displaystyle\sum_{n=2}^{\infty} \frac{1}{(\ln n)^{\ln(\ln n)}}.$

24. $\displaystyle\sum_{n=1}^{\infty} \frac{1}{(\tan^{-1} n)^n}.$

25. Let

$$s = \sum_{n=1}^{\infty} \frac{1}{2^{n+1/n}}.$$

(a) Compute the partial sum

$$s_{10} = \sum_{n=1}^{10} \frac{1}{2^{n+1/n}}.$$

(b) Show that the error in the approximation $s \approx s_{10}$ is at most 0.001. (Hint: Overestimate the tail t_{10} with a series that you can evaluate.)

26. Let

$$s = \sum_{n=1}^{\infty} \frac{1}{n^3 + 1}.$$

(a) Compute the partial sum

$$s_{10} = \sum_{n=1}^{10} \frac{1}{n^3 + 1}.$$

(b) Show that the error in the approximation $s \approx s_{10}$ is at most 0.005. (Hint: You could try to overestimate the tail t_{10} with an improper integral, as in the integral test, but the integral is difficult to evaluate. Try first comparing the tail to another series that is easier to overestimate.)

27. Let

$$s = \sum_{n=1}^{\infty} \frac{1}{3^n \sqrt{n}}.$$

(a) Compute the partial sum

$$s_3 = \sum_{n=1}^{3} \frac{1}{3^n \sqrt{n}}.$$

(b) Estimate the error in the approximation $s \approx s_3$. That is, find a small number c so that you can show that the error is at most c. (The smaller c is, the better.)

28. Approximate the value of the series

$$\sum_{n=1}^{\infty} \frac{1}{(3 + \sin n)^n}$$

with error at most 0.001.

29. Show that for every $n \geq 1$,

$$0 \leq \frac{1}{4\sqrt{n}} \leq \frac{1}{\sqrt{n} + 3},$$

and then apply the comparison test to show that $\sum_{n=1}^{\infty} \frac{1}{\sqrt{n} + 3}$ diverges.

30. Suppose that for all integers $n \geq k$, $a_n > 0$ and $b_n > 0$.

    (a) Show that if $\lim_{n \to \infty} a_n/b_n = 0$ and $\sum_{n=k}^{\infty} b_n$ converges, then $\sum_{n=k}^{\infty} a_n$ converges.

    (b) Show that if $\lim_{n \to \infty} a_n/b_n = \infty$ and $\sum_{n=k}^{\infty} a_n$ converges, then $\sum_{n=k}^{\infty} b_n$ converges.

31. Consider a number written in decimal notation as

    $$0.d_1 d_2 d_3 \cdots,$$

    where each d_n is one of the digits 0, 1, 2, ..., 9. As we have discussed before, this notation represents the infinite series

    $$\frac{d_1}{10} + \frac{d_2}{100} + \frac{d_3}{1000} + \cdots = \sum_{n=1}^{\infty} \frac{d_n}{10^n}.$$

    Show that this series converges. (This shows that decimal notation always succeeds in specifying a number.)

10.4 The Ratio and Root Tests

If a series $\sum_{n=1}^{\infty} a_n$ is geometric, then each term after the first is some fixed number r times the previous term. In other words, for every positive integer n, $a_{n+1}/a_n = r$.

Now consider a series $\sum_{n=1}^{\infty} a_n$ in which the ratios a_{n+1}/a_n are close to some number L, but not exactly equal to it. It is reasonable to think that such a series might be comparable to a geometric series with a ratio r that is close to L. That is the motivation for our next convergence test.

Theorem 10.4.1 (Ratio Test). *Suppose $\sum_{n=k}^{\infty} a_n$ is an infinite series, and for all $n \geq k$, $a_n > 0$. Suppose also that either $\lim_{n \to \infty} a_{n+1}/a_n$ is defined or $\lim_{n \to \infty} a_{n+1}/a_n = \infty$, and let $L = \lim_{n \to \infty} a_{n+1}/a_n$. (Thus, either L is a number or $L = \infty$.)*

1. *If $L < 1$, then $\sum_{n=k}^{\infty} a_n$ converges.*

2. *If $L > 1$ (including the case $L = \infty$), then $\sum_{n=k}^{\infty} a_n$ diverges.*

Proof. Since the terms of the series are all positive, it is clear that $L \geq 0$. To prove part 1, suppose that $L < 1$. Let r be any number such that $L < r < 1$. Then by the sequence version of Theorem 2.8.8, there must be some integer $N \geq k$ such that for every integer $n \geq N$, $a_{n+1}/a_n \leq r$, and therefore $a_{n+1} \leq a_n r$. Since this is true for *every* integer $n \geq N$, we have

$$a_{N+1} \leq a_N r,$$

$$a_{N+2} \leq a_{N+1} r \leq (a_N r) r = a_N r^2,$$

$$a_{N+3} \leq a_{N+2} r \leq (a_N r^2) r = a_N r^3,$$

and in general, for every $n \geq N$,

$$0 \leq a_n \leq a_N r^{n-N}.$$

We can now apply the comparison test to the series

$$\sum_{n=N}^{\infty} a_n \quad \text{and} \quad \sum_{n=N}^{\infty} a_N r^{n-N} = a_N + a_N r + a_N r^2 + \cdots.$$

The second series above is a geometric series, with $0 < r < 1$, so it converges. It follows, by the comparison test, that $\sum_{n=N}^{\infty} a_n$ converges, and therefore $\sum_{n=k}^{\infty} a_n$ converges.

The proof of part 2 is similar. If $L > 1$, then we can choose r such that $L > r > 1$, and then we can find an integer $N \geq k$ such that for every integer $n \geq N$, $a_n \geq a_N r^{n-N}$. Since $a_N > 0$ and $r > 1$, $\lim_{n \to \infty} a_N r^{n-N} = \infty$, and therefore $\lim_{n \to \infty} a_n = \infty$. It follows, by the test for divergence, that $\sum_{n=k}^{\infty} a_n$ diverges. $\qquad \square$

Note that if $L = 1$, then the ratio test doesn't tell us anything. The reason for this is clear from the proof: In the proof, we compared the ratio a_{n+1}/a_n to numbers that were close to L, but not equal to L. We needed some "wiggle room" between L and 1 to be able to make appropriate comparisons, so the idea of the proof wouldn't work if $L = 1$. In fact, if $L = 1$ then the series may converge and it may diverge; see Exercise 17.

The ratio test works best when the ratio a_{n+1}/a_n can be simplified easily. For example, the ratio test is often useful when the formula for a_n involves exponential functions or factorials.

Example 10.4.2. Determine whether the following series converge or diverge:

$$\text{(a)} \sum_{n=0}^{\infty} \frac{2^n}{n!}, \quad \text{(b)} \sum_{n=1}^{\infty} \frac{n^2}{2^n}, \quad \text{(c)} \sum_{n=1}^{\infty} \frac{n^n}{n!}.$$

Solution. In series (a), the nth term is $a_n = 2^n/n!$. Since the formula for a_n involves both 2^n and $n!$, the ratio test seems like a good bet. When computing the limit of a_{n+1}/a_n, it is often helpful to group together corresponding parts of the formulas for a_n and a_{n+1}. In this case, we group together the exponential terms and the factorial terms before simplifying:

$$L = \lim_{n \to \infty} \frac{a_{n+1}}{a_n} = \lim_{n \to \infty} \frac{2^{n+1}/(n+1)!}{2^n/n!} = \lim_{n \to \infty} \left[\frac{2^{n+1}}{(n+1)!} \cdot \frac{n!}{2^n} \right]$$

$$= \lim_{n \to \infty} \left[\frac{2^{n+1}}{2^n} \cdot \frac{n!}{(n+1)!} \right] = \lim_{n \to \infty} \left[2 \cdot \frac{1}{n+1} \right] = 2 \cdot 0 = 0 < 1.$$

By part 1 of the ratio test, series (a) converges.

Similarly, for series (b) we compute

$$L = \lim_{n\to\infty} \frac{(n+1)^2/2^{n+1}}{n^2/2^n} = \lim_{n\to\infty}\left[\frac{(n+1)^2}{2^{n+1}}\cdot\frac{2^n}{n^2}\right] = \lim_{n\to\infty}\left[\left(\frac{n+1}{n}\right)^2\cdot\frac{2^n}{2^{n+1}}\right]$$

$$= \lim_{n\to\infty}\left[\left(1+\frac{1}{n}\right)^2\cdot\frac{1}{2}\right] = 1^2\cdot\frac{1}{2} = \frac{1}{2} < 1.$$

Once again, we conclude that the series converges.

The computation for series (c) is a bit more difficult:

$$L = \lim_{n\to\infty} \frac{(n+1)^{n+1}/(n+1)!}{n^n/n!} = \lim_{n\to\infty}\left[\frac{(n+1)^{n+1}}{(n+1)!}\cdot\frac{n!}{n^n}\right]$$

$$= \lim_{n\to\infty}\left[\frac{(n+1)^{n+1}}{n^n}\cdot\frac{n!}{(n+1)!}\right] = \lim_{n\to\infty}\left[\frac{(n+1)^{n+1}}{n^n}\cdot\frac{1}{n+1}\right]$$

$$= \lim_{n\to\infty}\frac{(n+1)^n}{n^n} = \lim_{n\to\infty}\left(\frac{n+1}{n}\right)^n = \lim_{n\to\infty}\left(1+\frac{1}{n}\right)^n = e.$$

Note that in the final step we have used limit (c) from Example 7.6.3. Since $e \approx 2.718 > 1$, part 2 of the ratio test tells us that series (c) diverges.   □

As usual, finding the exact values of the convergent series in Example 10.4.2 is much harder than showing that they converge, but we can get approximate answers by computing partial sums. For series (a), using a computer for the arithmetic, we find that

$$s_{10} = \sum_{n=0}^{10} \frac{2^n}{n!} \approx 7.388995.$$

If we let s denote the exact value, then the error in the approximation $s \approx s_{10}$ is the tail

$$s - s_{10} = t_{10} = \sum_{n=11}^{\infty} \frac{2^n}{n!} = \frac{2^{11}}{11!} + \frac{2^{12}}{12!} + \cdots.$$

In Exercise 18 we ask you to show, by comparing this tail to a geometric series, that $0 \le t_{10} \le 0.000062$. Thus, the exact value of the series is between about 7.388995 and 7.389057. Is it just a coincidence that $e^2 \approx 7.389056$? No, it turns out that the exact value of the series is e^2, but you'll have to wait until Section 10.8 to find out why.

Series (b) in Example 10.4.2 converges a little more slowly, but the 20th partial sum turns out to be a very good approximation:

$$s_{20} = \sum_{n=1}^{20} \frac{n^2}{2^n} \approx 5.99954.$$

This time, the easiest way to estimate the error is to compare the tail to an improper integral. In Exercise 19, we ask you to show that $0 \le t_{20} \le 0.00064$, so the value of series (b) is between about 5.99954 and 6.00018. You can probably guess that the exact value is 6, but again we're not ready yet to explain why.

Sometimes it is useful to combine the ratio test with other tests. For example, consider the series

$$\sum_{n=1}^{\infty} \frac{n^5 + 3}{n! - 7}. \tag{10.13}$$

The presence of $n!$ in the denominator suggests that the ratio test might be helpful, but the -7 in the denominator will interfere with the cancellation that would allow us to simplify a_{n+1}/a_n. Furthermore, the hypotheses of the ratio test are not satisfied: the terms of this series are negative for $n \leq 3$, but the ratio test requires positive terms.

We therefore proceed as follows. First, we drop the first three terms and study the series

$$\sum_{n=4}^{\infty} \frac{n^5 + 3}{n! - 7}; \tag{10.14}$$

that takes care of the problem of negative terms. If we can determine whether or not series (10.14) converges, then the answer will be the same for series (10.13). Next, we compare series (10.14) to the similar but simpler series

$$\sum_{n=4}^{\infty} \frac{n^5}{n!}. \tag{10.15}$$

For this last series, the ratio test works well:

$$L = \lim_{n \to \infty} \frac{(n+1)^5/(n+1)!}{n^5/n!} = \lim_{n \to \infty} \left[\frac{(n+1)^5}{n^5} \cdot \frac{n!}{(n+1)!} \right]$$

$$= \lim_{n \to \infty} \left[\left(1 + \frac{1}{n} \right)^5 \cdot \frac{1}{n+1} \right] = 0.$$

Since $L < 1$, by the ratio test, series (10.15) converges.

Unfortunately, the terms of series (10.14) are larger, so the comparison test doesn't apply. We therefore try the limit comparison test, which requires that we compute the limit

$$L = \lim_{n \to \infty} \frac{(n^5 + 3)/(n! - 7)}{n^5/n!} = \lim_{n \to \infty} \left[\frac{n^5 + 3}{n^5} \cdot \frac{n!}{n! - 7} \right]$$

$$= \lim_{n \to \infty} \left[\left(1 + \frac{3}{n^5} \right) \cdot \frac{1}{1 - 7/n!} \right] = 1.$$

Since $L > 0$, we can conclude by the limit comparison test that series (10.14) converges, and therefore (10.13) converges as well.

The intuition behind the ratio test is that it allows us to tell if a series is comparable to a geometric series. Another convergence test that is based on a similar intuitive idea is the root test.

Theorem 10.4.3 (Root Test). *Suppose $\sum_{n=k}^{\infty} a_k$ is an infinite series, and for all $n \geq k$, $a_n \geq 0$. Suppose also that either $\lim_{n\to\infty} \sqrt[n]{a_n}$ is defined or $\lim_{n\to\infty} \sqrt[n]{a_n} = \infty$, and let $L = \lim_{n\to\infty} \sqrt[n]{a_n}$.*

1. *If $L < 1$, then $\sum_{n=k}^{\infty} a_n$ converges.*

2. *If $L > 1$ (including the case $L = \infty$), then $\sum_{n=k}^{\infty} a_n$ diverges.*

Proof. The proof is similar to the proof of the ratio test. For part 1, suppose that $L < 1$. Choose a number r such that $L < r < 1$, and then choose an integer $N \geq k$ such that if n is an integer and $n \geq N$, then $\sqrt[n]{a_n} \leq r$. Then for every integer $n \geq N$,

$$0 \leq a_n \leq r^n.$$

Since $0 < r < 1$, the geometric series $\sum_{n=N}^{\infty} r^n$ converges. By the comparison test, $\sum_{n=N}^{\infty} a_n$ converges, and therefore $\sum_{n=k}^{\infty} a_n$ converges.

If $L > 1$, then we choose r such that $L > r > 1$, and we choose an integer $N \geq k$ such that for every integer $n \geq N$, $a_n \geq r^n$. Since $r > 1$, $\lim_{n\to\infty} r^n = \infty$. Therefore $\lim_{n\to\infty} a_n = \infty$, so $\sum_{n=k}^{\infty} a_n$ diverges by the test for divergence. $\qquad\square$

Example 10.4.4. Determine whether the following series converge or diverge:

$$\text{(a) } \sum_{n=1}^{\infty} (\tan^{-1} n)^n, \qquad \text{(b) } \sum_{n=0}^{\infty} \frac{1}{2^{(n+1+(-1)^n)}}.$$

Solution. In series (a), it is easy to compute the nth root of the nth term, so we use the root test. We begin by computing

$$L = \lim_{n\to\infty} \sqrt[n]{(\tan^{-1} n)^n} = \lim_{n\to\infty} \tan^{-1} n = \frac{\pi}{2} \approx 1.57 > 1.$$

By part 2 of the root test, we conclude that series (a) diverges. You can reach this same conclusion by using the ratio test, but the calculations are much harder; see Exercise 20.

Series (b) looks similar to the geometric series $\sum_{n=1}^{\infty} 1/2^n$, so it seems reasonable to try using either the ratio test or the root test. It turns out that the ratio test doesn't work for this series—see Exercise 21. But the root test does work:

$$L = \lim_{n\to\infty} \sqrt[n]{\frac{1}{2^{(n+1+(-1)^n)}}} = \lim_{n\to\infty} \left(\left(\frac{1}{2} \right)^{(n+1+(-1)^n)} \right)^{1/n}$$

$$= \lim_{n\to\infty} \left(\frac{1}{2} \right)^{(1+1/n+(-1)^n/n)} = \left(\frac{1}{2} \right)^1 = \frac{1}{2} < 1,$$

so by part 1 of the root test, series (b) converges. $\qquad\square$

It is interesting to compare series (b) in the last example to the geometric series $\sum_{n=1}^{\infty} 1/2^n$:

$$\sum_{n=0}^{\infty} \frac{1}{2^{(n+1+(-1)^n)}} = \frac{1}{4} + \frac{1}{2} + \frac{1}{16} + \frac{1}{8} + \cdots,$$

$$\sum_{n=1}^{\infty} \frac{1}{2^n} = \frac{1}{2} + \frac{1}{4} + \frac{1}{8} + \frac{1}{16} + \cdots.$$

The terms in the two series are the same, but they appear in different orders. We say that one series is a *rearrangement* of the other. Notice that since the orders of the terms in the two series are different, the partial sums are also different. Are the values of the series the same? See part (b) of Exercise 21. We will have more to say about rearrangements in the next section.

Exercises 10.4

1–16: Determine whether the series converges or diverges.

1. $\displaystyle\sum_{n=1}^{\infty} \frac{n2^{n+1}}{3^{n-1}}$.

2. $\displaystyle\sum_{n=1}^{\infty} \frac{3^{2n}}{n2^{3n}}$.

3. $\displaystyle\sum_{n=1}^{\infty} \frac{n^3 + 1}{n!}$.

4. $\displaystyle\sum_{n=1}^{\infty} \frac{3^n + 1}{n!}$.

5. $\displaystyle\sum_{n=1}^{\infty} \frac{n!}{3^n n^3}$.

6. $\displaystyle\sum_{n=1}^{\infty} \frac{3^{(n^2)}}{9^n}$.

7. $\displaystyle\sum_{n=1}^{\infty} \frac{2^n \sqrt{n}}{9^n}$.

8. $\displaystyle\sum_{n=1}^{\infty} \frac{n!}{e^{(n^2)}}$.

9. $\displaystyle\sum_{n=1}^{\infty} \frac{1}{2^n - \sqrt{n}}$.

10. $\displaystyle\sum_{n=1}^{\infty} \frac{(n!)^2}{(2n)!}$.

11. $\displaystyle\sum_{n=1}^{\infty} \frac{n^n n!}{(2n)!}$.

12. $\displaystyle\sum_{n=1}^{\infty} \frac{n^{2n}}{(2n)!}$.

13. $\displaystyle\sum_{n=1}^{\infty} \frac{n^{3n}}{(3n)!}$.

14. $\displaystyle\sum_{n=1}^{\infty} \left(\frac{2n+3}{3n+2}\right)^n$.

15. $\displaystyle\sum_{n=1}^{\infty} \left(1 - \frac{1}{n}\right)^n$.

16. $\displaystyle\sum_{n=1}^{\infty} \left(1 - \frac{1}{n}\right)^{(n^2)}$.

17. Show that if $\sum_{n=1}^{\infty} a_n$ is a p-series, then $\lim_{n\to\infty} a_{n+1}/a_n = 1$. Since some p-series converge and some diverge, this illustrates why the ratio test is inconclusive when $L = 1$.

18. In the text, we approximated the value of series (a) from Example 10.4.2 with the 10th partial sum. The error in this approximation is the tail

$$t_{10} = \sum_{n=11}^{\infty} \frac{2^n}{n!}.$$

By comparing t_{10} to a geometric series, show that $0 \le t_{10} \le 0.000062$.

19. In the text, we approximated the value of series (b) from Example 10.4.2 with the 20th partial sum. The error in this approximation is the tail

$$t_{20} = \sum_{n=21}^{\infty} \frac{n^2}{2^n}.$$

By comparing t_{20} to an improper integral, show that $0 \le t_{20} \le 0.00064$.

20. Use the ratio test to show that series (a) in Example 10.4.4 diverges.

21. In this exercise you will study series (b) from Example 10.4.4, which was the following rearrangement of a geometric series:

$$\sum_{n=0}^{\infty} \frac{1}{2^{(n+1+(-1)^n)}} = \frac{1}{4} + \frac{1}{2} + \frac{1}{16} + \frac{1}{8} + \cdots.$$

   (a) Show that the ratio test cannot be used to test this series for convergence.

   (b) Find the value of the series. Is it the same as the value of the series $\sum_{n=1}^{\infty} 1/2^n$? (Hint: Compute the first few partial sums, and then determine in general what the Nth partial sum is.)

10.5 Absolute Convergence and the Alternating Series Test

So far, almost all of our convergence tests have applied only to series whose terms are nonnegative. How can we test a series for convergence if some of its terms are negative? Often, the best thing to try first is our next theorem.

Theorem 10.5.1 (Absolute Convergence Test). *If $\sum_{n=k}^{\infty} |a_n|$ converges, then $\sum_{n=k}^{\infty} a_n$ converges.*

Proof. We will use the fact that for every integer $n \geq k$, $-|a_n| \leq a_n \leq |a_n|$, and therefore

$$0 \leq a_n + |a_n| \leq 2|a_n|.$$

Suppose that $\sum_{n=k}^{\infty} |a_n|$ converges. Then $\sum_{n=k}^{\infty} 2|a_n|$ converges, and by the comparison test it follows that $\sum_{n=k}^{\infty} (a_n + |a_n|)$ converges. To show that $\sum_{n=k}^{\infty} a_n$ converges, we rewrite it in the form

$$\sum_{n=k}^{\infty} a_n = \sum_{n=k}^{\infty} ((a_n + |a_n|) + (-1)|a_n|).$$

We have just shown that $\sum_{n=k}^{\infty} (a_n + |a_n|)$ converges, and by part 2 of Theorem 10.1.4, $\sum_{n=k}^{\infty} (-1)|a_n|$ converges, so by part 1 of the same theorem, $\sum_{n=k}^{\infty} a_n$ converges. $\square$

If $\sum_{n=k}^{\infty} |a_n|$ converges, then we say that $\sum_{n=k}^{\infty} a_n$ *converges absolutely*. In this terminology, what Theorem 10.5.1 says is that if a series converges absolutely, then it converges. What makes this fact useful is that the terms of the series $\sum_{n=k}^{\infty} |a_n|$ are nonnegative, and therefore we can apply the convergence tests we have developed over the last few sections to it.

Example 10.5.2. Determine whether the following series converge or diverge:

(a) $\displaystyle\sum_{n=1}^{\infty} \frac{(-1)^{n-1}}{n^2}$, (b) $\displaystyle\sum_{n=1}^{\infty} \frac{\sin n}{2^n}$, (c) $\displaystyle\sum_{n=1}^{\infty} \frac{(2n)!}{(-2)^n n!}$.

Solution. We have

$$\sum_{n=1}^{\infty} \left| \frac{(-1)^{n-1}}{n^2} \right| = \sum_{n=1}^{\infty} \frac{1}{n^2},$$

which converges, since it is a p-series with $p = 2 > 1$. Therefore series (a) converges absolutely, so it converges. For the value of the series, see Exercise 18.

For series (b), we use the fact that for every positive integer n, $-1 \leq \sin n \leq 1$, and therefore

$$0 \leq \left| \frac{\sin n}{2^n} \right| \leq \frac{1}{2^n}.$$

Since $\sum_{n=1}^{\infty} 1/2^n$ is a convergent geometric series, $\sum_{n=1}^{\infty} |(\sin n)/2^n|$ converges by the comparison test, so series (b) converges by the absolute convergence test.

Finally, for series (c), we test

$$\sum_{n=1}^{\infty} \left| \frac{(2n)!}{(-2)^n n!} \right| = \sum_{n=1}^{\infty} \frac{(2n)!}{2^n n!} \tag{10.16}$$

for convergence by using the ratio test. We begin by computing

$$L = \lim_{n \to \infty} \frac{(2(n+1))!/(2^{n+1}(n+1)!)}{(2n)!/(2^n n!)} = \lim_{n \to \infty} \left[\frac{(2n+2)!}{(2n)!} \cdot \frac{2^n}{2^{n+1}} \cdot \frac{n!}{(n+1)!} \right]$$

$$= \lim_{n \to \infty} \left[(2n+2) \cdot (2n+1) \cdot \frac{1}{2} \cdot \frac{1}{n+1} \right] = \lim_{n \to \infty} (2n+1) = \infty.$$

By part 2 of the ratio test, we conclude that series (10.16) diverges. But what about the original series (c)? The absolute convergence test says that if (10.16) converges, then (c) converges; but it *doesn't* say that if (10.16) diverges, then (c) diverges.

Nevertheless, we can use the calculations we have done to complete this example. To see how, look back at the proof of the ratio test. The proof of part 2 of the test showed that if $\lim_{n \to \infty} a_{n+1}/a_n > 1$, then $\lim_{n \to \infty} a_n = \infty$. In this case, since we computed that $L = \infty > 1$, we can conclude that $\lim_{n \to \infty} (2n)!/(2^n n!) = \infty$. But then

$$\lim_{n \to \infty} \frac{(2n)!}{(-2)^n n!} = \lim_{n \to \infty} (-1)^n \frac{(2n)!}{2^n n!}$$

is undefined, because the terms oscillate between large magnitude positive and negative numbers. We conclude, by the test for divergence, that series (c) diverges. □

The reasoning we used in part (c) of the last example can be applied whenever we use the ratio test to show that a series does not converge absolutely, and similar reasoning works for the root test as well. These facts are important enough that we record them as new versions of the ratio and root tests.

Theorem 10.5.3 (Ratio Test for Absolute Convergence). *Suppose $\sum_{n=k}^{\infty} a_n$ is an infinite series, and for all $n \geq k$, $a_n \neq 0$. Suppose also that either $\lim_{n \to \infty} |a_{n+1}|/|a_n|$ is defined or $\lim_{n \to \infty} |a_{n+1}|/|a_n| = \infty$, and let $L = \lim_{n \to \infty} |a_{n+1}|/|a_n|$.*

1. *If $L < 1$, then $\sum_{n=k}^{\infty} a_n$ converges absolutely.*

2. *If $L > 1$, then $\sum_{n=k}^{\infty} a_n$ diverges.*

Theorem 10.5.4 (Root Test for Absolute Convergence). *Suppose $\sum_{n=k}^{\infty} a_n$ is an infinite series. Suppose also that either $\lim_{n\to\infty} \sqrt[n]{|a_n|}$ is defined or $\lim_{n\to\infty} \sqrt[n]{|a_n|} = \infty$, and let $L = \lim_{n\to\infty} \sqrt[n]{|a_n|}$.*

1. *If $L < 1$, then $\sum_{n=k}^{\infty} a_n$ converges absolutely.*

2. *If $L > 1$, then $\sum_{n=k}^{\infty} a_n$ diverges.*

Proofs. The first parts of both theorems follow immediately from the original ratio and root tests: if $L < 1$ in either theorem, then $\sum_{n=k}^{\infty} |a_n|$ converges, which means that $\sum_{n=k}^{\infty} a_n$ converges absolutely.

If $L > 1$ in either theorem, then the proofs of the original ratio and root tests show that $\lim_{n\to\infty} |a_n| = \infty$. It follows that $\lim_{n\to\infty} a_n$ cannot be 0, so $\sum_{n=k}^{\infty} a_n$ diverges by the test for divergence. This proves part 2 of both theorems. $\qquad\square$

Unfortunately, our other convergence tests do not have similar versions for absolute convergence. Furthermore, it is not true in general that if $\sum_{n=k}^{\infty} |a_n|$ diverges, then $\sum_{n=k}^{\infty} a_n$ diverges. For example, consider the series

$$\sum_{n=1}^{\infty} \frac{(-1)^{n-1}}{n} = 1 - \frac{1}{2} + \frac{1}{3} - \frac{1}{4} + \cdots . \tag{10.17}$$

This series is called the *alternating harmonic series*, because the terms are the same as those of the harmonic series, except that they alternate positive and negative. Of course, if we take the absolute values of the terms of series (10.17) then we get the harmonic series, which diverges, so the alternating harmonic series does not converge absolutely. But it turns out that it does converge.

To see why, let's look at the first few partial sums of the alternating harmonic series:

$$s_1 = 1, \qquad\qquad s_2 = 1 - \frac{1}{2} = 0.5,$$

$$s_3 = 1 - \frac{1}{2} + \frac{1}{3} \approx 0.833, \qquad s_4 = 1 - \frac{1}{2} + \frac{1}{3} - \frac{1}{4} \approx 0.583,$$

$$s_5 = 1 - \frac{1}{2} + \frac{1}{3} - \frac{1}{4} + \frac{1}{5} \approx 0.783, \quad s_6 = 1 - \frac{1}{2} + \frac{1}{3} - \frac{1}{4} + \frac{1}{5} - \frac{1}{6} \approx 0.617.$$

If we mark these partial sums on a number line, as in Figure 10.3, then we see that they jump back and forth, with the jumps getting smaller and smaller. In the figure, it looks plausible that the partial sums are zeroing in on a limiting value somewhere between $s_6 \approx 0.617$ and $s_5 \approx 0.783$.

To confirm that the alternating harmonic series converges, we pick out the crucial properties of the series that are responsible for the behavior of the partial sums that we see in Figure 10.3, and use them to formulate another convergence test.

Theorem 10.5.5 (Alternating Series Test). *Suppose $\sum_{n=k}^{\infty} a_n$ is a series whose terms alternate positive and negative. Suppose also that $|a_k| \geq |a_{k+1}| \geq |a_{k+2}| \geq \cdots$ and $\lim_{n\to\infty} |a_n| = 0$. Then $\sum_{n=k}^{\infty} a_n$ converges.*

Proof. For each integer $n \geq k$, let $b_n = |a_n|$. Then the hypotheses of the theorem tell us that $b_k \geq b_{k+1} \geq b_{k+2} \geq \cdots$ and $\lim_{n \to \infty} b_n = 0$. We will assume that the series starts with a positive term; the proof for a series that starts with a negative term is similar. Since the terms alternate positive and negative, this means that $a_k = b_k$, $a_{k+1} = -b_{k+1}$, $a_{k+2} = b_{k+2}$, and so on. Thus, the series is

$$\sum_{n=k}^{\infty} a_n = b_k - b_{k+1} + b_{k+2} - b_{k+3} + \cdots.$$

For $N \geq k$, let $s_N = \sum_{n=k}^{N} a_n$, the Nth partial sum. The key idea in the proof is to determine the order of these partial sums. Since $0 \leq b_{k+2} \leq b_{k+1}$,

$$b_k - b_{k+1} \leq b_k - b_{k+1} + b_{k+2} \leq b_k - b_{k+1} + b_{k+1} = b_k;$$

in other words,

$$s_{k+1} \leq s_{k+2} \leq s_k.$$

Similar reasoning shows that every partial sum is between the previous two partial sums. This means that $s_{k+1} \leq s_{k+3} \leq s_{k+2}$, $s_{k+3} \leq s_{k+4} \leq s_{k+2}$, and so on. Therefore the partial sums are ordered as follows:

$$s_{k+1} \leq s_{k+3} \leq s_{k+5} \leq \cdots \leq s_{k+4} \leq s_{k+2} \leq s_k. \tag{10.18}$$

This ordering matches the way the partial sums are arranged in Figure 10.3.

Another way to describe the ordering (10.18) is to say that the intervals $[s_{k+1}, s_k]$, $[s_{k+3}, s_{k+2}]$, $[s_{k+5}, s_{k+4}]$, ... form a nested sequence of intervals. The widths of the intervals are

$$s_k - s_{k+1} = b_k - (b_k - b_{k+1}) = b_{k+1},$$
$$s_{k+2} - s_{k+3} = (b_k - b_{k+1} + b_{k+2}) - (b_k - b_{k+1} + b_{k+2} - b_{k+3}) = b_{k+3},$$
$$s_{k+4} - s_{k+5} = (b_k - b_{k+1} + \cdots + b_{k+4}) - (b_k - b_{k+1} + \cdots + b_{k+4} - b_{k+5}) = b_{k+5},$$

and so on. Since $\lim_{n \to \infty} b_n = 0$, the widths of the intervals approach 0, and therefore we can apply the nested interval theorem (Theorem 2.8.7) to conclude that there is a unique number s that belongs to all of the intervals. This means that

$$s_{k+1} \leq s_{k+3} \leq s_{k+5} \leq \cdots \leq s \leq \cdots \leq s_{k+4} \leq s_{k+2} \leq s_k. \tag{10.19}$$

Furthermore, both the left and right endpoints of the nested intervals approach s, which implies that $\lim_{N \to \infty} s_N = s$. In other words, the series $\sum_{n=k}^{\infty} a_n$ converges to s. $\square$

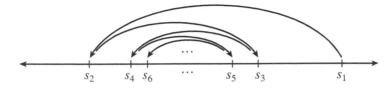

Figure 10.3: The first six partial sums of the alternating harmonic series.

A useful observation from the proof of the alternating series test is that according to (10.19), the value s of the series is between any two successive partial sums. It follows that, in the notation of the proof, for any $N \geq k$,

$$|s - s_N| \leq |s_{N+1} - s_N| = |(a_k + \cdots + a_N + a_{N+1}) - (a_k + \cdots + a_N)|$$
$$= |a_{N+1}| = b_{N+1}.$$

In other words, the magnitude of the error in the approximation $s \approx s_N$ is at most the absolute value of the first term missing from the partial sum s_N. Keep in mind that these conclusions are based on the proof of the alternating series test, and therefore *they apply only to series that satisfy the hypotheses of that test.*

Returning to the alternating harmonic series (10.17), we have $a_n = (-1)^{n-1}/n$, and therefore $b_n = |a_n| = 1/n$. Clearly $1 \geq 1/2 \geq 1/3 \geq \cdots$ and $\lim_{n\to\infty} 1/n = 0$, so the alternating series test tells us that the series converges. Furthermore, if we want to approximate the value s of the series with error at most, say, $1/200 = 0.005$, then it will suffice to compute the 199th partial sum s_{199}, since the magnitude of the error in this approximation will be at most the absolute value of the first missing term:

$$|s - s_{199}| \leq b_{200} = \frac{1}{200}.$$

Computation leads to the values

$$s_{199} \approx 0.6957, \qquad s_{200} \approx 0.6907.$$

The value of the series must be between these two partial sums. You may recall that in Section 7.3 we found that $\ln 2 \approx 0.693$. It turns out that this is the exact value of the alternating harmonic series. One proof of this can be found in Exercise 19, but we will give a more satisfying explanation in Section 10.7.

A series that converges, but does not converge absolutely, is said to converge *conditionally.* The alternating harmonic series is our first example of a series that converges conditionally. Let's try a more challenging example. Consider the series

$$\sum_{n=0}^{\infty} \frac{(-1)^n (2n)!}{4^n (n!)^2} = 1 - \frac{1}{2} + \frac{3}{8} - \frac{5}{16} + \cdots . \tag{10.20}$$

As in the proof of the alternating series test, we will use the notation

$$a_n = \frac{(-1)^n (2n)!}{4^n (n!)^2}, \qquad b_n = |a_n| = \frac{(2n)!}{4^n (n!)^2}.$$

To see if series (10.20) converges absolutely, we must determine whether or not the series

$$\sum_{n=0}^{\infty} b_n = \sum_{n=0}^{\infty} \frac{(2n)!}{4^n (n!)^2} \tag{10.21}$$

converges. A natural test to try is the ratio test. Unfortunately, we find that

$$\frac{b_{n+1}}{b_n} = \frac{(2n+2)!/(4^{n+1}((n+1)!)^2)}{(2n)!/(4^n(n!)^2)}$$

$$= \frac{(2n+2)!}{(2n)!} \cdot \frac{4^n}{4^{n+1}} \cdot \frac{(n!)^2}{((n+1)!)^2} = \frac{(2n+2)(2n+1)}{4(n+1)^2} = \frac{2n+1}{2n+2}, \quad (10.22)$$

and therefore

$$\lim_{n\to\infty} \frac{b_{n+1}}{b_n} = \lim_{n\to\infty} \frac{2n+1}{2n+2} = \lim_{n\to\infty} \frac{2+1/n}{2+2/n} = 1.$$

Thus, the ratio test is inconclusive; we'll need to try something else.

We leave it as an exercise for you (see Exercise 23) to check that for all $n \geq 1$,

$$\frac{1}{2\sqrt{n}} \leq b_n \leq \frac{1}{\sqrt{2n}}. \quad (10.23)$$

The series

$$\sum_{n=1}^{\infty} \frac{1}{2\sqrt{n}} = \sum_{n=1}^{\infty} \frac{1}{2} \cdot \frac{1}{n^{1/2}}$$

is 1/2 times a p-series with $p = 1/2 < 1$, so it diverges. Therefore, by (10.23) and the comparison test, series (10.21) diverges, which means that series (10.20) does not converge absolutely.

However, (10.20) does satisfy the hypotheses of the alternating series test: Clearly the terms alternate positive and negative. According to (10.22), for every n,

$$\frac{b_{n+1}}{b_n} = \frac{2n+1}{2n+2} \leq 1,$$

so $b_{n+1} \leq b_n$. And (10.23) implies that $\lim_{n\to\infty} b_n = 0$. Therefore the series converges conditionally. As usual, if we let s denote the value of the series, then we can approximate s by computing partial sums. For this series we have

$$s_{199} \approx 0.687, \quad s_{200} \approx 0.727,$$

so s must be between these two numbers. It turns out that the exact value of series (10.20) is $\sqrt{2}/2 \approx 0.707$, but, yet again, we will have to ask you to wait to find out why.

If a series converges, why should we care whether the convergence is absolute or conditional? We give one example to illustrate the importance of this distinction. Recall that we say that one series is a *rearrangement* of another if the two series consist of the same terms, but added up in a different order. We saw one example of a rearrangement at the end of the last section. Here is another example: Let $\sum_{n=1}^{\infty} a_n$ be the series

$$\sum_{n=1}^{\infty} a_n = 1 + \frac{1}{3} - \frac{1}{2} + \frac{1}{5} + \frac{1}{7} - \frac{1}{4} + \frac{1}{9} + \frac{1}{11} - \frac{1}{6} + \cdots. \quad (10.24)$$

This series is a rearrangement of the alternating harmonic series; the terms of both series consist of all numbers of the form $(-1)^{n-1}/n$. In the alternating harmonic series, the terms alternate positive and negative, whereas in series (10.24), we have two positive terms, then one negative, then two positive, then one negative, and so on. Surprisingly, in this case changing the order of the terms changes the value of the series! The value of the alternating harmonic series is $\ln 2 \approx 0.693$, but computation shows that the 100th partial sum of series (10.24) is about 1.03971. It turns out that the value of the series is $(3/2)\ln 2 \approx 1.03972$; for a proof, see Exercise 21.

How can changing the order of the terms change the value of a series? The value of a series is, by definition, the limit of its partial sums, and changing the order of the terms certainly changes the partial sums. But that's only because partial sums are *partial*: which terms are included in any partial sum depends on the order of the terms. Shouldn't the effect of these changes disappear when we take the limit of the partial sums, which takes *all* of the terms into account? The answer, it turns out, is that it depends on whether the series converges absolutely or conditionally.

Theorem 10.5.6.

1. *If a series converges absolutely, then all rearrangements of the series converge to the same value.*

2. *If a series converges conditionally, then for every real number c there is a rearrangement of the series that converges to c.*

Part 2 of the theorem is sometimes called the Riemann rearrangement theorem. For proofs of both parts, see Exercises 24 and 25.

Exercises 10.5

1–17: Determine whether the series converges absolutely, converges conditionally, or diverges.

1. $\displaystyle\sum_{n=1}^{\infty} \frac{(-1)^n}{5n+3}$.

2. $\displaystyle\sum_{n=1}^{\infty} \frac{(-1)^n}{5n^2+3}$.

3. $\displaystyle\sum_{n=1}^{\infty} \frac{\sin n}{n^2+\sqrt{n}}$.

4. $\displaystyle\sum_{n=0}^{\infty} \frac{(1+2\sin n)^n}{n!}$.

5. $\displaystyle\sum_{n=1}^{\infty}(-1)^n \frac{n+1}{3n+1}$.

6. $\displaystyle\sum_{n=1}^{\infty}(-1)^n \frac{n+1}{3n^2+1}$.

7. $\displaystyle\sum_{n=1}^{\infty}(-1)^n\frac{n+1}{3n^3+1}$.

8. $\displaystyle\sum_{n=1}^{\infty}\left(\frac{\cos(3n\pi/2)}{2}\right)^n$.

9. $\displaystyle\sum_{n=1}^{\infty}\left(\frac{3\cos(n\pi/2)}{2}\right)^n$.

10. $\displaystyle\sum_{n=1}^{\infty}(-1)^n\frac{n+100}{100n}$.

11. $\displaystyle\sum_{n=1}^{\infty}\frac{(-2)^n}{2^n+n}$.

12. $\displaystyle\sum_{n=1}^{\infty}(-1)^n\frac{n}{n^2+1}$.

13. $\displaystyle\sum_{n=1}^{\infty}(-1)^n\frac{n}{n^2+2^n}$.

14. $\displaystyle\sum_{n=1}^{\infty}(-1)^n\frac{n2^n}{n^2+2^n}$.

15. $\displaystyle\sum_{n=1}^{\infty}(-1)^n\frac{n2^n}{n^22^n+1}$.

16. $\displaystyle\sum_{n=1}^{\infty}\frac{1}{(1+3(-1)^n)^n}$.

17. $\displaystyle\sum_{n=1}^{\infty}\frac{1}{(1+2(-1)^n)^n}$.

18. In this exercise, you will find the value of series (a) in Example 10.5.2.

   (a) Show that for every positive integer N,

$$\sum_{n=1}^{2N}\frac{(-1)^{n-1}}{n^2}=\sum_{n=1}^{2N}\frac{1}{n^2}-\frac{1}{2}\sum_{n=1}^{N}\frac{1}{n^2}.$$

   (Hint: One approach is to use mathematical induction. Another is to think about the even- and odd-numbered terms separately.)

   (b) Use part (a), together with the fact that $\sum_{n=1}^{\infty}1/n^2=\pi^2/6$ (see Exercises 26 and 27 in Section 10.2), to find the value of $\sum_{n=1}^{\infty}(-1)^{n-1}/n^2$.

19. In this exercise, you will find the value of series (10.17), the alternating harmonic series.

    (a) Show that for every positive integer N,

    $$\sum_{n=1}^{2N} \frac{(-1)^{n-1}}{n} = \frac{1}{N+1} + \frac{1}{N+2} + \cdots + \frac{1}{2N} = \sum_{i=1}^{N} \frac{1}{N+i}.$$

    (Hint: This is similar to part (a) of Exercise 18.)

    (b) Show that the sum on the right-hand side of the equation in part (a) is a Riemann N-sum for the integral $\int_1^2 (1/x)\, dx$.

    (c) Use parts (a) and (b) to show that the alternating harmonic series converges to $\ln 2$.

20. (a) Show that $\displaystyle\sum_{n=0}^{\infty} \frac{4(-1)^n}{2n+1}$ converges conditionally.

    (b) Use a partial sum to estimate the value of the series in part (a) with error at most 0.01. (You will probably want to use a computer to do the arithmetic.)

    (c) Do you have a guess at what the exact value of the series is? (We will find the exact value in Section 10.7.)

21. In this exercise, you will find the value of series (10.24), which is the following rearrangement of the alternating harmonic series:

    $$\sum_{n=1}^{\infty} a_n = 1 + \frac{1}{3} - \frac{1}{2} + \frac{1}{5} + \frac{1}{7} - \frac{1}{4} + \frac{1}{9} + \frac{1}{11} - \frac{1}{6} + \cdots .$$

    We will make use of the following series in our solution:

    $$\sum_{n=1}^{\infty} b_n = \sum_{n=1}^{\infty} \frac{(-1)^{n-1}}{n} = 1 - \frac{1}{2} + \frac{1}{3} - \frac{1}{4} + \cdots = \ln 2,$$

    $$\sum_{n=1}^{\infty} c_n = \sum_{n=1}^{\infty} \frac{\sin((n-1)\pi/2)}{n} = 0 + \frac{1}{2} + 0 - \frac{1}{4} + 0 + \frac{1}{6} + \cdots .$$

    (a) Show that $\sum_{n=1}^{\infty} c_n = (1/2)\ln 2$. (Hint: What is the relationship between the partial sums of this series and the partial sums of the alternating harmonic series?)

    (b) Show that

    $$\sum_{n=1}^{\infty} a_n = \sum_{n=1}^{\infty} (b_n + c_n).$$

    (Hint: Write out the first few terms of $\sum_{n=1}^{\infty} (b_n + c_n)$, and then imitate part (a).)

    (c) Combine parts (a) and (b) to show that $\sum_{n=1}^{\infty} a_n = (3/2)\ln 2$.

22. Evaluate the series

$$1 + \frac{1}{2} - 1 + \frac{1}{3} + \frac{1}{4} - \frac{1}{2} + \frac{1}{5} + \frac{1}{6} - \frac{1}{3} + \cdots . \tag{10.25}$$

(Hint: One approach is to evaluate the following series, and then add them together:

$$1 + 0 - \frac{1}{2} + \frac{1}{3} + 0 - \frac{1}{4} + \frac{1}{5} + 0 - \frac{1}{6} + \cdots$$

$$0 + \frac{1}{2} - \frac{1}{2} + 0 + \frac{1}{4} - \frac{1}{4} + 0 + \frac{1}{6} - \frac{1}{6} + \cdots .$$

Notice that in series (10.25), every number of the form $1/n$ occurs twice, once positive and once negative. You might therefore expect the value to be 0, but it's not.)

23. In this exercise, you will confirm inequality (10.23). As in the inequality, we use the notation

$$b_n = \frac{(2n)!}{4^n (n!)^2}.$$

(a) Use mathematical induction to show that for every positive integer n,

$$\frac{1}{4n} \le b_n^2 \le \frac{2n-1}{4n^2}.$$

(b) Use part (a) to prove inequality (10.23).

24. In this exercise, you will prove part 1 of Theorem 10.5.6. Suppose that $\sum_{n=1}^{\infty} a_n$ converges absolutely. Let $s = \sum_{n=1}^{\infty} a_n$, and suppose that $\sum_{n=1}^{\infty} b_n$ is a rearrangement of $\sum_{n=1}^{\infty} a_n$. We must prove that $\sum_{n=1}^{\infty} b_n = s$.

Suppose $\epsilon > 0$. To complete the proof, we will find a number M such that for every positive integer $N > M$, $\left| \sum_{n=1}^{N} b_n - s \right| < \epsilon$.

(a) Show that there is some positive integer L such that $\sum_{n=L+1}^{\infty} |a_n| < \epsilon/2$. We will use this number L in the remaining parts of this exercise.

(b) Show that if $n_1, n_2, \ldots, n_k$ are distinct integers larger than L, then

$$|a_{n_1} + a_{n_2} + \cdots + a_{n_k}| \le \sum_{n=L+1}^{\infty} |a_n|.$$

(Hint: First try working this out for $k = 1$, $k = 2$, and $k = 3$.)

(c) Show that

$$\left| \sum_{n=1}^{L} a_n - s \right| \le \sum_{n=L+1}^{\infty} |a_n|.$$

(d) Show that there is some positive integer M such that all of the terms a_1, $a_2, \ldots, a_L$ occur among $b_1, b_2, \ldots, b_M$.

(e) Show that for every positive integer $N > M$,

$$\left| \sum_{n=1}^{N} b_n - s \right| < \epsilon.$$

25. Let c be any real number. In this exercise you will show that there is some rearrangement of the alternating harmonic series that converges to c. Similar reasoning could be applied to any conditionally convergent series to prove part 2 of Theorem 10.5.6.

    For each positive integer n, let $p_n = 1/(2n - 1)$ and $q_n = 1/(2n)$. Thus, p_1, p_2, p_3, ... are the positive terms of the alternating harmonic series, and $-q_1, -q_2, -q_3, \ldots$ are the negative terms.

    (a) Show that there is some positive integer M_1 such that

    $$\sum_{n=1}^{M_1} p_n > c.$$

    For the rest of this exercise, let M_1 be the smallest such integer.

    (b) Show that there is some positive integer N_1 such that

    $$\sum_{n=1}^{M_1} p_n - \sum_{n=1}^{N_1} q_n < c.$$

    Let N_1 be the smallest such integer.

    (c) Show that we can similarly choose the smallest M_2 such that $M_2 > M_1$ and

    $$\sum_{n=1}^{M_1} p_n - \sum_{n=1}^{N_1} q_n + \sum_{n=M_1+1}^{M_2} p_n > c,$$

    and then the smallest N_2 such that $N_2 > N_1$ and

    $$\sum_{n=1}^{M_1} p_n - \sum_{n=1}^{N_1} q_n + \sum_{n=M_1+1}^{M_2} p_n - \sum_{n=N_1+1}^{N_2} q_n < c,$$

    and so on.

    (d) Let $(b_n)_{n=1}^{\infty}$ be the sequence

    $$p_1, p_2, \ldots, p_{M_1}, -q_1, -q_2, \ldots, -q_{N_1}, p_{M_1+1}, \ldots, p_{M_2},$$
    $$-q_{N_1+1}, \ldots, q_{N_2}, p_{M_2+1}, \ldots.$$

    Show that the series $\sum_{n=1}^{\infty} b_n$ is a rearrangement of the alternating harmonic series that converges to c.

10.6 Power Series

We have now developed methods that allow us to tell in many cases whether or not a series converges. But often, after showing that a series converges, we have been unable to find the exact value of the series. For example, here are four series whose convergence we have verified in the last few sections:

$$\sum_{n=0}^{\infty} \frac{2^n}{n!} = 1 + 2 + 2 + \frac{4}{3} + \cdots, \tag{10.26a}$$

$$\sum_{n=1}^{\infty} \frac{n^2}{2^n} = \frac{1}{2} + 1 + \frac{9}{8} + 1 + \cdots, \tag{10.26b}$$

$$\sum_{n=1}^{\infty} \frac{(-1)^{n-1}}{n} = 1 - \frac{1}{2} + \frac{1}{3} - \frac{1}{4} + \cdots, \tag{10.26c}$$

$$\sum_{n=0}^{\infty} \frac{(-1)^n (2n)!}{4^n (n!)^2} = 1 - \frac{1}{2} + \frac{3}{8} - \frac{5}{16} + \cdots. \tag{10.26d}$$

The first two are series (a) and (b) from Example 10.4.2, and the third and fourth are series (10.17) and (10.20) from Section 10.5. We have told you that the values of these series are, respectively, e^2, 6, $\ln 2$, and $\sqrt{2}/2$, but we have not yet explained why these values are correct.

To give these explanations, we will turn to an idea that has helped us out several times in the past: the paradox of generalization. Instead of studying the series in (10.26), we will study these series:

$$\sum_{n=0}^{\infty} \frac{x^n}{n!} = 1 + x + \frac{x^2}{2} + \frac{x^3}{6} + \cdots, \tag{10.27a}$$

$$\sum_{n=1}^{\infty} n^2 x^n = x + 4x^2 + 9x^3 + 16x^4 + \cdots, \tag{10.27b}$$

$$\sum_{n=1}^{\infty} \frac{(-1)^{n-1} x^n}{n} = x - \frac{x^2}{2} + \frac{x^3}{3} - \frac{x^4}{4} + \cdots, \tag{10.27c}$$

$$\sum_{n=0}^{\infty} \frac{(2n)! x^n}{(n!)^2} = 1 + 2x + 6x^2 + 20x^3 + \cdots. \tag{10.27d}$$

Of course, these four series contain an unspecified number x, so their values will depend on the value of x. In other words, we can think of them as defining functions. If we can determine what these functions are, then we will be able to find the values of series (10.26a–d) by substituting 2 for x in series (10.27a), 1/2 for x in series (10.27b), 1 for x in series (10.27c), and $-1/4$ for x in series (10.27d).

Series like the ones in (10.27) are called *power series*. Here is the general definition.

Definition 10.6.1. Suppose $(c_n)_{n=0}^{\infty}$ is a sequence of numbers. Then the series

$$\sum_{n=0}^{\infty} c_n x^n = c_0 + c_1 x + c_2 x^2 + c_3 x^3 + \cdots$$

is called a *power series*. The number c_n is called the *coefficient* of x^n in the power series.

For example, consider the power series in which all of the coefficients are 1:

$$\sum_{n=0}^{\infty} 1 \cdot x^n = 1 + x + x^2 + x^3 + \cdots .$$

This is a geometric series with first term $a = 1$ and ratio $r = x$, so if $-1 < x < 1$ then it converges to $a/(1-r) = 1/(1-x)$, and for other values of x it diverges. In other words, we have

$$\sum_{n=0}^{\infty} x^n = 1 + x + x^2 + x^3 + \cdots = \frac{1}{1-x}, \qquad -1 < x < 1. \qquad (10.28)$$

We could think of equation (10.28) as saying that, although the function $f(x) = 1/(1-x)$ is not a polynomial, the restriction of this function to the interval $(-1, 1)$ is almost a polynomial; it can be written as a power series, which is just like a polynomial except that it has infinitely many terms. Polynomials are among the easiest functions to work with in calculus: they are continuous everywhere, and their derivatives and integrals are easy to compute. It is natural, then, to wonder what will happen if we apply the concepts of calculus to power series. Are functions defined by power series continuous, and can we differentiate and integrate them? As we will see in the next section, the calculus of power series is almost as easy as the calculus of polynomials.

Notice that the set of values of x for which the power series in (10.28) converges is an interval centered at 0, the interval $(-1, 1)$. We call this the *interval of convergence* of the power series. It turns out that the set of values of x for which a power series converges is always either just the number 0 or an interval centered at 0.

Theorem 10.6.2. *For any power series*

$$\sum_{n=0}^{\infty} c_n x^n,$$

one of the following is true:

1. *The power series converges only when $x = 0$.*

2. *There is some positive number R such that the power series converges absolutely if $|x| < R$ and diverges if $|x| > R$.*

3. *The power series converges absolutely for all values of x.*

If statement 2 of the theorem is true, then the series may converge absolutely, converge conditionally, or diverge when $x = \pm R$. Thus, the interval of convergence of the power series may be $(-R, R)$, $[-R, R)$, $(-R, R]$, or $[-R, R]$. The number R is called the *radius of convergence* of the power series. It is customary to say that if statement 1 is true then the radius of convergence of the power series is 0, and if statement 3 is true then the radius of convergence is infinite.

The key idea for the proof of Theorem 10.6.2 is the following lemma.

Lemma 10.6.3. *Suppose a and b are real numbers such that $|a| < |b|$. If the power series $\sum_{n=0}^{\infty} c_n x^n$ converges when $x = b$, then it converges absolutely when $x = a$. If it diverges when $x = a$, then it diverges when $x = b$.*

Proof. Suppose that the power series converges when $x = b$; in other words, the series $\sum_{n=0}^{\infty} c_n b^n$ converges. Then by Theorem 10.2.1, $\lim_{n\to\infty} c_n b^n = 0$. Setting $\epsilon = 1$ in the definition of limits, we see that there is some positive integer N such that for every integer $n > N$, $|c_n b^n| < 1$. It follows that for every $n > N$ we have

$$0 \leq |c_n a^n| = \left| c_n b^n \cdot \left(\frac{a}{b}\right)^n \right| = |c_n b^n| \cdot \left|\frac{a}{b}\right|^n \leq \left|\frac{a}{b}\right|^n.$$

We now apply the comparison test to the series

$$\text{(a)} \sum_{n=N+1}^{\infty} |c_n a^n|, \qquad \text{(b)} \sum_{n=N+1}^{\infty} \left|\frac{a}{b}\right|^n.$$

Series (b) is a geometric series, and since $|a| < |b|$, $|a/b| < 1$. Therefore series (b) converges. By the comparison test, series (a) converges as well. Therefore $\sum_{n=0}^{\infty} |c_n a^n|$ converges, which means that $\sum_{n=0}^{\infty} c_n a^n$ converges absolutely. This proves the first statement in the lemma. The second statement follows immediately from the first. $\square$

Proof of Theorem 10.6.2. Let A be the set of all values of x for which the power series converges. If $x = 0$, then the value of the power series is

$$\sum_{n=0}^{\infty} c_n \cdot 0^n = c_0 + 0 + 0 + \cdots = c_0.$$

(Recall our convention that $0^0 = 1$; see the footnote on page 165.) Thus, the series converges when $x = 0$, so $0 \in A$.

If A has no upper bound, then for every real number r, there is some $b \in A$ such that $b > |r|$. Since $b \in A$, the series converges when $x = b$, and therefore, by the lemma, it converges absolutely when $x = r$. Thus, the series converges absolutely for every value of x, so statement 3 of the theorem is true.

Now suppose that A has an upper bound. Then by the completeness of the real numbers, we can let R be the least upper bound of A. Notice that since $0 \in A$, $R \geq 0$.

Consider any number r. If $|r| < R$, then since R is the *least* upper bound of A, $|r|$ is not an upper bound, so there must be some $b \in A$ such that $b > |r|$. As before, since $b \in A$, the series converges when $x = b$, and we conclude by the lemma that the series converges absolutely when $x = r$. On the other hand, if $|r| > R$, then we can find a number a such that $R < a < |r|$. Since R is an upper bound for A, it cannot be the case that $a \in A$, so the series diverges when $x = a$. By the last statement in the lemma, it follows that the series diverges when $x = r$.

To summarize, we have shown that for every number r, if $|r| < R$ then the power series converges absolutely when $x = r$, and if $|r| > R$ then it diverges when $x = r$. It follows that if $R > 0$ then statement 2 of the theorem is true, and if $R = 0$ then statement 1 is true. □

Example 10.6.4. Determine the values of x for which the following power series converge:

$$\text{(a) } \sum_{n=0}^{\infty} \frac{x^n}{(n+1)3^n}, \qquad \text{(b) } \sum_{n=0}^{\infty} \frac{x^n}{n!},$$

$$\text{(c) } \sum_{n=0}^{\infty} \frac{(2n)!x^n}{(n!)^2}, \qquad \text{(d) } \sum_{n=0}^{\infty} \frac{x^n}{2^n + n^2}.$$

Solution. In each series, we treat x as a fixed but unknown number and test the series for convergence using the methods we have developed in the previous sections. Every power series converges when $x = 0$, so we only need to test the series for $x \neq 0$. Often a good first step is to use either the ratio test or the root test for absolute convergence. (Note that, since x could be negative, the series could have both positive and negative terms even if all of the coefficients are positive.)

Applying the ratio test for absolute convergence to series (a), we begin by computing

$$L = \lim_{n \to \infty} \frac{|x^{n+1}/((n+2)3^{n+1})|}{|x^n/((n+1)3^n)|} = \lim_{n \to \infty} \left[\frac{|x|^{n+1}}{(n+2)3^{n+1}} \cdot \frac{(n+1)3^n}{|x|^n} \right]$$

$$= \lim_{n \to \infty} \left[\frac{|x|^{n+1}}{|x|^n} \cdot \frac{n+1}{n+2} \cdot \frac{3^n}{3^{n+1}} \right] = \lim_{n \to \infty} \left[|x| \cdot \frac{1+1/n}{1+2/n} \cdot \frac{1}{3} \right] = \frac{|x|}{3}.$$

By the ratio test for absolute convergence, if $|x|/3 < 1$ then the series converges absolutely, and if $|x|/3 > 1$ then it diverges. In other words, the series converges absolutely if $|x| < 3$ and diverges if $|x| > 3$. (We leave it as an exercise for you to check that the root test for absolute convergence would lead to the same conclusion.) This is case 2 of Theorem 10.6.2, and the radius of convergence is 3. If $x = \pm 3$ then the ratio test is inconclusive, and we need to use a different test. We check these two values of x separately.

If $x = 3$, then the series takes the form

$$\sum_{n=0}^{\infty} \frac{3^n}{(n+1)3^n} = \sum_{n=0}^{\infty} \frac{1}{n+1} = 1 + \frac{1}{2} + \frac{1}{3} + \cdots.$$

This is the harmonic series, and as we already know, it diverges.

On the other hand, if $x = -3$ then the series is

$$\sum_{n=0}^{\infty} \frac{(-3)^n}{(n+1)3^n} = \sum_{n=0}^{\infty} \frac{(-1)^n}{n+1} = 1 - \frac{1}{2} + \frac{1}{3} - \cdots .$$

This time we have the alternating harmonic series, which converges. Thus, the interval of convergence of series (a) is $[-3, 3)$.

Notice that series (b) is the same as (10.27a). To test it for absolute convergence, we again use the ratio test for absolute convergence, so we compute

$$L = \lim_{n \to \infty} \frac{|x^{n+1}/(n+1)!|}{|x^n/n!|} = \lim_{n \to \infty} \left[\frac{|x|^{n+1}}{|x|^n} \cdot \frac{n!}{(n+1)!} \right] = \lim_{n \to \infty} \left[|x| \cdot \frac{1}{n+1} \right] = 0.$$

This shows that no matter what x is, $L = 0 < 1$, and therefore the series converges absolutely for all values of x. This is case 3 of Theorem 10.6.2. The interval of convergence of series (b) is $(-\infty, \infty)$, and the radius of convergence is infinite.

Series (c) is the same as (10.27d). As before, a good first step is the ratio test for absolute convergence:

$$L = \lim_{n \to \infty} \frac{|(2n+2)!x^{n+1}/((n+1)!)^2|}{|(2n)!x^n/(n!)^2|} = \lim_{n \to \infty} \left[\frac{|x|^{n+1}}{|x|^n} \cdot \frac{(2n+2)!}{(2n)!} \cdot \frac{(n!)^2}{((n+1)!)^2} \right]$$

$$= \lim_{n \to \infty} \left[|x| \cdot \frac{(2n+2)(2n+1)}{(n+1)^2} \right] = \lim_{n \to \infty} \left[|x| \cdot \frac{4n+2}{n+1} \right]$$

$$= \lim_{n \to \infty} \left[|x| \cdot \frac{4+2/n}{1+1/n} \right] = 4|x|.$$

By the ratio test for absolute convergence, the series converges absolutely if $4|x| < 1$, or in other words $|x| < 1/4$, and it diverges if $4|x| > 1$, or $|x| > 1/4$. The interval of convergence extends from $-1/4$ to $1/4$, but once again we must do separate tests to see if the endpoints are included.

At $x = 1/4$ and $x = -1/4$, the series is

$$x = \frac{1}{4}: \sum_{n=0}^{\infty} \frac{(2n)!}{4^n (n!)^2}, \qquad x = -\frac{1}{4}: \sum_{n=0}^{\infty} \frac{(-1)^n (2n)!}{4^n (n!)^2}.$$

These are series (10.21) and (10.20), and we showed in the last section that the first diverges and the second converges conditionally. Thus, the interval of convergence for series (c) is $[-1/4, 1/4)$.

Finally, we turn to series (d). We could use the ratio test or root test for absolute convergence, but in this case, it seems simpler to start with the comparison test. For all values of x we have

$$0 \le \left| \frac{x^n}{2^n + n^2} \right| = \frac{|x|^n}{2^n + n^2} \le \frac{|x|^n}{2^n} = \left(\frac{|x|}{2} \right)^n .$$

The series $\sum_{n=0}^{\infty} (|x|/2)^n$ is a geometric series, and it converges if $|x|/2 < 1$. Thus, by the comparison test, series (d) converges absolutely if $|x| < 2$. On the other hand, if $x = \pm 2$ then the series becomes

$$x = 2: \sum_{n=0}^{\infty} \frac{2^n}{2^n + n^2}, \qquad x = -2: \sum_{n=0}^{\infty} \frac{(-1)^n 2^n}{2^n + n^2}.$$

We leave it to you to verify that $\lim_{n \to \infty} n^2 / 2^n = 0$ (try l'Hôpital's rule on $\lim_{x \to \infty} x^2 / 2^x = \lim_{x \to \infty} x^2 / e^{x \ln 2}$, or apply Theorem 7.6.2), and therefore

$$\lim_{n \to \infty} \frac{2^n}{2^n + n^2} = \lim_{n \to \infty} \frac{1}{1 + n^2/2^n} = \frac{1}{1 + 0} = 1 \neq 0.$$

By the test for divergence, we conclude that series (d) diverges when $x = 2$. Similarly, when $x = -2$ the terms of the series oscillate between values close to 1 and -1, and therefore the series diverges by the test for divergence. There is no need to check values of x with $|x| > 2$: examining the possibilities in Theorem 10.6.2, we see that the radius of convergence of series (d) must be 2, and therefore the interval of convergence is $(-2, 2)$. □

As a consequence of part (b) of this example, we get the following fact, which will be useful to us later:

Proposition 10.6.5. *For every real number* x,

$$\lim_{n \to \infty} \frac{x^n}{n!} = 0.$$

Proof. As we showed in the last example, for every value of x, $\sum_{n=0}^{\infty} x^n / n!$ converges. It follows, by Theorem 10.2.1, that the terms approach 0. (Notice that you can also deduce this proposition from Exercise 19 in Section 2.9.) □

So far, the only power series whose value we know is $\sum_{n=0}^{\infty} x^n$; as we saw in (10.28), for $-1 < x < 1$ this series converges to $1/(1-x)$, and for other values of x it diverges. We can find the values of some other power series by substituting other expressions for x in (10.28). For example, substituting $-3x$ for x, we can say that if $-1 < -3x < 1$ then

$$\sum_{n=0}^{\infty} (-3x)^n = \frac{1}{1 - (-3x)} = \frac{1}{1 + 3x},$$

and for other values of x this series diverges. To solve the inequality $-1 < -3x < 1$, we divide through by -3 (and reverse the direction of the inequalities) to get $1/3 > x > -1/3$. Thus, we can say that

$$\frac{1}{1 + 3x} = \sum_{n=0}^{\infty} (-3x)^n = \sum_{n=0}^{\infty} (-3)^n x^n = 1 - 3x + 9x^2 - \cdots, \qquad -\frac{1}{3} < x < \frac{1}{3}.$$

Let's try another example. Suppose we want a power series that converges to $x/(x^2 + 4)$. We begin by writing

$$\frac{x}{x^2 + 4} = \frac{x}{4} \cdot \frac{1}{x^2/4 + 1} = \frac{x}{4} \cdot \frac{1}{1 - (-x^2/4)}.$$

Now we substitute $-x^2/4$ for x in (10.28) to get

$$\frac{1}{1-(-x^2/4)} = \sum_{n=0}^{\infty}\left(-\frac{x^2}{4}\right)^n = \sum_{n=0}^{\infty}\frac{(-1)^n x^{2n}}{4^n}, \qquad -1 < -\frac{x^2}{4} < 1.$$

The inequality $-1 < -x^2/4 < 1$ is equivalent to $4 > x^2 > -4$, and the solution to this inequality is $-2 < x < 2$. Thus, we have

$$\frac{x}{x^2+4} = \frac{x}{4}\cdot\frac{1}{1-(-x^2/4)} = \frac{x}{4}\cdot\sum_{n=0}^{\infty}\frac{(-1)^n x^{2n}}{4^n}$$

$$= \sum_{n=0}^{\infty}\frac{(-1)^n x^{2n+1}}{4^{n+1}} = \frac{x}{4} - \frac{x^3}{16} + \frac{x^5}{64} - \cdots, \qquad -2 < x < 2, \qquad (10.29)$$

where we have used part 2 of Theorem 10.1.4 in going from the first line to the second. Notice that this series has only odd powers of x, but in Definition 10.6.1 we defined a power series to be a series that has a term of the form $c_n x^n$ for every integer $n \geq 0$. To view (10.29) as a power series, we think of the even powers of x as having coefficients of 0. In other words, to express $x/(x^2+4)$ as a series that matches our definition of a power series, we would write

$$\frac{x}{x^2+4} = 0 + \frac{1}{4}\cdot x + 0\cdot x^2 + \left(-\frac{1}{16}\right)\cdot x^3 + 0\cdot x^4 + \frac{1}{64}\cdot x^5 + \cdots. \qquad (10.30)$$

However, we usually won't bother to include terms with coefficient 0 when we write power series.[3]

We close this section with one more example. This time we will find a power series representation for the function

$$\frac{x-10}{x^2+x-2}.$$

We begin with partial fractions. Using the methods of Section 8.1, you should be able to derive the partial fractions decomposition

$$\frac{x-10}{x^2+x-2} = \frac{4}{x+2} - \frac{3}{x-1}.$$

Next, we find a power series representation for each fraction on the right-hand side. For the first of these fractions, we have

$$\frac{4}{x+2} = \frac{4}{2}\cdot\frac{1}{x/2+1} = 2\cdot\frac{1}{1-(-x/2)}.$$

---

[3]Strictly speaking, series (10.29) and (10.30) are different; for example, their sequences of partial sums are different. However, their values, which are by definition the limits of their sequences of partial sums, are the same.

Substituting $-x/2$ for x in (10.28), we see that if $-1 < -x/2 < 1$ then

$$\frac{1}{1-(-x/2)} = \sum_{n=0}^{\infty} \left(-\frac{x}{2}\right)^n = \sum_{n=0}^{\infty} \frac{(-1)^n x^n}{2^n}.$$

Since the inequality $-1 < -x/2 < 1$ is equivalent to $-2 < x < 2$, we conclude that

$$\frac{4}{x+2} = 2 \cdot \frac{1}{1-(-x/2)} = 2 \cdot \sum_{n=0}^{\infty} \frac{(-1)^n x^n}{2^n} = \sum_{n=0}^{\infty} \frac{(-1)^n x^n}{2^{n-1}}, \qquad -2 < x < 2.$$

$$(10.31)$$

Similarly, we have

$$-\frac{3}{x-1} = 3 \cdot \frac{1}{1-x} = 3 \cdot \sum_{n=0}^{\infty} x^n = \sum_{n=0}^{\infty} 3x^n, \qquad -1 < x < 1.$$

$$(10.32)$$

We now want to combine (10.31) and (10.32), but we can do this only for values of x that work in both equations—that is, only for $-1 < x < 1$. For these values of x, we use part 1 of Theorem 10.1.4 to deduce that

$$\frac{x-10}{x^2+x-2} = \frac{4}{x+2} - \frac{3}{x-1} = \sum_{n=0}^{\infty} \frac{(-1)^n x^n}{2^{n-1}} + \sum_{n=0}^{\infty} 3x^n$$

$$= \sum_{n=0}^{\infty} \left(\frac{(-1)^n}{2^{n-1}} + 3\right) x^n = 5 + 2x + \frac{7}{2}x^2 + \cdots, \qquad -1 < x < 1.$$

We leave it to you to verify, by the test for divergence, that this power series diverges for all other values of x.

Exercises 10.6

1–15: Find the interval of convergence of the power series.

1. $\displaystyle\sum_{n=0}^{\infty} \frac{x^n}{4^n(n+4)}.$

2. $\displaystyle\sum_{n=2}^{\infty} \frac{x^n}{3^n \ln n}.$

3. $\displaystyle\sum_{n=1}^{\infty} \frac{(-2)^n x^n}{\sqrt{n}}.$

4. $\displaystyle\sum_{n=1}^{\infty} \frac{x^n}{n^2 2^n}.$

5. $\displaystyle\sum_{n=0}^{\infty} \frac{3^n x^n}{n!}.$

6. $\displaystyle\sum_{n=0}^{\infty} \frac{3^{(n^2)} x^n}{n!}.$

7. $\displaystyle\sum_{n=0}^{\infty} \frac{x^n}{2^n + 3^n}.$

8. $\displaystyle\sum_{n=1}^{\infty} \frac{x^n}{n2^n + n^2 3^n}.$

9. $\displaystyle\sum_{n=1}^{\infty} \frac{x^n}{n^2 2^n + n 3^n}.$

10. $\displaystyle\sum_{n=0}^{\infty} \frac{(2 + \sin n) x^n}{5^n}.$

11. $\displaystyle\sum_{n=1}^{\infty} \frac{x^n}{2^{\ln n}}.$

12. $\displaystyle\sum_{n=1}^{\infty} \frac{x^n}{3^{\ln n}}.$

13. $\displaystyle\sum_{n=0}^{\infty} \frac{x^n}{e^{\sqrt{n}}}.$

14. $\displaystyle\sum_{n=0}^{\infty} \frac{x^n}{2^{n+1} \sqrt{n^2 + 1}}.$

15. $\displaystyle\sum_{n=0}^{\infty} \frac{x^n}{2^{n^2+1} \sqrt{n + 1}}.$

16. Find a power series that converges to the function

$$f(x) = \frac{x^2}{x + 3}$$

on some interval. Specify the interval on which your series converges to $f(x)$.

17. Find a power series that converges to the function

$$f(x) = \frac{1}{x^2 + x - 6}$$

on some interval. Specify the interval on which your series converges to $f(x)$.

18. Let

$$f(x) = \frac{x}{1 - x^2}.$$

(a) Find a power series converging to $f(x)$ on some interval by substituting x^2 for x in (10.28) and then multiplying by x.

(b) Find a power series converging to $f(x)$ on some interval by applying partial fractions to write $f(x)$ as a sum of simpler fractions and then writing each of these simpler fractions as a power series.

(c) Are your answers to parts (a) and (b) the same?

19. (a) Find the interval of convergence of the power series

$$\sum_{n=0}^{\infty} \frac{x^{2n+1}}{3^n}.$$

(b) Evaluate the series in part (a) for all x in the interval of convergence.

20. (a) Find the interval of convergence of the power series

$$\sum_{n=0}^{\infty} (2^{n+1} - 1)x^n.$$

(b) Evaluate the series in part (a) for all x in the interval of convergence.

21. (a) Find the radius of convergence of the power series

$$\sum_{n=0}^{\infty} \frac{n^n x^n}{n!}.$$

(b) Show that for all positive integers n,

$$\left(1 + \frac{1}{n}\right)^n < e.$$

(Hint: First show that $\ln(n+1) - \ln n < 1/n$.)

(c) It can be shown (see the next exercise) that there are two positive numbers c_1 and c_2 such that for all positive integers n,

$$\frac{c_1 n^n \sqrt{n}}{e^n} \le n! \le \frac{c_2 n^n \sqrt{n}}{e^n}.$$

Use this fact and part (b) to determine whether or not the series in part (a) converges at the endpoints of the interval of convergence.

22. In this exercise you will establish the bounds on $n!$ that were used in the last exercise. Let n be any positive integer.

(a) Evaluate the integral

$$\int_1^n \ln x \, dx.$$

(b) Show that

$$\int_1^n \ln x \, dx \geq \ln(n!) - \frac{\ln n}{2}.$$

(Hint: Approximate the integral with the trapezoid rule (see Section 8.6), using trapezoids of width 1. Then explain why the trapezoid rule approximation is an underestimate of the integral.)

(c) Use parts (a) and (b) to find a constant c_2 such that

$$n! \leq \frac{c_2 n^n \sqrt{n}}{e^n}.$$

(d) Show that for $2 \leq k \leq n - 1$,

$$\int_{k-1/2}^{k+1/2} \ln x \, dx \leq \ln k.$$

(Hint: Approximate the integral with the area under the tangent line to the curve $y = \ln x$ at $(k, \ln k)$. Then explain why this approximation is an overestimate of the integral.)

(e) Show that

$$\int_1^{3/2} \ln x \, dx \leq \frac{\ln 3 - \ln 2}{2}, \qquad \int_{n-1/2}^n \ln x \, dx \leq \frac{\ln n}{2}.$$

(f) By adding up the inequalities in parts (d) and (e), show that

$$\int_1^n \ln x \, dx \leq \ln(n!) - \frac{\ln n}{2} + \frac{\ln 3 - \ln 2}{2}.$$

(g) Use parts (a) and (f) to find a constant c_1 such that

$$\frac{c_1 n^n \sqrt{n}}{e^n} \leq n!.$$

10.7 Calculus with Power Series

In the last section we introduced the idea of power series, but we have not yet fully explained why this idea will be useful to us. What makes power series so useful is the possibility of applying to them the fundamental concepts of calculus: limits, derivatives, and integrals.

Theorem 10.7.1. *Suppose the power series $\sum_{n=0}^{\infty} c_n x^n$ has a positive radius of convergence, and let I be its interval of convergence. For $x \in I$ let*

$$f(x) = \sum_{n=0}^{\infty} c_n x^n = c_0 + c_1 x + c_2 x^2 + \cdots .$$

Then:

1. *The function f is continuous on I.*

2. *The function f is differentiable on the interior of I, and for all x in the interior of I,*

$$f'(x) = \sum_{n=1}^{\infty} n c_n x^{n-1} = c_1 + 2c_2 x + 3c_3 x^2 + \cdots .$$

3. *On the interval I,*

$$\int f(x) \, dx = \left(\sum_{n=0}^{\infty} \frac{c_n x^{n+1}}{n+1} \right) + C = \left(c_0 x + \frac{c_1 x^2}{2} + \frac{c_2 x^3}{3} + \cdots \right) + C.$$

For the most part, the theorem says that the application of calculus concepts to power series works exactly as one might expect. In particular, derivatives and integrals of power series can be computed by differentiating or integrating the individual terms of the series, using the usual power rules for derivatives and integrals, exactly as one would when differentiating or integrating a polynomial. This is sometimes called "term-by-term differentiation and integration." Notice that the series for $f'(x)$ starts with $n = 1$ rather than $n = 0$, because the $n = 0$ term of the series for $f(x)$ is the constant c_0, whose derivative is 0, and therefore this term drops out when we differentiate. The only surprise in the theorem is the restriction of the differentiation formula in part 2 to the *interior* of the interval of convergence. As we will see, this restriction is necessary; sometimes things go wrong with the differentiability of f, or the formula for $f'(x)$, at the endpoints of I.

The proof of Theorem 10.7.1 does not require any concepts beyond those that we have already introduced. However, the arguments are somewhat long and technical. We therefore put off the proof until the end of this chapter, and turn now to applications of the theorem.

We begin with the familiar geometric series

$$\frac{1}{1-x} = \sum_{n=0}^{\infty} x^n = 1 + x + x^2 + \cdots , \qquad -1 < x < 1. \qquad (10.33)$$

Substituting $-x$ for x, and noting that $-1 < -x < 1$ if and only if $-1 < x < 1$, we see that

$$\frac{1}{1+x} = \sum_{n=0}^{\infty} (-x)^n = \sum_{n=0}^{\infty} (-1)^n x^n = 1 - x + x^2 - \cdots , \qquad -1 < x < 1. \qquad (10.34)$$

Now we're going to use part 3 of Theorem 10.7.1, term-by-term integration. Integrating equation (10.34), we get

$$\int \frac{dx}{1+x} = \left(\sum_{n=0}^{\infty} \frac{(-1)^n x^{n+1}}{n+1} \right) + C = \left(x - \frac{x^2}{2} + \frac{x^3}{3} - \cdots \right) + C, \quad -1 < x < 1.$$

$$(10.35)$$

In other words, the right-hand side of (10.35) gives a general formula for all antiderivatives of $1/(1+x)$ on the interval $(-1, 1)$. But clearly $\ln(1+x)$ is one such antiderivative. It follows that for some number C,

$$\ln(1+x) = \left(\sum_{n=0}^{\infty} \frac{(-1)^n x^{n+1}}{n+1} \right) + C, \qquad -1 < x < 1. \qquad (10.36)$$

To evaluate C, we set $x = 0$ in (10.36). On the left-hand side we get $\ln(1+0) = 0$, and on the right we get

$$\left(\sum_{n=0}^{\infty} \frac{(-1)^n \cdot 0^{n+1}}{n+1} \right) + C = (0+0+0+\cdots) + C = 0 + C = C.$$

Thus, $C = 0$, and substituting this into (10.36) we see that

$$\ln(1+x) = \sum_{n=0}^{\infty} \frac{(-1)^n x^{n+1}}{n+1} = x - \frac{x^2}{2} + \frac{x^3}{3} - \cdots, \qquad -1 < x < 1. \qquad (10.37)$$

Notice that the series in (10.37) is series (10.27c) from the last section, written using slightly different Σ-notation. Rewriting the series in the notation of (10.27c), we can say that

$$\ln(1+x) = \sum_{n=1}^{\infty} \frac{(-1)^{n-1} x^n}{n} = x - \frac{x^2}{2} + \frac{x^3}{3} - \cdots, \qquad -1 < x < 1. \qquad (10.38)$$

We are not quite done with series (10.27c). We have found its value on the interval $(-1, 1)$, but that is not the entire interval of convergence of the series. At the endpoints $x = \pm 1$, the series becomes:

$$x = 1: \sum_{n=1}^{\infty} \frac{(-1)^{n-1}}{n}, \qquad x = -1: \sum_{n=1}^{\infty} \frac{(-1)^{n-1}(-1)^n}{n} = \sum_{n=1}^{\infty} -\frac{1}{n}.$$

Thus, when $x = 1$ the series is the alternating harmonic series, which converges, but when $x = -1$ it is -1 times the harmonic series, and therefore it diverges. It follows that the radius of convergence must be 1, and the interval of convergence is $(-1, 1]$.

Here's another way of describing the situation. We can define a function g, with domain $(-1, 1]$, by the equation

$$g(x) = \sum_{n=1}^{\infty} \frac{(-1)^{n-1}x^n}{n}.$$

What equation (10.38) says is that, for $-1 < x < 1$, $g(x) = \ln(1+x)$. But we have not yet determined the value of $g(1)$.

To find $g(1)$, we apply Theorem 10.7.1 to g. According to part 1 of the theorem, g is continuous on its domain $(-1, 1]$, and in particular this means that it is continuous from the left at 1. Therefore

$$g(1) = \lim_{x \to 1^-} g(x) = \lim_{x \to 1^-} \ln(1+x) = \ln 2.$$

Thus, the equation $g(x) = \ln(1+x)$ holds on the entire interval $(-1, 1]$:

$$\ln(1+x) = \sum_{n=1}^{\infty} \frac{(-1)^{n-1}x^n}{n} = x - \frac{x^2}{2} + \frac{x^3}{3} - \cdots, \qquad -1 < x \leq 1. \qquad (10.39)$$

This, finally, is the complete determination of the value of the power series (10.27c); for values of x outside of the interval $(-1, 1]$, the series diverges.

Now that we have a complete understanding of the power series (10.27c), we can use it to find the value of series (10.26c), as suggested at the beginning of the last section. Setting $x = 1$ in equation (10.39), we find that

$$\ln 2 = \sum_{n=1}^{\infty} \frac{(-1)^{n-1}}{n} = 1 - \frac{1}{2} + \frac{1}{3} - \cdots.$$

Thus, we have finally discovered why the alternating harmonic series converges to $\ln 2$.

Notice that we have found a power series representation for the function $g(x) = \ln(1+x)$ on the interval $(-1, 1]$, and g is differentiable on this interval, but the power series for $g'(x) = 1/(1+x)$ only works on the interval $(-1, 1)$; it diverges when $x = 1$. This illustrates why it was necessary to restrict part 2 of Theorem 10.7.1 to the interior of the interval of convergence.

We can also use term-by-term integration to find a power series representation for the inverse tangent function. We begin by substituting x^2 for x in (10.34). Since the solution of the inequality $-1 < x^2 < 1$ is $-1 < x < 1$, this gives us

$$\frac{1}{1+x^2} = \sum_{n=0}^{\infty} (-1)^n (x^2)^n = \sum_{n=0}^{\infty} (-1)^n x^{2n} = 1 - x^2 + x^4 - \cdots, \qquad -1 < x < 1.$$

Now we integrate term-by-term. Since we know that $\tan^{-1} x$ is an antiderivative of $1/(1+x^2)$, we conclude that for some number C,

$$\tan^{-1} x = \left(\sum_{n=0}^{\infty} \frac{(-1)^n x^{2n+1}}{2n+1} \right) + C = \left(x - \frac{x^3}{3} + \frac{x^5}{5} - \cdots \right) + C, \qquad -1 < x < 1.$$

By substituting 0 for x in this equation we discover that $C = \tan^{-1} 0 = 0$, and therefore

$$\tan^{-1} x = \sum_{n=0}^{\infty} \frac{(-1)^n x^{2n+1}}{2n+1} = x - \frac{x^3}{3} + \frac{x^5}{5} - \cdots, \qquad -1 < x < 1. \qquad (10.40)$$

Once again, it turns out that the interval of convergence of the series in (10.40) is larger than $(-1, 1)$. We leave it to you to verify that the interval of convergence in this case is $[-1, 1]$. As in our last example, to find the values of the series at the endpoints of this interval we use continuity. Let h be the function with domain $[-1, 1]$ defined by the equation

$$h(x) = \sum_{n=0}^{\infty} \frac{(-1)^n x^{2n+1}}{2n+1}.$$

Equation (10.40) says that for $-1 < x < 1$, $h(x) = \tan^{-1} x$, and by part 1 of Theorem 10.7.1, h is continuous on $[-1, 1]$. We can therefore find $h(\pm 1)$ by taking limits:

$$h(1) = \lim_{x \to 1^-} h(x) = \lim_{x \to 1^-} \tan^{-1} x = \tan^{-1} 1 = \frac{\pi}{4},$$

$$h(-1) = \lim_{x \to -1^+} h(x) = \lim_{x \to -1^+} \tan^{-1} x = \tan^{-1}(-1) = -\frac{\pi}{4}.$$

This means that we can extend equation (10.40) to the entire interval of convergence:

$$\tan^{-1} x = \sum_{n=0}^{\infty} \frac{(-1)^n x^{2n+1}}{2n+1} = x - \frac{x^3}{3} + \frac{x^5}{5} - \cdots, \qquad -1 \le x \le 1. \qquad (10.41)$$

For example, setting $x = 1$ in equation (10.41), we find that

$$\sum_{n=0}^{\infty} \frac{(-1)^n}{2n+1} = 1 - \frac{1}{3} + \frac{1}{5} - \cdots = \tan^{-1} 1 = \frac{\pi}{4}.$$

This explains the result of Exercise 20 in Section 10.5.

Example 10.7.2. Compute

$$\int_0^1 \tan^{-1}(x^2)\, dx$$

with error at most 0.01.

Solution. It is possible to evaluate this integral by using the methods of Chapter 8, but it is very difficult. It is easier to write the integrand as a power series and then use term-by-term integration. Substituting x^2 for x in (10.41), we get

$$\tan^{-1}(x^2) = \sum_{n=0}^{\infty} \frac{(-1)^n (x^2)^{2n+1}}{2n+1} = \sum_{n=0}^{\infty} \frac{(-1)^n x^{4n+2}}{2n+1}, \qquad -1 \le x^2 \le 1.$$

The solution to the inequality $-1 \le x^2 \le 1$ is $-1 \le x \le 1$, so using term-by-term integration we find that on the interval $[-1, 1]$,

$$\int \tan^{-1}(x^2)\, dx = \left(\sum_{n=0}^{\infty} \frac{(-1)^n x^{4n+3}}{(4n+3)(2n+1)} \right) + C.$$

Since the limits of integration in the requested definite integral belong to the interval $[-1, 1]$, we can use this antiderivative in the fundamental theorem of calculus:

$$\int_0^1 \tan^{-1}(x^2)\, dx = \left[\sum_{n=0}^{\infty} \frac{(-1)^n x^{4n+3}}{(4n+3)(2n+1)} \right]_0^1$$

$$= \sum_{n=0}^{\infty} \frac{(-1)^n}{(4n+3)(2n+1)} - 0$$

$$= \frac{1}{3} - \frac{1}{21} + \frac{1}{55} - \frac{1}{105} + \cdots.$$

Clearly this last series satisfies the hypotheses of the alternating series test, so if we approximate it with the sum of the first three terms then the magnitude of the error will be at most $1/105 < 0.01$. Thus, we can achieve the requested accuracy with the approximation

$$\int_0^1 \tan^{-1}(x^2)\, dx \approx \frac{1}{3} - \frac{1}{21} + \frac{1}{55} = \frac{117}{385} \approx 0.304. \qquad \square$$

Example 10.7.3. Evaluate the power series

$$\sum_{n=1}^{\infty} \frac{(-1)^n x^{2n}}{2n}.$$

Solution. This series looks like the series for $\tan^{-1} x$, except that it involves even powers of x rather than odd powers. However, if we factor a $-1/2$ out of the series, then we see that it is actually more closely related to the series for $\ln(1+x)$. Applying (10.39), with x^2 substituted for x, we get

$$\sum_{n=1}^{\infty} \frac{(-1)^n x^{2n}}{2n} = -\frac{1}{2} \cdot \sum_{n=1}^{\infty} \frac{(-1)^{n-1}(x^2)^n}{n} = -\frac{1}{2} \ln(1+x^2), \qquad -1 < x^2 \le 1.$$

Since the solution to the inequality $-1 < x^2 \le 1$ is $-1 \le x \le 1$, we conclude that

$$\sum_{n=1}^{\infty} \frac{(-1)^n x^{2n}}{2n} = -\frac{1}{2} \ln(1+x^2), \qquad -1 \le x \le 1. \qquad \square$$

We have now derived two new power series formulas, equations (10.39) and (10.41), by using term-by-term integration. Let's try working out another by using term-by-term differentiation. We're going to figure out what function is represented by the power series $\sum_{n=1}^{\infty} n^2 x^n$, which was series (10.27b) in the last section.

We start with our series for $1/(1-x)$, equation (10.33), and differentiate both sides:

$$\frac{d}{dx}\left(\frac{1}{1-x}\right) = \frac{1}{(1-x)^2} = \sum_{n=1}^{\infty} nx^{n-1} = 1 + 2x + 3x^2 + \cdots, \quad -1 < x < 1.$$

$$(10.42)$$

The coefficient in the nth term of this series is n, and the coefficient in the nth term of series (10.27b) is n^2. We'll need another application of term-by-term differentiation to get another factor of n. It takes just a few more steps to complete the derivation.

Multiplying equation (10.42) by x, we get

$$\frac{x}{(1-x)^2} = x \cdot \sum_{n=1}^{\infty} nx^{n-1} = \sum_{n=1}^{\infty} nx^n = x + 2x^2 + 3x^3 + \cdots, \quad -1 < x < 1,$$

$$(10.43)$$

and then differentiating again gives us

$$\frac{d}{dx}\left(\frac{x}{(1-x)^2}\right) = \sum_{n=1}^{\infty} n^2 x^{n-1} = 1 + 4x + 9x^2 + \cdots, \quad -1 < x < 1. \quad (10.44)$$

By the quotient rule,

$$\frac{d}{dx}\left(\frac{x}{(1-x)^2}\right) = \frac{(1-x)^2 \cdot 1 - x \cdot 2(1-x)(-1)}{(1-x)^4} = \frac{(1-x) + 2x}{(1-x)^3} = \frac{1+x}{(1-x)^3},$$

and substituting this into equation (10.44) leads to

$$\frac{1+x}{(1-x)^3} = \sum_{n=1}^{\infty} n^2 x^{n-1} = 1 + 4x + 9x^2 + \cdots, \quad -1 < x < 1.$$

Finally, to get a formula for power series (10.27b), we multiply through by x again:

$$\frac{x+x^2}{(1-x)^3} = \sum_{n=1}^{\infty} n^2 x^n = x + 4x^2 + 9x^3 + \cdots, \quad -1 < x < 1. \quad (10.45)$$

Once again, as suggested at the beginning of the last section, we can use this formula to clear up a mystery left over from a previous section. Substituting $1/2$ for x in (10.45), we get

$$\sum_{n=1}^{\infty} n^2 \left(\frac{1}{2}\right)^n = \sum_{n=1}^{\infty} \frac{n^2}{2^n} = \frac{1}{2} + 1 + \frac{9}{8} + \cdots = \frac{(1/2) + (1/2)^2}{(1-1/2)^3} = 6.$$

This gives us the value of series (10.26b), which originally appeared as series (b) in Example 10.4.2. When we first studied this series in Section 10.4, we computed some partial sums and showed that the value of the series was between 5.99954 and 6.00018, and we guessed that the value was 6. Now we know that our guess was correct.

We have now cleared up two mysteries from previous sections: we have found the values of two of the power series in equations (10.27), and we have used these to find the values of two of the series in (10.26). We will find the values of the other two in Sections 10.8 and 10.9. But we have also discovered that, surprisingly, some familiar functions can be expressed as power series. In particular, in equations (10.39) and (10.41) we have expressed $\ln(1+x)$ and $\tan^{-1} x$, at least for certain ranges of values of x, as power series.

You may be wondering at this point if we can modify our formula for $\ln(1+x)$ in order to express $\ln x$ as a power series. One possibility is to substitute $x-1$ for x in equation (10.39), which gives us

$$\ln(1+(x-1)) = \sum_{n=1}^{\infty} \frac{(-1)^{n-1}(x-1)^n}{n}, \qquad -1 < x-1 \le 1.$$

Simplifying, we get

$$\ln x = \sum_{n=1}^{\infty} \frac{(-1)^{n-1}(x-1)^n}{n} = (x-1) - \frac{(x-1)^2}{2} + \frac{(x-1)^3}{3} - \cdots, \quad 0 < x \le 2.$$

$$(10.46)$$

Equation (10.46) expresses $\ln x$ as an infinite series, but the series doesn't quite fit our definition of a power series, since it involves powers of $x-1$ rather than powers of x. The series in (10.46) is called a *power series in $x-1$*. More generally, we extend our definition of power series as follows.

Definition 10.7.4. Suppose $(c_n)_{n=0}^{\infty}$ is a sequence of numbers and a is a number. Then the series

$$\sum_{n=0}^{\infty} c_n(x-a)^n = c_0 + c_1(x-a) + c_2(x-a)^2 + \cdots$$

is called a *power series in $x-a$*, or a *power series centered at a*.

From now on we will use the term "power series" to include power series in $x-a$. Notice that if $a=0$ then we get back our original definition of power series. Thus, the power series we have studied up until now are power series centered at 0.

Many of our previous results for power series centered at 0 can easily be extended to power series centered at any number. For example, notice that equation (10.46) holds for values of x in an interval centered at 1. In general, we have the following version of Theorem 10.6.2.

Theorem 10.7.5. *For any power series*

$$\sum_{n=0}^{\infty} c_n(x-a)^n,$$

one of the following is true:

1. *The series converges only when $x = a$.*

2. *There is some positive number R such that the series converges absolutely if $|x - a| < R$ and diverges if $|x - a| > R$.*

3. *The series converges absolutely for all values of x.*

Theorem 10.7.5 follows immediately from Theorem 10.6.2 simply by substituting $x - a$ for x. As before, we refer to R as the radius of convergence of the power series. We can also generalize Theorem 10.7.1 to get the following theorem.

Theorem 10.7.6. *Suppose the power series $\sum_{n=0}^{\infty} c_n(x - a)^n$ has a positive radius of convergence, and let I be its interval of convergence. For $x \in I$ let*

$$f(x) = \sum_{n=0}^{\infty} c_n(x - a)^n = c_0 + c_1(x - a) + c_2(x - a)^2 + \cdots .$$

Then:

1. *The function f is continuous on I.*

2. *The function f is differentiable on the interior of I, and for all x in the interior of I,*

$$f'(x) = \sum_{n=1}^{\infty} nc_n(x - a)^{n-1} = c_1 + 2c_2(x - a) + 3c_3(x - a)^2 + \cdots .$$

3. *On the interval I,*

$$\int f(x)\, dx = \left(\sum_{n=0}^{\infty} \frac{c_n(x - a)^{n+1}}{n + 1} \right) + C$$

$$= \left(c_0(x - a) + \frac{c_1(x - a)^2}{2} + \frac{c_2(x - a)^3}{3} + \cdots \right) + C.$$

To illustrate how Theorem 10.7.6 follows from Theorem 10.7.1, we sketch the idea behind the proof of part 2. Let g be the function defined by the equation

$$g(x) = \sum_{n=0}^{\infty} c_n x^n.$$

Then $f(x) = g(x - a)$. By Theorem 10.7.1, g is differentiable on the interior of its interval of convergence, and

$$g'(x) = \sum_{n=1}^{\infty} nc_n x^{n-1}.$$

We can now find $f'(x)$ by the chain rule:

$$f'(x) = g'(x-a) \cdot \frac{d}{dx}(x-a) = g'(x-a) \cdot 1 = \sum_{n=1}^{\infty} n c_n (x-a)^{n-1}.$$

Let's give one more example to illustrate these new ideas. We're going to write $1/(1-x)$ as a power series again, but this time we'll write it as a power series in $x+1$. We begin with the reexpression

$$\frac{1}{1-x} = \frac{1}{2-(x+1)} = \frac{1}{2} \cdot \frac{1}{1-(x+1)/2}.$$

Next, we substitute $(x+1)/2$ for x in our original power series for $1/(1-x)$, equation (10.33):

$$\frac{1}{1-(x+1)/2} = \sum_{n=0}^{\infty} \left(\frac{x+1}{2}\right)^n = \sum_{n=0}^{\infty} \frac{(x+1)^n}{2^n}, \qquad -1 < \frac{x+1}{2} < 1.$$

Since the solution of the inequality $-1 < (x+1)/2 < 1$ is $-3 < x < 1$, our final answer is

$$\frac{1}{1-x} = \frac{1}{2} \cdot \frac{1}{1-(x+1)/2} = \frac{1}{2} \cdot \sum_{n=0}^{\infty} \frac{(x+1)^n}{2^n} = \sum_{n=0}^{\infty} \frac{(x+1)^n}{2^{n+1}}$$

$$= \frac{1}{2} + \frac{x+1}{4} + \frac{(x+1)^2}{8} + \cdots, \qquad -3 < x < 1.$$

Exercises 10.7

1–10: Find the interval of convergence of the power series.

1. $\displaystyle\sum_{n=0}^{\infty} \frac{(x-4)^n}{3^n}$.

2. $\displaystyle\sum_{n=1}^{\infty} \frac{(x-4)^n}{n3^n}$.

3. $\displaystyle\sum_{n=1}^{\infty} \frac{(x-4)^n}{n^2 3^n}$.

4. $\displaystyle\sum_{n=1}^{\infty} \frac{(x-4)^n}{\ln(n^2+1)3^n}$.

5. $\displaystyle\sum_{n=1}^{\infty} \frac{(x+3)^n}{n4^{2n+1}}$.

6. $\displaystyle\sum_{n=1}^{\infty} \frac{(x+3)^{2n+1}}{n4^n}$.

7. $\displaystyle\sum_{n=0}^{\infty} \frac{e^n(x-2)^n}{n!}$.

8. $\displaystyle\sum_{n=0}^{\infty} \frac{e^{n\sqrt{n}}(x-2)^n}{n!}$.

9. $\displaystyle\sum_{n=0}^{\infty} \left(\frac{1}{2^n+3^n}\right)(x-1)^n$.

10. $\displaystyle\sum_{n=0}^{\infty} \left(\frac{1}{2^n}+\frac{1}{3^n}\right)(x-1)^n$.

11. Find a power series centered at 3 that converges to $1/x$ on some interval around 3.

12. Find a power series centered at 2 that converges to $\ln x$ on some interval around 2. (Hint: Begin with the equation

$$\ln x = \ln(2+x-2) = \ln\left(2\cdot\left(1+\frac{x-2}{2}\right)\right).$$

Then use algebraic properties of the natural logarithm and equation (10.39).)

13. Use equation (10.41) to approximate $\tan^{-1}(1/2)$ with error at most 0.001.

14. Use equation (10.43) to evaluate the series

$$\sum_{n=1}^{\infty} \frac{n}{2^n}.$$

Compare your answer to the one you found in Exercise 17 in Section 10.1.

15. Evaluate the series

$$\sum_{n=1}^{\infty} \frac{(-2)^n}{n3^n}.$$

(Hint: Use equation (10.39).)

16. (a) Show that

$$\sum_{n=1}^{\infty} \frac{1}{n2^n} = \ln 2.$$

(Hint: Begin by substituting $-1/2$ for x in (10.39).)

(b) Approximate $\ln 2$ by computing s_7, the 7th partial sum of the series in part (a).

(c) Show that the error in the approximation in part (b) is at most 0.001.

17. Evaluate

$$\sum_{n=0}^{\infty} \frac{(-1)^n}{(2n+1)3^n}.$$

(Hint: Begin by substituting some value for x in equation (10.41) to get a series that is similar to the one above.)

18. (a) Find the interval of convergence of the power series

$$\sum_{n=0}^{\infty} \frac{x^{2n+1}}{2n+1}.$$

(b) Evaluate the series in part (a) for all x in the interval of convergence. (Hint: Integrate another power series to get this one. Compare your answer to Exercise 35 in Section 7.5.)

(c) Evaluate

$$\sum_{n=0}^{\infty} \frac{1}{(2n+1)3^{2n+1}}.$$

(d) Evaluate

$$\sum_{n=0}^{\infty} \frac{1}{(2n+1)3^n}.$$

19. (a) Find a power series centered at 0 that converges to $1/(x^3+1)$ on some interval. What is the interval?

(b) Use term-by-term integration on your answer to part (a) to find an infinite series that converges to

$$\int_0^{1/2} \frac{dx}{x^3+1}.$$

(c) Use a partial sum to approximate the value of the definite integral in part (b) with error at most 0.0001.

(d) Find the exact value of the definite integral in part (b), and verify the accuracy of your approximation in part (c).

20. (a) Use term-by-term integration to find an infinite series that converges to

$$\int_0^1 \ln(1+x^2)\,dx.$$

(b) Use a partial sum to approximate the value of the definite integral in part (a) with error at most 0.01.

(c) Find the exact value of the definite integral in part (a), and verify the accuracy of your approximation in part (c).

21. (a) Find a power series centered at 0 that converges to

$$\frac{1}{(1+x)^3}$$

on some interval. (Hint: Differentiate a series for $1/(1+x)$ twice.)

(b) Find a power series centered at 0 that converges to

$$\frac{1}{(1+x)^4}$$

on some interval.

(c) Generalize: For any positive integer p, can you find a power series centered at 0 that converges to

$$\frac{1}{(1+x)^p}$$

on some interval?

10.8 Taylor Series

In the last section, we discovered power series representations for the functions $\ln x$ and $\tan^{-1} x$, but we found these power series mostly by luck. It turns out that many familiar functions can be written as power series, but to show this we'll need a more systematic method of finding such series. We will begin to develop such a method in this section.

The method is based on a formula for the coefficients of any power series representing a function f. Suppose that

$$f(x) = \sum_{n=0}^{\infty} c_n (x-a)^n = c_0 + c_1(x-a) + c_2(x-a)^2 + \cdots, \quad a - R < x < a + R,$$

$$(10.47)$$

where $R > 0$. Then in particular equation (10.47) holds when $x = a$, so

$$f(a) = \sum_{n=0}^{\infty} c_n \cdot 0^n = c_0 + 0 + 0 + \cdots = c_0.$$

Thus, we have a formula for the first coefficient of the power series: $c_0 = f(a)$.

It turns out we can find the next coefficient by taking the derivative of f. Differentiating equation (10.47), we get

$$f'(x) = \sum_{n=1}^{\infty} n c_n (x-a)^{n-1}$$

$$= 1 \cdot c_1 + 2 \cdot c_2(x-a) + 3 \cdot c_3(x-a)^2 + \cdots, \quad a - R < x < a + R. \quad (10.48)$$

Evaluating again at $x = a$ gives us

$$f'(a) = \sum_{n=1}^{\infty} nc_n \cdot 0^{n-1} = 1 \cdot c_1 + 0 + 0 + \cdots = c_1,$$

so we have found a formula for the next coefficient: $c_1 = f'(a)$.

Of course, to get the next coefficient we differentiate again. Differentiating (10.48) leads to

$$f''(x) = \sum_{n=2}^{\infty} n(n-1)c_n(x-a)^{n-2}$$
$$= 2 \cdot 1 \cdot c_2 + 3 \cdot 2 \cdot c_3(x-a) + 4 \cdot 3 \cdot c_4(x-a)^2 + \cdots, \quad a - R < x < a + R.$$
$$(10.49)$$

We have chosen not to multiply out the coefficients in (10.49), because we want to see the pattern in the formula for these coefficients. Evaluating again at $x = a$, we get

$$f''(a) = \sum_{n=2}^{\infty} n(n-1)c_n \cdot 0^{n-2} = 2 \cdot 1 \cdot c_2 + 0 + 0 + \cdots = 2! \cdot c_2,$$

and therefore $c_2 = f''(a)/2!$.

Let's work out one more coefficient. Differentiating (10.49) gives us

$$f'''(x) = \sum_{n=3}^{\infty} n(n-1)(n-2)c_n(x-a)^{n-3}$$
$$= 3 \cdot 2 \cdot 1 \cdot c_3 + 4 \cdot 3 \cdot 2 \cdot c_4(x-a)$$
$$+ 5 \cdot 4 \cdot 3 \cdot c_5(x-a)^2 + \cdots, \quad a - R < x < a + R.$$

When $x = a$ this says that

$$f'''(a) = \sum_{n=3}^{\infty} n(n-1)(n-2)c_n \cdot 0^{n-3} = 3 \cdot 2 \cdot 1 \cdot c_3 + 0 + 0 + \cdots = 3! \cdot c_3,$$

so $c_3 = f'''(a)/3! = f^{(3)}(a)/3!$.

Here are the coefficients we have computed so far:

$$c_0 = f(a), \quad c_1 = f'(a), \quad c_2 = \frac{f''(a)}{2!}, \quad c_3 = \frac{f'''(a)}{3!} = \frac{f^{(3)}(a)}{3!}.$$

You can probably guess that $c_4 = f^{(4)}(a)/4!$, $c_5 = f^{(5)}(a)/5!$, and, in general,

$$c_n = \frac{f^{(n)}(a)}{n!}.$$

And, indeed, it is not hard to extend the reasoning we have given to show that this general formula for c_n is correct. Notice that this formula works even in the case $n = 0$ if we interpret $f^{(0)}$, the "0th derivative" of f, to be f itself.

We summarize our conclusions in a theorem:

Theorem 10.8.1. *Suppose that*

$$f(x) = \sum_{n=0}^{\infty} c_n(x-a)^n = c_0 + c_1(x-a) + c_2(x-a)^2 + \cdots, \quad a-R < x < a+R,$$

where $R > 0$. *Then for every* $n \geq 0$,

$$c_n = \frac{f^{(n)}(a)}{n!}.$$

Filling in the formula for c_n in the power series in Theorem 10.8.1, we see that the series has the form

$$\sum_{n=0}^{\infty} c_n(x-a)^n = \sum_{n=0}^{\infty} \frac{f^{(n)}(a)}{n!}(x-a)^n.$$

There is a name for this series:

Definition 10.8.2. Suppose that for every positive integer n, $f^{(n)}(a)$ is defined. Then the power series

$$\sum_{n=0}^{\infty} \frac{f^{(n)}(a)}{n!}(x-a)^n = f(a) + f'(a)(x-a) + \frac{f''(a)}{2!}(x-a)^2 + \frac{f'''(a)}{3!}(x-a)^3 + \cdots$$

is called the *Taylor series for f centered at a*. When $a = 0$, the series takes the simpler form

$$\sum_{n=0}^{\infty} \frac{f^{(n)}(0)}{n!}x^n = f(0) + f'(0)x + \frac{f''(0)}{2!}x^2 + \frac{f'''(0)}{3!}x^3 + \cdots.$$

This is the Taylor series for f centered at 0, but it is also often called the *Maclaurin series for f*.

Taylor and Maclaurin series are named for Brook Taylor (1685–1731) and Colin Maclaurin (1698–1746), who were among a number of mathematicians who studied power series representations of functions in the late 17th and early 18th centuries.

It is important to understand what Theorem 10.8.1 says and what it doesn't say. In the terminology of Definition 10.8.2, what the theorem says is that if any power series centered at a converges to $f(x)$ in an interval around a, then that series must be the Taylor series for f centered at a. However, it does *not* say that the Taylor series is guaranteed to converge to $f(x)$. To put it another way, the theorem says that if you are looking for a power series in $x - a$ that converges to $f(x)$ in some interval around a, then you may as well focus your attention on the Taylor series, because no other series has any hope of working. The Taylor series may or may not work, but if it doesn't work, then nothing does.

For example, consider our series for $\ln x$ from the last section:

$$\ln x = \sum_{n=1}^{\infty} \frac{(-1)^{n-1}}{n}(x-1)^n = (x-1) - \frac{(x-1)^2}{2} + \frac{(x-1)^3}{3} - \cdots , \quad 0 < x \le 2.$$

(10.50)

In this case, $f(x) = \ln x$ and $a = 1$. According to Theorem 10.8.1, this series must be the Taylor series for f centered at 1. To verify this, we need to compute $f^{(n)}(1)$ for all $n \ge 0$. Here are the first few values:

$$f(x) = \ln x, \qquad\qquad\qquad\qquad\qquad f(1) = \ln 1 = 0,$$

$$f'(x) = \frac{1}{x} = x^{-1}, \qquad\qquad\qquad\qquad f'(1) = 1,$$

$$f''(x) = (-1) \cdot x^{-2} = -\frac{1}{x^2}, \qquad\qquad\quad f''(1) = (-1) \cdot 1,$$

$$f'''(x) = (-1)(-2) \cdot x^{-3} = \frac{(-1)^2 \cdot 2!}{x^3}, \qquad f'''(1) = (-1)^2 \cdot 2!,$$

$$f^{(4)}(x) = (-1)(-2)(-3) \cdot x^{-4} = \frac{(-1)^3 \cdot 3!}{x^4}, \qquad f^{(4)}(1) = (-1)^3 \cdot 3!,$$

$$\vdots \qquad\qquad\qquad\qquad\qquad\qquad\qquad \vdots$$

The pattern is probably clear at this point: $f(1) = 0$, and for every $n \ge 1$, $f^{(n)}(x) = (-1)^{n-1}(n-1)!/x^n$ and $f^{(n)}(1) = (-1)^{n-1}(n-1)!$. Thus, the coefficients of the Taylor series are $c_0 = 0$ and, for $n \ge 1$,

$$c_n = \frac{f^{(n)}(1)}{n!} = \frac{(-1)^{n-1}(n-1)!}{n!} = \frac{(-1)^{n-1}}{n}.$$

Looking back at equation (10.50), we see that these are precisely the coefficients of the power series in that equation. Thus, series (10.50) is, as expected, the Taylor series for $\ln x$ centered at 1.

While it is nice to see Theorem 10.8.1 confirmed in the case of the natural logarithm function, the real usefulness of the theorem comes from applying it to functions for which we don't yet know a power series representation. For example, suppose we want to find a power series centered at 0 that converges to the function $f(x) = e^x$ in some interval around 0. Theorem 10.8.1 suggests that we try the Maclaurin series, so we need to compute $f^{(n)}(0)$ for all $n \ge 0$. As soon as we compute the first few values, the pattern becomes clear:

$$f(x) = e^x, \qquad\qquad\qquad f(0) = e^0 = 1,$$

$$f'(x) = e^x, \qquad\qquad\qquad f'(0) = e^0 = 1,$$

$$f''(x) = e^x, \qquad\qquad\qquad f''(0) = e^0 = 1,$$

$$\vdots \qquad\qquad\qquad\qquad\qquad \vdots$$

Evidently, $f^{(n)}(x) = e^x$ and $f^{(n)}(0) = e^0 = 1$ for every $n \geq 0$, and therefore the Maclaurin series is

$$\sum_{n=0}^{\infty} \frac{f^{(n)}(0)}{n!} x^n = \sum_{n=0}^{\infty} \frac{1}{n!} x^n = \sum_{n=0}^{\infty} \frac{x^n}{n!} = 1 + x + \frac{x^2}{2!} + \frac{x^3}{3!} + \cdots . \qquad (10.51)$$

Again, we want to stress that we don't yet know if series (10.51) converges to e^x; what we do know is that if any power series centered at 0 works, then it has to be this one. Of course, we're hoping that this series converges to e^x, but there are two things that could go wrong. The series could diverge, at least for some values of x. And even if the series converges, it might converge to something other than e^x. (For an example of how a Taylor series for a function f can converge to something other than $f(x)$, see Exercise 25.) We need to address both of these worries.

Fortunately, the first worry has already been taken care of. Series (10.51) is the same as series (10.27a) from Section 10.6. We studied this series in part (b) of Example 10.6.4, and found that it converges for all values of x. We can therefore define a function g with domain $\mathbb{R}$ by the equation

$$g(x) = \sum_{n=0}^{\infty} \frac{x^n}{n!}.$$

The question we need to address now is whether or not $g(x) = e^x$.

From the point of view of calculus, the most distinctive property of the function $f(x) = e^x$ is that it is equal to its own derivative: $f'(x) = f(x)$. As a first step toward seeing if $g(x) = e^x$, let's see if g has this property. By term-by-term differentiation, we see that

$$g'(x) = \sum_{n=1}^{\infty} \frac{n x^{n-1}}{n!} = \sum_{n=1}^{\infty} \frac{x^{n-1}}{(n-1)!} = 1 + x + \frac{x^2}{2!} + \frac{x^3}{3!} + \cdots = \sum_{n=0}^{\infty} \frac{x^n}{n!} = g(x).$$

So g at least shares this property with f: it is equal to its own derivative. If you did Exercise 25 in Section 7.4, then you know that it follows that there must be some constant C such that $g(x) = Ce^x$. For those who didn't do that exercise, we give a solution here. Our plan is to show that $g(x)/e^x$ is equal to a constant (note that e^x is never 0, so division by e^x is acceptable). Often in calculus the easiest way to show that something is constant is to show that its derivative is 0, so we use the quotient rule to compute

$$\frac{d}{dx}\left(\frac{g(x)}{e^x}\right) = \frac{e^x g'(x) - g(x)e^x}{(e^x)^2} = \frac{e^x g(x) - e^x g(x)}{e^{2x}} = 0.$$

Thus, there is some number C such that for every x, $g(x)/e^x = C$, and therefore $g(x) = Ce^x$.

All that remains is to compute C, which we do by plugging in a convenient value for x. Setting $x = 0$, we get

$$C = Ce^0 = g(0) = \sum_{n=0}^{\infty} \frac{0^n}{n!} = 1 + 0 + 0 + \cdots = 1.$$

Thus, $g(x) = Ce^x = e^x$. In other words, we have shown that the Maclaurin series for e^x converges to e^x for all values of x:

$$e^x = \sum_{n=0}^{\infty} \frac{x^n}{n!} = 1 + x + \frac{x^2}{2!} + \frac{x^3}{3!} + \cdots, \qquad -\infty < x < \infty. \qquad (10.52)$$

Using this power series representation for e^x, we can clear up one more mystery from before. Setting $x = 2$ in equation (10.52), we get

$$\sum_{n=0}^{\infty} \frac{2^n}{n!} = e^2.$$

This confirms the guess at the value of series (10.26a) that we made all the way back in Example 10.4.2.

We could also compute the Taylor series for e^x centered at some other number. For example, to find the Taylor series centered at 3, we first observe that for all $n \geq 0$, $f^{(n)}(x) = e^x$, and therefore $f^{(n)}(3) = e^3$. Therefore the Taylor series is

$$\sum_{n=0}^{\infty} \frac{f^{(n)}(3)}{n!} (x - 3)^n = \sum_{n=0}^{\infty} \frac{e^3}{n!} (x - 3)^n.$$

We ask you to verify in Exercise 19 that this series also converges to e^x for all values of x.

Let's try another example. We're going to see if we can find a power series centered at 0 that converges to the function $f(x) = \sin x$. Once again we begin by computing the Maclaurin series, which means that we must find $f^{(n)}(0)$ for $n \geq 0$:

$$
\begin{aligned}
f(x) &= \sin x, & f(0) &= \sin 0 = 0, \\
f'(x) &= \cos x, & f'(0) &= \cos 0 = 1, \\
f''(x) &= -\sin x, & f''(0) &= -\sin 0 = 0, \\
f^{(3)}(x) &= -\cos x, & f^{(3)}(0) &= -\cos 0 = -1, \\
f^{(4)}(x) &= \sin x, & f^{(4)}(0) &= \sin 0 = 0, \\
f^{(5)}(x) &= \cos x, & f^{(5)}(0) &= \cos 0 = 1, \\
&\;\vdots & &\;\vdots
\end{aligned}
$$

This time there is a pattern of four numbers that repeats: $0, 1, 0, -1$. Thus, the Maclaurin series is

$$\sum_{n=0}^{\infty} \frac{f^{(n)}(0)}{n!} x^n = 0 + \frac{1}{1} x + 0 + \frac{-1}{3!} x^3 + 0 + \frac{1}{5!} x^5 + 0 + \frac{-1}{7!} x^7 + \cdots .$$

Leaving out the terms that are 0, we have alternating positive and negative terms with odd powers of x:

$$x - \frac{x^3}{3!} + \frac{x^5}{5!} - \frac{x^7}{7!} + \cdots = \sum_{n=0}^{\infty} \frac{(-1)^n x^{2n+1}}{(2n+1)!} .$$

As before, there is more work to do to determine whether or not this power series converges to $\sin x$. We leave it as an exercise for you (see Exercise 20) to check that the series converges for all values of x. We can therefore let g be the function with domain $\mathbb{R}$ defined by the equation

$$g(x) = \sum_{n=0}^{\infty} \frac{(-1)^n x^{2n+1}}{(2n+1)!} . \tag{10.53}$$

We must now investigate whether or not $g(x) = \sin x$.

Imitating the reasoning we used in the case of e^x, we ask: Is there a distinctive property of the function $f(x) = \sin x$? A reasonable candidate is the fact that $f''(x) = -\sin x = -f(x)$. To see if g has the same property, we compute:

$$g(x) = \sum_{n=0}^{\infty} \frac{(-1)^n x^{2n+1}}{(2n+1)!} = x - \frac{x^3}{3!} + \frac{x^5}{5!} - \frac{x^7}{7!} + \cdots ,$$

$$g'(x) = \sum_{n=0}^{\infty} \frac{(-1)^n (2n+1) x^{2n}}{(2n+1)!} = \sum_{n=0}^{\infty} \frac{(-1)^n x^{2n}}{(2n)!} = 1 - \frac{x^2}{2!} + \frac{x^4}{4!} - \frac{x^6}{6!} + \cdots ,$$

$$\tag{10.54}$$

$$g''(x) = \sum_{n=1}^{\infty} \frac{(-1)^n (2n) x^{2n-1}}{(2n)!} = \sum_{n=1}^{\infty} \frac{(-1)^n x^{2n-1}}{(2n-1)!} = -x + \frac{x^3}{3!} - \frac{x^5}{5!} + \frac{x^7}{7!} - \cdots$$

$$= -\sum_{n=0}^{\infty} \frac{(-1)^n x^{2n+1}}{(2n+1)!} = -g(x).$$

Yes, g has the same property.

Once again, our next step is to show that some quantity is constant by showing that its derivative is equal to 0. We are going to show simultaneously that $g(x) = \sin x$ and $g'(x) = \cos x$, which motivates us to compute

$$\frac{d}{dx} \Big[(g(x) - \sin x)^2 + (g'(x) - \cos x)^2 \Big]$$

$$= 2(g(x) - \sin x)(g'(x) - \cos x) + 2(g'(x) - \cos x)(g''(x) + \sin x)$$

$$= 2(g'(x) - \cos x)(g(x) - \sin x + g''(x) + \sin x)$$

$$= 2(g'(x) - \cos x)(g(x) - g(x)) = 0.$$

We conclude that there is some number C such that for every x,

$$(g(x) - \sin x)^2 + (g'(x) - \cos x)^2 = C.$$

As usual, we evaluate C by setting $x = 0$. Using equations (10.53) and (10.54), we find that

$$g(0) = \sum_{n=0}^{\infty} \frac{(-1)^n \cdot 0^{2n+1}}{(2n+1)!} = 0 + 0 + 0 + \cdots = 0,$$

$$g'(0) = \sum_{n=0}^{\infty} \frac{(-1)^n \cdot 0^{2n}}{(2n)!} = 1 + 0 + 0 + \cdots = 1,$$

and therefore

$$C = (g(0) - \sin 0)^2 + (g'(0) - \cos 0)^2 = (0 - 0)^2 + (1 - 1)^2 = 0.$$

We conclude that for all x,

$$(g(x) - \sin x)^2 + (g'(x) - \cos x)^2 = 0.$$

But the only way this can happen is if $g(x) = \sin x$ and $g'(x) = \cos x$. In other words, we have simultaneously found two power series formulas:

$$\sin x = \sum_{n=0}^{\infty} \frac{(-1)^n x^{2n+1}}{(2n+1)!} = x - \frac{x^3}{3!} + \frac{x^5}{5!} - \cdots, \qquad -\infty < x < \infty, \qquad (10.55a)$$

$$\cos x = \sum_{n=0}^{\infty} \frac{(-1)^n x^{2n}}{(2n)!} = 1 - \frac{x^2}{2!} + \frac{x^4}{4!} - \cdots, \qquad -\infty < x < \infty. \qquad (10.55b)$$

Example 10.8.3. Find the following series:

(a)   The Maclaurin series for the function $f(x) = e^{2-x^2}$.
(b)   The Taylor series centered at $\pi/6$ for the function $g(x) = \sin x$.

Solution. Although we could use the definition of Maclaurin series for part (a), it is easier to use other methods to find a series representation for f. Using exponent laws and (10.52), we see that for all x,

$$e^{2-x^2} = e^2 \cdot e^{-x^2} = e^2 \cdot \sum_{n=0}^{\infty} \frac{(-x^2)^n}{n!} = \sum_{n=0}^{\infty} \frac{(-1)^n e^2 x^{2n}}{n!}.$$

By Theorem 10.8.1, this must be the Maclaurin series for f.

Similarly, by the sum rule for the sine function and equations (10.55), for all x we have

$$
\begin{aligned}
\sin x &= \sin\left(\frac{\pi}{6} + \left(x - \frac{\pi}{6}\right)\right) \\
&= \sin\left(\frac{\pi}{6}\right)\cos\left(x - \frac{\pi}{6}\right) + \cos\left(\frac{\pi}{6}\right)\sin\left(x - \frac{\pi}{6}\right) \\
&= \frac{1}{2}\cdot\cos\left(x - \frac{\pi}{6}\right) + \frac{\sqrt{3}}{2}\cdot\sin\left(x - \frac{\pi}{6}\right) \\
&= \frac{1}{2}\cdot\sum_{n=0}^{\infty}\frac{(-1)^n(x - \pi/6)^{2n}}{(2n)!} + \frac{\sqrt{3}}{2}\cdot\sum_{n=0}^{\infty}\frac{(-1)^n(x - \pi/6)^{2n+1}}{(2n+1)!} \\
&= \left(\frac{1}{2} + 0 - \frac{1}{2\cdot 2!}\left(x - \frac{\pi}{6}\right)^2 + 0 + \cdots\right) \\
&\quad + \left(0 + \frac{\sqrt{3}}{2}\left(x - \frac{\pi}{6}\right) + 0 - \frac{\sqrt{3}}{2\cdot 3!}\left(x - \frac{\pi}{6}\right)^3 + \cdots\right) \\
&= \frac{1}{2} + \frac{\sqrt{3}}{2}\left(x - \frac{\pi}{6}\right) - \frac{1}{2\cdot 2!}\left(x - \frac{\pi}{6}\right)^2 - \frac{\sqrt{3}}{2\cdot 3!}\left(x - \frac{\pi}{6}\right)^3 + \cdots.
\end{aligned}
$$

Notice that we have included the terms with coefficient 0 in the power series for $\cos(x - \pi/6)$ and $\sin(x - \pi/6)$, to make it easier to add these two series in the last step. Again, by Theorem 10.8.1, we conclude that this is the Taylor series for g centered at $\pi/6$. In Exercise 24 we ask you to verify this directly from the definition of Taylor series. $\square$

Example 10.8.4. Evaluate the following series:

$$
\text{(a)}\ \sum_{n=0}^{\infty}\frac{(-1)^n}{n!}, \qquad \text{(b)}\ \sum_{n=0}^{\infty}\frac{(-1)^n\pi^{2n}}{9^n(2n)!}.
$$

Solution. These series are easy to evaluate if you recognize how they are related to Maclaurin series we have studied. Series (a) is the result of substituting -1 for x in the Maclaurin series for e^x, and therefore, by (10.52),

$$
\sum_{n=0}^{\infty}\frac{(-1)^n}{n!} = e^{-1} = \frac{1}{e} \approx 0.368.
$$

Similarly, after a little rewriting we can use (10.55b) to evaluate series (b):

$$
\sum_{n=0}^{\infty}\frac{(-1)^n\pi^{2n}}{9^n(2n)!} = \sum_{n=0}^{\infty}\frac{(-1)^n(\pi/3)^{2n}}{(2n)!} = \cos\left(\frac{\pi}{3}\right) = \frac{1}{2}. \qquad \square
$$

Power series have many applications. The most straightforward use of power series is to find particular values of functions, as in our next example.

Example 10.8.5. Compute $\sin(20°)$ with error at most 0.0001.

Solution. First, we must convert $20°$ to radians. Since $360°$ is 2π radians, $20°$ is $20\pi/180 = \pi/9$ radians. Therefore, by (10.55a),

$$\sin(20°) = \sin(\pi/9) = \sum_{n=0}^{\infty} \frac{(-1)^n (\pi/9)^{2n+1}}{(2n+1)!} = (\pi/9) - \frac{(\pi/9)^3}{3!} + \frac{(\pi/9)^5}{5!} - \cdots .$$

The terms of this series alternate positive and negative, and, since $\pi/9 < 1$, it is not hard to see that the terms decrease in magnitude. Also, by Proposition 10.6.5, the terms approach 0. Therefore all of the hypotheses of the alternating series test hold, and that implies that if we approximate the value of the series with a partial sum, then the magnitude of the error will be at most the absolute value of the next term. Calculation shows that $(\pi/9)^5/5! \approx 0.00004 < 0.0001$. Thus, we can say that, with error at most 0.0001,

$$\sin(20°) \approx (\pi/9) - \frac{(\pi/9)^3}{3!} \approx 0.34198.$$

In fact, as your calculator will tell you, $\sin(20°) \approx 0.34202$. □

   Power series can also be used to solve calculus problems that might be difficult to solve by other methods. Our next two examples illustrate this.

Example 10.8.6. Compute $\int_{-1}^{2} e^{-x^2} \, dx$ with error at most 0.01.

Solution. We used the midpoint and trapezoid rules to approximate this integral in Section 8.6, and found that the value was approximately 1.6289. Now we'll evaluate the integral by expressing the integrand as a power series. Substituting $-x^2$ for x in (10.52) and then integrating term-by-term, we get

$$\int_{-1}^{2} e^{-x^2} \, dx = \int_{-1}^{2} \sum_{n=0}^{\infty} \frac{(-x^2)^n}{n!} \, dx = \int_{-1}^{2} \sum_{n=0}^{\infty} \frac{(-1)^n x^{2n}}{n!} \, dx = \left[\sum_{n=0}^{\infty} \frac{(-1)^n x^{2n+1}}{(2n+1)n!} \right]_{-1}^{2}$$

$$= \sum_{n=0}^{\infty} \frac{(-1)^n 2^{2n+1}}{(2n+1)n!} - \sum_{n=0}^{\infty} \frac{(-1)^n (-1)^{2n+1}}{(2n+1)n!} = \sum_{n=0}^{\infty} \frac{(-1)^n (2^{2n+1} + 1)}{(2n+1)n!}.$$

$$\text{(10.56)}$$

We leave it as an exercise for you (see Exercise 22) to show that if we approximate the last series in (10.56) with its 10th partial sum, then the error in the approximation will be at most 0.01. Using a computer to do the arithmetic, we find that

$$\int_{-1}^{2} e^{-x^2} \, dx \approx \sum_{n=0}^{10} \frac{(-1)^n (2^{2n+1} + 1)}{(2n+1)n!} \approx 1.6359. \qquad □$$

Example 10.8.7. Find

$$\lim_{x \to 0} \frac{\sin x - x}{\tan^{-1} x - x}.$$

Solution. This limit is an indeterminate form of type $0/0$. It can be solved by l'Hôpital's rule (see Exercise 23), but it may be easier to proceed by expressing numerator and denominator as power series. By (10.41) and (10.55a), if $-1 \le x \le 1$ and $x \ne 0$ then

$$\frac{\sin x - x}{\tan^{-1} x - x} = \frac{(x - x^3/3! + x^5/5! - \cdots) - x}{(x - x^3/3 + x^5/5 - \cdots) - x} = \frac{-x^3/6 + x^5/120 - \cdots}{-x^3/3 + x^5/5 - \cdots}$$

$$= \frac{x^3(-1/6 + x^2/120 - \cdots)}{x^3(-1/3 + x^2/5 - \cdots)} = \frac{-1/6 + x^2/120 - \cdots}{-1/3 + x^2/5 - \cdots}.$$

By the continuity of power series (part 1 of Theorem 10.7.1), we can find the limits as $x \to 0$ of the numerator and denominator of the last fraction above by substituting 0 for x. Therefore

$$\lim_{x \to 0} \frac{\sin x - x}{\tan^{-1} x - x} = \lim_{x \to 0} \frac{-1/6 + x^2/120 - \cdots}{-1/3 + x^2/5 - \cdots} = \frac{-1/6 + 0 + 0 + \cdots}{-1/3 + 0 + 0 + \cdots} = \frac{-1/6}{-1/3} = \frac{1}{2}.$$

$\square$

Exercises 10.8

1–7: Find the Taylor series for f centered at a.

1. $f(x) = 3/x^2, a = 2$.

2. $f(x) = \ln x, a = e$.

3. $f(x) = \ln(3x + 1), a = 1$.

4. $f(x) = \cos x, a = \pi/4$.

5. $f(x) = \sin(2x), a = \pi/4$.

6. $f(x) = 2^x, a = 0$.

7. $f(x) = 1/\sqrt{x}, a = 1$.

8. (a) Find the Taylor series for the function $f(x) = x^3$ centered at 2.

   (b) Show that the Taylor series converges to $f(x)$ for all values of x.

9. Let $f(x) = \tan^{-1} x$. Find $f^{(10)}(0)$ and $f^{(11)}(0)$. (Hint: Computing the 11th derivative of $\tan^{-1} x$ would be very messy. It is easier to use the fact that series (10.41) must be the Maclaurin series for f.)

10. Compute $\cos(18°)$ with error at most 0.001.

11. Setting $x = 1$ in equation (10.52) gives us the equation

$$e = \sum_{n=0}^{\infty} \frac{1}{n!} = 1 + 1 + \frac{1}{2} + \frac{1}{6} + \cdots. \tag{10.57}$$

For any positive integer N, let t_N be the Nth tail of this series:

$$t_N = \sum_{n=N+1}^{\infty} \frac{1}{n!} = \frac{1}{(N+1)!} + \frac{1}{(N+2)!} + \cdots .$$

(a) Show that for every N,

$$0 \le t_N \le \frac{N+2}{(N+1)(N+1)!}.$$

(Hint: Overestimate t_N with a geometric series.)

(b) Use a partial sum of series (10.57) to compute the value of e with error at most 0.001.

12. Compute $\sqrt{e}$ with error at most 0.001.

13. (a) Find a power series that converges to

$$\int_{-1}^{0} e^{(x^3)}\, dx.$$

(b) Use a partial sum of this series to approximate the value of the integral with error at most 0.001.

14. The hyperbolic cosine and sine functions are the functions cosh and sinh defined as follows:

$$\cosh x = \frac{e^x + e^{-x}}{2}, \qquad \sinh x = \frac{e^x - e^{-x}}{2}.$$

Find power series centered at 0 that converge to $\cosh x$ and $\sinh x$ for all values of x. (By Theorem 10.8.1, these are the Maclaurin series for cosh and sinh. For more on the hyperbolic cosine and sine functions, see Exercises 29–35 in Section 7.5.)

15. The function

$$S(x) = \int_0^x \sin(t^2)\, dt$$

is called a *Fresnel integral*; it has applications in optics.

(a) Find the Maclaurin series for S. (Hint: Use the Maclaurin series for the sine function and term-by-term integration to find a power series centered at 0 that converges to $S(x)$. By Theorem 10.8.1, it must be the Maclaurin series for S.)

(b) Compute $S(1)$ with error at most 0.001.

16. The function

$$\mathrm{sinc}(x) = \begin{cases} \frac{\sin x}{x}, & \text{if } x \ne 0, \\ 1, & \text{if } x = 0 \end{cases}$$

is called the *cardinal sine function*. It plays an important role in digital signal processing.

(a) Find the Maclaurin series for sinc.

The function

$$\text{Si}(x) = \int_0^x \text{sinc}(t)\, dt$$

is called the *sine integral*; it is also important in digital signal processing.

(b) Find the Maclaurin series for Si.

(c) Compute Si(1) with error at most 0.001.

17. (a) Find

$$\lim_{x \to 0} \frac{\cos x - 1}{e^{(x^2)} - 1}$$

by writing the numerator and denominator as power series.

(b) Check your answer to part (a) by using l'Hôpital's rule.

18. Find

$$\lim_{x \to 0} \frac{\ln(1 + x^3) - x^3}{\sin(x^2) - x^2}.$$

19. In the text, we showed that the Taylor series for e^x centered at 3 is

$$\sum_{n=0}^{\infty} \frac{e^3}{n!}(x - 3)^n.$$

Show that this series converges to e^x for all values of x.

20. In the text, we showed that the Maclaurin series for $\sin x$ is

$$\sum_{n=0}^{\infty} \frac{(-1)^n x^{2n+1}}{(2n + 1)!}.$$

Show that the interval of convergence for this power series is $(-\infty, \infty)$.

21. Show that if $f''(x) = -f(x)$, then there are numbers a and b such that $f(x) = a \sin x + b \cos x$. (Hint: First show that for any a and b,

$$(f(x) - (a \sin x + b \cos x))^2 + (f'(x) - (a \cos x - b \sin x))^2$$

is equal to a constant. Then find values of a and b for which this constant is 0.)

22. Let

$$s = \sum_{n=0}^{\infty} \frac{(-1)^n(2^{2n+1}+1)}{(2n+1)n!}, \qquad s_{10} = \sum_{n=0}^{10} \frac{(-1)^n(2^{2n+1}+1)}{(2n+1)n!}.$$

Show that the magnitude of the error in the approximation $s \approx s_{10}$ is at most 0.01.

23. Use l'Hôpital's rule to find

$$\lim_{x \to 0} \frac{\sin x - x}{\tan^{-1} x - x},$$

and check that your answer agrees with the one we found in Example 10.8.7.

24. Use the definition of Taylor series to verify that the series we found in part (b) of Example 10.8.3 is the Taylor series for $g(x) = \sin x$ centered at $\pi/6$.

25. Let

$$f(x) = \begin{cases} e^{(-1/x^2)}, & \text{if } x \neq 0, \\ 0, & \text{if } x = 0. \end{cases}$$

In this problem you will study the Maclaurin series for f.

(a) Show that for any polynomial $p(x)$,

$$\lim_{x \to \infty} \frac{p(x)}{e^{(x^2)}} = 0, \qquad \lim_{x \to -\infty} \frac{p(x)}{e^{(x^2)}} = 0.$$

(b) Show that for any polynomial $p(x)$,

$$\lim_{h \to 0} p(1/h)e^{(-1/h^2)} = 0.$$

(Hint: Use part (a).)

(c) Show that $f'(0) = 0$. (Hint: Use the definition of derivative and part (b).)

(d) Show that for every $n \geq 0$, there is some polynomial $p_n(x)$ such that

$$f^{(n)}(x) = \begin{cases} p_n(1/x)e^{(-1/x^2)}, & \text{if } x \neq 0, \\ 0, & \text{if } x = 0. \end{cases}$$

(Hint: Use mathematical induction. You may find it helpful to work out the first few derivatives first.)

(e) Find the Maclaurin series for f.

(f) Show that for every value of x, the Maclaurin series converges, but if $x \neq 0$ then it does *not* converge to $f(x)$.

10.9   The Binomial Series

It is straightforward to compute that

$$(1+x)^2 = 1+2x+x^2,$$
$$(1+x)^3 = 1+3x+3x^2+x^3,$$

and if you know the binomial theorem then you can generalize and give a formula for $(1+x)^n$ for any nonnegative integer n (and if not, then see Exercise 1). But in this section, we will see that we can even generalize to $(1+x)^p$ for any real number p. To do this, we will find a power series representation for the function $f(x) = (1+x)^p$. We begin by finding the Maclaurin series for f.

As usual, we must compute $f^{(n)}(0)$ for all $n \geq 0$:

$$
\begin{aligned}
f(x) &= (1+x)^p, & f(0) &= 1, \\
f'(x) &= p(1+x)^{p-1}, & f'(0) &= p, \\
f''(x) &= p(p-1)(1+x)^{p-2}, & f''(0) &= p(p-1), \\
f'''(x) &= p(p-1)(p-2)(1+x)^{p-3}, & f'''(0) &= p(p-1)(p-2), \\
&\ \ \vdots & &\ \ \vdots
\end{aligned}
$$

It will be convenient to have notation for the values of $f^{(n)}(0)$, so we define $(p)_0 = 1$, $(p)_1 = p$, $(p)_2 = p(p-1)$, $(p)_3 = p(p-1)(p-2)$, and, in general,

$$(p)_n = p(p-1)(p-2)\cdots(p-(n-1)).$$

These numbers are called *falling factorials*, because, like factorials, they are products of decreasing lists of numbers. (Indeed, it is not hard to see that for any integer $n \geq 0$, $(n)_n = n!$.) Falling factorials could also be defined by the following recursive definition:

$$(p)_0 = 1, \text{ and for every integer } n \geq 0,\ (p)_{n+1} = (p)_n \cdot (p-n).$$

Using falling factorial notation, we have $f^{(n)}(0) = (p)_n$, and therefore the Maclaurin series for f is

$$\sum_{n=0}^{\infty} \frac{f^{(n)}(0)}{n!} x^n = \sum_{n=0}^{\infty} \frac{(p)_n}{n!} x^n = 1 + px + \frac{p(p-1)}{2!}x^2 + \frac{p(p-1)(p-2)}{3!}x^3 + \cdots.$$

This power series is called the *binomial series with exponent p*.

If p is a nonnegative integer, then for $n > p$ the nth term of the series will be 0. For example, if $p = 2$ then the series takes the form

$$\sum_{n=0}^{\infty} \frac{(2)_n}{n!} x^n = 1 + 2x + \frac{2 \cdot 1}{2!} x^2 + \frac{2 \cdot 1 \cdot 0}{3!} x^3 + \frac{2 \cdot 1 \cdot 0 \cdot (-1)}{4!} x^4 + \cdots$$

$$= 1 + 2x + x^2 + 0 + 0 + \cdots.$$

Of course, for all x this series converges to $1 + 2x + x^2 = (1 + x)^2$. In Exercise 1 we ask you to show that if p is any nonnegative integer, then the binomial series with exponent p turns into a finite sum that is equal to $(1 + x)^p$ for all values of x.

However, we are most interested in seeing what happens if p is not a nonnegative integer. In that case, all of the coefficients in the Maclaurin series are nonzero, and we can use the ratio test for absolute convergence to determine the radius of convergence. Applying the recursive definition of falling factorials to help us simplify, we get

$$L = \lim_{n\to\infty} \frac{|(p)_{n+1} x^{n+1}/(n+1)!|}{|(p)_n x^n/n!|} = \lim_{n\to\infty} \left[\frac{|x|^{n+1}}{|x|^n} \cdot \frac{|(p)_n| \cdot |p-n|}{|(p)_n|} \cdot \frac{n!}{(n+1)!} \right]$$

$$= \lim_{n\to\infty} \left[|x| \cdot \frac{n-p}{n+1} \right] = \lim_{n\to\infty} \left[|x| \cdot \frac{1 - p/n}{1 + 1/n} \right] = |x|.$$

Thus, the series converges absolutely if $|x| < 1$ and diverges if $|x| > 1$. (We'll worry about the endpoints later.) We can therefore define $g(x)$, for $-1 < x < 1$, by the formula

$$g(x) = \sum_{n=0}^{\infty} \frac{(p)_n}{n!} x^n.$$

We now want to see if $g(x) = (1 + x)^p$.

As in the last section, we begin by looking for a distinctive property of $f(x) = (1 + x)^p$. Since $f'(x) = p(1 + x)^{p-1}$, we have

$$(1 + x)f'(x) = (1 + x)p(1 + x)^{p-1} = p(1 + x)^p = pf(x).$$

To see if g has the same property, we first compute

$$g'(x) = \sum_{n=1}^{\infty} \frac{n(p)_n}{n!} x^{n-1} = \sum_{n=1}^{\infty} \frac{(p)_n}{(n-1)!} x^{n-1}$$

$$= (p)_1 + (p)_2 x + \frac{(p)_3}{2!} x^2 + \frac{(p)_4}{3!} x^3 + \cdots. \tag{10.58}$$

This is just like the binomial series, except that the subscripts on the falling factorials are one larger than the exponents on x. Thus, we can rewrite this series with different

Σ-notation and then apply the recursive definition of falling factorials to get

$$g'(x) = \sum_{n=0}^{\infty} \frac{(p)_{n+1}}{n!} x^n = \sum_{n=0}^{\infty} \frac{(p)_n \cdot (p-n)}{n!} x^n = p \cdot \sum_{n=0}^{\infty} \frac{(p)_n}{n!} x^n - \sum_{n=0}^{\infty} \frac{n(p)_n}{n!} x^n.$$

$$(10.59)$$

Now, look at the two series on the right-hand side of (10.59). The first is just the binomial series again, so it is equal to $g(x)$. And the second is very similar to the first step in (10.58). In fact, rewriting it to see how it relates to $g'(x)$, we see that

$$\sum_{n=0}^{\infty} \frac{n(p)_n}{n!} x^n = 0 + \sum_{n=1}^{\infty} \frac{n(p)_n}{n!} x^n = x \cdot \sum_{n=1}^{\infty} \frac{n(p)_n}{n!} x^{n-1} = xg'(x).$$

Substituting into the right-hand side of (10.59), we get

$$g'(x) = pg(x) - xg'(x),$$

and therefore

$$(1+x)g'(x) = pg(x). \qquad (10.60)$$

In other words, g' is related to g in exactly the same way that f' is related to f.

Imitating the approach we used with e^x, we now check that $g(x)/(1+x)^p$ is a constant. (Note that, since we are only considering x in the open interval $(-1, 1)$, the denominator here is not 0.) Applying (10.60), we see that

$$\frac{d}{dx}\left(\frac{g(x)}{(1+x)^p}\right) = \frac{(1+x)^p g'(x) - g(x)p(1+x)^{p-1}}{((1+x)^p)^2}$$

$$= \frac{(1+x)^{p-1}((1+x)g'(x) - pg(x))}{(1+x)^{2p}} = \frac{pg(x) - pg(x)}{(1+x)^{p+1}} = 0.$$

Thus, there is some number C such that for all $x \in (-1, 1)$, $g(x)/(1+x)^p = C$, and therefore $g(x) = C(1+x)^p$. Finally, we set $x = 0$ to find C:

$$C = C(1+0)^p = g(0) = \sum_{n=0}^{\infty} \frac{(p)_n}{n!} \cdot 0^n = 1 + 0 + 0 + \cdots = 1.$$

We conclude that $g(x) = (1+x)^p$. In other words, we have proven the following theorem.

Theorem 10.9.1. *For every real number* p,

$$(1+x)^p = \sum_{n=0}^{\infty} \frac{(p)_n}{n!} x^n$$

$$= 1 + px + \frac{p(p-1)}{2!} x^2 + \frac{p(p-1)(p-2)}{3!} x^3 + \cdots, \qquad -1 < x < 1.$$

Let's try the theorem out with a few different values of p. If $p = -1$, then the theorem says that

$$\frac{1}{1+x} = 1 + (-1)x + \frac{(-1)(-2)}{2!}x^2 + \frac{(-1)(-2)(-3)}{3!}x^3 + \cdots$$

$$= 1 - x + x^2 - x^3 + \cdots, \qquad\qquad -1 < x < 1.$$

We already knew this; it is equation (10.34) from Section 10.7. In this case, the series diverges when $x = \pm 1$, so $(-1, 1)$ is the entire interval of convergence.

But we can discover a new power series formula by setting $p = -1/2$ to get

$$\frac{1}{\sqrt{1+x}} = \sum_{n=0}^{\infty} \frac{(-1/2)_n}{n!} x^n, \qquad\qquad -1 < x < 1. \qquad (10.61)$$

Now, it turns out that we can work out a formula for the falling factorial $(-1/2)_n$ that appears in this series. We start with

$$\left(-\frac{1}{2}\right)_n = \left(-\frac{1}{2}\right)\left(-\frac{3}{2}\right)\left(-\frac{5}{2}\right)\cdots\left(-\frac{1}{2}-(n-1)\right)$$

$$= \left(-\frac{1}{2}\right)\left(-\frac{3}{2}\right)\left(-\frac{5}{2}\right)\cdots\left(-\frac{2n-1}{2}\right) = \frac{(-1)^n \cdot 1 \cdot 3 \cdot 5 \cdots (2n-1)}{2^n}.$$

To simplify the product of odd integers in the numerator, we multiply and divide by all even integers from 2 to $2n$:

$$1 \cdot 3 \cdot 5 \cdots (2n-1) = \frac{1 \cdot 2 \cdot 3 \cdots (2n)}{2 \cdot 4 \cdots (2n)} = \frac{(2n)!}{(2 \cdot 1)(2 \cdot 2)\cdots(2 \cdot n)} = \frac{(2n)!}{2^n n!}.$$

Thus, we have

$$\left(-\frac{1}{2}\right)_n = \frac{(-1)^n \cdot 1 \cdot 3 \cdot 5 \cdots (2n-1)}{2^n} = \frac{(-1)^n (2n)!/(2^n n!)}{2^n} = \frac{(-1)^n (2n)!}{4^n n!}.$$

Finally, plugging this into series (10.61) gives us

$$\frac{1}{\sqrt{1+x}} = \sum_{n=0}^{\infty} \frac{(-1)^n (2n)!/(4^n n!)}{n!} x^n = \sum_{n=0}^{\infty} \frac{(-1)^n (2n)!}{4^n (n!)^2} x^n, \quad -1 < x < 1. \quad (10.62)$$

This time, it turns out the power series converges at one of the endpoints. Setting $x = \pm 1$ in (10.62) gives us two series we have already studied. When $x = 1$ we get series (10.20), and when $x = -1$ we get (10.21). We showed in Section 10.5 that the first of these converges and the second diverges, so the interval of convergence for series (10.62) is $(-1, 1]$. Imitating the reasoning we used in Section 10.7, we can use continuity to extend (10.62) to this interval:

$$\frac{1}{\sqrt{1+x}} = \sum_{n=0}^{\infty} \frac{(-1)^n (2n)!}{4^n (n!)^2} x^n = 1 - \frac{x}{2} + \frac{3x^2}{8} - \frac{5x^3}{16} + \cdots, \qquad -1 < x \le 1.$$

$$(10.63)$$

We can use (10.63) to resolve one remaining unsolved problem from before. Setting $x = 1$ in equation (10.63), we find that

$$\sum_{n=0}^{\infty} \frac{(-1)^n (2n)!}{4^n (n!)^2} = 1 - \frac{1}{2} + \frac{3}{8} - \frac{5}{16} + \cdots = \frac{1}{\sqrt{1+1}} = \frac{\sqrt{2}}{2} \approx 0.707.$$

This confirms our stated value for series (10.20), which was the same as (10.26d). We have now verified the values of all four of the series in (10.26). (For the value of series (10.27d), see Exercise 12.)

Since $\frac{d}{dx}(\sin^{-1} x) = 1/\sqrt{1 - x^2}$, we can also use (10.63) to find a power series for $\sin^{-1} x$. First we substitute $-x^2$ for x in (10.63), which gives us

$$\frac{1}{\sqrt{1 - x^2}} = \sum_{n=0}^{\infty} \frac{(-1)^n (2n)!}{4^n (n!)^2} (-x^2)^n = \sum_{n=0}^{\infty} \frac{(2n)! x^{2n}}{4^n (n!)^2}, \qquad -1 < -x^2 \le 1.$$

The solution to the inequality $-1 < -x^2 \le 1$ is $-1 < x < 1$, so we can integrate term-by-term to conclude that

$$\sin^{-1} x = \sum_{n=0}^{\infty} \frac{(2n)! x^{2n+1}}{4^n (n!)^2 (2n + 1)} + C, \qquad -1 < x < 1.$$

When $x = 0$ this means that

$$C = \sin^{-1} 0 = 0,$$

so we have

$$\sin^{-1} x = \sum_{n=0}^{\infty} \frac{(2n)! x^{2n+1}}{4^n (n!)^2 (2n + 1)}, \qquad -1 < x < 1. \qquad (10.64)$$

Finally, it follows from (10.23) that

$$0 \le \frac{(2n)!}{4^n (n!)^2 (2n + 1)} \le \frac{1}{\sqrt{2n}(2n + 1)} \le \frac{1}{n^{3/2}},$$

and the series $\sum_{n=1}^{\infty}(1/n^{3/2})$ is a convergent p-series, so by the comparison test, the series in (10.64) converges absolutely when $= \pm 1$. Therefore we can extend (10.64) by continuity to the interval $[-1, 1]$:

$$\sin^{-1} x = \sum_{n=0}^{\infty} \frac{(2n)! x^{2n+1}}{4^n (n!)^2 (2n+1)} = x + \frac{x^3}{6} + \frac{3x^5}{40} + \cdots, \quad -1 \le x \le 1. \quad (10.65)$$

For example, we have

$$\sum_{n=0}^{\infty} \frac{(2n)!}{4^n (n!)^2 (2n+1)} = \sin^{-1}(1) = \frac{\pi}{2}.$$

We have now found power series for many familiar functions. We summarize the most important of these in Table 10.1. We end this section with two more examples of applications of power series.

$$\frac{1}{1-x} = \sum_{n=0}^{\infty} x^n = 1 + x + x^2 + \cdots, \qquad -1 < x < 1$$

$$\ln(1+x) = \sum_{n=1}^{\infty} \frac{(-1)^{n-1} x^n}{n} = x - \frac{x^2}{2} + \frac{x^3}{3} - \cdots, \qquad -1 < x \le 1$$

$$\tan^{-1} x = \sum_{n=0}^{\infty} \frac{(-1)^n x^{2n+1}}{2n+1} = x - \frac{x^3}{3} + \frac{x^5}{5} - \cdots, \qquad -1 \le x \le 1$$

$$e^x = \sum_{n=0}^{\infty} \frac{x^n}{n!} = 1 + x + \frac{x^2}{2!} + \cdots, \qquad -\infty < x < \infty$$

$$\sin x = \sum_{n=0}^{\infty} \frac{(-1)^n x^{2n+1}}{(2n+1)!} = x - \frac{x^3}{3!} + \frac{x^5}{5!} - \cdots, \qquad -\infty < x < \infty$$

$$\cos x = \sum_{n=0}^{\infty} \frac{(-1)^n x^{2n}}{(2n)!} = 1 - \frac{x^2}{2!} + \frac{x^4}{4!} - \cdots, \qquad -\infty < x < \infty$$

$$(1+x)^p = \sum_{n=0}^{\infty} \frac{(p)_n}{n!} x^n = 1 + px + \frac{p(p-1)}{2!} x^2 + \cdots, \qquad -1 < x < 1$$

$$\sin^{-1} x = \sum_{n=0}^{\infty} \frac{(2n)! x^{2n+1}}{4^n (n!)^2 (2n+1)} = x + \frac{x^3}{6} + \frac{3x^5}{40} + \cdots, \qquad -1 \le x \le 1$$

Table 10.1: Power series representations of some familiar functions.

Example 10.9.2. Compute $\sqrt[3]{10}$ with error at most 0.001.

Solution. While there are other ways to do this computation, one possibility is to use the binomial series. Since $\sqrt[3]{8} = 2$, $\sqrt[3]{10}$ must be a little more than 2. But how much more? We start by rewriting the problem as

$$\sqrt[3]{10} = \sqrt[3]{8 + 2} = \sqrt[3]{8\left(1 + \frac{1}{4}\right)} = 2\left(1 + \frac{1}{4}\right)^{1/3}.$$

Next, we turn to the binomial series. By Theorem 10.9.1,

$$(1 + x)^{1/3} = \sum_{n=0}^{\infty} \frac{(1/3)_n}{n!} x^n, \qquad -1 < x < 1,$$

and therefore

$$\sqrt[3]{10} = 2\left(1 + \frac{1}{4}\right)^{1/3} = 2 \cdot \sum_{n=0}^{\infty} \frac{(1/3)_n}{n!}\left(\frac{1}{4}\right)^n = \sum_{n=0}^{\infty} \frac{(1/3)_n}{2^{2n-1}n!} = 2 + \frac{1}{6} - \frac{1}{72} + \cdots.$$

We leave it to you to check (see Exercise 11) that, after the first term, this series satisfies the hypotheses of the alternating series test, and therefore we can achieve the desired accuracy by adding terms until we reach a term whose absolute value is at most 0.001. Since the $n = 4$ term is about -0.0003, we can use the approximation

$$\sqrt[3]{10} \approx \sum_{n=0}^{3} \frac{(1/3)_n}{2^{2n-1}n!} \approx 2.1547. \qquad \square$$

Example 10.9.3. Compute

$$\int_0^1 \sin^{-1}(x^3)\, dx$$

with error at most 0.001.

Solution. We substitute x^3 for x in (10.64) and then integrate term-by-term:

$$\int_0^1 \sin^{-1}(x^3)\, dx = \int_0^1 \sum_{n=0}^{\infty} \frac{(2n)!\, x^{6n+3}}{4^n (n!)^2 (2n+1)}\, dx$$

$$= \left[\sum_{n=0}^{\infty} \frac{(2n)!\, x^{6n+4}}{4^n (n!)^2 (2n+1)(6n+4)}\right]_0^1 = \sum_{n=0}^{\infty} \frac{(2n)!}{4^n (n!)^2 (2n+1)(6n+4)}.$$

If we approximate the value s of this series with a partial sum s_N, then the error will be the tail

$$s - s_N = t_N = \sum_{n=N+1}^{\infty} \frac{(2n)!}{4^n (n!)^2 (2n+1)(6n+4)}.$$

To overestimate this tail, we first apply (10.23) to see that

$$0 \leq \frac{(2n)!}{4^n (n!)^2 (2n+1)(6n+4)} \leq \frac{1}{\sqrt{2n}(2n)(6n)} = \frac{1}{12\sqrt{2}n^{5/2}}.$$

Now, as in the proof of the integral test, we can overestimate the tail t_N with an improper integral:

$$t_N \leq \sum_{n=N+1}^{\infty} \frac{1}{12\sqrt{2}n^{5/2}} \leq \int_N^{\infty} \frac{x^{-5/2}}{12\sqrt{2}} \, dx = \lim_{t \to \infty} \left[-\frac{x^{-3/2}}{18\sqrt{2}} \right]_N^t$$

$$= \lim_{t \to \infty} \left(-\frac{1}{18\sqrt{2}t^{3/2}} + \frac{1}{18\sqrt{2}N^{3/2}} \right) = \frac{1}{18\sqrt{2}N^{3/2}}.$$

To achieve an error of at most $0.001 = 1/1000$, we choose N so that $18\sqrt{2}N^{3/2} \geq 1000$, which means

$$N \geq \left(\frac{1000}{18\sqrt{2}} \right)^{2/3} \approx 11.57.$$

Thus, we can get the desired accuracy by using the approximation

$$\int_0^1 \sin^{-1}(x^3) \, dx \approx s_{12} = \sum_{n=0}^{12} \frac{(2n)!}{4^n (n!)^2 (2n+1)(6n+4)} \approx 0.2766. \qquad \square$$

Exercises 10.9

1. If p is a nonnegative integer, then $(1+x)^p$ can be expanded as a polynomial. This polynomial can be thought of as a power series in which all terms beyond some point are 0, and by Theorem 10.8.1, this series must be the Maclaurin series for $(1+x)^p$. In other words, the binomial series with exponent p must be equal to $(1+x)^p$. In this problem you will verify this fact directly.

   (a) Show that if p is a nonnegative integer, then for all $n > p$, $(p)_n = 0$, and therefore

   $$\sum_{n=0}^{\infty} \frac{(p)_n}{n!} x^n = \sum_{n=0}^{p} \frac{(p)_n}{n!} x^n.$$

   (b) Use mathematical induction to show that for every nonnegative integer p,

   $$(1+x)^p = \sum_{n=0}^{p} \frac{(p)_n}{n!} x^n.$$

2. For any positive integer p, use the binomial series to find a power series centered at 0 that converges to

   $$\frac{1}{(1+x)^p}$$

   on some interval. Compare your answer to your answer to Exercise 21 in Section 10.7. Are the two series the same?

3. Compute $\sin^{-1}(1/4)$ with error at most 0.001. (Hint: The tail of the series is hard to overestimate. Replace it with something larger that is easier to overestimate.)

4. (a) Find an infinite series that converges to

$$\int_0^{1/2} \sin^{-1} x \, dx.$$

   (b) Use a partial sum to approximate the value of the integral in part (a) with error at most 0.001.

   (c) Find the exact value of the integral in part (a), and verify the accuracy of your approximation in part (b).

5. Use the binomial series to compute $\sqrt{10}$ with error at most 0.001.

6. (a) Show that

$$\sqrt{15} = \sum_{n=0}^{\infty} \frac{15(2n)!}{4^{3n+1}(n!)^2}.$$

   (Hint: Start with the calculation

$$\sqrt{15} = \frac{15}{\sqrt{15}} = \frac{15}{\sqrt{16-1}} = \frac{15}{\sqrt{16 \cdot (1-1/16)}} = \frac{15}{4} \cdot \frac{1}{\sqrt{1-1/16}},$$

   and then apply (10.63).)

   (b) Use part (a) to compute $\sqrt{15}$ with error at most 0.001.

7. (a) Find an infinite series that converges to

$$\int_0^{1/2} \frac{dx}{\sqrt{1+x^3}}.$$

   (Hint: Use (10.63) and term-by-term integration.)

   (b) Compute the integral in part (a) with error at most 0.001.

8. In this problem you will use three methods to find the Maclaurin series for the function $f(x) = \sqrt{1+x}$.

   (a) Find a power series representation for f by setting $p = 1/2$ in the binomial series.

   (b) Find a power series representation for f by multiplying series (10.63) by $1+x$.

   (c) Find a power series representation for f by integrating series (10.63) term-by-term.

   (d) Verify that all three series in parts (a)–(c) are the same.

9. (a) Find the interval of convergence of the power series

$$\sum_{n=0}^{\infty} \frac{(-1)^n (2n)! x^{2n+1}}{4^n (n!)^2 (2n+1)}.$$

   (b) Evaluate the series in part (a) for all x in the interval of convergence. (Hint: Notice that the series is the same as series (10.65) for $\sin^{-1} x$, except for the factor of $(-1)^n$. Imitate the derivation of (10.65). Compare your answer to Exercise 32 in Section 7.5.)

10. (a) Show that

$$\sum_{n=0}^{\infty} \frac{3(2n)!}{16^n (n!)^2 (2n+1)} = \pi.$$

   (Hint: Begin with a series for $\sin^{-1}(1/2)$.)

   (b) Find a similar series converging to π by starting with a series for $\sin^{-1}(\sqrt{2}/2)$.

11. Show that the series

$$\sum_{n=1}^{\infty} \frac{(1/3)_n}{2^{2n-1} n!}$$

satisfies the hypotheses of the alternating series test, as stated in our solution to Example 10.9.2.

12. Series (10.27d) in Section 10.6 was the power series

$$\sum_{n=0}^{\infty} \frac{(2n)! x^n}{(n!)^2}.$$

In part (c) of Example 10.6.4 we showed that its interval of convergence is $[-1/4, 1/4)$. For values of x in this interval, find the value of the series. (Hint: Substitute something for x in (10.63).)

13. Show that if $|x| \geq 1$ then

$$\sec^{-1} x = \frac{\pi}{2} - \sum_{n=0}^{\infty} \frac{(2n)!}{4^n (n!)^2 (2n+1) x^{2n+1}}.$$

(Hint: Use Exercise 26 in Section 7.5. Note that the series above is not a power series, because the powers of x are in the denominators of the terms. It is called a *Laurent series*, after Pierre Alphonse Laurent (1813–1854).)

10.10 Taylor Polynomials and the Taylor Remainder

In the last few sections we have studied functions defined by power series. In this section, we will see that it can also be useful to treat the partial sums of a power series as functions.

Consider the Taylor series for a function f centered at a:

$$\sum_{n=0}^{\infty} \frac{f^{(n)}(a)}{n!}(x-a)^n.$$

The Nth partial sum of this series is the function

$$T_N(x) = \sum_{n=0}^{N} \frac{f^{(n)}(a)}{n!}(x-a)^n$$

$$= f(a) + f'(a)(x-a) + \frac{f''(a)}{2!}(x-a)^2 + \cdots + \frac{f^{(N)}(a)}{N!}(x-a)^N.$$

Notice that T_N is a polynomial of degree at most N. (The formula above can be put in the form in which we usually write polynomials by multiplying out all of the powers of $x-a$.). The function T_N is called the *N*th-order Taylor polynomial for f centered at a. If $a = 0$, then it is also called the *N*th-order Maclaurin polynomial for f. The formula for $T_N(x)$ consists of those terms of the Taylor series that involve powers of $x-a$ up to N. If the Taylor series converges to $f(x)$ for some value of x, then

$$f(x) = \sum_{n=0}^{\infty} \frac{f^{(n)}(a)}{n!}(x-a)^n = \lim_{N \to \infty} \sum_{n=0}^{N} \frac{f^{(n)}(a)}{n!}(x-a)^n = \lim_{N \to \infty} T_N(x).$$

In other words, for large N, $T_N(x)$ should be close to $f(x)$.

For example, in part (b) of Example 10.8.3 we showed that the Taylor series for the function $f(x) = \sin x$ centered at $a = \pi/6$ is

$$\frac{1}{2} + \frac{\sqrt{3}}{2}\left(x - \frac{\pi}{6}\right) - \frac{1}{2 \cdot 2!}\left(x - \frac{\pi}{6}\right)^2 - \frac{\sqrt{3}}{2 \cdot 3!}\left(x - \frac{\pi}{6}\right)^3 + \cdots,$$

and this series converges to $\sin x$ for all values of x. The first few Taylor polynomials for this function are

$$T_0(x) = \frac{1}{2},$$

$$T_1(x) = \frac{1}{2} + \frac{\sqrt{3}}{2}\left(x - \frac{\pi}{6}\right),$$

$$T_2(x) = \frac{1}{2} + \frac{\sqrt{3}}{2}\left(x - \frac{\pi}{6}\right) - \frac{1}{2 \cdot 2!}\left(x - \frac{\pi}{6}\right)^2,$$

$$T_3(x) = \frac{1}{2} + \frac{\sqrt{3}}{2}\left(x - \frac{\pi}{6}\right) - \frac{1}{2 \cdot 2!}\left(x - \frac{\pi}{6}\right)^2 - \frac{\sqrt{3}}{2 \cdot 3!}\left(x - \frac{\pi}{6}\right)^3.$$

Figure 10.4 shows the graphs of f and these four Taylor polynomials. All of the curves pass through the point $(\pi/6, 1/2)$, and it appears that as N increases, the graph of T_N follows the shape of the graph of f more and more closely near this point. The graph of $y = T_3(x)$ stays quite close to the graph of $y = f(x)$ over the interval $-\pi/2 \leq x \leq \pi/2$.

Will this trend continue? Figure 10.5 shows the graphs of f and T_{25}. Remarkably, the graph of T_{25} is almost indistinguishable from the graph of f over the entire range $-3\pi \leq x \leq 3\pi$.

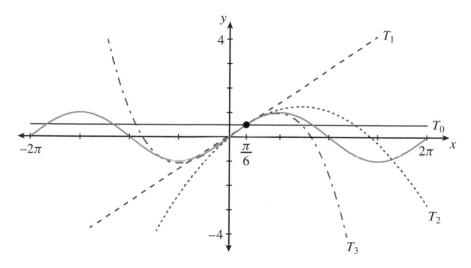

Figure 10.4: The graphs of the function $f(x) = \sin x$ (blue), and the Taylor polynomials T_0, T_1, T_2, and T_3 (red). The black dot is the point $(\pi/6, 1/2)$.

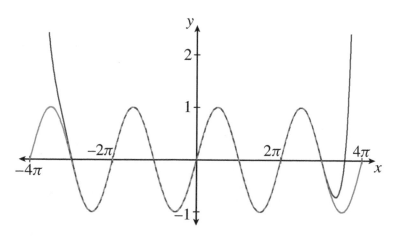

Figure 10.5: The graphs of the function $f(x) = \sin x$ (blue) and the Taylor polynomial T_{25} (red).

Let's try another example. As we saw in Section 10.8, the Taylor series for the function $f(x) = \ln x$ centered at $a = 1$ is

$$\sum_{n=1}^{\infty} \frac{(-1)^{n-1}}{n}(x-1)^n = (x-1) - \frac{(x-1)^2}{2} + \frac{(x-1)^3}{3} - \cdots ,$$

and this series converges to $\ln x$ for $0 < x \le 2$. In Figure 10.6 we see that the graph of the Taylor polynomial T_{25} is very close to the graph of f on the interval $(0, 2]$, but it deviates sharply beyond that interval. In the rest of this section, we investigate this phenomenon.

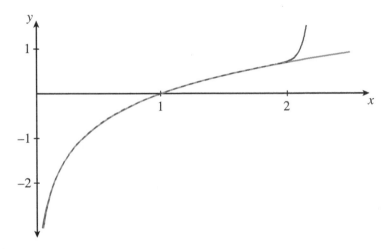

Figure 10.6: The graphs of the function $f(x) = \ln x$ (blue) and the Taylor polynomial T_{25} (red).

To understand the relationship between a function and its Taylor polynomials, let's look at the Taylor polynomials in order. Consider a function f and its Taylor polynomials T_N centered at some number a. The Taylor polynomial T_0 is the constant function $T_0(x) = f(a)$, so its graph is the horizontal line $y = f(a)$. This line intersects the graph of f at the point $(a, f(a))$. The graph of T_1 is the graph of the equation $y = T_1(x) = f(a) + f'(a)(x-a)$, which we can rewrite in the form

$$y - f(a) = f'(a)(x-a).$$

This is the point-slope form for the line through the point $(a, f(a))$ with slope $f'(a)$. But that means that the graph of T_1 is the line tangent to the graph of f at the point $(a, f(a))$! If you look back at Figure 10.4, you'll see that the graph of T_0 is a horizontal line through the point $(\pi/6, 1/2)$, and the graph of T_1 is the line tangent to the curve $y = \sin x$ at that point.

Another way to describe the relationship between f and T_1 is to say that $T_1(a) = f(a)$ and $T_1'(a) = f'(a)$. Since $T_1(a) = f(a)$, the graph of T_1 touches the graph of f

at the point $(a, f(a))$, and since $T_1'(a) = f'(a)$, the two graphs have the same slope at that point. But T_1 is a linear function, and therefore $T_1''(x) = 0$.

Similar results hold for T_2:

$$T_2(x) = f(a) + f'(a)(x - a) + \frac{f''(a)}{2!}(x - a)^2, \qquad T_2(a) = f(a),$$
$$T_2'(x) = f'(a) + f''(a)(x - a), \qquad\qquad\qquad\quad T_2'(a) = f'(a),$$
$$T_2''(x) = f''(a), \qquad\qquad\qquad\qquad\qquad\qquad\quad T_2''(a) = f''(a),$$
$$T_2'''(x) = 0.$$

As with T_1, the fact that $T_2(a) = f(a)$ and $T_2'(a) = f'(a)$ means that the graphs of T_2 and f touch at the point $(a, f(a))$, and they have the same slope at that point. But now we also have $T_2''(a) = f''(a)$, which means that the concavity of T_2 matches that of f at $(a, f(a))$. The graph of T_2 is a parabola that matches the shape of the graph of f near $(a, f(a))$ more closely than the tangent line does, because it curves the same way as f at $x = a$. Once again, this is visible in Figure 10.4.

We'll let you check that $T_3(a) = f(a)$, $T_3'(a) = f'(a)$, $T_3''(a) = f''(a)$, $T_3'''(a) = f'''(a)$, and $T_3^{(4)}(x) = 0$. The pattern is probably clear now. In general,

$$T_N^{(n)}(a) = f^{(n)}(a) \quad \text{for } n \le N,$$
$$T_N^{(N+1)}(x) = 0. \tag{10.66}$$

It is not hard to verify this by extending the calculations we did above for T_2. But there is another way of looking at it that makes equations (10.66) make intuitive sense. We can think of the Taylor polynomial $T_N(x)$ as a power series in which all terms beyond a certain point have coefficient 0:

$$T_N(x) = f(a) + f'(a)(x - a) + \cdots + \frac{f^{(N)}(a)}{N!}(x - a)^N + 0 \cdot (x - a)^{N+1}$$
$$+ 0 \cdot (x - a)^{N+2} + \cdots .$$

By Theorem 10.8.1, this must be the Taylor series centered at a for T_N. Thus, the Taylor series centered at a for T_N is the same as the series for f up to term number N, and it follows, by the formula for the coefficients of the Taylor series, that the derivatives of T_N and f at a match up to the Nth derivative. But T_N is a polynomial of degree at most N, and therefore $T_N^{(N+1)}(x) = 0$. That is precisely what equations (10.66) say.

To summarize what we have discovered so far, we can say that the reason that the shape of the graph of T_N is similar to the shape of the graph of f near the point $(a, f(a))$ is that T_N and its first N derivatives agree with f and its derivatives at a. What is remarkable is that, although this matching of derivatives holds only at the number a, the similarity in the shapes of the graphs often extends over a large interval around a.

The similarity between the graphs of f and T_N means that in many cases, if N is large, then $f(x) \approx T_N(x)$. It turns out that we can use (10.66) to analyze the accuracy of this approximation. As usual, we define the error of the approximation to be the difference between $f(x)$ and $T_N(x)$. This difference is called the *Nth Taylor remainder*, and it is denoted $R_N(x)$. Thus,

$$R_N(x) = f(x) - T_N(x).$$

Since $T_N(a) = f(a)$, $R_N(a) = 0$. For $x \neq a$, the following theorem gives a formula for $R_N(x)$.

Theorem 10.10.1 (Taylor's Theorem). *Suppose that $x \neq a$, and let I be the closed interval with endpoints a and x. (Thus, either $I = [a, x]$ or $I = [x, a]$, depending on whether $x > a$ or $x < a$.) Suppose also that $f^{(N)}$ is defined and continuous on I, and $f^{(N+1)}$ is defined on the interior of I. Then there is some number c in the interior of I such that*

$$R_N(x) = \frac{f^{(N+1)}(c)}{(N+1)!}(x - a)^{N+1}.$$

The Nth-order Taylor polynomial $T_N(x)$ includes all terms of the Taylor series up to $(f^{(N)}(a)/N!)(x - a)^N$, so the first term missing from $T_N(x)$ is $(f^{(N+1)}(a)/(N + 1)!)$ $(x - a)^{N+1}$. Notice that this is almost the same as the formula for $R_N(x)$ in Taylor's theorem. Thus, we could describe Taylor's theorem as saying that the Nth Taylor remainder is the same as the first term of the Taylor series that is missing from $T_N(x)$, except that the derivative $f^{(N+1)}$ is evaluated at some point c between a and x, rather than at a.

Proof of Taylor's Theorem. We will assume that $x > a$, so that $I = [a, x]$; the proof when $x < a$ is similar. It may be worthwhile to first work out the case $N = 0$. In that case, the hypothesis of the theorem says that f is continuous on $[a, x]$ and differentiable on (a, x). We have $R_0(x) = f(x) - T_0(x) = f(x) - f(a)$, so what we must prove is that for some c in the interval (a, x), $f(x) - f(a) = f'(c)(x - a)$, or equivalently

$$\frac{f(x) - f(a)}{x - a} = f'(c).$$

But we already know that this is true: it is just the mean value theorem!

To prove the theorem for $N > 0$, we will find a number c such that $a < c < x$ and

$$\frac{R_N(x)}{(x - a)^{N+1}} = \frac{f^{(N+1)}(c)}{(N+1)!}.$$

Our method will be to apply Cauchy's mean value theorem, Theorem 4.8.8, $N + 1$ times to the fraction on the left-hand side above. To do this, we will need to know something about the first $N + 1$ derivatives of the numerator and denominator of the fraction.

For the numerator, note that by (10.66),

$$R_N(x) = f(x) - T_N(x), \qquad\qquad R_N(a) = f(a) - T_N(a) = 0,$$
$$R_N'(x) = f'(x) - T_N'(x), \qquad\qquad R_N'(a) = f'(a) - T_N'(a) = 0,$$

$$\vdots$$

$$(10.67)$$

$$R_N^{(N)}(x) = f^{(N)}(x) - T_N^{(N)}(x), \qquad R_N^{(N)}(a) = f^{(N)}(a) - T_N^{(N)}(a) = 0,$$
$$R_N^{(N+1)}(x) = f^{(N+1)}(x) - T_N^{(N+1)}(x)$$
$$= f^{(N+1)}(x) - 0 = f^{(N+1)}(x).$$

To work out the derivatives of the denominator, we let $g(x) = (x - a)^{N+1}$. Then

$$
\begin{aligned}
g(x) &= (x - a)^{N+1}, & g(a) &= 0, \\
g'(x) &= (N+1) \cdot (x - a)^N, & g'(a) &= 0, \\
g''(x) &= (N+1) \cdot N \cdot (x - a)^{N-1}, & g''(a) &= 0, \\
&\;\;\vdots & & \\
g^{(N)}(x) &= (N+1) \cdot N \cdot (N-1) \cdots 2 \cdot (x - a), & g^{(N)}(a) &= 0, \\
g^{(N+1)}(x) &= (N+1)!.
\end{aligned}
\tag{10.68}
$$

We are now ready for the application of Cauchy's mean value theorem. Since $R_N(a) = g(a) = 0$, we have

$$\frac{R_N(x)}{(x-a)^{N+1}} = \frac{R_N(x)}{g(x)} = \frac{R_N(x) - R_N(a)}{g(x) - g(a)}.$$

By Cauchy's mean value theorem, there must be some number c_1 such that $a < c_1 < x$ and

$$\frac{R_N(x)}{(x-a)^{N+1}} = \frac{R_N(x) - R_N(a)}{g(x) - g(a)} = \frac{R'_N(c_1)}{g'(c_1)}.$$

But now by (10.67) and (10.68) we also have $R'_N(a) = g'(a) = 0$, so we can rewrite this in the form

$$\frac{R_N(x)}{(x-a)^{N+1}} = \frac{R'_N(c_1)}{g'(c_1)} = \frac{R'_N(c_1) - R'_N(a)}{g'(c_1) - g'(a)},$$

and then apply Cauchy's mean value theorem again to conclude that for some number c_2, $a < c_2 < c_1 < x$ and

$$\frac{R_N(x)}{(x-a)^{N+1}} = \frac{R'_N(c_1) - R'_N(a)}{g'(c_1) - g'(a)} = \frac{R''_N(c_2)}{g''(c_2)}.$$

Continuing in this way, after $N + 1$ applications of Cauchy's mean value theorem we will have a number c_{N+1} such that $a < c_{N+1} < x$ and

$$\frac{R_N(x)}{(x-a)^{N+1}} = \frac{R_N^{(N+1)}(c_{N+1})}{g^{(N+1)}(c_{N+1})}.$$

According to (10.67) and (10.68), this means that

$$\frac{R_N(x)}{(x-a)^{N+1}} = \frac{f^{(N+1)}(c_{N+1})}{(N+1)!},$$

and setting $c = c_{N+1}$, it follows that

$$R_N(x) = \frac{f^{(N+1)}(c)}{(N+1)!}(x-a)^{N+1},$$

which is what we had to prove.   □

Example 10.10.2. Compute $\sin(20°)$ with error at most 0.0001.

Solution. We have already done this computation in Example 10.8.5, but we redo it now in a slightly different way, using Taylor's theorem to guarantee the accuracy of our approximation. Let $f(x) = \sin x$. As we showed in Example 10.8.5, $20°$ is $\pi/9$ radians, so we must compute $f(\pi/9)$. We will approximate this value by using the Maclaurin series representation for f:

$$f(x) = \sin x = x - \frac{x^3}{3!} + \frac{x^5}{5!} - \cdots .$$

Let T_N be the Nth-order Maclaurin polynomial for f; thus, $T_N(x)$ consists of all terms of the Maclaurin series up to the x^N term. Our plan is to compute $T_N(\pi/9)$ for some N that is large enough to achieve the desired accuracy. Since the error in the approximation is given by $R_N(\pi/9)$, we must choose N so that $|R_N(\pi/9)| \leq 0.0001$. Setting $a = 0$ and $x = \pi/9$ in Taylor's theorem, we get

$$R_N(\pi/9) = \frac{f^{(N+1)}(c)}{(N+1)!}\left(\frac{\pi}{9}\right)^{N+1}.$$

Unfortunately, this formula involves an unknown number c between 0 and $\pi/9$. As usual, we deal with this problem by overestimating $|R_N(\pi/9)|$, rather than trying to compute it exactly. Since $f(x) = \sin x$, $f^{(N+1)}(c)$ is either $\pm\sin c$ or $\pm\cos c$, and it follows that $|f^{(N+1)}(c)| \leq 1$. Therefore

$$|R_N(\pi/9)| = \frac{|f^{(N+1)}(c)|}{(N+1)!}\left(\frac{\pi}{9}\right)^{N+1} \leq \frac{(\pi/9)^{N+1}}{(N+1)!}.$$

If we try out a few values of N, we discover that

$$|R_4(\pi/9)| \leq \frac{(\pi/9)^5}{5!} \approx 0.00004 < 0.001,$$

so we can achieve the desired accuracy with the approximation $\sin(20°) \approx T_4(\pi/9)$.

Taking all terms of the Maclaurin series up to the x^4 term, we see that

$$T_4(x) = x - \frac{x^3}{3!}.$$

(Notice that the coefficient of x^4 in the Maclaurin series is 0, so $T_4(x)$ turns out to be a polynomial of degree 3.) Our answer is therefore

$$\sin(20°) \approx T_4(\pi/9) = (\pi/9) - \frac{(\pi/9)^3}{3!} \approx 0.34198.$$ □

According to the definition of the Taylor remainder, $R_N(x) = f(x) - T_N(x)$. One enlightening way of rewriting this, using the formula for the remainder from Taylor's theorem, is

$$f(x) = T_N(x) + R_N(x)$$

$$= f(a) + f'(a)(x - a) + \cdots + \frac{f^{(N)}(a)}{N!}(x - a)^N + \frac{f^{(N+1)}(c)}{(N+1)!}(x - a)^{N+1}.$$

Thus, Taylor's theorem says that we can get the exact value of $f(x)$ by writing down the Taylor polynomial $T_{N+1}(x)$, and then, in the last term, replacing $f^{(N+1)}(a)$ with $f^{(N+1)}(c)$, for some number c between a and x.

Another helpful way of rewriting the definition of the remainder is $T_N(x) = f(x) - R_N(x)$. This leads immediately to the following observation.

Theorem 10.10.3. *For every $N \geq 0$, let T_N be the Nth-order Taylor polynomial for f centered at a, and let R_N be the Nth Taylor remainder. For any value of x, if $\lim_{N \to \infty} R_N(x) = 0$, then the Taylor series converges to $f(x)$.*

Proof. Suppose that for some value of x, $\lim_{N \to \infty} R_N(x) = 0$. Then the value of the Taylor series is

$$\sum_{n=0}^{\infty} \frac{f^{(n)}(a)}{n!}(x - a)^n = \lim_{N \to \infty} \sum_{n=0}^{N} \frac{f^{(n)}(a)}{n!}(x - a)^n = \lim_{N \to \infty} T_N(x)$$

$$= \lim_{N \to \infty} (f(x) - R_N(x)) = f(x) - 0 = f(x). \qquad \square$$

As an example of the use of this theorem, consider again the Taylor series for the function $f(x) = \ln x$ centered at $a = 1$:

$$\sum_{n=1}^{\infty} \frac{(-1)^{n-1}}{n}(x - 1)^n = (x - 1) - \frac{(x - 1)^2}{2} + \frac{(x - 1)^3}{3} - \cdots .$$

We already know that this series converges to $\ln x$ when $x = 2$, but let's see how we can confirm this fact by using Theorem 10.10.3. As we saw in Section 10.8, for any $N \geq 1$, $f^{(N)}(x) = (-1)^{N-1}(N - 1)!/x^N$. Applying Taylor's theorem with $a = 1$ and $x = 2$, we conclude that for some c in the interval $(1, 2)$,

$$R_N(2) = \frac{f^{(N+1)}(c)}{(N+1)!}(2 - 1)^{N+1} = \frac{(-1)^N N!/c^N}{(N+1)!} = \frac{(-1)^N}{(N+1)c^N}.$$

According to Theorem 10.10.3, if we can show that $\lim_{N \to \infty} R_N(2) = 0$, then we can conclude that the Taylor series converges to $f(x) = \ln x$ when $x = 2$. Unfortunately, our formula for $R_N(2)$ contains an unknown number c. But as usual, we can deal with this problem by overestimating $|R_N(2)|$. Since $1 < c < 2$, we have

$$0 \le |R_N(2)| = \frac{1}{(N+1)c^N} < \frac{1}{N+1}.$$

Since $\lim_{N \to \infty} 1/(N+1) = 0$, we conclude by the squeeze theorem that $\lim_{N \to \infty} |R_N(2)| = 0$, and therefore $\lim_{N \to \infty} R_N(2) = 0$. Theorem 10.10.3 now tells us that the Taylor series converges to $\ln x$ when $x = 2$.

Example 10.10.4. Use Theorem 10.10.3 to show that the Maclaurin series for $\sin x$ converges to $\sin x$ for all values of x.

Solution. Let $f(x) = \sin x$. It is easy to see that when $x = 0$, the Maclaurin series converges to $f(0) = \sin 0 = 0$. For any $x \ne 0$, we have

$$R_N(x) = \frac{f^{(N+1)}(c)}{(N+1)!} x^{N+1},$$

for some number c between 0 and x. Since $f^{(N+1)}(c)$ is either $\pm \sin c$ or $\pm \cos c$, $|f^{(N+1)}(c)| \le 1$, and therefore

$$0 \le |R_N(x)| = \frac{|f^{(N+1)}(c)|}{(N+1)!} |x|^{N+1} \le \frac{|x|^{N+1}}{(N+1)!}.$$

By Proposition 10.6.5, $\lim_{N \to \infty} |x|^{N+1}/(N+1)! = 0$. Therefore $\lim_{N \to \infty} |R_N(x)| = 0$ by the squeeze theorem, so $\lim_{N \to \infty} R_N(x) = 0$. It follows, by Theorem 10.10.3, that the Maclaurin series converges to $\sin x$. □

Exercises 10.10

1–10: Find the Nth-order Taylor polynomial for f centered at a.

1. $f(x) = e^x, a = 0, N = 3$.

2. $f(x) = \cos x, a = 0, N = 5$.

3. $f(x) = e^{x^2+1}, a = 2, N = 3$.

4. $f(x) = \sin(x^2 + \pi/3), a = 0, N = 4$.

5. $f(x) = \ln(x^2 + e^2), a = e, N = 3$

6. $f(x) = x^3 + 2x^2 - 3x + 5, a = 0, N = 2$.

7. $f(x) = x^3 + 2x^2 - 3x + 5, a = 0, N = 3$.

8. $f(x) = x^3 + 2x^2 - 3x + 5, a = 0, N = 4$.

9. $f(x) = \tan x, a = 0, N = 5$.

10. $f(x) = \sec^{-1} x, a = 2, N = 2$.

11. Let $f(x) = e^x$. For each positive integer N, let $T_N(x)$ be the Nth-order Maclaurin polynomial for f, and let $R_N(x)$ be the Nth remainder.

    (a) Show that if $x < 0$ then for all N,

$$|R_N(x)| \leq \frac{|x|^{N+1}}{(N+1)!}.$$

    (b) Show that if $x > 0$ then for all N,

$$|R_N(x)| \leq \frac{e^x x^{n+1}}{(N+1)!}.$$

    (c) Use parts (a) and (b) to give another proof that for every x, the Maclaurin series for f converges to $f(x)$.

    (d) Use part (b) and the fact that $e \leq 3$ to show that the error in the approximation $e \approx T_5(1)$ is at most $1/240 \approx 0.0042$.

    (e) Compute $T_5(1)$, and verify that $|e - T_5(1)| \leq 1/240$.

12. Let $f(x) = \sqrt{x}$ and $a = 9$.

    (a) Find $T_2(x)$, the 2nd-order Taylor polynomial for f centered at a.

    (b) Show that $|R_2(10)| \leq 1/3888 \approx 0.00026$.

    (c) Compute $T_2(10)$, and verify that $|f(10) - T_2(10)| \leq 0.00026$.

13. Let $f(x) = \cos(\ln x)$ and $a = 1$.

    (a) Find $T_3(x)$, the 3rd-order Taylor polynomial for f centered at a.

    (b) Show that $|R_3(3/2)| \leq 5/192 \approx 0.026$.

    (c) Compute $f(3/2)$ and $T_3(3/2)$, and verify that $|f(3/2) - T_3(3/2)| \leq 0.026$.

14. Let $f(x) = e^{\sin x}$.

    (a) Find $T_3(x)$, the 3rd-order Maclaurin polynomial for f.

    (b) Use the fact that $e \leq 3$ to show that $|R_3(1/4)| \leq 15/2048 \approx 0.0073$.

    (c) Compute $f(1/4)$ and $T_3(1/4)$, and verify that $|f(1/4) - T_3(1/4)| \leq 0.0073$.

10.11 Proof of Theorem 10.7.1

Most of what we have learned about power series in the last few sections has relied on Theorem 10.7.1, but we have not yet proven that theorem. In this section we present the proof. For the convenience of the reader, we first restate the theorem.

Theorem 10.7.1 (restated). Suppose the power series $\sum_{n=0}^{\infty} c_n x^n$ has a positive radius of convergence, and let I be its interval of convergence. For $x \in I$ let

$$f(x) = \sum_{n=0}^{\infty} c_n x^n = c_0 + c_1 x + c_2 x^2 + \cdots.$$

Then:

1. The function f is continuous on I.

2. The function f is differentiable on the interior of I, and for all x in the interior of I,

$$f'(x) = \sum_{n=1}^{\infty} n c_n x^{n-1} = c_1 + 2c_2 x + 3c_3 x^2 + \cdots.$$

3. On the interval I,

$$\int f(x)\,dx = \left(\sum_{n=0}^{\infty} \frac{c_n x^{n+1}}{n+1} \right) + C = \left(c_0 x + \frac{c_1 x^2}{2} + \frac{c_2 x^3}{3} + \cdots \right) + C.$$

Of course, it is important that we not use in the proof any results from the last few sections that depend on this theorem, since this would result in circular reasoning. We will use only ideas that were discussed before our original statement of the theorem.

One fact that we will use several times is the triangle inequality. Recall that the triangle inequality says that for any two numbers a and b, $|a + b| \le |a| + |b|$. In fact, the triangle inequality holds more generally for sums of more than two numbers: for any numbers $a_1, a_2, \ldots, a_N$,

$$|a_1 + a_2 + \cdots + a_N| \le |a_1| + |a_2| + \cdots + |a_N|,$$

or equivalently, in Σ-notation,

$$\left| \sum_{n=1}^{N} a_n \right| \le \sum_{n=1}^{N} |a_n|.$$

If $\sum_{n=1}^{\infty} |a_n|$ converges, then we can let $N \to \infty$ to conclude that

$$\left| \sum_{n=1}^{\infty} a_n \right| \le \sum_{n=1}^{\infty} |a_n|.$$

We ask you to verify these extended versions of the triangle inequality in Exercise 1.

Proof of Theorem 10.7.1. For $N \ge 0$, let

$$s_N(x) = \sum_{n=0}^{N} c_n x^n = c_0 + c_1 x + \cdots + c_N x^N.$$

Then s_N is a polynomial of degree at most N, and for all $x \in I$,

$$\lim_{N \to \infty} s_N(x) = \lim_{N \to \infty} \sum_{n=0}^{N} c_n x^n = \sum_{n=0}^{\infty} c_n x^n = f(x).$$

Thus, $(s_N(x))_{N=0}^{\infty}$ converges to $f(x)$ for every number x in the interval I. One of the key ideas in the proofs of parts 1 and 3 of the theorem is the following fact about the nature of this convergence.

Lemma 10.11.1. *Suppose that $b > 0$ and b is in the interior of I. Suppose also that $\epsilon > 0$. Then there is some positive integer M such that for every $N > M$ and every $x \in [-b, b]$, $|f(x) - s_N(x)| < \epsilon$.*

Note that since I is centered at 0, $[-b, b] \subseteq I$. It follows that for each $x \in [-b, b]$, $\lim_{N \to \infty} s_N(x) = f(x)$, which means that there must be some M such that for every $N > M$, $|f(x) - s_N(x)| < \epsilon$. But in this statement, for different values of x, the choice of M might be different. The significance of the lemma is that it says that there is a single choice of M that will work for all values of x in $[-b, b]$. In other words, we can find an M such that for every $N > M$, $s_N(x)$ is within ϵ of $f(x)$ over the entire interval $[-b, b]$, not merely at one point in that interval. We say that the functions s_N converge *uniformly* to f on the interval $[-b, b]$.

Proof of Lemma 10.11.1. Since b is in the interior of I, according to Theorem 10.6.2, the power series converges absolutely when $x = b$. Thus, we can choose a positive integer M such that for every $N > M$,

$$\sum_{n=0}^{\infty} |c_n b^n| - \sum_{n=0}^{N} |c_n b^n| < \epsilon.$$

The difference on the left-hand side above is equal to the Nth tail of the series $\sum_{n=0}^{\infty} |c_n b^n|$. If we denote this tail by t_N, then for every $N > M$ we have

$$t_N = \sum_{n=N+1}^{\infty} |c_n b^n| = \sum_{n=0}^{\infty} |c_n b^n| - \sum_{n=0}^{N} |c_n b^n| < \epsilon.$$

Now consider any $N > M$ and $x \in [-b, b]$. Then $|x| \le b$, and therefore $|c_n x^n| \le |c_n b^n|$ for every n. Combining this with our extended version of the triangle inequality, we find that

$$|f(x) - s_N(x)| = \left| \sum_{n=0}^{\infty} c_n x^n - \sum_{n=0}^{N} c_n x^n \right| = \left| \sum_{n=N+1}^{\infty} c_n x^n \right|$$

$$\le \sum_{n=N+1}^{\infty} |c_n x^n| \le \sum_{n=N+1}^{\infty} |c_n b^n| = t_N < \epsilon,$$

which is what we needed to prove. $\square$

We now turn to the proof of part 1 of the theorem. The hardest part of the proof is the verification of one-sided continuity of f at the endpoints of I, if those endpoints belong to I. We will put off this part of the proof until the end, and begin by proving that f is continuous on the interior of I.

Suppose that a is in the interior of I; we must prove that $\lim_{x \to a} f(x) = f(a)$. We will use the ϵ-δ definition of this limit statement, so we begin by assuming that $\epsilon > 0$.

Since a is in the interior of I, and I is an interval centered at 0, we can choose some $b > 0$ such that b is also in the interior of I and $-b < a < b$. Let $\delta_1 = \min(b - a, a - (-b)) > 0$, so that $-b \le a - \delta_1 < a < a + \delta_1 \le b$. By Lemma 10.11.1, we can choose some N such that for all $x \in [-b, b]$, $|f(x) - s_N(x)| < \epsilon/3$. Since s_N is a polynomial, it is continuous at a, so we can choose $\delta_2 > 0$ so that if $0 < |x - a| < \delta_2$ then $|s_N(x) - s_N(a)| < \epsilon/3$. Let $\delta = \min(\delta_1, \delta_2) > 0$. We verify now that δ fulfills the requirements in the ϵ-δ limit definition.

Suppose that $0 < |x - a| < \delta$. Then $0 < |x - a| < \delta_2$, so

$$|s_N(x) - s_N(a)| < \frac{\epsilon}{3}. \tag{10.69}$$

Also, $|x - a| < \delta_1$, so $a - \delta_1 < x < a + \delta_1$, and therefore $-b < x < b$. Thus, a and x both belong to $[-b, b]$, and therefore by the choice of N,

$$|f(x) - s_N(x)| < \frac{\epsilon}{3}, \qquad |f(a) - s_N(a)| < \frac{\epsilon}{3}. \tag{10.70}$$

Combining (10.69) and (10.70) and applying the triangle inequality, we can now compute

$$
\begin{aligned}
|f(x) - f(a)| &= |(f(x) - s_N(x)) + (s_N(x) - s_N(a)) + (s_N(a) - f(a))| \\
&\le |f(x) - s_N(x)| + |s_N(x) - s_N(a)| + |s_N(a) - f(a)| \\
&< \frac{\epsilon}{3} + \frac{\epsilon}{3} + \frac{\epsilon}{3} = \epsilon.
\end{aligned}
$$

This shows that $\lim_{x \to a} f(x) = f(a)$. Since this conclusion holds for every a in the interior of I, we have shown that f is continuous on the interior of I.

Now we turn to the proof of part 3 of the theorem. Our plan is to prove that for all $x \in I$,

$$\int_0^x f(t)\, dt = \sum_{n=0}^{\infty} \frac{c_n x^{n+1}}{n+1}. \tag{10.71}$$

From this it will follow, by the fundamental theorem of calculus, that

$$\frac{d}{dx}\left(\sum_{n=0}^{\infty} \frac{c_n x^{n+1}}{n+1} \right) = f(x).$$

In other words, on I,

$$\int f(x)\, dx = \left(\sum_{n=0}^{\infty} \frac{c_n x^{n+1}}{n+1} \right) + C = \left(c_0 x + \frac{c_1 x^2}{2} + \frac{c_2 x^3}{3} + \cdots \right) + C,$$

as claimed in part 3 of the theorem.

Once again, we leave consideration of the endpoints of I for later; for the moment, we will focus on proving that (10.71) holds for x in the interior of I, and hence that part 3 holds on the interior of I. It is clear that (10.71) holds when $x = 0$, since both sides of the equation are 0. We now prove (10.71) for $x > 0$; a similar argument would work for $x < 0$.

Notice that for any N, it is easy to integrate s_N, since it is a polynomial. We have

$$\int_0^x s_N(t)\,dt = \int_0^x \sum_{n=0}^N c_n t^n\,dt = \left[\sum_{n=0}^N \frac{c_n t^{n+1}}{n+1}\right]_0^x = \sum_{n=0}^N \frac{c_n x^{n+1}}{n+1}.$$

Thus, by the definition of infinite series,

$$\sum_{n=0}^\infty \frac{c_n x^{n+1}}{n+1} = \lim_{N\to\infty} \sum_{n=0}^N \frac{c_n x^{n+1}}{n+1} = \lim_{N\to\infty} \int_0^x s_N(t)\,dt.$$

Substituting into (10.71), we see that what we must prove is that

$$\lim_{N\to\infty} \int_0^x s_N(t)\,dt = \int_0^x f(t)\,dt. \tag{10.72}$$

We prove this by using the definition of limits.

Suppose $\epsilon > 0$. By Lemma 10.11.1, there is some positive integer M such that for all $N > M$ and all $t \in [-x, x]$, $|f(t) - s_N(t)| < \epsilon/(2x)$, and therefore

$$-\frac{\epsilon}{2x} \le f(t) - s_N(t) \le \frac{\epsilon}{2x}.$$

Applying Corollary 5.3.8 to the right half of this inequality, we conclude that

$$\int_0^x f(t) - s_N(t)\,dt \le \int_0^x \frac{\epsilon}{2x}\,dt = \frac{\epsilon}{2x}\cdot\int_0^x 1\,dt = \frac{\epsilon}{2x}\cdot x = \frac{\epsilon}{2}.$$

Similarly, the left half implies that $\int_0^x f(t) - s_N(t)\,dt \ge -\epsilon/2$. Therefore

$$\left|\int_0^x f(t)\,dt - \int_0^x s_N(t)\,dt\right| = \left|\int_0^x f(t) - s_N(t)\,dt\right| \le \frac{\epsilon}{2} < \epsilon.$$

This completes the proof of (10.72), which establishes that part 3 of the theorem holds on the interior of I.

Using part 3 for the interior of I, we can now prove part 2. We begin by showing that the proposed series for $f'(x)$ converges on the interior of I. Our method is similar to the one used in the proof of Lemma 10.6.3.

Suppose that x is in the interior of I. Then we can choose a number b such that $b > |x|$ and b is also in I, so $\sum_{n=0}^\infty c_n b^n$ converges. By Theorem 10.2.1, it follows that $\lim_{n\to\infty} c_n b^n = 0$, so there is some positive integer N such that for all $n > N$, $|c_n b^n| < 1$. Therefore for all $n > N$,

$$0 \le |nc_n x^{n-1}| = \left|c_n b^n \cdot \frac{n}{b}\cdot\left(\frac{x}{b}\right)^{n-1}\right| \le \frac{n}{b}\cdot\left|\frac{x}{b}\right|^{n-1}.$$

We leave it to you to check, using the ratio test, that $\sum_{n=N+1}^{\infty}(n/b)\cdot|x/b|^{n-1}$ converges. By the comparison test, it follows that $\sum_{n=N+1}^{\infty}|nc_nx^{n-1}|$ converges, and therefore $\sum_{n=1}^{\infty}nc_nx^{n-1}$ converges. In other words, the series in part 2 of the theorem converges on the interior of I.

Let

$$g(x) = \sum_{n=1}^{\infty} nc_nx^{n-1}.$$

As we have just seen, the interval of convergence for this series includes the interior of I. Applying part 3 for the interior of the interval of convergence, we can conclude that on the interior of I,

$$\int g(x)\,dx = \left(\sum_{n=1}^{\infty}\frac{nc_nx^n}{n}\right) + C = \left(\sum_{n=1}^{\infty}c_nx^n\right) + C.$$

In other words, if we assign any value to C in the formula on the right above, then we get an antiderivative of g on the interior of I. Assigning the value $C = c_0$, we get

$$\left(\sum_{n=1}^{\infty}c_nx^n\right) + c_0 = \sum_{n=0}^{\infty}c_nx^n = f(x).$$

Thus, f is an antiderivative of g on the interior of I, which means that for x in the interior of I,

$$f'(x) = g(x) = \sum_{n=1}^{\infty}nc_nx^{n-1}.$$

This proves part 2 of the theorem.

All that remains is to prove parts 1 and 3 at the endpoints of I, if those endpoints belong to I. For these proofs we will use a method called *summation by parts*. Recall the integration by parts formula for definite integrals: $\int_a^b u\,dv = [uv]_a^b - \int_a^b v\,du$. The formula for summation by parts looks very similar:

Lemma 10.11.2 (Summation by Parts). *For any numbers* $u_0,\ u_1,\ \ldots,\ u_N,\ v_0,\ v_1,\ \ldots,$ $v_{N+1},$

$$\sum_{n=0}^{N}u_n(v_{n+1} - v_n) = u_Nv_{N+1} - u_0v_0 - \sum_{n=1}^{N}v_n(u_n - u_{n-1}).$$

Proof. The proof is just simple algebra:

$$\sum_{n=0}^{N}u_n(v_{n+1} - v_n) = u_0(v_1 - v_0) + u_1(v_2 - v_1) + \cdots + u_N(v_{N+1} - v_N)$$

$$= -u_0v_0 + u_0v_1 - u_1v_1 + u_1v_2 - \cdots - u_Nv_N + u_Nv_{N+1}$$

$$= u_Nv_{N+1} - u_0v_0$$

$$- (-u_0v_1 + u_1v_1 - u_1v_2 + u_2v_2 - \cdots - u_{N-1}v_N + u_Nv_N)$$

$$= u_N v_{N+1} - u_0 v_0$$
$$- (v_1(u_1 - u_0) + v_2(u_2 - u_1) + \cdots + v_N(u_N - u_{N-1}))$$
$$= u_N v_{N+1} - u_0 v_0 - \sum_{n=1}^{N} v_n(u_n - u_{n-1}). \qquad \square$$

Now suppose that b is the right endpoint of I, and $b \in I$. We will prove that parts 1 and 3 of the theorem can be extended to b; the proof for the left endpoint of I would be similar.

For part 1, we must prove that $\lim_{x \to b^-} f(x) = f(b)$. This is known as Abel's theorem, after Niels Henrik Abel (1802–1829). It will simplify the proof if we assume initially that $f(b) = 0$. Later we will show how to eliminate this assumption.

For $0 < x < b$, we have

$$f(x) = \sum_{n=0}^{\infty} c_n x^n = \sum_{n=0}^{\infty} \left(\frac{x}{b}\right)^n \cdot c_n b^n = \lim_{N \to \infty} \sum_{n=0}^{N} \left(\frac{x}{b}\right)^n \cdot c_n b^n. \qquad (10.73)$$

We now apply summation by parts, with

$$u_n = \left(\frac{x}{b}\right)^n, \qquad v_0 = 0, \qquad v_1 = c_0, \qquad v_2 = c_0 + c_1 b, \qquad v_3 = c_0 + c_1 b + c_2 b^2,$$

and in general, for $n \geq 1$,

$$v_n = c_0 + c_1 b + \cdots + c_{n-1} b^{n-1} = \sum_{i=0}^{n-1} c_i b^i.$$

With this choice of notation,

$$v_{n+1} - v_n = (c_0 + c_1 b + \cdots + c_n b^n) - (c_0 + c_1 b + \cdots + c_{n-1} b^{n-1}) = c_n b^n.$$

Thus, by the summation by parts lemma,

$$\sum_{n=0}^{N} \left(\frac{x}{b}\right)^n \cdot c_n b^n = \sum_{n=0}^{N} u_n(v_{n+1} - v_n)$$

$$= u_N v_{N+1} - u_0 v_0 - \sum_{n=1}^{N} v_n(u_n - u_{n-1})$$

$$= \left(\frac{x}{b}\right)^N v_{N+1} - 1 \cdot 0 - \sum_{n=1}^{N} v_n \left[\left(\frac{x}{b}\right)^n - \left(\frac{x}{b}\right)^{n-1}\right]$$

$$= \left(\frac{x}{b}\right)^N v_{N+1} - \left(\frac{x}{b} - 1\right) \sum_{n=1}^{N} v_n \left(\frac{x}{b}\right)^{n-1}.$$

Rearranging this equation, we can say that

$$\sum_{n=1}^{N} v_n \left(\frac{x}{b}\right)^{n-1} = \frac{1}{x/b - 1} \left[\left(\frac{x}{b}\right)^{N} v_{N+1} - \sum_{n=0}^{N} \left(\frac{x}{b}\right)^{n} \cdot c_n b^n \right]. \tag{10.74}$$

We now let $N \to \infty$ in (10.74). Notice that since $0 < x < b$, $\lim_{N \to \infty} (x/b)^N = 0$, and

$$\lim_{N \to \infty} v_{N+1} = \lim_{N \to \infty} \sum_{n=0}^{N} c_n b^n = \sum_{n=0}^{\infty} c_n b^n = f(b) = 0.$$

Applying these observations and equation (10.73), we see that when $N \to \infty$ in (10.74), we get

$$\sum_{n=1}^{\infty} v_n \left(\frac{x}{b}\right)^{n-1} = \frac{1}{x/b - 1} [0 \cdot 0 - f(x)].$$

In other words, if $0 < x < b$ then

$$f(x) = \left(1 - \frac{x}{b}\right) \sum_{n=1}^{\infty} v_n \left(\frac{x}{b}\right)^{n-1}. \tag{10.75}$$

We can now use equation (10.75), rather than the original power series representation of $f(x)$, to prove that $\lim_{x \to b^-} f(x) = 0 = f(b)$. Suppose $\epsilon > 0$. Since $\lim_{n \to \infty} v_n = f(b) = 0$, we can choose a positive integer N such that for all $n > N$, $|v_n| < \epsilon/2$, and then we can rewrite equation (10.75) in the form

$$f(x) = \left(1 - \frac{x}{b}\right) \sum_{n=1}^{N} v_n \left(\frac{x}{b}\right)^{n-1} + \left(1 - \frac{x}{b}\right) \sum_{n=N+1}^{\infty} v_n \left(\frac{x}{b}\right)^{n-1}. \tag{10.76}$$

Our plan now is to show that, if x is close to b and less than b, then both terms on the right in (10.76) are small.

The second of these terms will be small as long as $0 < x < b$. To see why, we use the extended triangle inequality, the fact that $|v_n| < \epsilon/2$ for $n > N$, and the formula for the sum of a geometric series:

$$\left| \left(1 - \frac{x}{b}\right) \sum_{n=N+1}^{\infty} v_n \left(\frac{x}{b}\right)^{n-1} \right|$$

$$= \left(1 - \frac{x}{b}\right) \left| \sum_{n=N+1}^{\infty} v_n \left(\frac{x}{b}\right)^{n-1} \right|$$

$$\leq \left(1 - \frac{x}{b}\right) \sum_{n=N+1}^{\infty} \left| v_n \left(\frac{x}{b}\right)^{n-1} \right| \qquad \text{(by triangle inequality)}$$

$$\leq \left(1 - \frac{x}{b}\right) \sum_{n=N+1}^{\infty} \frac{\epsilon}{2} \cdot \left(\frac{x}{b}\right)^{n-1} \qquad \text{(since } |v_n| \leq \epsilon/2\text{)}$$

$$= \left(1 - \frac{x}{b}\right) \cdot \frac{(\epsilon/2)(x/b)^N}{1 - x/b} \qquad \text{(by geometric series formula)}$$

$$= \frac{\epsilon}{2} \cdot \left(\frac{x}{b}\right)^N < \frac{\epsilon}{2}. \tag{10.77}$$

The first term in (10.76) is a polynomial, so it is easy to see that

$$\lim_{x \to b^-} \left(1 - \frac{x}{b}\right) \sum_{n=1}^{N} v_n \left(\frac{x}{b}\right)^{n-1} = (1 - 1) \sum_{n=1}^{N} v_n = 0.$$

Thus, we can choose some $\delta_1 > 0$ such that if $b - \delta_1 < x < b$ then

$$\left| \left(1 - \frac{x}{b}\right) \sum_{n=1}^{N} v_n \left(\frac{x}{b}\right)^{n-1} \right| < \frac{\epsilon}{2}. \tag{10.78}$$

Now let $\delta = \min(\delta_1, b) > 0$, and suppose that $b - \delta < x < b$. Since $\delta \leq b$, this implies that $0 < x < b$, so (10.77) holds. And also $\delta \leq \delta_1$, so $b - \delta_1 < x < b$, and therefore (10.78) holds. Combining these with (10.76), we conclude that

$$|f(x) - 0| = |f(x)| = \left| \left(1 - \frac{x}{b}\right) \sum_{n=1}^{N} v_n \left(\frac{x}{b}\right)^{n-1} + \left(1 - \frac{x}{b}\right) \sum_{n=N+1}^{\infty} v_n \left(\frac{x}{b}\right)^{n-1} \right|$$

$$\leq \left| \left(1 - \frac{x}{b}\right) \sum_{n=1}^{N} v_n \left(\frac{x}{b}\right)^{n-1} \right| + \left| \left(1 - \frac{x}{b}\right) \sum_{n=N+1}^{\infty} v_n \left(\frac{x}{b}\right)^{n-1} \right|$$

$$< \frac{\epsilon}{2} + \frac{\epsilon}{2} = \epsilon.$$

This completes the proof that $\lim_{x \to b^-} f(x) = 0 = f(b)$.

To finish the proof of part 1 of the theorem, we must explain how to eliminate the assumption that $f(b) = 0$. If $f(b) \neq 0$, then we define $g(x) = f(x) - f(b)$. Clearly g has the power series representation

$$g(x) = f(x) - f(b) = (c_0 + c_1 x + c_2 x^2 + \cdots) - f(b)$$
$$= (c_0 - f(b)) + c_1 x + c_2 x^2 + \cdots,$$

and this representation has the same interval of convergence as the series for f, namely I. But $g(b) = f(b) - f(b) = 0$, so by what we have just proven, g must be continuous from the left at b. Therefore

$$\lim_{x \to b^-} f(x) = \lim_{x \to b^-} (g(x) + f(b)) = 0 + f(b) = f(b),$$

so f is also continuous from the left at b.

Finally, we extend part 3 of the theorem to the endpoint b. To do this, we must prove that (10.71) holds when $x = b$. In other words, we must show that

$$\int_0^b f(t)\,dt = \sum_{n=0}^{\infty} \frac{c_n b^{n+1}}{n+1}. \tag{10.79}$$

The hardest part of the proof is showing that the series on the right-hand side above converges. For this we will again use summation by parts.

As before, we let $v_0 = 0$ and $v_n = c_0 + c_1 b + \cdots + c_{n-1}b^{n-1}$ for $n \geq 1$, so that $v_{n+1} - v_n = c_n b^n$. This time we let $u_n = b/(n+1)$. Then for every positive integer N, the summation by parts lemma gives us

$$\sum_{n=0}^{N} \frac{c_n b^{n+1}}{n+1} = \sum_{n=0}^{N} \frac{b}{n+1} \cdot c_n b^n = \sum_{n=0}^{N} u_n(v_{n+1} - v_n)$$

$$= u_N v_{N+1} - u_0 v_0 - \sum_{n=1}^{N} v_n(u_n - u_{n-1})$$

$$= \frac{b}{N+1} \cdot v_{N+1} - b \cdot 0 - \sum_{n=1}^{N} v_n \left(\frac{b}{n+1} - \frac{b}{n} \right). \tag{10.80}$$

To prove convergence of the series in (10.79), we now want to let $N \to \infty$.

As we observed before, $\lim_{n\to\infty} v_n = f(b)$. It follows that there is some positive integer N such that for all $n > N$, $|v_n - f(b)| < 1$, and therefore

$$0 \leq \left| v_n \left(\frac{b}{n+1} - \frac{b}{n} \right) \right| = |f(b) + (v_n - f(b))| \cdot \left| \frac{b}{n+1} - \frac{b}{n} \right|$$

$$\leq (|f(b)| + 1) \left(\frac{b}{n} - \frac{b}{n+1} \right).$$

We ask you in Exercise 2 to verify that the series

$$\sum_{n=N+1}^{\infty} (|f(b)| + 1) \left(\frac{b}{n} - \frac{b}{n+1} \right)$$

is a telescoping series that converges. We can then use the comparison test to conclude that

$$\sum_{n=1}^{\infty} v_n \left(\frac{b}{n+1} - \frac{b}{n} \right)$$

converges absolutely, and therefore, letting $N \to \infty$ in (10.80), we get

$$\sum_{n=0}^{\infty} \frac{c_n b^{n+1}}{n+1} = -\sum_{n=1}^{\infty} v_n \left(\frac{b}{n+1} - \frac{b}{n} \right).$$

Thus, the series on the right-hand side of (10.79) converges.

To complete the proof of (10.79), let

$$g(x) = \int_0^x f(t)\,dt, \qquad h(x) = \sum_{n=0}^{\infty} \frac{c_n x^{n+1}}{n+1}.$$

We have already shown that $g(x) = h(x)$ for x in the interior of I. We now want to use continuity to extend this equation to b.

Since we have already extended part 1 of the theorem to b, we know that f is continuous from the left at b, so $g(b)$ is defined. Furthermore, by the first fundamental theorem of calculus, g is differentiable from the left at b, and therefore also continuous from the left at b. We have also now shown that $h(b)$ is defined, and applying part 1 of the theorem to h we know that h is also continuous from the left at b. Therefore

$$g(b) = \lim_{x \to b^-} g(x) = \lim_{x \to b^-} h(x) = h(b).$$

In other words, we have established equation (10.79), which completes the extension of part 3 of the theorem to b. □

Exercises 10.11

1. (a) Show that for any positive integer N and any numbers $a_1, a_2, \ldots, a_N$,

$$|a_1 + a_2 + \cdots + a_N| \le |a_1| + |a_2| + \cdots + |a_N|.$$

   (Hint: Use the triangle inequality (Theorem 1.1.5) and mathematical induction.)

   (b) Show that if a series $\sum_{n=1}^{\infty} a_n$ converges absolutely, then

$$\left| \sum_{n=1}^{\infty} a_n \right| \le \sum_{n=1}^{\infty} |a_n|.$$

   (Hint: Use part (a) and the preservation of weak inequalities by limits (Corollary 2.8.9).)

2. Show that for any positive integer N and any numbers K and b, the series

$$\sum_{n=N+1}^{\infty} K \left(\frac{b}{n} - \frac{b}{n+1} \right)$$

   converges. (Hint: Show that it is a telescoping series.)

3. For each positive integer N, let s_N be the function with domain $(-\infty, \infty)$ defined by the formula $s_N(x) = x/N$. Let f be the function with domain $(-\infty, \infty)$ defined by the formula $f(x) = 0$.

   (a) Show that for every real number x, $\lim_{N \to \infty} s_N(x) = f(x)$.

(b) Show that the functions s_N converge uniformly to f on the interval $[-1, 1]$. In other words, show that for every $\epsilon > 0$ there is some positive integer M such that for every $N > M$ and every $x \in [-1, 1]$, $|f(x) - s_N(x)| < \epsilon$.

(c) Show that the functions s_N do not converge uniformly to f on the interval $(-\infty, \infty)$. In other words, show that it is not the case that for every $\epsilon > 0$ there is some positive integer M such that for every $N > M$ and every $x \in (-\infty, \infty)$, $|f(x) - s_N(x)| < \epsilon$. (Hint: Suppose the functions s_N do converge uniformly to f on $(-\infty, \infty)$. Then in particular, letting $\epsilon = 1$, there must be some positive integer M such that for every $N > M$ and every $x \in (-\infty, \infty)$, $|f(x) - s_N(x)| < 1$. Now derive a contradiction by finding an $N > M$ and an $x \in (-\infty, \infty)$ such that $|f(x) - s_N(x)| \geq 1$.)

4. For each positive integer N, let s_N be the function with domain $[0, 1]$ defined by the formula $s_N(x) = x^N$.

(a) Find a function f with domain $[0, 1]$ such that for every $x \in [0, 1]$, $\lim_{N \to \infty} s_N(x) = f(x)$.

(b) Show that the functions s_N do not converge uniformly to f on $[0, 1]$.

5. Suppose that I is an interval, and for every positive integer N, s_N is a function that is continuous on I. Suppose also that the functions s_N converge uniformly on I to a function f.

(a) Show that f is continuous on I. (Hint: Imitate the proof of part 1 of Theorem 10.7.1.)

(b) Use part (a) to give another solution to part (b) of Exercise 4.

6. In this exercise you will use summation by parts to evaluate $\sum_{n=1}^{N} n2^n$.

(a) Show that if we use the formulas $u_n = n$ and $v_n = 2^n$, then

$$\sum_{n=1}^{N} n2^n = \sum_{n=0}^{N} n2^n = \sum_{n=0}^{N} u_n(v_{n+1} - v_n).$$

(b) Use the summation by parts lemma to conclude that

$$\sum_{n=1}^{N} n2^n = N2^{N+1} - \sum_{n=1}^{N} 2^n.$$

(c) Show that

$$\sum_{n=1}^{N} n2^n = (N - 1)2^{N+1} + 2.$$

Compare this answer to Exercise 14 from Section 5.1.

7. (a) Use summation by parts to find a formula for

$$\sum_{n=1}^{N} \frac{n}{2^n}.$$

(Hint: Let $u_n = n$ and $v_n = -1/2^{n-1}$.) Compare your answer to part (a) of Exercise 17 from Section 10.1.

(b) Find a formula for

$$\sum_{n=1}^{N} \frac{n^2}{2^n}.$$

(Hint: Use summation by parts with $u_n = n^2$ and $v_n = -1/2^{n-1}$, and then use your answer to part (a).)

(c) Use your answer to part (b) to evaluate the infinite series

$$\sum_{n=1}^{\infty} \frac{n^2}{2^n}.$$

Notice that this is series (10.26b), whose value we found in Section 10.7. Compare your answer to the answer we found in that section.

8. Show that for every positive integer N,

$$\sum_{n=1}^{N} n \ln\left(1 + \frac{1}{n}\right) = \ln\left(\frac{(N+1)^N}{N!}\right).$$

(Hint: Use summation by parts with $u_n = n$ and $v_n = \ln n$.)

Appendix

Answers to Odd-Numbered Exercises

Chapter 1

Section 1.1

1. $\{x : |x - 5| < 7\} = \{x : -2 < x < 12\} = (-2, 12)$.

3. $\{t : |5 - 2t| \geq 4\} = \{t : \text{either } t \leq 1/2 \text{ or } t \geq 9/2\} = (-\infty, 1/2] \cup [9/2, \infty)$.

5. $\{x : 2|x - 4| < x\} = \{x : 8/3 < x < 8\} = (8/3, 8)$.

7. $\{x : |x + 4| < 2x\} = \{x : x > 4\} = (4, \infty)$.

9. $\{y : |3y| \leq |y| + 10\} = \{y : -5 \leq y \leq 5\} = [-5, 5]$.

11. $\{x : |2x - 3| \geq |x| - 1\} = \{x : \text{either } x \geq 2 \text{ or } x \leq 4/3\} = (-\infty, 4/3] \cup [2, \infty)$.

13. $\{x : |3x - 6| \leq |3 - 6x|\} = \{x : \text{either } x \leq -1 \text{ or } x \geq 1\} = (-\infty, -1] \cup [1, \infty)$.

Section 1.2

1.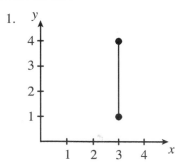

3.

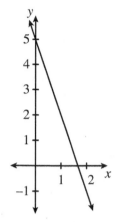

5.

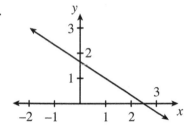

7.

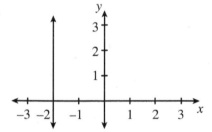

9.

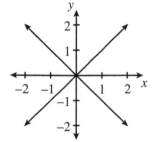

11.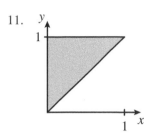

13.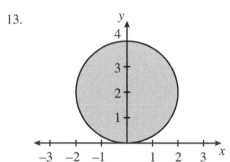

15. Slope 2, y-intercept $-9/2$.

17. $y - 1 = 2(x + 1)$.

19. $y = 7$.

21. $y - 3 = (1/3)(x - 2)$.

23. $y - 3 = (2/3)(x - 1)$.

25. $(3/5, -1/5)$ and $(-1, 3)$.

27. Area of gray triangle $= 30$, area of striped triangle $= 15$.

Section 1.3

1. 5.

3. 11.

5. $\sqrt{4x^2 + 4x + 6}$.

7. $x^2 + 1$.

9. $5\lfloor 5x \rfloor - 12$.

11. $(-\infty, -3) \cup (-3, 2) \cup (2, \infty)$.

13. $[-4, 4]$.

15. $[-4, 2) \cup [4, \infty)$.

17. $(-\infty, 3)$.

19.   (b) and (c).

21.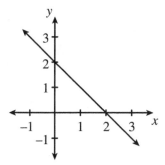

23.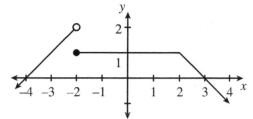

25.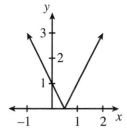

27.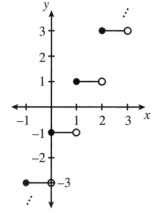

29.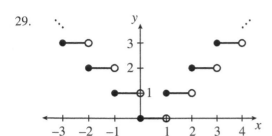

31. (a) $(-1, 6]$.

(b) $[-1, 4)$.

(c) 0.

(d) 2.

(e) $(-1, 1) \cup (4, 6]$.

(f) 3.

(g) None.

Section 1.4

1. $1 - 2/x + x$; domain $= (-\infty, -2) \cup (-2, 0) \cup (0, \infty)$.

3. $(x^2 - 2x)/(3x - 2)$; domain $= (-\infty, -2) \cup (-2, 0) \cup (0, 2/3) \cup (2/3, \infty)$.

5. $1 - 2/x$; domain $= (-\infty, 0) \cup (0, 2/5) \cup (2/5, \infty)$.

7. $\sqrt{3 - x} - 2$; domain $= (-\infty, 3]$.

9. $\cot x - \pi/x$; domain $= \{x : x \neq n\pi, \text{ for any integer } n\}$.

11. $\pi/\lfloor x \rfloor$; domain $= (-\infty, 0) \cup [1, \infty)$.

13. x; domain $= (-\infty, 0) \cup (0, \infty)$.

15. $f = f_1 + f_2$, where $f_1(x) = \sin x$ and $f_2(x) = \cos x$.

17. $h = h_1 \circ h_2$, where $h_1(x) = \sqrt{x}$ and $h_2(x) = 9 - x^2$.

19. $g = g_1 \circ (g_2/g_3)$, where $g_1(x) = \sqrt{x}$, $g_2(x) = x + 7$, and $g_3(x) = \cos x$.

21. (a) $((f + g) \circ h)(x) = f(h(x)) + g(h(x)) = ((f \circ h) + (g \circ h))(x)$.

(b) No.

(c) Yes.

23.

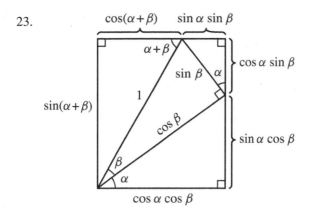

Chapter 2

Section 2.1

1. (a) 7 m/sec.

(b) $f(x) = x + 4 \ (x \neq 0)$.

(c)

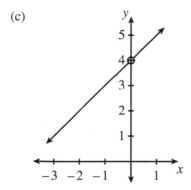

(d) 4 m/sec.

3. $-1/5$.

5. $1/2$.

7. $1/3$.

9. $-1/4$.

11. 0.

Section 2.2

1. (b) $1/30$.

3. (b) 3.

5. (b) and (c).

7. (a) and (c).

9.

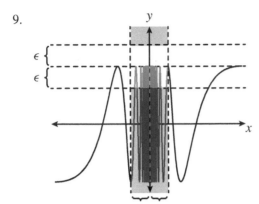

11. (a) Yes.

    (b) No.

    (c) No.

    (d) Yes.

Section 2.3

We don't give the proofs, but we give an acceptable choice for δ for those exercises in which a formula for δ was not given.

1. $\delta = \epsilon/2$.

3. $\delta = \epsilon$.

5. $\delta = 2\epsilon/3$.

7. $\delta = \epsilon/2$.

15. $\delta = \min(1, \epsilon/4)$.

17. $\delta = \min(1, \epsilon/3)$.

19. $\delta = \min(1, 2\epsilon)$.

21. $\delta = \min(1, \epsilon/4)$.

23. $\delta = \min(3, 6\epsilon)$.

Section 2.4

1. In the proof, let $\delta = \epsilon$.

3. 5.

5. 1.

7. 7/5.

9. 5/3.

11. Undefined.

13. 2.

15. 6/7.

17. −1.

19. −1/4.

21. 5.

23. 5/3.

25. 1.

27. 0.

Section 2.5

5. 0.

7. 3.

9. −1/2.

11. −∞.

13. ∞.

15. 0.

17. ∞.

19. 2/7.

21. −2.

23. 4.

25. 2.

27. −2/7.

Section 2.6

3. ∞.

5. −∞.

7. Undefined.

9. 9.

11. 7.

13. 0.

15. 0.

17. ∞.

19. (a) 1.

    (b) -2.

    (c) 3.

    (d) 1.

Section 2.7

1. 1.

3. 1/4.

5. 5/4.

7. 0.

9. 1/12.

11. 1/3.

13. $-1/2$.

15. 1/2.

17. 2.

19. $\sqrt{2}/2$.

21. 1/2.

23. 1.

25. $-5/3$.

29. $\pi/180$.

Section 2.8

1. 5, 20, 75, 200, …. Converges to 0.

3. 0, 0, 0, 0, …. Converges to 0.

5. -1.

7. 0.

9. 0.

11. 0.

13. ∞.

15. −1.

19. Converges to 1/2.

23. One example: $f(x) = x^2, a = 0$.

Section 2.9

1. Strictly increasing.

3. Strictly increasing.

5. Not monotone.

7. Strictly decreasing.

9. Not monotone.

11. (d) $L = 0$.

13. (d) $L = 0$.

15. (c) 6.

17. (a) $1, -3, 9, 5$.

    (c) 3.

Chapter 3

Section 3.1

1. 11.

3. 12.

5. −1/32.

7. 1/2.

9. $-\sqrt{2}/8$.

11. $y = 2x + 3$.

13. −2/5 amps/ohm; current decreases as resistance increases.

15. 3 years/au. P increases as r increases.

Section 3.2

1. $-2x$.

3. $-1/(2x\sqrt{x})$.

5. $2 - 1/(2\sqrt{x})$.

7. $1/(4(\sqrt[4]{x})^3)$.

9. $3/(2x+1)^2$.

11. $-1/(2\sqrt{t}(\sqrt{t}+1)^2)$.

13. $1 + 3/\sqrt{x}$.

15. Yes.

17. Yes.

Section 3.3

1. $12x^5 - 12x^3 + 3x^2 - 14x + 4$.

3. $\dfrac{-14x^2 + 32x - 22}{(3x^2 - 5x + 1)^2}$.

5. $\dfrac{-4x^3 - 9x^2 + 4}{(x^3 + 2)^2}$.

7. $\dfrac{-8x^2\sqrt{x} + 4x\sqrt{x} + 4x + 1}{2\sqrt{x}(x^2 + \sqrt{x})^2}$.

9. $\dfrac{\sqrt{x} + 3\sqrt[6]{x} + 2}{6(\sqrt[3]{x})^2(\sqrt[3]{x} + 1)^2}$.

11. $2x^{-1/3}/3$.

13. $\dfrac{2x(\cos^2 x - \sin^2 x) + \sin x \cos x}{2\sqrt{x}}$.

15. $\dfrac{\cos x - x}{1 + \sin x}$.

17. $y = (7/2)x - 2$.

19. $(2, 1/2)$.

21. $y = -2x + 17$.

23. $6x^2 - 10x + 6$.

27. $f(x)g(x)h'(x) + f(x)h(x)g'(x) + g(x)h(x)f'(x)$.

Section 3.4

1. $60(5x - 8)^{11}$.

3. $\dfrac{x}{\sqrt{x^2 + 4}}$.

5. $-\dfrac{2}{(3x + 4)\sqrt[3]{3x + 4}}$.

7. $-6\cos(3x)\sin(3x)$.

9. $\dfrac{5x + 6}{2(\sqrt[4]{2x + 3})^3}$.

11. $-3\cos(\cos(\tan(3x)))\sin(\tan(3x))\sec^2(3x)$.

13. $\dfrac{203(2x + 5)^6}{(2 - 5x)^8}$.

15. $\dfrac{\sec^2 x}{2\sqrt{\tan x}}$.

17. $g'(3) = 9,\ g''(3) = -3$.

19. $5670(3x - 2)^4$.

21. $2^{100}\sin(2x)$.

23. $\cos(2x) = \cos^2 x - \sin^2 x$.

Section 3.5

3. $-\dfrac{y^{2/3}}{x^{2/3}}$.

5. $-\dfrac{2y^2\sqrt{x + y} + 1}{4xy\sqrt{x + y} + 1}$.

7. $\dfrac{y\cos(xy)}{1 - x\cos(xy)}$.

9. $\dfrac{3\cos(3x - y) - y^2}{2xy + \cos(3x - y)}$.

11. $\dfrac{8x - y^3}{3xy^2 + 6y}$.

13. $-\dfrac{2}{(2y + 1)^3}$.

15. $\dfrac{1}{3x^{4/3}y^{1/3}}$.

17. $\dfrac{2(y-1)(y^2+3)}{4x^2}$.

19. (a) $3\pi/8$.

 (b) $y = -(\pi/2)x + 9\pi/8$.

Chapter 4

Section 4.1

1. Power is increasing 400 watts/min.

3. Radius is increasing $5/(36\pi)$ in/sec.

5. Shadow is shrinking 4 ft/min.

7. Top of ladder is sliding down 3/2 ft/sec.

9. Rocket is rising 1/4 km/sec.

11. Height is increasing $2/(25\pi)$ ft/sec.

13. (a) Distance is increasing $2\sqrt{10}$ ft/sec.

 (b) Distance is decreasing $2\sqrt{10}$ ft/sec.

15. He must turn the camera 8/25 rad/sec.

17. Area is increasing 36π in^2/sec.

19. Depth is increasing $845/705672 \approx 0.0012$ ft/min.

Section 4.2

1. 5/2.

3. 4.

5. $-\sqrt[3]{32/5}$.

Section 4.3

1. (a) Strictly increasing on $(p, q]$, strictly decreasing on $[q, r]$, constant on $[r, s]$, strictly decreasing on $[s, t]$, strictly increasing on $[t, u)$, strictly increasing on $(u, v]$. Also: weakly decreasing on $[q, t]$. Attains maximum value at v; no minimum value. Local maximum at q, local minimum at t. Also, both a local maximum and a local minimum at every number in the interval (r, s).

 (b) Strictly increasing on $[p, r]$, strictly decreasing on $[r, s]$, strictly decreasing on $(s, t]$, strictly increasing on $[t, v]$. Attains maximum value at r, minimum value at p. Local maximum at r, local minima at s and t.

3. Critical number 3/2. Strictly increasing on $(-\infty, 3/2]$, strictly decreasing on $[3/2, \infty)$. Local maximum at 3/2.

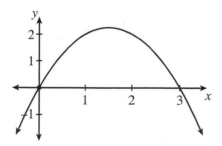

5. Critical numbers $-1, 0, 1$. Strictly decreasing on $(-\infty, -1]$, strictly increasing on $[-1, 0]$, strictly decreasing on $[0, 1]$, strictly increasing on $[1, \infty)$. Local minima at -1 and 1, local maximum at 0.

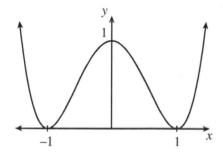

7. Critical numbers 0, 1, 5. Strictly decreasing on $(-\infty, 0]$, strictly increasing on $[0, 1]$, strictly decreasing on $[1, 5]$, strictly increasing on $[5, \infty)$. Local minima at 0 and 5, local maximum at 1.

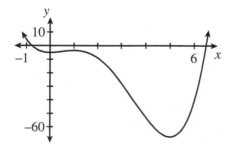

9. Domain $[0, \infty)$. Critical numbers 0, 1. Strictly decreasing on $[0, 1]$, strictly increasing on $[1, \infty)$. Local minimum at 1.

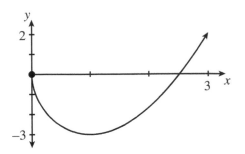

11. Critical numbers -1, 1. Strictly decreasing on $(-\infty, -1]$, strictly increasing on $[-1, 1]$, strictly decreasing on $[1, \infty)$. Local minimum at -1, local maximum at 1. $y = 0$ is an asymptote.

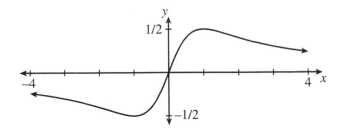

13. Strictly decreasing.

15. Not monotone.

17. Not monotone.

19. Strictly decreasing.

21. $a = -3, b = -4$.

Section 4.4

1. (a) Concave down on $[p, r]$, concave up on $[r, t]$, concave down on $[t, u]$, concave up on $[u, v]$. Inflection points at $x = r, t, u$.

   (b) Concave down on $[p, r]$, concave up on $[r, t]$, concave up on $[t, v]$, concave down on $[v, w]$. Inflection points at $x = r, v$.

3. Strictly increasing on $(-\infty, -5]$, strictly decreasing on $[-5, 1]$, strictly increasing on $[1, \infty)$. Concave down on $(-\infty, -2]$, concave up on $[-2, \infty)$. Local maximum at -5, local minimum at 1, inflection point at $(-2, 26)$.

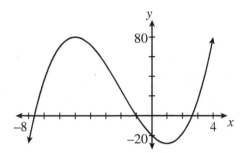

5. Strictly increasing on $(-\infty, 0]$, strictly decreasing on $[0, 4]$, strictly increasing on $[4, \infty)$. Concave down on $(-\infty, 0]$ and $[0, 3]$ (in fact, it can be shown that it is concave down on $(-\infty, 3]$), concave up on $[3, \infty)$. Local maximum at 0, local minimum at 4, inflection point at $(3, -62/5)$.

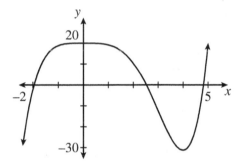

7. Strictly increasing $(-\infty, 0]$ and $[0, \infty)$ (in fact, it is strictly increasing on $(-\infty, \infty)$). Concave up on $(-\infty, 0]$, concave down on $[0, \infty)$. Inflection point at $(0, 0)$.

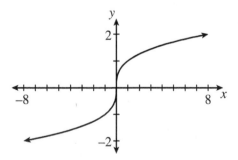

9. Strictly increasing on $(-\infty, -1]$, strictly decreasing on $[-1, 1]$, strictly increasing on $[1, \infty)$. Concave down on $(-\infty, 0]$, concave up on $[0, \infty)$. Local maximum at -1, local minimum at 1, inflection point at $(0, 0)$.

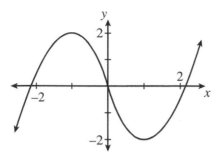

11. Strictly decreasing on $(-\infty, -\sqrt{3}]$, strictly increasing on $[-\sqrt{3}, \sqrt{3}]$, strictly decreasing on $[\sqrt{3}, \infty)$. Concave down on $(-\infty, -3]$, concave up on $[-3, 0]$, concave down on $[0, 3]$, concave up on $[3, \infty)$. Local minimum at $-\sqrt{3}$, local maximum at $\sqrt{3}$, inflection points at $(-3, -1/4)$, $(0, 0)$, and $(3, 1/4)$. $y = 0$ is an asymptote.

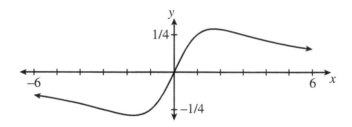

13. Strictly increasing on $(-\infty, 0]$ and $[0, \infty)$ (in fact, it is strictly increasing on $(-\infty, \infty)$). Concave up on $(-\infty, -3]$, concave down on $[-3, 0]$, concave up on $[0, 3]$, concave down on $[3, \infty)$. Inflection points at $(-3, -9/4)$, $(0, 0)$, and $(3, 9/4)$.

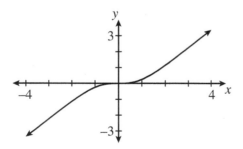

15. Strictly increasing on $[0, 2\pi/3]$, strictly decreasing on $[2\pi/3, 4\pi/3]$, strictly increasing on $[4\pi/3, 2\pi]$. Concave down on $[0, \pi]$, concave up on $[\pi, 2\pi]$. Local maximum at $2\pi/3$, local minimum at $4\pi/3$, inflection point at (π, π).

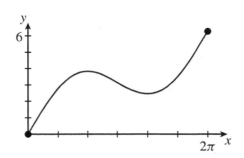

17. Strictly increasing on $(-\pi/2, -\pi/6]$, strictly decreasing on $[-\pi/6, \pi/6]$, strictly increasing on $[\pi/6, \pi/2)$. Concave down on $(-\pi/2, 0]$, concave up on $[0, \pi/2)$. Local maximum at $-\pi/6$, local minimum at $\pi/6$, inflection point at $(0, 0)$.

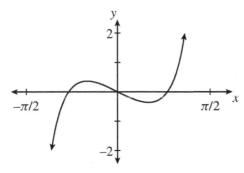

Section 4.5

1.

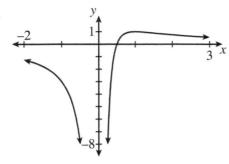

3.

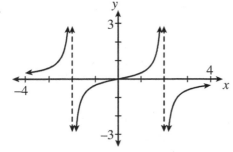

5.

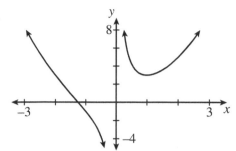

7.

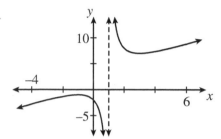

9.

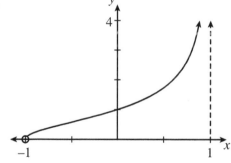

11.

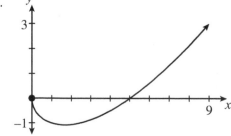

13.

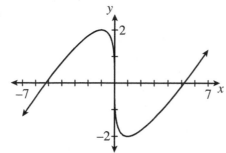

15.

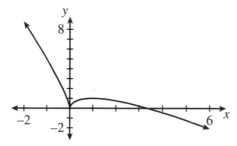

Section 4.6

1. 500.

3. 125/2.

5. 20 feet wide, 30 feet deep.

7. 8 feet wide, 10 feet high.

9. Radius 1 inch, height 3 inches.

11. Radius 3 inches, height 9 inches.

13. $5\sqrt{5}$ meters.

15. Radius 2 inches, height 8 inches.

Section 4.7

1. $(0, \sqrt{3})$.

3. 20 inches wide, 30 inches high.

5. Radius of semicircular top part of window and height of rectangular bottom part are both $P/(\pi + 4)$.

7. Row to the point 45 feet west of the point directly across the river.

9. (a) 60 feet by 60 feet.

    (b) 65 feet by 70 feet, with one 70 foot side coinciding with back of barn.

Section 4.8

1. 4/5.
3. 12.
5. 12/13.
7. $\pi/4$.
9. 3/4.
11. 1/6.
13. -2.
15. -1.
19. -1.

Section 4.9

1. $x^4/2 - 3x^3 + 3x^2/2 - x$, on $(-\infty, \infty)$.
3. $-\cos x + \sin x$, on $(-\infty, \infty)$.
5. $x^3/3 - 1/x$, on $(-\infty, 0)$ and $(0, \infty)$.
7. $x^7/7 + (3/2)x^4 + 9x$, on $(-\infty, \infty)$.
9. $(1/3)(2x + 3)^{3/2}$, on $[-3/2, \infty)$.
11. $g(x) = (2/3)x^{3/2} - \cos x + 2$.
13. $g(x) = (3x^3 \sqrt[3]{x} + 17)/10$.
15. 50 ft/sec.
17. 5 sec.

Chapter 5

Section 5.1

1. (a) $(2n^3 - 3n^2 + 13n)/6$.
   (b) 40.
3. $(n^4 + 2n^3 - n)/2$.
5. $(n^4 + 6n^3 + 13n^2 + 12n)/4$.
7. $(n^3 + 9n^2 + 8n)/3$.
13. (a) $r(r^n - 1)$.

(b) $r(r^n - 1)/(r - 1)$.

(c) $1 - 1/2^n$.

Section 5.2

1. (a) $72936 \, \text{m}^3$.

  (b) $68184 \, \text{m}^3$.

3. (a) 72 kWh.

  (b) 84 kWh.

5. (a)

  (b) 1.097.

7. 500 gal.

9. 3975 ft.

Section 5.3

1. 12.

3. 64/3.

5. 8/3.

7. 3.

9. **Theorem**. *Suppose that $a > b$ and f is continuous on $[b, a]$. Let m be the minimum value of f on $[b, a]$, and let M be the maximum value. Then*

$$M(b - a) \le \int_a^b f(x) \, dx \le m(b - a).$$

Section 5.4

1. 10.

3. $-21/2$.

5. 16/3.

7. 2π.

9. 119/6.

11. 668/15.

13. −9/4.

17. $-1/(x^3+1)$.

19. $2x/(x^6+1) - 1/(x^3+1)$.

21. $\sin x/(2\sqrt{x})$.

Section 5.5

1. $-\cos(x^2+3)+C$.

3. $(1/3)\sin(x^3)+C$.

5. $-\cos^4(x^3)/12+C$.

7. $-2/(3(\sqrt{x}+1)^3)+C$.

9. $\tan^3 x/3+C$.

11. 2.

13. 7/3.

15. $\tan x - x + C$.

Chapter 6

Section 6.1

1. 12.

3. 1/12.

5. $2\sqrt{2}$.

7. 4/3.

9. 32/3.

11. 1/3.

13. 9/2.

Section 6.2

1. 8π.

3. 18π.

5. $2\pi/3$.

7. $1088\pi/15$.

9. $64\pi/15$.

11. $19\pi/15$.

13.   (a) $4\pi b \int_{-a}^{a} \sqrt{a^2 - y^2}\, dy$.

 (b) $2\pi^2 a^2 b$.

15. $16r^3/3$.

Section 6.3

1. 12π.

3. $68\pi/3$.

5. $8\pi/3$.

7. $8\pi/3$.

9. $8\pi/3$.

11. $\pi r^2 h/3$.

Section 6.4

1. $236/3$.

3.   (a) $[1, 5]$. Force $= (x - 1)^2/4$.

 (b) $16/3$.

 (c) $\sum_{i=1}^{n} 2(t_i^*)^2 \Delta t$.

 (d) $16/3$.

5.   4375 ft-lb.

7. $5/2$ inch-lb.

9.   326.7 ft-lb.

Section 6.5

1. $-7/8$.

3. $2338/915$.

5. $(0, 12/5)$.

7. $(0, 4r/(3\pi))$.

9.   (a) $(r/3, h/3)$.

 (b) $\pi r^2 h/3$.

Chapter 7

Section 7.1

3. One-to-one.

5. Not one-to-one.

7. Not one-to-one.

9. $f^{-1}(x) = (5-x)/2$. Domain: $(-\infty, \infty)$. Range: $(-\infty, \infty)$.

11. $f^{-1}(x) = (x-2)^3 + 8$. Domain: $(-\infty, \infty)$. Range: $(-\infty, \infty)$.

13. $f^{-1}(x) = x^4 - 2x^2 + 1$. Domain: $[1, \infty)$. Range: $[0, \infty)$.

15. $f^{-1}(x) = x/\sqrt{1-x^2}$. Domain: $(-1, 1)$. Range: $(-\infty, \infty)$.

Section 7.2

1. $(-\infty, \infty)$.

3. $[0, 3]$.

5. $(-\infty, 0) \cup (0, \infty)$. The domain of f is not an interval.

7. -3.

9. (a) $1/(2\sqrt{x}) = (1/2)x^{-1/2}$.

    (b) $1/(n(\sqrt[n]{x})^{n-1}) = (1/n)x^{1/n-1}$.

Section 7.3

1. $2/(2x-3)$.

3. $\sec^2(\ln x)/x$.

5. $x/(x^2+1)$.

7. $\cos x \ln(\sin x)$.

9. $-\ln|\cos x| + C$.

11. $x + 2\ln|x-1| + C$.

13. $\ln|\tan x| + C$.

15. $-(1/2)\ln 5$.

17. 0.

19. 1.

23.

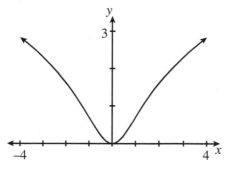

25. $\pi \ln 4$.

27. (a) 1200π in-lb.

   (c) $600\pi(\ln 4 - 1)$ in-lb.

Section 7.4

3. $3e^{3x-2}$.

5. $\cos(xe^{2x})e^{2x}(2x+1)$.

7. $(x+1)^{x-1} \cdot ((x-1)/(x+1) + \ln(x+1))$.

9. $e^{\tan x} + C$.

11. $-3/(2e^{2x}) + C$.

13. $e^{e^x} + C$.

17. ∞.

19. 3.

23.

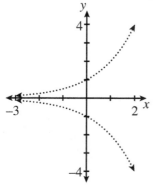

27. (a) $f(t) = 0.8e^{t\ln(1.5)/3}$.

   (b) 1.8 mg.

29. (d) $f(y) = \sqrt{w/(\pi s)}e^{dy/(2s)}$.

31. $\pi(e^2 - 1)/2$.

Section 7.5

3. $\pi/3$.

5. $\pi/7$.

7. $-3/4$.

9. $\dfrac{3\sqrt{x}}{2\sqrt{1-x^3}}$.

11. $\dfrac{1}{x((\ln x)^2 + 1)}$.

13. $\dfrac{2e^{\sin^{-1}(2x)}}{\sqrt{1-4x^2}}$.

15. $\ln(x^2 - 2x + 2) + 2\tan^{-1}(x - 1) + C$.

17. $\sin^{-1}(e^x) + C$.

19. $\sqrt{x^2 - 2x} + C$.

21. $-\pi/2$.

23. $2\sqrt{3}$.

27. $2\ln 2/\pi$.

31.

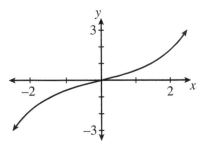

$y = \sinh x$

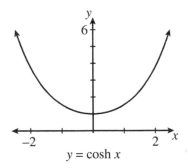

$y = \cosh x$

35. (a)

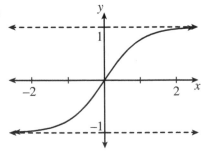

(b) $\tanh^{-1} x = (\ln(1+x) - \ln(1-x))/2$.

(c) $\frac{d}{dx}(\tanh^{-1} x) = 1/(1-x^2)$.

Section 7.6

1. 1.

3. ∞.

5. $\pi^2/2$.

7. $\sqrt{3}/3$.

9. 0.

11. 1/2.

13. 0.

15. 1.

17. e^2.

19. 1.

21. ∞.

23. $e^{-2/\pi}$.

25. 1.

27. (b) 0.

(c) One example is $f(x) = e^{1/x^2}$, $g(x) = x^2 \ln c$.

Chapter 8

Section 8.1

1. $\ln |x| - 2/x + C$.

3. $3 \ln |x - 4| + 2 \ln |x + 3| + C$.

5. $-\ln |x| + 2 \ln |x - 2| + C$.

7. $-\dfrac{1}{2} \ln |x| + \dfrac{1}{x} + \dfrac{3}{2} \ln |x - 2| + C$.

9. $-\ln |t| + +\dfrac{1}{2} \ln(4t^2 + 1) + \dfrac{1}{2} \tan^{-1}(2t) + C$.

11. $-\dfrac{1}{2} \ln |x| + \dfrac{1}{3} \ln |x - 1| + \dfrac{1}{6} \ln |x + 2| + C$.

13. $u^3/3 + 4u + 4 \ln |u - 2| - 4 \ln |u + 2| + C$.

15. $\frac{3}{4}\ln|x-2| - \frac{1}{4}\ln|x+2| + \frac{1}{4}\ln(x^2+4) + \tan^{-1}(x/2) + C.$

17. $-\frac{1}{4}\ln(2-\sin\theta) + \frac{1}{4}\ln(2+\sin\theta) + C.$

19. $\pi\ln 6.$

Section 8.2

1. $-\frac{x}{3}\cos(3x+1) + \frac{1}{9}\sin(3x+1) + C.$

3. $x\ln(x^3-8) - 3x - 2\ln(x-2) + \ln(x^2+2x+4) + 2\sqrt{3}\tan^{-1}\left(\frac{x+1}{\sqrt{3}}\right) + C.$

5. $(x-1)e^{x-2} + C.$

7. $(x^2-2x+2)e^{x-2} + C.$

9. $\frac{x^3}{3}\sin^{-1}x + \frac{1}{3}\sqrt{1-x^2} - \frac{1}{9}(1-x^2)^{3/2} + C.$

11. $-\frac{\tan^{-1}x}{x} + \ln|x| - \frac{1}{2}\ln(x^2+1) + C.$

13. $\theta\tan\theta + \ln|\cos\theta| + C.$

15. $x\tan^{-1}(x^2) - \frac{\sqrt{2}}{4}\ln(x^2-\sqrt{2}x+1) - \frac{\sqrt{2}}{2}\tan^{-1}(\sqrt{2}x-1)$

$+\frac{\sqrt{2}}{4}\ln(x^2+\sqrt{2}x+1) - \frac{\sqrt{2}}{2}\tan^{-1}(\sqrt{2}x+1) + C.$

17. $\frac{x^2(2\sin(\ln x) - \cos(\ln x))}{5} + C.$

19. The constant of integration has been left out.

21. $2\pi^2.$

23. $2\pi(\ln 2)^2 - 4\pi\ln 2 + 2\pi.$

Section 8.3

1. $\frac{\cos^3\theta}{3} - \cos\theta + C.$

3. $-\frac{1}{2\sin(2\theta)} - \frac{\sin(2\theta)}{2} + C.$

5. $\frac{4\sin^3\theta}{3} + C.$

7. $\frac{9\theta}{2} - 4\cos\theta - \frac{\sin(2\theta)}{4} + C.$

9. $\sin x - 2\ln|\sin x| + C.$

11. $\dfrac{\sec^5\theta}{5} - \dfrac{\sec^3\theta}{3} + C.$

13. $-\dfrac{\cos^3\theta}{3} + C.$

15. $\ln|\sec\theta + \tan\theta| - \sin\theta + C.$

17. $\ln|\sec\theta + \tan\theta| + \dfrac{\sec^3\theta}{3} - \sec\theta + C.$

19. $\tan x \ln(\cos x) + \tan x - x + C.$

21. $\displaystyle\int \cos^n\theta\,d\theta = \dfrac{\sin\theta\cos^{n-1}\theta}{n} + \dfrac{n-1}{n}\int \cos^{n-2}\theta\,d\theta.$

25. $\pi^2/2.$

27. (c) $\dfrac{1-u^2}{u^2+1}.$

(d) $\pi/4.$

Section 8.4

1. $2(\sqrt{x}-1)e^{\sqrt{x}} + C.$

3. $3(\sqrt[3]{x}-1)e^{\sqrt[3]{x}} + C.$

5. $3(\sqrt[3]{x})^2/2 - 3\sqrt[3]{x} + 3\ln|\sqrt[3]{x}+1| + C.$

7. $2\ln(\sqrt{x}+1) + C.$

9. $2\sqrt{x} - 4\sqrt[4]{x} + (32/3)\ln(\sqrt[4]{x}+2) + (4/3)\ln|\sqrt[4]{x}-1| + C.$

11. $(6/5)(\sqrt{x}+1)^{5/3} - 3(\sqrt{x}+1)^{2/3} + C.$

13. $2\sqrt{1+e^x} + \ln(\sqrt{1+e^x}-1) - \ln(\sqrt{1+e^x}+1) + C.$

15. $(6 - 3(\sqrt[3]{x})^2)\cos(\sqrt[3]{x}) + 6\sqrt[3]{x}\sin(\sqrt[3]{x}) + C.$

17. $\sec(\tan^{-1}x) + C = \sqrt{x^2+1} + C.$

21. $\pi(2\ln 2 - 1).$

Section 8.5

1. $-\dfrac{\sqrt{1-x^2}}{x} + C.$

3. $\sqrt{x^2+1} + (1/2)\ln(\sqrt{x^2+1}-1) - (1/2)\ln(\sqrt{x^2+1}+1) + C.$

5. $x\sqrt{x^2-9}/2 - (9/2)\ln\left|x+\sqrt{x^2-9}\right| + C.$

7. $\sqrt{3}/12$.

9. $\ln(2+\sqrt{3})-\sqrt{3}/2$.

11. $\sqrt{x^2-4x}+2\ln\left|x-2+\sqrt{x^2-4x}\right|+C$.

13. $3\sin^{-1}\left(\dfrac{x-1}{2}\right)-\dfrac{(x+3)\sqrt{3+2x-x^2}}{2}+C$.

15. $\dfrac{(2x^2-1)\sin^{-1}x+x\sqrt{1-x^2}}{4}+C$.

17. $\sin^{-1}(\sqrt{x})+\sqrt{x}\sqrt{1-x}+C$.

19. (a) $\ln\left|\dfrac{\sqrt{4x^2+1}-1}{2x}\right|+C$.

21. $(\pi^2+2\pi)/8$.

23. (a) 5.96 in.

   (b) No.

Section 8.6

1. $M_4=21/8$, $\int_0^2 x^2\,dx=8/3$, error $=1/24$.

3. $M_5=20$, $\int_1^6 2x-3\,dx=20$, error $=0$.

5. $M_6\approx1.3322$, $\int_0^\pi \sin^3 x\,dx=4/3$, error ≈0.0011.

7. $S_4=20$, $\int_{-1}^3 x^3\,dx=20$, error $=0$.

9. $S_6\approx2.0009$, $\int_0^\pi \sin x\,dx=2$, error ≈-0.0009.

11. $M_5\approx1.3730$.

17. $S_6\approx4.4547$.

Section 8.7

1. 1/2.

3. 1/3.

5. Diverges.

7. 1/4.

9. 2.

11. Diverges.

13. Diverges.

15. $\ln(3/4)$.

17. $\pi/6$.

19. Diverges.

21. $2\ln 2$.

23. $3\ln 3$.

25. $\pi/2$.

27. (a) Diverges.

    (b) Diverges.

    (c) $(1/2)\ln 2$.

29. (b) 0.

    (c) 1.

Chapter 9

Section 9.1

1.

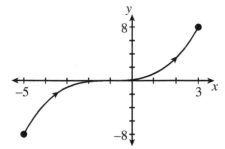

3.

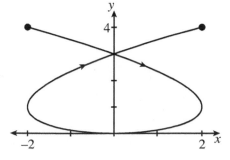

5.

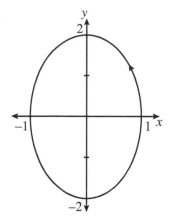

7.

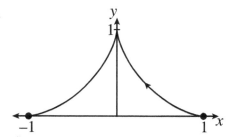

9.

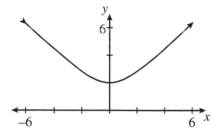

11. $y = -2x + 3\sqrt{3}/2.$

13. $y = \pm(\sqrt{3}/3)x + 3.$

15. One branch of a hyperbola.

19. (a) $x = r\theta - a \sin \theta, \; y = r - a \cos \theta.$

(b)

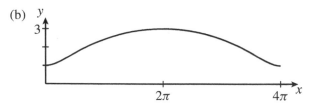

(c)
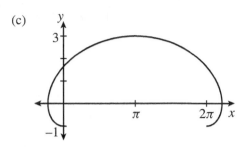

23. $\lim\limits_{t\to1^-}\dfrac{g(t)-g(1)}{f(t)-f(1)}=-\infty,\ \lim\limits_{t\to1^+}\dfrac{g(t)-g(1)}{f(t)-f(1)}=\infty,$

$\lim\limits_{t\to3^-}\dfrac{g(t)-g(3)}{f(t)-f(3)}=\infty,\ \lim\limits_{t\to3^+}\dfrac{g(t)-g(3)}{f(t)-f(3)}=-\infty.$

This shows that the tangent lines at $t=1$ and $t=3$ are vertical.

Section 9.2

1. 14.

3. $\pi/4$.

5. $\ln(2+\sqrt{3})$.

7. $2e-2/e$.

9. $\sqrt{5}+(1/2)\ln(\sqrt{5}-1)-(1/2)\ln(\sqrt{5}+1)-\sqrt{2}-(1/2)\ln(\sqrt{2}-1)$
    $+(1/2)\ln(\sqrt{2}+1)$.

11. $\pi^2/4$.

13. $2\ln(\sqrt{2}+1)$.

Section 9.3

1. $816\pi/5$.

3. $37\pi/12$.

5. $(\pi^2+2\pi)/4$.

7. $(10\sqrt{10}-1)\pi/27$.

9. $(64-16\sqrt{2})\pi/3$.

11. $(2800\sqrt{10}-172)\pi/1215$.

13. $(2\sqrt{2}+\ln(\sqrt{2}+1)-\ln(\sqrt{2}-1))\pi$.

15. $20\pi-30$.

Section 9.4

1.

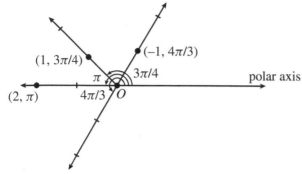

3. (a) $(3, \pi + 2\pi n), (-3, 2\pi n)$.

  (b) $(2\sqrt{2}, 5\pi/4 + 2\pi n), (-2\sqrt{2}, \pi/4 + 2\pi n)$.

  (c) $(2, -\pi/6 + 2\pi n), (-2, 5\pi/6 + 2\pi n)$.

5. $x = 2$:

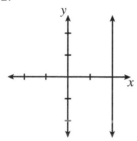

7. $x^2 + (y - 2)^2 = 4$:

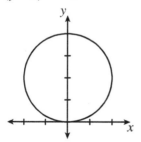

9. $y = x^2$:

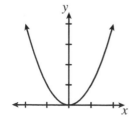

11.

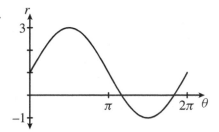

13.

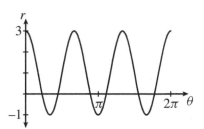

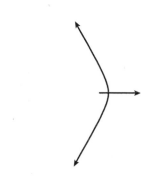

15.

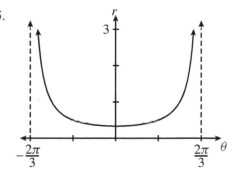

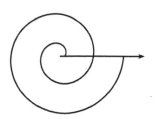

17.

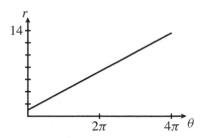

19. One branch of a hyperbola.

23. (b) $8((\pi^2 + 1)^{3/2} - 1)/3$.

Section 9.5

1. $\pi^5/60$.

3. $(4 - \pi)/8$.

5. 1/2.

7. $9\pi/2$.

9. $\pi/4$.

11. $(5\pi - 6\sqrt{3})/24$.

Chapter 10

Section 10.1

1. 1.

3. Diverges.

5. 10.

7. 14/15.

9. $\sqrt{3}/2$.

11. 2/3.

13. 1.

15. 3/2.

17. (b) 2.

19. (a) $s_2 = 3/4$, $s_3 = 8/9$, $s_4 = 15/16$.

    (b) $s_N = (N^2 - 1)/N^2$.

    (c) 1.

Section 10.2

1. Diverges.

3. Converges.

5. Converges.

7. Converges.

9. Converges.

11. Converges.

13. Converges.

15. Converges.

17. Diverges.

19. Converges.

21. (a) $s_{10} \approx 1.19753$.

23. (a) $s_4 \approx 0.57132$.

    (c) $s = 1/(e-1) \approx 0.58198$.

25. (c) $1.64488 \leq s \leq 1.64498$.

Section 10.3

1. Converges.

3. Converges.

5. Converges.

7. Converges.

9. Diverges.

11. Converges.

13. Diverges.

15. Converges.

17. Converges.

19. Diverges.

21. Converges.

23. Diverges.

25. (a) $s_{10} \approx 0.633048$.

27. (a) $s_3 \approx 0.381370$.

    (b) Error $\leq 1/5832 \approx 0.00017$.

Section 10.4

1. Converges.

3. Converges.

5. Diverges.

7. Diverges.

9. Converges.

11. Converges.

13. Converges.

15. Diverges.

21. (b) 1, which is the same as the value of $\sum_{n=1}^{\infty} 1/2^n$.

Section 10.5

1. Converges conditionally.

3. Converges absolutely.

5. Diverges.

7. Converges absolutely.

9. Diverges.

11. Diverges.

13. Converges absolutely.

15. Converges conditionally.

17. Diverges.

Section 10.6

1. $[-4, 4)$.

3. $(-1/2, 1/2]$.

5. $(-\infty, \infty)$.

7. $(-3, 3)$.

9. $[-3, 3)$.

11. $[-1, 1)$.

13. $[-1, 1]$.

15. $(-\infty, \infty)$.

17. $\dfrac{1}{x^2 + x - 6} = \sum_{n=0}^{\infty} \left(\dfrac{(-1)^{n+1}}{10 \cdot 2^n} - \dfrac{1}{15 \cdot 3^n} \right) x^n, \qquad -2 < x < 2.$

19. (a) $(-\sqrt{3}, \sqrt{3})$.

   (b) $3x/(3 - x^2)$.

21. (a) $1/e$.

   (c) Converges when $x = -1/e$, diverges when $x = 1/e$.

Section 10.7

1. $(1, 7)$.

3. $[1, 7]$.

5. $[-19, 13)$.

7. $(-\infty, \infty)$.

9. $(-2, 4)$.

11. $\dfrac{1}{x} = \displaystyle\sum_{n=0}^{\infty} \dfrac{(-1)^n (x-3)^n}{3^{n+1}}, \qquad 0 < x < 6.$

13. $6229/13440 \approx 0.4635$.

15. $\ln(3/5)$.

17. $\sqrt{3}\pi/6$.

19. (a) $\dfrac{1}{x^3 + 1} = \displaystyle\sum_{n=0}^{\infty} (-1)^n x^{3n}, \qquad -1 < x < 1.$

    (b) $\displaystyle\int_0^{1/2} \dfrac{dx}{x^3 + 1} = \sum_{n=0}^{\infty} \dfrac{(-1)^n}{(3n+1)2^{3n+1}}.$

    (c) $435/896 \approx 0.48549$.

    (d) $\ln 3/6 + \sqrt{3}\pi/18 \approx 0.48540$.

21. (a) $\dfrac{1}{(1+x)^3} = \displaystyle\sum_{n=0}^{\infty} \dfrac{(-1)^n (n+2)(n+1) x^n}{2}, \qquad -1 < x < 1.$

    (b) $\dfrac{1}{(1+x)^4} = \displaystyle\sum_{n=0}^{\infty} \dfrac{(-1)^n (n+3)(n+2)(n+1) x^n}{6}, \qquad -1 < x < 1.$

    (c) $\dfrac{1}{(1+x)^p} = \displaystyle\sum_{n=0}^{\infty} \dfrac{(-1)^n (n+p-1)! x^n}{(p-1)! n!}, \qquad -1 < x < 1.$

Section 10.8

1. $\displaystyle\sum_{n=0}^{\infty} \dfrac{(-1)^n 3(n+1)}{2^{n+2}} (x-2)^n.$

3. $\ln 4 + \displaystyle\sum_{n=1}^{\infty} \dfrac{(-1)^{n-1} 3^n}{n 4^n} (x-1)^n.$

5. $\displaystyle\sum_{n=0}^{\infty} \frac{(-1)^n 4^n}{(2n)!} \left(x - \frac{\pi}{4}\right)^{2n}$.

7. $\displaystyle\sum_{n=0}^{\infty} \frac{(-1)^n (2n)!}{4^n (n!)^2}(x - 1)^n$.

9. $f^{(10)}(0) = 0$, $f^{(11)}(0) = -10!$.

11. (b) $e \approx 1957/720 \approx 2.7181$.

13.
   (a) $\displaystyle\int_{-1}^{0} e^{(x^3)}\, dx = \sum_{n=0}^{\infty} \frac{(-1)^n}{(3n+1)n!}$.

   (b) $\displaystyle\int_{-1}^{0} e^{(x^3)}\, dx \approx \frac{2941}{3640} \approx 0.8080$.

15.
   (a) For all x, $\displaystyle S(x) = \sum_{n=0}^{\infty} \frac{(-1)^n x^{4n+3}}{(4n+3)(2n+1)!}$.

   (b) $S(1) \approx 13/42 \approx 0.3095$.

17. Both parts: $-1/2$.

23. $1/2$.

25. (e) $0 + 0 \cdot x + 0 \cdot x^2 + \cdots$.

Section 10.9

3. $\sin^{-1}(1/4) \approx 97/384 \approx 0.2526$.

5. $\sqrt{10} \approx 683/216 \approx 3.1620$.

7.
   (a) $\displaystyle\int_{0}^{1/2} \frac{dx}{\sqrt{1+x^3}} = \sum_{n=0}^{\infty} \frac{(-1)^n (2n)!}{2^{5n+1}(n!)^2(3n+1)}$.

   (b) $\displaystyle\int_{0}^{1/2} \frac{dx}{\sqrt{1+x^3}} \approx \frac{63}{128} \approx 0.4922$.

9. (a) $[-1, 1]$.
   (b) $\ln(\sqrt{1+x^2} + x)$.

Section 10.10

1. $T_3(x) = 1 + x + x^2/2 + x^3/6$.

3. $T_3(x) = e^5 + 4e^5(x - 2) + 9e^5(x - 2)^2 + (44e^5/3)(x - 2)^3$.

5. $T_3(x) = 2 + \ln 2 + (x - e)/e - (x - e)^3/(6e^3)$.

7. $T_3(x) = 5 - 3x + 2x^2 + x^3$.

9. $T_5(x) = x + x^3/3 + 2x^5/15$.

11. (e) $T_5(1) = 163/60 \approx 2.7167, |e - T_5(1)| \approx 0.0016$.

13. (a) $T_3(x) = 1 - (x - 1)^2/2 + (x - 1)^3/2$.

(c) $f(3/2) \approx 0.9189, T_3(3/2) = 15/16 \approx 0.9375, |f(3/2) - T_3(3/2)| \approx 0.0186$.

Section 10.11

7. (a) $2 - (N + 2)/2^N$.

(b) $6 - (N^2 + 4N + 6)/2^N$.

(c) 6.

Index

A CATALOG OF SELECTED
DOVER BOOKS
IN SCIENCE AND MATHEMATICS

Mathematics–Bestsellers

HANDBOOK OF MATHEMATICAL FUNCTIONS: with Formulas, Graphs, and Mathematical Tables, Edited by Milton Abramowitz and Irene A. Stegun. A classic resource for working with special functions, standard trig, and exponential logarithmic definitions and extensions, it features 29 sets of tables, some to as high as 20 places. 1046pp. 8 x 10 1/2.   0-486-61272-4

ABSTRACT AND CONCRETE CATEGORIES: The Joy of Cats, Jiri Adamek, Horst Herrlich, and George E. Strecker. This up-to-date introductory treatment employs category theory to explore the theory of structures. Its unique approach stresses concrete categories and presents a systematic view of factorization structures. Numerous examples. 1990 edition, updated 2004. 528pp. 6 1/8 x 9 1/4.   0-486-46934-4

MATHEMATICS: Its Content, Methods and Meaning, A. D. Aleksandrov, A. N. Kolmogorov, and M. A. Lavrent'ev. Major survey offers comprehensive, coherent discussions of analytic geometry, algebra, differential equations, calculus of variations, functions of a complex variable, prime numbers, linear and non-Euclidean geometry, topology, functional analysis, more. 1963 edition. 1120pp. 5 3/8 x 8 1/2.   0-486-40916-3

INTRODUCTION TO VECTORS AND TENSORS: Second Edition–Two Volumes Bound as One, Ray M. Bowen and C.-C. Wang. Convenient single-volume compilation of two texts offers both introduction and in-depth survey. Geared toward engineering and science students rather than mathematicians, it focuses on physics and engineering applications. 1976 edition. 560pp. 6 1/2 x 9 1/4.   0-486-46914-X

AN INTRODUCTION TO ORTHOGONAL POLYNOMIALS, Theodore S. Chihara. Concise introduction covers general elementary theory, including the representation theorem and distribution functions, continued fractions and chain sequences, the recurrence formula, special functions, and some specific systems. 1978 edition. 272pp. 5 3/8 x 8 1/2.   0-486-47929-3

ADVANCED MATHEMATICS FOR ENGINEERS AND SCIENTISTS, Paul DuChateau. This primary text and supplemental reference focuses on linear algebra, calculus, and ordinary differential equations. Additional topics include partial differential equations and approximation methods. Includes solved problems. 1992 edition. 400pp. 7 1/2 x 9 1/4.   0-486-47930-7

PARTIAL DIFFERENTIAL EQUATIONS FOR SCIENTISTS AND ENGINEERS, Stanley J. Farlow. Practical text shows how to formulate and solve partial differential equations. Coverage of diffusion-type problems, hyperbolic-type problems, elliptic-type problems, numerical and approximate methods. Solution guide available upon request. 1982 edition. 414pp. 6 1/8 x 9 1/4.   0-486-67620-X

VARIATIONAL PRINCIPLES AND FREE-BOUNDARY PROBLEMS, Avner Friedman. Advanced graduate-level text examines variational methods in partial differential equations and illustrates their applications to free-boundary problems. Features detailed statements of standard theory of elliptic and parabolic operators. 1982 edition. 720pp. 6 1/8 x 9 1/4.   0-486-47853-X

LINEAR ANALYSIS AND REPRESENTATION THEORY, Steven A. Gaal. Unified treatment covers topics from the theory of operators and operator algebras on Hilbert spaces; integration and representation theory for topological groups; and the theory of Lie algebras, Lie groups, and transform groups. 1973 edition. 704pp. 6 1/8 x 9 1/4.   0-486-47851-3

Browse over 9,000 books at www.doverpublications.com

A SURVEY OF INDUSTRIAL MATHEMATICS, Charles R. MacCluer. Students learn how to solve problems they'll encounter in their professional lives with this concise single-volume treatment. It employs MATLAB and other strategies to explore typical industrial problems. 2000 edition. 384pp. 5 3/8 x 8 1/2. 0-486-47702-9

NUMBER SYSTEMS AND THE FOUNDATIONS OF ANALYSIS, Elliott Mendelson. Geared toward undergraduate and beginning graduate students, this study explores natural numbers, integers, rational numbers, real numbers, and complex numbers. Numerous exercises and appendixes supplement the text. 1973 edition. 368pp. 5 3/8 x 8 1/2. 0-486-45792-3

A FIRST LOOK AT NUMERICAL FUNCTIONAL ANALYSIS, W. W. Sawyer. Text by renowned educator shows how problems in numerical analysis lead to concepts of functional analysis. Topics include Banach and Hilbert spaces, contraction mappings, convergence, differentiation and integration, and Euclidean space. 1978 edition. 208pp. 5 3/8 x 8 1/2. 0-486-47882-3

FRACTALS, CHAOS, POWER LAWS: Minutes from an Infinite Paradise, Manfred Schroeder. A fascinating exploration of the connections between chaos theory, physics, biology, and mathematics, this book abounds in award-winning computer graphics, optical illusions, and games that clarify memorable insights into self-similarity. 1992 edition. 448pp. 6 1/8 x 9 1/4. 0-486-47204-3

SET THEORY AND THE CONTINUUM PROBLEM, Raymond M. Smullyan and Melvin Fitting. A lucid, elegant, and complete survey of set theory, this three-part treatment explores axiomatic set theory, the consistency of the continuum hypothesis, and forcing and independence results. 1996 edition. 336pp. 6 x 9. 0-486-47484-4

DYNAMICAL SYSTEMS, Shlomo Sternberg. A pioneer in the field of dynamical systems discusses one-dimensional dynamics, differential equations, random walks, iterated function systems, symbolic dynamics, and Markov chains. Supplementary materials include PowerPoint slides and MATLAB exercises. 2010 edition. 272pp. 6 1/8 x 9 1/4. 0-486-47705-3

ORDINARY DIFFERENTIAL EQUATIONS, Morris Tenenbaum and Harry Pollard. Skillfully organized introductory text examines origin of differential equations, then defines basic terms and outlines general solution of a differential equation. Explores integrating factors; dilution and accretion problems; Laplace Transforms; Newton's Interpolation Formulas, more. 818pp. 5 3/8 x 8 1/2. 0-486-64940-7

MATROID THEORY, D. J. A. Welsh. Text by a noted expert describes standard examples and investigation results, using elementary proofs to develop basic matroid properties before advancing to a more sophisticated treatment. Includes numerous exercises. 1976 edition. 448pp. 5 3/8 x 8 1/2. 0-486-47439-9

THE CONCEPT OF A RIEMANN SURFACE, Hermann Weyl. This classic on the general history of functions combines function theory and geometry, forming the basis of the modern approach to analysis, geometry, and topology. 1955 edition. 208pp. 5 3/8 x 8 1/2. 0-486-47004-0

THE LAPLACE TRANSFORM, David Vernon Widder. This volume focuses on the Laplace and Stieltjes transforms, offering a highly theoretical treatment. Topics include fundamental formulas, the moment problem, monotonic functions, and Tauberian theorems. 1941 edition. 416pp. 5 3/8 x 8 1/2. 0-486-47755-X

Browse over 9,000 books at www.doverpublications.com

Mathematics–Logic and Problem Solving

PERPLEXING PUZZLES AND TANTALIZING TEASERS, Martin Gardner. Ninety-three riddles, mazes, illusions, tricky questions, word and picture puzzles, and other challenges offer hours of entertainment for youngsters. Filled with rib-tickling drawings. Solutions. 224pp. 5 3/8 x 8 1/2. 0-486-25637-5

MY BEST MATHEMATICAL AND LOGIC PUZZLES, Martin Gardner. The noted expert selects 70 of his favorite "short" puzzles. Includes The Returning Explorer, The Mutilated Chessboard, Scrambled Box Tops, and dozens more. Complete solutions included. 96pp. 5 3/8 x 8 1/2. 0-486-28152-3

THE LADY OR THE TIGER?: and Other Logic Puzzles, Raymond M. Smullyan. Created by a renowned puzzle master, these whimsically themed challenges involve paradoxes about probability, time, and change; metapuzzles; and self-referentiality. Nineteen chapters advance in difficulty from relatively simple to highly complex. 1982 edition. 240pp. 5 3/8 x 8 1/2. 0-486-47027-X

SATAN, CANTOR AND INFINITY: Mind-Boggling Puzzles, Raymond M. Smullyan. A renowned mathematician tells stories of knights and knaves in an entertaining look at the logical precepts behind infinity, probability, time, and change. Requires a strong background in mathematics. Complete solutions. 288pp. 5 3/8 x 8 1/2.

0-486-47036-9

THE RED BOOK OF MATHEMATICAL PROBLEMS, Kenneth S. Williams and Kenneth Hardy. Handy compilation of 100 practice problems, hints and solutions indispensable for students preparing for the William Lowell Putnam and other mathematical competitions. Preface to the First Edition. Sources. 1988 edition. 192pp. 5 3/8 x 8 1/2. 0-486-69415-1

KING ARTHUR IN SEARCH OF HIS DOG AND OTHER CURIOUS PUZZLES, Raymond M. Smullyan. This fanciful, original collection for readers of all ages features arithmetic puzzles, logic problems related to crime detection, and logic and arithmetic puzzles involving King Arthur and his Dogs of the Round Table. 160pp. 5 3/8 x 8 1/2. 0-486-47435-6

UNDECIDABLE THEORIES: Studies in Logic and the Foundation of Mathematics, Alfred Tarski in collaboration with Andrzej Mostowski and Raphael M. Robinson. This well-known book by the famed logician consists of three treatises: "A General Method in Proofs of Undecidability," "Undecidability and Essential Undecidability in Mathematics," and "Undecidability of the Elementary Theory of Groups." 1953 edition. 112pp. 5 3/8 x 8 1/2. 0-486-47703-7

LOGIC FOR MATHEMATICIANS, J. Barkley Rosser. Examination of essential topics and theorems assumes no background in logic. "Undoubtedly a major addition to the literature of mathematical logic." – Bulletin of the American Mathematical Society. 1978 edition. 592pp. 6 1/8 x 9 1/4. 0-486-46898-4

INTRODUCTION TO PROOF IN ABSTRACT MATHEMATICS, Andrew Wohlgemuth. This undergraduate text teaches students what constitutes an acceptable proof, and it develops their ability to do proofs of routine problems as well as those requiring creative insights. 1990 edition. 384pp. 6 1/2 x 9 1/4. 0-486-47854-8

FIRST COURSE IN MATHEMATICAL LOGIC, Patrick Suppes and Shirley Hill. Rigorous introduction is simple enough in presentation and context for wide range of students. Symbolizing sentences; logical inference; truth and validity; truth tables; terms, predicates, universal quantifiers; universal specification and laws of identity; more. 288pp. 5 3/8 x 8 1/2. 0-486-42259-3

Mathematics–Algebra and Calculus

VECTOR CALCULUS, Peter Baxandall and Hans Liebeck. This introductory text offers a rigorous, comprehensive treatment. Classical theorems of vector calculus are amply illustrated with figures, worked examples, physical applications, and exercises with hints and answers. 1986 edition. 560pp. 5 3/8 x 8 1/2.   0-486-46620-5

ADVANCED CALCULUS: An Introduction to Classical Analysis, Louis Brand. A course in analysis that focuses on the functions of a real variable, this text introduces the basic concepts in their simplest setting and illustrates its teachings with numerous examples, theorems, and proofs. 1955 edition. 592pp. 5 3/8 x 8 1/2.   0-486-44548-8

ADVANCED CALCULUS, Avner Friedman. Intended for students who have already completed a one-year course in elementary calculus, this two-part treatment advances from functions of one variable to those of several variables. Solutions. 1971 edition. 432pp. 5 3/8 x 8 1/2.   0-486-45795-8

METHODS OF MATHEMATICS APPLIED TO CALCULUS, PROBABILITY, AND STATISTICS, Richard W. Hamming. This 4-part treatment begins with algebra and analytic geometry and proceeds to an exploration of the calculus of algebraic functions and transcendental functions and applications. 1985 edition. Includes 310 figures and 18 tables. 880pp. 6 1/2 x 9 1/4.   0-486-43945-3

BASIC ALGEBRA I: Second Edition, Nathan Jacobson. A classic text and standard reference for a generation, this volume covers all undergraduate algebra topics, including groups, rings, modules, Galois theory, polynomials, linear algebra, and associative algebra. 1985 edition. 528pp. 6 1/8 x 9 1/4.   0-486-47189-6

BASIC ALGEBRA II: Second Edition, Nathan Jacobson. This classic text and standard reference comprises all subjects of a first-year graduate-level course, including in-depth coverage of groups and polynomials and extensive use of categories and functors. 1989 edition. 704pp. 6 1/8 x 9 1/4.   0-486-47187-X

CALCULUS: An Intuitive and Physical Approach (Second Edition), Morris Kline. Application-oriented introduction relates the subject as closely as possible to science with explorations of the derivative; differentiation and integration of the powers of x; theorems on differentiation, antidifferentiation; the chain rule; trigonometric functions; more. Examples. 1967 edition. 960pp. 6 1/2 x 9 1/4.   0-486-40453-6

ABSTRACT ALGEBRA AND SOLUTION BY RADICALS, John E. Maxfield and Margaret W. Maxfield. Accessible advanced undergraduate-level text starts with groups, rings, fields, and polynomials and advances to Galois theory, radicals and roots of unity, and solution by radicals. Numerous examples, illustrations, exercises, appendixes. 1971 edition. 224pp. 6 1/8 x 9 1/4.   0-486-47723-1

AN INTRODUCTION TO THE THEORY OF LINEAR SPACES, Georgi E. Shilov. Translated by Richard A. Silverman. Introductory treatment offers a clear exposition of algebra, geometry, and analysis as parts of an integrated whole rather than separate subjects. Numerous examples illustrate many different fields, and problems include hints or answers. 1961 edition. 320pp. 5 3/8 x 8 1/2.   0-486-63070-6

LINEAR ALGEBRA, Georgi E. Shilov. Covers determinants, linear spaces, systems of linear equations, linear functions of a vector argument, coordinate transformations, the canonical form of the matrix of a linear operator, bilinear and quadratic forms, and more. 387pp. 5 3/8 x 8 1/2.   0-486-63518-X

Mathematics–Probability and Statistics

BASIC PROBABILITY THEORY, Robert B. Ash. This text emphasizes the probabilistic way of thinking, rather than measure-theoretic concepts. Geared toward advanced undergraduates and graduate students, it features solutions to some of the problems. 1970 edition. 352pp. 5 3/8 x 8 1/2. 0-486-46628-0

PRINCIPLES OF STATISTICS, M. G. Bulmer. Concise description of classical statistics, from basic dice probabilities to modern regression analysis. Equal stress on theory and applications. Moderate difficulty; only basic calculus required. Includes problems with answers. 252pp. 5 5/8 x 8 1/4. 0-486-63760-3

OUTLINE OF BASIC STATISTICS: Dictionary and Formulas, John E. Freund and Frank J. Williams. Handy guide includes a 70-page outline of essential statistical formulas covering grouped and ungrouped data, finite populations, probability, and more, plus over 1,000 clear, concise definitions of statistical terms. 1966 edition. 208pp. 5 3/8 x 8 1/2. 0-486-47769-X

GOOD THINKING: The Foundations of Probability and Its Applications, Irving J. Good. This in-depth treatment of probability theory by a famous British statistician explores Keynesian principles and surveys such topics as Bayesian rationality, corroboration, hypothesis testing, and mathematical tools for induction and simplicity. 1983 edition. 352pp. 5 3/8 x 8 1/2. 0-486-47438-0

INTRODUCTION TO PROBABILITY THEORY WITH CONTEMPORARY APPLICATIONS, Lester L. Helms. Extensive discussions and clear examples, written in plain language, expose students to the rules and methods of probability. Exercises foster problem-solving skills, and all problems feature step-by-step solutions. 1997 edition. 368pp. 6 1/2 x 9 1/4. 0-486-47418-6

CHANCE, LUCK, AND STATISTICS, Horace C. Levinson. In simple, non-technical language, this volume explores the fundamentals governing chance and applies them to sports, government, and business. "Clear and lively ... remarkably accurate." – *Scientific Monthly*. 384pp. 5 3/8 x 8 1/2. 0-486-41997-5

FIFTY CHALLENGING PROBLEMS IN PROBABILITY WITH SOLUTIONS, Frederick Mosteller. Remarkable puzzlers, graded in difficulty, illustrate elementary and advanced aspects of probability. These problems were selected for originality, general interest, or because they demonstrate valuable techniques. Also includes detailed solutions. 88pp. 5 3/8 x 8 1/2. 0-486-65355-2

EXPERIMENTAL STATISTICS, Mary Gibbons Natrella. A handbook for those seeking engineering information and quantitative data for designing, developing, constructing, and testing equipment. Covers the planning of experiments, the analyzing of extreme-value data; and more. 1966 edition. Index. Includes 52 figures and 76 tables. 560pp. 8 3/8 x 11. 0-486-43937-2

STOCHASTIC MODELING: Analysis and Simulation, Barry L. Nelson. Coherent introduction to techniques also offers a guide to the mathematical, numerical, and simulation tools of systems analysis. Includes formulation of models, analysis, and interpretation of results. 1995 edition. 336pp. 6 1/8 x 9 1/4. 0-486-47770-3

INTRODUCTION TO BIOSTATISTICS: Second Edition, Robert R. Sokal and F. James Rohlf. Suitable for undergraduates with a minimal background in mathematics, this introduction ranges from descriptive statistics to fundamental distributions and the testing of hypotheses. Includes numerous worked-out problems and examples. 1987 edition. 384pp. 6 1/8 x 9 1/4. 0-486-46961-1

Browse over 9,000 books at www.doverpublications.com

Mathematics–Geometry and Topology

PROBLEMS AND SOLUTIONS IN EUCLIDEAN GEOMETRY, M. N. Aref and William Wernick. Based on classical principles, this book is intended for a second course in Euclidean geometry and can be used as a refresher. More than 200 problems include hints and solutions. 1968 edition. 272pp. 5 3/8 x 8 1/2. 0-486-47720-7

TOPOLOGY OF 3-MANIFOLDS AND RELATED TOPICS, Edited by M. K. Fort, Jr. With a New Introduction by Daniel Silver. Summaries and full reports from a 1961 conference discuss decompositions and subsets of 3-space; n-manifolds; knot theory; the Poincaré conjecture; and periodic maps and isotopies. Familiarity with algebraic topology required. 1962 edition. 272pp. 6 1/8 x 9 1/4. 0-486-47753-3

POINT SET TOPOLOGY, Steven A. Gaal. Suitable for a complete course in topology, this text also functions as a self-contained treatment for independent study. Additional enrichment materials make it equally valuable as a reference. 1964 edition. 336pp. 5 3/8 x 8 1/2. 0-486-47222-1

INVITATION TO GEOMETRY, Z. A. Melzak. Intended for students of many different backgrounds with only a modest knowledge of mathematics, this text features self-contained chapters that can be adapted to several types of geometry courses. 1983 edition. 240pp. 5 3/8 x 8 1/2. 0-486-46626-4

TOPOLOGY AND GEOMETRY FOR PHYSICISTS, Charles Nash and Siddhartha Sen. Written by physicists for physics students, this text assumes no detailed background in topology or geometry. Topics include differential forms, homotopy, homology, cohomology, fiber bundles, connection and covariant derivatives, and Morse theory. 1983 edition. 320pp. 5 3/8 x 8 1/2. 0-486-47852-1

BEYOND GEOMETRY: Classic Papers from Riemann to Einstein, Edited with an Introduction and Notes by Peter Pesic. This is the only English-language collection of these 8 accessible essays. They trace seminal ideas about the foundations of geometry that led to Einstein's general theory of relativity. 224pp. 6 1/8 x 9 1/4. 0-486-45350-2

GEOMETRY FROM EUCLID TO KNOTS, Saul Stahl. This text provides a historical perspective on plane geometry and covers non-neutral Euclidean geometry, circles and regular polygons, projective geometry, symmetries, inversions, informal topology, and more. Includes 1,000 practice problems. Solutions available. 2003 edition. 480pp. 6 1/8 x 9 1/4. 0-486-47459-3

TOPOLOGICAL VECTOR SPACES, DISTRIBUTIONS AND KERNELS, François Trèves. Extending beyond the boundaries of Hilbert and Banach space theory, this text focuses on key aspects of functional analysis, particularly in regard to solving partial differential equations. 1967 edition. 592pp. 5 3/8 x 8 1/2.
0-486-45352-9

INTRODUCTION TO PROJECTIVE GEOMETRY, C. R. Wylie, Jr. This introductory volume offers strong reinforcement for its teachings, with detailed examples and numerous theorems, proofs, and exercises, plus complete answers to all odd-numbered end-of-chapter problems. 1970 edition. 576pp. 6 1/8 x 9 1/4. 0-486-46895-X

FOUNDATIONS OF GEOMETRY, C. R. Wylie, Jr. Geared toward students preparing to teach high school mathematics, this text explores the principles of Euclidean and non-Euclidean geometry and covers both generalities and specifics of the axiomatic method. 1964 edition. 352pp. 6 x 9. 0-486-47214-0

Browse over 9,000 books at www.doverpublications.com

Mathematics–History

THE WORKS OF ARCHIMEDES, Archimedes. Translated by Sir Thomas Heath. Complete works of ancient geometer feature such topics as the famous problems of the ratio of the areas of a cylinder and an inscribed sphere; the properties of conoids, spheroids, and spirals; more. 326pp. 5 3/8 x 8 1/2. 0-486-42084-1

THE HISTORICAL ROOTS OF ELEMENTARY MATHEMATICS, Lucas N. H. Bunt, Phillip S. Jones, and Jack D. Bedient. Exciting, hands-on approach to understanding fundamental underpinnings of modern arithmetic, algebra, geometry and number systems examines their origins in early Egyptian, Babylonian, and Greek sources. 336pp. 5 3/8 x 8 1/2. 0-486-25563-8

THE THIRTEEN BOOKS OF EUCLID'S ELEMENTS, Euclid. Contains complete English text of all 13 books of the Elements plus critical apparatus analyzing each definition, postulate, and proposition in great detail. Covers textual and linguistic matters; mathematical analyses of Euclid's ideas; classical, medieval, Renaissance and modern commentators; refutations, supports, extrapolations, reinterpretations and historical notes. 995 figures. Total of 1,425pp. All books 5 3/8 x 8 1/2.

Vol. I: 443pp. 0-486-60088-2
Vol. II: 464pp. 0-486-60089-0
Vol. III: 546pp. 0-486-60090-4

A HISTORY OF GREEK MATHEMATICS, Sir Thomas Heath. This authoritative two-volume set that covers the essentials of mathematics and features every landmark innovation and every important figure, including Euclid, Apollonius, and others. 5 3/8 x 8 1/2.

Vol. I: 461pp. 0-486-24073-8
Vol. II: 597pp. 0-486-24074-6

A MANUAL OF GREEK MATHEMATICS, Sir Thomas L. Heath. This concise but thorough history encompasses the enduring contributions of the ancient Greek mathematicians whose works form the basis of most modern mathematics. Discusses Pythagorean arithmetic, Plato, Euclid, more. 1931 edition. 576pp. 5 3/8 x 8 1/2.

0-486-43231-9

CHINESE MATHEMATICS IN THE THIRTEENTH CENTURY, Ulrich Libbrecht. An exploration of the 13th-century mathematician Ch'in, this fascinating book combines what is known of the mathematician's life with a history of his only extant work, the Shu-shu chiu-chang. 1973 edition. 592pp. 5 3/8 x 8 1/2.

0-486-44619-0

PHILOSOPHY OF MATHEMATICS AND DEDUCTIVE STRUCTURE IN EUCLID'S ELEMENTS, Ian Mueller. This text provides an understanding of the classical Greek conception of mathematics as expressed in Euclid's Elements. It focuses on philosophical, foundational, and logical questions and features helpful appendixes. 400pp. 6 1/2 x 9 1/4. 0-486-45300-6

BEYOND GEOMETRY: Classic Papers from Riemann to Einstein, Edited with an Introduction and Notes by Peter Pesic. This is the only English-language collection of these 8 accessible essays. They trace seminal ideas about the foundations of geometry that led to Einstein's general theory of relativity. 224pp. 6 1/8 x 9 1/4. 0-486-45350-2

HISTORY OF MATHEMATICS, David E. Smith. Two-volume history – from Egyptian papyri and medieval maps to modern graphs and diagrams. Non-technical chronological survey with thousands of biographical notes, critical evaluations, and contemporary opinions on over 1,100 mathematicians. 5 3/8 x 8 1/2.

Vol. I: 618pp. 0-486-20429-4
Vol. II: 736pp. 0-486-20430-8

Browse over 9,000 books at www.doverpublications.com